Cancer: An Integrative Approach

Doru Paul

Cancer: An Integrative Approach

Principles and Strategies - from Precision Targeting to Process Oncology

Springer

Doru Paul
Department of Hematology and Oncology
NewYork–Presbyterian Hospital
NY, NY, USA

ISBN 978-3-032-08807-9 ISBN 978-3-032-08808-6 (eBook)
https://doi.org/10.1007/978-3-032-08808-6

This Springer imprint is published by the registered company Springer Nature Switzerland AG
The registered company address is: Gewerbestrasse 11, 6330 Cham, Switzerland

This book is dedicated to my patients

Acknowledgments

I would like to express my heartfelt gratitude to all those who have inspired me, as well as to my friends, colleagues, and collaborators: David Lyden, Aurora Nedelcu, Radu Minea, Irina Matei, Florica Paul, Florin Popescu, Ilana Vale, Seta Balmanoukian, Mihai Dinu, Zhu Rong, Natalya Komarova, Mircea Ivan, Frédéric Thomas, Florinel Bădulescu, Adina Croitoru, Steve Jackowicz, Livia Kohn, Daniela Anton, Kavitha Chinnaiyan, Robert Gatenby, Mark Vincent, Maurice Israël, Nevena Lucic, Elizabeta Popa, Cristina Antonescu, Pavel Chirilă, Elena Udrişte, Eduard Nedea, Florin Băcanu, Alexandru Buturugă, Daniela Matei, George Călin, Velisar Rill, Jim Gordon, Lucia Debeleac, Marie-France Poupon, Cristian Calude, Mihaela van der Schaar, Mircea Leabu, John-William Sidhom, and Charles Kinzig. I would also like to thank Ana Tourian and Laura Pătraş for their artistry in designing the figures for the book. Matthew Calhoun's expertise with editing was invaluable and I would like to express my sincere thanks and appreciation to him here. Gregory Sutorius, Executive Editor at Springer Nature Group, provided me several books from Springer's brand-new collection, and I thank him warmly for his generosity. I would also like to thank Pinky Sathiskumar, Springer Production Editor, for ensuring I stayed on track and for checking in periodically on the progress of the manuscript. A very special thanks goes to my family—and especially to my brother, Bogdan Paul, with whom I shared countless late-night conversations about the key themes of this book, and whose unwavering support sustained me through the many challenges of writing it. Last but not the least, I wish to express my deepest gratitude to the late Anna Velcich whose mentorship, friendship, and wisdom helped me through the whole process of shaping my ideas into a book format. Though no longer with us due to her premature death from pancreatic cancer, Anna's guidance continues to resonate with me. This book is, in part, a tribute to her enduring legacy.

The original version of the book has been revised. A correction to this book can be found at https://doi.org/10.1007/978-3-032-08808-6_16

Contents

About the Author

Doru Paul is a seasoned medical oncologist who has provided care to over 10,000 cancer patients, with particular clinical expertise in the treatment of head and neck as well as lung cancers. He has a track record of developing successful multidisciplinary approaches to treat these conditions. A graduate of the esteemed "Carol Davila" Faculty of Medicine in Bucharest, Romania, he pursued advanced training in Paris, France, in molecular biology, chronobiology, and aerospace medicine. He also holds a PhD in personalized oncology. He has served since 2002 as Principal Investigator or co-Principal Investigator on more than 70 oncology clinical trials. He has designed and conducted multiple investigator-initiated studies and initiated three FDA-approved investigational new drug studies. Among his pioneering contributions is the design and execution of the first-in-human pilot study using positrons for targeted cancer therapy—a world premiere. Currently, he is an Associate Professor of Clinical Medicine at Weill Cornell Medical College in New York and serves as an Associate Attending Physician in Hematology and Medical Oncology at New York-Presbyterian Hospital.

1 Introduction

Once, someone asked me whether I believed in miracles and I answered that miracles are not a question of belief—they are about opening your eyes and seeing what is right in front of you.

Consider this: a single cell develops into a human being composed of over 37 trillion cells—each communicating, adapting, and evolving. That is a miracle. We are rightly awed by the vastness of the cosmos, yet the human body contains approximately 35,000 times more atoms than the number of stars in the visible universe. This internal universe is no less wondrous than the one we see in the sky. To answer the question more directly, yes, patients I have managed with what is widely considered to be incurable cancer have achieved complete remission. I have met and cared for individuals with metastatic small-cell and non-small-cell lung cancer, metastatic breast, prostate, and colorectal cancers—even patients with metastatic pancreatic cancer—who are alive and well more than a decade after treatment and completely free of disease. These cases challenge conventional expectations. Were they miracles? If by miracle we mean events without a clear explanation, then yes, I have witnessed such phenomena. To me, these cures are not supernatural—they are part of nature's extraordinary logic. Life itself is the ultimate miracle, and within that context, exceptional cancer recoveries are completely natural. I don't merely believe in cancer cures; I have witnessed firsthand. I know it. I hold an unshakable certainty that metastatic cancer can be cured. The pressing question is: how?

This book is unusual in its structure and intention. Rather than beginning with a predefined model of cancer, it opens by asking fundamental questions about the nature of life itself. From that inquiry, it builds a new integrative framework for understanding cancer, bottom-up and top-down, as a novel life form, with its own dynamic but, at the same time, following the same laws and rooted in the same principles as all living systems. Within this framework, several strategies for cancer management have been proposed.

There are currently three strategies that could significantly impact the oncology field. The first is building-up cancer interception and prevention programs that may detect both the propensity for cancer and cancer in early stages and curb the cancer incidence by as much as 40%. The second is developing early interception and prevention of the metastatic process. The third is designing cures for metastatic disease, which is currently responsible for the vast majority of cancer deaths.

At its core, biological organization is not solely spatial but fundamentally temporal: spatial patterns arise from underlying temporal dynamics, and living systems can be understood as ensembles of interwoven narratives inscribed within organismal forms. As Galileo's statement "E pur si muove" reminds us, movement lies both at the heart of the Universe and at the heart of life. Thus, biology must look beyond static forms like molecules, cells, tissues and organisms and embrace the *processes*, *transitions*, and *rhythms* that animate them.

This perspective is particularly vital in cancer research. Cancer is not a fixed entity but a heterogeneous, evolving system—a dynamic constellation of interacting processes and shifting targets. A framework centered on time, change, and action is essential for grasping its complexity and for developing therapies that can adapt alongside it. In this light, emerging paradigms such as *process biology* offer a new foundation for rethinking the nature of life and disease. By prioritizing temporal structure and system dynamics, they pave the way for a more accurate understanding of carcinogenesis and for new treatment strategies.

A second characteristic of biology is *scaling*: organisms are nested entities. Each level is organized into integrated modules and introduces new functions progressively, from the nanoscale, up to the emergent organism level up to that interacts with the environment. On Earth, there are three levels of more or less autonomous life forms: a) *the molecular level*: viruses depend entirely on the host for transmission; b) *the cellular level*: autonomous unicellular life forms exist; however, in multicellular organisms, cells are only partially autonomous, and their autonomy can fluctuate, as demon-

D. Paul, *Cancer: An Integrative Approach*, https://doi.org/10.1007/978-3-032-08808-6_1

strated by the cancer phenomenon; and c) *the multicellular organismic level*: multicellular organisms are part of an ecosystem, and autotrophs like Plantae and Protista, or heterotrophs like animals and fungi, depend on their modality of harvesting energy. In multicellular organisms, bottom-up and top-down communication and their particular design glues into a functional whole different scales, from molecules to cells and then to tissues and organs. An important trans-scale connector is the *internal body climate*, a novel concept described in Chap. 9. Its importance is crucial for the malignant process, being both directly involved in cancer inception and, reversely, profoundly disturbed by cancer progression.

Finally, another key characteristic is *complexity*. The complexity of life forms is both *quantitative* (life is a group phenomenon) and *qualitative* (each level of life is organized into integrated modules and introduces new functions up to the emergent organism level that interact with the environment as a separate entity). This is where system biology tools like network theory or dynamic graph models come into the picture. As opposed to the classic formulation one gene-one disease, it became apparent that the majority of chronic diseases are multigenic. The majority of cancers involve aneuploidies related to qualitative and quantitative chromosomal changes that impact the expression of hundreds of genes. *Cancer is a network disease and involves both intracellular networks, cell-cell interaction networks, organism-wide networks, and translevel interactions.* At the genetic level, the majority of cancers are not related to alterations in one gene but rather alteration in a large number of genes organized in networks. The particular network composition varies not only across different cancer locations but also within the same cancer type, and, significantly, this is associated with very different prognostics. At the cellular level, in cancer, the communication network between the nucleus and the mitochondria is disturbed. At the tissue level, the cell-cell communication between cancer cells and non-cancer cells leads to the formation of a cancerized stroma that supports cancer development. At the organismic level, new cancer-induced pathogenic networks are formed that are responsible for the systemic hallmarks of cancer: metabolic changes, increase in the risk of thrombosis, increase in systemic inflammation, etc.

A wider biological framework that takes into consideration the *interactions and communication processes* between the cancer system—understood as a complex, dynamic, and evolving network—and the human organism may serve as a foundation for integrative oncology, a new holistic model that could open new research avenues and lead to more effective cancer treatments.

This book is not about a single "integration" but rather several different "integrations." Three integrations represent the major themes in the book.

The *first* integration is between cancer, which is considered a life form, and the traditional categorization of cancer as a disease.

Biologists typically approach cancer through a dualistic lens, differentiating malignant from nonmalignant processes by focusing on the intrinsic properties of cancer cells and their interactions with surrounding cellular components. Their emphasis lies in dissecting the cellular and molecular mechanisms that drive oncogenesis and tumor progression. In contrast, medical doctors adopt a holistic perspective, treating cancer within the broader context of the patient's overall condition. Clinicians address not only the tumor but also the systemic, physiological, and psychosocial dimensions of the individual, recognizing that effective therapy must consider the patient as an integrated whole rather than a collection of isolated cellular phenomena.

The *second* integration is between an "object-oriented" understanding of biological processes and a "process-oriented" approach.

In the study of life, two distinct yet complementary epistemological frameworks shape our understanding of biological phenomena: object-oriented and process-oriented approaches. The object-oriented perspective, which is deeply rooted in classical biology and molecular reductionism, emphasizes discrete entities—genes, proteins, organelles, and cells—as the primary units of analysis. This framework excels in mapping structural features, identifying genetic mutations, cataloging molecular interactions, and describing anatomical architectures. It has been foundational in developing modern molecular biology, genomics, and targeted therapeutics, enabling precise interventions aimed at well-defined biological "objects," such as oncogenes or mutated receptors. However, life cannot be fully captured through its material constituents alone. The process-oriented approach shifts the analytical focus from static entities to dynamic interactions, emphasizing fluxes, rhythms, transformations, and emergent behaviors across time and scale. In this view, biological phenomena are seen not as properties of objects but as outcomes of ongoing processes—signaling cascades, metabolic fluxes, immune responses, morphogenetic movements, or the diachronic evolution of cell states. The process-oriented model aligns more closely with systems biology, developmental dynamics, and ecological or organismal perspectives, where the emphasis is on relationships, temporal unfolding, and feedback loops rather than on isolated components. This distinction becomes particularly salient in oncology. An object-oriented lens might focus on the presence of a KRAS mutation or the overexpression of HER2, treating these features as targets for intervention. A process-oriented view, in contrast, considers how these molecular events participate in evolving networks—how cancer cells modulate their microenvironment, adapt metabolically, evade immune surveillance, or rewire their epigenetic machinery over time. Such temporal and relational complexity often escapes static, object-based models but is critical for understanding cancer resilience, adaptability, and systemic entanglement with the host.

The *third* integration is between the molecular, cellular and tissue hallmarks of cancer and the *systemic hallmarks of cancer*. Cancer is not merely a consequence of genetic mutations, cellular dysregulation, or disordered tissue architecture. While the hallmarks of cancer at the molecular and cellular levels, such as sustained proliferative signaling, evasion of growth suppressors, and resistance to cell death, provide essential insights into its pathogenesis, they offer only a partial account of the disease. Equally important, yet often underappreciated, are the *systemic hallmarks of cancer*, which reflect the profound disruption of organism-wide homeostasis. Cancer should be viewed not simply as a localized pathology but also as a dynamic systemic condition involving complex interactions between the tumor and the host's physiological, metabolic, immune, neuroendocrine, and circadian regulatory networks. From its inception, cancer has engaged with the body as an integrated whole. Tumor growth and progression perturb systemic parameters such as energy balance, inflammation, oxidative stress, hormonal rhythms, and immune surveillance. Conversely, systemic factors, chronic stress, metabolic syndrome, circadian disruption, inflammation, and dysbiosis can facilitate the emergence and expansion of malignant clones. The cancerized body is a *biosphere* composed of interlinked ecosystems: tumor niches, immune microenvironments, endocrine loops, vascular territories, and neural networks, all modulated by a shared *internal body climate*. When this climate becomes dysregulated, it not only fosters tumor growth but also contributes to an array of cancer-associated symptoms, such as fatigue, cachexia, cognitive decline, and emotional distress. Integrating cellular and tissue-level hallmarks with systemic hallmarks is essential for a comprehensive understanding of cancer as a *multiscale disease*. The tumor phenotype emerges from an interplay between molecular alterations, cellular behaviors, the tissue microenvironment, and organism-level physiological states. For instance, the tumor-induced reprogramming of the immune system alters not only local immune cell infiltration but also systemic inflammatory states and neuroimmune communication, feeding into the chronic nature of the disease. Therefore, effective cancer treatment must move beyond the molecular targeting of oncogenic drivers toward *modulating the systemic context* in which tumors arise and thrive. Interventions that aim to restore organismal homeostasis by targeting the hypothalamic-pituitary axis, synchronizing circadian rhythms, reducing systemic inflammation, and reestablishing metabolic equilibrium may not only curb tumor progression but also improve resilience, quality of life, and therapeutic responsiveness. In this integrative framework, cancer is understood not as an alien invader but as a maladaptive process within the organismal biosphere, an adaptive response to internal systemic perturbations.

This work introduces a unifying theoretical framework based on the concept of dynamic and creative persistence (DCP), redefining life—and cancer—as an emergent, homeodynamic interplay of bioenergy, bioinformation, and biostructure. Homeodynamics theory is proposed as a new foundational biology model, and, *more geometrico*, a comprehensive formalism consisting of an axiom, three laws, and eight heuristic principles is introduced as universal rules governing the living process. Under this framework, cancer is not simply a collection of genetic mutations or cellular malfunctions but also a cyclical, self-organizing process that coevolves with the host, sustained by its ability to maintain internal coherence while adaptively reshaping its ecological niche within the organism. At the heart of this model lies the insight that disruptions in organismal homeodynamics, particularly those affecting the internal body climate, are not passive preconditions but active drivers of tumorigenesis and cancer progression. Oncogenesis emerges as a systemic disturbance rooted in imbalances in metabolic, immune, neuroendocrine, and circadian regulation. The DCP framework emphasizes early interception by identifying preclinical shifts in metabolomic signatures, systemic vulnerabilities, and niche formation. Integrating multiomics platforms, liquid biopsy, and AI-driven analytics, enables individualized, predictive diagnostics. Preventive strategies, such as exercise, dietary modulation, circadian alignment, and stress reduction, aim to recalibrate systemic balance and reinforce resilience before malignancy becomes entrenched. Therapeutically, this model promotes strategies of *dynamic modulation,* aiming to influence the underlying generative processes of cancer rather than merely targeting discrete oncogenic features. This includes influencing metabolic control nodes (e.g., acetyl-CoA), controlling phenotypic plasticity, regulating exosomal communication, and fostering ecological containment to constrain cancer within manageable bounds. As traditional precision oncology faces limitations, especially in metastatic settings, the homeodynamic approach offers a complementary path—uniting systemic interventions, AI-personalized treatment, and organism-wide level strategies. Here, cancer care becomes not a battle but an adaptive stewardship of the body's intrinsic capacity for self-organized balance and resilience.

The reality of the oncology ward is very different from the world of theoretical biology or basic research. Navigating the complex world of oncology is difficult for both doctors and patients. For a patient diagnosed with cancer, this condition involves life or death decisions and represents an extremely serious existential threat. Arguably, *the most important integration* in oncology happens between patients with their educational and cultural values and healthcare specialists who have their own biases and understanding of oncologic processes.

The lived reality within the oncology clinic starkly contrasts with the abstractions of theoretical biology or the controlled environment of basic research laboratories. While molecular models and mechanistic frameworks offer invaluable insights into tumorigenesis, they often fall short of capturing the profound complexity, urgency, and emotional gravity that define real-world cancer care. In clinical oncology, both physicians and patients must navigate an intricate and often overwhelming landscape of diagnostic uncertainty, therapeutic trade-offs, and rapidly evolving scientific knowledge—an experience that is as much existential as it is medical.

For the patient, a cancer diagnosis is not merely a clinical label; it is an existential fracture—an intrusion that redefines identity, disrupts life trajectories, and imposes choices that carry the weight of life and death. Each decision, from treatment selection to participation in clinical trials, unfolds against a backdrop of fear, hope, cultural beliefs, and personal values. For clinicians, delivering care requires far more than scientific expertise—it demands empathy, communication, and the ability to guide individuals through uncertainty while translating complex biological phenomena into humane and actionable choices.

In this context, probably the most important yet underrecognized integration in oncology is not between signaling pathways or therapeutic agents but between two epistemic worlds: the patient's world, shaped by personal history, cultural frameworks, and lived experience; and the clinician's world, shaped by biomedical training, clinical guidelines, and ingrained professional paradigms. These worlds often operate on different logics: subjective versus objective, narrative versus analytical, intuitive versus evidence-based, and their intersection defines the therapeutic alliance. The quality of this integration can profoundly influence outcomes: it determines trust, adherence, psychological resilience, and ultimately the ability to build up together a path through the disease.

Thus, oncology must evolve not only toward precision medicine but also toward *precision communication* and *relational integration*. This demands that healthcare professionals remain aware of their own cognitive frameworks and biases while remaining open to the values, fears, and meaning systems of their patients. It also calls for educational models that bridge the gap between bench science and bedside empathy, fostering a more holistic understanding of what it means to care for a person with cancer. In doing so, we can create a truly integrative oncology, one that unites molecular insight with human understanding and clinical rigor with existential care.

This book is structured as a conceptual and scientific exploration of cancer, aiming to integrate current knowledge with new theoretical perspectives, including the potential benefit of AIs in assisting medical oncologists in patient care.

Below is an overview of the chapters:

Chapter 2 offers an overview of the entire book and introduces its main themes, which will be subsequently be expanded.

Chapter 3 is a reflection on some aspects of cancer in the context of the COVID-19 pandemic experience.

Chapter 4 presents a concise overview of contemporary cancer theories, each offering a partial understanding of the disease. These theories are pieces of a larger, more complex puzzle, which, in my view, in order to be solved requires the integration of several key conceptual components.

Chapter 5 provides a detailed analysis of the current scientific understanding of cancer, at the molecular, cellular, and tissue levels, respectively.

Chapter 6 explores the idea of biology as the science of both forms and processes and discusses the features needed for a comprehensive, integrative model of living systems.

Chapter 7 discusses various theories about the nature of life itself, drawn from biology, physics and systems theory.

Chapter 8 introduces homeodynamics, a novel perspective on the nature of life, and proposes a new interpretative framework based on the concepts of cyclic causality and dynamic and creative persistence (DCP).

Chapter 9 presents three fundamental laws and eight guiding principles of living systems. These are discussed from both the perspective of the healthy organism and that of cancer, which is considered at the same time a divergent expression of life's organizing logic and as a distinct, semi-autonomous form of life itself.

Chapter 10 explores the concept of biological time—its structure, properties, and role in carcinogenesis.

Chapter 11 traces the temporal trajectory of cancer, identifying key landmarks and transitions in its evolution within the host.

Chapter 12 applies homeodynamic theory specifically to cancer, offering reinterpretations of its origins, behavior, and dynamics through the lens of the proposed new model of life.

Chapter 13 proposes various therapeutic strategies and introduces a novel, AI-assisted integrated cancer treatment plan that aligns with the author's systemic understanding of cancer.

Chapter 14 focuses on oncological patients and discusses the personal and existential impact of cancer, including shifts in values, goals, and worldviews.

Chapter 15 offers a conclusion and suggests future directions for both research and clinical applications.

One Flew Over the Cancer's Nest

2

2.1 The Elephant in the Room

You may be familiar with that anecdote of a group of blind men getting hold of an elephant, an animal they have never seen before, none of them aware of its shape and form. "Out of curiosity, they said: "We must inspect and know it by touch, of which we are capable." So, they sought it out, and when they found it, they groped about it. The first one, whose hand landed on the trunk, said, "This being is like a thick snake." For another one whose hand reached its ear, it seemed like some fan. As for another person, whose hand was upon its leg, the elephant was a pillar like a tree trunk. The blind man, who placed his hand upon its side, said "This is a wall." Another, who felt its tail, described it as a rope. The last felt its tusk, stating the elephant is that which is hard, smooth and like a spear" (Goldstein 2010) (see Fig. 2.1). Like the elephant, cancer has also been considered a mysterious condition, for which a wealth of perspectives and theories regarding its nature have been put forward by the oncology community. The reason the cancer puzzle has not yet been solved is thought to be either the absence of some critical pieces, or, as Swanton et al. and Huang et al. suggested recently (Swanton et al. 2024; Huang et al. 2025a, b), cancer may be a complicated zoo of animals of all colors, shapes, and sizes with different behaviors that are difficult to pin down.

D. Paul, *Cancer: An Integrative Approach*, https://doi.org/10.1007/978-3-032-08808-6_2

Fig. 2.1 Holistic view and functional perceptions of an elephant. (Courtesy of Ana Tourian)

2.2 Why Do Patients with Cancer Die? Insights from the COVID-19 Pandemic

Since the 1914 publication of Theodor Boveri's book *The origin of malignant tumors*, a large amount of data has accumulated demonstrating the roles that chromosomes and genes play in carcinogenesis, and, in the last decades of the twentieth century, the somatic mutation theory (SMT) of cancer has become the dominant narrative in oncology. But cancer is not solely a disorder of genetic alterations and chromosomal abnormalities. DNA alterations alone cannot explain key cancer features like preprimary tumor or premetastatic niche formations that do not necessitate any DNA alterations. For several decades, different explanatory maps have been drawn, highlighting the roles of other cellular processes besides DNA alterations like epigenetic processes and metabolism, and, more recently, over the last two decades or so, cell-cell and cell-stromal interactions that have led to the extraordinary success have also been reported by immunotherapy. At present, cancer is considered both a cellular disease and a tissue disease. Importantly, it appears that, in general, cancer represents a perturbation of genetic networks, and individual genetic alterations play a secondary role (Huang et al. 2025a, b). The focus on the microenvironment has expanded our explanatory framework, but it is still incomplete.

2.3 Current Cancer Landscape

The March 28, 2014, cover of *Newsweek* magazine read, in reference to cancer, "You can't cure what you can't understand." In more than a decade since then, we still have not made significant progress in our understanding of cancer, and as of today, the cure for cancer seems as remote as ever. According to the International Agency for Research on Cancer (IARC), the global cancer burden is steadily increasing, and as a result, it is projected that in 2025, there will be more than 20 million new cancer cases and approximately ten million deaths from cancer. About 1 in 5 people develop cancer in their lifetime, and approximately 1 in 9 men and 1 in 12 women die from the disease (WHO website, March 2025). By 2050, the number of cancer cases is predicted to increase to 35 million based solely on projected population growth (ACS Global Cancer Facts & Figures n.d.).

What exactly is needed in order to cure cancer? *First, we should aim for a cure. Neither current research studies nor treatment protocols aim for a cure.* As a clinical oncologist, at the beginning of treatment for metastatic disease, I ask patients to certify that they understand that the purpose of the treatment they are going to receive is not curative and to sign a consent form, certifying that they understand that the treatment they are going to receive may lead to unforeseen side effects, including death. The wording of the consent form is an acknowledgment of the unlikelihood of current oncologic treatments to cure advanced and metastatic cancers.

Using the surrogate endpoint of long-term survival (LTS), cure should be included as a valuable purpose in our metastatic cancer research trials, which currently include, as goals, more or less durable responses—partial and complete response (PR and CR), progression-free survival (PFS), median survival (MFS), and overall survival (OS). As an important corollary, a recent study analyzed global LTS and specific causes of death in large longitudinal cohorts of patients diagnosed with cancer in the United States (Cheng et al. 2022). Overall, patients diagnosed with lung and bronchial cancer had poor short- and long-term survival rates and mostly died from their primary cancer, as opposed to patients diagnosed with other common cancers, who were more likely to die from causes other than primary cancers. Two different basic trends were observed for cumulative *cancer-specific mortality*. The first trend was found in prostate and breast cancer, which presented a low but excess risk after cancer diagnosis, that persisted over time pointing to the need for continuous and long-term monitoring for disease recurrence. The second trend was found in lung, colorectal, urinary bladder, melanoma, uterine corpus, and thyroid cancer, which were associated with a greatly diminished risk of index cancer-specific mortality 10 years or more following diagnosis. It appears that non-breast cancer in women and non-prostate cancer in men do not need the long-term surveillance needed in breast and prostate cancer survivors. An important limitation noted by the authors of the study themselves is that the study did not analyze specifically the survival of patients with metastatic disease, the cancer subtype known to be associated with the highest mortality. Recently, a population-based study of the *Norway Cancer Registry* (Dillekås et al. 2019) reported that approximately 66.7% of patients diagnosed with metastatic disease die from their cancers, as opposed to the 90% commonly cited figure. A study done in the Netherlands that analyzed data from almost two million patients diagnosed with solid cancer between January 1, 1989, and December 31, 2018, compared the 1- and 5-year survival for de novo metastatic cancer between the two time periods of 1989–1993 and 2014–2018 and demonstrated improvements in most de novo metastatic cancer subtypes, but this improvement was found to be unevenly distributed among cancers (Luyendijk et al. 2023). For example, a net improvement in survival was observed in metastatic cancers of the breast, prostate, GIST, and NETs between the two time periods, as opposed to hepatocellular cancers or cancers of the pancreas, where almost no improvement in survival was found. The natural history of different cancer subtypes is also significant. Approximately 50% of lung or pancreatic cancers are diagnosed de novo as metastatic, and only 1% of head and neck cancers are diagnosed in the metastatic stage (Niu et al. 2016, Luyendijk et al. 2023), which is a significant difference. An interesting recent article compared the survival trajectories of patients with incurable cancers receiving different systemic therapies and concluded that immunotherapy is the only systemic treatment that has historically improved significantly LTS, but only in certain cancer subtypes like melanoma, for example. Unfortunately, sooner or later, immunotherapy and other systemic treatments like targeted agents lead to therapeutic resistance (Geijteman et al. 2024).

The best strategy to cure cancer is to prevent it from occurring in the first place. Furthermore, discovered early, most cancers can be cured with local treatments like surgery or radiotherapy. Prevention strategies and early detection have already helped curb cancer epidemics significantly. Designing effective means of primary and secondary prevention and early diagnosis would represent tremendous steps forward and potentially decrease cancer mortality by at least 40%. With the recent technological advancements in omics and AIs, it is conceivable that soon we will also be able to predict cancer inception before its actual clinical diagnosis.

As a proof of principle, a Japanese study reported that an increase in plasma branched chain amino acid levels was associated with an increased risk of pancreatic cancer up to 10 years before diagnosis (Katagiri et al. 2018). Last year, a longitudinal proteomics study (Carrasco-Zanini et al. 2024) reported that higher plasma levels of TNFRSF17 and lower plasma levels of TNFSF13B are strong, specific predictors of an increased risk of multiple myeloma (MM) up to 10 years before the actual diagnosis. Also, in 2024, a deep learning AI tool using medical records data identified people at the highest risk for pancreatic cancer up to 3 years before the actual diagnosis (Placido et al. 2023).

The real problem of cancer starts when it is diagnosed at advanced or metastatic stages, at a point when local treatments are of limited benefit, and we are obliged to use systemic treatments with proven limited efficacy. In the best-case scenario, all the cases considered, taken together with the three currently used treatment modalities (chemotherapy, targeted agents, and immunotherapy) may lead to the cure of approximately 30% of patients with metastatic solid cancers. Of course, as noted above, the natural history of locally metastatic cancer varies tremendously between different cancer histologies and locations. As an example, well-differentiated metastatic thyroid cancer has a 5-year survival of approximately 70%, compared to the close to zero 5-year survival of metastatic adenocarcinoma of the pancreas, advanced small cell lung cancer or glioblastoma multiforme. Therefore, it is extremely important for an oncologist to properly explain the prognosis of metastatic disease to an individual patient while avoiding words like "incurable." Occurrences of spontaneous regression (SR) have been reported for multiple cancer types, and although this phenomenon is extremely rare (Cole and Everson 1956; Jessy 2011; Radha and Lopus 2021) and appears to be more frequently associated with certain cancer types (melanomas, lymphoma, leukemia, neuroblastoma, renal cell carcinoma, and germ cell cancers) (Papac 1998), oncologists should be aware of it. The cause of SR of cancer is considered to be the stimulation of the immune system (Radha and Lopus 2021), but evidence for this proposed mechanism is lacking. Medicine, as opposed to mathematics or physics, is the science of particulars and idiosyncrasies and, above all, involves interaction with fellow human beings; therefore, a dose of cautionary hope is always recommended.

2.4 How Are Cancer Cells Different from Normal Cells?

As Harold Varmus pointed out, "cancer cells are a distorted version of ourselves" (Varmus 1989 Nobel banquet speech). It is therefore not surprising that their properties are very similar to those of normal cells. Recently, João Pedro de Magalhães, a researcher at the University of Liverpool, conducted a literature analysis of the journal database PubMed and found that 17,371 different human genes (87.7%) are mentioned in at least one paper related to cancer (de Magalhães 2022).

The 14 hallmarks of cancer described so far (Hanahan and Weinberg 2000, 2011; Hanahan, 2022) are not unique to malignant tumors. As noticed by Yuri Lazebnik, benign neoplasms may present the same hallmarks as cancer (Lazebnik 2010), including cellular immortality and induction of angiogenesis, with the notable exception of invasion and metastasis (Gomes 2022). Clinically, the diagnosis of malignancy is made by pathologists based on structural abnormalities of cells and tissues, i.e., cellular atypia and tissue invasion, and by clinicians by detection during direct examination or imaging of macroscopic metastases.

Senescent cells and polymorphic microbes are also present in normal tissues and organisms. Some of the hallmarks of cancer like phenotypic plasticity, immortality, and inhibition of apoptosis are shared with normal stem cells (Quesenberry et al. 2002; Rando 2006; Iannolo et al. 2008). Others, like inflammation and sustained angiogenesis, are also present in wound healing (Koh and DiPietro 2011; Han et al. 2017). The hypermutator phenotype associated with some forms of cancer is also present in normal B and T lymphocytes. Aerobic glycolysis (the Warburg effect) is the "hallmark of rapid proliferation" and not only of cancer. Aerobic glycolysis is the preferred metabolic program of normal cells when robust transient responses are needed (Abdel-Haleem et al. 2017). Chromosomal instability (CIN), which is present in many solid tumors, is also present in several congenital CIN syndromes in which affected patients are predisposed to cancer although only a minority of them develop cancer during their life (Taylor et al. 2019). Aneuploidies associated with cancer by Thomas Boveri more than a century ago and cancer-driving gene mutations have been found in samples of normal tissues (Knouse et al. 2014; Fiala and Diamandis 2020). Like cancer cells, macrophages can travel to distant sites and invade tissues. Fusion between macrophages and cancer cells has been observed in vivo and has been proposed as an explanation for the invasion and metastatic properties of cancer cells (Manjunath et al. 2020). Epigenetic reprogramming is present in normal development (Feng et al. 2010; Singh et al. 2023). Similarly, clonal evolution, which has been the focus of many articles in recent years, is not restricted to cancer cells. Clonal hematopoiesis, for example, appears to be a universal phenomenon associated with aging, and clonality is also present in normal cells (Jaiswal and Ebert 2019).

The similarity between normal and cancer cells is not surprising, as, in most cases, the original founder of the first cancer cell is a normal cell. Like multicellular organisms, with few exceptions, cancers come from a unique, first cell. Cancer inception is a slow process and appears to have developed over the years and does not resemble the fast-paced process of embryonal development. Research in oncology has focused mainly on the abnormalities present in cancer cells as opposed to normal cells, disregarding the fact that, biologically speaking, cancer cells are very much alive and, fundamentally, cancer is a form of life. As Gloria Heppner suggested, cancer is not solely a "metaphor for organismal life," but rather an "exemplar, of life itself [...] a microcosm of fundamental biologic principles played out over a time course of years rather than millennia" (Heppner 1984).

So, how are cancer cells different from normal cells? The initial diagnosis of cancer is typically made by a pathologist. A pathologist will immediately note that, morphologically, cancer cells differ from their normal counterparts in several ways: they vary in size and shape, exhibit increased fragility, and may be found in abnormal anatomical locations (e.g., epithelial cells appearing in a bone marrow biopsy). These cells often display enlarged, variably sized, or multiple nuclei, prominent nucleoli (greater than 5 nm), an increased nuclear-to-cytoplasmic ratio, and abnormal chromatin distribution. The cytoplasm may show basophilia and vacuolization. Basophilia in cancer cells is thought to reflect their high content of ribosomes and RNA, as these cells are actively engaged in protein synthesis. It is worth noting that basophilic and eosinophilic designations do not necessarily indicate that a structure is chemically acidic or basic but rather how it stains under microscopic analysis.

On cytology, epithelial cancer cells tend to appear in clusters, while hematopoietic malignancies and melanomas more commonly present as individual cells. Mesenchymal tumors may show either pattern. At the tissue level, the architectural organization of cancer is often distorted, and the basement membrane is typically disrupted. Although metastatic tumors may be visualized through imaging modalities, histopathological confirmation is always necessary to establish the nature of any abnormal growth.

Thus, at both the cellular and tissue levels, cancers exhibit distinct morphological differences compared to normal cells and tissues. Molecular analyses further corroborate these findings by identifying genetic and epigenetic alterations consistent with malignancy. These changes, whether at the molecular, cellular, or tissue scale, reflect underlying pathological processes. But which comes first: genetic alterations or homeostatic imbalances?

The current oncological paradigm predominantly considers genetic mutations to be the primary drivers of cancer. However, throughout this book, I argue that *homeostatic imbalances* are probably more important. These imbalances prompt cells to adopt altered internal parameters, distinguishing them from their normal counterparts. For instance, the plasma membrane of cancer cells is typically depolarized compared to that of normal cells. Additionally, the intracellular pH of cancer cells is more alkaline, whereas, in general, the extracellular environment—or tumor microenvironment—is more acidic than that of normal tissues. These altered biophysical parameters may serve both as hallmarks and drivers of the malignant process.

2.5 Why Is There No Cure for Metastatic Cancer?

The primary obstacle to curing metastatic cancer lies in its profound spatial and temporal heterogeneity (Fidler and Kripke 2015). No two cancers are the same. Also, cancer cells are different from each other, and they evolve. Once cancer spreads, tumor cells must adapt to diverse microenvironments across the body, resulting in considerable variation in their genetic, epigenetic, metabolic, and antigen profiles. This heterogeneity leads to differential responses to systemic therapies, especially those aimed at targeting proliferative mechanisms or specific protein products of oncogenic drivers. In order to apply precision medicine strategies, identifying specific targets is crucial. Everything in oncology is about the "target rush": finding, for therapeutic purposes, a target that is differentially expressed in cancer or cancer tissues compared to normal cells or tissues that can be used for therapeutic purpose. The history of oncology therapeutics is, literally, the history of finding a good target. Currently, there are no cancer treatments *specifically designed* to prevent or inhibit metastases. We are not able to prevent or target metastases. Also, our systemic treatments lack efficacy for treating cancers that grow in different locations.

Since cancer is understood primarily as a disease of uncontrolled cell growth, cancer treatments have been directed toward cell growth. Historically, chemotherapy was the first systemic approach employed to treat *already established* metastatic disease. While chemotherapy remains curative in certain contexts—most notably in pediatric leukemias—some adult hematologic malignancies, metastatic testicular cancer, and choriocarcinoma are rare exceptions. For most advanced solid tumors, chemotherapy is primarily palliative, offering transient symptom relief and modest survival extension. In exceptional cancers where cures are achieved, such as testicular cancer, tumor cells exhibit *uniform* and *extreme chemosensitivity*, which is a rare biological feature. In contrast, the general population of metastatic tumors demonstrates variable and often limited sensitivity, which differs depending on the site of metastasis and tends to be short-lived due to the inevitable emergence of resistance.

The second major systemic approach involves targeted therapies. The introduction of imatinib, a tyrosine kinase inhibitor (TKI), for chronic myeloid leukemia (CML) was heralded as a revolution, and the cover of *Time* magazine claiming that "magic bullets" able to cure cancer has been discovered. While TKIs have indeed transformed the prognosis of diseases like CML and certain melanomas, durable responses in other cancers such as lung adenocarcinomas with specific genetic alterations are far less common. Moreover, most targeted agents are cytostatic rather than cytotoxic, halting cell proliferation without inducing cell death. This, coupled with eventual resistance, limits their curative potential in the metastatic setting.

Finally, over the last decade, immunotherapy has been hailed as a major game changer and has replaced chemotherapy as a first-line treatment for the majority of cancers. Immunotherapy comes in three main flavors: checkpoint inhibitors, chimeric antigen receptor T cells (CAR-Ts), and adoptive T cell therapy (ACT). Checkpoint inhibition leads to a cure in a significant number of patients with melanoma, long-term remission in some patients with MSI-high metastatic colorectal cancers, and, occasionally, in patients with other types of metastatic cancers (e.g., lung cancer). Considering all forms of metastatic cancer together, *responses* using checkpoints inhibitor agents are observed only in approximately 20% of cancer patients. Also, apart from melanoma, in approximately 15% of patients with metastatic non-small cell lung cancer (NSCLC), checkpoint inhibition leads to *cure* in only a minority of cases. In certain areas, advances in immunology are substantial, and CAR-Ts have demonstrated significant benefits in patients with heavily pretreated hematological malignancies. Approximately, 50% of the 43 patients with hematological malignancies treated with CAR-T cells were disease free at 3 years, and durable responses lasting 8–9 years have been documented (Cappell et al. 2020). Importantly, CAR-Ts work mainly in hematological malignancies. So far, the results of clinical trials with CAR-Ts in metastatic solid malignancies have been negative. Tumor-infiltrating lymphocyte (TIL) is another treatment modality that was introduced and championed by Steven Rosenberg for more than half a century, and, anecdotally, it has demonstrated responses in heavily pretreated metastatic cancer patients. As an example, in 2011, Rosenberg et al. reported durable 5-year responses in 19 out of 20 patients who achieved a complete response after treatment with TILs (Rosenberg et al. 2011). More recently, durable responses were also demonstrated for lifileucel (ACTs using autologous ex vivo-expanded TIL) in patients with heavily pretreated advanced melanoma. Overall, 48 (31.5%) patients in the cohort responded with 8 complete responses and 40 with partial responses, and 41.7% of the responders maintained the response without any evidence of their cancer progressing for ≥18 months (Chesney et al. 2022).

Compared to chemotherapy, immunotherapy, despite occasionally causing significant side effects, is generally better tolerated. However, when considering all types of metastatic cancers collectively, immune checkpoint inhibitors elicit objective responses in only about 20% of patients, and sustained remissions—potential cures—are achieved in just a small subset of these patients. Notably, cellular therapies such as CAR-T cells and tumor-infiltrating lymphocytes (TILs) exert cytocidal—not merely cytostatic—effects, which may explain their higher efficacy.

Another underappreciated limitation of systemic treatments for metastatic disease lies in their dosing schedule. Most agents are administered continuously until disease progression, which may not always be optimal. Evidence from prostate cancer, for instance, suggests that intermittent treatment schedules can yield superior outcomes in certain contexts (Zhang et al. 2022a).

The uncomfortable reality is that most patients diagnosed with metastatic cancer who receive systemic therapy ultimately succumb to their disease. While some may experience temporary improvements in quality of life or modest survival gains, long-term survival is rarely achieved. Given the limited efficacy of current therapies, all patients with metastatic cancer should be considered for clinical trial enrollment. Encouragingly, progress in oncology drug development has been substantial. From 2000 to 2019, phase 1 trial response rates across solid tumors nearly doubled, from 9.6% to 18%, and complete response rates rose from 2.5% to 4.3% (Chihara et al. 2022). Importantly, combination therapies outperformed monotherapies (15.8% vs 3.5%). Despite these gains, trial participation remains low, with only 7.1% of patients enrolled in therapeutic studies (Unger et al. 2024)—a modest improvement from historical levels (~5%) but still inadequate. Crucially, durable remissions remain limited and do not always equate to cure.

Metastatic cancer continues to pose one of the greatest challenges in oncology. *An efficient way to manage metastasis would be to prevent it from occurring in the first place.* Adjuvant treatments should not be directed solely against cell growth but also against preventing cell migration. Also, novel treatments directed against already established metastases need to be developed.

While significant progress has been made, much work remains. Answering the fundamental question: "What is cancer?" through a reassessment of its systemic, homeostatic, developmental, evolutionary, and ecological dimensions may be the key to devising curative strategies, which are currently lacking.

2.6 What Is Cancer Exactly?

The official answers found on the National Cancer Institute (NCI) website are not very helpful:

A. "A tumor, or mass of cells, formed of these abnormal cells may remain within the tissue in which it originated (a condition called in situ cancer), or it may begin to invade nearby tissues (a condition called invasive cancer). An invasive tumor is said to be malignant, and cells shed into the blood or lymph from a malignant tumor are likely to establish new tumors (metastases) throughout the body. Tumors threaten an individual's life when their growth disrupts the tissues and organs needed for survival.
B. Cancer is a group of more than 100 diseases that develop across time and involve the uncontrolled division of the body's cells. Although cancer can develop in virtually any of the body's tissues, and each type of cancer has its unique features, the basic processes that produce cancer are quite similar in all forms of the disease.
C. Today, we know that cancer is a disease of molecules and genes, and we even know many of the molecules and genes involved."

These answers are not incorrect; they are incomplete. I propose to expand our perspective by considering the following points:

A. Cancer is not solely defined by uncontrolled proliferation, local invasion, and distant dissemination of malignant cells. Equally critical to the oncogenic process are the roles of the local stroma, the host organism, and the broader tissue and systemic microenvironments. *The mere presence of malignant cells is not inherently pathological*; rather, it is the physiological disruption caused by macroscopic tumors and the dynamic interplay between the cancer-associated stroma and the host drive disease manifestations. Inflammatory responses, present both in the cancerized stroma and systemically, constitute a major source of cancer-related symptoms. Conventional oncological models tend to dichotomize cancerous and noncancerous cells, malignant and normal stroma, and affected versus unaffected organs. However, such distinctions obscure the fact that stromal and organismal elements are essential for both the initiation and progression of malignancy. In this context, both the local microenvironment and the whole organism undergo a sort of "cancerization" and actively contribute to tumor biology. Cancer is fundamentally a systemic disease of the entire organism rather than one confined to genetic, epigenetic, metabolic, or stromal abnormalities. Analogous to the way maternal physiology sustains a pregnancy, the host organism supports the initiation, maintenance, and clinical progression of cancer. Cancer is literally, a sort of malignant embryo nurtured by the host organism.
B. Cancer is not only a "group of more than 100 diseases." Of course, no two cancers are *exactly* the same, but, fundamentally, there are both universal characteristics of the malignant process and unique idiosyncratic features.
C. Originally, it was thought that cancer behavior is the result of mutations, and only recently *has it become clear that the oncogenic process is not driven by genetic mutations only*. Identical driver mutations with those present in cancer have been described in normal tissues and benign conditions (Martincorena et al. 2015, 2016; Kato et al. 2016; Yizhak et al. 2019; Gomes et al. 2022). As an example, in pulmonary artery hypertension (PAH), driver mutations also present in cancer have been found. Several publications have shown a parallel between the two conditions (Boucherat et al. 2017; Pullamsetti et al. 2017, 2020; D'Alessandro et al. 2018; Cool et al. 2020). Due to its similarity with cancer, tyrosine kinase inhibition (TKI) has been proposed as a therapeutic intervention for PAH (Pitsiou et al. 2014; Dhoble and Patravale 2019), but important questions remain before TKIs become part of the routine armamentarium for this disease (Szulcek et al. 2020; Le Vely et al. 2022). Sometimes genetic alterations are even completely absent from cancer cells, and in these cases, it has been thought that the malignant process may be driven by epigenetic changes. As an example, a comprehensive genomic analysis of ependymoma brain tumors revealed the presence of three distinct subtypes: the first subtype is driven by an intrachromosomal translocation; the second is driven by aberrant epigenetic modifications; and *the third subtype shows neither gene mutations nor epigenetic alterations* (Parker et al. 2014; Mack et al. 2014). Importantly, as described previously (Witt et al. 2011), chromosomal gains or losses are present in all three subtypes. It appears that the inception of the cancer phenotype may occur through different mechanisms related to genetic or epigenetic alterations, chromosomal aneuploidies, or a combination of the three. Also, mechanistically, at least in some cases, the metastatic process seems to be uncoupled from tumorigenesis, and epithelial cells can disseminate through the circulation, seed, and differentiate in response to distal local environments, *in the absence of oncogenic driver mutations* (Rahrmann et al. 2022).

At its core, cancer is not solely defined by genetic or epigenetic abnormalities although these remain crucial therapeutic targets but also by distinct phenotypes and dysregulated biological processes, both of which present additional opportunities for intervention. Precision oncology, which traditionally focuses on molecular alterations and pathogenic interactions between cancer cells and the surrounding stroma, can be meaningfully expanded to incorporate phenotype-based and process-oriented therapeutic strategies beyond genetic mutations equal the cancer paradigm.

This expanded view entails three key conceptual shifts: (a) integrating the broader dimensions of the organism, along with its cellular and local environments, into cancer models; (b) addressing not only genetic and epigenetic changes but also the emergent phenotypes and dynamic processes driving malignancy; and (c) moving beyond the conventional disease-centric paradigm to consider *cancer as a novel, adaptive life form*—one that arises from, coevolves with, and continuously influences the host organism. These questions naturally prompt a deeper, foundational question: "What is life?"

While this inquiry might appear overly philosophical or complex at first glance, its relevance becomes clear when one acknowledges cancer as a life form in its own right—a novel form of life that disrupts and engages with the life of the host. Exploring the essence of life is thus not a theoretical detour but a necessary step toward a unified understanding of cancer. Such an approach holds transformative potential, reframing oncology's goal from winning a war against malignant cells to a broader project to restore the coherence and integrity of life across all levels of the organism.

2.7 What Is Life Exactly?

Defining the nature of life is not easy. There have been literally hundreds of life definitions, none of which are completely satisfactory (Popa 2004; Trifonov 2011; Mix 2015). As Carlo Mariscal and W Ford Doolittle noticed, "For any set of defining traits, one can either show obviously living cases that are left out of the definition or unexpected nonliving cases included in the definition. Life is organized. So are geological formations. Life processes energy, so does fire. Life evolves, so do star systems. Life is at thermodynamic disequilibrium, so is a car. Life is made of complex biochemistry, so are prions. Life exhibits functional complexity, so does a clock. Living creatures embody contained, functionally integrated programs, but so do robots. Life reproduces. Mules don't, bubbles do. Life is self-sustaining. Parasites are not, the carbon cycle is." (Mariscal and Doolittle 2018).

Life on Earth can be defined as the happy marriage between abiotic and biotic components forming the fourth geosphere (Smith and Morowitz 2016). Arguably, the most important symbiosis in the history of life occurred more than four billion years ago, when abiotic and biotic processes were confined together by membranes within the same space and became interconnected. Their coexistence within the proto-cellular space and their codependence led to the formation of the first cell, the last universal common ancestor (LUCA), which is considered to represent the origin of all life on Earth. In metabolism first, the life-origin theory, the *primum movens* of life may have been an abiotic process represented by cyclic autocatalytic chemical reactions (Kauffman 1971; Xavier et al. 2020a, b) that lead to the formation of the first metabolism (Trefil et al. 2009; Xavier et al. 2020a, b) and, subsequently or in parallel, to the constitution of DNA-RNA-protein coding reactions, which may have been the first purely biotic process. It is yet unclear whether the coding reactions are derived directly from metabolic reactions or if both are the outcome of an as yet uncharacterized process (Freeland 2022). Importantly, in all organisms, abiotic and biotic components are *seamlessly integrated*. Perturbations in abiotic components or parameters significantly impact biological functions and behavior and vice versa. As an example, hypoxia, acidic pH and mechanical pressure are related to cellular vitality and behavior, and in malignant tissues, they are associated with increases in the cellular division rate, metabolic switches, and metastasis. These correlations represent opportunities to control cancer behavior by targeting abiotic components in cancer cells and tissues.

Given its remarkable persistence throughout Earth's history, life can be conceptualized as a *quasi-perpetuum* mobile—a self-sustaining process that continuously harnesses energy, guided by bioinformational control, within membrane-bound or otherwise compartmentalized, spatially organized systems. Unlike a true perpetuum mobile, which is thermodynamically impossible, the living state is up and running due to its thermodynamic openness, requiring constant exchange of matter, energy, and information with the external environment to sustain itself. To exist and maintain structure, living systems must import and transform external inputs into organized biostructures. Fundamentally, all living organisms, regardless of their activity level, depend on energy to persist. Even in states of reduced metabolic activity, such as spore germination, hibernation, or cancer cell dormancy, life continues to require minimal but essential inputs of matter and energy, underscoring the continuous yet variable demand for environmental exchange across all forms of life.

Living forms appear and disappear, but the living process persists. The dynamics of the living process may vary with time and may be different in different classes of organisms, but their fundamental properties are the same. Life follows a *homeologic* (similar to itself) dynamic and not a *tautologic* (identical to itself) dynamic. In order to perpetuate their living state, all organisms need to maintain the union of three key processes (bioenergy transformation, bioinformation computing, and biostructure recycling), and continuously adjust these processes to environmental challenges. Regardless of the particularities of these three processes in different living systems (bacteria, for example, are able to process energy without mitochondria; some bacteria like *Mycoplasma*, do not have a cell wall but maintain a cytoplasmic membrane, etc.), this triad is always present, and the interconnectivity of bioinformation/bioenergy/biostructure makes life possible.

The dynamic and evolving unity of bioinformation, bioenergy, and biostructure surrounded by a membrane is able to control the exchange of energy and information with the outside environment, and the continuous interaction with this precise environment represents the key component of life's secret recipe. The life process is autopoietic, i.e., copies and continuously creates/recreates itself, and is sustained by multiple feedback mechanisms that have been under continuing selective pressure through evolution. Life forms and the environment coevolve together. Therefore, all three, i.e., bioinformation, biostructure and bioenergy, in conjunction with the environment and several mechanisms maintaining their union (i.e., autopoiesis and homeostasis), are essential; none of them are primordial in relationship to the other, each of them being a primus inter pares (Otto 2022).

2.8 What Is a Cell?

The National Cancer Institute defines the cell as "the smallest unit that can live on its own and that makes up all living organisms and the tissues of the body" (Wright and Poulsom

2012). Yet, this definition, while technically sound, barely touches the mystery and dynamism of what a cell truly is. A cell is not merely the smallest unit of life; it is the living process unfolding, sculpting, and *singing itself into form*. From a single cell arises the intricate choreography of the human body, just as cancer, too, is born of a cell gone awry.

Two centuries ago, François-Vincent Raspail coined the enduring principle, *omnis cellula e cellula*—every cell arises from another cell (Harris 1999), an aphorism subsequently popularized by Robert Virchow (Wright and Poulsom 2012). Raspail also provided one of the first definitions of a cell as being "a type of laboratory [...] that organize themselves and develop within its innermost substance; its imperforate walls, to judge from our strongest magnifying instruments, have the property of drawing out by aspiration from the ambient liquid the elements necessary for its elaboration. They thus have the property of acting as a sorter, of admitting certain substances and preventing the passage of others, and consequently of separating the elements of certain combinations in order to admit only a portion of them" (Klein 1975).

After decades of gene-centric dogma following the discovery of DNA's double helix, attention is once again returning to the cell—not as a passive executor of genetic scripts but as the true architect of life. Alfonso Martínez Arias, in *The Master Builder*, reframes the hierarchy of biology by asserting that it is the cell, not the DNA, that orchestrates development (Martinez Arias 2023). In his opinion, the cell is the builder, the designer, and the master engineer, while genes represent just toolboxes. Echoing this shift, Siddhartha Mukherjee, who once chronicled cancer through its genetic mutations in *The Emperor of All Maladies* (Mukherjee 2011) and *Genes* (Mukherjee 2016), also changed his narrative and focused on the cell in his most recent book *The Song of the Cell* (Mukherjee 2022).

All three, Raspail, Arias, and Mukherjee, emphasize the cell as an entity, a "thing" that sorts, builds, and "sings." Their ontology, although richly evocative, remains rooted in an object-based worldview. Can we change this perspective entirely and move to a *processual* understanding of life? In such a view, *the cell is not an object that builds, but the very act of building*. It does not sing, *it is the song*. The cell is not a static node in a network but a flow, a happening, rhythm of emergence, and dissolution. It is not merely alive, it is *aliveness itself*.

To embrace this perspective is to let go of the view that life is made of things. Instead, we can see life as a recursive dynamic, continuous process. The cell, then, is understood as a verb, not a noun. It is not the container of action but the action itself.

In oncology, this shift in perspective could be transformative. Cancer is not simply a cluster of abnormal cells; it is a process gone awry, a distortion of life's rhythm. Understanding the cancer cell as a pattern of becoming, not merely as a unit of being, may allow us to conceive new strategies for intervention—not by solely targeting static molecules but also by retuning the processes of cellular communication, differentiation, and coherence.

2.9 What Is an Organism?

An organism can be defined in at least three ways: an individual (a society of cells), a whole with different parts (a holobiome), or the material embodiment of an individual.

Beginning in the latter half of the twentieth century, a biophysical paradigm grounded in self-organization and dynamic systems theory enabled a deeper exploration of the quantitative and qualitative dimensions of the organized complexity inherent to living systems. This integrative framework, reinforced by advances in experimental techniques, revealed that while feedback mechanisms governing steady states can confer stability, such homeostasis represents only one of several operational modes—and perhaps not the main one.

More than three decades ago, in an attempt to develop a generalized mechanics of biology, the term *homeodynamics* was introduced by Francis Eugene Yates in order to better describe any "fluid-plastic-elastic complex system, from molecular ensembles and networks to galactic clusters, in which bulk/shear viscosity ratios (X/,u) are large, with the consequence that interactions among parts at any level lead to internalization of action (energy z time) with variable and often long process delays between external inputs and outputs" (Yates 1994). Subsequently, Lloyd et al. used the term *homeodynamics* as a more comprehensive and dynamic term for biological regulation than homeostasis (Lloyd et al. 2001). The authors noted that rather than striving for static equilibrium, living systems continuously reorganize themselves near critical thresholds—bifurcation points—where stability is transiently lost and novel functional states can emerge. These transitions give rise to diverse and often nonlinear behaviors, including bistability, threshold effects, oscillations, waves, gradients, mutual entrainment, and even chaotic dynamics. Such emergent properties are observable in cellular processes like supramolecular organization, metabolic flux coordination, and signal transduction. Crucially, these behaviors unfold across a wide range of spatial and temporal scales, from submolecular interactions in membranes to long-term evolutionary adaptations. Similar to Lloyd et al., I use *homeodynamics* as a general concept that refers to different control mechanisms responsible for the capacity for dynamic self-organization, underlying the remarkably structured complexity and adaptability of life (Chap. 8).

From the perspective of *homeodynamics*, an organism can be understood as a multiscale, self-organizing, and adaptive system that sustains its functional integrity and identity through ongoing dynamic interactions among its components and the continuous modulation of internal states across temporal and spatial scales. The organism maintains coherence not by static equilibrium but through *homeodynamics*—a flexible, nonlinear integration of bioenergetic, biochemical, informational, and structural processes. This dynamic regulation enables the system to adapt to internal perturbations and external environmental fluctuations while preserving its overall integrity and functional unity.

At its core, a *multicellular organism* is characterized by the coordinated activity of multiple interacting subsystems that span the molecular, cellular, tissue, organ, and organismal levels. Unlike machines, organisms do not merely operate through fixed hierarchical instructions but through *relational integration*, where feedback, feedforward, and cross-scale signaling ensure adaptability and resilience. Communication between components—via cytokines, metabolites, extracellular vesicles, and neuroendocrine circuits—enables distributed yet coherent regulation across spatial and temporal domains.

While some living systems display partial autonomy (e.g., cancer subsystems), true organisms maintain a stable identity through internally coordinated self-regulation—the hallmark of homeodynamic function.

This homeodynamic nature implies that organisms are open, nested entities linked internally by recursive loops and externally by exchange with their environment. Energy and information cascade across levels, from quantum biological reactions (such as photosynthesis) to circadian rhythms and immune-endocrine regulation. As articulated by thinkers like Nicolas Rashevsky and James Miller, the *unity of biological systems* lies in their *structural-functional recursion*: similar principles of organization reappear at different scales, forming *relational fractals* of process and function (Rashevsky 1960; Miller 1978). This functional fractal organization, spanning through several biological scales, represents a fundamental property of living systems.

Cancer, in contrast, may be viewed using Miller's classification as a *parti-potential living system*—a parasitic or symbiotic subsystem embedded within the host's homeodynamic whole. It lacks full autonomy, draws on host resources, coopts communication networks, and disrupts the organism's systemic coherence. From this vantage, cancer is not merely a cellular aberration but a *breakdown in the organism's scale-linking dynamics*, reflecting a loss of coordinated homeodynamics across levels of biological organization. A suggestive metaphor for cancer is that of a malignant embryo produced and nurtured by the organism itself, which sabotages and harms its own creator.

In sum, an organism is best understood as a functionally integrated, scale-linking network that sustains itself through continuous adaptation, recursive communication, and systemic coherence. Health, within this model, is not a property of isolated parts but rather the emergent consequence of the relational balance and the dynamic order across the entire living organism.

2.10 The Elephant Vanishes

Creating more reductionistic theories of cancer will not lead us anywhere. Cancer is not a static object to be isolated and studied in parts; it is a temporal process that is deeply entangled with the organism in which it emerges. Time, integral to this process, is not a substance to be grasped—*time cannot be touched*—and it cannot be disentangled from the biological system that enacts it.

The aggressiveness of cancer can stem not only from intrinsic features of the cancer cells—some malignancies seem inherently "primed" for high virulence (some cancers are "born to be bad") but also from contextual factors such as the surrounding stroma and systemic properties of the host organism. Consequently, in addition to targeting tumor cells directly, it becomes imperative to modulate cancer-host interfaces, for example, by attenuating cancer-induced inflammation or thrombosis. Historically, chemotherapy regimens have been modeled on the infectious disease paradigm, treating cancer as a pathogenic invader to be eliminated. However, pathogenicity in microbes is context dependent. As seen in *E. coli*, a commensal strain can adopt pathogenic traits through the acquisition of pathoadaptive mutations, driven by sustained immune pressures like macrophage phagocytosis (Proença et al. 2017). Similarly, the concept of immunoediting posits that the immune system not only combats but also sculpts malignancy (Vesely and Schreiber 2013). The immunoediting framework includes three sequential stages: (a) elimination—immune-mediated clearance of tumor cells; (b) equilibrium—a period of selection in which resistant tumor clones survive; and (c) escape—whereby resistant cancer cells proliferate unrestrained (Dunn et al. 2002). These dynamics influence both the emergence and trajectory of cancer, underscoring the organismal context of malignancy.

Life begins at the molecular scale, emerging through stochastic events such as Brownian motion, wherein random molecular collisions drive the dynamic behavior of particles suspended in fluid. Remarkably, certain cells harbor intracellular vortices that stir the cytoplasm into turbulent, ocean-like flows (Dutta et al. 2023). Also, a decade ago, a subtler form of molecular movement—nanomotion—was detected within living cells. Unlike Brownian motion, nanomotion reflects active metabolic processes. This phenomenon is now being explored as a powerful diagnostic marker, enabling rapid bacterial susceptibility testing and offering a potential biosignature for life in extreme or extraterrestrial environments (Kasas et al. 2015; Venturelli et al. 2021; Walther-Antonio and Schulze-Makuch 2023). Because nanomotion directly signals metabolic activity (Pleskova et al. 2023), its detection may also serve as a fast, sensitive screening method for assessing the cytotoxicity of anticancer compounds, identifying those that selectively target cancer cells while sparing normal tissue (Stupar et al. 2021).

Philosophers such as Margarida Hermida and, more recently, Carlo Mariscal and W Ford Doolittle have suggested that life on Earth—from its origins over four billion years ago to the present—constitutes a singular, continuous process called *life* (Hermida 2016; Mariscal and Doolittle 2018). In Hermida's provocative framing, *life* is not a class or a phenomenon but an *individual*. This interpretation shifts the emphasis away from discrete organisms and species and toward the temporal continuity of living processes. Moreover, the emerging view of organisms as multispecies collectives-holobionts further supports Hermida's thesis (Gilbert et al. 2012). Life, seen through this lens, is less about bounded forms and more about enduring, evolving dynamics.

Recognizing that every cancer is a unique, temporally embedded process, therapeutic strategies must address more than isolated genomic mutations or neoantigens. They must consider the specific features of the cancer process itself, as well as concurrent pathological conditions that might exacerbate malignancy. Patient age and comorbidities, quantified by tools such as the Charlson Comorbidity Index (Charlson et al. 1987), can profoundly influence clinical outcomes—sometimes even more than the cancer itself (Gould et al. 2017). Thus, a truly effective cancer therapy must be as integrative, individualized, and temporally aware as the process it aims to modulate.

2.11 Temporal Configurations

Unlike in relativistic physics, where time is embedded as a dimension within a four-dimensional space-time continuum, in biology, time plays a key separate role from space. The living process self-generates and self-maintains itself, and, in order to do that, it literally creates its own temporal domain. In contrast to the static temporal coordinates of the inanimate world, biological systems instantiate a distinct, intrinsic temporality. Organisms are not merely entities in time; they *are* temporal configurations. It is important to note that while all organisms are biological systems, not all biological systems qualify as organisms—ecosystems and the biosphere are biological wholes that function without being unitary organisms.

Crucially, temporal structures in biology exhibit both vertical and horizontal organization. Vertically, the human organism functions as a dynamic temporal hierarchy (Baedke and McManus 2018), with processes unfolding across nested timescales—from nanosecond-scale light perception to the multidecade span of human life. Horizontally, biological time is strategically fragmented and modulated. *In living systems, the passage of time is slowing down.* Metabolic pathways, for example, operate through stepwise progression that introduces purposeful delays (Ho 1998). Developmental processes unfold in precise sequences, with each stage following the other with *algorithmic* fidelity, reflecting a noncommutative structure of time (Letsou and Cai 2016). Also, living processes are rhythmic, occurring in loops called *biorhythms*. Biorhythms—circadian, ultradian, and infradian—serve as integrative cycles that synchronize cellular activities across tissues and organs. Biorhythms act as temporal adhesives, ensuring systemic coherence. Distributed pacemakers—from the suprachiasmatic nucleus to Cajal cells in the gut and Purkinje fibers in the heart—govern these rhythms. Furthermore, life processes are embedded within *life cycles* that adaptively coevolve with environmental constraints. Biological time is also characterized by properties such as *simultaneity*, *duration* (completed distorted in cancer) and a sort of, *actualized historicity*, and the ability of an organism to memorize past events and use solutions from the past to survive present stressors. Last but not the least, two key qualities of time, the *flow* and the *present*, are key qualities of biological time.

Within organisms, events do not progress uniformly or metronomically. Temporal dynamics are under active regulation. For instance, cell growth may accelerate, decelerate (as in dormancy), or abruptly change trajectory through events like fusion or meiosis. Cancer, arguably, presents the most perplexing disruption of this regulatory machinery. What fuels the relentless ticking of the cancer clock? For a tumor to grow beyond the diffusion-limited size of 1–2 mm^3, at least four conditions must be met: (1) suppression of negative regulators of the cell cycle (e.g., telomere erosion evasion), (2) activation of proliferative signals via oncogenic drivers (e.g., KRAS), (3) inactivation of genome stability guardians (e.g., TP53 loss), and (4) initiation of angiogenesis (Saman et al. 2020). Recent experimental work has shown that the malignant phenotype can be reversed by silencing just three master regulatory genes, *MYB*, *HDAC2*, and *FOXA2*, highlighting a potential route for reprogramming cancer (Gong et al. 2025). Whether such interventions will translate into viable clinical strategies remains to be seen.

Yet, cancer's unregulated proliferation is only one side of the coin. The microenvironment and systemic organismal context also play active roles in sustaining malignancy. Cancer does not evolve in isolation: it recruits and reprograms its surroundings to support its growth. Traditional models—the somatic mutation theory (SMT) and the tissue organization field theory (TOFT)—fail to capture this bidirectionality. SMT views cancer as a cell-autonomous consequence of accumulating mutations. TOFT, in contrast, posits that proliferation and motility are the default states of all cells and that cancer arises when regulatory constraints from the stroma are lifted (Sonnenschein and Soto 2016). However, both models overlook the reciprocal influence of cancer on its micro- and macroenvironments. Cancer is not simply the unleashing of dormant proliferative potential, nor merely development run amok. It initiates stromal reprogramming and systemic alterations that create a permissive landscape for its own expansion. The role of tumor-secreted factors in modifying the organismal milieu has been well documented (McAllister and Weinberg 2014), emphasizing that malignancy is a coconstructed state involving both the tumor and its host (McAllister and Weinberg 2010; Alečković et al. 2019).

Normal cells in a multicellular organism are harmoniously interconnected, and thousands of simultaneous events are finely tuned in a coherent and harmonious flow, resulting in a kind of global organismal concert (Noble, *The Music of Life*). Ultimately, cancer is not only a spatial pathology—defined by rogue genes, aberrant cells, and disordered tissues—but also a *temporal disease, following its own rhythms, out of tune with the rest of the organism* (Verschuren 2021). Disruptions in circadian regulation, cell cycle timing, developmental trajectories, and evolutionary adaptation all converge to reveal malignancy as a breakdown in temporal coherence. A time-sensitive view allows oncology to move beyond static, spatial models and toward dynamic, process-oriented frameworks. Emerging strategies, such as chronotherapeutics, circadian gene modulation, and temporally informed biomarkers, offer the promise of reestablishing synchrony and combating cancer not only in space but also in time itself.

2.12 Forms and Processes

Both quantum physics and the life sciences operate within frameworks of complementarity—wave and particle duality in the former and form and process in the latter. Notably, as in quantum theory, the primacy in biological systems belongs not to static entities but to dynamic processes (Heylighen and Busseniers 2023). The existence of different types of programmed death, i.e., *apoptosis*, which supports life, points to the preeminence in living systems of process versus form. In this context, biocausality emerges as the key mediator between biological form, structure, and syntactic information on one side and process, function, and semantic information on the other. Through the embedding of syntactic content within causal networks, living systems are able to generate meaning, anticipate future states, and carry out intentional actions. The twentieth century focus of molecular biology on concrete components—DNA and proteins, cells—reflected a predominantly object-oriented paradigm. Yet, contemporary findings increasingly illuminate that these so-called entities are inseparable from the intricate, time-dependent processes in which they are entangled. According to process ontology, reality is best understood as being composed of interwoven processes rather than discrete substances (Nicholson and Dupré 2018). In my opinion, a more rigorous characterization of biocausality will lead to a better understanding of the complementarity between forms and processes, an interplay that is central to deciphering the logic of life.

Across the biosphere, from bacteria to whales, the living process switches from process to form: in bacteria, life is pure transience and process; in a whale, it becomes a macroscopic form that changes slowly over time. Life sculpts time, modulating its flow, stretching or compressing it, and making time its playground.

Even the most seemingly immutable structures, from stars to stones, are in fact expressions of slow transformations—forms—as stabilized patterns of action. At its core, life is kinetic, defined by recursive, layered motions. Cells are never static objects; their identity lies in perpetual motion, internally, and in interaction with their environment. Yet, microscopy and medical imaging fix them into deceptive stasis. These captured forms are but shadows of unseen dynamics. All natural systems—abiotic and biotic—are in continuous flux. We are habituated to calling dead cells under a light microscope "cells," although in truth, we observe remnants—lifeless shapes that once pulsed with motion. Living cells dance, whirl and swirl, move and change shape continuously. Notably, coherent angular motion has been documented in normal mammary epithelial cells, in stark contrast to the erratic behavior of breast cancer cells (Tanner et al. 2012). Darwin's emphasis on the organism may reflect the observational constraints of his time: macroscopic organisms yield to the eye, while microscopic processes, such as bacterial division, reveal life as a temporal flow rather than a discrete form.

Nearly a century ago, Alfred North Whitehead urged a radical shift: "there is urgency in coming to see the world as a web of interrelated processes" (cited by Dupré 1998). Yet, his process ontology (Whitehead 1929), outlined in *Process and Reality*, while influential in philosophical circles, remains absent from most biomedical curricula. The transient nature of organisms stands in contrast to the durability of life as process—a continuity that has persisted across more than four billion years of Earth's history. Organisms are ephemeral—they age, deteriorate, and perish—while the process of life endures. *Apoptosis* exemplifies a sacrifice in service of the greater continuity of the living system. *Life is fundamentally a verb, not a noun*. Philosophers of biology have argued persuasively that core concepts such as autopoiesis and biological autonomy become intelligible only within a process-based ontological framework (Meincke 2018). As

Daniel Nicholson and John Dupré assert, the most coherent way to understand living systems—at every scale—is as a nested hierarchy of processes, not as an assembly of objects (Nicholson and Dupré 2018). Cancer, often conceptualized as an entity, or a cellular ensemble, can also be reframed as a process—emergent, relational, and deeply entangled with the organism from which it arises.

2.13 Biocognition

An intriguing and underexplored feature of biological time is the phenomenon of the *present*. In living systems, *presence* is not merely an epiphenomenon of cognition but is also a core adaptive mechanism honed by evolution. *The capacity to inhabit the present* enables organisms to perceive coherently, make rapid decisions, and execute actions that ensure survival. The construction of a functional present—through tightly coupled signal-response dynamics—permits organisms to engage with temporality in a meaningful and actionable way. Far from a metaphysical abstraction, the biological present has a quantifiable duration—a measurable "thickness"—defined by the minimal time required for environmental signals to be detected, processed, and acted upon. The present, in this view, is a creative synthesis that is evolutionarily sculpted to maximize responsiveness and optimize survival under changing conditions.

This discussion naturally leads to the concept of *biocognition*—the capacity of living systems to perceive, interpret, and react to information in *real time*. *Biocognition* encompasses sentience, the ability to sense and feel, and presence, the power to instantiate *the now*. These capabilities are unique to living systems and are foundational for their survival. To persist, organisms must maintain a continuous state of alertness to changes in both internal and external environments and must immediately adapt their physiological states accordingly. *Biocognition* facilitates tight integration between the organism and the environment. In complex organisms like humans, the brain, our principal cognitive organ, performs this function by processing signals arriving from the body through the autonomic nervous system and from the external world via sensory inputs relayed to central processing hubs.

At the cellular level, *biocognition* is mediated through the interaction of three fundamental components: (a) biosensors capable of detecting stimuli; (b) integrative processing networks that decode diverse informational streams; and (c) effector mechanisms that adapt cellular behavior in response. Cells are not passive recipients of external input; they are active agents within a feedback-rich, predictive system. The perception-action loop, as proposed in active inference theory (Friston 2010), posits that organisms modulate their sensory experience through movement and interaction, effectively transforming perception into a hypothesis-testing function within a closed causal loop.

Within the human brain, cognition is processed through networks of varying complexity: the archicortex (three-layered), paleocortex (three to five layers), and neocortex (six-layered), each associated with distinct forms of information processing. Analogously, cellular cognition involves back-and-forth bidirectional information flowing between the membrane and the nucleus. The architecture and complexity of cellular membranes vary across evolutionary levels, from prokaryotic cells to unicellular and multicellular eukaryotes. Likewise, the number and intricacy of transcriptional regulators, signaling proteins, noncoding RNAs, and genomic response pathways differ dramatically between bacteria and eukaryotic cells. One may speculate that, similar to cerebral architecture, cells exhibit a sort of cellular *cognitive network architecture*—hypothetical "archi-," "paleo," and "neo"—cellular cortices. The basic "cognitive layers", i.e., the hypothetical "archi" cortex and hypothetical "paleo" cortex, may be specific to prokaryotes and unicellular eukaryotes, respectively, and only the hypothetical "neo" cortex layer may be specific to cells of multicellular eukaryotes (Gaddam and Ogas 2023), which are uniquely able to respond flexibly with high sophistication to complex signals. Also, as it has been suggested by Stephen Grossberg in his magnum opus (Grossberg 2021), similar to brains, cells may have integrated short-, medium-, and long-term *memory mechanisms* related to different patterns of intracellular calcium waves (Trewavas 1999).

The experience of the *present moment*, sometimes termed the *specious present*, is central to both perception and biological action. Recent hypotheses suggest that cell membranes may function as active components in cellular biocognitive processing (Baluška et al. 2021), contributing to the organism's capacity to perceive and act in real time.

Cancer cells, derived from eukaryotic lineages that are able to engender multicellular organisms, retain and, even possibly, refine the *biocognitive* abilities of eukaryotes. They exhibit adaptive behaviors that allow rapid sensing and integration of environmental threats, including therapeutic assaults. These behaviors are coordinated by intracellular signaling networks with varying architectures and processing speeds. Importantly, beyond individual cellular cognition, within the tumor stroma, cancer cells can form multicellular communication networks, enabling emergent group intelligence. Also, within the cancer system established between the primary tumor and distal metastases, a distributive type of cognition exists akin to the distributive cognition of the immune system. The capacity of cancer cells and groups of cells to toggle between unicellular and multicellular states grants tumors with a flexible mode of existence, supporting the emergence of collective, adaptive behaviors (Nicolazzo et al. 2023). Tumors may exhibit heterogeneous compositions, harboring cell types with diverse specialization levels (Hausser and Alon 2020), contributing to a collective system that resembles a primitive social organization. These intercellular networks may enhance the *bio-cognitive* capabilities of tumors compared to solitary cells, representing a novel form of evolutionary adaptation (Levin 2019). The cognitive-like flexibility of cancer—its ability to sense, learn, and adapt—hinges on its manipulation of temporal structure, particularly its capacity to operate within a distributed, *specious present*. Disrupting this temporal coherence may represent a new frontier in therapeutic intervention.

The same principles underlying Stephen Grossberg's category learning, expectation, attention, resonance, and synchrony (CLEARS) framework originally developed to elucidate how neural networks in the brain process information, learn, and adapt (Grossberg 2017, 2020, 2021), may be conceptually extended to understand the adaptive behaviors and communication strategies of cancer cells within primary tumors *and* between primary tumors and metastases.

Cancer cells exhibit remarkable plasticity, enabling them to adapt to diverse microenvironmental conditions. This adaptability mirrors the *category learning* aspect of CLEARS, where cells classify environmental cues to make survival decisions. For instance, tumor cells can undergo epithelial-to-mesenchymal transition (EMT), altering their phenotype in response to stressors such as hypoxia or chemotherapy, leading to invasion and metastasis. At the same time, primary tumors release exosomes and cytokines that modulate distant microenvironments, making them more receptive to incoming tumor cells. *Resonance,* i.e., the amplification of signals through feedback loops, is present in both autocrine and paracrine signaling within tumors, where cancer cells and stromal components mutually reinforce protumorigenic signals and within the interaction between tumor-derived factors and the target organ's microenvironment, which facilitates metastatic colonization. These processes may be both reactive and anticipatory akin to the *expectation* component, allowing cancer cells sometimes to preemptively activate survival pathways.

The *attention* mechanism in CLEARS, which prioritizes certain stimuli over others, can be likened to how cancer cells focus on specific signaling pathways to promote growth and survival. For example, cancer-associated fibroblasts (CAFs) in the tumor microenvironment secrete factors that selectively enhance tumor cell proliferation and angiogenesis (Maia et al. 2018).

Finally, *synchrony* in tumors manifests as the collective behavior of cancer cells, such as synchronized proliferation and migration. Recently, a research group from Heidelberg coordinated by Frank Winkler described pacemaker-like autonomous rhythmic activity in glioma networks driving brain tumor growth (Hausmann et al. 2023). Also, circulating tumor cell (CTC) clusters exhibit synchronized behavior through intercellular connections, allowing them to resist shear stress in circulation and evade immune detection. Upon reaching a distant site, their synchronized activity aids in successful extravasation and colonization. Exosomes and other extracellular vesicles facilitate this coordination by transferring oncogenic signals between cells, promoting uniform responses to environmental stimuli.

Understanding cancer through the lens of the CLEARS framework highlights potential therapeutic targets. Disrupting the *resonance* and *synchrony* among cancer cells by targeting intercellular communication (Su et al. 2024), exosome release pathways (Irep et al. 2024), premetastatic niche formation (Pătraş et al. 2023a, b; Wang et al. 2024a), or, specifically, the KCa3.1 channel responsible for pacemaker activity in glioblastoma networks (Hausmann et al. 2022) could impair their collective synergy, disrupting the cancer-induced pathologic networks (CISPN) present both at the level of the cancerized stroma and at the level of the entire organism (Paul 2020; Paul and Komarova 2021). Moreover, targeting the mechanisms underlying *Category Learning* and *Expectation* by inhibiting the EMT process may prevent cancer cells from effectively responding to therapeutic pressures (Davis et al. 2014; Antony et al. 2019; Zhong and Sun 2023 Vol I and Vol II).

Finally, whether related or not to cellular cognition, cytosolic calcium (Ca^{2+}) may represent one of the most important, if not the most important cell signal in biology (Bruce and James, 2020), as it is involved in essential cellular process important for both cellular life and programmed death. Cytosolic Ca^{2+} and the components of the Ca^{2+} signaling machinery are involved in all key cancer hallmarks, including cell proliferation, invasion, and migration; resistance to apoptosis; and tumor vascularization and metastasis. The reason why no cancer therapies that interfere with intracellular Ca^{2+} pathways approved at present is related to the ubiquitous nature of many Ca^{2+} channels and transporters and their pleiotropic involvement in cellular functions and the high probability of side effects related to their targeting. The search for Ca^{2+} channels, regulatory proteins, or processes Ca^{2+} channels uniquely expressed in cancer continues (Bruce and James 2020).

2.14 Cancer as Dyschronia

"There can't be a single thing which is not a cloud," Jorge Luis Borges famously wrote, evoking the amorphous, shifting nature of reality as it appears in the mind (Borges 2000). The key paradigm shift that this book is advocating is to integrate the static object-oriented paradigm of precision oncology with a complementary, dynamic process-oriented explanatory frame. Cells and organisms are not static objects;

they are on-going processes; they change continuously like clouds forming and dissolving on a summer sky. *Forms support processes and processes support forms. Understanding their unity is the key to understanding cancer.* Under this paradigm, in order to cure it, we need to be able to change trajectories, modulate processes, and induce transitions not only in cancer cells and tissues but also in normal cells and at the level of the whole organism. Cancer has been increasingly recognized as a disease of cellular plasticity, where malignant cells transition dynamically between phenotypes to adapt, survive, and resist treatment. Among these transitions, the interplay between cancer stem cells (CSCs) and non-stem cancer cells, the epithelial-mesenchymal transition (EMT) and its reverse process (MET), and lineage transitions such as neuroendocrine transdifferentiation are critical to tumor heterogeneity, metastasis, and drug resistance. Other transitions relevant to cancer are a) the mesenchymal-amoeboid transition (MAT) (Panková et al. 2010), in which cancer cells can switch between mesenchymal and amoeboid migration modes to invade tissues; b) the metabolic transition back and forth between glycolysis (Warburg effect) and oxidative phosphorylation supporting cancer survival in fluctuating environments; and c) immune evasion transitions, in which cancer cells can upregulate PD-L1 and other immune checkpoint molecules, facilitating immune escape.

Organisms are finely tuned orchestras playing in unison, in which singers and songs form a unique functional unit. The extraordinary organization of organisms is due to tightly controlled processes that are codependent on their architecture. Bioorganization is the integration of bioprocesses and biostructures. Cancer disrupts this unity and introduces different local and global dynamics. The hypothesis that cancer represents dyschronia and that it can also be controlled through biorhythms manipulation has been present for more than half a century (Gilbert 1968).

The hypothalamic suprachiasmatic nuclei host a master circadian clock that coordinates several physiologic parameters, i.e., core body temperature, sleep and wakefulness rhythms, and hormone secretion (e.g., cortisol and melatonin). The master circadian clock is synchronized to Earth's rotation, primarily by light-dark cycles. Circadian rhythms are also generated at the cellular level by a molecular clock system that involves 15 specific clock genes (Karaboué et al. 2024). At least 40% of protein-coding genes show daily rhythms in expression in a tissue-specific manner, and many of the organism's physiological and metabolic activities, including immune responses and xenobiotic detoxification as well as renal and cardiovascular functions, are coordinated by cellular oscillators (Wittenbrink et al. 2018). Disruption of the cellular clock is increasingly recognized as a significant factor in cancer development and progression (Verlande and Masri 2019; Zhu et al. 2024). The core clock genes exert tumor-promotive functions, depending on the cancer cell status or type. It has been shown, for example, that in the setting of breast cancer, BMAL1 gene overexpression promotes breast cancer cell invasion and metastasis, and in other settings, enhanced levels or activity of circadian clock genes (i.e., CLOCK/BMAL1) inhibit tumor proliferation and growth by promoting the degradation of oncoproteins (i.e., cMYC, E2F, and TLK2) and induce cell cycle arrest and apoptotic cell death (Lee 2021). Significantly, in some cancers, the circadian rhythms of host organisms appear to dictate tumor growth and proliferation (Kinouchi and Sassone-Corsi 2020), and in these cases, restoring the balance of metabolic rhythms with chemical, hormonal and behavioral interventions is a potential strategy for cancer treatment (Su et al. 2025).

Cancer itself represents both local and systemic processes. The structural disorder observed in cancer cells and tissues by pathologists and the supposedly random growth and distribution of tumors in distal organs may just represent the cancer's strategy of increasing entropy as a survival mechanism. Importantly, this entropy increase is *both* internal and external. This observation points to the fact that there are some essential processes inside cancer cells that are important for cancer survival and are still under tight homeostatic control as opposed to other processes that are considered not essential. The presence of a core of conserved processes is a vulnerability that can be explored for cancer treatment.

When cancer is diagnosed, it has already been progressing silently for years or even decades. The eco-evo-devo models developed so far are only partially capturing cancer's complexity that not only is located spatially and temporally at a single scale but also extends through multiple spatiotemporal micro- and macroscales. At present, we have detailed static descriptions of molecular, cellular and tissular alterations and of various metastatic patterns present in cancer, but we lack a comprehensive description of the transitions from a normal cell to a cancer cell that divides uncontrollably, creating clones that invade the local stroma and are able to metastasize to one or several distant tissues, interfering with the microenvironment and the whole organism homeostasis. Time and timing are of essence in cancer. The large temporal gap between cancer inception and clinical symptoms makes prevention and interception strategies in the preclinical stages the most logical and effective ways to curb the cancer epidemic.

Swarm-like collective migration of cancer cells leading to cluster metastases appears to be more efficient compared to single-cell metastasis and is associated with aggressive disease (Cheung and Ewald 2016; Yamamoto et al. 2023).

Under this dynamic perspective, the variety of afflictions and diseases present in an organism can be considered blockages, turbulences, and rhythm distortions. As compared to physics, where time has been either integrated into space in the general relativity model or has even been considered an illusion (Barbour 1999), in biology, time is fundamental. Life persists through time by continuously and actively supporting its own dynamic persistence.

2.15 The Oncosphere

The living process is present both within organisms and across biological systems, i.e., within ecosystems and biospheres (Smith and Morowitz, 2016). The distinction between an organism and a biological system is not as sharp as originally thought (Baedke 2019). Organisms have been recently described as holobionts (Aldana and Robeva 2021; Bordenstein et al. 2024). As Scott Gilbert formulated it suggestively: "If organisms are to be perceived as individuals, we are individual teams. And just like teams, organisms are formed through a mixture of cooperative and competitive processes" (Gilbert 2024). The term "holobiont" was introduced originally by Adolf Meyer-Abich in 1943 (Baedke et al. 2021) and was subsequently used by Lynn Margulis in an article from a book that she coedited in 1991 (Margulis 1961). The "holobiont" used by Margulis describes an integrated composite organism that contains other species besides the eukaryotic host (bacteria, fungi, viruses, etc.). "Holobiont" here is used with the original meaning coined by Meyer-Abich, who defined the "holobiont" as a "specific biont *within* a strongly integrated whole, a holobiome ("holobiom") created through holobiosis, rather than as the whole itself" (Baedke et al. 2021).

As recent research on the microbiota has demonstrated (Sepich-Poore et al. 2021), cancer itself can be considered a holobiont (an onco-holobiont), and the microbiota may contribute to all cancer's hallmarks (Fulbright et al. 2017). It can also be seen as a complex ecosystem (Basanta and Anderson 2013; de Visser and Joyce 2023), and an innovative cancer classification based on ecological and evolutional cancer features has been described (Maley et al. 2017). Both the holobiont and the different ecosystem models of cancer, the swamp model, for example (Amend and Pienta 2015), are related to the microenvironment of the primary tumor. A more comprehensive model than the holobiont model is needed for metastatic cancer.

Taken as a whole, metastatic cancer, with its different ecosystems spread across the organism, is a multilevel system that can also be seen as an *oncosphere*, developed within the *biosphere* of the organism (Pienta et al. 2025), influenced by it, and, in turn, influencing the whole body (Paul 2020). Three components are important in the context of cancer's *oncosphere*.

1. *First*, the different ecosystems. Cancer metastases in different locations may represent very different ecosystems. On the one hand, cancer cells may exhibit different genetic and epigenetic alterations and different metabolisms, and on the other hand, the stroma components of different metastases may be very different.
2. *Second*, the various ecosystems are linked. The union of the human organism is established by its anatomical structure, by the circulatory system and by the neuroendocrine system. This union is also established through a dynamic network of interorgan communication. Body organs exchange cellular, microvesicular, metabolic, or protein signals. Cancer is able to hack into the interorgan communication system, prompting the entire organism to support cancer growth, invasion, and metastasis through an instigation process previously described by Robert Weinberg and his team (McAllister et al. 2008). As pointed out by Micheal Sporn almost three decades ago, "carcinoma is not a disease of an individual cell" (Sporn 1996). Cancer is also establishing its own wireless communication system (Pătraş and Paul 2024) through pathologic cancer networks between primary tumor and metastases (Paul 2020; Lieff 2020). Exosomes have been described as key components of the cancer metastatic network (Hoshino et al. 2015; Wortzel et al. 2019; Dai et al. 2020). To dismantle such a complex multilayered network as cancer, multiscale targeted approaches are needed that simultaneously target key elements of the cellular, tissular and systemic cancer networks. Top-down therapeutic approaches, targeting the tissular or the organismic level, may be less affected by cancer genetic instability and heterogeneity than treatments targeting the cancer cells themselves and represent a "new wave" of cancer treatments (Paul 2020; Paul and Komarova 2021).
3. *Third*, the membrane of the first cell separated the internal and external environments, and this asymmetry is fundamental for a cell's functioning. It created a biological border between the climate of the outside world and the internal climate of the cell. The story of life is also the story of multiple codependent climates: one external, i.e., the Earth climate, and several internal: the *nanoclimates* at the molecular level, the *miniclimates* present inside every cell, and then, in multicellular organisms, the *microclimates* of the different tissues and, finally, at the highest scale, the global internal *macroclimate* of the body's biosphere.

"The internal body climate" amply discussed in Chap. 9 is a fundamental characteristic resulting from spatial and temporal interactions between internal body components themselves and with the external environment. It significantly impacts both the human organism and cancer and their different nested levels, and it contextualizes a novel heuristic tool that can be used both to design cancer prevention interventions and treatment protocols by manipulating or restoring the internal climate of the body and obstructing cancer inception, progression, and metastasis.

The molecular biology research projects developed over the last half a century have focused mostly on the molecular aspects of the cell and, more recently, on the interaction between cells and the stroma, grossly neglecting the organism in which all these processes occur. It is time to zoom out and expand our cancer models and treatment strategies to the whole-body scale and focus not only on cancer itself but also on cancer-organism interactions and the well-being of normal body organs and systems and the organism itself.

Cancer's behavior as a systemic disease is undeniable. Far from being isolated pathologies, tumors communicate extensively with distant organs and systems, reprogramming the host environment to facilitate their growth and spread. Exosomes stand at the center of this interorgan dialogue, shaping immune responses, metabolic states, and even neurological outcomes. Understanding and harnessing this complex web of communication is critical for developing more effective, personalized, and comprehensive cancer therapies that address not only the tumor but also the whole patient.

Reconceptualizing the human organism as a biosphere and cancer as a biological system within it (the *oncosphere*) offers a powerful systems biology paradigm. It integrates ecological principles, evolutionary theory, and complex systems thinking to better understand and treat cancer. Future therapeutic strategies may benefit from targeting not only cancer cells but also the ecosystems that sustain them, reestablishing internal ecological balance within the biosphere of the body. Disentangling the complex interplay between tumor cell-intrinsic, cell-extrinsic, and systemic mediators of cancer cell initiation, promotion and progression is critical for the design of effective anticancer treatments.

In the first quarter of the twenty-first century, the human organism still remains an uncharted territory, and the next frontier in oncology will be an integration of the cellular, tissular and the whole biosphere of the human organism into a broader, multilevel encompassing perspective that includes interactions with cancer's own biosphere. Only through integrative approaches can a cancer cure become a reality.

2.16 Cancer as a Systemic Disease

The idea that cancer occurs at the molecular level has led cancer research to focus mainly on molecular biology trying to solve the root problem of cancer at the molecular level. The relative success of immunotherapy has demonstrated that effective anticancer therapies can be designed, targeting cancer cell interactions at the tissue level. In the near future, we need to take into consideration the whole organism, as causality circuits in an organism are both bottom-up and top-bottom, and novel therapies acting at the whole organism level can be designed (see Fig. 2.2).

This represents a fundamental conceptual leap. Superficially, the body appears to behave like a democratic society of cells in which every cell can communicate freely with every other cell. In fact, the organism is at the same time a horizontal holarchy and a tightly controlled vertical hierarchy. The design of the nervous, endocrine, immune and vascular systems allows both hierarchical control and cell-to-cell communication. Communication between cells allows the state -which is the organism-, through its government-like entities (i.e. organs and body systems), to closely control the interaction and behavior of its members (i.e. cells themselves). The hypothalamus is the central agency that surveys

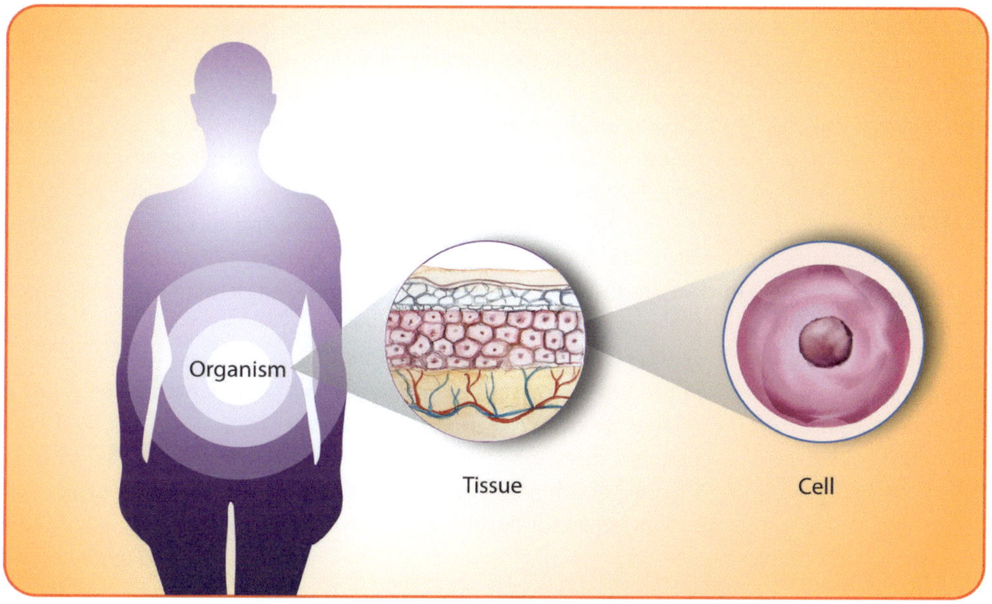

Fig. 2.2 Three levels of organization: cell/tissue/organism. (Courtesy of Ana Tourian)

closely and adjusts the entire body homeostasis. The pancreas secretes insulin, allowing glucose to enter cells only at certain times. The kidneys continuously filter urine, and the heart incessantly contracts and relaxes, keeping the organism alive. Cancer taps into the global communication systems of the organism (vascular and lymphatic, neurohormonal and immune) and manipulates them for its own benefit.

The concept of drivers of the cancer state can be further extended progressively, from the genetic level to the tissue level and then to the level of the whole organism. As an example, non-cancer cells present in stroma-like cancer-induced fibroblasts (Sahai et al. 2020) or macrophages may function as cellular drivers of the oncogenic process (Gargiulo et al. 2024). Also, the *group phenotypical composition (GPC) of the primary tumor* may be *functionally* selected and represent, as a whole, a driver of tumor progression or metastasis (Thomas et al. 2024). Given the complex interactions present between the primary tumor and its various metastases, sometimes *the entire primary tumor represents the driver of the malignant state*, and resecting it may lead to spontaneous regression of distal metastases (Heng et al. 2014; Marquina et al. 2021; Shaashua et al. 2020). Also, it has been shown that age-induced accumulation of methylmalonic acid may lead to a systemic environment that favors tumor progression and increases aggressiveness of cancerous tumors (Gomes et al. 2020).

Recently, in 25% of patients who had both clonal hematopoiesis of indeterminate potential (CHIP) mutations and cancer—6% of them with solid tumors—CHIP was detected inside the tumors—a phenomenon called tumor-infiltrating clonal hematopoiesis (TI-CH) (Pich et al. 2025). Importantly, TI-CH was associated with adverse patient outcomes. The incidence of TI-CH varies across cancer types, being more frequently present in non-small cell lung cancer (NSCLC), head and neck cancer, pancreatic cancer, and mesothelioma. Pich et al. reported a remarkable finding: age-related somatic clonal expansions in one tissue, such as the hematopoietic system, may impact malignant progression in distant tissues, a novel phenomenon that points to an interplay between non-malignant somatic clones, aging, and malignant disease and offers a novel, systemic perspective on cancer biology (Pich et al. 2025).

A profound unity underlies all biologic processes, both in health and in disease. In contrast to the dichotomous world portrayed in the Matrix movie series, in biology, matter, energy, and information are collapsed into a single integrated reality; *a protein, for example, may represent in different contexts a signal, a type of fuel or a cell building block*. Deciphering the interrelations and mechanisms of the control of bioenergy, bioinformation, and biostructure both of cancer and of the whole organism is key for developing more effective cancer treatments.

Integrated biology and integrated oncology are seamlessly related. A holistic understanding of living processes may also clarify some of the mechanisms and trajectories of oncologic conditions and, importantly, may help us design novel and more effective cancer treatment strategies.

2.17 Consistent Patterns of Cancer Progression

"Unlike laws of nature, causal relations between initiating processes of tumor development are not anchored in an invariance of nature. Therefore, molecular and cytogenetic aberrations at initial diagnosis are generally heterogeneous in both tumors and single tumor types. *Invariance within the tumor process may be observed during tumor progression.* In interaction with normal human tissue, tumor cells use processes *according to laws of nature* to build up a favorable infrastructure for proliferation." (Reichle and Vogt 2008).

Albrecht Reichle and Thomas Vogt capture two key features of cancer cells. The *relative invariance of the cancer process* is the first cancer characteristic. Even if, as mentioned priorly, some types of normal cells may share the same hallmarks with cancer cells, normal cells do not grow uncontrollably, do not invade locally and do not metastasize. On the other hand, despite their extreme genetic, epigenetic and metabolic heterogeneity, despite arising from different cells and growing in different local environments, it is striking that cancer cells across individuals and tumor types consistently converge upon a common set of core phenotypic traits, i.e., they divide uncontrollably, invade the local stroma, and spread at a distance.

Cancer cells continue to interact with normal cells and the tissue surrounding them following their own dynamic and instigating them to support the cancer process, *according to the laws of nature*, a second characteristic of cancer cells. *Cancer does not invent anything*. It uses preexistent cellular programs like wound healing, germ cells, embryonal or fetal programs, or even placental programs. The open demethylated cancer genome allows cancer cells to respond to stressors in real time and switch their genetic/epigenetic programs, leading to invasion or metastatic phenotypes. Cancer's phenotypes open new communication channels with the neighboring cells of the microenvironment and with distal organs.

2.18 Cancer as a Stress Response

This "*bottleneck*" behavior is the basis of the TNM (tumor-nodes-metastasis) cancer classification, which is currently used in oncology for both treatment and prognosis. Mark Vincent referred to this convergence as a remarkable instance of evolutionary parallelism (Vincent 2012). As an explanation for this, I suggested that the diversity of mutations and chromosomal rearrangements observed in cancer may serve as triggers for a limited set of conserved cellular programs (Paul 2020), which in turn give rise to these shared traits. Supporting this concept, it has been observed that chromosomal instability (CIN) frequently results in the generation of cytosolic double-stranded DNA (dsDNA), which activates the cGAS-cGAMP-STING signaling pathway (Kwon et al. 2008). Notably, *this activation occurs independently of the specific DNA sequences present in the cytosol*. To adapt, tumor cells suppress downstream type I interferon responses while maintaining NF-κB-dependent signaling pathways that facilitate metastasis (Bakhoum and Landau 2017). Concurrently, cancer cells can evade immune surveillance by degrading extracellular cGAMP, a key immunostimulatory molecule (Li et al. 2021a). Another example is aneuploidy. Dysfunctional short telomeres lead to a telomere crisis that results in transient genomic instability that may lead to aneuploidies and chromothripsis. Chromothripsis is a mutagenic process in which one or a few chromosome segments are affected by genomic rearrangements. Some aneuploidies and chromosome rearrangements may be useful for cancer cell survival, but others may lead to lethal variants. Telomerase activation represents the way out of the telomere crisis, and an increase in telomerase activity stabilizes the genome and reestablishes fitness (Maciejowski and de Lange 2017).

The theory of cancer as the activation of a stress response program was previously proposed by Douglas Brash and Junmei Cairns in 2009. The two authors noticed that confronted by environmental stressful conditions (bacterial, yeast and mammalian cells) *activated a program that raised their mutation rate in the absence of any preexisting mutations* (Brash and Cairns 2009). Also, as proposed by Luo et al., cancer may represent a *universal stress response*, like senescence (Luo et al. 2009). The plasticity of cancer cells and their dedifferentiation (Hanahan, 2022) may be a stress-coping mechanism (Bensellam et al. 2018). All roads of cancer cell behavior lead to stress reactions and the *bottleneck* list of hallmarks may thus be explained by the commonality of the toolkit of stress response mechanisms that lead to a limited variety of phenotypes. The mutator phenotype, for example, leading to widespread genetic mutations, may be favored when the generation of new variants is needed as a cellular survival mechanism. Some new mutations may allow cancer cells to persist in the face of new external or internal stressors.

Recently, Armando Aranda-Anzaldo and Myrna Dent presented evidence that cancer may originate in the epigenetic processes associated with cellular adaptation to stressful conditions, and so it could be a direct consequence of stress-buffering mechanisms that allow cells with aberrant phenotypes (not necessarily associated with genetic mutations) to survive and propagate within the organism (Aranda-Anzaldo and Dent 2021). The authors advanced the hypothesis that there would be an inverse correlation between the activation threshold of the cellular stress responses (CSRs) and the risk of cancer so that species or individuals with low-threshold CSRs would display a higher incidence or risk of cancer. In this context, Thomas Seyfried and colleagues proposed the 'Press–Pulse' therapeutic strategy, which aims to push cancer cells beyond their adaptive limits by combining chronic metabolic stress ('press'), targeting tumor energy metabolism, with intermittent acute stressors ('pulse') that restrict glucose and glutamine availability while inducing cancer-specific oxidative stress (Seyfried et al. 2017).

As a variation to the stress *adaptation* cancer theory, João Carvallho also put forward the cell *reversal* theory (CRT) as a reversal of a cell to a stem cell-like phenotype due to a perturbation (a potential carcinogenic event) in the cell and/or its environment (Carvalho 2020). A decade ago, Friedmann-Morvinski and Verma advanced a theory similar to CRT for the origin of cancer stem cells (CSC), but they considered the CSC phenotype to be a consequence of tumor progression and not at its origin (Friedmann-Morvinski and Verma 2014). In parallel, many other authors advocated the concept of cancer as a reversal to early embryonic programs leading to dedifferentiation (Naxerova et al. 2008; Manzo 2019; Oginuma et al. 2020) or, as briefly mentioned above, to a stem-cell-like phenotype. The common denominator of all these reversal theories may be the plasticity of epigenetic plasticity and increased chromatin accessibility of cancer cells (Corces et al. 2018), particularly at enhancer regions and promoters of proliferation-related genes.

2.19 The Story of Two Life Forms

Rather than representing a mere opportunistic parasitic process, cancer may be conceptualized as a novel life form, one that emerges and is supported by the host organism from its earliest stages—functioning in a manner reminiscent of the placenta during pregnancy (Costanzo et al. 2018; Khorami Sarvestani et al. 2023; Doria-Borrell and Pérez-García 2024). This analogy, central to the long-standing *placentation cancer theory*, first proposed over a century ago, has gained renewed scientific support in recent years (Wagner et al. 2022). Both cancer and placental development share key features, such as the induction of neovascularization and the invasion of cells into the surrounding stroma (Rekowska et al. 2023). Emerging research has highlighted the striking molecular and cellular parallels between carcinogenesis and placentation (Khorami Sarvestani et al. 2023; Rekowska et al. 2023). For instance, the tumor suppressor gene product BAP1 (BRCA1-associated protein 1), known to regulate cellular proliferation, apoptosis, metabolic activity, and hypoxia responses, plays pivotal roles in both tumor development and placental function (Doria-Borrell and Pérez-García 2024). Additionally, a shared mechanism of immune evasion underlies both cancer progression and placental protection of the fetus from maternal immune attack, involving dynamic immune-editing processes (Bronchud et al. 2016; Bronchud 2018).

Cancer originates from a normal cell, and the organism itself facilitates its development and dissemination; in this regard, cancer represents an intrinsic condition that is inherently more challenging to treat than exogenous threats like infectious diseases. No cell exists in isolation, and cancer cells are no exception. Both the local tissue environment and the systemic physiological milieu are implicated from the onset of carcinogenesis and continue to play active roles throughout progression and metastasis. Supporting this, a murine study revealed that the *transcriptomes of surrounding normal cells* undergo consistent changes in response to metastatic invasion, while cancer cells themselves exhibit either transient transcriptional regulation or permanent alterations driven by clonal selection (Alzubi et al. 2019). Furthermore, pathological stromal remodeling observed in malignancies such as breast cancer, non-small cell lung cancer, hepatocellular carcinoma, and cholangiocarcinoma, has been shown to arise from aberrant interactions between fibroblasts and macrophages in the tumor microenvironment (Baer et al. 2023). Extracellular vesicles (EVPs) are also increasingly recognized as key mediators of cancer's systemic impact. In pancreatic cancer, for example, tumor-derived EVPs have been shown to induce a proinflammatory microenvironment and promote hepatic steatosis (Wang et al. 2023a). Additionally, distal organs actively contribute to cancer progression: reprogrammed interstitial macrophages in the lungs of cancer patients release pro-thrombotic EVPs, thereby facilitating thrombosis and enhancing pulmonary metastasis (Lucotti et al. 2025).

At the heart of cancer's virulence lies the interaction between two biologically discordant yet coexisting living entities—the host organism and the cancer itself. This bidirectional interplay is responsible for many of cancer's clinical manifestations and helps determine its trajectory and prognosis from the earliest stages. Hallmarks such as inflammation, metabolic dysregulation, and both local and systemic disruptions stem directly from this interaction.

Although cancer cells may fulfill all Hanahan and Weinberg's widely cited hallmarks of malignancy, they remain clinically silent until their activities start disrupting tissue architecture and impairing organ function. Cancer becomes a clinical concern only when it manifests through local and systemic interference with physiological processes. Importantly, it is not always the cancer cells themselves that lead to mortality; rather, morbidity and mortality often arise from inflammation, systemic derangements, and organ dysfunction provoked by the tumor-host interaction. This distinction between subclinical, indolent cancers—such as incidental breast, prostate, or thyroid tumors found during autopsies—and overt, clinically impactful malignancies highlights the necessity of understanding the organism's response as a codeterminant of disease progression and symptomatology. *The contribution of the organism to cancer progression is as important as the effect of cancer itself on the organism.*

Adopting a paradigm that regards cancer and the host as mutually interactive systems, rather than as isolated or antagonistic entities, can help reorient therapeutic strategies toward targeting not only tumor-intrinsic properties but also the broader cancer-organism interface. Such a perspective encourages the development of integrative treatments that address both malignant processes and the systemic context in which they arise. Given this codependence, effective cancer control must involve not only suppressing cancer's adaptive mechanisms but also reinforcing the organism's homeostatic responses and resilience.

2.20 Cancer and Artificial Intelligence

In addition to inducing profound structural, energetic, and temporal alterations across various levels of biological organization, cancer orchestrates the formation of aberrant communication networks both within individual cells and across tissues and the entire organism. It can be viewed as a disorder of intercellular signaling, wherein malignant cells hijack developmental pathways and manipulate both the local microenvironment and distant physiological systems to promote their own survival and proliferation. These pathological communication circuits can be systematically investigated using tools derived from systems biology, including artificial intelligence (AI).

Recently, advanced AI technologies have become instrumental in helping researchers integrate and analyze data from diverse high-throughput platforms, such as genomics, epigenomics, transcriptomics, proteomics, and metabolomics, to enhance diagnostic accuracy, refine prognostic models, and increase their clinical utility. However, a common misconception is that AI performance is driven solely by data volume; in reality, both the quality and quantity of data are critical. Designing effective AI architectures requires not only skilled computer scientists capable of evaluating data fidelity but also access to immense computational resources to process and interpret massive datasets. Deep learning models, including large language models (LLMs) like ChatGPT, or causal machine learning architectures, rely on similar optimization strategies to navigate high-dimensional solution spaces. These nonlinear dynamics may yield unexpected (Longo 2023), or even erroneous outputs—commonly referred to as AI—"hallucinations"—underscoring their unpredictability. Nevertheless, the promise of generative AI lies in its potential to accelerate biomedical discovery through the creation of novel hypotheses, molecular entities, and biological insights.

Biological systems do not operate within either strictly deterministic or entirely stochastic frameworks; rather, they exist along a continuum in which structured mechanisms interact with inherent randomness to shape outcomes to external stressors such as therapeutic response. As our tools for measuring and modeling biology grow more sophisticated, integrating deterministic models with stochastic variability will be essential for refining therapeutic predictions and moving closer to the realization of truly personalized oncology. In this context, AI holds significant potential for identifying new drug targets, designing safer and more effective molecules, uncovering hidden patterns in multi-omics data, and predicting protein structures, which are key to understanding functions and informing drug development. Additionally, AI can simulate intricate biological phenomena such as tumor progression and metastasis and play a pivotal role in precision oncology by analyzing patient-specific datasets not only to match them with FDA-approved, off-label (Zarkavelis et al. 2023), or repurposed therapies (Hijazi et al. 2023) but also to predict which treatment options are most likely to benefit everyone.

Currently, several metastatic cancers remain essentially incurable, highlighting the urgent need for transformative strategies and paradigms. While the AI "genie" has been released from its lamp (Van der Schaar 2025), it has not yet granted our wish for a cancer cure. Nevertheless, with human support and innovative applications, future AI systems may serve as navigational tools guiding us toward that goal. Notably, rudimentary cognitive processes are present even in unicellular life forms, suggesting that intelligence, in its most basic form, is embedded in biology itself. The strategic mastery demonstrated by AI in complex games such as chess and Go—exemplified by Deep Blue's victory over Kasparov in 1997 and AlphaGo's triumph over Lee Sedol in 2016—reflects an evolving capacity for long-term planning, contextual learning, and adaptive behavior. These victories were grounded in predictive modeling and dynamic strategies. Cancer, like a master opponent in a game, evolves, adapts, and counters the host's defensive and therapeutic strategies. By conceptualizing cancer as a strategic game between tumor and host, new paradigms in treatment can be developed where organismal intelligence and therapeutic algorithms may be combined in strong joint strategies designed and assisted by AI, which is able to contain and defeat cancer (Gatenby et al. 2009; Stanková et al. 2019).

Looking ahead, it is conceivable that the oncology toolkit will be significantly augmented by increasingly powerful AI systems capable of confronting the complex, adaptive intelligence of cancer and equipping clinicians and researchers with a resourceful arsenal in the ongoing battle against this multifaceted disease (Sharma and Sarode 2023).

2.21 Human Being: The Forgotten Dimension

Zooming out beyond cells and tissues and looking at the whole organism under the lens of biology is important, but a significant dimension needs to be added. A human being is much more than an organism, and cancer is not only a complex biological problem waiting to be solved but also a threatening condition with a profound psychological impact on the life of an individual. Every single cancer is a unique disease affecting a unique person. We need to personalize our approaches not only toward the idiosyncrasies of each distinct form of cancer but also to individualize and humanize our interactions with our suffering human fellows. Human beings have bodies and minds, and suffering is not only physical but also mental. Cancer brutally reminds us of mortality. The fear of death is the blind spot, the unvisited zone that needs to be added to all the oncology maps. "Doctor, how long I have to live?" This is one of the most sensitive questions oncologists confront almost daily. Alleviating the impact of cancer threat to human existence and reducing the heavy load of anxiety associated with this disease may significantly improve the quality of life of patients diagnosed with cancer and, sometimes, may help them live not only better but also longer lives. While current research does not definitively link stress reduction to a direct impact on survival rates, managing stress and anxiety may help restore the immune system (Witek Janusek et al. 2019) and improve cancer patient fatigue, insomnia, quality of life, and biological, psychological, and behavioral markers of health (Gosain et al. 2020). Recognizing the profound impact of cancer on patients' inner lives, healthcare providers play a crucial role in supporting patients all through this transformative process. Engaging in open, empathetic conversations about patients' values and goals can inform treatment decisions and enhance the quality of care.

Cancer is a multiscale disease that transcends the sum of its cellular abnormalities. An integrative approach that spans the molecular intricacies of cancer cells, the complex ecologies of tumor tissues, and the broader psychosocial and physiological systems of the human organism is essential for effective, enduring, and humane care. Addressing cancer's multifaceted nature requires an integrative approach that considers the biological, psychological, existential, and spiritual needs of patients. A global model of cancer treatment that embraces these interwoven dimensions will move oncology from a battlefield to a healing ecosystem, where strategy meets sensitivity and science is aligned with the whole person.

Integrating these layers calls for a paradigm shift: from seeing cancer merely as a pathological event to understanding it as a deeply human experience. *Treatments must be biologically precise and personally meaningful*. Support must be technologically advanced *and* emotionally attuned. An oncologist becomes not just a prescriber of drugs but also a steward of life transitions. A hospital becomes not just a site of cure but also a space of connection, reflection, and support. Healthcare providers can collaborate across disciplines to offer comprehensive care plans that include medical treatment, mental health support, and spiritual counseling. Such an approach aims not only to treat the disease but also to support the whole person, enhancing quality of life and overall well-being.

Empowering cancer patients to be active participants in their treatments is a vital aspect of modern oncology care. It enhances psychological resilience, improves clinical outcomes, fosters better communication, and ultimately leads to more patient-centered care. Healthcare providers must prioritize empowerment by fostering supportive environments, enhancing health literacy, leveraging technology, and embracing shared decision-making. In doing so, they not only improve the quality of care but also affirm the dignity and autonomy of those facing one of life's most challenging diagnoses.

The integration of a cancer-centered and patient-centered model of care in oncology represents an evolution toward holistic, ethical, and compassionate care. It aligns cutting-edge science with humanism, ensuring that patients not only live longer but also live better. As cancer becomes increasingly treatable yet complex, such integration is not only beneficial—it is essential. Oncology professionals, health systems, and policymakers must commit to reshaping care models to serve not only the disease but also the whole person who lives with it.

Why Do Patients with Cancer Die? Insights from the COVID-19 Pandemic

3

3.1 Disease of the Century

In the twenty-first century, cancer is still a leading cause of death. Currently, every year, more than 20 million people in the world are diagnosed with cancer, and more than ten million die from the disease, accounting for one in six deaths globally. Two-thirds of these deaths occur in lower- and middle-income countries where cancer is often diagnosed late and access to treatment is limited, but even in higher-income countries, more than one-third of patients diagnosed with cancer present with late-stage disease and a significantly increased risk of dying within a year (Sung et al. 2021). Also, in recent years, the incidence of cancer has been rising in adults younger than 50 years, a trend seen in many countries, including the US (https://seer.cancer.gov/report_to_nation/statistics.html). This recent trend, a real early-cancer-onset epidemic (Ugai et al. 2022), is probably related both to exposure in utero to reproductive factors, smoking, diet, alcohol consumption, lifestyle components and comorbidities of the mother and to an increase in the consumption of highly processed or westernized foods together with changes in lifestyles, the environment, and morbidities in early life and/or young adulthood (Ugai et al. 2022).

Approximately one in five people develop cancer before age 74 (slightly more men than women), and although cancer is currently the second leading cause of death worldwide after ischemic heart disease, by 2060, it is expected to become the leading cause of death worldwide (Mattiuzzi and Lippi 2019).

3.2 Reasons for Optimism

According to the January 2023 ACS report (https://www.cancer.org/content/dam/cancer-org/research/cancer-facts-and-statistics/annual-cancer-facts-and-figures/2023/2023-cancer-facts-and-figures.pdf), there was a 33% decrease in cancer death rates from 1991 at 215.1 cancer deaths per 100,000 people to 2020 at 143.8 per 100,000 people. An estimate of 3.8 million lives (2.6 million men and 1.2 million women) have been saved over the last three decades, which would have been lost to cancer if the death rate had not dropped significantly since 1991. The steepest declines have been observed in metastatic melanoma (−6.4%) and lung cancer (−4.3%), largely owing to the transformative impact of immunotherapy (Ganesh and Massagué 2021). Despite this progress, lung cancer still tops the lives lost to cancer statistics for the years covered (Radkiewicz et al. 2023). A significant improvement in survival has also been noted in breast cancer: currently, 5-year survival rates are in the range of 90%, and 10-year survival is about 80% (Nardin et al. 2020). Despite these significant improvements, as a result of population growth and a continuous increase in the aging population (WHO estimates), cancer incidence continues to rise.

3.3 Cancer Deaths

What is the most common cause of death from cancer? The short answer is metastatic disease. The majority of cancers are solid tumors (approximately 90%), and approximately 67% of cancer deaths are attributed to metastasis (Dillekås et al. 2019). It has been estimated that in 2025, approximately 693,452 patients will be living with metastatic cancer in the United States, and this number has been increasing by 70,000 since 2018 (Gallicchio et al. 2022). So, why do metastases lead to death? The classical answer found in the literature is that death may be due to functional failure of a vital organ involved (for example, brain metastases) or death to a generalized inflammation syndrome related to systemic cytokine overproduction and subsequent multiple organ failure (i.e., cachexia) (Loberg et al. 2007; Boire et al. 2024). In other words, metastases may cause death either by directly interfering with a critical functional structure of the organism or indirectly by inducing systemic inflammation. Another hypothesis raised recently is that cancer, together with other chronic conditions, may be considered a sort of "accelerant" of aging process occurring in

D. Paul, *Cancer: An Integrative Approach*, https://doi.org/10.1007/978-3-032-08808-6_3

healthy individuals. This hypothesis has clinical implications, since targeting senescent cells with antiaging therapies may curb the systemic effects of cancer (Boire et al. 2024). Sometimes, metastases do not alter either the local or the global functions of the organism, and patients may live for many years with metastatic disease. As an example, 5-year survival of patients with well-differentiated thyroid cancer is in average 67% (Chen et al. 2024), and most of these patients are alive at 10 years, the main risk factor for patients with thyroid cancer being their age and not the stage of their disease (Ganly et al. 2015). So, what drives the lethality of metastases? Is it something related to the metastatic cells, the metastatic stroma, the organism of the patient as a whole or all these together? Importantly, only up to 10% of patients die from direct failure of a local organ from the growth of the primary tumor or its metastases. The vast majority of patients die due to systemic effects of various products (cytokines, chemokines, thrombogenic factors, reactive oxygen species, lactic acid, etc.) released by cancer or the organism itself in the bloodstream that is literally "poisoning" the body's internal environment (Pienta et al. 2025).

As an oncology physician with several decades of experience, I have been treating thousands of patients with metastatic disease, and finding answers to these questions, far from being theoretical dilemmas, is crucial in order to improve the prognostic of cancer patients. The hospital where I currently practice is in New York City, and during the 2020 COVID-19 pandemic, I treated many cancer patients infected with COVID-19. Being exposed to both the heterogeneity of the clinical presentations of patients with metastatic disease and the variety of symptoms and outcomes of patients treated for COVID-19, I noticed some similarities between these two conditions, and I looked for explanations for these similarities.

3.4 Lessons from the COVID-19 Pandemic

By the end of 2023, the COVID-19 epidemic had killed approximately seven million people worldwide. One of the striking mysteries related to the pandemic was that although, as of the summer of 2023, more than 664 million people infected with the virus had cleared it within a few days without major complications, a lot of infected people had died. Why this difference? What is the explanation for different disease evolution and outcomes? In the COVID-19 literature, old age (Rosenthal et al. 2020), comorbidities (Docherty et al. 2020), or both, leading to reduced organ reserve (Marmiere et al. 2021), have been the most frequent reasons incriminated as causing death from COVID-19. Indeed, as studies have shown, a significant number of patients with severe presentations of COVID-19 were old, had cardiovascular diseases, diabetes, and hypertension, and were overweight or obese (Makhoul et al. 2022), and patients with at least one comorbidity accounted for the vast majority (94.9%) of COVID-19 deaths (Rosenthal et al. 2020). A recent study analyzed in detail the influence of age on COVID-19 mortality and found that aging leads to greater activation of the coagulation system and the endothelium and, importantly, to stronger elevation of inflammation and organ damage markers, which, importantly, is independent of an increase in age-related comorbidities (Michels et al. 2023).

A German series also found that in the ICU setting, the majority of patients infected with COVID-19 died of septic shock and multiorgan failure related to purulent lung infection (Elezkurtaj et al. 2021). Importantly, in the German series, with one exception, all the deceased patients also had various comorbidities (arterial hypertension, chronic kidney or heart disease, and chronic pulmonary disease). Another important cause of death described in patients infected with SARS-CoV-2 has been a disproportionate inflammatory response leading to acute respiratory distress syndrome and thromboembolic events. A minority of patients died of rare acute clinical manifestations (i.e., myocarditis) (Marmiere et al. 2021).

It is important to note that COVID-19 illness has two phases, an upper respiratory phase and a pulmonary phase, and death may occur in both instances (Huang et al. 2020). In the acute upper respiratory phase, death from COVID-19 appears to be related to the virulence of the virus itself replicating in the upper respiratory tract, and subsequently, although the virulence of the COVID-19 virus remains an important factor, the host reaction initiating inflammatory lung injury (Russell et al. 2023) plays a major role. Russell et al. found that the impact of comorbidities on patients infected with COVID-19 was even more significant than age and, comorbidities affected the outcome through a combination of host factors: defective resistance, defective tolerance, and the generic effect of frailty and reduced functional reserve (Russell et al. 2023).

In conclusion, both age and comorbidities have been described as predisposing conditions for severe presentation and death from COVID-19. But what is the immediate cause of death? Many patients with chronic conditions also have systemic inflammation that can be exacerbated by COVID-19 infection. Infection with COVID-19 may be considered a global inflammatory disease that, in severe form, leads to a fulminant "cytokine storm," resulting in a septic shock-like clinical picture associated with multisystemic end-organ damage (Buicu et al. 2021).

3.5 Mortality Trends for COVID-19 and Cancer

Basically, on the one hand, patients with decreased immunity due to age or comorbidities or both have a severe presentation of COVID-19. On the other hand, younger patients sometimes respond to the virus with an inappropriate overblown inflammatory response that damages their body systems. Who has the highest mortality from COVID-19 infection?

Analyzing the US mortality from COVID-19 and cancer data, we see a similar trend. For both diseases, approximately 80% of deaths occur in patients aged 65 years and older. But is age, by itself, a risk factor for death from COVID-19 and from cancer?

3.6 Impact of Comorbidities as Opposed to Aging on Mortality from COVID-19 Infection

Recently, a multivariate analysis investigating the impact of age versus the impact of comorbidities on COVID-19 mortality showed that the effect size of age decreased with the number of diagnoses considered and that comorbidities played a much more significant role. When adjusted for comorbidities, age per se had a minimum role, and the comorbidity burden at admission was responsible for 98.2% (95% CI 84.1–117.1%) of mortalities (Valero-Bover et al. 2023).

3.7 Comorbidities Play a Significant Role in Cancer Incidence and Mortality Independent of Age

What about cancer? Chronological age is considered the most significant carcinogen, as cancer incidence increases dramatically after age 55, and it has been proposed that the cancer-prone phenotype of older humans might reflect the combined effects of cumulative mutational load, increased epigenetic gene silencing, telomere dysfunction, and an altered stromal environment (De Pinho 2000). Age-induced tissue abnormalities and malfunctioning of the immune system are currently considered prominent factors associated with cancer in the elderly as opposed to genetic mutation accumulation (Laconi et al. 2020; De Gregori et al. 2020).

The incidence of comorbidities also increases with age and may predispose to cancer. For example, hypertension is linked with a higher incidence of breast cancer, especially for postmenopausal women (Han et al. 2017), renal cancer (Weikert et al. 2008; Radišauskas et al. 2016), and, intriguingly, colorectal cancer only in men (Xuan et al. 2021). Diabetes is associated with an increased risk of several cancers, including pancreatic, liver, colon, breast, and endometrial cancer (Tao et al. 2020; Shahid et al. 2021). Comorbidities appear to be more common in certain cancers than in others. They are the most common in lung cancer (46.7%) and colorectal cancer (40.0%), followed by prostate cancer (28.5%), melanoma (28.3%), and breast cancer (22.4%). The most common types of comorbidities were hypertension (29.7%), pulmonary diseases (15.9%), and diabetes (13.5%) (Vrinzen et al. 2023).

More recently, a large study enrolled 128,999 individuals aged 55—74 years over approximately 20 years without a history of cancer (Lavery et al. 2025). Overall, 38.4% of patients had cardiovascular conditions, 27.6% had metabolic conditions, 16.4% had gastrointestinal conditions, 6.2% had respiratory conditions, and 3.5% had liver conditions. After a median follow-up of 20 years, 32,857 cancer cases were diagnosed across 20 malignancy types. Across the cohort, the cancer incidence was significantly increased in patients with respiratory conditions (HR = 1.07; 95% CI, 1.02–1.12) and cardiovascular conditions (HR = 1.02; 95% CI, 1–1.05). Importantly, *each* comorbidity classification significantly increased the risk for at least one specific cancer type. Interestingly, some comorbidities are also associated with a decrease in the incidence of certain cancer types.

The highest associations included liver conditions with liver cancer (HR = 5.57; 95% CI, 4.03–7.71) and decreased risk of endometrial cancer (HR, 0.52; 95% CI, 0.31–0.87); metabolic conditions with liver cancer (HR = 2.04; 95% CI, 1.58–2.65), endometrial cancer (HR = 1.87; 95% CI, 1.62–2.17), kidney cancer (HR = 1.54; 95% CI, 1.33–1.79), biliary (HR, 1.48; 95% CI, 1.07–2.05), thyroid (HR, 1.31; 95% CI, 1.01–1.70), rectal (HR, 1.28; 95% CI, 1.06–1.55), colon (HR, 1.22; 95% CI, 1.10–1.36), pancreas (HR, 1.20; 95% CI, 1.03–1.40), hematopoietic cancer (HR, 1.14 [95% CI, 1.05–1.23]), and decreased risk of lung cancer (HR, 0.75; 95% CI, 0.69–0.81) and head and neck cancer (HR, 0.82; 95% CI, 0.68–0.98); respiratory conditions with lung cancer (HR = 1.8; 95% CI, 1.63–1.98), pancreatic cancer (HR, 1.33; 95% CI, 1.02–1.73), and decreased risk of prostate cancer (HR, 0.70; 95% CI, 0.63–0.78); gastrointestinal conditions with thyroid cancer (HR = 1.5; 95% CI, 1.13–1.99), breast cancer (HR = 1.46; 95% CI, 1.37–1.57), kidney cancer (HR, 1.39; 95% CI, 1.17–1.66), ovarian cancer (HR, 1.25; 95% CI, 1.01–1.56), and decreased risk of prostate cancer (HR, 0.60; 95% CI, 0.56–0.64); and cardiovascular conditions with kidney cancer (HR = 1.47; 95% CI, 1.27–1.7) and biliary cancer (HR = 1.42; 95% CI, 1.04–1.94).

Chronological age (CA) does not appropriately describe the heterogeneity between older individuals. Biological age (BA) is considered to be a better reflection of a person's physiology and cancer incidence. In a recent study (Mak et al. 2023), BA was computed from 18 biological biomarkers that were considered important for an individual's health: forced expiratory volume (FEV1), systolic blood pressure, blood urea nitrogen, glycated hemoglobin (HbA1c), total cholesterol, creatinine, serum glucose, waist circumference, red cell distribution width, albumin, alkaline phosphatase, triglyceride, mean cell volume, uric acid, lymphocyte percent, red blood cell count, C-reactive protein, and diastolic blood pressure. Three different algorithms were used for computing the BA, and five cancer types were analyzed: breast, prostate, lung, colorectal, and melanoma skin cancer. Remarkably, after adjusting for sociodemographic and lifestyle factors, all BA measures were associated with cancer risk, in general. Also, specifically, all BA measures were associated with increased risks of lung and colorectal cancers. Taken together, these findings may indicate that multiple aging processes captured by BA measures (i.e., inflammation and metabolic changes) and, importantly, comorbidities and lifestyles may together play a role in cancer development.

In terms of survival, most studies found that patients with comorbidities have worse survival, which, in most cases, is independent of treatment (Lee et al. 2011). Comorbidities may directly impact cancer progression and metastasis and, ultimately, survival (Panigrahi and Ambs 2021). On the other hand, as shown recently, approximately 80% of patients living with metastatic cancer die of their diagnosed cancer, while 20% die of their comorbidities (heart disease, COPD, and stroke). The impact of comorbidities on survival depends on the cancer type. Patients with pancreatic, lung, stomach and esophageal tumors are the most likely to die of metastatic cancer, while the mortality of patients with prostate and breast cancer, in many cases, is mostly related to comorbidities (Mani et al. 2024).

Similar to COVID-19, for the majority of adults, cancer is associated with modifiable risk behaviors (smoking, alcohol consumption, and obesity), preventable chronic conditions (heart conditions, diabetes, and COPD), and avoidable exposures, which increase coincidentally with age.

3.8 Cancer Incidence and Mortality in the Oldest-Old

An interesting observation is that cancer incidence *does not increase with age indefinitely*. The data shows that cancer prevalence and cancer incidence increase until the age of 85–89, after which the rates decrease into the oldest-old, presenting a threefold decrease after the age of 90 years and reaching 0–4% above the age of 100 years.

The National Institute on Aging classifies elderly persons into young-old (aged 65–75), old (aged 76–85) and oldest-old (older than 85 years), but there is no general agreement on the age at which a person becomes old. The oldest-old is the fastest growing age group in the United States and some other nations.

In addition, the number of metastatic sites in the oldest-old were found to be remarkably less and incidental malignant tumors or multiple primary cancers were more frequent, pointing to a more indolent behavior of cancer in this population. Also, mortality from cancer was found to be 20–25% in nonagenarians and 0–4% in centenarians (Pavlidis et al. 2012). This observation may be related to innate protective genetic polymorphism, including immune system-related genes—the oldest-old having an increase with age of CD8 and CD4 cells bearing NK features and secreting a special pattern of cytokines (IL-4 and INF-gamma), creating an immunologically hostile environment for cancer cells (Bonafè et al. 2002), and to increased tumor screening, which may have resulted in the removal of precancerous diseases (De Santis et al. 2019). Another factor may be related to the local tissular environment and the global decline, which may have a protective role in the oldest-old (Cui et al. 2024). Lastly, differences in environmental health and public health policy in different geographical regions may be contributing factors (Nolen et al. 2017).

3.9 Why Do Patients with Cancer Die?

For more than half a century, clinical oncology has focused mainly on cancer cells, "the seed" in Paget's terms. From the data presented before, the host characteristics play an essential role in the mortality of people infected with COVID-19. How about cancer? On the NCI website, cancer is still defined as a disease of the "seed": "cancer is a disease in which some of the body's cells grow uncontrollably and spread to other parts of the body—that is, it is caused by changes to genes that control the way our cells function, especially how they grow and divide." The importance of the "soil" is disregarded by this definition. Historically, the first observations between the tumors and the stroma were made by Robert Virchow in 1863, but the importance of the host factors, or the "soil," has started to be acknowledged progressively less than a century ago following the seminal work of Judah Folkman and Mina Bissell (Maman and Witz 2018). By the "soil," most authors refer to the local microenvironment. The role of the whole organism or the "macroenvironment" of the host has only started recently to be acknowledged. In a recent publication, together with Aurora Nedelcu, we expanded Paget's classical "seed" (cancer cell) and "soil" (cancer stroma) components by also including "trees" (normal organs) and the "internal body climate" (environment of the whole organism) (Paul and Nedelcu 2022) (see Chap. 9).

So, returning to our initial question: how are COVID-19 and cancer related in terms of morbidity and mortality? First, cancer incidence and mortality are also higher in patients with chronic comorbidities, and, similar to patients infected with COVID-19, 80% of patients with cancer are over 55 years of age. Similar to patients infected with COVID-19, patients with cancer may sometimes die due to factors related to the intrinsic aggressivity of cancer cells (the "seed"), but this is a rare occurrence. More often, patients with cancer die due to comorbidities present in other organs and systems (i.e., the "trees") and conditions of the whole organism (the "internal body climate") related, for example, to aging. In some severe cases, the COVID-19 virus triggers a domino-like effect involving multiple organ systems, ultimately leading to multiple organ failure. This clinical picture has been called "systemic inflammatory response syndrome" (SIRS), which precedes "acute shock syndrome" (ACS). In patients with advanced and metastatic cancer, a similar slower domino-like effect occurs not evolving over hours and days like SIRS and ACS but rather over weeks and months, involving progressively multiple organs—a sort of "chronic strain syndrome" (CSS). In some cases, CSS is associated with anorexia and glucose intolerance without significant weight loss (pre-cachexia), but in most cases, this preliminary phase is followed by full-blown cachexia, which is defined as more than 5% involuntary weight loss over 6 months (Fearon et al. 2011). According to the cancer type, cancer cachexia may affect 50–85% of patients with advanced cancer and is a major cause of mortality in patients with metastatic disease (Ferrara et al. 2022). Similar to the cytokine storm present in severe COVID-19 cases, cachexia is also closely related to global inflammation leading to multiple organ failure, but as opposed to the fulminant progression of the cytokine storm in COVID-19 patients, it develops much slower over weeks and months and is a sort of "slow storm" caused by cytokines, exosomes, and other circulating factors induced by systemic inflammation.

3.10 Cancer Symptoms as an Overreaction of the Organism

In both fulminant cases of COVID-19 and cancer cachexia, the clinical symptoms are related not as much to the virulence of the triggering agent (the COVID-19 virus or the cancer tumor(s)) but significantly more by the body's response to the offending agent. Cytokine storm damage is caused by an overreaction of the immune system. Similarly, in cancer, symptoms are often caused by the inflammatory response of the organism and *not by the cancer cells themselves.* Also, the most common clinical complications of cancer (organs malfunctioning, airway blockages, GI, biliary or urinary tract obstructions, thromboses, pain) are mainly related to the organism responses to cancer (i.e., inflammation) but not by cancer per se. Metastatic disease, which is currently largely incurable, is closely related to inflammation – metastases are triggered by inflammation, and represent an inflammation trigger themselves.

In a recent analysis of 1856 elderly participants in the prospective REGARDS cohort who were cancer free at baseline and analyzed in relation to cancer mortality prospectively, several inflammatory markers (IL-6, IL-8, and IL-10) considered to be related to an increased risk of mortality (Akinyemiju et al. 2019) were measured. Elevated IL-6 increases the risk of cancer incidence as high as 12 times. Several studies have reported that global inflammation is associated with increased cancer incidence and mortality (Mantovani et al. 2008; Balkwill and Mantovani 2012; Greten and Grivennikov 2019; Turizo-Smith et al. 2024). Recently, a review of two meta-analyses totaling more than 12,000 cancers of patients with inoperable cancers found that 29% to 79% of these patients were systemically inflamed (Dolan and McMillan 2020).

Cancer Theories

4

4.1 Cancer: A Disease in Search of a Paradigm

Working in unison, cellular, tissular, and organismal factors are responsible for tissue homeostasis. Cancer as a disease is related to both the disruption of tissue homeostasis, leading to an excess of cancer cells and the local invasion and distal spread of cancer cells. Cancer theories try to stitch together different aspects of the cancer phenotype and come up with a narrative that explains both cancer inception and cancer progression. For this purpose, a particular mechanism or mechanisms linking the observed molecular and microscopic morphological and processual changes in cancer cells and tissues are used as explanations. The dominant cancer theories focus on morphological changes (chromosome abnormalities; genetic, epigenetic, and abnormal stem cells; and metabolic or tissue alterations). There is also a plethora of other theories that focus on a particular process or processes that may lead to cancer (immune system dysfunction, regression to a primitive state, stress responses, evolutionary dynamics, etc.).

4.2 Oncology: The Science of Growth

Since antiquity, growth has been considered the main cancer characteristic. The Latin term "oncos" means "swelling, growth" and was introduced by Galen in the second century to refer to tumors—another word for swelling—and cancer, in general. The current dominant paradigm in oncology, popularized by bestseller books like *The Emperor of All Maladies* considers cancer a disease of uncontrolled growth caused by random genetic alterations thought to be related to either cell-intrinsic (i.e., errors during replication) or exogenous (i.e., inhaling cigarette smoke) factors.

Chemotherapy agents are drugs that have been designed to stop or slow the growth of cancer cells. On March 15, 1949, Mechloretamine was the first FDA-approved chemical agent for the treatment of cancer. Its effect on cancer was mainly through growth inhibition, and for the following 43 years, up to the FDA's approval of Interleukin-2 in 1992, chemotherapy agents were the only systemic therapeutic options available for cancer treatment.

Chemotherapy benefits should not be dismissed. Among the four different classes of systemic agents used in cancer treatment (chemotherapy, hormonal therapy, targeted agents, and immunotherapy), chemotherapy is, by far, the most commonly used treatment modality. A new study predicts that between 2018 and 2040, the number of patients needing chemotherapy each year will rise from 9.8 million to 15 million globally—a 53% increase (Wilson et al. 2019). Chemotherapy is able to offer a cure for the majority of children with advanced cancers; in adults, it also cures the majority of testicular cancers in men and the majority of choriocarcinomas in women, and it also improves the quality of life and survival of many patients with other forms of cancer. Anecdotally, there are also several reports of various types of metastatic cancers considered incurable and cured by standard chemotherapy (see, for example, Paul et al. 2014). Why are some cancer types resistant to chemotherapy and others so sensitive? There are many factors involved in chemotherapy resistance including inactivation of drugs through enhanced drug metabolism, enhanced elimination of chemotherapy drugs that are controlled by drug transporters and drug-metabolizing enzymes, changes in drug targets, DNA methylation, DNA repair, and avoidance of apoptosis (Prihantono and Faruk 2021).

A group from Cornell University has shown that the spectacular responses to chemotherapy seen in testicular cancers, as opposed to other solid tumors, may be related to the fact that cancer stem cells, which are usually resistant to chemotherapy, are extremely sensitive to chemotherapy in testicular tumors (Pierpont et al. 2017).

The main issue with chemotherapy agents given either alone or in combination, though, is that except for children and adults with testicular cancers or choriocarcinoma, they do not work as well for most solid tumor cancers. Their use in metastatic disease is considered palliative, sometimes

D. Paul, *Cancer: An Integrative Approach*, https://doi.org/10.1007/978-3-032-08808-6_4

improving the quality and prolonging the duration of life of patients but rarely curative. Also, unfortunately, as it is becoming more and more apparent, chemotherapy may sometimes accelerate cancer progression (Middleton et al. 2018) and may have a prometastatic effect (Karagiannis et al. 2018; D'Alterio et al. 2020). Recently, it has been found that DNA released into the environment by breast cancer dying cells in the environment triggers growth-promoting signaling in neighboring tumor cells and enhances their metastatic outgrowth in the lung (Park et al. 2023).

4.3 Quest for a Unified Cancer Theory

A unified theory of cancer has been a sort of Holy Grail in the oncology field, and dozens of theories have been proposed over the years, some of which so farfetched that can be used as sci-fi movie scripts. On the other hand, some theories have withstood the test of time. For example, more than a century ago, Theodor Boveri proposed that cancer is caused by an abnormal chromosomal number (i.e., aneuploidy) and/or morphology. Aneuploidy is still acknowledged to be a key cancer feature. More recently, Peter Duesberg and his collaborators proposed the speciation theory of cancer. According to this theory, carcinogens initiate cancer by aneuploidization, which automatically distorts the expression of thousands of genes and subsequently induces chain reactions of additional aneuploidization. Subsequently, the karyotypes of new autonomous cancer cells are stabilized and immortalized, by clonal selection for autonomy and immortality (Hirpara et al. 2018). As a corollary, each cancer represents a new species (Duesberg et al. 2011), a proposal made also priorly by Mark Vincent (2010, 2011).

In the 1930s, Otto Warburg discovered a switch to aerobic metabolism in hepatoma cells and thought that malfunctioning mitochondria leading to aerobic glycolysis represents the *primum movens* of *all* cancers. Subsequently, at a 1966 conference, he explicitly expressed his doubts regarding Rous' cancer viral origin (which turned out to be true and paved the way for the discovery of oncogenes) and continued to blow the horn of his own "unifying cancer theory" related to mitochondrial dysfunction.

More recently, in 1998, after several outstanding responses to angiostatin in a lung cancer mouse model, James Watson thought angiogenesis was cancer's key characteristic and declared for *The New York Times* that "Judah Folkman will cure cancer in two years." Unfortunately, the clinical trials using angiostatin in human lung cancer were not as successful as in a mouse model, and angiostatin was never FDA approved in the United States, although subsequently it was approved in China, where it is currently still in use.

For several decades, the somatic mutational theory (SMT) initially proposed by Robert Weinberg in 1999 (Weinberg 1999) and then expanded through several editions of his *magnum opus* textbook (Weinberg 2006) has occupied center stage. The contender theory also proposed in 1999 by Anna Soto and Carlos Sonnenschein is the tissue organization field theory (TOFT), in which the disorganized stroma and not genetic alterations represent the cause of cancer (Sonnenschein and Soto 1998). The remarkable success of immunotherapy over the past decade has underscored the critical role of cancer-stromal interactions. Although TOFT emphasized tissue architecture rather than stromal dynamics per se, these advances have shifted the focus of cancer research from purely genetic alterations toward the immune system.

Cancer is an extremely heterogeneous phenomenon, and researchers have been struggling to build a cancer theory that could explain all its aspects and lead to a cancer cure. In the last few years, almost every month, we read in the news about a seminal discovery that may lead to a cancer cure. Under the large umbrella of "cancer," there are literally hundreds of diseases with different origins, behaviors, treatments and prognoses. If we read the articles cited by the media, most of these "seminal discoveries" refer to only one out of hundreds of cancer conditions, and several years later, the work that was advertised as leading to a potential cancer cure turned out to apply just to a fraction of cancer or be just another blind alley. Sometimes, however, a real insight is gained like the discovery by Judah Folkman of the crucial role played by angiogenesis or the discovery of the immune checkpoint inhibitor system for whom two researchers, Tasuku Honjo and James Allison, received the Nobel prize in 2018.

4.4 Hallmarks of Cancer

How can we make sense of this tremendous cancer heterogeneity? In 2000, Douglas Hanahan and Robert Weinberg described the core features of cancers in a paper that became one of the most cited articles in the oncology field. In the original article they described six characteristics that they considered universal cancer hallmarks. In 2011, they increased the list to ten. In 2022, in a single author article, Hanahan added four new hallmarks to the list (Hanahan 2022). Here is the 2022 updated list of 14 cancer hallmarks:

– Sufficiency in growth signals (original 2000 paper) -Insensitivity to growth inhibitory (antigrowth) signals (original 2000 paper) -Evasion of programmed cell death (apoptosis) (original 2000 paper) -Limitless replicative potential (original 2000 paper) -Sustained angiogenesis (original 2000 paper) -Tissue invasion and metastasis (original 2000 paper) -Genetic instability and mutation (2011 paper) -Tumor promoting inflammation (2011

paper) -Deregulated energetics (2011 paper) -Avoidance of immune destruction (2011 paper) -Nonmutational epigenetic reprogramming (2022 paper).
– Phenotypic plasticity (2022 paper) -Presence of senescent cells (2022 paper) -Polymorphic microbes (2022 paper).

In an article written in collaboration with Douglas Hanahan in 2023, Michelle Monje proposed that the interaction between the nervous system and cancer may represent a 15-th hallmark (Hanahan and Monje 2023). As a constituent of the tumor environment, both innervation and/or co-opted neuronal signaling in cancer cells can modulate six (sustaining proliferative signaling, resisting cell death, avoiding immune destruction, tumor-promoting inflammation, inducing/accessing vasculature, activating invasion, and metastasis) of the ten hallmarks described in the 2011 paper by Hanahan and Weinberg. It is important to note that the hallmarks proposal is not only descriptive but also prescriptive in the sense that in this model, genetic instability is considered the unifying cause of underlying human cancers.

Twelve of the fifteen hallmarks refer directly to *cancer cells* (sufficiency in growth signals, insensitivity to growth inhibitory (antigrowth) signals, evasion of programmed cell death (apoptosis), limitless replicative potential, genetic instability and mutation, tissue invasion and metastasis, deregulated energetics, avoidance of immune destruction, non-mutational epigenetic reprogramming, tumor-promoting inflammation, unlocking phenotypic plasticity and sustained angiogenesis), one refers to the cancerous tumor as a whole (tumor-promoting inflammation), and only two are not directly cancer related (senescent cells and polymorphic microbes).

In the original 2000 paper, the authors presciently wrote, "looking forward in time, we believe that important new inroads will come from regarding tumors as complex tissues in which mutant cancer cells have conscripted and subverted normal cell types to serve as active collaborators in their neoplastic agenda. The interactions between the genetically altered malignant cells and these supporting coconspirators will prove critical to understanding cancer pathogenesis and to the development of novel, effective therapies" (Hanahan and Weinberg 2000).

In their 2011 updated hallmarks article, they noted that "the biology of tumors can no longer be understood simply by enumerating the traits of the cancer cells but instead must encompass the contributions of the 'tumor microenvironment' to tumorigenesis." They also described tumor progression as a result of "reciprocal signaling interactions between cancer cells of the parenchyma and stromal cells that convey the increasingly aggressive phenotypes that underlie growth, invasion, and metastatic dissemination." As examples of stromal cells co-opted by cancer cells, they mentioned stem cells, epithelial cells, pericytes, cancer-induced fibroblasts and immune inflammatory cells. The inclusion of cancer-promoting inflammation, the presence of senescent cells, and polymorphic microbes is a recognition of the important role played by stromal components in the cancer process. The 2011 hallmarks article also introduced the idea that metastasis is supported by a "repertoire of cell types originating in the bone marrow that play crucial roles in pathological angiogenesis" and, in general, by a permissive microenvironment.

The foundational theory underlying the Hanahan and Weinberg articles is the somatic mutational theory (SMT). In 2003, in a seminal paper published in the *New England Journal of Medicine*, William Hahn and Robert Weinberg proposed that four or five mutations are sufficient for malignant transformation (Hahn and Weinberg 2002). Four years before, Weinberg's team was able to transform in vitro normal fibroblast cells by artificially expressing a sequence of several mutated genes and subsequently obtaining tumors by introducing artificially created cells into nude mice (Hahn et al. 1999). The main problem with the experiment, noted by the authors themselves, was its artificiality. Although on average, spontaneously developed human cancers have four driver mutations, cancers with less or more than four, or *no driver mutations* have also been described. Also, the tumors obtained in the Hahn et al. mouse model *did not form spontaneous metastases and only grew locally* (Hahn et al. 1999). On close examination, not only SMT but also the different variants of stem cell theory (stem cells have not been described in all cancer types), Warburg's metabolic theory of cancer (in many instances, OXPHOS is required for malignant transformation and progression), and tissue cancer theories (TOFT) are incomplete. Both stromal *and* cellular changes are required for cancer to develop.

Since 2010, several authors have proposed that cancer is primarily a disease of the whole organism (Zajicek 1978, 1996; Egeblad et al. 2010; McAllister and Weinberg 2010; Castaño et al. 2011; Redig and McAllister 2013; Al-Zoughbi et al. 2014, McAllister and Weinberg 2014; Paul 2015, 2020; Borniger 2019; Alečković et al. 2019; Venerin et al. 2020). This is not a novel idea. In 1931, Willy Meyer, a physician practicing in New York, presciently wrote: "it appears more likely that all cells, and consequently all cancer cells as well, are the creatures of the conditions which surround them; that they must adapt themselves to these conditions, or else perish; that cancer cells have no energy of their own; that the energy they seem to possess is imparted to them by the vis-a-tergo exerted through the growing cell multiplication and the demand for space of the daughter cells, and that therefore tumors of every description are not active, but passive growths. That is to say, they do not grow like a green leaf through inside forces, but rather like a heap of dry leaves through outside forces without any contribution to the growth on the part of the heap. The views referred to as erroneous regard the tumor as the primary phenomenon and know systemic involvement only after the malignancy has become generalized. On the other hand, the here advocated view starts with systemic disturbances as the primary disease, on the basis of which malignancy sets in as a secondary development. For those sharing the first-mentioned views, cancer is a cell problem; for those supporting the standpoint mentioned in the second place, cancer becomes primarily a problem of the physicochemical condition of the humors of the body" (Meyer 1931). Also, since 1978, Gershom Zajicek, a researcher from Israel, wrote a series of articles (Zajicek 1978, 1996) reasoning that "cancer is a chronic systemic disease with local manifestations like arteriosclerosis, which is also systemic and manifested solely by its local manifestations, e.g. stroke and myocardial infarction. In the same way as treatment of an ailing heart does not cure the underlying arteriosclerosis, tumor removal does not cure cancer, as it is 'metabolically' systemic" (Zajicek 1996). In 2010, Mikala Egeblad and her collaborators introduced a model of a tumor as an organ that may influence the immunity, metabolism and coagulation status of the host (Egeblad et al. 2010). In a paper published the same year, Sandra McAllister and Robert Weinberg suggested that tumor-host interactions extend well beyond the local tissue microenvironment and introduced the tumor "instigation" concept, in which primary tumors perturb normal host organs and support the growth of metastatic tumors at distant anatomic sites (McAllister and Weinberg 2010). The two authors mentioned several factors secreted by tumors with systemic effects: vascular endothelial growth factor (VEGF), interleukin (IL)-6, IL-8, stromal cell-derived factor-1 (SDF1), fibroblast growth factor, growth-related oncogene-α (CXCL1), platelet-derived growth factor (PDGF), angiopoietins, transforming growth factor β (TGF-β), hepatocyte growth factor (HGF), angiogenin, leptin, and sonic hedgehog homolog, regulated upon activation normal T cell expressed (CCL5), and osteopontin (McAllister and Weinberg 2010). They also proposed a novel treatment for metastases by blocking their access to supporting stromal cells derived from the bone marrow. In a subsequent publication, the same authors (McAllister and Weinberg 2014) expanded their original model bringing additional support for the complementary idea that tumors can also be significantly influenced by systemic processes. In 2014, a research team from Austria introduced the concept of the tumor macroenvironment and described the global metabolic changes that tumors exert on the whole organism (Al-Zoughbi et al. 2014). In three consecutive papers published between 2018 and 2021, Jeremy Borniger, a neuroscientist from the Cold Spring Harbor Laboratory, investigated brain-tumor interactions, emphasizing the relationship between subcortical neural populations and cancer (Borniger et al. 2018; Borniger 2019; Francis and Borniger 2021). Specifically, he and his team discovered that nonmetastatic breast tumors may influence the endocrine and immune systems of the host. Using a breast cancer mouse model, they demonstrated that by inhibiting the signaling of lateral hypothalamic orexin/hypocretin neurons, the sleep quality and metabolic dysregulation induced by tumors were improved (Borninger et al. 2018). Michelle Monje also proposed that cross talk between the nervous system, immune system, and cancer—both local and systemic—regulates protumor inflammation and anti-cancer immunity and wrote recently two influential papers on this subject (Mancusi and Monje 2023; Hanahan and Monje 2023). Also, recently, Inamura et al. highlighted the relationship between the exposome (i.e., the totality of exposure), which encompasses diets, supplements, smoking, alcohol, other lifestyle factors, medications, etc., the body microbiome (including bacteria, viruses, archaea, fungi, and parasites), and the immune system at various body sites and the induction of different tumor phenotypes (Inamura et al. 2022).

I have also previously proposed the need for a broader, systemic perspective on cancer and introduced the idea of the *cancer system* in several papers (Paul 2015, 2020; Paul and Komarova 2021). The concept of the *cancer system*, although closely related to the idea of the cancer-systemic disease, is different in a fundamental way. The meaning of the word "systemic" refers to "affecting the body generally." Metastatic and, sometimes, localized cancers influence the whole organism and are therefore classified as systemic diseases. On the other hand, macroscopically, cancer behaves like a parasitic system within an organism, and metastasis appears to be a finely orchestrated process. The concept of a *cancer system* tries to capture precisely this deterministic behavior. In 2022, in a paper written in collaboration with Aurora Nedelcu, we introduced the concept of the internal body climate serving as an integrated framework for studying all aspects of cancer as a systemic disease (Paul and Nedelcu 2022). We present this framework in Chap. 9.

The concept of cancer as a systemic disease became the most recent installment in the Cancer Hallmarks series. On March 28, 2024, an impressive team of 33 cancer researchers and medical oncologists led by Douglas Hanahan published a paper titled "Embracing cancer complexity: Hallmarks of systemic disease," an article that, by serendipity, mirrors in its title "The systemic hallmarks of cancer" paper published by me in 2020. In their article, almost apologetically, Swanton et al. noted that "this modern synthesis of cancer biology (i.e., the cancer hallmarks) does not fully consider the broader interactions of an evolving tumor with distant organs of the host, nor the impact of host pathophysiology, germline genetic diversity, and environmental exposures, on cancer initiation and evolution. The reassuring simplicity of the hallmarks is clearly insufficient to fully understand the multifarious manifestations of disease mechanisms" (Swanton et al. 2024).

In this paper, instead of expanding the list of hallmarks, as the prior four installments had, Swanton et al. had a radically different approach: they postulated thirteen "clouds of complexity" related to cancer and, as the authors confessed, incompletely understood:

- Inception and promotion.
- Macro- and microtumor environments.
- Aging systems and tissues.
- Senescent cell states.
- Metabolic effects.
- Cancer cachexia.
- Obesity and physical activity.
- Dietary influences.
- Physiological prevention.
- Circadian clocks.
- Neural-cancer crosstalk.
- Thromboinflammation.

and, at the same time, six "mechanistic effectors—the building blocks of cancer" are also incompletely understood:

- Mutation caused by environmental agents.
- Mutations from misrepair of aging DNA and replication errors.
- Nonmutational epigenetic reprogramming.
- Polymorphic germline susceptibility.
- Environmental and physiological promoters of tumorigenesis.
- Phenotypic plasticity (Swanton et al. 2024).

Importantly, the accent of the Swanton et al. article is on *prevention*. The authors ended their review by saying that their intention was to "to provide a logical framework for initiatives aimed to address the clouds of complexity described herein, aiming to increasingly prevent the inception of cancer, to reduce the incidence of symptomatic cancer, as well to diagnose cancer earlier" (Swanton et al. 2024).

I cannot emphasize enough the importance of the Swanton et al. article. In my view, this marks a major paradigm shift in oncology—from a cell- and gene-centric model of cancer to a broader framework that includes not only genetic and cellular factors but also local stromal elements and organism-wide components. Cancer cells have traditionally been considered *individuals,* and their behavior is self-determined by internal processes; thus, oncology is mainly a cell-centric discipline. Swanton et al.'s article proposes a more nuanced view in which the *environment* may play different roles in the cancer process. It also opens the possibility that cancer individuality is not fixed and can switch between the different classes of individualities.

A decade ago, Marusyk et al. demonstrated in a mouse model that tumor growth can be driven by subclones that stimulate whole tumor growth by inducing tumor-promoting changes in the microenvironment and focusing solely on the dominant clone in tumor biopsies to inform targeted therapies may be misleading (Marusyk et al. 2014). Recent data have clearly demonstrated that cancer progression and metastasis are not autonomous events solely due to genetic alterations. The tumor stroma induces cancer progression and metastasis. In turn, cancer cells transform the biology and composition of the stromal microenvironment by attracting immune cells, triggering angiogenesis and inducing cancer-associated fibroblast (CAF) activation (Hamidi and Ivaska 2018; Barbazán et al. 2019).

4.5 Cancer as a Wound that Does Not Heal

Historically, cancer has been compared to wounds that do not heal; however, importantly, this analogy refers to the primary tumor (Dvorak 1986) and not to the metastatic process. Another important analogy is between infectious diseases and cancers. Cancer as a disease resembles, in more than one way, a chronic infectious process like tuberculosis, for example (Richmond and Roehner 2019). Daunorubicin and doxorubicin are both antibiotics developed in the 1960s (Dimarco et al. 1964; Arcamone et al. 1969) from different strains of the bacterium *Streptomyces peucetius*.

4.6 Cancer Complexity and Somatic Mutation Theory

There are multiple sources of heterogeneity in cancer genetic, epigenetic, metabolic and microenvironmental (reviewed in Basanta and Anderson 2017). Importantly, this heterogeneity is dynamic (Fisher et al. 2013), which means that both the nature of the genetic/epigenetic/metabolic alterations and the timing of these alterations are equally important (Swanton 2012; Sottoriva et al. 2015, 2017).

As clearly stated before by Serena Nik-Zainal and Benjamin Hall, "additional extrinsic factors, including cellular interactions and the microenvironment, seem to be required to create a selective environment that triggers a cell to become a dominant cancer clone. It is likely that the mutated genome of a cell contributes to the potential for malignant transformation, but it is not deterministic of it. This concept is not unfamiliar: Inherited mutations in DNA repair genes, such as BRCA1, BRCA2, TP53, and MLH1 (mutL homolog 1), cause familial cancer predisposition. They confer an increased lifetime risk of cancer, but cancer does not always arise. Additional factors are required" (Nik-Zainal and Hall 2019).

Oncogenic mutations, epigenetic changes and metabolic switches are codependent. As an example, oncogenic mutations may promote an increase in glucose uptake, which may itself dictate distinct phenotypes inside the tumor, allowing the coexistence of cancer stem and migratory cells as well as highly proliferative cells (García-Jiménez et al. 2013).

SMT is incomplete, a fact that has been noticed for more than a decade by even one of its main promoters (Weinberg 2014). Cancer is an extremely complex disease that cannot be reduced to genes, genetic mutations, or even cancer cells or the cancer stroma. The original hallmarks of cancer need to be supplemented by multiple other perspectives: the *epigenetic hallmarks of cancer* (Esteller et al. 2024), *metabolic cancer hallmarks* (Pavlova and Thompson 2016; Pavlova et al. 2022), *immune hallmarks* (Cavallo et al. 2011) including *hallmarks of immune evasion* (Galassi et al. 2024), *hallmarks of immunologic interactions between stromal cells, and cancer cells* (Turley et al. 2015), the *cancer microenvironment hallmarks* (Pietras and Ostman 2010; Naser et al. 2023), the microenvironment role in metastasis (de Visser and Joyce 2023), the *hallmarks of metastatic cancer* (Welch and Hurst 2019; Fares et al. 2020), the *systemic hallmarks of cancer* (Paul 2020) including the *internal body environment* (Paul and Nedelcu 2022), and of cancer understood as a systemic disease (Paul 2020; Swanton et al. 2024). Recently, several research teams pointed to the fact that all cancer hallmarks can be explained by *cell-to-cell communication* (Becker et al. 2016; Dominiak et al. 2020; Lazar et al. 2022), in general, and communication through tumor-secreted microvesicles (Xavier et al. 2020; Clancy and D'Souza-Schorey 2023), especially exosomes, in particular (Wortzel et al. 2019; Lucotti et al. 2022), play a prominent role. Piecing the cancer puzzle together is a complex endeavor that needs extensive collaborative efforts (Alcolea et al. 2024).

Two key perspectives should also be included: the *hallmarks of aging* (López-Otín et al. 2023) and, importantly, the *hallmarks of health* (López-Otín and Kroemer 2021). Recently, a computational approach using machine learning was applied to DNA methylation and pathway/ontology data and was able to generate quantifiable hallmarks of aging and cancer (Xiong 2024). Dependency network analysis also identified "superregulators," such as genomic stability for colon cancer, described as the main enabling characteristic in the 2011 Hanahan and Weinberg hallmarks article (Hanahan and Weinberg 2011). Interestingly, for breast cancer, mitochondrial metabolism, and not one or several genes, was found to represent the main superregulator. In contrast, in both breast cancer and colon cancer, tissue remodeling was picked as the key disease controller. The intriguing observations reported by Jianghui Xiong are preliminary and require independent validation.

In 2000, in the frequently cited hallmarks of cancer article, Hanahan and Weinberg described cancer as comprising cells with *"defects in regulatory circuits that govern normal cell proliferation and homeostasis"* with the accent on *"growth,"* a detailed analysis of homeostatic mechanism being omitted, though, in both versions of their article (Hanahan and Weinberg 2000, 2011). The focus on growth is, indeed, important: many cancers kill patients by excessive tumor growth, and if we are able to better control both local and metastatic cancer growth with our therapeutic armamentarium, cancer can become a chronic disease, like diabetes. Unfortunately, with few notable exceptions, in most cancer cases, there are no drugs that can control cancer growth for a prolonged period of time. Metastatic disease is considered incurable mainly because, although systemic treatments may prolong the life of some patients, most of them will succumb due to a combination of uncontrolled cancer growth and both cancer- and therapy-induced systemic complications.

4.7 Five Main Cancer Theories

There are five principal oncogenesis theories, each of them placing a different emphasis on genetic and cellular factors as opposed to tissular factors.

First, the currently "dominant" theory, *called the somatic mutation theory* (SMT) accentuates the role of oncogenes and tumor suppressor gene mutations in causing uncontrolled proliferation and cancer (Nordling 1953; Armitage and Doll 1954; Weinberg 2006). This theory led to the development of targeted agents (small molecules and antibodies) and to the creation of the field of personalized/precision oncology.

Second, the *"bad luck" theory is* a variant of SMT proposing that cancer is caused by random mutations appearing in self-renewing stem cells that subsequently generate malignant clones (Tomasetti and Vogelstein 2015; Tomasetti et al. 2017). Robert Weinberg has also been a proponent of the cancer stem cell (CSC) theory for more than a decade (Pattabiraman and Weinberg 2014; Chaffer and Weinberg 2015). More recently, it has been shown that the dichotomy between "stemness" versus "non-stemness" is artificial, and stemness is an "emergent, contextual property of interactions with the microenvironment and other cell types, within both the cancer and normal hierarchy" (Prager et al. 2019), rather than an independent cellular state. As pointed by Weinberg before, CSCs are unlikely to represent the "object" of selection during cancer progression, as they do not divide often, and epithelial stem cells have a quasimesenchymal phenotype (Lambert and Weinberg 2021). The malignant phenotype is protean, and cancer cells have the capacity to shift dynamically between a differentiated state, with limited tumorigenic potential, and an undifferentiated or CSC state, which is responsible for long-term tumor growth, resistance, and recurrence (da Silva-Diz et al. 2018).

Third, *epigenetic theories* of cancer have also been proposed by several authors (Israel and Schwartz 2006; Esteller 2008; Baylin and Jones 2016; Esteller et al. 2024). Genes are considered passive agents compared to the action of different epigenetic mechanisms (methylation, histone acetylation, different types of RNA messengers, etc.) that are able to coordinate hundreds of genes simultaneously.

Fourth, the *metabolic theory* was initiated by Otto Warburg almost a century ago (Warburg 1956). He firmly believed that mitochondrial abnormalities are responsible for the cancer phenotype. There are several other contemporary metabolism-first theories (Seyfried 2012; Pavlova and Thompson 2016; Pavlova et al. 2022; Lane 2022).

Fifth, *tissue theories.* The tissue organization field theory (TOFT) advocates that malignancies are driven at the tissue level by abnormal interactions between different stromal components (Soto and Sonnenschein 2011). The significance of cancer cell-host tissue interactions and, in general, the importance of the "soil," as opposed to the "seed" in cancer development, has been strongly supported by the work of Isaiah Fidler (Fidler and Kripke 1977; Price et al. 1988). The work of Mina Bissell (Bissell and Radisky 2001) and Robert Weinberg (Orimo and Weinberg 2006) brought additional experimental evidence to the theory of cancer as caused by an abnormal microenvironment. The tissue organization field theory proposed by Carlos Sonnenschein and Ana Soto (Soto and Sonnenschein 2011) also suggests that cancer appears as a failure of the stroma to keep under control individual cells. Björn Brücher and Ijaz Jamall also proposed a microenvironment-centric theory in which genetic mutations are considered epiphenomena. They postulated that cancer originates following a sequence of six steps that include *(1)* a pathogenic stimulus (biological or chemical), followed by *(2)* subclinical chronic inflammation, from which develops *(3)* fibrosis with associated changes in the cellular microenvironment. For the authors, the distinction between "benign" inflammation that promotes the development of normal cells and "promalignant" inflammation that leads to cancer lies in the ability of inflammation to cause the onset of fibrosis. From these changes, *(4)* a precancerous niche (PCN) develops, which triggers *(5)* deployment of a chronic stress escape strategy (CSES) induced by TGFβ and continuous LOX activation, inducing remodeling of the ECM, which forms the precancerous niche. Subsequently, the formation of cell-cell complexes is deregulated, the stability of adherens junctions decreases, and the apical-basal polarity and reorganization of the cytoskeletal architecture are lost. From this, follows (6) a normal cell to a cancer cell transition (NCCCT) (Brücher and Jamall 2014, 2016).

Taking into consideration the tissue component of cancer opens the possibility of treating cancer indirectly, i.e., treating the cancer stroma and the cancer vasculature and enhancing the immune response against cancer. Indeed, tissular-directed approaches, alone or in combination with other treatment modalities, have been demonstrated to be successful in different clinical setups. Of note, immunotherapy did not develop as a result of TOFT; however, the fact that the action of checkpoint inhibitors and CAR-T cells occurs at the tissular level, and not inside cells, like chemotherapies or targeted agents, demonstrates the fundamental role of the stromal environment and cellular interactions in oncology.

The five theories presented above are not mutually exclusive, and theories combining elements of several theories have been proposed, for example, a theory that combines ideas from stem cell theory with epigenetic theory (Wainwright and Scaffidi 2017). As pointed by Zhu et al., there are multiple possible etiologies and mechanisms involved in cancer: some mutational, some non-mutational, some intracellular, and other extracellular (Zhu et al. 2022), and these mechanisms may be more or less prominent in different cancer types.

Of note, SMT and TOFT may not be incompatible (Bedessem and Ruphy 2015), and several authors have tried to reconciliate them (Rosenfeld 2013; Prokop 2021). Recently, *the ground state theory* proposed by Gilbertson and his collaborators uses elements from the SMT, the "bad luck theory," and TOFT, suggesting that both extrinsic and intrinsic cellular factors may be responsible for cancer transformation, adding also epigenetic factors as important contributors (Jassim et al. 2023). According to this theory, different stem cell populations are transformed by different mutations, even within the same tissue, and stem cells appear to show age-related differences in their susceptibility to transformation independent of proliferation; thus, additional forces (i.e., microenvironmental influences) must be at work to determine the susceptibility of specific cells to specific mutations (Jassim et al. 2023).

4.8 Other Cancer Theories

Besides these "orthodox" cancer theories, many other alternative, more or less plausible, cancer theories have been formulated.

4.8.1 Cancer as a Return to a Primitive Life Form

The hypothesis that malignancy may represent a reactivation of ancestral genetic programs has been explored since the late twentieth century. In 1978, the Romanian physician Octavian Udrişte proposed that cancer results from the derepression of ancestral prokaryotic genes present in all multicellular organisms. He referred to these as "anaerobic genes," suggesting that they originated during the anoxic conditions of early Earth (Udrişte 1978). In the second edition of his book, published in 1982, Udrişte included an English-language summary asserting that the "phenotypical expression of the cancerous cells is controlled by the ancestral genetic program" (Udrişte 1982).

Along similar lines, physicist Rafael Sorkin posited that cancer could be understood as the breakdown of the regulatory mechanisms that evolved during the transition from unicellularity to multicellularity (Sorkin 2000). He suggested that the comparison of unicellular and multicellular genomes could reveal potential intervention targets to restore these damaged control mechanisms and halt cancer progression.

Expanding on this evolutionary framework, a series of publications by Charles Lineweaver, Paul Davies, and Mark Vincent (Davies and Lineweaver 2011; Vincent 2010, 2011, 2012; Fernandes et al. 2012; Davila and Zamorano 2013; Lineweaver et al. 2014, 2021; Vinogradov and Anatskaya 2023) formulated different versions of the "atavistic theory of cancer." This theory posits that cancer represents a regression to a primitive, unicellular-like state. In this view, tumor cells express robust, ancient genetic programs that are typically suppressed in differentiated multicellular organisms. Building on this, Han Chen et al. argued that cancer progression and metastasis are driven by a selective loss-of-function of genes essential for multicellular cooperation, thereby reverting cells to a unicellular "ground state" (Chen et al. 2015).

Further empirical support for the atavistic model was provided through phylostratigraphic analyses by Zhou et al. and Anna Trigos et al., who applied this evolutionary gene-dating approach to transcriptomic data from The Cancer Genome Atlas (TCGA) across seven solid tumors (Trigos et al. 2017; Zhou et al. 2018). Their findings demonstrated that genes of unicellular origin were disproportionately upregulated in cancer, while genes associated with multicellular complexity were frequently suppressed.

In several publications, Mark Vincent characterized cancer as a "protozoan-like" biological entity, emerging through the derepression of an ancient survival program embedded within the genome (Vincent 2010, 2011). He framed this regression as a reawakening of a "Precambrian organism"in competition with the host's metazoan physiology. According to Vincent, the metabolic hallmark of this regression is the Warburg effect—an increased reliance on aerobic glycolysis.

In line with this perspective, I conducted a pilot study (Paul et al. 2020a) exploring the therapeutic potential of targeting cancer cell glucose avidity using escalating doses of 18F-fluorodeoxyglucose (18F-FDG) in patients with hypermetabolic tumors. Preliminary results suggested that some clinical effects at doses lacking any toxicities. However, after an initial response, all tumors progressed, suggesting that monotherapies targeting singular metabolic features are likely to encounter resistance, as tumor cells may adapt by utilizing alternative energy sources, such as glutamine, reflecting metabolic heterogeneity and plasticity.

Another innovative therapeutic strategy proposed by Lineweaver, Davies, and Vincent involved targeting tumor-associated macrophages with non-attenuated strains of *Listeria monocytogenes* (Lineweaver et al. 2014). While this specific approach has not yet demonstrated clinical efficacy, alternative Listeria-based immunotherapeutic interventions are under active investigation (Morrow et al. 2019).

Critiques of atavistic theory have emerged. Notably, Aurora Nedelcu contended that cancer does not constitute a reversion to a unicellular lifestyle per se. Instead, she described cancer cells as "opportunistic cheaters"—mutant derivatives within multicellular systems that disrupt cooperative cellular dynamics (Nedelcu 2020). She argued that the loss of cooperative gene function, while offering short-term survival advantages, could also compromise cellular fitness, such as increasing sensitivity to environmental stress, an exploitable vulnerability in therapeutic contexts.

A recent variant of the atavistic model, termed the systemic evolutionary theory of cancer (SETOC), has also been proposed, emphasizing the systemic evolutionary dynamics that underlie cancer progression and treatment resistance (Mazzocca 2019).

4.8.2 Cancer as a Developmental Disease

The conceptualization of cancer as a dysregulated embryonic developmental program has been proposed for over four decades (Udrişte 1978; Pierce 1993). As outlined by Anushka Dongre and Robert Weinberg, numerous genes implicated in embryogenesis are also central to oncogenesis (Dongre and Weinberg 2019). Analogous to neural crest cell migration during development (Gallik et al. 2017), the metastatic cascade in cancer represents a reactivation and pathological deployment of evolutionarily conserved cellular programs. Metastasis is orchestrated through a complex interplay between tumor cells and their microenvironment, including the secretion of signaling molecules such as transforming growth factor-beta (TGF-β), hepatocyte growth factor (HGF), tumor necrosis factor-alpha (TNF-α), Wnt, and platelet-derived growth factor (PDGF) by the surrounding stroma. These signals, in turn, activate key embryonic transcriptional regulators—Twist, Snail, Slug, Zeb1, and Zeb2—that are fundamental to the epithelial-to-mesenchymal transition (EMT), a process essential for both embryogenesis and cancer metastasis (Dongre and Weinberg 2019; Lambert et al. 2017).

Between 1989 and 1993, Vladimir Vinnitsky introduced the "oncogerminative cell theory," which conceptualizes tumorigenesis as a self-organizing and dynamic process that recapitulates early embryonic development (Vinnitsky 1989, 1993). He proposed five developmental stages of tumor progression originating from an "oncogerminative cell":

1. *Proliferation of oncogerminative cells,* akin to cleavage-stage embryos.
2. *Aggregation into a tumor germ,* resembling the morula stage.
3. *Transformation into a heterogeneous tumor spheroid,* analogous to the avascular blastocyst.
4. *Vascularization of the tumor spheroid,* mimicking the postimplantation embryo.
5. *Dissemination of oncogerminative cells* via *metastasis,* reflecting primordial germ cell migration (Vinnitsky 2014).

Building on a similar developmental perspective, Jingsong Liu and colleagues hypothesized that polyploid giant cancer cells (PGCCs) may have derepressed latent embryonic programs present within somatic tumor cells, thereby contributing to therapy resistance and disease relapse. Analysis of 38 postchemotherapy human ovarian cancer tissue samples revealed an upregulation of embryonic stem cell markers, supporting this hypothesis (Niu et al. 2017).

While cancer and embryonic development share regulatory networks and structural analogies, the comparison is inherently limited. Unlike the linear, teleological nature of embryogenesis, tumorigenesis often follows a pattern better described by deterministic chaos—nonlinear, unpredictable, yet governed by underlying constraints (Calin et al. 2003; Sharma 2016).

Neither classical developmental biology nor conventional evolutionary frameworks fully capture the complexity of cancer's causal dynamics. Integrative approaches such as evolutionary developmental biology (Evo-Devo) and developmental evolutionary biology (Devo-Evo) have been proposed as more holistic frameworks to reconcile these perspectives.

4.8.3 Cancer as Polyploidization

The hypothesis that cell fusion contributes to cancer progression has a historical basis extending over a century (Rachkovsky et al. 1998). Notably, the analogy of cancer as a form of "*somatic cell pregnancy*" has appeared intermittently in the oncology literature (Old 2001). In 1911, Professor Otto Aichel proposed that malignancy might originate from the fusion of leukocytes with somatic cells, producing hybrid cells with oncogenic potential (Aichel 1911 cited by Shabo et al. 2020). Subsequent work suggested that the fusion of tumor cells with immune cells, such as lymphocytes, could contribute to metastatic competency (Mekler 1971 cited by Rachkovsky et al. 1998). The origin of cancer stem cells has similarly been linked to cell fusion, with one model proposing that fusion between stem cells and somatically mutated cells could initiate tumorigenesis (Bjerkvig et al. 2005). Similarly, Thomas Seyfried and Leanne Huysentruyt proposed that metastatic cells may arise from epithelial-macrophage fusion hybrids (Seyfried and Huysentruyt 2013).

Cell fusion can give rise to aneuploid hybrid cells capable of altering the transcriptome and inducing adaptive phenotypic and genotypic changes. While the exact contribution of this mechanism to oncogenesis remains under investigation, cell fusion is increasingly recognized as a process that may facilitate tumor heterogeneity, promote aggressive traits, and enhance metastatic potential.

Polyploid cells, likewise, have been known for more than a century; however, their significance in oncology has only recently gained attention. Jekaterina Erenpreisa and colleagues described the capacity of these polyploid cells to revert to diploid or quasi-diploid states, potentially contributing to tumor initiation (Erenpreisa et al. 2000; Illidge et al. 2000). Polyploidization events have been hypothesized to have occurred early in vertebrate evolution, potentially accounting for gene family expansions such as Hox gene clusters (Davoli and de Lange 2011). Also, more recently, Amend et al. characterized polyploid giant cancer cells (PGCCs) as potential drivers of tumorigenesis, metastasis, and therapeutic resistance (Amend et al. 2019).

In vitro, tetraploidization may be induced by prolonged DNA damage responses, a phenomenon reproducibly observed upon exposing tumor cells to genotoxic chemotherapy agents (Davoli and de Lange 2011; Pienta et al. 2020). Experimental data from multiple laboratories, including those of de Lange, Liu, and Pienta, support the induction of polyploidy in cancer cells following DNA damage.

Jinsong Liu formalized a mechanistic framework for PGCC behavior through the polyploid giant cancer cell (PGCC) cycle (Liu 2018). This model includes several distinct phases: (1) *Initiation,* wherein failed mitosis or cytokinesis leads to endoreplication, allowing cells to escape senescence and apoptosis; (2) *self-renewal,* during which PGCCs expand autonomously and generate embryoid-like structures; (3) *termination,* characterized by the budding of small, stem-like progeny from the PGCCs; and (4) *stability,* during which these progeny cells stabilize their karyotype, acquire mitotic competence, and give rise to tumors of varying differentiation grades, including germ cell tumors. Notably, Liu argues that genetic mutations are not strictly necessary for PGCC formation. Instead, embryonic reprogramming through the activation of developmental transcription factors may suffice, although oncogenic mutations (e.g., TP53, RB1, BRCA1/2, KRAS, and BRAF) can facilitate dedifferentiation and uncouple proliferation from differentiation (Liu 2018).

Physiologically, the conversion from diploidy to polyploidy is a feature of normal mammalian development, occurring in specialized cell types such as trophoblasts (8 N–64 N), megakaryocytes (16 N–128 N), and hepatocytes (4 N, 8 N). Fusion-mediated polyploidization also occurs in skeletal muscle fibers and osteoclasts. Polyploidy has additionally been sporadically reported in other tissues, including the lactating mammary gland, urothelium, mesothelium, and Purkinje neurons (Davoli and de Lange 2011). Three primary mechanisms for tetraploidization in cancer have been identified: (1) cell fusion, (2) cytokinesis failure or mitotic errors, and (3) endoreduplication (Davoli and de Lange 2011).

In cancer, polyploidy is proposed to buffer deleterious chromosomal mutations and support clonal evolution under selective pressure (Sansregret and Swanton 2017). De Lange proposed that polyploidy and chromosomal instability may arise from telomere dysfunction in early tumorigenesis (De Lange 2005), similar to Heng's model of "genomic chaos," who also proposed a transient phase of karyotypic instability followed by telomerase-mediated stabilization (Heng 2019). De Lange posited that a transient episode of genomic instability, driven by telomere attrition, might be followed by telomerase reactivation and karyotypic stabilization, enabling selection of advantageous mutations while minimizing mutational burden: "A brief episode of high mutation rate followed by return to a more stable genome would avoid a potential mutational load that might hamper proliferation. In this regard, telomere dysfunction is different from other sources of genome instability, since it is reversible through the upregulation of telomerase. Upon acquisition of sufficient telomerase levels, this period of telomere-related scrambling will end, resulting in more stable, yet altered, genomes. The notion that tumors develop through a brief period of telomere dysfunction that generates extensive genetic diversity is borne out by data on genomic alterations during the development of breast cancer" (De Lange 2005).

Polyploid cancer cells often exhibit supernumerary centrosomes, necessitating centrosome clustering for survival. This dependency provides a therapeutic opportunity: HSET/KIFC1, a kinesin motor essential for centrosome clustering, is overexpressed in multiple tumor types, including glioblastoma, lung, breast, colon, and cervical cancers, and dispensable in normal cells, making it an attractive drug target (Pannu et al. 2015). HSET inhibitors and centrosome-declustering agents such as the antiparasitic drug griseofulvin have shown efficacy in preclinical models (Kwon et al. 2008; Pannu et al. 2014).

Future cancer therapies may aim to block polyploidization or redirect PGCC differentiation toward benign lineages. Moreover, embryoid-like structures derived from PGCCs have been proposed as tumor vaccine candidates. Supporting this idea, Kooreman et al. demonstrated that induced pluripotent stem cells (iPSCs) can elicit antitumor immunity in murine models. In prophylactic settings, iPSC-based vaccines have shown efficacy against syngeneic models of breast cancer, mesothelioma, and melanoma (Kooreman et al. 2018).

Metastatic cancer cells exhibit behaviors, such as migration, immune evasion, and dormancy, that parallel those of immune cells like macrophages. This observation prompted the theory that fusion events between macrophages and tumor cells can generate hybrid cells with metastatic properties (Huysentruyt and Seyfried 2010). Notably, such behaviors are not exclusive to cancer; similar phenomena occur in infectious diseases such as tuberculosis, where *Mycobacterium tuberculosis* survives intracellularly and manipulates host immunity (Richmond and Roehner 2019).

While PGCCs display aggressive phenotypes in vitro, especially under chemotherapy-induced stress, these behaviors may not fully reflect the natural history of solid tumors, which often evolve over extended periods. As such, the PGCC model may be more relevant to therapy-induced tumor evolution rather than spontaneous cancer initiation and progression.

4.8.4 Cancer as Evolution

The connection between cancer and evolution was first formally articulated by John Cairns and Peter Nowell in their landmark papers published in *Nature* (Cairns 1975) and *Science* (Nowell 1976), respectively. Both authors proposed a stepwise model of tumor evolution driven by accumulating genetic variation. This evolutionary lens has become foundational in understanding therapy resistance in solid tumors. A well-known example is EGFR-mutant lung adenocarcinoma, where after 10–16 months, initial responses to targeted therapies like erlotinib often give way to resistance due to intratumoral clonal evolution (Gillies et al. 2012). When resistant clones preexist treatment, standard cytotoxic strategies inadvertently select for these variants, hastening disease progression (Gatenby et al. 2009). As such, acquired resistance to targeted agents seems inevitable, and future therapies need to address spatial and temporal tumor heterogeneity more effectively (Gerlinger et al. 2012; Marusyk et al. 2014).

It has been shown that multiple models of evolution (see Fig. 4.1) may operate at different stages of progression or concurrently during tumor evolution (Davis et al. 2017). Marticorena et al. demonstrated that the vast number of cancer mutations are neutral and that only a small number (~5%) are positively selected (Martincorena et al. 2017); thus, in Fig. 4.1, the left lower branch is best represented. Importantly,

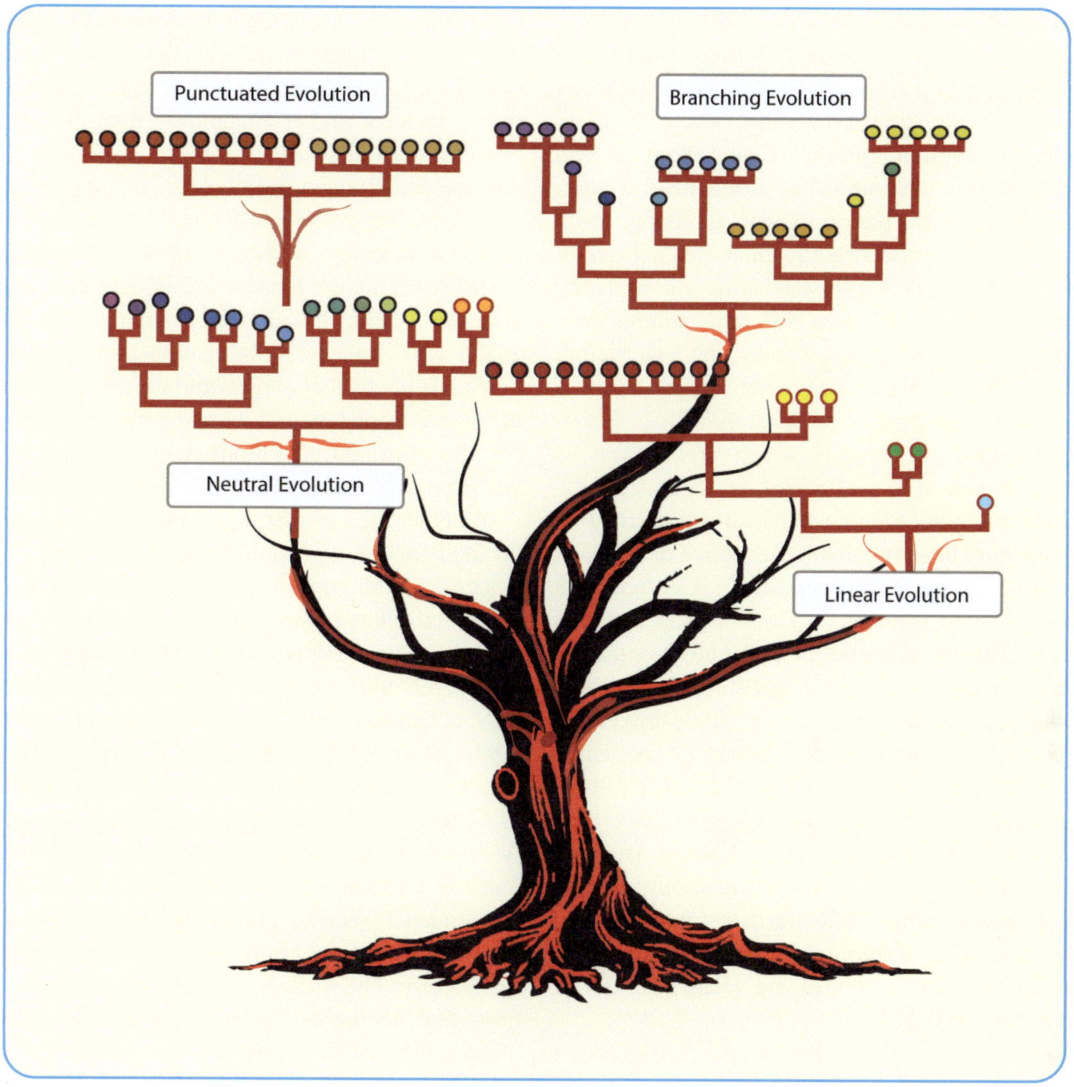

Fig. 4.1 Evolutionary mechanisms: punctuated evolution, branching evolution, linear evolution, neutral evolution. (Courtesy of Ana Tourian)

negative selection, which is commonly present in the germ line, is almost absent in cancer cells, pointing to the extreme resilience of cancer cells to an increased mutational burden. The increase in mutational tumor burden may come to yet uncharacterized fitness advantage, a phenomenon called Muller's ratchet (Bakhoum and Landau 2017).

Cancer is a robust system that has the ability to maintain stable functioning despite various perturbations. The essential robustness of cancer is maintained through heterogeneous redundancy, i.e., the cancer tissue contains a heterogeneous distribution of genetically different cancer cells maintained by genetic instability (Kitano 2004).

Genetic mutations responsible for intra- and intertumoral heterogeneity have been described by Charles Swanton and his team in several seminal papers (Gerlinger et al. 2012; Jamal-Hanjani et al. 2017; Watkins et al. 2020); Swanton made popular the suggestive notions of "trunk" mutations present in the majority of cancer clones versus "branch" mutations present in cancer subclones only.

Expanding the evolutionary framework, Henry Heng introduced the concept of *genome chaos*; his theory proposes that, under stress, cells undergo a burst of chromosomal reorganization (macroevolution), followed by clonal stabilization (microevolution) (Heng 2019). His Li-Fraumeni fibroblast model supports this hypothesis, revealing fluctuating karyotypes that eventually stabilize over passages. Heng proposes that cancer transitions—primary tumor formation, metastasis, and drug resistance—are driven by chromosomal system-level changes: "the entire set of chromosomes of a given species is a new genomic coding, which defines the network structure (how individual genes interact) and serves as a blue print or system inheritance" and, also, "the physical position of the gene (or the address of the gene in the nucleus) is important for networks dynamics" (Heng 2016). This view is reminiscent of Richard Goldschmidt's early twentieth century concept of repatterned genomes driving evolutionary leaps: "A repatterning of a chromosome may have the exact same effect as an accumulation of mutations. And even more, a complete repatterning might produce a new chemical system which as such, i.e., as a unit, has a definite and completely divergent action upon development, an action which can be conceived of as surpassing the combined actions of numerous individual changes by establishing a completely new chemical system. Model: two different pictures produced with the same set of mosaic blocks; the new picture 'emerging' only when all blocks are in their place. It is certainly most remarkable that the new developments in genetics lead to the same conclusions which are derived as postulates from an unbiased analysis of evolutionary field. This encourages me to believe that the dead end reached by neo-Darwinian theory based upon the conception of classical genetics can now be passed successfully" (Goldschmitdt 1940). A similar theory of genome system architecture was also proposed in 2005 by James Shapiro: "From an organizational point of view, distant effects of repetitive element dosage tell us that the whole genome is a single integrated system, regulated both in *cis*- and *trans*- by networks employing DNA repeats" (Shapiro 2005). Also, other authors (Lin et al. 2012; Bloomfield and Duesberg 2016) have likewise emphasized the systemic nature of genome architecture, suggesting that *chromosomal translocations are not random* but structured events embedded in transcriptional and epigenetic regulation. Karyotypic complexity appears to be conserved in metastases, reinforcing Heng's model of one-step metastatic seeding.

Supporting this idea, single-nucleus sequencing of breast tumors revealed early, stable aneuploid rearrangements and gradual point mutations (Wang et al. 2014). Chromothripsis, a single catastrophic genomic event, has been detected in ~2–3% of cancers (Stephens et al. 2011). These findings suggest that carcinomas may follow branching evolutionary models for mutations but punctuated evolution for chromosomal alterations (Davis et al. 2017; Zack et al. 2013). TracerX studies from Swanton's group (Turajlic et al. 2018) further advanced this model, showing that renal cell carcinomas with high chromosomal complexity but low clonal diversity progressed more aggressively than heterogeneous tumors, highlighting the dual impact of macro- and microevolution on prognosis and therapy decisions.

Chromosomal instability (CIN) gives rise to both somatic copy number alterations (SCNAs) and structural changes, with implications for heterogeneity, metastasis, and resistance. Studies by Swanton and others have charted the landscape of SCNAs, identifying both selection and neutrality as the driving forces behind subclonal variation (Watkins et al. 2020). A prior large study (Taylor et al. 2018) that analyzed aneuploidy across 10,522 tissue samples spanning 33 cancer types from the Cancer Genome Atlas (TCGA) dataset found that whole chromosome or chromosome-arm imbalance occurs in 88% of cancers. Some types of cancer have been shown to have higher aneuploidy than others: virtually all glioblastomas and testicular germ cell tumors have at least one aneuploidy event. In colon cancer, most tumors (85%) are aneuploid, likely due to CIN, and a minority of them (15%) have microsatellite instability (MSI). In general, these two mechanisms are not present within the same tumor (Dunican et al. 2002), but this rule is not universal (Bolhaqueiro et al. 2019). This is an oversimplification, as the aneuploidy tumors may be further subclassified into tumors with low copy number chromosomal alterations and tumors with high copy number chromosomal alterations, with different histopathology and gene expression profile (Barresi et al. 2018).

Whether aneuploidy is a driver or a passenger has been debated for a long time. In prostate cancer, its presence is associated with poor prognosis (Stopsack et al. 2019), yet in colorectal cancer, single-chromosome gains can paradoxically promote or suppress metastasis (Vasudevan et al. 2020). CIN levels also influence prognosis nonlinearly: moderate CIN correlates with poor outcomes and drug resistance, while extreme CIN may impair tumor viability (Sansregret and Swanton 2017; Bakhoum and Compton 2012). A recent in vitro study (Hoevenaar et al. 2020) performed in mice predisposed to intestinal cancer (ApcMin/+) showed that certain levels of CIN can have contrasting effects on distinct tissues. Moderate CIN causes a remarkable increase in adenoma burden throughout the entire intestinal tract, especially in the distal colon but, strikingly, a higher level of CIN although promoted adenoma formation in the distal colon, has no effect on the small intestine.

In a recent analysis of over 10,000 tumors using computational tools, a research team from Boston demonstrated that aneuploidy functions as a driver of cancer progression. They identified multiple chromosomal regions whose duplication or deletion either promoted or hindered tumor development, thereby providing strong validation—more than a century later—of Boveri's original hypothesis (Shih et al. 2023).

Efforts to therapeutically target CIN are underway, with trials focusing on key regulators like ATR, CHK1, and WEE1 kinases, alongside numerous preclinical agents targeting replication stress and mitotic dysfunction (Baillie and Stirling 2021).

A central critique of cancer evolution theory is its inability to fully explain metastasis, a complex, coordinated phenotype difficult to attribute solely to random mutation and clonal selection. Evolutionary frameworks must be expanded to include insights from ecology (eco-evo) and developmental biology (evo-devo). For instance, metastatic potential may preexist in the primary tumor and do not appear to be selected by an evolutionary process (LaFave et al. 2020), and tissue aging and the decline of defense mechanisms may represent alternative explanations for cancer development (De Gregori 2018).

In conclusion, cancer evolution is far more nuanced than Darwinian models suggest. Integrating chromosomal dynamics, systemic inheritance, environmental context, and cellular plasticity yields a richer, multilayered understanding that is better suited to confronting the heterogeneity, adaptability, and resilience of cancer.

4.8.5 Cancer as Organization Dissolution

More than 60 years ago, David Waldron Smithers articulated a foundational perspective on biological organization, emphasizing that the life of a multicellular organism is not merely a summation of the functions of its individual cells. He noted that "the life of a complex organism is not merely the sum of the individualities of its millions of component cells; it has a structure all its own which is a by-product of their actions [...]. Organismal organization and its disturbances can no more be tied down solely to the laws of individual cell action than everything which happens in society can be attributed to the direct design of individuals within that society, or even of powerful groups [...]. The organization of cells into organs and of organs into organisms is something more than individual cell performance: it exists in its own and performs the most important and fundamental functions of life. The behavior of individual cells is, in fact, the result and not the origin of organized living." (Smithers 1962).

Subsequent scholars have reinforced this paradigm. Wallace Clark, for instance, emphasized that a defining feature of cancer cells is their detachment from the organismal organizational framework (Clark 1995). In this view, malignancy arises not from intrinsic cellular anomalies alone but from the dissociation of the cell from the hierarchical control systems that govern normal tissue function.

Moreover, cancer can be conceptualized as a nested hierarchy of organizational levels—cellular, tissue, and organismal—each containing complex, interdependent networks. While malignant growth may superficially appear chaotic, it may in fact follow emergent patterns of organization analogous to those observed in normal developmental and physiological processes. As Cliff Hooker noted, elucidating causality in complex systems such as cancer is inherently challenging, as these systems are characterized by context-dependent dynamics (Hooker 2011).

4.8.6 Cancer as Eco-Evo: Cancer and the Tissue Environment

If ecology is also taken into picture with evolution, a complex picture appears in which tumors interact continuously with the micro- and macroenvironment of the organism. Cancer cells phenotypic heterogeneity, epigenetic plasticity, and phenomena like dormancy can be understood as nongenetic adaptation strategies to environmental pressures (Pienta et al. 2025). Following a consensus conference convened by experts in cancer evolution and cancer ecology, a novel framework for tumor classification was proposed, culminating in a 16-type tumor taxonomy grounded in two indices: the Evo-index and the Eco-index (Maley et al. 2017). The Evo-index reflects intratumoral heterogeneity and temporal dynamics, capturing the evolutionary potential of neoplastic cells. The Eco-index quantifies the environmental pressures—both resource availability and hazards—that influence tumor cell survival. Together, these indices aim to guide clinical decision-making by offering prognostic insight into tumor evolvability and informing the aggressiveness of therapeutic interventions. Furthermore, this framework opens the door to dynamic treatment monitoring through serial biopsies, enabling more responsive and tailored oncologic care.

Building on similar eco-evolutionary principles, Robert Gatenby pioneered "adaptive therapy," an innovative treatment strategy that has already shown promise in prostate cancer (West et al. 2020; Zhang et al. 2022a). Rather than aiming for complete eradication of cancer cells, *adaptive therapy* seeks to maintain a stable tumor burden by deliberately preserving a population of treatment-sensitive cells. This approach reduces selective pressure for resistant clones by applying lower-than-conventional drug doses in a cyclic manner. For over a decade, Robert Gatenby and his colleagues at Moffitt Cancer Center have refined this protocol through extensive preclinical and clinical work (Gatenby and Gillies 2004, 2008; Gatenby et al. 2009; Orlando et al. 2012; Enriquez-Navas et al. 2015). Recently, they expanded the adaptive therapy framework by integrating three key principles: frequency-dependent treatment cycles, drug selection sequencing, and evolutionary velocity modulation (West et al. 2020). This strategy reframes cancer treatment not as a war of annihilation but as a controlled equilibrium. The new strategy of playing a "game of clones" (Arney 2020) under the banner "Make peace with cancer, not war," may work for certain cancers.

Nonetheless, caution is warranted. The success of adaptive therapy has thus far been demonstrated primarily in prostate cancer, a typically indolent malignancy. Many elderly patients with prostate cancer ultimately die *with*, rather than *from*, the disease. Translating this paradigm to more aggressive malignancies such as pancreatic or small cell lung cancer presents considerable challenges, requiring further adaptation of evolutionary control principles to higher-velocity tumor dynamics. Ecological strategies directed to the tumor microenvironment like anti-antiangiogenic agents or hypoxia induced factor inhibitors, and targeted nutrient depletion (decreasing adenosine levels, for example) or restoration (increasing tryprofan levels, for example) can curb, in some cases, cancer aggressivity and improve immune responses (Pienta et al. 2025).

In 2021, a collaborative initiative led by a consortium of eco-evo oncology researchers produced a comprehensive list of 84 key research questions organized under nine broad themes: *cancer initiation*, *metastasis*, *tumor microenvironment*, *holobiont dynamics*, *immune interactions*, *mathematical modeling*, *eco-evo therapies*, *links to species ecology and evolution*, and *cancer in wild-type animal populations* (Dujon et al. 2021). This ambitious undertaking aims to expand the portfolio of cancer research at the ecology-evolution-oncology interface and foster interdisciplinary collaboration. Importantly, also underscores the enduring gaps in our understanding of cancer's fundamental nature, reminding us that many of the most critical questions in oncology remain unanswered.

4.8.7 Cancer as Evo-Devo: Evolution and Aging

James DeGregori brought forth a tissue theory of cancer, advancing the hypothesis that cancer represents a maladaptation to aging, leading to a transformed microenvironment in which the state of the tissue environment, and not genetic mutations, is the key determinant for cancer development, i.e., healthy youthful tissues maintaining normal cells over cells with cancer-causing mutations, while old or damaged tissues being more likely to favor cells with cancer-causing mutations (De Gregori 2018; Montégut et al. 2024).

Evo-Devo-based treatment approaches were recently proposed by James DeGregori in his book *Adaptive Oncogenesis*. Starting from the observation that approximately 80% of cancers occur after the age of 55 years and considering that recent data indicate the presence of mutations in various normal tissues from young individuals without cancer, DeGregori suggested a cell extrinsic origin of oncogenesis due to the physiologic decline of tissular control over cellular activity associated with aging. So, he brings forth the idea that cancer is an adaptation to a modified environment and he proposes that therapeutic approaches are directed to the microenvironment, i.e., the oxygen level or the pH of the environment, instead of the cancer cell itself (De Gregori 2018; Mazzone et al. 2009; Ibrahim-Hashim et al. 2017).

4.8.8 Cancer as an Immune System Failure

The theory of cancer immune surveillance, first proposed independently by Burnet and Thomas, postulates that the immune system plays a fundamental role in monitoring and eliminating nascent malignant cells, thereby preventing tumor initiation and preserving tissue homeostasis. This concept lays the groundwork for the modern understanding of the immune system's dual role in tumor suppression and progression (Burnet 1957; Thomas 1959). It is now well recognized that an intricate and dynamic interplay exists between immune effector mechanisms and evolving tumor cell populations (Kim et al. 2007; Schreiber et al. 2011).

Building upon the immune surveillance hypothesis, the concept of cancer immunoediting provides a more comprehensive framework encompassing three distinct phases: *elimination*, *equilibrium*, and *escape* (Dunn et al. 2002; Mittal et al. 2014). During the elimination phase, cytotoxic T lymphocytes (CTLs), natural killer (NK) cells, and other immune components target and destroy immunogenic tumor cells through the recognition of tumor-specific or tumor-associated antigens presented via major histocompatibility complex (MHC) molecules (van der Burg et al. 2016). However, selective pressure exerted by the immune system can promote the survival of tumor variants with reduced immunogenicity—either through loss of antigen presentation, MHC downregulation, or epigenetic silencing of immunogenic genes—leading to immune escape and tumor progression (Muenst et al. 2016; Tumeh et al. 2014).

A key mechanism through which tumors evade immune detection is the exploitation of immune checkpoint pathways. These regulatory pathways, such as PD-1/PD-L1 and CTLA-4/CD80/86, normally function to maintain peripheral tolerance and prevent autoimmunity but are co-opted by tumor cells to suppress T cell activation and effector functions (Pardoll 2012; Ribas and Wolchok 2018). The *immune synapse*, initially described as the structured interface between T cells and antigen-presenting cells (Norcross 1984), has since been extended to encompass tumor-immune cell interactions, providing a structural and functional framework for understanding checkpoint-mediated immune modulation (Dustin and Long 2010; Dustin 2014).

Immune checkpoint blockade, which disrupts these inhibitory pathways, has revolutionized the treatment landscape of various cancers by restoring T cell-mediated antitumor activity (Sharma and Allison 2015; Ribas and Wolchok 2018). Unlike traditional cytotoxic or targeted therapies, checkpoint inhibitor therapies can induce durable, sometimes curative, responses, even in metastatic settings. The efficacy of these agents is notably enhanced in tumors with a high tumor mutational burden (TMB), which correlates with increased neoantigen generation and immune recognition (Rizvi et al. 2015; Yarchoan et al. 2017). This biological rationale underpinned the FDA's landmark approval of pembrolizumab as a tumor-agnostic therapy for microsatellite instability-high (MSI-H) or TMB-high malignancies (Marabelle et al. 2020).

Importantly, immune checkpoint blockade is less susceptible to intratumoral genetic heterogeneity compared to molecularly targeted therapies, as it relies on broader immune recognition rather than targeting specific oncogenic drivers (McGranahan et al. 2016). Clinical trials have validated the transformative potential of immune checkpoint blockade across a range of cancers. In metastatic melanoma, the combination of anti-CTLA-4 (ipilimumab) and anti-PD-1 (nivolumab) antibodies has extended 10-year overall survival rates to more than 50%, a remarkable improvement over historical norms. In addition, among patients who were alive and progression-free at 3 years, the 7.5-year melanoma-specific survival was 96% with nivolumab plus ipilimumab, 97% with nivolumab, and 88% with ipilimumab, which is remarkable (Wolchok et al. 2025). Similarly, in non-small cell lung cancer (NSCLC), the 5-year survival rates in non-small cell lung cancer have more than doubled after the introduction of checkpoint inhibitors (nivolumab plus ipilimumab and pembrolizumab) into frontline treatment protocols (Gettinger et al. 2018; Garon et al. 2019).

As research advances, combinatorial strategies involving checkpoint inhibitors, personalized neoantigen vaccines, and adoptive T cell therapies offer promising avenues for further improving outcomes, particularly in immunologically "cold" tumors. However, immune-related adverse events (irAEs), biomarker heterogeneity, and resistance mechanisms remain key challenges in optimizing immunotherapy for broader clinical benefit (Postow et al. 2018).

4.8.9 Cancer as a Bioelectric Disease

Since the landmark study published in 1947, in which Louis Langman and H.S. Burr demonstrated striking differences in voltage readings between women with cervical cancers and normal subjects (Langman and Burr 1947), several studies have demonstrated the role of bioelectric signaling in cancer cell proliferation and tumor growth (reviewed by Michael Levin in 2020).

The membrane depolarization cancer theory is a variant of TOFT in which cancer is considered a disease of tissue organization (Moore et al. 2017), and several theories of cancer caused by membrane depolarization have been proposed (Yang and Brackenbury 2012; Levin 2014; Levin et al. 2019; Carvalho 2021). In this model, a stressful event overwhelms homeostatic tissue control systems, leading to global changes in the bioelectric properties of cells normally coupled via gap junctions and intercellular electric fields. The importance of cell-cell interaction both in promoting and inhibiting tumorigenesis has been demonstrated by the observation of oncogenic transformation of cells isolated from their neighbors (Moore et al. 2017). Experiments led by Michael Levin demonstrated the role of global tissular bioelectricity in regulating cancer cell metastasis (Payne et al. 2019). As a proof of principle, Levin and his team demonstrated that the electrical state overrules the genetic state by inducing, in a tadpole model, metastatic melanoma behavior in the absence of genetic mutations simply by disrupting the electrical communication between cells (Blackiston et al. 2009).

Membrane depolarization may influence cells at both the genetic and epigenetic levels. As an example, depolarization leads to changes in the activity of genes such as Notch, BMP, Sox10, Nurr1, Slug, Fos, Jun, NPY, and Wnt and can also regulate chromatin acetylation through butyrate signaling (Levin 2014). Of note, this model points to the importance of neighboring cells in inducing a particular cell state, and cancer is seen as the outcome of cellular community dysregulation and not a process of cancerization of a unique cell. In this model, the depolarized state corresponds to a more proliferative and motile cell, and the polarized state to quiescent behavior (Carvalho 2021). Notably, membrane depolarization may be important for the emergence and maintenance of a cancer stem cell state (Yang and Brackenbury 2013; Sundelacruz et al. 2013).

In recent years, the idea of using electricity to treat cancer has led to some interesting results. Tumor-treating fields (TTF) are low-intensity, intermediate-frequency, alternating electric fields delivered externally through noninvasive transducers placed locoregionally around the anatomic region of the tumor (Mun et al.2018). TTF has been FDA approved for the treatment of brain cancer (Ghiaseddin et al. 2020), mesothelioma (Ceresoli et al. 2019), metastatic non-small cell lung cancer (Leal et al. 2023), and, recently, metastatic pancreatic cancer (Babiker et al. 2025). Another device using tumor-specific amplitude-modulated radiofrequency electromagnetic fields (AM RF EMF) has also been approved in Europe since 2018 for the treatment of liver cancer (Jimenez et al. 2019). Prior proposals have been made to consider cancer a "channelopathy" (Prevarskaya et al. 2018), and the modulation of the bioelectricity of cancer cells using targeted agents directed to membrane ion channels, as suggested by Michael Levin, may represent a novel way of cancer treatment. Interestingly, both TTF and AM RF EMF treatments work by activating specific calcium channels and result in increased levels of cytosolic Ca2+ (Jimenez et al. 2019), so it is plausible that similar effects can be obtained by directly manipulating cancer cell membrane channels.

4.8.10 Cancer as an Intercellular Communication Disorder

The importance of communication for understanding life and, by extension, cancer has been discussed in a series of papers by Guenther Witzany (2010; Baluška and Witzany 2014; Witzany 2020). The importance and diversity of the intercellular communication modalities of cancer cells have been reviewed recently (Ebrahim et al. 2024). Miroslav Radman introduced the term "cellular parabiosis" to describe "the compensation of functional deficiencies of individual cells in tissues by intercellular communication via molecular traffic" (Radman 2019) and proposed that cancer promotion is due to cell separation from their neighbors. During inflammation, macrophages secrete matrix metalloproteinases (MMPs) and other proteases that may disrupt the stroma architecture, leading to the isolation of some cells from their neighbors. Senescent cells express a senescence-associated secretory phenotype (SASP) and secrete cytokines and proteases that may lead to stroma architecture disruption and cellular isolation. In his model, phenotypic suppression occurs between neighboring cells through the transfer of functional and informational molecules, including RNAs and proteins (including prions), as well as entire organelles (mitochondria, lysosomes, and the Golgi apparatus) and even viruses. Transfer occurs mainly through gap junctions but also through tunneling nanotubes, extracellular vesicles and/or exosomes. In his opinion, cellular parabiosis is the mechanism underlying tissue homeostasis, and disruption of cellular parabiosis is the cause of cancer and other chronic conditions. Parabiosis among cancer cells (Osswald et al. 2015) and between cancer cells and other cells of the cancer stroma (Zhang and Zhang 2015) has been well described in brain cancer. Cellular parabiosis can contribute to cancer cell fitness (Radman 2019). Through cellular parabiosis, cancer cells can parasitize the neighboring cells by depleting them of nutrients, can instigate them to support the oncogenic process (inducing them to destroy the stroma, for example), and can avoid being killed by the immune system through checkpoint inhibition.

4.8.11 Cancer as an Intelligent Process

Biological systems represent a distinct class of systems that, as highlighted in numerous discussions since Schrödinger's seminal inquiry *What is Life*, cannot be fully explained by their physical, chemical, or even informational properties alone (Schrödinger 1944).

A growing body of literature suggests that cognition may be an intrinsic characteristic of all living cells (Dennett 2017; Baluška and Levin 2016; Miller and Torday 2018; Reber 2018; Shapiro 2021; Marshall 2021; Levin 2022; Torday 2024). These authors argue that the ability of cells to perceive environmental stimuli, adaptively respond, and engage in goal-directed behaviors constitutes a rudimentary form of cognition. For example, James Shapiro posits that prokaryotic cells exhibit cognitive capabilities by identifying and discriminating among nutrients, directing their movement toward nutrient-rich environments, engaging in chemical communication, forming biofilms, undertaking coordinated group migrations, and activating genome repair mechanisms in response to DNA damage. Notably, many of these behaviors are also observed in cancer cells (Shapiro 2021).

Building on these arguments, Perry Marshall proposed the model of the cell as a "volitional Turing machine," capable of information processing and decision-making. His broader thesis suggests that life, cognition, cancer, evolution, and artificial intelligence may all converge upon the fundamental problem of cognition. He specifically references cancer cells as systems capable of self/non-self-discrimination, therapy resistance, and angiogenesis (Marshall 2021).

The notion of cellular intelligence has historical roots; in 1916, Nels Quevli asserted that evolution is guided by cellular intelligence (Quevli 1916). More recently, Ford argued that intelligence is a defining cellular trait (Ford 2017). The concept of intelligence may offer a more comprehensive framework than cognition when analyzing cancer, especially if intelligence is understood as adaptive capacity. Nicolazzo et al. defined biological adaptation across four interrelated layers: (1) morphological flexibility to fit environmental constraints; (2) metabolic reprogramming to enhance fitness; (3) environmental modification; and (4) engagement in collective adaptation. They cite metastatic cancer cells as capable of all four adaptive strategies (Nicolazzo et al. 2023).

Intelligence is multifaceted and shaped by both genetic and environmental inputs (Neisser et al. 1996). Consequently, the diverse genetic makeup and microenvironmental context of different cancers may yield varying types and degrees of cellular intelligence.

Daniel Dennett proposed a classification of intelligence into four hierarchical levels: Darwinian, Skinnerian, Popperian, and Gregorian (Dennett 2017). *Darwinian intelligence* involves selection through heritable variation; *Skinnerian intelligence* incorporates learning via reinforcement; *Popperian intelligence* entails the capacity to simulate hypothetical scenarios; and finally, *Gregorian intelligence* encompasses the use of external symbolic tools (e.g., mathematics and language) and social learning.

Cancer cell behavior has been extensively analyzed in Darwinian terms, although not all tumors conform strictly to Darwinian evolution. Cancer cells also exhibit Skinnerian traits, adjusting their phenotype in response to environmental feedback (Frost 2020). It remains uncertain whether they possess Popperian or Gregorian intelligence, that is, whether they can simulate alternative future scenarios, such as remaining quiescent or migrating to a different organ site based on anticipated environmental conditions.

Athena Aktipis and colleagues proposed that cancer cells "cheat" by violating the cooperative rules of multicellular life (Aktipis et al. 2015; Aktipis 2020). These rules include regulated proliferation, task-specific function, limited resource usage, cellular housekeeping, and timely apoptosis.

Cancer cells defy these norms, leading to unchecked division, tissue invasion, immortality, and metastasis. Athena Aktipis described them as "cheaters," while Kat Arney labeled them "rebels" (Arney 2020). These anthropomorphic descriptions reflect a broader epistemological challenge: whether it is scientifically valid to attribute intention or goal-directedness to malignant cells. Michael Levin argues that such metaphors should be evaluated based on their heuristic utility rather than dismissed outright. He suggests that behaviors traditionally attributed to cognition, such as memory, anticipation, context-sensitive decision-making, and learning, can be modeled as feedback systems that seek to stabilize attractor states in dynamic systems (Levin 2019).

In his technological approach to mind everywhere (TAME) theory, Michael Levin redefined intelligence as the capacity to navigate various conceptual spaces (e.g., morphospace, transcriptional space, and physiological space) toward favorable outcomes while avoiding local minima. He proposed that cognition is not exclusive to living systems but can also be present in artificial intelligence, framing it as a continuum rather than a binary property (Levin 2022). His earlier work introduced the concept of "scale-free cognition," suggesting that higher-order cognitive structures (Selves) emerge from the interaction and communication of simpler cognitive units (selves), thereby enhancing computational capacities (Levin 2019).

Building upon these ideas, Michael Levin also proposed a theory of cancer in which cells, losing integration with the organism through disruption of physiological signaling (e.g., bioelectric cues), revert to a unicellular behavioral mode (Levin 2019). This perspective aligns with earlier theories such as Radman's "cellular parabiosis" and the "reprimitivization" theories of Octavian Udriște, Mark Vincent, and Paul Davies.

The question arises: is cancer merely a stress-induced reactivation of primitive survival strategies—what Arney termed a "safe mode" (Arney 2020) or a "ground state" (Jassim et al. 2023)—or does it exhibit a more complex form of adaptive intelligence than a simple switch to a "default mode"?

4.8.12 Cancer as Organismal Programmed Self-Destruction

Given the association between cancer and aging, can cancer represent the outcome of a programmed kamikaze-like behavior of the aging organism that self-destroys itself? Russian researchers have introduced the concept of *phenoptosis*, a genetically regulated process of organismal death that purportedly evolved to eliminate aged or damaged individuals from the population (Skulachev 1999; Lichtenstein 2005; Skulachev and Skulachev 2014). Kirkwood has made several strong arguments against programmed aging, pointing to the fact, for example, that the theory is circular because it starts from the premise that old individuals are worn out when exactly the fact that they are worn out needs an explanation (Kirkwood 2005). Mikhail Blagosklonny also contested this view (Blagosklonny 2012, 2013). He proposed the *hyperfunction theory*, which postulates that aging and age-related diseases (ARD), including cancer, are driven by the continued activity of developmental programs beyond their optimal time frame (Blagosklonny 2021). According to this model, functional genetic programs, such as those mediated by mTOR, persist past their developmental utility, ultimately leading to senescence and pathology. In his opinion, aging and ARD are neither programmed per se (in Blagosklonny's terminology, they are *quasiprogramed*) nor entirely random. He proposes the following universal temporal sequence responsible for aging and ARD: function → hyperfunction(aging) → malfunction (ARD). His argument is that genetic programs determine developmental growth and the onset of reproduction, but when these programs are completed, they are not switched off. Blagosklonny considers aging and ARD, a sort of "residue" of a completed program that is no longer beneficial for the organism. He offers as an example, hyperfunctional mTOR, which is useful in development, may lead to senescence once the developmental program ends (Blagosklonny 2013).

Nevertheless, cancer is not exclusively a disease of aging, complicating purely age-related theories. Additionally, systemic stress signals, such as mitochondrial DNA (mtDNA) and reactive oxygen species, may act as markers rather than mechanisms of programmed decline. Circulating mtDNA has been correlated with increased mortality in critical care settings (Nakahira et al. 2013) and is associated with frailty and cognitive decline in the elderly (Fan et al. 2022; Nidadavolu et al. 2023). Elevated levels of mtDNA have also been observed in various cancers and are linked to poor prognosis (Meng et al. 2019).

4.9 Personalized Oncology

Over the past century, numerous theories of cancer have emerged—each emphasizing a particular facet of tumor biology such as genetic mutations, metabolic deregulation, epigenetic modifications, immune evasion, or disordered tissue architecture—and extrapolating this singular lens across all cancer types. While each theory has contributed valuable insights, their limitations become apparent in light of the immense heterogeneity that characterizes cancer. We now recognize that each malignancy is a unique entity shaped not only by its intrinsic molecular and cellular properties but also by the broader context of the host's physiological status, comorbid conditions, immune competence, and complex bidirectional tumor-host interactions (Hanahan 2022; McAllister and Weinberg 2010). This diversity of biological and contextual factors has undermined efforts to construct a universal theory of cancer pathogenesis and has complicated the development of standardized therapeutic protocols applicable across all tumor types.

The field of personalized oncology has been developed precisely in order to address this heterogeneity, with the guiding principle of delivering "the right treatment, at the right dose, at the right time, to the right patient" (Obama 2015). Importantly, personalized medicine has been defined as "the ability to classify individuals into subpopulations that differ in their susceptibility to a particular disease, in the biology and/or prognosis of those diseases they may develop, or in their response to a specific treatment" and not as we may think, as "the creation of drugs or medical devices that are unique to a patient" (National Research Council 2011). This nuanced interpretation underpins many current efforts in clinical oncology.

At present, for metastatic solid tumor treatment, the U.S. Food and Drug Administration (FDA) has approved only two therapies that are fully "personalized" at the cellular level using autologous immune cells modified to target the patient's own malignancy. Sipuleucel-T, an autologous dendritic cell-based vaccine, has shown relatively modest benefits in metastatic prostate cancer (Kantoff et al. 2010). More recently, lifileucel, a tumor-infiltrating lymphocyte (TIL) therapy, was approved for metastatic melanoma (Sarnaiket al. 2021). In contrast, chimeric antigen receptor (CAR) T cell therapies, though created using a patient's T or NK cells, are engineered to target shared tumor antigens rather than individual-specific neoantigens (Neelapu et al. 2017). While these approaches represent significant advances, truly personalized immunotherapies remain technically challenging and economically prohibitive for routine use in most clinical settings (Phillips et al. 2014; Mills 2017; Schleidgen et al. 2013).

Nonetheless, the field of functional precision medicine is rapidly expanding. Innovative molecular profiling platforms, including next-generation sequencing (NGS), transcriptomic analyses, phosphoproteomics, and patient-derived organoid testing, are being integrated into clinical trials to identify the most effective therapies based on tumor-specific vulnerabilities (Malani et al. 2022; Kornauth et al. 2022; Wang et al. 2023b; Li et al. 2023; Acanda de la Rocha et al. 2025). Despite these technological strides, precision oncology still faces considerable methodological hurdles, including tumor heterogeneity, sampling bias, temporal evolution, and discrepancies between in vitro drug sensitivity and clinical efficacy. Cost and accessibility further limit the widespread implementation of these technologies in standard oncology care (Vogenberg et al. 2010; Schork 2015).

In reality, despite the conceptual advances in precision oncology, often considered a rebranding of personalized medicine, *la même Jeannette mais autrement coiffée* (same Janet, just with a different hairstyle), most treatment regimens for solid tumors, with some notable exceptions (i.e., targeted treatments of breast and prostate cancer, advanced melanoma and lung adenocarcinoma), remain largely empirical. Even immunotherapy, which has transformed care for select malignancies, demonstrates limited predictive reliability. Biomarkers such as PD-L1 expression, high tumor mutational burden (TMB), or inflamed tumor microenvironments do not consistently predict the response to immune checkpoint inhibitors, which generally achieve objective response rates of only ~20–25% across most solid tumors (Topalian et al. 2012; Yarchoan et al. 2017). Moreover, long-term survival remains an exception rather than the rule and is observed mainly in certain cancer types (i.e. melanoma).

Thus, while personalized and precision oncology represent powerful paradigms, their practical impact is still constrained by biological complexity, technical limitations, and systemic barriers. Future progress will likely depend on integrating multilayered biomarker data (Chang et al. 2025b), AI-guided decision algorithms (Wang et al. 2025b), and organism-level contextual factors into a truly holistic oncology framework.

The Main Cancer Theories in Focus

5

5.1 Biological Basis of Somatic Mutation Theory

Proto-oncogenes are defined as normal genes involved in normal cell growth and proliferation. They also have the potential to contribute to cancer development if their expression is altered through genetic mutations or genetic amplifications transforming into oncogenes (Cline 1987). The first proto-oncogene discovered was *src*, a gene of the Rous sarcoma virus (RSV), identified by Peyton Rous as a tumor-causing agent in 1911 (Rous 1911).

The similarity between viral-src (v-src), a gene coding a nonreceptor tyrosine kinase present in the chicken sarcoma virus genome, and a similar gene (c-src), present in the normal chicken genome, was discovered by Dominique Stehelin, J. Michael Bishop, and Harold E. Varmus in 1976. The authors demonstrated that because of genetic alterations (mutations, amplifications, etc.), proto-oncogenes can transform into oncogenes. The finding of the c-*src proto-oncogene* in normal cells suggested that mutations in genes with similar functions may induce the formation of tumors unrelated to viral infections, leading to the subsequent discovery of *oncogenes* in cancer. Importantly, the Rous sarcoma virus itself is not capable of infecting human cells. Humans have their own version of the c-src gene called SRC, and high levels of SRC expression have been strongly implicated in the development, maintenance, progression, and metastatic spread of several human cancers, such as prostate, lung, breast, and colorectal (Wheeler et al. 2009). Importantly, as opposed to *tumor repressor genes*, which require both copies of the allele to become inactivated and lose their protective effects against unregulated cell growth, *oncogenes* require only a single mutation on one allele to become activated and act in an autosomal dominant fashion.

Of note, in humans, only the human T lymphotropic virus (HTLV-1) has been demonstrated to *directly* transform human cells (an "acute" or transducing mechanism through which viral oncogenes are expressed using the cellular machinery) as opposed to other retroviruses, like the avian myelocytomatosis virus MC29, which inserts in the vicinity of the cellular avian Myc gene, altering its expression (a "slow" or cis-acting mechanism). It is estimated that approximately 15%–20% of human cancers worldwide are related to viruses (zur Hausen 1991) that may be either DNA (i.e., HPV, EBV, HepB, Kaposi sarcoma herpes virus) or RNA viruses (i.e., HTLV-1 and Hep C). Besides these six viruses, HIV also, by infecting the CD4 cells and depressing immunity, indirectly increases the risk of several cancers (Kaposi sarcomas, non-Hodgkin lymphomas, cervical cancers, anal and lung cancers).

Oncoproteins are proteins coded by oncogenes that play an important role in oncogenesis across the animal kingdom (Stehelin et al. 1976). Most oncoproteins act as growth factors, growth factor receptors (the class of tyrosine kinase receptors is particularly important for cancer therapeutics), transducers in signaling cascades, and transcription factors regulating cell growth, proliferation, and differentiation in the nucleus or play, for example, various other roles like programmed cell death regulators. A special category of transcription factors is *lineage-survival oncogenes*, which act as master regulators of a specific cell lineage and, when deregulated, may support the proliferation and survival of cancers derived from that lineage, like microphthalmia-associated transcription factor (MITF) (melanoma), the androgen receptor (AR) (prostate cancer), and ESR1 (luminal breast cancer) (Garraway and Sellers 2006; Hegde et al. 2019). More commonly than viral insertions, point mutations, amplifications, or chromosomal translocations may alter the structure or activity of oncogenes, resulting in modified proteins contributing to the malignant process.

The discovery of the *tumor suppressor genes* came as a result of the work of Alfred Knudson (1971, 2001) who speculated that in the familial forms of retinoblastoma (Rb), a second mutated gene was required to trigger tumorigenesis since this gene had already been inherited in the mutated form from a parent of the affected child. In 1983, the nature of the second mutation was discovered: it involved the loss

D. Paul, *Cancer: An Integrative Approach*, https://doi.org/10.1007/978-3-032-08808-6_5

of the second, hitherto intact, copy of the Rb gene (Cavenee et al. 1983).

The proponents of the somatic mutation theory (SMT), the main explanatory framework of the gene-centric theory, argue that carcinogenesis is a multistep process caused by genome-altering mutations, translocations, or chromosomal aberrations (Pecorino 2021) that disrupt cellular activities by the corroborated action of *three* broad classes of genes: *oncogenes*, which lead to activation of the neoplastic process; *tumor-suppressor genes*, which curtail cells' ability to inhibit abnormal cell proliferation; and *stability genes*, which hamper cell's damage repair mechanisms (Kinzler and Vogelstein 1997; Vogelstein and Kinzler 2004).

It has been proposed that alterations in one or several oncogenes, tumor-suppressor genes, or stability genes may initiate and then further promote cancer progression by providing a selective advantage to cancer cells through positive selection (Bozic et al. 2010; Vogelstein et al. 2013; Vogelstein and Kinzler 2004; Wood et al. 2007).

There are several genetic mechanisms associated with oncogene activation (Chial 2008):

- Point mutations, deletions, or insertions leading to a hyperactive gene product.
- Point mutations, deletions, or insertions in the promoter region of a proto-oncogene leading to increased transcription.
- Gene amplification events leading to extrachromosomal copies of a proto-oncogene.
- Chromosomal translocation events relocating a proto-oncogene to a new chromosomal site that leads to higher expression of the proto-oncogene.
- Chromosomal translocations leading to a fusion between a proto-oncogene and a second gene, resulting in a fusion protein with oncogenic activity.

Genome alterations present in cancer cells have been carefully analyzed, and both the timing of cancer inception and the nature of internal and external cancer factors and processes associated with different mutational signatures have been described. A classic example of step-by-step tumorigenesis is the model of Fearon and Vogelstein (1990).

The SMT model gained traction after the initial experiments done by Robert Weinberg and his collaborators, who were able to induce a malignant phenotype in normal fibroblasts—but, importantly, *not* a metastatic phenotype—that were transfected serially with a small number of oncogenes (Hahn et al. 1999; Hahn and Weinberg 2002). The same group demonstrated that the metastatic potential of cancer cells may be related to the cell of origin. They reported that an identical set of oncogenes introduced in melanoma cells and epithelial cells led to a much higher metastatic potential in transformed melanoma cells versus the epithelial cells (Gupta et al. 2005).

These experiments provided the foundation for the SMT theory by demonstrating that cancer can be induced in normal cells through a relatively small number of genetic alterations (three or four).

As opposed to epigenetic transformations, which are historically considered transitory, mutations are considered permanent in the cell's life. Mutations can be transmitted from generation to generation as part of inherited cancer syndromes. This sort of dogged persistence over time partially explains why it is considered essential for cancer etiology. Mutations are thought to rewire key intracellular signaling pathways, leading to the "addiction" of cancer cells to the (hyper)activity of a mutated protein belonging to one of these pathways. The term "*oncogene addiction*" was initially introduced by Bernard Weinstein in 2000 (Weinstein 2000). In 2002, he proposed a slightly modified definition (Weinstein 2002). The "*oncogene addiction*" concept postulates that some tumors rely on "one single dominant oncogene for growth and survival, so that inhibition of this specific oncogene is sufficient to halt the neoplastic phenotype" (Torti and Trusolino 2011). Cancer cells are considered dependent on or "addicted" to oncogenes that were initially acquired during multistep tumorigenesis and remain critical to the ongoing proliferation and viability of these cells long after they have progressed to a fully neoplastic state (Weinstein and Joe 2006). A classic example is VHL gene alterations in renal cell carcinoma (Gerlinger et al. 2012). Subsequently, the concept of a single-molecule "*oncogene addiction*" has been extended to the idea of a "*signaling pathway addiction*," and cancer has been viewed as a "*signaling pathway disease*" (Sever and Brugge 2015). Importantly, the "oncogenicity" of a gene versus its "tumor suppression" qualities are highly context-dependent properties that are influenced not only by the particular intracellular genetic and epigenetic context but also by the characteristics of the microenvironment (Monti et al. 2022).

5.2 Driver Gene Alterations

The genes responsible for this "bizarre circuitry" present inside the cancer cells (Torti and Trusolino 2011) have been called "*driver*" genes, in order to point to their relevance for the cancer process (they are "driving" the cancer process), as opposed to "*passenger*" genes, which are also genes mutated in the cancer cells but are considered to lack oncologic pathophysiologic relevance. "*Driver*" gene mutations have been defined as mutations that confer a fitness advantage to somatic cells in their microenvironment, i.e., leading the cells to more frequent divisions and therefore to positive selection during cancer evolution, in contrast to "*passenger*" mutations, which are defined as mutations that provide no such proliferative benefit and therefore are not selected (Vogelstein et al. 2013).

The vast majority (97%) of mutations in cancer are passengers as opposed to drivers (Lawrence et al. 2014). This traditional paradigm of the neutrality of passenger mutations has been challenged by in vitro experiments suggesting that the passenger load may reduce the fitness of cancer cells and slow or prevent the progression of both primary and metastatic disease (McFarland et al. 2017). The dichotomy between oncogenes and tumor repressors is an oversimplification since genes with both functions have been described in which the oncogenic versus the tumor suppressor properties vary according to the physiologic context; for example, some genes function as tumor repressors in normal cells and oncogenes in cancer cells (Shen et al. 2018).

Some driver mutations are closely related to only one cancer type (the oncogenic fusion BCR-ABL in chronic myeloid leukemia), while other driver mutations belong to one or several of the pathways described by Vogelstein's team (Vogelstein et al. 2013) and may be present in multiple cancer types. Finally, some driver mutations fall in between, being shared by a restricted type of cancer (Iranzo et al. 2018).

In general, searches for driver gene mutations have been limited to the coding regions of the genome, with driver mutations found in approximately two-thirds of the protein-coding genome (Sholl et al. 2016). The study of Scholl et al. demonstrated that potentially actionable alterations may be found in rare tumors, suggesting the utility of a "cancer-agnostic" approach in genomic profiling.

It has been previously estimated that about 60% of the variation in protein expression arises from posttranscriptional regulation (Schwanhäusser et al. 2013); thus, posttranscriptional changes are an important contributing factor in oncogenesis. Although cancer is thought to be driven primarily by changes in DNA, some driver alterations may manifest themselves via changes in RNA rather than DNA sequence mutations (PCAWG Transcriptome Core Group et al. 2020).

Recently, a computational method using neural networks has been used to find driving elements anywhere in the genome, including coding and noncoding areas. Driver elements found in noncoding cryptic splice sites, 5′ UTRs, and rarely mutated genes are estimated to collectively contribute to the disease pathology of up to 10% of tumors (Sherman et al. 2022).

Also, it has been recently demonstrated that driver gene mutations can be present in a significant number of the 3′ UTR noncoding areas of the genome (Fu et al. 2024), including nine of the ten canonical oncogenic pathways (Sanchez-Vega et al. 2018). The 3′ UTR genetic locus harbors *cis*-regulatory elements that interact with *trans*-factors, including microRNAs (miRNAs) and RNA-binding proteins (RBPs), affecting many critical processes, such as RNA stability and abundance, as well as RNA localization and translation. Fu et al. analyzed the frequency of rare 3' UTR variants and found that they were present in 155 driver genes and represented 33% of the 11,929 somatic mutations analyzed, leading to an increase in mRNA abundance (in the case of oncogenes) or to a decrease in mRNA abundance (in the case of suppressor genes) (Fu et al. 2024).

Clearly, binary driver/passenger classification is an oversimplification. As it has been suggested by Creixell et al., on the one hand, different tumors can develop similar phenotypes by acquiring mutations in different proteins while, on the other hand, two different mutations unable by themselves to drive cancer can do so when both are present within the same cell or even within two neighboring cells (Wu et al. 2010). This has been described as two passengers becoming drivers, referred to as "synthetic oncogenes" (Wu et al. 2010; Creixell et al. 2012).

A comprehensive integrative molecular analysis of the complete set of tumors present in The Cancer Genome Atlas (TCGA), consisting of approximately 10,000 specimens and representing 33 types of cancer, was performed by integrating data from exome mutation analysis, chromosome-arm-level aneuploidy, DNA hypermethylation, mRNA, and miRNA expression levels, and reverse-phase protein arrays. Except for aneuploidy, tumors can be organized by histology, tissue type, or anatomic origin into 28 clusters. A key finding was that anatomically homogeneous clusters showed mixed types of immune infiltration and variable degrees of stemness, attesting to the high levels of intertumor heterogeneity present not only at the molecular level but also at the tissue level (Hoadley et al. 2014).

The influence of cell type was predominant across all tumors and was evident in DNA methylation-based clustering, even after excluding sites with known preexisting tissue type-specific methylation (Hoadley et al. 2018). The same group previously noted some exceptions. In the Pan-cancer-12 collection, for example, breast basal-like cancers were found to be different from luminal/ER+ breast cancers as they were from lung cancers (Hoadley et al. 2014).

Melanoma and lung cancer are considered the cancers with the highest number of genetic mutations (Alexandrov et al. 2013), melanoma bearing roughly 200 protein function-altering mutations per tumor, with only a minority of these considered to be "driver" mutations directly involved in the oncogenic process (Vogelstein et al. 2013). Vogelstein et al. pointed out that "more than 99.9% of the alterations in tumors (including point mutations, copy number alterations, translocations, and epigenetic changes distributed throughout the genome, not just in the coding regions) are simply passenger changes that mark the time that has elapsed between successive clonal expansions" (Vogelstein et al. 2013). Confirming the earlier experiments of Weinberg's team, Tomasetti et al. estimated that only three driver genes are required for the development of lung and colorectal cancers (Tomasetti et al. 2015). Of note, this analysis suggested that three mutations are, in general, sufficient for a lethal cancer to develop; however, the authors clearly specified that this number may be higher in different tumors due to additional mutations that can develop subsequently. A recent study analysis done on 7664 tumors across 29 cancer types (Marticorena et al. 2017) found that, on average, tumors carry approximately 4 driver mutations ranging from <1/tumor in thyroid and testicular cancers to >10/tumor in endometrial and colorectal cancers. Interestingly, the pattern of mutations in tumors and normal tissues is similar, pointing to common mechanisms, and the number of mutations per tumor does not vary in early versus advanced cancer stages (Martincorena et al. 2017).

Driver mutations appear throughout the history of a given malignant tumor from early to advanced stage, and, interestingly, the overall number of late driver mutations is equal to or greater than that of early driver mutations in 44% of patients with metastatic tumors (Gomez et al. 2018). In 2020, an in-depth analysis of ~4600 whole-genome-sequenced samples and ~19,000 whole-exome-sequenced samples revealed 81 distinct patterns of genetic alterations or "mutational signatures": 49 single-base-substitution, 11 doublet-base-substitution, 4 clustered-base-substitution, and 17 small insertion-and-deletion signatures (Alexandrov et al. 2020). The authors proposed specific etiologies for each signature, both exogenous (i.e., tobacco-related exposure, UV light exposure, aflatoxin exposure, and exposure to various chemotherapies) and endogenous (defective DNA mismatch repair, defective base excision repair, reactive oxygen species, and deamination by apolipoprotein B mRNA-editing-APOBEC-enzymes) (Alexandrov et al. 2020). The repertoire likely represents an oversimplification as "a mutational signature may be the primary, direct outcome of a particular process; it could also be an indirect, secondary consequence of multiple processes" and "even endogenous signatures can be brought on or amplified by exogenous exposures" (Koh et al. 2021).

5.3 Cancer Pathways, Networks, and Modules

In a landmark article published in 2013 in Science by Vogelstein et al., 136 "driver" mutations present across different cancer types (Vogelstein et al. 2013) were listed. The "driver" genes were classified as belonging to one or more of the 12 pathways that confer the cancer cell a selective growth advantage:

1. RAS pathway.
2. MAPK pathway.
3. STAT pathway.
4. PIK3CA pathway.
5. TGF beta pathway.
6. Cell cycle/apoptosis pathway.
7. Notch pathway.
8. Hedgehog (HH) pathway.
9. APC pathway.
10. Chromatin modification pathway.
11. Transcription regulation pathway.
12. DNA damage control pathway.

The functions of these "driver" genes were further grouped into three categories (see Fig. 5.1):

1. Cell survival: RAS, MAPK, STAT, PIK3CA, TGF beta, and cell cycle pathways.
2. Cell fate: Notch, HH, APC, chromatin modification, and transcription regulation pathways.
3. Genome maintenance: DNA damage control pathway.

The number of driver genes subsequently extended to 299 (Bailey et al. 2018). It has been discovered that, in general, both the gain of function of oncogenes and the loss of function of tumor repressors are needed for cancer to develop (Pecorino 2021), but there are tumors that contain alterations only in either oncogenes or tumor suppressor genes.

This division into 12 distinct pathways may represent an oversimplification. For example, other researchers (Perrimon et al. 2012, for example) have reported that the signaling pathways involved in cell fate fall into ten classes and not five. These ten pathways are as follows:

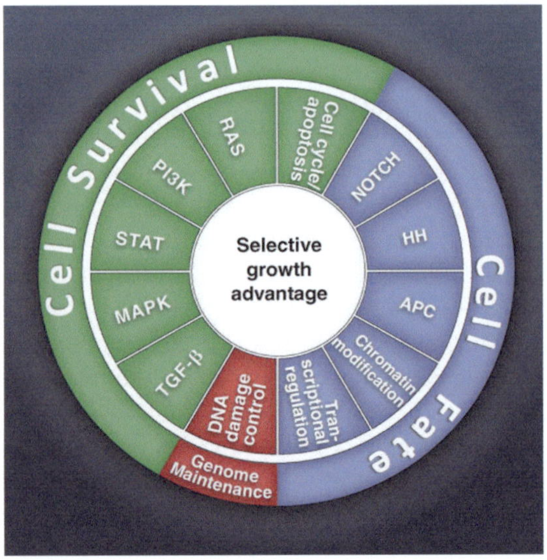

Fig. 5.1 Cancer cell signaling pathways (inner ring) and the processes they regulate (outer ring). (Courtesy Nature Publications Vogelstein B et al. Science (2013) 339, 1546–1558)

1. Notch.
2. Receptor tyrosine kinase (EGFR, FGFR, etc.)
3. Wnt/Wingless (Wg).
4. Hedgehog (HH).
5. Transforming growth factor beta (TGFb)/BMPs.
6. Cytokine/nonreceptor tyrosine kinase JAK-STAT (signal transducers and activators of transcription) pathway.
7. Hippo.
8. Jun kinase (JNK).
9. NF-kB.
10. Nuclear receptors like the retinoic acid receptor (RAR).

An important observation of Perrimon et al. is that developmental signaling elicited by ligand-receptor binding appears to be transmitted in a linear fashion within the cell, leading to the induction of target genes, which is different from typical signaling schemes in which multiple converging and diverging links are observed. This signaling linearity probably also applies to cancers driven by *one dominant driver oncogene* (BCR-ABL-driven CML; mutated EGFR-driven adenocarcinoma of the lung; BRAF V600E mutated melanomas, etc.) that respond well to the inhibition of the respective oncogene, as opposed to *cancers driven by multiple oncogenes* in which two or more signaling pathways can act in parallel and converge to regulate the activity of the same target (Perrimon et al. 2012). In the second type of signaling architecture, the inhibition of only one driver oncogene or signaling pathway is not sufficient to curb cancer cell behavior.

In the 2013 *Science* paper, the FGF, EGF, and JAK2 pathway alterations are listed within the cell survival pathways and not within the cell fate pathways, so some of the mutated genetic "drivers" may simultaneously serve multiple cellular functions.

A subsequent cancer classification described 10 pathways altered in 64 cancer subtypes (Sanchez-Vega et al. 2018). Using mutations, copy number changes, mRNA expression, gene fusions, and DNA methylation in 9125 tumors profiled by The Cancer Genome Atlas (TCGA), the authors analyzed the mechanisms and patterns of somatic alterations in ten canonical pathways almost completely overlapping with the ten signaling pathways reported by Perrimon et al.: the cell cycle, Hippo, Myc, Notch, Nrf2, PI-3-kinase/Akt, RTK-RAS, TGFβ signaling, p53, and β-catenin/Wnt. A detailed landscape of pathway alterations was analyzed in 33 cancer types, and 64 subtypes were described. The extent, mechanisms, and co-occurrence of alterations in these pathways differed between individual tumors and tumor types, but some patterns were noted. About 89% of tumors had at least one driver alteration in these pathways, and, significantly, 57% of tumors had at least one alteration potentially targetable by currently available drugs. Thirty percent of tumors harbored multiple targetable alterations, highlighting the need for combination therapy.

Another more recent report published described 760 drivers distributed across 21 pathways (Dietlein et al. 2020).

The most comprehensive analysis published to date has analyzed 2658 whole-cancer genomes and their matching normal tissues across 38 tumor types from the pan-cancer analysis of whole genomes (PCAWG), the International Cancer Genome Consortium (ICGC), and The Cancer Genome Atlas (TCGA) (ICGC/TCGA Pan-Cancer Analysis of Whole Genomes Consortium, 2020). The PCAWG driver list identified a total of 722 protein-coding genes as cancer driver genes and 22 noncoding driver mutations. It was found that 91% of tumors had at least one identified driver mutation, with an average of 4–5 drivers per tumor identified when combining coding and noncoding genomic elements, showing extensive variation across cancer types. For coding point mutations, the average was 2.6 drivers per tumor. Twenty-five percent of PCAWG tumors bore at least one putative noncoding driver point mutation, and one-third (237 out of 785) affected the TERT promoter. Overall, noncoding driver point mutations were less frequent than coding driver mutations. Three types of single catastrophic genetic events have been identified with different frequencies across different tumors: (1) *chromoplexy,* in which chromosomal rearrangements appear as a result of the repair of co-occurring double-stranded DNA breaks; (2) *kataegis,* a focal hypermutation process that leads to locally clustered nucleotide substitutions biased toward a single DNA strand; and (3) *chromothripsis,* in which tens to hundreds of DNA breaks occur simultaneously, clustered on one or a few chromosomes, with near-random stitching together of the resulting fragments. Chromoplexy events were identified in 17.8% of the samples and were most common in prostate adenocarcinoma, lymphoid malignancies, and thyroid adenocarcinomas. Kataegis events were found in 60.5% of all cancers, with particularly high abundance in lung squamous cell carcinoma, bladder cancer, acral melanoma, and sarcomas. Finally, chromothripsis events were found in 22.3% of the samples and occurred most frequently among sarcoma, glioblastoma, lung squamous cell carcinoma, melanoma, and breast adenocarcinoma. Chromothripsis increased with whole-genome duplications in most cancer types and was found to be an early event in tumor evolution.

In an attempt to classify signaling networks within cancer cells, Creixell et al. proposed six types of network-attacking mutations (NAMs) (Creixell et al. 2015) divided into three fundamental classes: 1. dysregulation of phosphorylation sites, 2. network rewiring, and 3. dysregulation of network dynamics. As proposed by the authors:

> "1. The first is one that disrupts signaling network dynamics by constitutively activating or inactivating a protein kinase, thereby maintaining the information flow either "on" or "off" uninterrupted over time. Examples of such "on" mutations are those substitutions that mimic activation loop phosphorylations, whereas examples of "off" mutations include those that alter

catalytically essential residues of kinases, or residues in SH2 domains that are critical for phospho-tyrosine binding.
2. A second, largely undescribed type of NAMs are those mutations that shift the signaling network structure by "rewiring" upstream or downstream interactions (of the mutated protein or node). Upstream rewiring can be caused by mutations in a kinase substrate that disrupt the linear motif around a phosphorylation site, thus causing a new upstream kinase to phosphorylate the mutant substrate. Downstream rewiring, by contrast, can be caused by drifts in the peptide specificity upon mutation of the determinants of specificity (DoS) in kinase (or SH2) domains.
3. A third type of NAMs could exist where mutations would generate or destroy phosphorylation sites, effectively generating new molecular logic gates in cancer cells (Creixell et al. 2015)."

In order to support their theoretical model, the authors validated their data using the COSMIC database of somatic cancer variants. In addition, they performed NGS exome analysis and quantitative mass spectrometry (MS)-based (phospho-)proteomics on a set of five ovarian cancer cell lines. An interesting corollary of their study was the discovery that nonrecurrent and nonconserved mutations, classically considered passenger mutations, can also be functional NAMs, challenging the dogma of the neutrality in cancer of passenger mutations in cancer (Creixell et al. 2015).

Iranzo et al. also analyzed 7665 tumors from 30 cancer types and were able to classify 27% of them as belonging to one of twelve network modules of driver mutations in tissue-specific cancer genes, with the rest presenting mutations in nontissue-specific genes (Iranzo et al. 2018).

The authors proposed that a cell may acquire the cancer phenotype in two distinct ways: via mutations in a set of genes belonging to one of twelve modules (*the modular mode*) or via mutations in a broader set of genes, some of which have not yet been identified as cancer genes (*the diffuse mode*). Colorectal, pancreatic, endometrial, kidney (clear cell), breast, thyroid, and brain cancers and acute myeloid leukemia, sarcoma, mesothelioma, melanoma, and uterine carcinosarcoma were found to be significantly associated with mutations in tissue-specific sets of genes (*modular type*). Lung, gastric, and esophageal cancers were found to have a less specific mode of driver accumulation (*diffuse type*). Bladder, prostate, head and neck, and germ cell cancers of the testicle exhibited specific mutations in only a fraction of the samples. Interestingly, *modular and diffuse cancer types appear to be mutually exclusive,* and cancers arising from the same tissue may preferentially adopt one of these oncogenesis routes but not both (Iranzo et al. 2018).

We are nearing an almost-complete catalog of cancer driver genes, and we know to a large extent the types of genetic alterations associated with a particular cancer type. An important observation is that the mutational repertoire associated with a given cancer type may be very different, even if cancer cells come from the same organ (for example, adenocarcinoma of the lung does not share drivers with either squamous cell carcinoma of the lung or with small cell lung cancer). Also, even within the same type of tumor derived from the same type of cell, the mutational repertoire may be very different and associated with strikingly different prognostics. For example, patients with metastatic lung adenocarcinoma tumors bearing EGFR mutations or ALK, RET, or ROS1 fusions had a median overall survival (mOS) of approximately 33 months, while patients with metastatic lung adenocarcinoma tumors with KEAP1/STK11 co-mutations had a mOS of 7.3 months (Shen et al. 2019).

Synchronous multiple primary lung cancer (SMPLC), defined as two or more primary tumors simultaneously identified in ipsilateral or contralateral lungs with distinct drivers and genomic make-ups, has also been described, and in some series, the occurrence of this phenomenon is as high as 20% (Qu et al. 2022).

An explanation for this phenomenon is the impact of both the interaction between the driver genes themselves and between the germline genome and the driver genes (Porta-Pardo et al. 2020).

Besides mutations, there are other genetic factors that may contribute to cancer heterogeneity like the germline and the tissue of origin.

5.4 Importance of Germline in Cancer

Germline variants have traditionally been studied primarily in relation to cancer susceptibility, with landmark discoveries, such as pathogenic mutations in BRCA1, BRCA2, TP53, and MLH1, shaping our understanding of hereditary cancer syndromes and informing risk assessment, prevention strategies, and early detection guidelines (Mavaddat et al. 2013; Stadler et al. 2021). These studies have significantly advanced our knowledge of fundamental cellular processes such as DNA repair, cell cycle regulation, and apoptosis. However, the role of germline variation has largely been confined to prediagnostic risk stratification, with limited clinical application after cancer diagnosis. In contrast, somatic alterations, such as mutations in EGFR, BRAF, and KRAS, are routinely used to guide therapeutic decisions and stratify patients for targeted treatments (Horlings et al. 2015; Matias-Guiu et al. 2020).

The interesting hypotheses that specific germline variants impact a limited number of somatic events and that *mutations are not random* but rather generated and selected for in cancer cells during tumorigenesis have been investigated in depth by Ramroop et al. (2019). As the authors pointed out, EGFR mutations are much more common in lung tumors from patients of Asian ancestry than in lung tumors from patients of European ancestry. It is possible that different frequencies of somatic mutations observed in different racial and ethnic groups result from different germline variant frequencies between these groups.

Also, certain cancers are more frequent in women than in men (thyroid cancer being three times more common in women, for example) and vice versa (bladder cancer being twice as common in men). This phenomenon appears to be only partially dependent on differences in lifestyles (Ramroop et al. 2019). It has been noted that the frequency of certain somatic driver mutations is related to sex (Li et al. 2018a), and that a possible explanation for this is the presence of multiple oncogenes and tumor suppressors on chromosome X, which can contribute to sex bias and induce specific cancer phenotypes by escaping X-inactivation in women.

Emerging research now highlights a broader role for germline variants in postdiagnostic cancer biology, including tumor progression, metastasis, immune evasion, and response to therapy (Chatrath et al. 2020; Chao et al. 2022). For instance, by mapping more than 330,000 protein-coding germline variants, it has been discovered that germline differences may impact the expression of genetic drivers, alter protein activity, and modulate immune infiltration (Martins Rodrigues et al. 2025).

Some germline alleles have been shown to affect the rate of tumor evolution and prognostic outcomes independently of somatic mutations (Ramroop et al. 2019).

These insights suggest that incorporating germline information into cancer management frameworks may identify distinct biological subtypes with implications for treatment responsiveness and disease monitoring. For example, TP53 germline mutations in Li-Fraumeni syndrome not only predispose to early-onset malignancies but are also associated with distinct somatic mutation profiles and therapeutic vulnerabilities (Bougeard et al. 2015). Likewise, inherited polymorphisms in DNA damage response genes have been linked to variable sensitivity to chemotherapy and radiotherapy, influencing therapeutic efficacy and toxicity (Wang et al. 2018).

Despite these advances, germline variants remain underutilized in routine oncology practice beyond their role in risk assessment. Systematic integration of germline data into tumor profiling platforms and clinical decision-making algorithms may offer a more personalized and predictive approach to cancer care. As the field moves toward increasingly multiomics and precision-guided strategies, understanding how inherited genetic variation shapes tumor behavior and therapeutic outcomes will be critical to refining individualized treatment paradigms.

5.5 Significance of Tissue of Origin in Cancer

It has also been shown that the tissue of origin has a strong influence on the genetic pattern of most cancer types (Polak et al. 2015; Hoadley et al. 2018; Iranzo et al. 2018). Both pathways that control cell proliferation and cancer driver genes are tissue specific; therefore, the behavior of a particular cancer depends on the cellular context. For example, APC and KRAS mutations are closely related to colorectal cancer, while combined VHL-BAP1 mutations are related to clear cell renal cell carcinoma (Iranzo et al. 2018). Along similar lines, recent analysis of synonymous and nonsynonymous substitutions in cell-of-origin chromatin features is a much stronger determinant of cancer mutation profiles than chromatin features of matched cancer cell lines, chromatin accessibility and modification, or replication timing, explaining up to 86% of the variance in mutation rates along cancer genomes (Polak et al. 2015).

Recent studies have highlighted how the crosstalk between the cell type-specific rewiring of signaling pathways and the tissue type-specific microenvironment shapes the tumor type-specific mutational frequency of cancer driver genes, as well as the response to therapy (Porta-Pardo et al. 2020). Research has demonstrated that genes associated with cancer susceptibility loci exhibit tissue-specific coregulation patterns that converge on known somatic driver genes. This coexpression is most pronounced in noncancerous tissues, indicating that the tissue of origin plays a vital role in the functional dynamics of driver genes (Urzúa-Traslaviña et al. 2024).

The contribution of tissue of origin to the function of driver genes is currently a critical area of research in understanding cancer biology. Recent studies indicate that the tissue from which a tumor originates significantly influences the behavior of driver genes, impacting tumor growth, the immune response, and overall cancer progression. This relationship is underscored by several key findings.

Cancer driver genes are central to tumor development and progression. However, the biological and clinical impact of a driver gene is not uniform across all cancers; rather, it is profoundly influenced by the tissue of origin. The tissue context not only shapes the downstream signaling pathways and transcriptional programs activated by these genes but also determines cellular vulnerabilities, immune microenvironment interactions, and therapeutic responses.

Classic oncogenes such as KRAS, TP53, PIK3CA, and BRAF show varying effects across tissue types. For instance, KRAS mutations are highly oncogenic and early events in pancreatic adenocarcinoma, while in colorectal cancer, they arise later and are associated with therapeutic resistance (Prior et al. 2020). BRAF V600E mutations drive melanoma by activating the MAPK pathway, yet in colorectal cancer, the same mutation fails to yield a similar proliferative phenotype unless other cooperating alterations are present (Yaeger et al. 2018).

This context dependence is due to several factors, including lineage-specific transcription factors, chromatin accessibility, the epigenetic landscape, and existing cellular programs that modulate the function of the mutated driver gene (Hoadley et al. 2014). For example, TP53 mutations have distinct transcriptional consequences in lung versus breast cancer due to differential chromatin configurations and interacting protein networks.

Tumors exhibit distinct immune cell infiltrations based on their tissue of origin. For instance, lung cancers show higher infiltration of antitumoral immune cells compared to pancreatic cancers, which are characterized by a predominance of immunosuppressive cells. The differential immune response is linked to unique transcriptional regulations of immune cell-recruiting factors, suggesting that the tissue context shapes the tumor immune microenvironment, plays a critical role in cancer progression and influences therapy outcomes (Anari et al. 2018; Bianchi et al. 2020; de Visser and Joyce 2023).

5.6 Significance of Co-mutations in Cancer

Rather than attributing cancer to individual gene mutations, Huang et al. recently proposed a gene regulatory network (GRN) model, in which cancer emerges from the collective behavior of multiple genes (Huang et al. 2025a, b). These genes are coregulated by transcription factors (TFs), regulatory RNAs, and chromosomal configurations that bind genetic targets in a sequence-specific manner. Together, these components define the "wiring diagram" of the GRN—that is, which genes regulate others, how they do so, and under what conditions.

Unlike genome-, transcriptome-, or proteome-wide association studies (GWAS, TWAS, PWAS), which typically evaluate one genetic variant or molecular feature at a time, coexpression-wide association studies (COWAS) have recently been developed to detect gene or protein pairs whose genetically regulated coexpression is associated with complex traits. Notably, COWAS methods can identify biologically relevant interactions even when neither member of a protein pair individually shows an association with the trait in question (Malakhov and Pan 2025).

A compelling clinical example of the relevance of gene interactions comes from lung adenocarcinoma research (Shen et al. 2019). Co-mutations in STK11 and KEAP1 are associated with a particularly poor prognosis: patients harboring both mutations have a median overall survival (OS) of only 7.3 months, compared to 14.7 months in those with KEAP1 mutation alone.

Similarly, co-mutations in TP53 are linked to worse outcomes in patients with tumors harboring targetable alterations, such as EGFR mutations or ALK, ROS1, or RET fusions, who are treated with tyrosine kinase inhibitors (TKIs). Patients with targetable mutations but without TP53 co-mutations had a median OS of 32.8 months, compared to 24.3 months in those with both targetable alterations and TP53 co-mutations.

5.7 Essential Genes

Essential genes are defined as those necessary for cellular survival. They are involved in critical functions such as metabolism, DNA replication and translation, maintenance of the cellular structure, and transport processes. It is conceivable that the loss of heterozygosity (LOH) of an essential gene in a cancer cell could create cancer-specific vulnerability (Bányai et al. 2021). When a cancer cell loses one allele of an essential gene, further loss or inhibition of the remaining allele may lead to selective cell death while normal cells, retaining one functional copy, remain viable. This scenario exemplifies *synthetic lethality*, in which inactivation of a single allele maintains cell survival, but simultaneous inactivation of both alleles results in lethality.

An analysis of copy number variations and LOH across 9686 patient samples from 33 tumor types in The Cancer Genome Atlas (TCGA) revealed wide variability in genome-wide LOH rates, ranging from a median of 45% in adenoid cystic carcinoma to 0.01% in thyroid carcinoma. On average, tumors harbored LOH in 189 essential genes (Nichols et al. 2020). As a proof of concept, Nichols and colleagues demonstrated that targeting two essential genes exhibiting LOH, PRIM1 (rs2277339) and EXOSC8 (rs117135638), in vitro led to highly effective tumor cell inhibition. They estimated that therapeutic targeting of PRIM1 rs2277339 and EXOSC8 rs117135638 could potentially benefit approximately 22,470 and 5307 patients per year, respectively, in the United States.

Two other essential genes, AQP7 and ATP1A1, were recently identified as crucial for proper mitotic chromosome segregation. These genes encode membrane-bound ion transporters, which play essential roles in maintaining intracellular ion homeostasis during cell division (Funk et al. 2022).

Bányai et al. further investigated whether negative selection occurs in cancer, using applied statistical analysis of COSMIC and other publicly available genomic datasets. Based on mutation patterns, they identified three major gene categories: passenger genes, driver genes, and tumor essential genes. Tumor essential genes are defined as those necessary for cell division, metabolism, migration, and metastasis. Notably, tumor essential genes were found to *be negatively selected in cancer evolution,* suggesting a functional constraint that preserves their integrity during tumor progression (Bányai et al. 2021).

5.8 Oncogenic Mutations Are Present in Normal Tissues

A recent comprehensive RNA-sequencing analysis detected thousands of somatic mutations across all human tissues and, in almost all tested individuals, including mutations at cancer hotspots and other cancer genes (Yizhak et al. 2019).

Identical genetic alterations with those described in cancer have also been found in normal tissues or tissues related to several benign conditions (Martincorena et al. 2015; Martincorena 2019; Lee-Six et al. 2019; Moore et al. 2021; Poon et al. 2021; for a review Gomes 2022).

For example, V600E mutations, besides melanoma, are more frequently present in normal nevi (Pollock et al. 2003), and PIK3CA mutations are not only present in 35% of breast cancers but also more commonly found in several breast conditions of low malignant potential, including ductal hyperplasia, columnar cell change, and flat epithelial atypia. (Ang et al. 2014). PIK3CA mutations are also commonly found in atypical ductal hyperplasia and ductal carcinoma in situ, suggesting that PIK3CA is most likely a *predisposing* mutation.

A striking example is the KRAS G12V mutation, the second most prevalent KRAS mutation in pancreatic adenocarcinomas, a clinically aggressive disease that is also found in adenomatoid odontogenic tumor, a benign lesion with an indolent behavior (Gomes 2022). It is likely, as Gomes suggested, that the effect of mutations on malignant transformation is tissue- and context dependent. The genetic, epigenetic, micro-, and macroenvironmental contexts are all important. Malignant cells tend to have a higher tumor mutation burden, and in tumors with ultralow mutation burdens, like the majority of benign tumors, upregulation of some genes (ECRG4, OGN, and ATP1A2) may suppress TP53 gene inactivation, resulting in the inhibition of cancer cell migration and invasion (Hatakeyama et al. 2020). In conclusion, the traditional multistage model of carcinogenesis may not be really challenged by the presence of cancer gene mutations in normal tissues because the probability of a single cell carrying the right combination of multiple driver events leading to malignancy is low (Feunteun et al. 2022), explaining the rarity of cancer-initiating cells compared to the several billions of cells present in the human body. Nonmutagenic factors are also essential for carcinogenesis. Given the presence of the same driver genes in benign and malignant tumors, the distinction between *advantageous* and *predisposing* mutations in terms of carcinogenesis is crucial.

5.9 The "Bad Luck" Debate and the Main Cause of Cancer

In 2015 and 2017, Vogelstein and Tomasetti published two papers in *Science* in which they advanced the hypothesis that, based on a calculated stem cell mutation rate throughout a person's lifetime, approximately 66% of cancers are due to random stem cell mutations or "bad luck" that accumulate in various organs during life, 29% can be traced to environmental causes, and 5% are inherited (Tomasetti and Vogelstein 2015; Tomasetti et al. 2017).

These papers raised animated discussion among biologists, who vehemently criticized the Vogelstein and Tomasetti papers (Various authors 2015). For example, Rozhok et al. pointed to multiple incorrect assumptions in the paper, including a mixture of mouse and human data; the fact that tissue architecture was not taken into consideration; the lack of discrimination in the model between the likelihood of mutations happening anywhere in a tissue and the probability of multiple driver mutations happening in one cell; and, importantly, the fact that the model contradicts evolution (Rozhok et al. 2015).

Wu et al. noticed the fact that the correlation between stem-cell division and cancer risk does not distinguish between the effects of intrinsic and extrinsic factors and showed that the rates of endogenous mutation accumulation by intrinsic processes are not sufficient to account for the observed cancer risks (Wu et al. 2016; Wu and Hannun 2016). Wu et al. proposed that cancer risk is heavily influenced by extrinsic factors and that intrinsic risk factors contribute only modestly (less than ~10–30% of lifetime risk) to cancer development (Wu et al. 2016).

It is important to note that the discovery of a correlation between cancer incidence and stem cell divisions cannot be used to differentiate between contributions from replication, hereditary, and environmental factors to cancer risk. Importantly, Tomasetti and Vogelstein did not make this claim in their papers (Nowak and Waclaw 2017). In their second paper, Tomasetti et al. explained that "A cancer in which 50% of the mutations are due to R (replication) can still be preventable. The reason for this is that it generally requires more than one mutation to develop the disease. A cancer that required two mutations is still preventable if one of the mutations was due to R and the other due to an avoidable environmental factor." (Tomasetti et al. 2017).

Using the data from Nurses' Health Study and Health Professionals Follow-up Study, Song and Giovannucci performed a direct comparison of cancer rates between participants with distinct lifestyle profiles. A healthy lifestyle pattern was defined as never smoking or past smoking with pack years less than 5; no or moderate alcohol drinking (≤ 1 drink/day for women, ≤ 2 drinks/day for men); body mass index ≥ 18.5 and < 27.5 kg/m2; and weekly aerobic physical activity of at least 75 vigorous-intensity or 150 moderate-intensity minutes (Song and Giovanucci 2016). They found that, overall, 20–40% of carcinoma cases and about 50% of carcinoma deaths can potentially be prevented through lifestyle modification. The authors commented that these figures were conservative and may increase to 40–70% in the general US population of whites, which has a much worse lifestyle pattern than the two cohorts analyzed (Song and Giovannucci 2016).

Two Russian authors, Aleksey Belikov and Sergey Leonov, showed that the number of rate-limiting driver events per tumor can be estimated from the age distribution of cancer incidence. They classified cancers as *anthropogenic*—induced by factors that arose with human civilization, such as air pollution, occupational hazards, ionizing radiation, smoking, alcohol, poor diet, insufficient exercise, obesity, insufficient breastfeeding, postmenopausal hormone therapy and oral contraceptives—and *nonanthropogenic*—induced by UV radiation and viruses. For the *anthropogenic* cancer group (the majority of human cancers), they suggested that the more driver events a cancer type requires, the more cases are due to preventable anthropogenic risk factors (Belikov and Leonov 2022).

Anatoly Lichtenstein's comments to the "bad luck" article were that "a population with a high living standard (and therefore having a long lifespan and good environment, food and hygiene) will be characterized by a high lifetime risk of cancer and a predominant contribution of endogenous factors to morbidity. In contrast, a population with a low living standard (and, consequently, a shorter lifespan and an excess of adverse external factors) would be characterized by a lower life-time risk of cancer and, in this case, a greater contribution of exogenous factors to cancer incidence. Cancer can be [...] depending on circumstances, both bad luck (in favorable life conditions) and punishment (in adverse life conditions)" (Lichtenstein 2017). He proposes "to obtain a holistic view of mutual influences of the tumor and the host and the factors mediating these interactions" and offers as a solution therapeutic approaches that enhance a "peaceful coexistence" between cancer and its host.

The contribution of hereditary factors to cancer development has been demonstrated by twin studies. The first Nordic analysis of cancer published in *The New England Journal of Medicine* included approximately 90,000 twins and provided strong evidence for a genetic component for three of the most common cancers, the breast, prostate and colon (Lichtenstein et al. 2000). The larger NorTwinCan I included 200,000 twins, both mono- and dizygotic, who were followed over an average of 32 years between 1943 and 2010 (Mucci et al. 2016). In this series, cancer heritability was estimated at up to 33%, with environmental factors accounting for the remaining incidence. Subsequently, based on the NorTwinCan I results, the list of cancers with a strong genetic component was expanded to melanoma and nonmelanoma skin cancer and corpus uteri, ovary, and kidney cancers (Mucci et al. 2016).

Interestingly, among the pairs of twins in which both developed cancer, *more than two-thirds were diagnosed with a different malignancy*. This key, novel finding also held true for monozygotic pairs. As Harris et al. suggested, this may reflect at the same time "a general genetic proneness or resistance" to cancer in different individuals and the fact that "the actual tissue that develops cancer may be more likely to be determined by nongenetic factors and stochastic effects" (Harris et al. 2019). Lung cancer risk in twin pairs in which one was a smoker at baseline and the other had never smoked was also assessed. There were 35 such discordant monozygotic twin pairs with lung cancer diagnosed in one twin; among them, lung cancer was diagnosed in the current smoker for 31 pairs, in contrast to 4 pairs in which the never smoker had lung cancer. The hazard ratio was 6.0 (95% CI (2.1, 17.3), $p = 0.001$) (Harris et al. 2019). These results confirm the causal role of smoking in lung cancer independent of genetic factors described before by the same team (Hjelmborg et al. 2017). The independent role of smoking has also been demonstrated in other smoking-associated tumor types besides the lung (esophagus, kidney, larynx, liver, oral cavity, pancreas, pharynx, and urinary bladder). Among a cohort of 109 monozygotic twin pairs discordant for cancer and smoking, the HR was 1.85 (95% CI: 1.15–2.98) among current smokers and 1.69 (1.00–2.87) among former smokers when compared to their never-smoking co-twin (Korhonen et al. 2022).

Also, as an argument against the "bad luck" scenario, it has been demonstrated by Blokzijl et al. that mutations accumulate at steady similar rates over time (approximately 40 novel mutations per year) in human stem cell-derived small intestine, colon, and liver organoids despite the large variation in cancer incidence among these tissues (Blokzijl et al. 2016).

Finally, pediatric cancers are much rarer than adult cancers although juvenile tissues accumulate mutations at much greater rates than their adult counterparts do (Bae et al. 2018; Kuijk et al. 2019).

In conclusion, it is apparent that both external (chemical, physical, and biological-viral-carcinogenic) and internal factors (genetic predisposition and endogenous processes) are involved in cancer etiology. Importantly, the environment and inherited or acquired genetic alterations are interrelated, and both may contribute at the same time to cancer susceptibility and evolution (Brennan 2002; Mbemi et al. 2020).

Tissue-specific patterns of mutations in sporadic cancers and organ-restricted patterns of tumorigenesis in inherited cancer syndromes demonstrate that cells are not equally susceptible to transformation and that different tissues are transformed by different oncogenic mutations.

In 2018, a consensus paper was published signed by both Vogelstein and Tomasetti, the authors of the original "bad luck" article, and Song and Giocannucci, two of its critics. The conclusion of the paper was that "A much greater emphasis on cancer prevention research is needed to avoid, detect, and treat early cancers and premalignant lesions. With such further research, *we envision that cancer death rates could be reduced by 70% around the world, even without the development of any new therapies*" (Song et al. 2018). Song et al. proposed a sequence in which hereditary (H), random, replicative errors (R), and environmental factors (E) all play a role in the oncogenic process.

The sequence presented by Song et al. is not universal. In a mouse model of breast cancer, 80% of metastases were derived from early disseminated cancer cells (Hosseini et al. 2016), and parallel evolution of primary tumors and metastases seems to be typical in breast cancer (Klein 2009). Also, more recently, as shown by Christina Curtis' team, metastases can appear early, not only in breast cancer but also in cancers of the colon and lung (Hu et al. 2020).

5.10 Sequence of Genetic Alterations Causing Cancer

The two-hit hypothesis of Alfred Knudson applies to tumor suppressor genes (TSG), for example, the retinoblastoma gene (Knudson 1971). On the other hand, the clonal expansion model of oncogenesis developed by Bert Vogelstein and Kenneth Kinzler contends that in order for an individual mutation to be selected, it must cause an expansion of the clone derived from that mutant cell by increasing relative proliferation and survival (Vogelstein and Kinzler 1993). For this purpose, *the oncogene selected needs to be dominant.* As explained by Davoli et al., for a hemizygous mutation in a TSG to be selected in cancer, either the mutation needs to be dominant negative or the TSG needs to be haploinsufficient (Davoli et al. 2013). The results of the Davoli et al. analysis performed on 8700 cancers demonstrated that virtually all cases of sporadic TSGs are likely to be haploinsufficient. The loss of a second allele of a TSG or additional genetic alterations may amplify the clonal selective pressure imposed by the first hit. Davoli et al.'s haploinsufficiency model complements the clonal expansion model.

In typical patients with cancers of self-renewing tissues, it has been shown that a large part of the somatic mutations occur before tumor initiation. In chronic lymphoid leukemia (CLL), colorectal, and ovarian cancer patients, half or more (68%, 57%, and 51%, respectively) of the passenger somatic mutations appear to have occurred before the tumor-initiating event (Tomasetti et al. 2013).

In order to account for these conflicting findings, several explanations have been offered.

Peyton Rous distinguished between initiation, which was a necessary but not sufficient precondition for carcinogenesis, and promotion, which, given an initiated state, could end the latency period, resulting in the beginning of phenotypic transformation and, finally, the progressive malignant transformation of the cancerous tissue (Moss 2003).

Bert Vogelstein and Kenneth Kinzler proposed *three broad phases* of sequential genetic alterations: "In *the breakthrough phase*, a cell acquires a driver gene mutation and begins to proliferate abnormally. It takes many cell divisions, and many years, for the cells resulting from this proliferation to be observable clinically, if they ever are. For example, nevi on the skin and small adenomas in the intestine are detected only because they can be easily observed on visual inspection or colonoscopy. Similarly sized lesions in internal organs would generally be undetectable clinically. Examples of such mutations are BRAF for melanoma, KRAS for pancreatic ductal adenocarcinoma, TP53 and RB for cervical carcinoma and APC for colon cancer, are provided as examples of such mutations (Vogelstein and Kinzler 2015).

The expansion phase is driven by a second driver-gene mutation enabling the cell to thrive in its local environment, despite low concentrations of growth factors, nutrients, oxygen, and appropriate cell-to-cell contacts. Known cancer mutation rates suggest that this second mutation is unlikely to occur absent a large increase in cell number during the breakthrough phase. The likelihood of progression thus depends on both the mutation rate and the number of cells at risk for acquiring another mutation. Examples of secondary mutations are TERT for melanoma, CDK2NA for pancreatic ductal adenocarcinoma, PIK3CA for cervical carcinoma and KRAS for colorectal cancer. These first two mutations lead to the abnormal proliferation and disordered cellular architecture defining benign tumors."

In the third, *invasive phase* "subsequent mutations enable cells to invade normal tissues and grow in otherwise hostile environments; such cells are by definition malignant. The mutation initiating the breakthrough phase is often very specific—a limited number of growth-regulating pathways seem able to initiate neoplasia in a given cell type. As tumors progress, this specificity seems to be progressively lost, so a greater number of driver genes can transform a cell from the expansion phase to the invasive phase" (Vogelstein and Kinzler 2015). Examples of such mutations are CDK2NA, TP53, PIK3CA for melanoma, SMAD4, TP53, for pancreatic ductal adenocarcinoma, MAPK1, STK11, FBXW7, for cervical carcinoma, and SMAD4, TP53, PIK3CA, FBXW7 for colon cancer.

Similarly, Kuhner et al. classify mutations as *advantageous*, conferring a growth or survival advantage on their cell lineage but otherwise not contributing to tissue invasion or metastasis, yet *predisposing*, when it pushes the cell toward a cancer phenotype, tissue invasion, and/or metastasis (Kuhner et al. 2016). For cancer to develop, mutations with both properties or both types of mutations are needed. Therefore, in theory, only mutations that are both advantageous and predisposing to cancer are true driver mutations (Kuhner et al. 2016; Gomes 2022).

Importantly, in both Vogelstein and Kinsler and Kuhner et al., the theoretical frames for the process of cellular proliferation and of cellular "malignization" are separated. This important distinction opens the room for the consideration of other mechanisms besides genetic alterations that may be responsible for the different phenotypes of malignant cells.

A multistage carcinogenesis model was recently published by the team of Charles Swanton (Weeden et al. 2023). This model, which is supported by mouse tumor model data (Balmain 2020), points out that mutations alone are not sufficient for tumor development and the nonmutagenic tumor-promoting agents have an essential role in generating actively growing lesions. This model concurs with all the recent data describing mutations without pathogenic significance in normal tissues and emphasizes that the promotion and not the initiation stage is the essential, rate-limiting step in tumor formation. A clear separation of the *initiation* stage, where mutations, although present, are not tumor inducing, from the *promotion* stage, where internal and external nonmutagenic factors are involved in cancer development, may have tremendous implications for cancer prevention programs that may include interventions even after the initiation stage.

The detailed sequence of events proposed by Weeden et al. is the following: "1. Tumor *initiation* occurs when a cell gains an oncogenic driver mutation or other genetic alterations, either from endogenous or exogenous sources of DNA damage (smoking, viral infections, alcohol, UV radiation, etc.). These cells harboring genetic alterations ("initiated" cells) go through a process of selection and compete with other stromal cells for space and resources to either persist in the tissue or are removed (repair, apoptosis, or extrusion). This somatic evolution is a normal tissular homeostatic process, given that these mutant cells may accumulate over time and are found in healthy tissues with no evidence of disease. (2) Initiated cells can either be removed from the tissue via both cellular or extracellular processes ("outcompetition"), persist with no functional outcomes, or *gain a selective advantage* through additional genetic and nongenetic changes and proliferate. The fitness landscape of the tissue may be altered by cancer risk factors that may contribute to further growth of the "initiated" cell (inflammation, progeny generation). (3) By escaping 3D tissular constrains and immune evasion, the clonal expansions of oncogenic cells can progress to form *pre-invasive lesions* that distort the confines of the normal tissue architecture and evade immune detection. It is important to note that some precancerous lesions may remain indolent or regress invasive cancers under the influence of the microenvironment factors and immune surveillance. (4) The "*oncogenic switch*" refers to precancerous lesions that break the basal membrane to form invasive cancers and, importantly, constitutes an irreversible process. This switch may involve either genetic processes like chromosomal instability (CIN), leading to an accumulation of copy-number alterations (CNAs), or/and microenvironmental remodeling" (Weeden et al. 2023).

An important note is that the advancement of "initiated" cells through the 2–4 *promotion* stages may take several years or even decades, and most cancers have only one founding cell at their origin. Accordingly, the occurrence of mutant clonal expansions within normal tissue appears to be a common phenomenon, but their progression to invasive cancers is extremely rare. This temporal gap represents a window of opportunity for cancer prevention interventions. Weeden et al. also made the key observation that "cancer risk factors may affect tissues before the acquisition of oncogenic alterations and their non-mutagenic tumor-promoting roles can be imprinted upon the tissue." This observation opens the possibility that interception interventions can prevent the oncogenic process *before* the initiation stage.

5.11 Mutations and Aging

The lifetime probability of developing cancer in the United States is approximately 40% (Siegel et al. 2014). It seems likely that if people lived long enough, virtually everyone would get cancer; the cumulative risk to age 120 years for men is nearly 90% and over 70% for women (Ahmad et al. 2015). It has even been suggested that *all* elderly people develop covert cancer in various tissues.

Similar to cancer, the rate of aging is determined by both internal factors (mutation accumulation, epigenetic instability, and immune system dysregulation with age) and external stressors (such as exposure to UV light or smoking). Consequently, one should expect different mutational processes to provide interacting, synergistic contributions to the age-dependent mutation load and burden. The dynamics of aging, in turn, affect the age-dependent change in the sensitivity of an organism to stressors. In addition, stem cell (SC) division rates (through age-dependent SC niche depletion), as well as the probability of inducing mutation(s) during DNA replication (through the loss of transcription-induced repair efficacy), may also change with age. Deleterious mutations causing cancer can arise within this age-dependent background.

Podolskiy and Gladyshev pointed out that in Alexandrov's original analysis of mutational signatures, among the 30 known mutational signatures, only Signature 1 (due to spontaneous deamination of 5′-methylcytosine) and Signature 5 (of unknown etiology, possibly due to errors in transcription-induced repair) contribute consistently across cancer samples and cancer types to incidence and correlate with the patient age (Alexandrov et al. 2013; Podolskiy and Gladyshev 2016). Other signatures correlate very weakly (or not at all) with age. A simple conclusion is that the total age-dependent mutational load, which correlates with cancer incidence, might also comprise contributions from other mutational signatures and perhaps other mutational processes.

Helen Byrne designed a mathematical model and demonstrated that, for most choices of parameter values, the optimal strategy for a tumor would be a mutation rate that is initially high and subsequently reduces over time (Byrne 2010). This prediction is consistent with the behavior of many cancers that show a high degree of genetic instability during early growth that stabilizes as the disease progresses.

5.12 Aneuploidies

The somatic mutation theory has been the emperor of cancer theories for half a century, but it has become progressively clear that the gene-centric cancer theory is incomplete. As has been clearly stated, "additional extrinsic factors, including cellular interactions and the microenvironment, seem to be required to create a selective environment that triggers a cell to become a dominant cancer clone." It is likely that the mutated genome of a cell contributes to the potential for malignant transformation, but it is not deterministic. This concept is not unfamiliar: Inherited mutations in DNA repair genes, such as BRCA1, BRCA2, TP53, and MLH1 (mutL homolog 1), cause familial cancer predisposition. They confer an increased lifetime risk of cancer, but cancer does not always arise. Additional factors are required "to tip the balance toward cancer development" (Nik-Zainal and Hall 2019).

It has been observed for a long time that some chromosomal aneuploidies, like Down syndrome, for example, are associated with an increase in the incidence of certain malignancies (i.e., acute leukemias in patients with Down syndrome) and a decrease in the incidence of other cancers (i.e., a large variety of solid tumor types in the case of patients with Down syndrome) (Osuna-Marco et al. 2021). This clinical observation suggested that chromosomal number variation may be an important component of cancer prognostic.

Although the rate of point mutations in tumors is similar to that in normal cells, the rate of chromosomal changes in cancer is elevated, and most solid tumors display widespread changes in chromosome number (aneuploidy), as well as deletions, inversions, translocations, and other genetic abnormalities (Vogelstein et al. 2013). Homozygous deletions of a few genes frequently drive carcinogenesis, and the target genes involved in such deletions are tumor suppressor genes (Cheng et al. 2017). Oncogene amplification is one of the most common molecular alterations in cancer and often can be the result from double-strand break events, such as tandem duplications, breakage–fusion–bridge cycles, and chromothripsis, which may result in complex chromosomal rearrangements (Verhaak et al. 2019). Recently, oncogene-carrying extrachromosomal DNA (ecDNA) has been identified as a new mechanism by which oncogenes are amplified and may increase genomic heterogeneity during GBM evolution, independent of chromosomal DNA alterations (Turner et al. 2017; De Carvalho et al. 2018).

Chromosomal translocations may also lead to the formation of novel oncogenes (i.e., BCR-ABL translocation).

In the ICGC/TCGA Pan-Cancer Analysis of Whole Genomes Consortium, pan-cancer analysis of whole genomes revealed that two tumor types had a surprisingly high fraction of patients without any identified driver mutations: chromophobe renal cell carcinoma (44%; 19 out of 43) and pancreatic neuroendocrine cancers (22%; 18 out of 81). A notable feature of the missing driver cases in both tumor types was a remarkably consistent profile of chromosomal aneuploidy. The absence of other identified driver mutations in these patients raises the possibility that certain combinations of whole-chromosome gains and losses may be sufficient to initiate a cancer in the absence of more targeted driver events such as point mutations or fusion genes of focal copy number alterations (CNAs) (ICGC/TCGA Pan-Cancer Analysis of Whole Genomes Consortium 2020).

In 2013, Stephen J Elledge and his team published a paper describing a computational model called Tumor Suppressor and Oncogene (TUSON) Explorer that was used to analyze the patterns of mutational signatures in more than 8700 sporadic cancers (Davoli et al. 2013). TUSON Explorer was able to predict the likelihood that any individual gene would function as a tumor suppressor gene (TSG) or an oncogene (OG). More generally, TUSON Explorer was able to describe STOP genes, i.e., genes able to inhibit proliferation and tumorigenesis, and GO genes, i.e., either oncogenes or genes essential for proliferation in general. The researchers were also able to pair GO and STOP genes to their respective chromosomes and found that chromosomes or chromosome arms that had more STOP genes relative to GO genes were much more likely to be deleted in tumors and that chromosomes with the highest density of GO genes relative to STOP genes were the most likely to be amplified in tumors. These specific aneuploid patterns pointed to the fact that aneuploidies are not random and support the oncogenic process.

On average, the cancers in the dataset analyzed by Elledge's team had a mean number of ~1 OG mutation, ~3 TSG mutations (loss of function and damaging missense), ~3 chromosomal arm gains, ~5 chromosomal arm losses, ~2 whole chromosome gains, ~2 whole chromosome losses, ~12 focal deletions, and ~ 11 focal amplifications. Given the importance of DNA copy number changes in oncogenesis, Elledge suggested that they "represent an important copilot of cancer," adding another level of complexity to the driver/passenger dichotomy.

Analyzing a large cohort of cancer patients for whom both genetic information about their tumors and clinical outcomes were available, Joan Smith and Jason Sheltzer reported that genetic copy number alterations (CNA) and not mutations had the strongest impact on cancer survival (Smith and Sheltzer 2018). CNA are commonly associated with aneuploidies, and it has been known for a long time that abnormal chromosome copy numbers represent a poor prognosis in multiple cancer types (Merkel and McGuire 1990).

Recently, analyzing more than 10,000 cancer tumors with the help of a computation tool, a group of researchers from Boston have shown that aneuploidy is a cancer progression driver and identified multiple chromosome regions that, when duplicated or deleted, either helped or inhibited cancer progression (Shih et al. 2023), thus demonstrating Boveri's hypothesis, after more than a century. It is important to note that for a specific cancerous tumor, *karyotype evolution is not random,* i.e., despite random chromosomal instability, there is selection for specific chromosomal copy numbers in order to maintain the cancer phenotype (Wangsa et al. 2018).

Key Points

- Genetic and chromosomal alterations are present in the vast majority of cancers. They may contribute to cancer initiation, promotion, and progression, but they are neither necessary nor sufficient to cause cancer. It is unclear whether mutations represent the *primum movens* of oncogenic transformation or are bystanders. It is currently unknown why in some cancers, point mutations are prevalent, while in others, qualitative and quantitative chromosomal alterations, including genetic translocations, are more common. It appears that, depending on the tissue and the cell type where it originated, the malignant phenotype can be acquired in multiple ways that may be lineage dependent. Additional factors—some intrinsic to the cell, such as metabolic or epigenetic changes—and other extrinsic factors, including cellular interactions, the microenvironment, and the macroenvironment—are necessary to "tip the balance" toward cancer development.
- There are at least 12 distinct intracellular pathways that may be altered in different types of cancer cells based on the growth advantage benefits conferred by them.
- An average of four driver mutations is present per cancer cell. Cancer cells are wired differently, explaining why targeting a single abnormal protein product of a genetic alteration sometimes works in cancers "addicted" to a strong genetic driver (CML, EGFR mutated adenocarcinoma of the lung, etc.) and lacks efficiency in cancers "addicted" to multiple genetic drivers or with no genetic drivers.
- The dominant genetic signature across most cancer types is related to the cell of origin. The tissue of origin also plays a significant role.
- Fundamentally, cancer cannot be equated with either genetic alterations or excessive growth. Besides cell division, a full-blown cancer cell is also capable of local invasion and distal metastasis, which are strongly determined by micro- and macroenvironments.
- In order for cancer to develop, several changes, some at the genetic level and some nongenetic, are needed. Carcinogenesis is a slow process that takes years or sometimes even decades.
- SMT has dominated the oncology field for more than half a century, and it led to the development of more than 130 FDA-approved targeted therapies, a number that is steadily growing (NCI website).

5.13 Epigenetic Cancer Theories

The genomes of higher eukaryotes exhibit complex, multilevel hierarchical organization. At the broadest scale, chromosomes occupy distinct spatial territories within the nucleus. Within each territory, chromatin is further partitioned into topologically associating domains (TADs), which demarcate regions of active and inactive chromatin. These TADs facilitate the formation of DNA loops, enabling regulatory elements such as enhancers and silencers to physically interact with the promoters of their target genes. Although this structural hierarchy might suggest a relatively static genome architecture, growing evidence indicates that the genome is highly dynamic and adaptable. The 3D organization of the genome is dynamically reshaped in response to developmental signals and environmental stimuli. This dynamic behavior includes the coordinated interaction of coregulated genes and the assembly of enhancer hubs formed by regulatory elements located on different chromosomes to modulate the expression of a single gene (Rubio et al. 2024). The expression of specific genes at certain times is a fundamental feature of cells in multicellular organisms. DNA in eukaryotic cells is wrapped around proteins known as histones, which protect and regulate the packed DNA. This structure of DNA wrapped around histones is known as chromatin and plays a vital role in regulating gene expression. The ability of eukaryotic cells to maintain a distinct phenotype while containing identical genes is ensured by chromatin-associated proteins and the various chemical modifications occurring on DNA and histones.

The way the genetic blueprint of an organism is accessed during development and differentiation in distinct cells is regulated by epigenetics (Biswas and Rao 2018). Genes can change expression based on cues from epigenetic factors influenced by the environment (Li et al. 2021a).

"Epigenetics," a term originally coined by Conrad Waddington in 1942, is defined as a series of biological processes that regulate gene expression, independent of changes in the original DNA sequence (Biswas and Rao 2018; Singh et al. 2022; Esteller et al. 2024). Waddington introduced the *epigenetic landscape* as a visual metaphor that illustrates the genetic control of embryonic differentiation. Waddington famously illustrated the above concept as a marble rolling down a rugged hillside, where valleys of this "epigenetic landscape" were held in place by wires underneath attached to pegs in the ground representing genes (Waddington 1957).

Epigenetic processes refer to the activity of transcription factors that establish specific chemical modifications to both protein components of the nucleosome and the DNA itself. Epigenetic regulatory mechanisms are represented mainly by DNA methylation, histone acetylation, and noncoding RNAs (see Fig. 5.2).

Classically, epigenetic changes such as DNA methylation or histone phosphorylation, acetylation, and methylation can inhibit or activate gene expression. Protein complexes that control epigenetic modifications (including DNA methylation and covalent histone modification) can be divided into writers, readers, and erasers (Biswas and Rao 2018). Epigenetic "writers" add distinct epigenetic chemical modifications to DNA or histones, while "readers" are methyl-CpG-binding domain proteins (MBPs) that identify and interpret the specialized domains of modified proteins. Chromatin-modifying enzymes function as "erasers" by removing epigenetic markers (Yu et al. 2024; Esteller et al. 2024). More recently, an operational definition of an epigenetic trait, a "stably heritable phenotype resulting from changes in a chromosome without alterations in the DNA sequence," was established at a Cold Spring Harbor meeting in 2008 (Berger et al. 2009a, b).

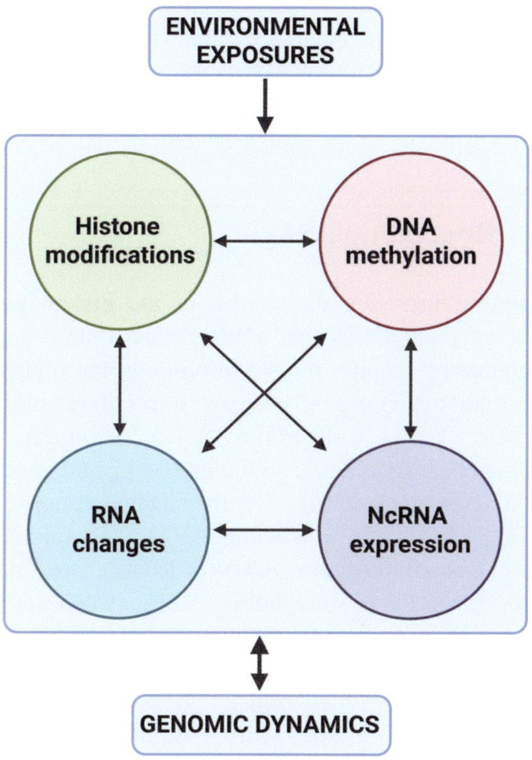

Fig. 5.2 Environmental exposures and influences may cause histone modifications, changes in RNA and ncRNA expression, and DNA methylation, leading to changes in gene expression. At their turn, gene expression may cause epigenetic changes (adapted from Li et al. 2021b, Science of the Total Environment, 776, 145998). Created in BioRender. (Courtesy of Laura Pătraş)

Epigenetic regulation involves a three-tiered signaling framework for establishing and maintaining heritable chromatin states. First, an *extracellular "epigenator"* signal—originating from the environment—triggers the pathway. This activates an *"epigenetic initiator,"* which specifies the chromatin locus or DNA region targeted for epigenetic modification. Finally, an *"epigenetic maintainer"* preserves the established chromatin (depicted in blue) configuration across cell divisions and generations. Sustained epigenetic memory often requires cooperative interaction between the initiator and maintainer (Berger et al. 2009b).

We now know that besides the external environment, signals for the initiator and the maintainer can also come from inside the cell, from metabolic cues, for example.

Historically, epigenetic changes have been considered transient and therefore less important for the cancer phenotype, and this is why, given the relative permanence of genetic mutations, SMT got more traction. Recently, their role in oncology has been formally acknowledged, and epigenetic changes have been included among the 14 recent updates of cancer hallmarks. Less than 10% of cancers are related to hereditary factors, and epigenetic mechanisms related to environmental and lifestyle changes are major contributors to cancer risk.

Over the last two decades, several theories of cancer as an epigenetic disease have been proposed (Israel and Schwartz 2006; Timp and Feinberg 2013), leading to ideas for therapeutic interventions (Fardi et al. 2018). At present, it has been clearly demonstrated that both genetic and epigenetic changes contribute to carcinogenesis: genes involved in shaping the epigenome are frequently mutated in common cancers (Yang and Brackenbury 2013; Di Domenico et al. 2017; Chatterjee et al. 2018).

A recent publication describes six epigenetic hallmarks shared among the majority of human malignancies and affecting DNA methylation, histone modifications, and chromatin remodeling:

Hallmark 1: Targeted DNA hypermethylation of gene regulatory regions.
Hallmark 2: DNA hypomethylation events at the global genome level.
Hallmark 3: Shift in histone modification patterns.
Hallmark 4: Acquisition of a new 3D chromatin architecture.
Hallmark 5: Flexible nature of the epigenetic setting.
Hallmark 6: Bidirectional impact of epigenetics and genetics (Esteller et al. 2024).

These 6 marks directly impact the 14 cancer hallmarks and can be considered to be sort of meta-hallmarks of the malignant process. Targeted hypermethylation of tumor suppressors may lead to their silencing and prevent them from curbing the oncogenic process. Global hypomethylation may result in an increase in mitotic recombination, causing translocations, deletions, chromosomal rearrangements, and aneuploidy. General DNA hypomethylation can also reactivate retrotransposons, which in turn may drive chromosomal structural and copy number changes and activate oncogenes. Chromatin accessibility is also modified in cancer cells, leading to increased flexibility of cancer cells switching between different phenotypes. This flexibility allows cancer cells to respond and adapt rapidly to different environmental stressors (hypoxia, a lack of nutrients, increased mechanical pressure, and insult induced by treatments) and change the phenotype, allowing them to survive. Epigenetic mechanisms also create a large diversity of cancer cell phenotypes that, together with genetic heterogeneity, confer fitness advantages to the tumor as a whole (Esteller et al. 2024).

Sometimes, mutations in genes encoding epigenetic modifiers alone are sufficient to drive the malignant phenotype. An analysis of 24 types of childhood cancer found several genes, including KMT2C, KMT2D and SMARCA4, to be the most commonly mutated genes across all molecular subtypes; mutations in epigenetic modifiers affected 25% of tumors (Chen et al. 2018).

In cancer, promoter hypermethylation events may lead to the silencing of tumor suppressor genes (Pfeifer 2018). Furthermore, promoter hypomethylation can promote carcinogenesis if it leads to increased expression of proto-oncogenes (Van Tongelen et al. 2017).

Histones also play a key role in cancer progression. The level of histone H1.0, for example, appears to correlate with cancer aggressivity, as demonstrated by measuring H1.0 in samples from six aggressive tumor samples: breast cancer, glioblastoma, liver cancer, kidney cancer, melanoma, and low-grade gliomas (Torres et al. 2016).

A recent study (Parreno et al. 2024) suggested that cancer can occur in the absence of genetic mutations after transient dysfunction of an epigenetic system of histone methylation. A prior study described transcription silencing driven by CpG methylation as the main driver of childhood ependymomas in the absence of genetic mutations (Mack et al. 2014). A more recent study suggested that temporary epigenetic changes arising from environmental influences such as certain diets, medications, or exposure to chemical agents may have long-term consequences (Classen 2024).

5.14 Epigenetic Regulation of the Metastatic Process

The majority of the "metastatic genes" proposed by Massagué and collaborators more than a decade ago (Nguyen and Massagué 2007; Chiang and Massagué 2008) in their step-by-step model of metastasis *are not mutated.* In addition, Vogelstein et al. also noticed that despite considerable effort, specific genetic alterations that distinguish cancers that metastasize from cancers that do not metastasize have not yet been identified (Vogelstein et al. 2013). The immediate conclusion, drawn by the Vogelstein team, is that there are no specific metastasis genes.

Two studies described widespread epigenetic reprogramming during the evolution of distant metastasis in the absence of metastasis-specific driver mutations. The first one was done on patients with pancreatic cancer (McDonald et al. 2017) and the second one performed on patients with prostate cancer (Aytes et al. 2018).

A comprehensive genomic analysis of primary and metastatic tumors from 25,000 patients did not find any significant genetic alterations between local and distal tumors (Nguyen et al. 2022). On the other hand, metastases from seven tumor types had significantly more whole-genome duplications versus the correspondent primary tumors; furthermore, an increase in tumor mutation burden was detected in metastases from ten tumor types versus the correspondent primary tumors.

5.15 Noncoding RNAs

One of the surprising discoveries of the human genome sequencing project was that although less than 3% of the human genome encodes translated proteins, more than 80% of the human genome is actively transcribed into RNA (ENCODE Project Consortium, 2012). Strikingly, some RNAs may be bifunctional, with both coding and noncoding potential (Nam et al. 2016). Over the last two decades, it has also been shown that noncoding RNAs (i.e., microRNAs (miRNAs), small interfering RNAs (siRNAs), small nucleolar RNAs (snoRNAs), small nuclear RNAs (snRNAs), transfer RNA (tRNA)-derived fragments, piwi protein-interacting RNAs (piRNAs), and long ncRNAs) play an important role in epigenetic control (Moazed 2009; Djebali et al. 2012; Slack and Chinnaiyan 2019; Liu et al. 2019; Statello et al. 2021). More recently, the family of ncRNA has expanded and now includes stable intronic sequence RNA (sisRNA), YRNA, miRNA-offset RNA (moRNA), telomerase RNA component (TERC), natural antisense transcript (NAT), transcribed ultraconserved regions (TUCR), and pseudogene transcripts with unclear pathogenic impact (Irimie et al. 2018). All the data accumulated over the last two decades points to the role of ncRNAs as master regulators of the genome. Importantly, tumor-specific promoter hypermethylation of ncRNAs also affects ncRNAs, and as a result of ncRNAs expression loss, all the classical hallmarks of cancer are impacted positively (Esteller et al. 2024).

An abundance of studies has shown that, based on complementary base pairing, ncRNAs may function both as oncogenes (by inhibiting the activity of tumor suppressor genes) or as tumor suppressors (by inhibiting the activity of oncogenes or tumor essential genes) (reviewed in Diamantopoulos et al. 2018; Anastasiadou et al. 2018; Yan and Bu 2021). Since it has been demonstrated that they can contribute to all classical hallmarks of cancer, the most widely studied classes of ncRNAs are miRNAs and lncRNAs (Ruan et al. 2009; Schmitt and Chang, 2016). miRNAs are small ncRNAs of ~22 nucleotides but regulate 60% of coding genes (Shu et al. 2017) by binding to mRNAs and are involved in proliferation, differentiation, apoptosis, and development. So far, according to different authors, 2300 to 2600 human mature miRNAs have been identified (Alles et al. 2019; Plotnikova et al. 2019). lncRNAs are considered to be longer than 200 nucleotides and make up about 80% of mammalian ncRNAs. A total of 95,243 lncRNA genes and 323,950 transcripts were described. lncRNA as a class has been relatively poorly conserved during evolution (30–40%) (Gibert et al. 2022). Evolutionary conservation is considered a marker for biological significance and, therefore, of particular interest are 481 lncRNA, 100% conserved among orthologous regions of the human, mouse, and rat genomes (Oliveira 2023). Due to this peculiarity, these lncRNAs are transcribed from ultraconserved regions (TUCRs). TUCRs are more frequently found in cancer-associated genomic regions (CAGRs), and the differential expression of T-UCRs has been described in several types of cancer (Zambalde et al. 2022). Out of 481 TUCRs, 297 were found to be differentially expressed in cancer samples. Like other lncRNAs, TUCR can have both oncogenic and tumor suppressor qualities (Oliveira 2023). The mechanisms through which some TUCRs contribute to cancer progression or to metastasis have not yet been elucidated. It is thought that an interaction with miRNAs is essential for their tumorigenicity (Terracciano et al. 2017).

ncRNAs are master regulators of all the classical hallmarks of cancer described by Hannahan and Weinberg (Ruan et al. 2009; Schmitt and Chang, 2016). As an example, epigenetic regulation, alternative splicing, chromatin modification, and translational control by miRNAs like miR-21, miR-155, and miR-200 (Ding 2014; Marima et al. 2021; Szczepanek and Tretyn 2023) or lncRNAs like MALAT1, HOTAIR, and TRE (Dhamija and Diederichs 2016) are some examples of ncRNAs that mediate control of cell migration and invasion, EMT, and metastasis (Dhamija and Diederichs 2016).

The picture of the unidirectional flow of information from genes to proteins through mRNA is incomplete. In cells, bio-information flows back and forth between genes and a complex network of miRNAs, proteins, lncRNAs, mRNAs, and other ncRNAs that regulate gene expression at the epigenetic, transcriptional, posttranscriptional, translational, and posttranslational levels in a variety of ways. On the other hand, as proposed by Hervé Seitz, although this regulatory network may affect many genes at the microscopic scale, it appears that only a few regulatory events propagate from the microscopic to the macroscopic scale; others appear to be censored by homeostatic mechanisms (Seitz 2024). Specifically, so far, most studies point to a single target, a one-to-one miRN/mRNA relationship of the mRNA target with the resulting phenotype (Pinzón et al. 2017). To date, only a single in vivo experiment reported in Drosophila has demonstrated that miRNAs can impact multiple targets simultaneously, resulting in phenotype changes (Verma and Cohen 2015). In this experiment, string and wingless genes were targets of miR-965, leading to a mutant phenotype.

5.16 Epigenetic Modulation of the Stroma

The significance of epigenetic mechanisms for the oncogenic process goes far beyond the cancer cells themselves. The morbidity of cancer in the human host is related to the cancerized local stroma and to the systemic effects of cancer on the whole organism. Studies spanning over two decades have demonstrated how epigenetic reprogramming alters the phenotypes of immune cells and tumor-associated stromal cells, including fibroblasts and endothelial cells. As a result, epigenetic dysregulation of these cells reshapes the tumor microenvironment, changing it from an antitumor environment to a tumor permissive environment (Hu et al. 2005; Pan and Zheng 2020).

It has been known since the last century that genes and signaling pathways normally expressed during embryonic development are also involved in cancer. These genes that govern cell division, growth, and differentiation in order to ensure normal development within each lineage (Perrimon et al. 2012) are often reactivated or dysregulated during tumorigenesis and metastasis, leading to aberrant proliferation, epithelial-to-mesenchymal transition (EMT), and stem cell potency, which affects the cellular lineage hierarchy (Lee 2022). Also, it was described that BMP/TGF-β, NOTCH, WNT/β-catenin, Hedgehog, and Hippo pathways, which are active during embryogenesis, are reexpressed in cancer cells during epithelial to mesenchymal transition (EMT) (Monk and Holding 2001). Downstream of these signaling pathways, EMT is regulated by transcription factors belonging to the Snail, ZEB, TWIST, and PRRX families (Fazilaty and Basler 2023).

Importantly, even after undergoing differentiation, cells preserve embryonal memory in the form of epigenetic imprints, reflected in DNA methylation patterns and histone marks (Plasschaert and Bartolomei 2014). These preexisting epigenetic landscapes, different in different tissues, determine how oncogenes and tumor suppressor genes can influence a tissue's response to a specific oncogenic signal, which depends on the permissiveness of its epigenetic state (Monk and Holding 2001; Peters et al. 2024). Epigenetic permissiveness also depends on the particularities of the cancer cell origin, which varies according to the location of the primary tumor. This may explain the idiosyncratic patterns of genetic alterations in cancers of different cellular lineages, i.e., adenocarcinoma of the lung, for example, bears genetic alterations in oncogenes like EGFR, ALK, ROS1, BRAF, HER2 neu, and NTRK versus small-cell lung cancers that have mutations in TP53 and Rb tumor suppressor genes.

Key Points

- Epigenetic changes have been universally described in cancer. The genome and epigenome thus represent two complex mechanisms that are tightly integrated in both oncogenesis and cancer progression. Signaling pathways often converge on transcription factors, and the activation of the same transcription factor may lead to different effects in multiple cell types, depending on their chromatin state (Flavahan 2020). If "driver" mutations in genes can be seen as "driving" the cellular cancerous process, the function of epigenetic alterations may be to "drive the drivers" themselves, acting as master regulators of switching between different cellular states. Environmental exposure can modulate genetic expression through different epigenetic mechanisms.
- Epigenetic changes appear to be consistently related to the malignant state. Several pan-cancer liquid biopsies using analysis of DNA methylation patterns have been developed.
- Epigenetic mechanisms are involved in cancer cells both hijacking and reactivating embryonic programs, and also in instigating stromal cells to create procancerous niches both locally and distally.
- Setback: The field of epigenetic treatments has been somewhat held back due to concerns over the pleiotropic effects of both drugs and their targets. In general, epigenetic agents alone lack efficacy in the treatment of solid tumors and need to be combined with other therapeutic interventions. The era is near when targeted epigenetics will be used with not only methylation- or histones-directed drugs but also agents able to target ncRNAs.

5.17 Metabolic Cancer Theory

Metabolism plays a key role in cancer throughout the oncogenic process, starting from cells with mutations through neoplasia, malignant transformation, distant metastasis, and dormant cancer cells. When they metastasize, in order to maintain proliferation and survive in various microenvironments, cancer cells' plasticity allows them to adapt their metabolism and their phenotype as they pass from differentiation to dedifferentiation and vice versa several times (Pacella et al. 2018). For example, whereas breast cancer primary tumors rely on glucose and glutamine metabolism, metastatic tumors in the lung use pyruvate and, in the brain, serine and acetate to sustain the TCA cycle (Elia and Haigis 2021).

Otto Warburg's groundbreaking observation that cancer cells preferentially use aerobic glycolysis over oxidative phosphorylation for energy production, even in the presence of oxygen (the Warburg effect), is now more than a century old (Warburg 1924, 1956; Warburg et al. 1924). Until the end of his life, Warburg was deeply convinced that the reason for the cancer cells' switch to aerobic glycolysis was a defect in mitochondrial respiration: "Just as there are many remote causes of plague-heat, insects, rats-but only one common cause, the plague bacillus, there are a great many remote causes of cancer-tar, rays, arsenic, pressure, urethane- but there is only one common cause into which all other causes of cancer merge, the irreversible injuring of respiration" (Warburg 1956). According to this mitochondrial metabolic theory of cancer, the large genomic heterogeneity seen in tumor cells arises as a consequence, rather than as a cause, of mitochondrial malfunction (Seyfried 2015).

Recent work supports an important role for respiration—and, more broadly, mitochondrial metabolism—in cancer cell growth. The Warburg effect's significance for human cancer has been put into perspective, and the mitochondrial theory of cancer is seen as an oversimplification (DeBerardinis and Chandel 2016). The transition to aerobic glycolysis may be related to both an increase in oncogene activity (KRAS, PI3K/AKT, and EGFR) and a decrease in tumor suppressor activity (TP53 and PTEN). The hypoxic environment may also raise the levels of HIF1-α, which induces an adaptive switch to glycolysis (Zheng 2012).

Cancer cells in a tumor may exhibit different types of metabolism, and metastases may also display different metabolic traits compared with the tumors from which they originate (Bergers and Fendt 2021). Metabolic rewiring is clearly needed for metastasis as the energy requirements increase (Wei et al. 2020b). For example, it has been shown recently that dysregulation of propionate metabolism produces a proaggressive signature in breast and lung cancer cells, increasing their metastatic potential (Gomes et al. 2022).

As proposed by Peiris-Pagès et al., the metabolism of cancer cells may also be related to their degree of differentiation (Peiris-Pagès et al. 2016). Cancer stem cells (CSCs) that are in a quiet state use oxidative phosphorylation (OXPHOS)-based metabolism, whereas rapidly dividing cells use aerobic glycolysis preferentially. More differentiate cancer cells may also cooperate with cancer-associated fibroblasts (CAFs) to obtain lactate (reverse Warburg effect). The quiescent state may be reversible, and CSCs can also enter the path of proliferation, accompanied by a change in metabolism. CSCs in this state can benefit from aerobic glycolysis and OXPHOS and rely on the reverse Warburg effect. This metabolic plasticity might allow CSCs to survive in multiple or changeable environments, such as distal metastatic sites. In support of this metabolic plasticity scenario, it has been shown that in some tumors, only by inhibiting both aerobic glycolysis and OXPHOS, cancer growth can be curbed (Cheong et al. 2011).

The Warburg phenomenon has also recently received another explanation (Luengo et al. 2021). Cells switch to aerobic glycolysis when the demand for NAD+ required for anabolic reactions is in excess of the demand for ATP. The main reason for cancer cells switching to an inefficient ATP production pathway is to regenerate NAD+, a molecule needed to synthesize DNA and other important molecules.

It is important to note that glycolysis is redox neutral and does not net regenerate NAD+; thus, this model still does not explain why increased glucose uptake and glycolysis are associated with proliferation. Cancer cells do not switch completely to aerobic glycolysis and rely on oxidative phosphorylation (OXPHOS) to obtain the ATP necessary for their survival. In fact, the only known output of glycolysis is ATP, and if excess ATP limits net NAD+ regeneration to promote glycolysis, it is unclear why glucose metabolism is increased during rapid proliferation. Nevertheless, a model in which aerobic glycolysis reflects a metabolic state in which the requirement for NAD+ supersedes the demand for ATP is consistent with the reality of this metabolic phenotype, which is best characterized by a relative increase in fermentation rather than a complete switch from mitochondrial respiration to glycolysis (Luengo et al. 2021).

5.18 Warburg's Hypothesis Revisited

At the cellular level, it appears that both metabolism and DNA changes contribute to the oncogenic process. Somatic mutation theory considers genetic changes to be more important than metabolism in oncogenesis. Supporters of the metabolic theory of cancer (Otto Warburg, Thomas Seyfried) herald the opposite view. A comparison of isogenic lung and pancreas tumor models demonstrated that the presence of a given metabolic pathway in a cancer cell may be defined by lineage rather than by driver mutations (Mayers and Vander Heiden 2017).

Otto Warburg's proposal was that cancer originates from irreversible injury to mitochondrial oxidative phosphorylation (OXPHOS), which leads to an increased rate of aerobic glycolysis in the majority of malignancies. As discussed above, his theory was dismissed in the 1970s with the discovery of oncogenes, which provided support for the SMT. More recently, the discovery that some OXPHOS proteins act as tumor suppressors renewed interest in Warburg's theory. For example, the OXPHOS enzyme succinate dehydrogenase (SDHD) has been shown to be mutated in hereditary paragangliomas and pheochromocytomas; hereditary mutations in the Krebs cycle enzyme fumarate hydratase (FH) have been found to be related to leiomyomas, uterine fibroids, and renal cell carcinoma; and somatic mutations affecting isocitrate dehydrogenase 1 and 2 (IDH1 and IDH2), which catalyze the conversion of isocitrate to the Krebs cycle-intermediate ketoglutarate, have been identified in brain tumors (Chandra and Singh 2011), leading to the development of specific inhibitors of the IDH enzymes currently in use in the clinic. Aging is associated with mutations in mitochondrial DNA, which encodes 13 essential components of the OXPHOS system. It has recently been shown that aging-related OXPHOS dysfunction may contribute to the development of intestinal cancer (Smith et al. 2020). Reduced mitochondrial copy number has also been reported in hepatocellular, breast, renal, gastric, ovarian, and lung cancer, but increased mitochondrial copy numbers have also been observed in esophageal, colorectal, and head and neck cancers (Sun et al. 2018), pointing to the fact that both a reduction in mitochondrial number and mitochondrial dysfunction may contribute to cancer development.

Given their multifunctional and malleable nature, the frequently used terms "mitochondrial function" and "mitochondrial dysfunction" are misleading misnomers that fail to capture the complexity of mitochondrial biology. In order to improve the conceptual and experimental specificity in mitochondrial science, Monzel et al. proposed a terminology system that distinguishes between (1) cell-dependent properties, (2) molecular features, (3) activities, (4) functions, and (5) behaviors (Monzel et al. 2023). These authors also suggested that mitochondrial function can be improved by a Mediterranean diet.

Reprogramming of cellular metabolism is a fundamental aspect of malignant transformation resulting from mutations in proto-oncogenes and tumor suppressors (Ward and Thompson 2012). As described recently, there are many metabolic alterations associated with the cancer state, including (1) a shift from oxidative phosphorylation to aerobic glycolysis to support the increased need for ATP; (2) increased glutaminolysis for NADPH regeneration; (3) altered flux through the pentose phosphate pathway and the tricarboxylic acid cycle for macromolecule generation; (4) increased lipid uptake, lipogenesis, and cholesterol synthesis; (5) upregulation of one-carbon metabolism for the production of ATP, NADH/NADPH, nucleotides, and glutathione; (6) altered amino acid metabolism; (7) metabolism-based regulation of apoptosis; and (8) the utilization of alternative substrates, such as lactate and acetate (Schiliro and Firestein 2021). The transformed metabolic flux of cancer cells is controlled both by genes (oncogenes and repressor genes) and intracellular epigenetic mechanisms (i.e., micro-RNAs) and by tumor-host interactions.

To this list, seven other hallmarks were added in 2022: (8) use of opportunistic modes of nutrient acquisition; (9) deregulated uptake of glucose and amino acids; (10) expanded needs for electron acceptors; (11) increased demand for nitrogen; (12) alterations in metabolite-driven gene regulation; (13) elevated reliance on protection mechanisms from oxidative stress; and (14) metabolic interactions with the microenvironment (Pavlova and Thompson, 2016; Pavlova et al. 2022).

Metabolic enzymes can act as tumor suppressors (e.g., fumarate hydratase and succinate dehydrogenase) or oncogenes (e.g., mutant isocitrate dehydrogenase 1 and 2—IDH1 and IDH2).

Cancer cells, by virtue of their high metabolic rate, outcompete stromal cells, deprive the microenvironment of nutrients and alter the tumor microenvironment by secreting lactate. Lactate in the microenvironment can, in turn, contribute to immunosuppression in tumors by affecting the behavior of stromal cells by infiltrating lymphocytes and macrophages. Beyond nutrient sharing, stromal cells in the microenvironment can serve as alternative sources of nutrients. In nutrient-deprived conditions, cancer-related macrophages and tumor cells can literally cannibalize other stromal cells through the process of *entosis* (Muir et al. 2018).

An interesting observation is that cancer cells tightly control mitochondrial reactive oxygen species (ROS) production (Martínez-Reyes and Chandel 2021), and they may either switch to aerobic glycolysis to protect themselves against excessive oxidative damage and cell death or, paradoxically, may increase mitochondrial superoxide production when they become metastatic (Porporato et al. 2014). Some cancers may thrive in the presence of antioxidants (Martínez-Reyes and Chandel 2021), or on the contrary, mitochondrial superoxide scavenging may prevent spontaneous human tumor metastasis (Porporato et al. 2014). An explanation for this phenomenon is that at the beginning of oncogenesis, limiting ROS production is necessary for initiation, whereas sustaining ROS levels promotes metastasis (Martínez-Reyes and Chandel 2021). A recent screening of two drug libraries with a total of 727 drugs revealed that many commonly used chemotherapy drugs increase oxidative stress (Haas et al. 2020), suggesting that their main use should be cancer prevention. Also, the largest category of drugs found through the same screening that may ameliorate oxidative stress were drugs used in cardiovascular medicine, such as statins, angiotensin receptor blockers, angiotensin-converting enzyme inhibitors, β-blockers, and spironolactone. Unfortunately, a recent meta-analysis demonstrated that studies of currently used cardiovascular or anti-inflammatory drugs repurposed in the hope of improving survival or decreasing cancer relapse were mostly negative (Benjamin et al. 2024).

Key Points

- Compared to the high number of genomic alterations, cancer cells metabolize a relatively small variety of substrates, i.e., glucose, lactate, pyruvate, hydroxybutyrate, acetate, glutamine, and fatty acids. At first glance, it seems that cancer metabolism may be a good therapeutic target; however, cancers exhibit high variability even within the same tumor. Also, the metabolic ecology of tumors is complex, as they usually contain multiple metabolic compartments linked by the transfer of various catabolites. Importantly, experimental evidence indicates that metabolic coupling between cell populations with different, complementary metabolic profiles can induce cancer progression. In cancer, metabolism has high plasticity both in space and in time. The metabolic heterogeneity of cancer cells within the primary tumor and between primary tumor and metastatic sites, along with their high flexibility, allows them to adapt rapidly to therapeutic stressors and switch quickly between anaerobic glycolysis and OXPHOS or to use them both at the same time.
- A common application of the abnormal metabolism of cancer cells is positron emission tomography (PET) scanning. PET scanning using the radiotracer [18F]-fluorodeoxyglucose (FDG) is a functional imaging technique widely used in clinical oncology for assessing both local and regional cancer spread. FDG is a glucose analog that is taken up by cells using glucose due to the increased number of glucose receptors on their membranes and the increased activity of hexokinase, the first enzyme in the glycolytic pathway. High uptake of the FDG tracer indicates increased tissue metabolic activity commonly associated with cancer but also with other inflammatory and infectious diseases, such as tuberculosis, for example. Immune cells are also FDG avid, and some cancer cells preferentially use glutamine as fuel;

therefore, high FDG uptake on an FDG-PET scan is not always related to high metabolism in cancer cells and should be interpreted taking into consideration the clinical context.

– Mutations in isocitrate dehydrogenase (IDH), an enzyme involved in the citrate cycle, have been described in approximately 4% of malignancies, including brain and cholangiocarcinomas. Ivosidenib, an IDH1 inhibitor, is currently FDA-approved for advanced cholangiocarcinoma, while vorasidenib, a dual IDH1/IDH2 inhibitor, is FDA-approved for the treatment of grade 2 astrocytoma or oligodendroglioma. Selective IDH inhibitors, such as enasidenib (IDH2), have also demonstrated clinical efficacy in acute myeloid leukemia (AML) and intrahepatic cholangiocarcinoma, representing a paradigmatic example of targeting mutant mitochondrial metabolism (DiNardo et al. 2018; Abou-Alfa et al. 2020; Stein et al. 2019).

  Metabolites like D-2HG, which are found at higher levels in cancer cells than in normal cells, have been presumed to fuel cancer and are called oncometabolites. Inhibition of this metabolic enzyme altered glucose metabolism in T cells. Notarangelo et al. showed that high concentrations of d-2HG may have immunosuppressive activity, acting as a direct inhibitor of lactate dehydrogenase in T cells and curbing their proliferation, cytokine production, and ability to kill target cells (Notarangelo et al. 2022).

– Abnormalities in mitochondrial morphology and function are commonly described in malignancies. Increased fission and reduced fusion of mitochondria have been described in several cancer types, and mitochondrial dynamics have been linked to cancer cell migration (Senft and Ronai 2016). Strikingly, cancer cells can also, literally, "steal" mitochondria from surrounding T cells through nanotubes in one stroke, increasing their ATP-producing capacity and impeding the activity of T cells (Zhang et al. 2023b). Also, recently, it was also found that neurons may "donate" mitochondria to cancer cells, and this transfer enhances their metastatic capabilities (Hoover et al. 2025). It seems that, at least in some cases, Otto Warburg has been proven right. Mitochondria and metabolism are key players in cancer.

5.19 Who Controls Who Inside and Outside Cancer Cells?

Implicitly, SMT postulates that genetic alterations are the master drivers of the oncogenic process and cell fate. Taking a contrary position, the authors of cancer epigenetic models attribute this role to various epigenetic mechanisms. Contemporary supporters of the primordial role played by metabolism in cancer, such as Thomas Seyfried, the author of the now classic *Cancer as a Metabolic Disease: On the Origin, Management, and Prevention of Cancer* (Seyfried 2012), consider both genetic and epigenetic alterations in cancer, epiphenomena.

Another possibility supported by multiple recent studies is that genetic and epigenetic programs reciprocally influence each other, and the direction of this influence depends on the context. For example, chromatin-modifying enzymes use metabolites as cofactors, while cellular metabolism can influence the capacity of the cell to write or erase chromatin marks, directly impacting gene expression (Paro et al. 2021) and thus influencing both cell identity and metabolism (Bernard and Goldstein 2024).

Epigenetic factors have been shown to control both gene expression and metabolism (Zong et al. 2023; Wang et al. 2024b; Bernard and Goldstein 2024) and are controlled at the same time by proteins expressed by genes and metabolic factors (Xu et al. 2008), including oncometabolites (Lanzetti, 2024). Also, oncogenes directly steer metabolism (Levine and Puzio-Kuter 2010; Tarrado-Castellarnau et al. 2016; Tufail et al. 2024), and mitochondria can control cellular division (Piergentili and Sechi 2024). Importantly, metabolism switches induced by oncogenes are necessary for the induction of cancer metastasis and the cancer cell phenotype (Lee et al. 2018).

An interesting aspect of the codependence between nuclear and cytoplasmic factors is the physical interaction between ncRNAs and proteins (Li 2023). ncRNAs can affect protein assembly and activity by forming complexes with proteins, participating in various biological processes such as signal transduction and enzyme activity. They interact with proteins through their complex tertiary structures, acting as scaffolds or molecular chaperones. These interactions play a pivotal role in the recruitment and assembly of protein complexes, as well as the regulation of their activities. Recently, lncRNAs have also been found to be anchors of phase liquid-liquid phase separation (LLPS) by creating molecular condensates both in the nucleus and the cytoplasm. LLPS is a major mechanism of both intracellular biomolecular interactions and homeostasis by creating membrane-less organelles (MLOs), which are ad hoc liquid compartments that selectively partition a specific set of macromolecules from others. LncRNAs are emerging as master regulators of the formation of biomolecular condensates (Somasundaram et al. 2022), which may represent an emerging mechanism of cancer progression (Qiu et al. 2024).

Genetic and epigenetic programs and metabolism are codependent, and, therefore, genetic, epigenetic, or metabolic cancer theories are all partially correct (see Fig. 5.3).

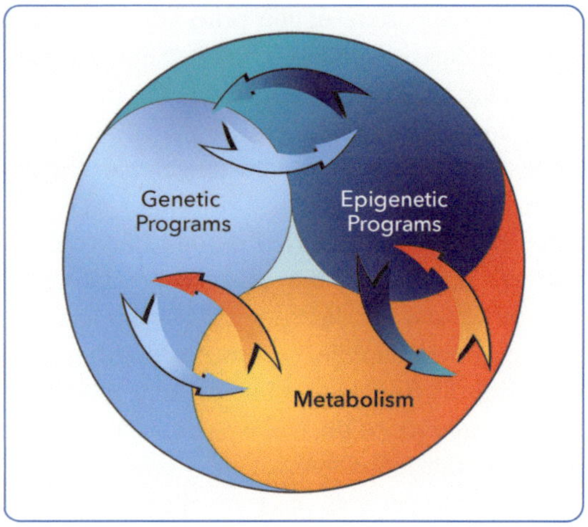

Fig. 5.3 Genetic programs, epigenetic programs, and metabolism codependency. (Courtesy of Ana Tourian)

5.20 Role of the Microenvironment in Cancer Initiation, Progression, and Metastasis

5.20.1 The Tissue Organization Field Theory and Beyond

Tissue organization field theory (TOFT) emphasizes the role of tissular changes in oncogenes and considers genetic alterations as epiphenomena. In the opinion of the two TOFT authors, Carlos Sonnenschein and Anna Soto, cancer arises when regulatory constraints from the stroma are lifted (Huang et al. 2025a, b). An interesting phenomenon known for several decades that brings support to the tissular theories is the phenomenon of donor-cell derived leukemia, which occurs in 0.2–1% of the recipients of allogeneic bone marrow transplants (Suárez-González et al. 2018). The inception of leukemia in bone marrow recipients has been explained by an abnormal bone marrow niche and/or a damaged microenvironment following intensive chemotherapy and/or radiotherapy that induces leukemic mutations in the transplanted cells (Torra and Loeb 2011).

Stromal cells that were initially considered bystanders of tumorigenesis have been shown to play critical roles in the pathogenesis of cancer. Therefore, as an extension of the "*omnigenic*" model, which proposes that all genes expressed in disease-relevant cells are liable to affect the functions of core disease-related genes and explains this phenomenon as a consequence of the high interconnectivity of gene regulatory networks (Boyle et al. 2017), an "*omnistromal*" cancer model can be envisioned in which all cellular and noncellular stromal components play a more or less central role in the dynamics of cancer progression (Hanahan and Coussens 2012; Lu et al. 2012; Quail and Joyce 2013; Kalluri 2016a, b; Neophytou et al. 2021; Hinshaw and Shevde 2019; de Visser and Joyce 2023). Importantly, the metabolic phenotype of cancer cells is also directly related to microenvironmental factors within the tumor, such as nutrient availability and interactions with stromal cells (Muir et al. 2018).

Intratumoral symbiosis, for example, has been proposed in the case of lactate, which can be shuttled between neighboring cancer cells, adjacent stroma, and vascular endothelial cells, inducing metabolic reprogramming (Doherty and Cleveland 2013; Nakajima and Van Houten 2013). In some nutrient-deprived conditions, stromal cells can even become nutrition for cancer cells and macrophages via *entosis* (Hamann et al. 2017; Krajcovic et al. 2013). The physical characteristics of tumors such as pressure (Tung et al. 2015) or acidosis can also affect metabolic gene expression. A strong lactic acidosis response signature was correlated with improved survival outcomes, possibly related to the role of lactic acidosis in directing energy generation toward aerobic respiration and the utilization of other energy sources via the inhibition of glycolysis (Chen et al. 2008).

The tumor microenvironment (TME) comprises biotic and abiotic components. The biotic component consists of various cell types, such as immune cells, including tumor-associated macrophages (TAMs), T cells, B cells, natural killer (NK) cells, myeloid-derived suppressor cells (MDSCs), tumor-associated neutrophils (TANs), and dendritic cells (DCs); stromal cells, including cancer-associated fibroblasts (CAFs), mesenchymal stromal cells (MSCs), and pericytes; and the extracellular matrix (ECM) and secreted molecules, including extracellular vesicles (ECVs), growth factors, hormones, cytokines, and chemokines. The abiotic component is represented by the interstitial fluid (Baronzio et al. 2014; Apiz Saab and Muir 2023), which has a different composition from blood (Sullivan et al. 2019). Sullivan et al. discovered that the composition of interstitial fluid can vary based on tumors' location in the body and at the site where the tumor originated. Both interstitial fluid composition in nutrients and its physical properties (i.e., mechanical pressure) have a significant impact on cancer progression.

Immunotherapy has emerged as a transformative approach in oncology, offering a new therapeutic paradigm that harnesses the patient's immune system to recognize and eliminate cancer cells. Unlike conventional treatments such as chemotherapy or radiation, which act directly on tumor cells, immunotherapy targets the regulatory mechanisms of the immune system itself, restoring or enhancing its capacity to mount effective antitumor responses. This strategy has significantly broadened the scope of cancer treatment, with applications across a growing array of malignancies.

Immunotherapy represents one of the most significant advances in cancer treatment in recent decades. For instance, immune checkpoint inhibitors targeting the CTLA-4 and PD-1/PD-L1 pathways have revolutionized the management of advanced melanoma, non-small cell lung cancer (NSCLC), renal cell carcinoma, and urothelial carcinoma. In some patients, these therapies have led to durable responses lasting years— a stark contrast to the transient benefits typically seen with cytotoxic therapies. Recently, 10-year melanoma-specific survival has increased to approximately 50% by using a combination of nivolumab and ipilimumab (Wolchok et al. 2025), and the overall 5-year survival of non-small cell lung cancer using checkpoint inhibitors has more than doubled compared to historical data (Garon et al. 2019; Gettinger et al. 2018), demonstrating the role of cell-cell stromal interactions.

Given these excellent results, a plethora of immunotherapy treatments, including novel checkpoint inhibitors, CAR-Ts, CAR-NKs, adoptive cell therapies (ACT), cancer vaccines, oncolytic viruses, and cytokine-based therapies, are currently in development (Hamdan and Cerullo 2023). Chimeric antigen receptor T cell (CAR-T) therapies have shown striking efficacy in B-cell leukemias and lymphomas, with ongoing trials exploring their application in solid tumors (Maude et al. 2018; June and Sadelain 2018). Additionally, immune agonists and personalized neoantigen vaccines are being developed to target tumor-specific antigens with high precision, thereby expanding the immunotherapy landscape.

As per the Cancer Research Institute (CRI) Cancer Immunotherapy Insights and Impact 2025 report, since 2011, the US Food and Drug Administration (FDA) has had 150 distinct regulatory approvals, encompassing immune checkpoint inhibitors (ICIs), adoptive cell therapies, cytokine agonists, bispecific T cell engagers (BiTEs), and novel immune-stimulatory agents. In the United States, there are currently 20 immunotherapy agents that are FDA-approved for use in solid tumors, and 17 immunotherapy agents are FDA-approved for use in hematological malignancies (https://www.cancerresearch.org/wp-content/uploads/2025/06/2025-CRI-Insights-Impact.pdf).

Despite its promise, immunotherapy is not universally effective. Response rates vary, and resistance mechanisms—both intrinsic and acquired—pose significant challenges. Factors such as tumor mutational burden, antigen presentation defects, and immunosuppressive tumor microenvironments influence outcomes (Binnewies et al. 2018). Importantly, despite having the same targets (i.e., PD1–1) or blocking the same interaction (i.e., PD1-PD-L1), checkpoint inhibitors have very different clinical efficacy. As an example, as opposed to Pembrolizumab, Nivolumab alone has limited efficacy in the frontline treatment of non-small cell lung cancer (NSCLC), and, reversely, as opposed to Nivolumab, Pembrolizumab did not demonstrate an improvement in the second line treatment of squamous head and neck cancers. Also, recently, a combination of Atezolizumab and dabrafenib/trametinib demonstrated an approximately 2 years improvement compared to a combination of Pembrolizumab and dabrafenib/trametinib in the same setting of V600E BRAF-mutated anaplastic thyroid cancer (CabAnnilas et al. 2024; Zafereo et al. 2025). Ongoing research and a rapidly expanding portfolio of clinical trials are refining patient selection criteria, exploring synergistic combinations (e.g., immunotherapy with targeted or metabolic agents), and identifying novel biomarkers of response.

Besides immune cells, stroma cells play a key role in cancer progression. The most recent generation of immune therapies is moving beyond direct tumor inhibition and targeting other cellular and noncellular components of the stroma that are important for a strong and durable immune response (Marin-Acevedo et al. 2021; Schaub and Tang 2025).

An active area of research is the targeting of key cellular components of the tumor stroma that enable cancer cell invasion and tumor progression. Therapies against cancer-associated fibroblasts (Jenkins et al. 2022; Chhabra and Weeraratna 2023; Broz et al. 2024), tumor-associated neutrophils (Zhang et al. 2020b), M2-tumor-associated macrophages (Raskov et al. 2021; Wang et al. 2024c) and T regulatory cells (T-Regs) (Shan et al. 2022) are under development. Targeting the noncellular components of the tumor stroma, i.e., the extracellular matrix and cytokines, is also a promising avenue (Xu et al. 2022; Schaub and Tang 2025).

Recent attention has centered on the bidirectional relationship between tumor growth and systemic metabolism, with recognition of tumor–host crosstalk giving rise to the idea that host metabolic pathways can be harnessed to limit cancer progression (Altea-Manzano et al. 2025).

5.21 Inflammation, an Unsolved Puzzle

Tumor-promoting inflammation, one of the key cancer hallmarks described by Hanahan and Weinberg in the 2011 installment of the hallmarks series, has been identified as a key player in cancer development. Given the presence of cell renewal and proliferation both in physiological wound healing and as a result of tumor-associated inflammation, tumors have been referred to as *wounds, which never heal* (Dvorak 1986). More recently, it has been speculated that carcinogenesis may be due to the failure of mechanisms that normally shut down inflammation at the appropriate times (Fishbein et al. 2021).

There is a complex relationship between cancer and inflammation; basically, inflammation may both promote and inhibit cancer growth (Roncati and Figueiredo 2023). Furthermore, cancer inflammation may be both a cause and a consequence of cancer (Rakoff-Nahoum and Medzhitov 2009), and importantly, cancer and inflammation have different mechanisms.

Several types of tumor-associated inflammation have been described with different causes, mechanisms, outcomes, and intensities (Rakoff-Nahoum and Medzhitov 2009; Grivennikov et al. 2010). In some situations (i.e., bladder cancer), acute inflammation may, in fact, fight cancer. Acute inflammation induced by a mixture of bacterial toxins from the Gram-positive bacterium *Streptococcus pneumoniae* and the Gram-negative bacterium *Serratia marcescens* was used by Coley at the beginning of the twentieth century, with some efficacy as an antitumor therapeutic method (Coley 1893).

Inflammation is a fundamental physiological response to tissue injury and infection and is aimed at restoring homeostasis. However, when dysregulated or chronic, inflammation can contribute to tumorigenesis, progression, and metastasis. Both local and systemic (global) inflammatory responses play critical roles in cancer development, influencing various stages of the disease, from initiation to immune evasion and resistance to therapy.

Reactive oxygen species (ROS) and reactive nitrogen species (RNS), which are produced during chronic inflammation, cause oxidative damage to DNA, proteins, and lipids. This can result in mutagenesis, genomic instability, and epigenetic alterations that predispose to malignant transformation (Kawanishi et al. 2017). For instance, inflammatory bowel disease (IBD) is associated with an increased risk of colorectal cancer, which is partly due to chronic inflammation-induced DNA damage (Itzkowitz and Yio 2004).

Tumor-associated macrophages (TAMs), in particular, have been implicated in promoting tumor growth. They often exhibit an M2-like phenotype characterized by anti-inflammatory and protumorigenic functions, such as enhancing angiogenesis, promoting tissue remodeling, and suppressing cytotoxic T cell activity (Mantovani et al. 2008; Noy and Pollard 2014).

Proinflammatory cytokines such as IL-6, TNF-α, and IL-1β are commonly elevated in the TME. IL-6, for example, activates the JAK/STAT3 signaling pathway, which promotes cell survival, proliferation, and resistance to apoptosis (Grivennikov and Karin 2010). Chronic NF-κB activation, often triggered by TNF-α and IL-1β, leads to the transcription of genes involved in inflammation, cell proliferation, and angiogenesis (Karin 2006).

Tumor-promoting inflammation may be modulated through several therapeutic strategies (Grivennikov et al. 2010): (1) inhibition of key signal transducers and transcription factors involved in the prosurvival and proliferative responses of malignant cells to inflammatory cytokines; (2) neutralization or sequestration of chemokines and cytokines responsible for the recruitment and maintenance of protumorigenic immune cells within the tumor microenvironment; (3) modulation of therapy-induced inflammation, either by attenuating excessive proinflammatory responses or enhancing beneficial inflammatory effects; (4) selective depletion of immune and inflammatory cell populations that facilitate tumor initiation and progression while preserving immunological subsets and effector mechanisms critical for antitumor immunity; and (5) targeted inhibition of protumorigenic cytokines with minimal impact on the expression or activity of cytokines with antitumorigenic functions.

A small clinical trial recently demonstrated the potential value of inhibiting inflammation. Twenty-four patients with chemo-resistant locally advanced and/or metastatic triple-negative breast received chemotherapy plus a pan-anti-nitric oxide inhibitor called NG-monomethyl-l-arginine (L-NMMA), which blocks the production of nitric oxide, a molecule involved in inflammation. The combination shrank the tumors in approximately half of the patients, which is significantly better than historical data (Chung et al. 2021).

Global inflammation is also associated with cancer incidence and progression. For example, in a large prospective cohort study of systemic inflammation and cancer risk, an investigation of samples from a cohort of approximately 440,000 participants, part of the UK Biobank, found that three inflammation biomarkers, the systemic immune-inflammation index (SII), neutrophil-to-lymphocyte ratio (NLR), and platelet-to-lymphocyte ratio (PLR), were positively associated with the risk of 7 out of 17 cancers, as opposed to the lymphocyte-to-monocyte ratio (LMR) biomarker, which had a negative association in these cancers. Among the inflammation markers evaluated, the SII generally displayed the strongest association with cancer risk and was associated with 6 of the 17 sites evaluated (colorectal, kidney, liver, lung, lymphoma and myeloma cancers). Of note, in this study, there were no associations of any of the blood cell ratios with breast or pancreatic cancer (Nøst et al. 2021).

Key Points

The tumor microenvironment is a dynamic and integral component of cancer biology, influencing every stage from initiation to metastasis. Therapeutic strategies that target TME components, such as CAFs, immune cells, and ECM elements, hold promise for disrupting the supportive niche that tumors exploit. Also, inflammation, both local and systemic, plays a critical role in the initiation, progression, and treatment response of cancer. A comprehensive understanding of TME interactions is essential for the development of effective cancer therapies. In particular, understanding the dual roles of inflammation in cancer biology may provide critical insights for developing targeted anti-inflammatory and immunomodulatory therapies.

6 From Being to Becoming: Philosophical Echoes in the Biology of Life and Cancer

6.1 Being and Becoming

Since the dawn of Western thought, humanity has wrestled with the tension between permanence and change—two views of our world proposed by the Greek philosophers Parmenides and Heraclitus. Parmenides saw reality as a singular, static, immutable being that was stable, self-sufficient, and unmoving. In contrast, Heraclitus envisioned the cosmos as a river in eternal flux: "We step and do not step into the same rivers, we are and we are not [...]. The world, the same for all, neither any god nor any man made; but it was always and is and will be, fire ever-living, kindling in measures and being extinguished in measures" (Graham 2023).

Permanence and change are also fundamental concepts in biology, and the three main biological processes, evolution, development, and homeostasis, are related to permanence and change.

The relationship between philosophy and biology, although distinct disciplines, can be traced back to ancient Greece. Philosophers like Aristotle (384—322 BC) made pioneering contributions to both philosophy and biology. He looked at living beings through the lenses of his teleological philosophy, which posited that natural phenomena have inherent purposes. His ideas, which dominated Western thought for almost 2000 years until the Renaissance, were challenged only by the Scientific Revolution, a drastic paradigm shift that started during the sixteenth century.

The enlightenment of the seventeenth and eighteenth centuries marked another paradigm shift in the philosophy of biology. The shift was started by René Descartes, who introduced mechanistic perspectives and considered organisms complex machines governed by physical laws. Although Descartes insisted on the reciprocal influences of the mind and body (Damasio 2005), he considered them separate entities and proposed a sharp demarcation between physical and mental processes, a separation that has continued until today. His mechanistic and dualistic views had a high impact on the development of modern biology, particularly in the study of anatomy and physiology. At the same time, philosophers such as Immanuel Kant analyzed the complexities of living systems, emphasizing the need to understand organisms as unified wholes rather than mere assemblies of parts.

In the nineteenth century, Charles Darwin's theory of evolution by natural selection revolutionized biology and challenged traditional views of design and purpose in nature, shifting the focus to natural processes. Darwin's vision reshaped not only biology but also philosophical understandings of human nature, ethics, and the interconnectedness of all life.

From the beginning of the twentieth century, through the discovery of the DNA and the DNA-RNA-protein code and transcription-translation mechanisms, molecular mechanisms have been the main explanandum of biology. Fundamentally, life occurs at the molecular level, and bottom-up models focus on understanding biological systems by starting at the smallest level—molecular or cellular processes—and building upward. This approach proved its efficacy by uncovering the basic mechanisms of life, providing insights into diseases at the molecular level, and prompting the development of targeted therapies, including more recent checkpoint inhibitors, which have significantly improved survival in several cancer types over the last decade.

In contrast, top-down holistic models work downward by focusing on the organization and functions of higher levels of biological systems, such as tissues, organs, and organisms. The top-down models are crucial for understanding phenomena that emerge from interactions between different components of the biological system and rely on networks, complexity, and emergence theories. Organisms are both hierarchies *and* holarchies: recursive systems where parts influence wholes and wholes influence parts. Cellular signals ripple upward to shape tissue and organismal behavior, while systemic cues flow downward to reprogram cells. The *hierarchical* architecture allows feedback and feedforward loops between levels, and communication within the same level allows *holarchic* integration between the bioenergetic, bioinformation, and biostructural elements, resulting in a whole integrated functioning process that we call the organism.

D. Paul, *Cancer: An Integrative Approach*, https://doi.org/10.1007/978-3-032-08808-6_6

Life is thus a dance of feedback and feedforward loops, structured through both bottom-up molecular logic and top-down organizational balance.

Therefore, in order to describe life phenomena, biology requires an integrated approach, blending both bottom-up and top-down models. Similarly, a thorough description of cancer needs to include not only the description of molecular processes but also the mechanisms through which cellular processes influence bottom-up behavior of tissues, organs, and the entire organism and vice versa, the top-down control of cells by the local and global environment of the organism. In the last decade, the rise of "omics" technologies has fostered a more holistic view, and more recently, with the help of AI, molecular data can be analyzed in the context of larger biological networks and systems, bottom-up and top-down being seamlessly blended, leading to a more realistic understanding of cells.

The twenty-first century saw further integration of mechanistic and holistic models both in philosophy and biology, particularly in areas like genetics, ecology, and cognitive science. Philosophers of biology now explore questions about the nature of species, the role of genes versus the environment, and the ethical implications of biotechnologies like cloning, gene editing and the use of AI in medicine. Studies of cognition and consciousness, hot topics that straddle both philosophy and biology, are currently a vibrant area of inquiry, with debates about the mechanisms of the cognition present in the simplest life forms, the origins of subjective experience, the mind-body relationship, and, last but not least, the potential threat to the human species of AI-embedded cognitive abilities.

Since the late nineteen/early twentieth century, theoretical biology has progressively become a sort of transdisciplinary science, an extended *metabiology* forum in which biology researchers and doctors, physicists, chemists, mathematicians, philosophers, and, more recently, computer scientists have collaborated in building more or less comprehensive maps of the biological world. Recently, this type of collaborative multidisciplinary effort has been extended to the oncology field. One example is the publication, *Rethinking Cancer: A New Paradigm for the Post Genomic Era*, edited by two philosophers of science (Marta Bertolaso and Bernhard Strauss) and two biologists (Mina Bissell and Ingemar Ernberg) (Strauss et al. 2021). Another example is "A Processual Perspective on Cancer," an insightful article written by two philosophers, Marta Bertolaso and John Dupré, and included in *Everything Flows: Towards a Processual Philosophy of Biology,* edited by Daniel Nicholson and John Dupré, both philosophers of science (Bertolaso and Dupré 2018). A few years ago, Anya Plutinsky, also a philosopher, presented in her book, *Explaining Cancer: Finding Order in Disorder*, a general description of the role of models and modeling in oncology and the nature of scientific evidence (Plutinsky 2018).

At its origin, clinical medicine was also heavily influenced by philosophy. In the East, traditional forms of medicine, like traditional Chinese medicine and Indian Ayurveda, rely heavily on philosophical models. In China, the dynamic relationship between Yin, an anabolic-like property associated with stasis and inertia, and Yang, a catabolic-like propensity for activity and change, both described in Huang Di Nei Jing's *Yellow's Emperor Inner Canon of Medicine*, are at the basis of all traditional Chinese medicine (TCM) schools. Importantly, Yin and Yang are considered complementary and not mutually exclusive. Yin and Yang represent a dialectic unity, at the same time opposing and controlling each other and supporting and transforming into each other. As opposed to the diads of Chinese philosophy, Indian metaphysics is based on triads. The foundation of Ayurvedic medicine describes three *gunas* (*sattva*, *rajas*, and *tamas*) and three *doshas* (*vatta*, *pitta*, and *kapha*) as fundamental properties of the universe and, respectively, of the human organism. Similar to yin, *tamas* represents a tendency toward inertia and rest, and, similar to yang, *rajas* represents a tendency toward activity and action. *Sattva* is understood in the Ayurvedic system as a third balancing, harmonizing agency between the two. According to Ayurveda, in the human body, the *tridoshas,* i.e., *vatta,* a sort of functional information force; *pitta*, related to energy and metabolism; and *kapha,* the structuring force of the organism, all of which interact continuously, and their imbalance leads to disease.

In the West, homeopathy, allopathy, and, more recently, German functional medicine have also been based on philosophical models. The principles of allopathy, *oppositorium currantum* (*opposites are treated by opposites*) and later homeopathy, *similia similibus curentur* (*likes are treated by like*), have been used by physicians, from Galen to Paracelsus to Hahneman well into the beginning of the twentieth century.

Contemporary evidence-based medicine has its own 5P mantra of *Personalized, Predictive, Preventive, Participatory, Population*-medical interventions endorsed by science and recently implemented using data-driven AI algorithms.

AI can and will certainly assist us to develop a deeper understanding of the cancer process and to navigate the multidimensional oncology space from cancer prevention to early diagnosis to prognosis and treatment.

On the other hand, AI cannot do the thinking for us in the quest for a better understanding of the cancer process. AI is a useful assistant, but it is limited; *it cannot think in our place.* It is my deep conviction that only through a deeper understanding of the living process will we get closer to finding a cure for cancer. To cure cancer—or any complex disease—we must first understand life. Not just in its chemistry, but in its wholeness. Not just in its structure but also in its unfolding. This requires a fusion of bottom-up mechanistic insights with top-down systemic understanding.

6.2 Do We Really Need a Better Understanding of Fundamental Biological Processes to Cure Cancer?

René Thom, Fields Medalist and founder of catastrophe theory, cautioned against this purely pragmatic approach to science, stating: "If we reduce science to a bunch of recipes that work, intellectually, we are not superior to a rat that when it presses on a pedal, food is delivered to his cage. The pragmatic theory of science brings us to the situation of the rat in his cage" (Thom 2009).

This cautionary view underscores a deeper issue: our persistent inability to cure cancer may stem from a fundamental gap in understanding life itself. We still lack a coherent theory of what life is, how its defining characteristics emerge, and how cells, tissues, and organisms preserve their structural and functional integrity amid internal and external perturbations. Cancer, in this context, represents not only a disease but also a unique and complex life form. To truly address it, we must begin by asking: What is life? How does it sustain coherence? My opinion is that true progress in the oncology field will come from addressing these foundational questions.

Mark Vincent proposed a fundamental classification of cancer therapies into two distinct categories: "causality inhibition strategies," which aim to intervene in the root causes of cancer, such as systemic, metabolic, or developmental disruptions, and "acausal" strategies, which focus on treating the disease by targeting specific molecular biomarkers or phenotypic signatures without necessarily addressing the initiating events (Vincent 2017). Unfortunately, the former remains largely aspirational, as current clinical oncology is still predominantly shaped by acausal approaches that focus on the visible manifestations of cancer rather than its deep etiological architecture.

In this view, cancer is not merely a disease but also a sort of *cryptobiont*—an elusive biological phenomenon whose ultimate logic remains only partially deciphered. But do we need to fully unravel the cancer enigma to treat it effectively? Not necessarily. History shows that major medical advances—such as the discovery of penicillin or cisplatin—did not emerge from a full understanding of disease mechanisms but from serendipitous observation coupled with bold experimentation. Likewise, future breakthroughs in oncology may stem not from solving the entire puzzle but from unexpected discoveries, such as the CR/SR mouse (see Sect. 12.8.3), which demonstrated spontaneous cancer resistance through unconventional mechanisms. *Both* sustained research and serendipitous insights are essential drivers of progress.

Many cancer patients are in urgent need of effective treatments, and for patients, the effectiveness of a treatment is much more important than the path to its discovery. The discovery of cisplatin in the 1960s was serendipitous (Monneret 2011). More recently, the unexpected observation that SARS-CoV-2 mRNA vaccination enhanced the efficacy of immunotherapy in patients with lung cancer and melanoma illustrates how chance findings can shape clinical practice (Grippin et al. 2025). While serendipity has sometimes accelerated discoveries in oncology, curing more patients with metastatic disease will require rationally designed strategies rather than relying on chance alone. In Chap. 13, I propose a theoretical framework of AI-assisted personalized cancer therapy that occupies an intermediate position between informed models, which rely on mechanistic or biologically grounded priors, and agnostic models, which depend purely on data-driven pattern recognition.

6.3 Building an Integrative Model

Biological phenomena present to us as undifferentiated wholes like a cloud (see Fig. 6.1). After performing a series of epistemic cuts (see Fig. 6.2) and then drafting theoretical models, science analyzes these analog, holistic processes as abstract maps of reality that can be experimentally tested. As a side note, the etymology of the word "abstract" is to "draw away," indicating a process of dragging or pulling something away from its original state or context. Any cut supposes a distinction. We should always remember that *epistemic cuts* are artificial demarcations and not real ontological distinctions. The cuts fragment the whole, and the attempt to restore the original wholeness is an act of "suturing" the fragments back together by some ad hoc defined connections. Precisely, because the *epistemic cuts* are not *ontological cuts* and the ad hoc connections between the fragments are more or less arbitrarily chosen, jumping back from the fragments to the whole does not reestablish the original analogic wholeness. However, if we start from interactions between individual

Fig. 6.1 An analog holistic cloud. (Courtesy of Ana Tourian)

Fig. 6.2 A digital reductionistic cloud. (Courtesy of Ana Tourian)

fragments, an emergent digitalized whole arises with emergent properties, different from both the properties of the fragments and the properties of the original whole.

Epistemic cuts are acts of division imposed by us. They are methodological strategies and artifices born of necessity. And because they are artificial, the reconstitution of the whole attempt to "suture" the fragments back together inevitably bears the limitations of our initial choice. The connections we draw to rejoin the severed pieces are just provisional constructs, tailored to fit the assumptions of our models rather than the seamless reality they aim to depict.

Every epistemic fragmentation made by science carves a distinction: dividing a biological continuum into digital separated parts. However, with each division, the complexity expands. As distinctions multiply, the model becomes increasingly intricate, and with this intricacy emerges a degree of randomness, a drift away from order. To reanchor the expanding complexity, science introduces structuring mechanisms—relational connections that attempt to rebind what has been fragmented. Yet these same connective efforts, far from simplifying, often deepen the model's entanglement. Complexity, like a double helix, folds back on itself.

Francis Heylighen articulated this interplay, distinguishing four types of complexity shaped by the tension between analytic cuts and integrative synthesis: *structural complexity* (spatial organization), *functional complexity* (temporal dynamics), *structural hierarchy* (nested spatial scales), and *functional hierarchy* (layered temporal processes) (Heylighen 1999). Complexity and emergence are closely related: complex systems cannot be understood by analyzing their parts in isolation, and when a system has many interacting parts with intricate relationships, it is more likely to display emergent properties. New properties arise not from the *multitude* of isolated parts but from their multiple, often unpredictable *interactions*.

Life is both a product and a process of emergence. It arises from nested interactions across scales: molecules interact within organelles; cells coordinate within tissues; tissues integrate and organize into organs; and organs orchestrate bodily functions.

We speak of life as complex, but in truth, we do not understand complexity—not at its core. Nor do we truly grasp emergence. These words, powerful as they are, serve as conceptual fingers pointing to phenomena we do not yet fully understand. And too often, we mistake the pointing for the thing itself. We ignore the rules that underscore the spectrum of scales that exist in the Universe both in the abiotic world, from the quanta to the whole cosmos, and in the living world, from viruses and unicellular organisms to multicellular organisms and the whole biosphere.

Our ignorance spans across all the scales of nature—from quanta to cosmos in the physical realm and from viruses to ecosystems in the biological one. We lack a *theory of scales* that explains how properties translate—or fail to—across these levels. Cancer, for instance, may not be a binary state but a continuum. The rigid division between "cancer" and "noncancer" may obscure important transitional states, leading to diagnostic errors and flawed interventions. The confusion between genetic mutations and functional disruptions is emblematic of a deeper problem: *we misidentify the seat of disease*. Genes alone do not dictate behavior; it is context and systemic impact that define malignancy. Reality also comes in degrees and spectra.

We navigate the world through a continuous stream of consciousness—a ceaseless flow of thoughts, sensations, and emotions that shape our experience from within. Yet the analytical intellect intervenes, slicing through this cloud of undifferentiated forms. It names, sorts, classifies, and fixes—transforming transient impressions into stable structures. Language itself enforces dichotomy: nouns and verbs; beings and doings.

Physics echoes this split with particles and fields—objects and actions—although these categories, too, are but human impositions. Science carves up the world into binary distinctions to craft a coherent map of reality, but even with our modern tools and theories, the map remains a crude approximation of a vast and still uncharted territory.

In quantum physics, this division is formalized through the taxonomy of particles. *Fermions*, the matter particles, comprise the scaffolding of the material world: quarks and leptons that form protons, neutrons, and electrons, the architecture of atoms. *Bosons*, by contrast, are the mediators—the silent messengers of force; photons carry electromagnetism; gluons hold nuclei together through the strong force; and W and Z bosons govern the weak nuclear interaction. These two categories differ in their spin: fermions with a half-integer spin resist cohabitation (the Pauli exclusion principle), while bosons with integer spin embrace indistinguishability, piling into the same quantum state.

Yet even these seemingly rigid categories are porous under the right conditions. In superconductivity and other quantum phenomena, fermions can pair and behave like bosons—emergent behaviors that defy simple classification. Still, the constituent particles remain true to their type: the essence of a fermion does not shift, even when its behavior appears to mimic that of its counterpart.

Theoretical physics has spent over a century refining this language of matter and interactions, weaving quantum mechanics and relativity into parallel, if not yet unified, frameworks. In contrast, theoretical biology lags behind. It lacks the same foundational architecture, the conceptual rigor, and mathematical formalism that might one day illuminate life itself with the same clarity. The cloud of the living world remains largely unparsed, waiting for its own language to emerge.

Let us return to the epistemic cuts. *The first cut: stasis versus movement.* We divide the biological world into *objects* (genes, proteins, carbohydrates, lipids, nuclei, cytoplasm and membranes, cell organelles, cells, tissues, organs and whole organisms) and *processes* (DNA transcription and RNA translation, protein maturation, DNA copying, autophagy, mitosis and meiosis, exosome secretion, metabolism, cellular division, apoptosis, autophagy, cellular communication, phagocytosis, and cell fusion). Dividing the world into objects and processes is of course dictated by practicality, but this split makes us forget about the original wholeness where objects and processes were indistinguishable. In biology, objects and processes are tightly linked and inseparable, like spacetime in Einstein's relativity, and they are artificially set apart as anatomical (object-oriented) and physiological (process-oriented) study subjects.

The second cut: nucleus, the cytoplasm, and the membrane. In reality, they represent one functional whole: the cell. In a cell, the cytoplasm, which chemically contains mostly water, is also structured and is able to hold and compute information. A cell is a crowded place where molecules and macromolecules are very close to each other and able to interact, shift shapes, and exchange energy and information. The dichotomy between the nucleus and the cytoplasm is misleading. Similarly, we divide the body into organs and interactions between them when, in fact, there is only one functional whole: the organism.

The third cut: scales of life. We slice between the microscopic and macroscopic, the cellular and the systemic, the human and its microbiota. But the body is not a homogeneous being—it is a *biosphere* composed of different ecosystems that contain coevolving life forms. Microbes shape our immunity, metabolism, and even mood. The boundaries between "human" and "other" are blurred, and the organism emerges from these interspecies dialogs.

The fourth cut: biotic and abiotic. We assume that life emerged from nonlife, but perhaps the Earth's abiotic conditions were never "dead" in the sense we imagined. They were life-ready-electrochemically fertile, structurally patterned, and rich in self-organizing potential. If so, it would follow that the first cell—i.e., the last universal common ancestor (LUCA)—did not emerge from a void but from a world already structured and woven with possibility. The biotic and abiotic worlds are seamlessly integrated from the get-go, life and the environment form a functional whole from the beginning, and the biotic world did not "evolve" from an abiotic world. The abiotic environment and life are one: life cannot function without being integrated into the abiotic environment. Life sculpts its environment—carving niches and oxidizing atmospheres—and the environment, in turn, nests itself inside organisms as intraorganelles *nanoenvironments*, intracellular *minienvironments*, tissue *microenvironments*, and systemic *macroenvironments*. On the one side, life creates hybrid biotic-abiotic niches, and on the other side, the abiotic environment is "invaginated" inside living organisms, forming hybrid biotic-abiotic environments. These hybrid spaces—biotic-abiotic interzones—are the very fabric of life's unity. Finally, besides the internal environments, there are also external, both biotic and abiotic, environments The human organism, for example, is influenced both by biotic microscopic (i.e., viruses, bacteria) and macroscopic environments: familial, social, cultural environments, etc. Both biotic and abiotic aspects of lifestyle have a direct impact on the organism and represent an important component of the external environment.

The fifth cut: body and mind. Descartes' dualism has reigned the Western world for centuries, but the schism between matter and thought is dissolving. Neuroscience, somatic psychology, and cognitive biology reveal that consciousness is embodied, and the body is mindful. Importantly, the fact that consciousness is embodied does not necessarily mean that matter and the mind *are the same thing*. They may possibly represent two faces of a single process looping in a continuous spiral like a Möbius strip.

Finally, in oncology, there is an additional, *sixth cut* between *cancer cells* and *normal cells* and between *cancer as a system* and *the whole organism*, and, very importantly, for oncology clinicians and researchers, between *cancer as a life form* and *cancer as a disease*. These epistemic cuts represent artificial distinctions. Cancer cells and the stroma are closely interrelated and form together the cancerized tissue. Also, in addition, the local and distal cancer tissues are interconnected in complex networks both between themselves and with the whole organism. Oncology has long treated cancer as an isolated invader. But cancer is not *other.* It is part of the organism—its cells, its matrix, and its signals. Cancer tissue is not merely a cellular mass but also a *system,* an ecology of cells, both malignant and stromal, interacting within a dynamically reconfigured niche. These cancerized niches do not exist in isolation; they are embedded in and shaped by the organism's systemic state.

Indeed, between a healthy and a malignant cell lies a continuum of transitional states—partially reprogrammed, context-sensitive, phenotype-shifting. Stemness, likewise, is not a fixed identity but a transient capability, i.e., a functional potential. As Joep Beumer and Hans Clevers pointed, "stemness" is an emergent behavior, not an ontological label (Beumer and Clevers 2024).

And so, our conceptual frame must shift. Cancer is not merely a *disease*; it is a *life form*, an evolving entity adapting to its niche. Only recently has oncology begun to see through the lenses of ecology and evolution (Maley et al. 2017), and in so doing, it reveals cancer not as an aberration but as an adaptive phenomenon. Health, too, has its hallmarks, and their exploration (López-Otín and Kroemer 2021) may prove more illuminating than the cancer hallmarks alone.

What we need is a framework that can "suture" these cuts—not by denying their analytic utility but by transcending them through integration. We need a holistic model that sees life as it presents itself: not as fragmented categories but as a living continuity of matter, form, energy, and meaning, a model where new perspectives of living systems, cells, multicellular organisms, and their interactions come into focus. Only then will we begin to grasp the essence of life and the enigmatic phenomenon we call cancer.

6.4 Living Systems

In general, a system can be defined as a dynamic entity of several interacting components that are codependent and function in an *integrated* way. A single cell, an organ, the entire human body are all living systems, which are subtypes of the general class of systems. *Functional integration* in living systems is achieved, on the one hand, through a *hierarchical architecture* and, on the other hand, through *communication*. For example, in the human body, there is communication between cells, tissues, and organs via secreted factors, such as cytokines, chemokines, metabolites, and hormones, as well as through extracellular vesicles and particles (EVPs). The distinction between "organism" and "living system"has been critically explored by David Krakauer and colleagues, who proposed that individuality—*organismality*—is not a binary state but a continuum (Krakauer et al. 2020). According to their framework, individuality can be understood as the degree to which a system preserves and propagates coherent information over time, and they identify three classes of individuality:

- *Colonial individuality,* where internal autonomy and external influence are in balance.
- *Organismal individuality*, dominated by internal self-regulation.
- *Environment-driven individuality,* where environmental forces override internal controls.

Classically, cancer is considered a disorder of individual cancer cells with behavior mainly determined by internal processes, and progress in clinical oncology has been mainly driven by molecular biology discoveries of intracellular and extracellular targets, which are more or less idiosyncratic, inside and outside the cancer cell. Observational and experimental data accumulated for more than a decade now point to the fact that cancer is likely to represent a living system comprising several complex interconnected networks of interactions located at different levels (cellular, tissular, and organismic). For example, at the organism level, we can look at metastatic cancer as a macroscopic system made of several geographically separated cancer tissues (primary tumor, local and distal metastases) that communicate with each other (Paul 2020). The recent Swanton et al.'s hallmarks article (Swanton et al. 2024) also pointed to the complexity of tumorigenesis and proposed a more nuanced view in which the local and distal environments play significant roles in the cancer process, opening the possibility of novel treatments directed to the local and distal environments and not directly to cancer cells.

Due to cancer conceptualization as a disorder of aberrant individual cells—rogue actors driven by intrinsic molecular defects—much of modern oncology has pursued intracellular targets, with therapies emerging from the identification of cancer cell-specific molecular signatures. Yet, accumulating empirical and clinical evidence over the past decade has challenged this reductionist perspective. Cancer now increasingly appears as a complex, emergent living system—a network of networks operating across cellular, tissue, and organismal scales. For instance, metastatic disease may be seen as a macrosystem composed of spatially dispersed but functionally connected tumor sites that communicate across the host's internal milieu (Paul 2020).

In light of Krakauer's taxonomy, cancer emerges as a shape-shifting process—its cellular elements able to transition fluidly between modes of individuality: autonomous, clonal, or environmentally entrained. This plasticity likely confers evolutionary advantage and survival, especially through group behaviors like invasion and metastasis. Interventions aimed at modulating these transitions—blocking the ability of cancer to oscillate between identity states—could offer a novel axis for therapeutic intervention.

Krakauer et al. further suggested that the individuality of a living system can be quantitatively assessed by plotting the system's behavior over time, revealing patterns of coherence or disintegration (Krakauer et al. 2020).

The fact that systems at different scales reflect similar principles has long been recognized. In 1960, Nicolas Rashevsky observed that biological systems could be mapped across levels through relational structures, proposing a principle of "*relational epimorphism*"to express the deep unity of the biological world: the higher organism mapped continuously onto the lower, not by form alone, but by shared functions and relational integrity. He wrote: "Any higher organism can be mapped in a many-to-one way onto a lower organism with continuous mappings of graphs or of topological spaces—or relation-preserving mappings of sets. In as much as the topological approach is now found to be only one of several more general possibilities—we propose here that instead of speaking of the principle of bio-topological mapping we speak of the principle of relational epimorphism in biology. Whatever its final precise formulation will be, this principle represents one of the most general biological facts and expresses the well-known unity of the biological world as a whole" (Rashevsky 1960).

The fact that living systems at different scales (i.e., the cell, organ, and organism) share similar features or subsystems was also noted by James Grier Miller more than four decades ago (Miller 1978). "Seven levels of living systems can be identified, each with characteristic structure and processes, and each composed of a system at the level below. The seven levels are: cell, organ, organism, group, organization, society, and supranational system. The structure, that is, the arrangement of their components in space, and the functional and historical processes of systems at each level are characteristically different from other levels, although the levels grade into each other in such a way that some systems may be difficult to classify. With some exceptions, systems at each higher level are larger than those lower in the hierarchy" (Miller 1985).

In his monumental work, *Living Systems*, Miller described and characterized 20 subsystems present across all scales in all living systems; 8 processing matter-energy only (ingestor, distributor, converter, producer, matter-energy storage, extruder, motor, and supporter); 10 processing energy only (input transducer, internal transducer, channel and net, decoder, associator, memory, decider, encoder, output transducer, and timer); and two that process both matter-energy and information: reproducer and boundary (Miller and Miller 1992; Bailey 2006). He also made the distinction between *toti-potential systems* like the human body, in which all 20 subsystems are fully functioning, with no help needed from external systems; *systems at different levels of the nested hierarchy* (Miller 1978); and *parti-potential systems* such as the cancer system, which may engage in a parasitic or symbiotic relationship with another system (the human body).

From this vantage point, cancer appears to be a system with *partial* individuality, embedded within the host's physiology, exploiting host resources while perturbing host integrity and health.

Fractal geometry, which was originally developed by Benoît Mandelbrot, has been successfully applied in the description of cellular membranes, lung's small airway tree-like structures, and other biological structures (Weibel 1994). Besides self-similar shapes, multicellular organisms also exhibit a sort of *functional fractalic-like self-similarity* among several temporal and spatial scales (Yates 1992; Jenkins et al. 2025).

In multicellular living systems, the organism level and the cellular level are not only interconnected and codependent, but they are also *functionally similar*. This may represent an evolutionary design logic: once viability is achieved at one scale, evolution reuses and adapts functional blueprints at larger ones. In Volume 2 of *Generelle Morphologie der Organismen* published in 1866, Haeckel introduced his "biogenetic law," which was later elaborated as follows: "Ontogenesis is the short and rapid recapitulation of phylogenesis, governed by the physiological functions of inheritance (reproduction) and adaptation (nutrition)" (cited by Kutschera 2011). Ernst Haeckel's dictum—"*Ontogeny recapitulates phylogeny*"—might be reimagined as a functional axiom extending from unicellular to multicellular organization.

Within the multicellular host, life forms of different scales coexist: somatic cells, germline cells, microbiota, viruses, fungi, and the host itself. As Sim Van der Ryn and Stuart Cowan wrote in *Ecological Design*, nature is scale-linking. Energy and information cascades across spatial and temporal domains, from quantum photosynthetic events to planetary cycles. "Nature's processes are inherently scale linking, for they inherently depend on the flow of energy and materials across scales. [...] Global cycles link organisms together in a highly effective recycling system crossing about seventeen tenfold jumps in scale, from the ten-billionth of a meter (the scale of photosynthesis) to ten thousand kilometers (the scale of the Earth itself). [...] Scale-linking systems imply a holism in which everything influences, or potentially influences everything else— because everything is in some sense constantly interacting with everything else. Nature is infused with the dynamical interpenetration of the vast and minute, an endless dervish mixing. Matter and energy continually flow across scales, the small informing the large and the large informing the small." (Van der Ryn and Cowan 1997). Multicellular organisms interact with the environment at the macroscopic level, but the energy needed for their existence is produced at the microscopic level.

As proposed by Carlos López-Otín and Guido Kroemer, the health of an organism resides in its overall organization, and it transcends any particular class of molecules, organelles, cell types, or anatomical structures. Health is not confined to DNA, proteins, mitochondria, or specific cell populations (López-Otín and Kroemer 2021). Instead, it encompasses supracellular units that form the functional building blocks of organs, entire organs, organ systems, systemic circuitries (like endocrine, neurological, or immune), and even the meta-organism that integrates the host and its microbiota. Health is inherently organizational, reflecting a series of dynamic features that maintain a precarious equilibrium across microscopic and macroscopic levels, preceding disease and infirmity.

Therefore, disturbances of the *hallmarks of health*, including disruptions of spatial compartmentalization (integrity of barriers and containment of local perturbations), loss of maintenance of homeostasis (recycling and turnover, integration of circuitries, and rhythmic oscillations), or erosions of the array of appropriate responses to stress (homeostatic resilience, hormetic regulation, and repair and regeneration), may originate not only at the microscopic level of molecules, organelles, and cells but also at mesoscopic levels—tissues—or at the macroscopic level—organs, organ systems, systemic circuitries and meta-organism (holobiont) (López-Otín and Kroemer 2021).

The nested levels of a multicellular organism are connected through complex circuitries like a gigantic organismal wide web, and perturbation at one level may lead to disruptions of circuitries within that level or between levels. Reversely, if the integration of circuitries is lost due, for example, to a deficiency of either intra- or extracellular communication systems, organismal health may be lost.

Taking into account these observations, *cancer may originate at any level in an organism*. The first perturbation may occur initially at the cellular, tissue, or organismal level, irrespectively, and as it progresses, reverberates through the whole organism and significantly alters all levels, leading to systemic multiorgan failure and death.

6.5 Biological Macroclimates, Microclimates, Miniclimates, and Nanoclimates

In the human organism, there is an internal *macroenvironment* present at the level of the whole body, as well as the *microenvironments* of different tissues and, finally, the *minienvironments* present in every cell and the *nanoenvironments* of the organelles. These environments located at different scales are codependent (Paul and Nedelcu 2022). Beyond the internal environments, there are also external microscopic (i.e., viruses, bacteria) and macroscopic environments: familial, social, and cultural environments. Lifestyle can also be considered part of the external environment and directly impacts the organism as a whole and should be part of an integrated model of the organism and its environment. This integrated understanding of life as having both biotic and abiotic components leads us to different models of cells and organisms and from that to a different medicine.

6.6 Active Matter, Biological Fields

Living organisms are made from materials unique to life: nucleic acids (DNA and RNA), proteins, carbohydrates, and lipids. They perform their biological functions in complex interface made of membranes, biopolymers, and aqueous solutions. The fifth type of matter, "active matter" (Needleman and Dogic 2017; Wu et al. 2017), is present in living cells and multicellular organisms. *Active matter*, or biological soft matter (colloids, gels, surfactants, polymers, liquid crystals, liquid glass, etc.), consists of structural units that are much larger than atoms; polymer molecules, for example, contain millions of atoms (Doi 2013).

The time of biological systems is also very different from ordinary time, and chronobiology—the study of biological time—has been steadily developing for more than six decades since Franz Halberg, an Austrian physiologist, started to study biorhythms and introduced the term "circadian." One of the particularities of biological time is synchronization. In an organism, synchronization has been related to various mechanisms, including direct physical interactions, chemical signaling, and electrical signals. Besides these mechanisms, synchronization is also a fundamental property of active matter that exhibits both static and dynamic synchronization, as described in the *swarmalator* model (O'Keeffe et al. 2017). Active/soft matter physics is a booming discipline that involves new theoretical tools like biostatistics and AI platforms as well as new experimental designs.

The importance of active/soft matter studies for oncology cannot be overstated. Different properties of the active/soft cancer present in cancer may explain some of its behavior, and therapeutic interventions targeting at these properties represent new avenues for cancer treatment.

The complexity and singularity of the biological world are at par with that of the quantum world. The properties of the quantum field were described by Heisenberg's and Schrödinger's mathematical formalisms. The living process is characterized by a unique set of *gradual qualitative changes* (Jaeger 2021). In order to dissect life's idiosyncratic behavior, new mathematical languages and, I would add, new AIS platforms need to be developed.

Biological events exhibit an eerie coherence and synchronization similar to the spooky actions at a distance related to quantum entanglement. This sort of "biological entanglement" occurring at macroscopic scales has been puzzling the human mind for millennia, and concepts like the Qi of traditional Chinese medicine (TCM) or the various types of pranas of Ayurvedic medicine were introduced precisely as an attempt to make sense of this phenomenon. Over the last century, similar concepts, like the morphogenetic field of Alexander Gurwitsch, the morphic field of Rupert Sheldrake and Harold Saxton Burr's life fields, were also developed by Western researchers. Biofield research is still in its infancy, and biofield properties need to be rigorously defined before they can be used on a larger scale. Recently, a series of carefully designed experiments performed by researchers at the MD Anderson Cancer Center reported intriguing data a human biofield reducing the growth of tumor cells in vitro and in vivo in a mouse model (Yang et al. 2019; Cohen et al. 2024). These preliminary results are difficult to reproduce, as they appear to be operator dependent, but they may represent a promising beginning in the study of human biofields.

A field is composed of both a medium and a wave (Strassler 2024). The active matter present in all living organisms is a medium fundamentally different from classical media (i.e., liquid, solid, etc.). The main component of cells is water, but besides the pressure and electromagnetic fields related to cellular water, fields of a different nature may also be present. Given the multiple properties of active matter, it is likely that one or more active yet uncharacterized fields are also present. A deeper understanding and characterization of *active biological fields* (i.e., biofields) has the potential to bring about dramatic changes in biology and medical fields, which are currently driven mainly by object-oriented ontologies.

On the other hand, the use of electromagnetic field for both diagnosis and treatment is currently a well-accepted modality. In 1974, Paul Lauterbur and Peter Mansfield independently published papers presenting a human body imaging technique based on magnetic fields that subsequently became known as magnetic resonance imaging (MRI) (Geva 2006). More recently, cancer treatment strategies using alternating electric fields, popularly known as tumor treatment fields (TTFs), have demonstrated their efficacy in several cancer types and are currently part of the standard of care for the treatment of glioblastoma, mesothelioma and lung cancer in the United States (Hong et al. 2022; Pohling et al. 2023; Li et al. 2023c; Heo et al. 2024). In parallel, amplitude-modulated radiofrequency electromagnetic fields are being developed for the treatment of hepatocellular carcinoma (Jimenez et al. 2019; Pasche et al. 2025). Interestingly, the use of both treatment modalities resulted in increased levels of cytosolic Ca^{2+}, but apparently, they activated different channel types. TTFs appear to exert their antitumor effects through activation of Cav 1.2 channels (Neuhaus et al. 2019) as opposed to the amplitude-modulated radiofrequency electromagnetic fields that activate Cav 3.2 channels (Jimenez et al. 2019). The role of Ca^{2+} and other ion channels in carcinogenesis is well known, and cancer has been described by several authors as a channelopathy (Prevarskaya et al. 2010; Litan and Langhans 2015; Prevarskaya et al. 2018; Tajada and Villalobos 2020; Bell et al. 2023).

6.7 Form, Process, and Elusiveness of Equations of Motion in Living Systems

In the biological sciences, the relationship between form and process has long been a subject of inquiry and debate. Charles Darwin viewed form as an evolutionary outcome, i.e., the result of adaptive pressures exerted over deep time. For him, morphology was a shadow of history, not a map of underlying mechanisms. D'Arcy Wentworth Thompson, in his seminal *On Growth and Form*, approached form from a complementary direction: he understood form as a "diagram of forces," implying that the visible structure of an organism could reveal the invisible physical laws and generative processes behind it (Thompson 1917). Von Bertalanffy, the father of systems theory, wrote, "What are called structures are slow processes of long duration; functions are quick processes of short duration" (Von Bertalanffy 1952). These thinkers saw process as ontologically primary as opposed to form considered a historical or mechanical trace.

The publication of René Thom's *Structural Stability and Morphogenesis: An Outline of a General Theory of Models* marked a seminal turning point in the modern synthesis of mathematics and biology, deepening and reconfiguring the foundational debate between form and process that had been catalyzed by D'Arcy Thompson's *On Growth and Form* (Thom 1975). While Thompson famously argued that biological form could be understood through the lens of physical forces and geometrical constraints, Thom advanced this intuition by proposing a mathematical theory of morphogenesis grounded in dynamical systems and singularity theory, known as *catastrophe theory*. Thom's work moved beyond the descriptive elegance of Thompson's morphogenetic transformations to offer a causal and topological framework for understanding how biological forms emerge, stabilize, and undergo sudden transitions. Central to his contribution is the idea of structural stability—the notion that certain patterns or forms persist under a range of perturbations—and the recognition that form is not merely a static outcome but a dynamical attractor within a space of possible developmental trajectories. In this view, morphogenesis is shaped by the underlying topology of the organism's internal and external constraints, giving rise to predictable yet context-sensitive patterns of biological development.

Thom's *catastrophe theory* formalized how qualitative changes in form—folds, cusps, and bifurcations—can arise from smooth variations in control parameters, thus offering a mathematical bridge between continuous processes and discrete morphological transformations. This framework revitalized the notion of form as a processual attractor, resisting the reduction of development to gene-centric linear causality and reintroducing geometry, topology, and dynamical systems into the heart of biological theory. In doing so, Thom implicitly challenged both mechanistic developmental models and rigid functionalism, aligning with a structuralist tradition that views biological organization as governed by intrinsic constraints and top-down global coherence rather than bottom-up accumulation alone.

Other modern formulations, such as Adrian Bejan's *constructal law* theory (Bejan 2000) and Albert-László Barabási's *network* theory (Barabási and Albert 1999), continue this tradition by linking the geometry of systems to their underlying flows and interactions. Bejan's constructal law posits that "for a finite-size flow system to persist in time (to live), it must evolve such that it provides greater and greater access to the currents that flow through it." (Bejan and Lorente 2010). Similarly, Barabási's scale-free network theory reveals how the architecture of a system reflects the probabilistic dynamics of its connectivity.

However, despite Thom's description of the "elementary catastrophes"—fundamental geometric shapes—that represent the basic ways discontinuities can occur in a system's structure or behavior, his and other theoretical frameworks, although they offered useful metaphors and local insights, they fall short of being universal. Besides catastrophe theory, other theories like fractal geometry, Fibonacci sequences, and small-world architectures, are seductive in their apparent universality. Yet in living systems, their applicability is contextual, partial, and often metaphorical. The lack of scale-invariant self-similarity across biological hierarchies undermines the hope of a unifying formalism. Thus, while forms can occasionally hint at underlying processes, they do not reliably encode the equations that drive them. Einstein spelled it out clearly: physics deals with rigid objects and takes into consideration only positional changes, ignoring state changes. In biology, state changes are fundamental.

The central challenge lies in the nature of time and scale in biological systems. Unlike mechanical systems governed by uniform time and global rules, biological time is non-monotonic and punctuated by rhythms, delays, feedback loops, and abrupt phase transitions. Moreover, life operates across nested hierarchies of scale, from molecules to ecosystems, each governed by different organizational principles. As a result, morphologies that appear mathematically tractable at one level (e.g., the logarithmic spirals in seashells or the branching of bronchial trees) may bear no predictive power at another.

The lack of an appropriate formalism for living systems is acutely relevant in cancer biology. Tumors may exhibit fractal vascular patterns (Baish and Jain 2000), but their behavior resists modeling by static geometries. Cancer is not simply a form—it is a process, a dynamic instability shaped by interactions across the cellular, tissular, and systemic levels. While the form in cancer (e.g., the histological grade or architectural disarray of tumors) provides diagnostic information, it rarely yields a deterministic "equation of motion" that can predict the progression, treatment response, or metastatic potential of the disease (Hanahan and Weinberg 2011). The emergent behaviors of tumor ecosystems—heterogeneity, adaptation, and immune evasion—are governed by contingent, context-dependent interactions, not fixed physical laws.

The form-process relationship in living systems, especially in complex pathologies like cancer, must be understood as historically contingent, context dependent, and irreducibly multiscalar. A truly predictive biology may arise not from decoding form, but from modeling the flows, feedbacks, and stochastics of life itself. This requires mathematical tools different from ordinary equations. AIs may offer partial solutions.

7 Life Theories

7.1 Life Definitions

Biology systems are complex, and as pointed out by Radu Popa, there is no single feature, principle, or parameter that can define life: "life is not one thing but a circumstance of many mechanisms acting together for the fulfillment of a common and more general goal." Popa classified the life definitions as "parametric (a collection of more general or particular attributes of life forms), cybernetic (a numerical analysis of the properties of life), cellularist/genetic (a small list of the most important features, generally resulting from an examination of life on Earth) and non-Earth centric" and indicated that there is no life definition that is generally accepted and that every single theory has been criticized as being "too narrow, too broad, too vague, too mathematical, too materialistic or too impractical" (Popa 2010).

Edward Trifonov searched for commonalities in 123 definitions of life and came up with the definition: "Life is self-reproduction with variations" (Trifonov 2011). More recently, Athel Cornish-Bowden and María Cárdenas also analyzed contemporary theories of life and found that most of them included the notion of *closure*. Closure refers to the fact that although organisms exchange energy and information with the external world, all the genes necessary for the production of proteins and all the enzymes necessary for functional cellular metabolism are synthesized by the organism itself: none are coming from outside (Cornish-Bowden and Cárdenas 2020). Basically, thermodynamically speaking, organisms are open systems, but organizationally, they are closed systems. Of note, closure refers to the circularity of life processes like the Krebs cycle, for example, where all the enzymes and metabolites are present in the cycle, which is *functionally* closed, creating an inward-turning coherence amidst outward flux.

Despite the lack of a universally accepted all-encompassing life definition, a heuristic set of life traits that can help our purpose of investigating the cancer process can be drafted. Instead of an abstract set of definitions, we can use an operational approach that integrates sets of contextually relevant principles that can be applied in cancer research and clinical practice (Bich and Green 2018).

It is important to notice that the levels described by James Grier Miller (1978, 1985) have different ontological properties. A cell has very different properties than an organ, and an organism is different from a community. A cell, an organ, or an organism each possesses distinct modes of qualitatively unique configurations of matter, energy, and information. Life, then, is not a singular entity to be pinned down but a multilevel, ever-evolving phenomenon—an open-ended symphony—played in recursive motifs across the nested scales of nature.

7.2 Searching for Life Laws

The functional fractal-like self-similarity (Mistriotis 2021) of all living systems both at different spatial scales (i.e., the cell, organ, organism, etc.) and all through the history of life on Earth since the last universal ancestor (LUCA), suggested the presence of universal laws.

Life on Earth is influenced by Earth's geologic and climate characteristics and appears at precise spatial scales, starting at nanometers (the size of viruses and DNA/RNA codons) and expanding to 30 m (the blue whale, weighing approximately 180,000 kg, the same as 30 Tyrannosaurus Rex or 40 elephants). On the other hand, cells have a relatively limited size ranging from approximately 100 nanometers (mycoplasmas; by comparison, *Mimivirus*, the largest existent virus, is 400–500 nanometers) to approximately 18 cm (the ostrich egg; the largest egg ever found belongs to the now extinct *elephant bird* whose eggs could measure 33 centimeters). Cell size is limited by the surface area-to-volume ratio of the cell. As the volume of a cell increases, the diffusion of molecules inside the cell takes longer, and a cell stops functioning due to starvation or waste accumulation.

In terms of the temporal scale, vision involves one of the fastest known physical processes: within just 200 femtoseconds, a photon is absorbed by a retinal molecule called

D. Paul, *Cancer: An Integrative Approach*, https://doi.org/10.1007/978-3-032-08808-6_7

rhodopsin in the eye, triggering a conformational change from the 11-cis to the all-trans isomer. This ultrafast molecular shift initiates a cascade of cellular signal transduction events that ultimately result in visual perception (Gruhl et al. 2023). At the other end of the spectrum, the oldest organism on Earth is a Great Basin bristlecone pine in California's White Mountains with a 5067-year-old core. Organisms' sizes and duration of biological processes are closely related, and it is safe to say that life on Earth starts at the quantum nanometric/nanosecond scale and expands into the macroscopic world at approximately 10 power and 10 metric/second scales. In terms of size, gravity may be the limiting factor. On other planets, life would be bigger or smaller according to gravitational pull. In terms of duration, life cycles are impacted by genetic and environmental factors and vary a lot.

Various biological relationships have been described across taxa, including, for example, the allometric law of mass-related metabolic rates. Several power laws have been described, including a 2/3-power "surface law" related to the Euclidean geometry of surface area and volume and a 3/4-power law based on the fractal-like geometry of resource-transport networks (Glazier 2023). Russian ecologists also described a size invariance (power law of 1) of mass-specific metabolic rates across taxa, with bacteria exhibiting the same range of values as vertebrates and even large mammals, from about 2–40 W kg^{-1} (Makarieva et al. 2005, 2008).

Douglas Glazier argued that the scaling slope varies mostly between 2/3 and 1 and that it is most likely related to the metabolic (activity) level (Glazier 2005, 2022). As opposed to *allometry*, which refers to space, Glazier recently proposed the concept of *allochrony*, which refers to temporal relationships between relatively short biological time periods and longer time periods. In his opinion, the "pace of life" across taxa may depend both on the fundamental time limits set by the duration of the life cycle and evolutionary patterns of the origination/extinction of large-bodied species during geological "deep" time. These concepts described for organisms may also apply to cancer.

7.3 Universal Principles of Living Systems

Evelyn Fox Keller considered the search for universal law in biology futile: "Even though we cannot expect to find any laws governing the search for generalities in biology, some rough, pragmatic guidelines could be very useful indeed" (Fox Keller 2007).

Ignoring Fox Keller's warning, different authors have proposed various life principles:

(a) Two; dual autocatalytic closure of structure (Jagers op Akkerhuis 2016, 2023, 2024).
(b) Three (Sabelli 2001, 2003, 2005; Yafremava et al. 2013; Trevors and Saier 2010).
(c) Six (Lane 2015).
(d) Seven (Koshland 2002).
(e) Fifteen (Rosslenbroich 2023).
(f) Eighteen (Hooker 2011; Çengel 2023).
(g) Twenty (see above Miller 1978).

The *operator theory* of Gerard Jagers op Akkerhuis proposes a unifying framework for understanding the emergence and classification of entities in the universe through a hierarchy based on dual closures. This theory seeks not only to rank types of objects (referred to as "operators") but also to describe the process of their emergence (ontogenesis). Each operator represents a new kind of object defined by a specific form of dual closure—an abstract construct in which two or more components of the previous level are linked structurally and functionally to create a higher-level entity. The sequence of such transformations constitutes the operator hierarchy, beginning with fundamental particles formed during the Big Bang.

A *dual closure* is the defining feature of each *operator* and serves as the generative principle for the next level of complexity. For example, atoms emerge through the dual closure of subatomic particles (nucleus and electron shells), and molecules arise through the bonding of atoms. Each operator level generalizes not individual objects but structure kinds, similar to how the term "brick structure" refers not to a specific building but to any assembly of bricks joined by cement. Thus, any configuration using the same kind of components and closures belongs to the same operator category, regardless of the specific form or function. The Operator Theory is grounded in two core principles: *operators* - abstract categories of entities defined by specific structural and functional organization and *dual closure* - the mechanism by which a new operator level emerges through the integration of components from the preceding level, enabling the ontological ascent toward increasing complexity. In the context of biology, the cell represents a foundational operator: the first entity in the biological domain to exhibit both structural and functional closure. *Structurally*, a cell is enclosed by a semipermeable membrane that defines its boundary and maintains internal conditions distinct from the external environment. *Functionally*, the cell exhibits closure through self-maintaining metabolic and genetic regulatory networks that enable autonomy, homeostasis, and reproduction. This dual closure—the membrane structure and internal regulation—marks the cell as a distinct level in the operator hierarchy. All living systems, as is currently known, are composed of or derived from such cellular operators. The transition from unicellular to multicellular life represents the emergence of a new operator through a novel form of dual closure between cells. In multicellular organisms, cells connect and coordinate via intercellular junctions (structural closure) and integrated physiological or developmental signaling networks (functional closure). These dual closures bind individual cellular operators into a higher-order system, with emergent properties such as differentiated tissues, coordinated development, and centralized control of growth and reproduction. This dual closure between cells does not merely aggregate individual cells; it transforms them into parts of a unified whole, giving rise to a new kind of operator: the multicellular organism. Just as atoms bound by covalent bonds become molecules, cells bound by structural adhesion and functional integration become organisms. The organism, as an operator, inherits and transcends the properties of its constituent cells.

Following the logic of operator theory, multicellularity can branch into different organizational forms—colonial organisms, simple differentiated tissues, or complex bodies with organ systems—all emerging from variations in the nature of cellular dual closure. Despite this diversity, each multicellular system that achieves coherent structural and functional integration qualifies as a new operator at the same hierarchical level. This reflects the theory's allowance for ontogenetic diversity within organizational levels. This framework provides a rigorous, hierarchical model of biological organization, unifying life complexity under a common ontogenetic logic (Jagers op Akkerhuis 2016, 2023, 2024).

Inspired by the work of Bourbaki, using a series of mathematical concepts, Héctor Sabelli developed *bios theory*, a comprehensive framework that seeks to explain the creative dynamics of natural processes through three universal generative principles: asymmetry, opposition, and continuous transformation. These principles, drawn from both mathematical and physical insights, are proposed as the foundational forms of creativity operating across all levels of organization in nature. Sabelli analogizes these creative principles to the mathematical structures of lattice, group, and topology—not as static abstractions but as generators of dynamic processes. In his view, these structures are necessary and sufficient to produce the richness of mathematical and natural order: "Asymmetry, opposition and continuous transformation *are found in natural processes at all levels of organization*." Specifically:

- *Lattice* reflects time order, a one-dimensional sequencing of events that introduces direction and asymmetry;
- *Group structures* represent informational complexity and oppositional interactions, capturing the two-dimensional interplay of complementary or antagonistic forces;
- *Topology* embodies continuous transformation and spatial extension, capturing the flexible, adaptive, and three-dimensional nature of physical and biological system.

Sabelli integrates these mathematical structures into a six-dimensional view of process organization, emphasizing that creativity in nature is not random but ordered yet open, unfolding through patterns of asymmetry, dialectical opposition, and transformation. This biosystemic framework seeks to bridge science and philosophy by grounding emergent complexity—namely life and consciousness—in formal generative structures that underlie both mathematics and biological reality (Sabelli 2001, 2003, 2005).

Liudmila Yafremava and colleagues proposed a conceptual framework known as the "persistence triangle" to describe the fundamental strategies that enable living systems to survive and persist in dynamic and often unpredictable environments. This model identifies three core, interdependent strategies—flexibility, robustness, and economy—that organisms deploy to maintain viability across changing conditions.

1. *Flexibility*

Flexibility refers to an organism's ability to adapt behaviorally, physiologically, or developmentally in response to information from its *umwelt*—its subjective perception of the environment. This adaptive responsiveness allows living systems to adjust to novel or shifting ecological conditions. Flexibility is crucial for short-term survival in uncertain environments and underlies traits such as learning, phenotypic plasticity, and signal transduction cascades.

2. *Robustness*

Robustness ensures stability and functional integrity in the face of disturbances. It is achieved through redundancy, modularity, and reliability in internal networks (genetic, metabolic, neural). Robustness enables living systems to absorb shocks, such as genetic mutations, environmental noise, or mechanical damage, without catastrophic failure. In biological systems, examples include fault-tolerant gene regulatory networks and homeostatic mechanisms.

3. *Economy*

Economy emphasizes the efficient use of energy and resources, minimizing unnecessary metabolic costs. Flexibility and robustness, while beneficial, are energetically expensive. Economy imposes selective pressure to optimize trade-offs, ensuring that energy and matter are directed toward reproduction, growth, or other fitness-enhancing outcomes. This principle is seen in streamlined genomes, frugal metabolic networks, and energy-efficient locomotion or neural computation.

The triangle metaphor highlights that these three strategies are not simultaneously maximizable; enhancing one often compromises another. For instance, high robustness via redundancy may incur metabolic costs that reduce the economy. Similarly, extreme flexibility may introduce instability or inefficiency. Thus, organisms balance these strategies based on ecological context, evolutionary history, and developmental constraints. The optimal configuration varies among species, life stages, and environmental conditions. Yafremava's persistence triangle provides a valuable framework for understanding evolutionary design principles, biological resilience, and even synthetic biology or bioinspired systems. It connects informational, structural, and energetic aspects of life, suggesting that the persistence of living systems is not defined by a single optimal state but by their capacity to navigate trade-offs dynamically. It frames biological existence not only as survival but also as a continual negotiation between adaptability, stability, and efficiency (Yafremava et al. 2013).

This theory emphasizes not static traits but processes of balance and modulation—a view increasingly relevant for understanding complex adaptive systems, from cells to ecosystems.

Jack Trevors and Milton Saier proposed three biological laws: 1. All living organisms obey natural laws. 2. All living organisms consist of membrane-encased cells. 3. All living organisms arose in an evolutionary process (Trevors and Saier 2010).

In *The Vital Question*, Nick Lane, a biochemist, attempts to explain both life's origin and evolution on a bioenergetical basis. He enumerates six basic properties shared, in his opinion, by all living cells:

> (i) a continuous supply of reactive carbon for synthesizing new organics;
> (ii) a supply of free energy to drive metabolic biochemistry—the formation of new proteins, DNA, and so on;
> (iii) catalysts to speed up and channel these metabolic reactions;
> (iv) excretion of waste, to pay the debt to the second principle of thermodynamics and drive chemical reactions in the right direction;
> (v) compartmentalization—a cell-like structure that separates the inside from the outside;
> (vi) hereditary material-RNA, DNA or an equivalent, to specify the detailed form and function (Lane 2015).

Daniel Koshland Jr. proposed seven pillars of life under the acronym PICERAS: P (rogram), I (mprovisation), C (ompartmentalization), E (nergy), R (egeneration), A (daptability), and S (eclusion).

Recently, a comprehensive list of fifteen life principles was proposed by Bernd Rosslenbroich in his book *Properties of Life. Towards a Theory of Biological Organisms*: (1) interdependencies, (2) integrative biological systems, (3) autonomy, (4) agency, (5) processing of molecules, (6) processing information, (7) processing energy, (8) processes of shape, (9) time autonomy, (10) sensitivity and affectability, (11) subjective experience, (12) the ability to evolve, (13) growth and development, (14) environmental interrelationship, (15) reproduction (Rosslenbroich 2023).

Cliff Hooker described 18 characteristics of complex systems: (1) nonlinear interactions, nonadditivity; (2) irreversibility; (3) constraints—holonomic and nonholonomic; (4) equilibria and stabilities—static and dynamic; (5) amplification, sensitivity to initial conditions; (6) finite deterministic unpredictability; (7) symmetry breaking, bifurcations, self-organization, emergence; (8) constraints—enabling and coordinated; (9) intrinsically global coherence; (10) order, organization; (11) modularity; hierarchy; (12) path-dependence and historicity; (13) constraint duality, supersystem formation; (14) coordinated spatial and temporal differentiation with functional organization; (15) multiscale and multiorder functional organization; (16) autonomy and adaptation, adaptiveness, learning, (17) model specificity/model plurality, model centeredness; and (18) condition-dependent laws (Hooker 2011).

More recently, Yunus Çengel also described 18 distinctive features separating living beings from the nonliving world: existence, subjectivity, agency, purposiveness and mission orientation, primacy and supremacy, naturality, field phenomenon, locality, transience, transcendence, simplicity, unicity, initiation, information processing, traits, code of conduct, hierarchy and nesting, and the aptitude to vanish (Çengel 2023).

Reviewing this list of principles, a question comes to mind: Are principles really needed? Christoper Kempes and David Krakauer argued that *yes*, for a broader understanding of life, we need to move from the *extant-centric* approaches, focused on characteristics and comparisons among existing life forms, or *history-centric* approaches focused on specific evolutionary trajectories leading to current life forms, to a *principle-centric* approach concentrated on describing the set of shared properties among all possible evolutionary trajectories and all possible origins of life. Starting from the assumption that life is adaptive information, they also proposed a general model of three nested levels: L1, a level of materials (e.g., the DNA or RNA genome and cells); L2, a level of constraints (e.g., energy requirements); and L3, a level of function optimization (e.g., fitness maximization) (Kempes and Krakauer 2021).

They specified that the three levels are "ontological" and not physico-spatial, and they provided as an example optimization that is ubiquitously present in living systems from the molecular level to ecosystems. An important note is that the three levels proposed by Kempes and Krakauer are *functional* and should not be confused with the seven levels of nested hierarchy listed by Miller. On the other hand, similar to Miller, the model proposed by the two authors is meant to apply not only to organismic life forms but also to other forms of organization, such as human culture, economic, and legal institutions, and even to human products like artificial intelligence. The model of Kempes and Krakauer describes, in fact, as the authors themselves acknowledge, the information-processing architecture of a computer system with hardware, software and three layers: "A computational or *functional level* that describes the computational problem. [...] A subvening algorithmic or *procedural level* that realizes iteratively the desired computation. [...] And a *foundational hardware implementation level* that supports the software realizing a computation" (Kempes and Krakauer 2021).

Developing a set of guiding principles for living systems holds significance not only for the foundational understanding of biology but also for advancing oncology. Since cancer can be viewed as a dynamic system in continuous interaction with the host organism—a separate dynamic system—applying the principles of living systems may provide valuable insights into cancer behavior and treatment. Unlike the hallmarks of cancer, which aim to define its unique and distinguishing features, the principles of living systems emphasize the fundamental similarities between cancer and the organism in which it arises. This perspective shifts the focus from what makes cancer different from the shared systemic properties that govern both malignant and healthy biological processes. In Chap. 9, we introduce eight living system principles that we discuss in the context of cancer treatment.

7.4 Life's Theory of Everything: Unifying Scales from Microcosm to Macrocosm

For over a century, both physics and biology have sought a "Theory of Everything" (TOE), capable of bridging the gap between the microcosmic world, governed by quantum mechanics, and the macrocosmic realm, guided by Einstein's theories of relativity. In physics, these two domains remain separated—each operating under its own distinct set of principles. Similarly, living systems span multiple scales, from the molecular to the organismal, each scale governed by its own dynamics and rules.

Biology faces a challenge parallel to physics: identifying trans-scale principles that can unify diverse organizational levels of life. Whether examining viruses, cells, or multicellular organisms, biology also seeks a framework capable of spanning these varied forms of life.

Since Miller's work, several other theories of living systems have been proposed focusing on other fundamental properties, such as Morowitz's ecology theory of life, examining the reciprocal relationship between a living system and its environment (Morowitz 1992).

There is a different relation between stochasticity and order in the biological world from the abiotic world. Biological systems are complex, and the solutions to equations of the dynamics of biological systems are heavily influenced by initial conditions, so, similar to chaotic systems fundamentally their trajectories are never completely predictable. On the other hand, usually, biological systems do not behave like weather systems, that can be disturbed by butterfly effects (Gleick 1987). Their homeostatic architecture grants them quasistability, and as a result, their behavior can be partially predicted.

Furthermore, the presence of cognition in living organisms bestows on them the ability of goal-directed actions that necessitate a different analysis from the mathematical formalism of physics. Of note, goal-directed actions do not necessarily imply the presence of a conscious "will." "Will" often implies conscious intention and effort, which may be a component of *some* goal-directed actions but not of *all*.

Biological systems are neither purely digital nor entirely analog. Also, they are neither entirely deterministic nor stochastic. They defy classification in terms commonly used to describe artificial systems. Artificial intelligence architectures can capture only fragments of biological behavior, reflecting the challenges posed by trying to model life in their full richness.

A parallel can be drawn between quantum systems and biological systems: just as the wave function in quantum mechanics can be computed via the Schrödinger equation without a comprehensive understanding of the underlying system, biological patterns can also be extracted from data without a full grasp of the intricate mechanisms at play. Nowadays, AI can identify these patterns; yet, AI remains incapable of deciphering the underlying biological intricacies and meaning.

The abiotic foundation of biology is represented by the quantum level of elementary particles combined into atoms that form molecules that interact between themselves, creating larger molecular assemblies. Importantly, quantum mechanisms are involved in fundamental biological processes like light sensing and olfaction, magnetoreception, enzyme catalysis, photosynthesis, and cellular respiration. A recent report suggested that quantum tunneling may also be involved in genetic mutations (Pusuluk et al. 2018; Srivastava 2019; Slocombe et al. 2021; Slocombe et al. 2022; Slocombe et al. 2023). It is possible that quantum phenomena are also present at the level of other molecules, like the water molecules, cytoplasmic proteins, different metabolites, and ions, like calcium ions.

A key point is that in cellular environments, unlike the spontaneous randomness of abiotic systems (such as Brownian motion), a new kind of process emerges—one exclusive to living organisms: communication. Supramolecular structures such as organelles, membranes, and nuclei *communicate* with each other through nonrandom channels, conveying semantic and functional information rather than mere energetic exchanges. This distinction is crucial: whereas basic interactions might only exchange syntactic information, biological communication conveys deeper meaning and functional significance. Communication represents directed interactions, i.e., *interactions with biological messages or meanings*. At higher levels, biological communication extends beyond the cell to tissues, organs, and across the organism. Cells communicate within tissues via gap junctions, and organs communicate with one another through chemical, electrical, and mechanical networks. Furthermore, communication transcends the individual organism, extending into the social realm where interactions with others also influence physiological responses.

In the human organism, these communication networks are not only horizontally integrated within a given scale but also vertically interlinked across scales. For instance, a drop in blood glucose triggers the release of hormones like glucagon and epinephrine, which not only induce hunger but also prompt the mobilization of stored glucose. Also, verbal exchanges between individuals may stir emotions that can lead to the release of molecules in the bloodstream that reverberate through the whole organism.

However, the challenge biology faces in distinguishing its systems is evident when comparing cancer cells with their normal progenitors. The success of antibiotics rests on the ability to target bacterial ribosomes, which differ from those in eukaryotic cells, allowing for effective treatment with minimal side effects. Cancer, by contrast, arises from normal cells, making it far more challenging to target without collateral damage to them.

Cells *as* unicellular organisms, cells *in* a multicellular organism and the multicellular organism itself, represent three distinct living entities. Despite these similarities, there are important differences between them. A multicellular organism can function without one or several of its cells; however, a cell inside a multicellular organism depends on the organism for its life. The structures (body organs) and functions (body systems) of the human body are similar to those of a cell but also have many unique emergent properties. For example, similar to the microtubules of a cell, the human body has two global transport systems (the cardiovascular and the lymphatic system); however, although a cell may be capable of basic cognitive mechanisms, it lacks higher cognitive abilities of a human being. Another important distinction refers to death and reproduction. Physiological mechanisms of cell death are important in multicellular organisms for development and morphogenesis, for cell number control mechanisms control cell number as a defensive strategy in order to remove infected, mutated, or damaged cells (Vaux and Korsmeyer 1999). Multicellularity has arisen independently at least twenty-five times during life history on Earth, but most multicellular groups are simple facultative aggregations. Each cell can survive and reproduce independently and can establish another multicellular group (Grosberg and Strathmann 2007). Obligate multicellularity observed in red algae, brown algae, fungi, and plants and in some ciliates and cyanobacteria animals is more rare (Howe et al. 2022). Under obligate multicellularity, cells cannot survive or reproduce independently of the multicellular group and become specialized into somatic (from "soma" body Gr.) and germline. This "fundamental division of labor" enables increased complexity (Howe et al. 2022), resulting in increased fitness of the multicellular organism.

Cancers can be either monoclonal (Yamamoto et al. 2003; Liu et al. 2009; Huang et al. 2014) or polyclonal (Parsons 2008; Gundem et al. 2015; Maddipati and Stanger 2015; Cheung et al. 2016). They exhibit characteristics of both unicellular life forms (i.e., forming colonies like bacteria) and multicellular organisms (i.e., the presence of distinct cells with stem cell properties, i.e., a form of cellular specialization). Importantly, cancer cells in a tumor are plastic, i.e., can switch between more or less differentiated cellular states. At a given time, some cells in a tumor exhibit stem cell-like properties and are able to recreate another tumor and/or acquire a mesenchymal-like phenotype and metastasize. The Weismann barrier, which traditionally separates somatic and germline cells, is often bypassed in cancer, allowing tumor cells to dedifferentiate and acquire the properties of germline or embryonic cells, further complicating the therapeutic landscape.

7.5 How About Viruses?

Regarding viruses, an animated debate has been going on in biology for many years regarding their nature. Are viruses alive or not? If we consider dynamic persistence the essential life characteristic, viruses are certainly alive since they are able to replicate and persist through time. Viruses are a parasitic intracellular form of life and depend on the host for their existence but, recently, viral factories have been described (Novoa et al. 2005; for a review, see Fernández de Castro et al. 2021), and an important distinction has been made between viral factories, which are large, complex life forms, and virions, which are simply the mechanisms used by viruses to spread from one cell to the other. According to Claverie and Forterre (Claverie 2006; Forterre 2006), viruses can also be considered capsid-encoding organisms, which are very different from unicellular and pluricellular organisms but organisms on their own. Claudiu Bandea proposed the term "*molecular organisms*" for viruses (Bandea 2009a, b). Separating viruses into a class of their own is supported by genomic data that have shown that the majority of viral genes have no cellular homologues, indicating that viral genes represent a unique pool of genetic diversity (Forterre 2010) that is larger than the eukaryotic genetic pool (Edwards and Rohwer 2005; Bandea 2009a, b).

Importantly, viruses may have appeared before the last universal cellular ancestor (LUCA), from which the three cellular domains diverged (Forterre 2006; Bandea 2009a, b).

Didier Raoult and Patrick Forterre proposed that the living world is comprised of two major groups of organisms: ribosome-encoding organisms (the descendants of LUCA, i.e., archaea, bacteria, and eukarya); and capsid-encoding organisms (viruses) (Raoult and Forterre 2008). Similar to the three realms of bacteria, prokaryota, and eukaryota, viruses can also be classified into three realms: archeoviruses, bacterioviruses, and eukaryoviruses (Forterre and Prangishvili 2009). Of note, viruses can also be classified according to their structure into three different classes (Bamford et al. 2005). Viruses have been actively involved in life evolution on Earth, and evolutionary novelties have been reinterpreted as side effects of a billion-year war between ribosome- and capsid-encoding organisms (Brosius 2003; Forterre and Prangishvili 2009; Bandea 2009a, b), with each organism struggling to maintain its integrity. At present, 42% of the human genome contains sequences whose origins can still be traced to retroposition (de Parseval and Heidmann 2005).

Based on the above, three different basic life forms have been described on planet Earth:

1. *Molecular* life forms—viruses.
2. *Microscopic* life forms—unicellular (bacteria) and multicellular (protists).
3. *Macroscopic* life forms—unicellular (*Brefeldia maxima*, *Stentor polyphemus*, etc.) and multicellular (mammals, etc.).

Multicellular organisms are also cell derived, so fundamentally, on planet Earth, there are only two life blueprints: a viral blueprint and a cellular blueprint. The largest viruses found (*Mimiviruses*) are 400–500 nm large, so they are still at the molecular level, and both microscopic and macroscopic natural viral life forms have been described.

Multicellular organisms are not simply society of cells; they are forms of life with distinct macroscopic physiologic properties. Human health is maintenance of homeostasis not only at the molecular, macroscopic, and microscopic levels but also in between the molecular, microscopic, and macroscopic levels. Disease may appear not only due to functional impairments of components of one particular level but also due to a lack of coordination or miscommunication between these levels.

7.6 Origin of Life-Origin of Cancer

The main cancer theories and the origin of life models are strikingly similar, and a similar debate regarding the *primum movens* factor causing cancer has been going on in oncology for almost a century.

The major scientific models of the origin of life are 1. cells first, metabolism second, and genes (information) third; 2. replication (information) first, metabolism second, and cells third; 3. metabolism first, cells second, and genetic mechanisms third; 4. order first (Popa 2004); and 5. ecosystem first (Morowitz 1992).

By comparison, the major scientific theories of cancer are 1. genes first (SMT); 2. metabolism first (Warburg, Seyfried); 3. epigenetics first (Schwartz, Feinberg); 4. cells first (the "bad luck" stem cell theory and the ground theory); and 5. tissue first (TOFT and tissue disorder, the hypothesis of D. W. Smithers).

The origin of life is not just one problem that requires an overarching solution but an enormous series of problems, including the origin of all the metabolic processes as well as self-replication that must each be solved, where no one model or computer program can provide the answer. Given the fact that at the origin of a cancer cell is another normal cell, at first glance, it seemed that finding a cancer's origin would be much easier than finding life origin. Surprisingly, finding cancer cell's origin has not been easy. Researchers in the cancer field realized many decades ago that there is no unique mechanism or explanation behind cancer inception. A "Big Bang" theory describing cancer inception as a process similar to fertilization has been recently proposed (Perry et al. 2023), as opposed to the gradual step-by-step "Small Fizzle" genetic theory sketched by Eric Fearon and Bert Vogelstein (Fearon and Vogelstein 1990) and, subsequently, in the microenvironment version described by Robert Gatenby and Robert Gillies (Gatenby and Gillies 2008).

Cancer is not one single problem waiting to be solved or one set of equations waiting to be discovered but rather a set of multiple interconnected problems forming a complex temporal puzzle from which some essential steps like the transformation of a cell with mutations to a malignant cell are missing. Computer simulation of certain cancer aspects may be more or less relevant, but a complete in silico simulation of the oncogenic process is still in its infancy. Similarly, basic research using cell cultures and, more recently, organoids and in vivo animal models may dissect some aspects of oncogenesis, but because essential information (i.e., the immune system and the vascular component) is also missing, they are incomplete and potentially misleading. Recently, it has been demonstrated that all eukaryotes originate from a unique ancestor.

The situation in oncology offers special challenges, since not all cancers develop from a single cell. Some cancers may be polyclonal, and an intriguing hypothesis that cancer may originate from the fusion of two (or several) cells has been proposed.

7.7 What Do We Talk About When We Talk About Life?

As noted by Yury Zhuravlev and Vladik Avetisov, the concept of "life" has three separate connotations:

- The first one relates to the *state* of being alive or the property of the aliveness of a system: "a specific state of matter (the living state) resulting from the interaction between matter and energy carriers. This interaction starts from the utilization of solar radiation by autotrophic organisms, and spreads over a diversity of organisms via numerous (bio)chemical cycles. A significant part of the utilized energy is retained in organisms by molecular carriers and "network channels" of high energy content; lessening of the utilized energy pool up to some critical level entails in death."
- The second one relates to the *structure* of living systems: "consisting of self-reproducing agents. These agents are the only reference matter of life and are often represented by organisms. They can sometimes be represented as more complex units: bisexual pair, beehive, etc. The agents being individuals can interact with each other and therefore the whole system can be considered as a fragmented and integral entity simultaneously. Different levels of the organization of agents correspond to different levels of life hierarchy. Life as a system shows its worth in the diversity of constraints, feedbacks and interconnections with surroundings."
- The third one relates to the living *process*: "It is expressed in transformations of surroundings (by agents) and in transmutations of the self-reproducing agents themselves. From the physico-chemical point of view, the living process has both dynamic and informational contents. It allows the agents to properly meet the changes in environment and to expand (spread) over a space, thus increasing the level of system complexity and differentiation" (Zhuravlev and Avetisov 2006).

So, life refers at the same time to the structure of a living system, to the process that maintains that ordered structure, and to the deployment in time of the same process through the life cycle of an organism (development) and through different generations (evolution). *These aspects are manifestations of the same phenomenon, but a concise definition that encompasses all three of them is currently lacking.* Similarly, cancer is simultaneously a *state* of the cell (cancer cell) and of the organism (organism with cancer), a specific *structure* (malignant cells and cancerized tissues), and a *process* (oncogenesis, invasion and metastasis).

7.8 So, What Is Life?

7.8.1 First Proposal: *Order from Order*

In his 1943 Dublin lectures, later published as *What Is Life?*, Erwin Schrödinger proposed the idea that life behaves like a "nonentropic phenomenon." He suggested that the complex nature of life could potentially necessitate a "*new type of physics*" beyond the known laws of physics and chemistry to fully understand how life functions at the molecular level, essentially arguing that life may not be entirely explained by existing physical laws (Schrödinger 1944). His most important insight was that "opposing the second principle of thermodynamics that leads to an inexorable entropy increase of all closed systems, life, being an open system, behaves like a "*non-entropic phenomenon*" that "*feeds upon negative entropy,*" attracting, as it were, a stream of negative entropy upon itself, to compensate the entropy increase it produces by living and thus to maintain itself on a stationary and fairly low entropy level" (Schrödinger 1944). "The awkward expression *'negative entropy'* can be replaced by a better one: entropy, taken with the negative sign, is itself a measure of order. Thus, the device by which an organism maintains itself stationary at a fairly high level of orderliness (i.e., fairly low level of entropy) really consists in continually sucking orderliness from its environment" (Schrödinger 1944). In the opinion of Ernst Bauer, one of the pioneers of theoretical biology, the anti-entropic activity of living systems is a universal life principle:

> The living and only the living systems are never in equilibrium, and, through their free energy, they are continuously working against that equilibrium which, within the given initial conditions, would be required by the physico-chemical laws. [...] We assert that the living system always converts its entire free energy content into work against the realization of the expected equilibrium. This is already a quantitative statement and can be experimentally verified by measurements (Bauer 1935 cited by Grandpierre 2008).

In order to resolve the apparent contradiction between the second law of thermodynamics (increasing entropy) and the highly ordered nature of living organisms, he also suggested that living organisms must contain a sort of "*aperiodic solid*" to help them maintain order against this entropic tendency: "An organism's astonishing gift of concentrating a 'stream of order' on itself and thus escaping the decay into atomic chaos - of 'drinking orderliness' from a suitable environment - seems to be connected with the presence of the 'aperiodic solids', the chromosome molecules, which doubtless represent the highest degree of well-ordered atomic association we know of - much higher than the ordinary periodic crystal—in virtue of the individual role every atom and every radical is playing here" (Schrödinger 1944). James Watson and Francis Crick were inspired by the chromosome idea and the "*aperiodic solid*," and when they discovered the DNA structure, they were sure that by the same stroke, they had also found the "aperiodic solid" predicted by Schrödinger a decade priorly. As a side note, Schrödinger considered the chromosomes as both "architect's plan and builder's craft," which is wrong because the cell itself is the builder and not the DNA. Turing introduced the stored-program computer, while von Neumann clarified that the program (description) is distinct from the universal constructor (executor). This distinction is crucial. The code provides only instructions for execution, not the execution itself (Brenner 2012).

Simply put *life, in order to maintain order, feeds on order*. By doing that, *only apparently*, it appears that it opposes the second principle of thermodynamics which states that in all closed systems, inexorably, entropy increases. I purposely wrote "only apparently" because living systems are open systems and, being so, they feed on order but at the expense of increasing the entropy of their environment. So, no trespassing of the second principle here. If we carefully do the books, we find that the overall entropy of living systems and their environment combined is *increasing*. Life maintains its inner orderly state by expelling disorder.

In recent years, a different perspective has emerged, proposing that kinetic reactions, rather than thermodynamic considerations, are responsible for the order observed in living systems. Addy Pross and Robert Pascal suggest that life's organization arises from dynamic processes that are not governed by classical thermodynamics but by the kinetics of chemical reactions that sustain life's ordered state (Pross and Pascal 2017).

A quick note: entropy has been defined by Rudolf Boltzmann as a measure of the number of possible microscopic arrangements or states of individual atoms and molecules of a thermodynamic system. It is important to emphasize that entropy is context and observatory dependent. The number of possible arrangements depends on the experimental setup that needs to be specified in advance. In other words, "entropy is an anthropomorphic concept. For it is a property not of the physical system, but of the particular experiments you or I choose to perform on it" (Jaynes 1965).

Importantly, the relationship between life and entropy is more complex than envisioned by Schrödinger. Life is not pure negentropic, and it does not create either order from order or order from disorder. Neither perfect order nor perfect disorder are compatible with life. *Life processes and forms are organized by maintaining a balance between order and disorder.* They are organized by controlling and not opposing entropy. In fact, in certain situations, life forms may benefit from temporary entropy increases and may promote entropy production for survival purposes.

Biological processes, situated between the quantum and cosmological scales, *do not require entirely new laws of physics* but rather new principles for their application to life's *mesoscale*. Life on Earth operates within a relatively narrow temperature range, dominated by entropy at these temperatures, with quantum effects being less significant. The unique materials of life, such as nucleic acids (DNA and RNA), proteins, carbohydrates, and lipids, form the basis of biological function. These molecules interact within a complex interface of membranes, biopolymers, and aqueous solutions, providing the foundation for life's complex behavior. The field of soft matter physics, which has emerged over the past four decades, studies the properties of such biomaterials—liquids, polymers, colloids, and biomaterials—using statistical theory and computational biology to understand how these molecules interact to sustain life.

Soft matter physics is particularly important in cancer, where stiffness is a key characteristic of cancerous tumors, promoting their progression.

Recently, Irun Cohen and Assaf Marron proposed *that entropy is the driving force in evolution*, selecting the organisms that best resist entropic destruction and introducing the "*survival-of-the-fitted*" concept as the opposing Darwin's "*survival-of-the fittest theory*" (Cohen and Marron 2020). But are biological systems fighting entropy? Let us continue our investigation.

7.8.2 Second Proposal: *Order from Disorder*

Four decades after Schrödinger, in a book written in collaboration with Isabelle Stengers, Ilya Prigogine, a distinguished physical chemist and a Nobel prize laureate, gave a different spin to the biological significance of the second principle of thermodynamics, proposing that "in far-from equilibrium conditions we may have transformation from disorder, from thermal chaos, into order. New dynamic state of matter may originate states that reflect the interaction of a given system with its surroundings. We have called these new structures dissipative structures to emphasize the constructive role of dissipative processes in their formation" (Prigogine and Stengers 1984).

Living organisms are considered dissipative structures because they are created and maintained by irreversible flows of energy and matter. The term "dissipative structures" introduced by Prigogine points to the fact that ordered structures are generated and maintained by irreversible macroscopic processes that are entropy and free-energy dissipating. (Note: free energy is energy capable of doing work.) An important mention is that the self-organizing behavior described by Prigogine takes place only in nonlinear open systems. In nonlinear systems, the speed of the reaction is not directly proportional to the concentration of the reactants since they may stimulate or inhibit the reaction itself. Compared to machines, entropy is used very differently by living organisms: "Entropy production in a machine is a measure of inefficiency, where energy is converted to lost heat rather than mechanical work. In dissipative structures, entropy producing processes maintain structure and function, as in the irreversible chemical transformations that drive spatiotemporal structuring in a chemical clock. Similarly, the endogenous chemical cycles of organisms (metabolic, behavioral) are driven by irreversible entropy-producing processes" (Kondepudi et al. 2020).

As explained by Deacon et al., life uses dissipation of energy gradients in order to maintain an organized state: "All dynamical processes, whether close to or far from equilibrium, are structured so that they dissipate any labile gradients and promote development toward ever higher global states of entropy. Self-organizing processes are special cases that increase the rate of gradient dissipation by developing increasingly more direct and therefore regular pathways for gradient reduction to occur. This local regularization is critical to generating order in living systems, but it is also the most efficient way to undermine the very conditions that produce it. In contrast, life must accomplish the opposite. It must be organized so that it preserves and reconstitutes the intrinsic and extrinsic conditions that support it, even in the face of influences that tend to undermine its continued existence" (Deacon et al. 2014).

In other words, under certain nonequilibrium conditions, "*entropy may produce*, rather than degrade, *order, organization—and therefore life*" (Alvin Toffler, in Prigogine and Stengers 1984). This is an important point, and we will stop for a moment to realize the importance of Prigogine insight. *Entropy is not necessary opposed to order.* In fact, sometimes it may create order: "*entropy increase generates macroscopic disorder in some cases, but macroscopic order of various kinds in others.* [...] the widespread idea that life is "*a local fight against entropy*", namely a trick to keep entropy locally low, is misleading" (Jeffery et al. 2019). Jeffrey et al. explain this using a suggestive example: "There are familiar cases where an increase in entropy does amount to an increase in disorder. If we put some wine in a glass of water, for example, the wine is initially concentrated: this is an ordered configuration because the wine is collected in the same region, rather than dispersed: the locations of the molecules are somewhat correlated. If we let the system evolve freely, the wine mixes with the water: the order is lost. This is an irreversible process and entropy increases. But consider the case where, instead of wine, we put some oil into the water and mix. Let the system evolve freely: the oil separates from the water and order is generated. But this is also an irreversible process, and again entropy increases. Therefore, increase of entropy generates disorder in some cases, but order in others" (Jeffery et al. 2019).

Again, we should not forget that the discussion of entropy takes place in the context of an open system. In an open system, the total number of possible trajectories is smaller than that in a closed system. "Statistically, this is equivalent to lower internal entropy" (Georgiev and Chatterjee 2016).In order to illustrate this, Georgi Yordanov Georgiev and Atanu Chatterjee use the following comparison: "This is similar to a situation where hot air will randomly distribute heat in a closed room, whereas the same heat will travel along a specific trajectory if the doors of the room are opened, reducing the number of possible states and the entropy of the air molecules. Heat can now enter the system through one door and leave it through the second door. While one door acts as a source and the other as a sink, the channeled flow lets the heat pass through the system and allows it to dissipate into the surrounding" (Georgiev and Chatterjee 2016).

This "principle of ordering" from disorder is also the foundation for the directional motion observed in ratcheting or rotating biomolecules present inside the cells and led to a recent definition of life as a "self-regulating process whereby matter undergoes cyclic, uniplanar conformation state changes that convert thermodynamic disequilibria into directed motion, performing work that locally reduces entropy" (Pierce 2023).

Living entities, as open systems, have been compared with fridges—they maintain internal order by pumping out entropy into the environment. But here is the catch: in order to function, a fridge requires not only energy (electricity) but also designed equipment that embodies instructions, i.e., information.

Also, in Arto Annila's opinion, the universal drive toward a free energy minimum, equivalent to an entropy maximum, supports and does not oppose biological processes (Annila 2024).

Life forms are not solely antientropic feeding on order like Schrödinger proposed, and they are not only able to harness order from disorder, as Prigogine demonstrated; they are also capable, when confronted with stress, of inducing stochastic events for survival purposes, temporarily creating randomness as a survival mechanism. Life is not only able to "feed on entropy"; *it can also harness entropy in order to persist and evolve*. The random creation of the antibody repertoire is such an example of entropy harnessing.

So, in the context of genetic mutations, a paradox emerges in which the increased division rates observed in cancer cells may serve as an anti-entropic strategy, fostering survival through cellular adaptation. The polyploidization present in cancer may be a similar mechanism. Also, as shown more than a decade ago by the lab of Angelika Amon, aneuploidy at the same time slows proliferation and induces genomic instability (Sheltzer et al. 2011; Replogle et al. 2020). Paradoxically, the newly acquired mutations that appear as a result of aneuploidy may be a way to counteract aneuploidy's deleterious effects (Sheltzer et al. 2011).

Taking into consideration the importance of entropy for living organisms, cancer resilience makes total sense. Both normal organism and cancer are dissipative structures, but cancer has an advantage against the organism, as it has a higher ability to produce and harness entropy, which increases its likelihood to adapt and persist.

Besides being a fundamental component of the living process, randomness is also essential for improving the performance of artificial intelligence systems. Randomness also represents a crucial component of many machine learning algorithms, particularly in training models, and in exploring different solutions, it helps them achieve better generalization and robustness.

Living processes are a class of their own, neither completely deterministic nor entirely random. Life is quasideterministic, and the difficulty of pinning down the nature of life is precisely this quasiness. Similarly, the cancer process is quasideterministic: some genetic mutations are random, while others occur through directed mutagenesis. Stress represents a source of randomness that pushes cells to rapidly adapt or die. Cancer may initially start as a cellular adaptation to stress. Uncontrolled division, local invasion, and migration may represent further unhinged stress responses.

7.8.3 Third Proposal: *Order from Complexity and Evolution*

Thus far, we have used the terms "order" and "organization" somewhat interchangeably, but it is crucial to acknowledge a fundamental distinction between the two, a distinction elucidated by the second principle of thermodynamics. This principle states that an isolated physical system inevitably tends toward its maximal state of entropy. While entropy is often equated with disorder, it is important to note that, as illustrated by the earlier example of the mixing of water and wine versus oil and water, *entropy and disorder are not synonymous*. "Order," as exemplified by the structure of a crystal or by the arrangement of cars in a parking lot, refers to a static configuration, a spatial pattern, or arrangement. In contrast, "organization" in biological systems represents a dynamic, self-sustaining process. It is maintained through communication, regulatory mechanisms, and the interplay of hierarchies and holarchies. Therefore, it is more accurate to describe the dynamic, self-regulating nature of living organisms as "organization" rather than static order.

In the *Critique of the Power of Judgment*, Immanuel Kant grasped the difference between biological systems and machines, using the example of a watch: "one wheel in the watch does not produce the other, and even less does one watch produce another, using for that purpose other matter (organizing it); hence, it also cannot by itself replace parts that have been taken from it, or make good defects in its original construction by the addition of other parts, or somehow repair itself when it has fallen into disorder: all of which, by contrast, we can expect from organized nature." Further, he explained that "an organized being is thus not a mere machine, for that has only a motive force, while the organized being possesses in itself a formative force (Bildungskraft), and indeed one that it communicates to the matter, which does not have it (it organizes the latter): thus, it has a self-propagating formative power, which cannot be explained through the capacity for movement alone (that is, mechanism)" (Kant 1790 cited in Mossio 2024).

In a thought-provoking book, Matteo Mossio proposed that biological organization cannot be equated with the self-organizing systems proposed by Prigogine (Mossio 2024). He boldly stated that biological systems, "unlike "self-organizing" dissipative structures, constrain and canalize the thermodynamic flow through the collective activity of their functional parts, which realize a specific form of mutual relationship, i.e., organization. Because of their distinctive functional complexity, furthermore, organized systems (usually) do not appear spontaneously when some specific boundary conditions are met, as self-organizing structures do. Rather, organized systems are the result of a long historical evolutionary process of increase and preservation of complexity, which means, in particular, that any individual biological system is generated through the reproduction of other biological systems. Thus, in spite of their common thermodynamic grounding, (biological) organization cannot be conflated with (physical) self-organization." He distinguished "organization" from "complexity" by specifying that organization designates a specific kind of complexity, a specific set of relations among elements: "to a first approximation, organization refers to the differentiation of functional roles (i.e., division of labor) among the parts of a system and, at the same time, to their integration and coordination as a whole. Furthermore, and crucially, organization involves a generative dimension in the form of a mutual dependence, such that the very activity and existence of each organized part depends on its mutual relationship with the others." He also distinguished "organization" from "organism," giving the example of an ecosystem that may be organized but is not an organism. In light of this distinction, the ten biological levels of organization of living things arranged from the simplest to most complex are organelle, cells, tissues, organs, organ systems, organisms, populations, communities, ecosystem, and biosphere. On the scale proposed by Mossio, only cells and organisms are considered organisms.

Importantly, he specified that "organization" in biological systems is closely related to purposiveness: "biological organization can be legitimately said to be the cause and effect of itself and thereby an intrinsically purposive regime."

Matteo Mossio proposed an "organization-first" model of life, suggesting that biological organization both explains the stability of living systems over time and accounts for the functional integration of systems that allows for both quantitative and qualitative innovations. He emphasized that organization explains not only the stability of biological systems at both the individual and evolutionary scales but also how innovations arise and are preserved and integrated through functional organization: "the notion of organization plays a twofold explanatory role: on the one hand, organization allows explaining the stability of biological systems (both at the individual and evolutionary scale), and the maintenance of their constitutive dynamics over time; on the other hand, it provides a ground to understand how quantitative and qualitative innovations can be produced, and then preserved though its functional integration" (Mossio 2024).

The idea that organization is a necessary ingredient of life has been recently brought to the forefront by a group of scientists who published two joint *biocosmos* articles (Cortês 1 et al. 2022; Cortês 2 et al. 2024). According to them, the world therefore consists of three distinct types of systems: ergodic systems (type I and type II) and nonergodic systems (type III), where an ergodic system is a system that "visits all its possible states over some 'reasonable' time period," while a nonergodic system is a system that "does not visit all its possible states." According to the authors, living systems are type III systems, different from the type II systems studied by Prigogine, or type I systems of classical physics, and they cannot be described by routine physical laws. The explanation for this difficulty resides in the fact that for open far-from-equilibrium systems, there are no physical laws capable of characterizing "a trajectory in a configuration space that is itself dynamically evolving." Instead of physical laws, ad hoc biological laws like evolution are introduced. In the authors' terms, "cells construct themselves." In one of their two *biocosmos* papers, the authors tried to provide a mathematical formalism for the theory of the "adjacent possible" (Cortês 1 et al. 2022).

By heritable variation and natural selection, cells continually evolve to construct boundary conditions that are new to the Universe, i.e., expansion in the "adjacent possible." The new boundary conditions create new phase spaces and hence new possibilities in the evolution of the Universe. An important note is that as specified before (Longo et al. 2012), "evolution is not optimization: "none of the issuing paths is "better" than the other, nor followed physical optimality criteria, even less biological ones: each is just a possible variation on an original common theme, just compatible with the coconstituted ecosystem that enabled them. In general, there is no way to define a real valued (Lagrangian) functional to be extremized as for biological functions, as this would require an ordered space (a real valued functional), where "this biological function (or phenotype)" could be said to be "better" than "that biological function (or phenotype)" and is unpredictable" (Longo et al. 2012).

The theory of the adjacent possible (TAP) was previously developed by Stuart Kauffman in a series of articles (for a critique, see van Der Merwe 2024). Basically, according to Kauffman, entities like hearts and kidneys exist "by virtue of the functional roles these systems and subsystems play in abetting the survival and further evolution of the organisms of which they are parts" (Kauffman 2016, 2019).

He explains emergence by using the example of a fish's swim bladder that evolved from ancestral lungs in lungfish. "With the swim bladder's emergence, a new function came to exist in the biosphere" (Kauffman 2019). He argues that the fact that a bacterium can adapt to live in such a swim bladder independently from natural selection that was not involved in specifically creating a niche for the bacterium implies that living systems are literally creating their own future by exploring adjacent possible functions. These novel trajectories are completely unpredictable.

Cells and organisms are "Kantian wholes," a term introduced by Stuart Kauffman to describe systems where the parts exist for and by means of the whole (Kauffman and Roli 2021). Of note, Arthur Koestler introduced the term "holon" to describe organisms: "biological holons are self-regulating open systems which display both the autonomous properties of wholes and the dependent properties of parts." The "Kantian whole" was first introduced by Stuart Kauffman and then given a mathematical formalism by Marina Cortês et al., which is the same concept as Koestler's holon.

Using both type III systems and Kantian wholes, Cortês et al. formulated a new definition of a living organism: "A living organism is a Kantian whole within a type III system that is a non-equilibrium self-reproducing system with a metabolism, an identity and a boundary, and which is capable of open-ended evolution by heritable variation and selection or drift" (Cortês et al. 2022).

In conclusion, the definition of life as a dynamic, self-organizing system, continuously evolving through the exploration of the "adjacent possible" highlights the complex interplay between organization, purpose, and evolution. Living systems do not simply act to minimize entropy; they construct themselves, creating new states and possibilities within the framework of a continually evolving and self-sustaining system.

Cancer Relevance

Why are thermodynamics and information theory important for cancer? For an organism to maintain its existence, it must actively control its internal entropy. One key characteristic of living systems is their capacity to modulate their entropy in a controlled manner, either increasing or decreasing it as needed. An increase in entropy expands the information space, allowing for greater complexity and adaptability. This control over entropy is not only a fundamental feature of life but also a crucial factor in the evolution of living organisms, enabling them to access new possibilities for survival and adaptation.

Life, in its very essence, unfolds at the boundary between order and disorder, necessitating control systems to prevent entropy from spiraling out of control. In this delicate balance, cancer can be viewed as the cellular response to an increase in entropy. Mutations contribute to this rise in entropy, potentially triggering an amplification in cellular division, which, in turn, generates further mutations. This process may be responsible for the transformation of a mutated cell into a cancerous one. Cancer cells are preprogrammed to deal with stress. As entropy increases, they often respond by dividing—thereby dissipating some of the excess entropy—or metastasizing to more favorable environments. The behavior of cancer cells—whether through division or metastasis—may therefore represent an adaptive response to the internal entropy that arises from various cellular and environmental stresses, such as activation of the STING pathway related to metastasis. In this framework, cancer primarily emerges as a *stress response*, with genetic alterations being secondary to the broader cellular need to mitigate stress, which may include chromosomal abnormalities, oxidative stress, and an accumulation of mutations (Luo et al. 2009; Gusev and Zotova 2019).

Noticed almost a century ago by Ervin Bauer but overlooked is the capacity of organisms to tinker with thermodynamic laws and use them to their own benefit: "In the living organism, the laws of physics and chemistry are valid, but they are subordinated to the higher laws of the biological unit, and therefore manifest themselves in a particular form" (Bauer 1930, cited by Grandpierre 2008).

Cancer, with its heterogeneous nature, provides a clear example of this principle. Some cancers contain cells that divide rapidly, while others grow more slowly. The entropy within these cells fluctuates and is influenced by both internal factors and the ever-changing external environment. As open thermodynamic systems, all cells are subject to entropy, which is shaped by both intrinsic and extrinsic influences. Importantly, cells are not isolated systems; their entropy is affected by direct contact and communication with their surroundings. When a cell divides, it temporarily increases its entropy, after which it settles into a state of slightly higher entropy than before. This "normalization" is not a return to the original state but rather a shift to a new behavioral state. Because cells are not isolated entities, processes such as epithelial-mesenchymal transition (EMT) or dormancy can occur, both of which are coinduced by the environment.

Plasticity, in this context, refers to a cell's ability to navigate different entropic phase spaces. A cell that is dividing or migrating consumes more energy, generates more heat, and experiences increased internal entropy as it performs its functions. This capacity for adaptive change is a hallmark of life, underscoring the dynamic nature of cellular behavior. It is this variability, driven by entropy, that sets biology apart from physics: while physics is concerned with statistical laws, biology is focused on individual organisms and their idiosyncratic behaviors. Cellular behavior is often characterized in statistical terms, but at its core, it reflects the unique responses of individual cells to environmental and internal conditions.

Strikingly, a research team from Princeton discovered that cancer cells from primary tumors were found to proliferate with decreased, rather than elevated, ATP production flux. Conversely, metastatic cancers appear to utilize more energy than primary tumors that have energy needs similar to normal tissues (Bartman et al. 2023). Part of the explanation may be the plasticity of cancer cells that are able to adapt to the relative lack of nutrients in their primary location and decrease their needs when food supplies are scarce and increase their metabolism when more abundant energy sources are found in distal tissues and organs. Cancer cells often rely on the glycolytic pathway (the Warburg effect) and avoid energetically expensive tissue-specific functions. Their focus is primarily on division, not on sustaining complex tissue functions. This behavior was previously postulated by Jeremy England: "The less durable or the less organized a self-replicator is, all things being equal, the less metabolic energy it must harvest at minimum in order to achieve a certain growth rate. Thus, in a competition among self-replicators to dominate the population of the future, one strategy for "success" is to be simpler in construction and more prone to spontaneous degradation. [..] In a race between competitors of similar form and construction, it is worthwhile to note that one strategy for reducing the minimal "metabolic" costs of growth for a replicator is to substitute in new components that are likely to "wear out" sooner" (England 2013). In other words, one evolutionary strategy for self-replicating systems is to *minimize* metabolic energy costs by adopting simpler, more prone-to-degradation structures, allowing for faster replication.

But are energy and information sufficient in the first place for life to exist? Energy and information are embedded into material structures and into physical units, so space and time components are also a must. An important ingredient of life is a boundary, a cellular membrane (Morowitz 1992; Gánti 2003; Nurse 2021) although it has been argued that boundaries may not have been required for life to appear in the first place (Pross 2012). Boundaries are trespassed and destroyed by cancer cells. A pathologist's definition of cancer is precisely the invasion of basal membranes, which differentiates benign tumors and in situ carcinomas from invasive cancers. Besides energy and information, a third essential component of life is structure and organization. Life is not merely a collection of matter and energy; it is a highly organized system governed by complex networks of interactions and processes that maintain stability and enable adaptability. This organization, from the molecular to the organismal level, is what allows life to persist and evolve.

Finally, life is not static; it unfolds over time. Organisms grow, develop, age, and reproduce, passing on their genetic material to future generations. Time, therefore, is the fourth essential ingredient of life. Paul Nurse, for instance, added the concept of evolution through natural selection as a defining characteristic of life (Nurse 2021). This idea resonates with NASA's definition of life: "Life is a self-sustaining chemical system capable of Darwinian evolution." Life, in this view, is not merely a series of biochemical reactions but a dynamic process that evolves and adapts over time.

Thus, thermodynamics, information theory, and the interplay of entropy and organization are foundational to our understanding of life, and cancer provides a compelling model for studying how these principles manifest in cellular behavior.

7.8.4 Fourth Proposal: *Order from Information*

7.8.4.1 Order from Syntactic Information

More recently, searching for explanations of life's nature outside the thermodynamic paradigms, Cara Walker and Paul Davies introduced the hypotheses of an algorithmic origin of life and of information as a top-down causal agency of life forms. In their opinion, life is a "transition in causal structure, and information management and control, whereby information gains causal efficacy over the matter it is instantiated in" (Walker and Davies 2012). They consider the origin of life a reversal in the causal architecture in which "an abstract and nonphysical systemic entity (algorithmic information) effectively becomes a causal agent capable of manipulating its material substance" (Walker and Davies 2012). But how did this reversal arise in the first place? How did the code appear in the first place from a material cause? This nontrivial problem has not yet been solved. Their hypothesis echoes Wheeler's "it from bit" concept introduced in 1989: In his opinion, reality has an immaterial information core: "It from Bit symbolizes the idea that every item of the physical world has at bottom—at a very deep bottom, in most instances—an immaterial source and explanation; that what we call reality arises in the last analysis from the posing of yes-no questions and the registering of equipment-evoked responses; in short, that all things physical are information-theoretic in origin and this is a participatory universe" (Wheeler 1989).

As briefly mentioned before, entropy is an anthropomorphic concept related to the uncertainty in the number of states of a system. Information is closely related to entropy. It measures the difference in uncertainty, the entropy before and after a measurement, and thus can never be absolute (Adami 2004). So, differences between initial and subsequent entropies or uncertainties are called information, which is a relative measure (Adami 2016). Information is always about something. "If it cannot be specified what the information is about, then we are dealing with entropy, not information" (Adami 2004). Regardless of the potential information stored in a DNA or a protein sequence, the actual meaning of any string of information is context dependent and has only a "relative (or conditional) one with respect to an environment" (Adami 2004). *No distinction between information on entropy can be inferred by looking at an isolated DNA sequence*, but the conservation of the given sequence in different genomes points to the significance for the cell of that particular sequence. Similarly, equating conserved sequence with biological significance, Vogelstein and his team developed the 20/20 rule: to be classified as an oncogene, >20% of the recorded mutations in the gene are required to be at recurrent positions and are missense, and to be classified as a tumor suppressor gene, the requirement is that >20% of the recorded mutations in the gene are inactivating and used to discover driver genes (Vogelstein et al. 2013).

Information has also been used to define and characterize life forms as individuals (Krakauer et al. 2020). The authors define individuals as "aggregates that propagate information from the past to the future and have temporal integrity." The relevance of this definition is that it encompasses different life forms, including organisms and colonies. In this context, the authors were able to separate organismal individuality from colonial individuality based on three parameters: "the information that is shared by system and environment (e.g., adaptive information); information that is unique to either the system or the environment (e.g., memory in each); and information that depends in some complicated way on both the system and the environment (e.g., regulatory information)."

In the cases of colonies as opposed to organisms, the information shared with the environment tends to be less. Similar to bacteria, cancer cells can organize in colonies or in pseudo-organisms, and these different forms of organization may be related to information parameters. The behavior of cancer may be closely dependent on communication and interactions with the environment.

Although it is tempting to relate the order present in living entities to an algorithm, life forms, in general, express a nonlinear, context dependent, variable degree of stochasticity. Their interaction with the environment does not follow either a sort of yes/no logical binary predictable script or a model of completely random behavior that can be computed using the formalism of quantum mechanics. Life literally lives in a world of its own creation and exploits the laws of physics and chemistry to its own benefit.

The thermodynamic model of life is closely related to the metaphor of the organism-machine that dominated the nineteenth century. In the second half of the twentieth century, the metaphor of the organism computer was also introduced. Both metaphors are useful, and cells maintain order both through a special implementation of the second principle of thermodynamics and through information computing.

The jury is still out regarding the priority of energy versus information both in terms of life origin and basic cellular functioning (Walker and Davies 2012). Are controlled fluxes of energy (i.e., metabolism) the driver of life, with information (DNA programs) just a passenger, or vice versa?

Recently, a team of French researchers argued that cancer is a *nanoscopic electronic disease* of modified energy flow and that the metabolic abnormalities of cancer cells are due to modifications of the electric currents in redox reactions. Their basic hypothesis is that "the flow of energy and matter drives the information flow along the signaling pathways, from the nanoscopic scale of electrons to the macroscopic scale of the tumor and of the organism." In their opinion, "energy dissipation drives information flow" and not vice versa (Attal et al. 2023).

This dichotomy between the primordiality of energy dissipation versus information computing may be artificial. It is possible to extend to all cellular functioning and life forms, including cancer, the double origin of life proposed by Freeman Dyson: "Either life began only once, with the functions of replication and metabolism already present in rudimentary form and linked together from the beginning, or life began twice, with two separate kinds of creatures, one kind capable of metabolism without exact replication and the other kind capable of replication without metabolism" (Dyson 1999). In other words, maybe part of life's secret formula is *exactly* this yin/yang synergy between energy and information.

Cells and Organisms Are Not Machines or Computers

The view of life as a computing process is common among computer scientists. Their narrative goes like this: "Does Nature compute? Indeed, we can model all the processes of Nature as information processes. For example, cell multiplication and DNA replication are seen as instances of text processing. A chemical reaction is simply an exchange of electrons, that is, an exchange of information between two molecules. The spin of an atom, whether spin up or spin down, is a binary process, the answer to a 'yes' or 'no' question. Information and computation are present in all natural occurrences, from the simplest to the most complex. From reproduction in ciliates to quorum sensing in bacterial colonies, from respiration and photosynthesis in plants to the migration of birds and butterflies, and from morphogenesis to foraging for food, all the way to human cognition, Nature appears to be continually processing information" (Adamatzky et al. 2017). Taking part in the same debate, Karl Svozil claimed that "Systems which compute can serve as, and even construct and produce, better interfaces for energy dissipation than systems without algorithmics" (Adamatzky et al. 2017). On the one hand, he reiterated that the common behavior of all living systems is to dissipate energy, but at the same stroke, he introduced the idea of optimization of dissipation by computing systems (Adamatzky et al. 2017). This argument is circular, and the fact that computing systems are able to construct and produce interfaces that dissipate better than systems without algorithmics begs the question: *Why is there a need for better energy dissipation in the first place*?

Scientific theories about the nature of life, evolved over the last two and a half centuries from machine analogies to computer analogies, mostly focused on the property of life's autopoiesis (for a review, Letelier et al. 2011). Yes, some functions of living entities are similar to machines or computers but only up to a point. Machines and computers are deterministic systems, and thus, their output is predictable. Ironically, Philippe Marliere points to the fact that for a computer scientist, *Darwin's Origin of Species* "just reads like a mechanistically sophisticated and brilliantly ornate description of a greedy algorithm, i.e., a procedure that iteratively optimizes a combinatorial configuration by choosing at the next step the variants of the prior configuration with equal or higher merit" and may lead to "poor optimization solutions to most combinatorial problems [...] which leads to the vexing notion that the whole process of spontaneous biological creation and evolution gave rise to the adoption of imperfect molecular devices and processes" (Marliere 2009).

Despite its similarity to a *greedy algorithm* or to other software programs that try to mimic evolution, none of them was able to replicate evolution in silico. Basically, all of them stop evolving after a certain time, and the forms generated tend to resemble each other. One explanation given by Florin Ghețu is that all computer simulations need to specify goals and build constraints from the beginning (Ghețu 2023), and these goals and constraints that guide evolution from the onset will also curb the intrinsic power of creativity.

As Maël Montévil pointed out, the constraints of living systems are continuously changing: "In physics, the theoretical definition of an object stems from its invariants and symmetries, and more generally, by a mathematized theoretical framework. It follows that the theoretical object is generic so that all electrons, for example, follow the same equations—they have no singularity in the philosophical sense. [...] In biology, there are no such theoretical constructs because, on the one hand, organisms embed a multiplicity of particular constraints that appeared over time, and, on the other hand, these constraints continue to change, even under standardized laboratory conditions" (Montévil 2022).

The difference between living organisms and machines has been noticed for a long time. Edwin Bauer in 1935 wrote, "The energy input of the machine comes from the outside world and used for relative motions of the components of the machines. In contrast, in living organism the energy of the nutrients is converted into internal work on the structure of living matter, changing the physical conditions within the 'components' of the organism in a directed way, to renew and transform the internal structure of the components themselves in a way increasing their capability to do work. Living organisms are not machines in the ordinary sense of the word. Living systems are neither thermodynamic nor chemodynamic machines. They obey their own special laws, which distinguish them from machines and inanimate systems" (Bauer 1935, cited by Müller and Igamberdiev 2024).

Cells exchange energy and information with the outside world and have structures similar to computers: hardware (DNA), software (genes) that can be hacked (viruses), sources of energy (mitochondria), and processors (organelles). Also, like computers, they can receive and transmit data (receptors, secondary messengers), and you can argue that they even have computer cases (cell membranes). But cells can also reproduce and respond in a meaningful way to their environments and exhibit cognitive abilities and behavior.

Is life a natural computing system? The jury is still out, but I will argue that it is not. Cells and organisms are complex systems and contain various degrees of randomness. Because of this very randomness, their behavior is much more unpredictable than a machine or computer software. Despite the suggestive analogy between cells and computers (or, more generally, Turing machines), cells and living organisms are not machines or computers. A cell is not a wetware computer, and the depiction of intracellular pathways as wires on a chip (Hanahan and Weinberg 2011) is misleading. The inside of a cell is closer to a colloid in which molecules can switch between different conformations and play pleiotropic roles and not a rigid computer motherboard. In the context of the constructor theory developed by David Deutsch and Chiara Marletto, organisms were considered universal von Neumann constructors (Marletto 2015).

Living systems can be viewed as *resembling* universal von Neumann constructors in that they generate complex structures from simple rules, with "imperfections" such as mutations, fluctuations, and symmetry breaking serving as drivers of creativity and evolution, *but they are not universal constructors*. As explained by Michael Conrad, "the organism is not programmable, and its program of development is not written into the states of its components and their pattern of connection according to a manageably small user's manual. If it could be so written, each component would have a functional specification that is independent of the number of components in the organism. But in the regime of multiple weak interactions this is not the case. Ultimately the only user's manual would be the equations of physics, and these in effect have an infinite number of entries (since they involve continuous dynamics). What about the DNA base sequence? This also cannot be a computer program in any way reminiscent of a digital computer program. Even if we adhere to the metaphor of a developmental program we cannot identify this with the genetic description of the organism. The genes are better viewed as knobs which are used to alter the developmental program. This involves continuous dynamical processes embedded in a sequence of bifurcations from one developmental stage to another. Some of the genotypic knobs control dynamic parameters in each stage (e.g. rate constants of various biochemical and cellular processes}, and others control the space-time order of the bifurcations (the bifurcation parameters}. Environmental factors also exert an influence, either on the expression of the genes or on their action" (Conrad 1990).

Biospheres—like the universe itself—are more accurately described as self-organizing, autopoietic dynamical systems, where novelty emerges from the interplay of natural laws, chance, and time. Embryonic development, for instance, is shaped not only by evolutionary inheritance but also by environmental constraints such as energy, entropy, and temporality. Life is therefore not apart from the universe but a nested expression of its ongoing creative unfolding on Earth. In this light, Haeckel's dictum that "ontogeny recapitulates phylogeny" may be extended toward a broader principle: "ontogeny recapitulates phylogeny recapitulates cosmogenesis."

Importantly, interoperability, which represents, as explained by Chiara Marchetto, a principle, which by definition is *a law about the law* of counterfactuals, allows us to make predictions about physical systems without knowing exactly the nature of their laws of motion (Marletto 2021). Together with causal machine learning methods (Weberpals et al. 2025), the principle of interoperability may be a useful heuristic tool that can be applied in life sciences, in general, and in oncology, in particular. A tumor may be understood as an imperfect universal von Neumann constructor: a self-replicating system whose very imperfection drives evolutionary novelty. This imperfect constructor is engaged in reciprocal feedback loops with the host's organism. It also generates emergent forms of "creativity," manifesting as new pathological states that progressively reshape the organism. Understanding cancer in this light opens the path to systemic therapies aimed at modulating the constructor, the genetic and epigenetic blueprints, or the environment.

Biology, as a science is a distinct field from thermodynamics or computer science, and therefore:

> the entire strategy associated with various attempts to reduce biological evolution and dynamics of living systems to the principles either of classical, or irreversible thermodynamics, or to statistical mechanics, or to statistical/structural information the-

ory - that is to say, to manifestations of simple, one level physical systems - is a theoretical cul-de-sac. Physics is highly relevant to biology, but its explanatory arsenal can deal only with a part of the multi-levelled, multi-faceted causal hierarchy that is found in living systems (Corning and Kline 1998).

Also, in Carl Woese's words, "If they are not machines, then what are organisms? A metaphor far more to my liking is this. Imagine a child playing in a woodland stream, poking a stick into an eddy in the flowing current, thereby disrupting it. But the eddy quickly reforms. The child disperses it again. Again it reforms, and the fascinating game goes on. There you have it! Organisms are resilient patterns in a turbulent flow- patterns in an energy flow. A simple flow metaphor, of course, fails to capture much of what the organism is. None of our representations of organism capture it in its entirety. But the flow metaphor does begin to show us the organism's (and biology's) essence. And it is becoming increasingly clear that to understand living systems in any deep sense, we must come to see them not materialistically, as machines, but as (stable) complex, dynamic organization" (Woese 2004).

As Spyridon Koutrofinis wrote, "Organisms do not simply react to external gradients of matter and energy but manipulate the inner conditions of the formation and maintenance of their morphological structure and their physiological and genetic dynamics in a way that enables the continuation and, if necessary, the purposeful modification of their metabolism. The autonomous manipulation of the inner-organismic conditions, which determine the internal functions and external actions of the organisms, identifies the distinctiveness of the organismic mode of being. This ability often allows organisms to modify their environment in a way that serves the preservation and steering of their metabolism. This means, however, that the autonomy and depth of organismic self-organization cannot even approximately be attained by the most complex examples of physical or chemical self-organization of inanimate matter. Thus, there are good reasons to expect that 21st century biology will be erected around the insight that the explanation of biological phenomena demands a sound theory of the organism" (Koutroufinis 2022).

7.8.4.2 Order from Semantic Information

Life itself is not a machine (Witzany and Baluška 2012; Witzany 2010, 2020), and as pointed out by Witzany, biological systems cannot be reduced to material (physical and chemical) or nonmaterial (informational) characteristics. The equation life = physics + chemistry + Shannon information is incomplete. Witzany proposed a different definition of life: "life is communicative interaction of three complementary levels in parallel: cell communication, RNA communication, and virus communication" (Witzany 2020). This communication "cannot be predicted or simulated by computing machines, because biocommunication does not function mechanistically and is not algorithm dependent" (Baluška and Witzany 2014). For Guenther Witzany, biocommunication is the mysterious ingredient that would infuse life into cellular automata (Witzany 2010).

Arnold De Loof proposed a new definition of information that can be applied to biological systems: "In essence, information is any propagation of cause and effect in a system. The many definitions that exist are often discipline-specific. My definition of information, intended for use in biology, reads: A message contains 'information' when, upon being decoded and amplified by a competent (= with matching receptor(s)) receiver, part of the stored energy in that receiver is mobilized sooner or later for doing some sort of 'work' (as defined in physics). This can be contraction of muscles, engaging in feedback, activation of the cognitive memory system, activation or inactivation of specific genes etc." (De Loof 2016).

De Loof defines life as the sum of all communication activities; for him life = communication. "What we call 'Life' is nothing other than the total sum of all acts of communication exerted by a given sender-receiver compartment at moment *t*, at all levels of its compartmental organization (cell organelle, cell, tissue, ..., whole organism, ..., population, community, Gaia level) [...]. The enabling role of communication follows, in part, from the fact that any act of communication is a problem-solving act. Indeed, any message is coded and needs to be decoded at its arrival at a target that has matching receptors. Idem for responding to the decoded message(s) sooner or later. This means that problem-solving is inherent to communication, and thus to life itself. No life without problem-solving. Problem-solving needs the presence of appropriate memory systems in which the decoding programs have to be stockpiled in advance. [...]. Life (as an activity) is an approximating synonym for communication activity. Through its inherent link with problem-solving, changes in communication drive evolution. Thus, life drives its own evolution" (De Loof 2023). He proposes three kinds of memories: genetic, electric, and cognitive. In his opinion, genetic memory is represented by the mechanisms of transcription and translation of genetic information, i.e., DNA → RNA → protein(s). Electric memory, which De Loof considers absolutely vital to "being alive, for functioning of epithelia, the nervous and muscular systems, osmoregulation, self-electrophoresis, some roles of the cytoskeleton," is maintained by the membrane electric activity, the "electrome of cells." His definition of death is "the absence of the ionic/voltage gradient, not by the disappearance of some inorganic ions or molecules."

Finally, De Loof also introduces the original idea of a cognitive memory: "The idea that the cognitive memory might (partially) be based upon "moving forth and back" electrical charges and fix an imprint at a specific location in the cytoskeleton is appealing. Such system is present in all prokaryotic and eukaryotic cells, it is a most versatile system, and it can, in principle, be used for short or for long term purposes. It is very fast, and it can be modulated during physiological changes and during development. Was perhaps, the original role of actin-like molecules primarily in storing cognitive information in electrical form and was a role in motility secondary?" (De Loof 2016).

For other authors, the missing ingredient that can bring cellular automata to life is *evolution,* as proposed by Carlo Rovelli, who stated that "meaning = information + evolution" (Rovelli 2016). As he explained, "the semantic notions of information and meaning are ultimately tied to their Darwinian evolutionary origin. The suggestion is that the notion of meaningful information serves as a ground for the foundation of meaning. That is, it could offer the link between the purely physical world and the world of meaning, purpose, intentionality and value. It could bridge the gap."

John Miller and John Torday introduced several years ago the model of cognition-based evolution as an explanatory framework, which focused on the reciprocal assessment and communication of information by self-referential cells in reaction to environmental stresses (Miller and Torday 2017). James Shapiro (Shapiro 2021) and Perry Marshall (2021) also proposed that cognition represents the key life ingredient. More recently, John Torday proposed a model of consciousness-based evolution (Torday 2025). In his model, consciousness is defined as "an awareness of the outside world" (Trewavas and Baluška 2011).

These different hypotheses regarding the missing essential ingredient of life are centered around the idea that besides the "structural information" present in the different structures present in a cell (DNA, RNA and protein molecules, the cellular compartment, etc.), there is also "functional information" related to the "functioning of the cell and hence the rate and number of chemical reactions taking place" (Davies et al. 2013). A whole field of biosemiotics has been developed around "functional information" and directed behavior (Hoffmeyer 2008).

In order to maintain entropy within a confined range, both structural information and functional information are under tight homeostatic control in a cell, since disruption of any of them can lead to disease and death.

Are all living systems capable of "cognition"? In the minimal sense, a bacterium is capable of simple behavior and therefore of cognition (for a review, see Baluška and Levin 2016). Other authors restrict cognition only to human beings capable of "symbol manipulating programmes in abstract contexts" (Moreno and Mossio 2015).

7.8.4.3 Order from Functional Information

More than a decade ago, Farnsworth et al. proposed that living represents functional information processing. The function for the authors had two layers of meaning: "The specific meaning of a function depends on both the functioning entity and the context in which it acts (function is relational, not inherent)" (Farnsworth et al. 2013). A cybernetic model of life was proposed: "Considering life as information processing (computation) where the subject of computation is life, we are faced with a 'program' running on itself, the function of which is to output itself. Such recursion is familiar and much exploited in computer science. It highlights the fact that for life, there is no distinction between the 'machine' and the program—both are information; they are the same information, ordering and re-ordering matter and energy so as to persist. It would not be right to think of life as a biochemical structure on which a program is run, because life is the program and the biochemical structure is its embodiment. This is why we say that information is not just in DNA, but it is in the whole biological system. The idea that 'life is information processing' brings reductionists and synthesists closer together as it shows life to supervene on chemistry strictly according to information content, but to also possess strictly emergent properties (at several levels) arising from the functions of the embodied information" (Farnsworth et al. 2013).

As will be further discussed in Chap. 8, the cyclic causality present in life forms is related not only to information but also to energy and organization. Life is not just computing itself; it also energizes and organizes itself within cyclical causality loops. Cyclic causality is needed in order to unveil the deeper principles of living systems.

Recently, Michael Wong et al. proposed a new model of life centered on evolution and a new definition of "functional information" (Wong et al. 2023). The authors started by enumerating different laws of physics (Newtonian laws of movement, Newtonian laws of gravitational attraction, laws of thermodynamics, etc.) in support of their proposal of a new universal law responsible for a "conceptual equivalence among disparate evolving systems," such as stellar evolution, nucleosynthesis, geological evolution, and biological evolution. They define an evolving system as a "collective phenomenon of many interacting components that displays a temporal increase in diversity, distribution, and patterned behavior." Basically, expanding Darwin's law of evolution to the abiotic world, they propose an all-encompassing law of evolution of natural systems, animate and inanimate.

The formulation of this new proposed law is as follows:

> Systems of many interacting agents display an increase in diversity, distribution, and/or patterned behavior when numerous configurations of the system are subject to selective pressure (Wong et al. 2023).

Subsequently, Wong et al. described three orders of selection, each ruled by a different selection law:

- The *first-order selection law* applies to static persistent systems like atoms of Earth's minerals. They stated that "Configurations of matter tend to persist unless kinetically favorable avenues exist for their incorporation into more stable configurations."

- The *second-order selection law* applies to dynamic persistent systems like stars, hurricanes, and organisms. They stated that "Insofar as processes have causal efficacy over the internal state of a system or its external environment, they can be referred to as functions. If a function promotes the system's persistence, it will be selected for."

The authors mention dynamic persistence as a key life characteristic that is "driven by chemical disequilibria in the environment and solar photons that are greatly out of equilibrium with the thermal background. [...] The fundamental process is the dissipation of free energy - without this function, no complex, dynamic entities could exist. Unlike static persistence, which only requires dissipation during formation, dynamic persistence requires active dissipation. Other functions—such as autocatalysis, homeostasis, and information processing—can emerge that prolong the act of dissipation through space and time" (Wong et al. 2023). For the authors, the core functions of life are represented by energy dissipation, autocatalysis, homeostasis, and information, but as stated above, they consider energy dissipation primordial.

In order to explain dynamic persistence, Wong et al. already introduced in the formulation of the second-selection law the concept of "function," defined as a process that has "causal efficacy over the internal state of a system or its external environment" and proposed the ad hoc principle "if a function promotes the system's persistence, it will be selected for."

Looking for a quantitative formulation of the law of selection for new functions, Wang et al. used the concept of "functional information," introduced earlier by Jack Szostak (2003). In their formulation "Functional information quantifies the state of a system that can adopt numerous different configurations in terms of the information necessary to achieve a specified "degree of function," where "function" may be as general as stability relative to other states or as specific as the efficiency of a particular enzymatic reaction" (Wong et al. 2023).

The law of increasing functional information reads:

> The functional information of a system will increase (i.e., the system will evolve) if many different configurations of the system are subjected to selection for one or more functions.

Of note, a similar law has been priorly proposed as the fourth law of thermodynamics by Cortês et al. (Cortês et al. 2022). In order to formalize Stuart Kauffman's idea of the adjacent possible, Cortês et al. postulated that the number of possible and actual functions of biological systems is continuously increasing.

It is important to note that in a prior paper, Stuart Bartlett and Michael Wong already proposed that *dissipation, autocatalysis, learning and homeostasis* are the four necessary and sufficient pillars of what they call *lyfe*, understood as a cosmic process, which is not uniquely related to planet Earth but can also be applied "to as yet undiscovered (or uninvented) systems that exist at myriad scales across the universe" (Bartlett and Wong 2020). According to Bartlett and Wong, as opposed to *life* on Earth described as "an autocatalytic network of organometallic chemicals in aqueous solution that records and processes information about its environment in molecular form and achieves dynamical order by dissipating any subset of the following disequilibria: redox gradients, chemiosmotic gradients, visible/thermal photons, the boundaries with which we use to refer to that system," *lyfe* is "any hypothetical phenomenon that maintains a low-entropy state via dissipation and disequilibria conversions, utilizes autocatalytic networks to achieve nonlinear growth and proliferation, employs homeostatic regulatory mechanisms to maintain stability and mitigate external perturbations, and acquires and processes functional information about its environment." An important detail of their proposal is the description of these four properties as pertaining to *systems*: "The pillars that can be ascribed to a certain system depend specifically on the boundaries with which we use to refer to that system" (Bartlett and Wong 2020).

The *third-order selection law* applies to "nested networks of smaller complex systems, each persisting and helping to maintain the persistence of the whole [...] "There exist selection pressures favoring systems that can open-endedly invent new functions—i.e., selection pressures for novelty generation" (Wang et al. 2023a, b). This law refers to systems capable of inventing new functions and is closely aligned with the Kantian notion of wholes introduced by Mossio and Kauffman.

From their description, it seems that, besides multicellular organisms, the third-order selection law also applies to supraorganismal living systems like ecosystems and human societies.

The functional selection theory of Michael Wong et al. gives selection the power of a natural law: "Selection based on static persistence, dynamic persistence, and novelty generation is a universal process that results in systems with increased functional information." Importantly, all three: static, dynamic persistence, and novelty generation are interdependent and overlapping (Wong et al. 2023). Wong's arguments for selection/evolution as a new natural law echo Alfred Lotka's century-old proposal of "natural selection, the principle of survival of the fittest, or … the principle of stable forms" as a fourth law of thermodynamics (Lotka 1922a). In his formulation, "Natural selection will so operate as to increase the total mass of the organic system, to increase the matter of circulation through the system and to increase the total energy flux through the system, so long as there is presented a utilized residue of matter and available energy. [...] Natural evolution tends to make the energy flux through the system a maximum, so far as compatible with the constraints to which the system is subjected" (Lotka 1922b). In other words, organisms tend to maximize power intake, i.e., energy per unit of time, as explained by Howard Odum, who applied Lotka's ideas to the ecology field and called Lotka's law the *maximum power principle* (Odum 1995). Of note, in order to maximize *useful* power (and not the energy that goes to heat), there should be a selection for an *intermediate* rate and *intermediate* efficiency (Hall 2004).

Of note, both in the theory of functional information and in NASA's definition of life, in order to exist, life needs to evolve (De Loof 2015, 2017).

As observed by Stephen Jay Gould, evolution does not proceed by following an optimal design like the principle of the minimum path of Le Chatelier, for example, or by continuously inventing new functions but by tinkering with what is already there (Gould 1981). The concept of "tinkering" also perfectly fits cancer, the great opportunist, since cancer does not invent new forms or processes and develops by hacking into the cell's and organism's already existing programs. Cancers compete for matter and energy with local tissues and the whole organism not by maximizing their energy flow or their efficiency of energy use, but by turbo-amplifying the energy intake pathways and mechanisms, i.e., maximizing their intake power (power defined as the energy transferred or converted per unit of time). A finding that supports the maximization of power exhibited by cancer cells is the positivity of FDG-PET scans described in many types of cancer tissues pointing to their higher intake of energy in the form of glucose molecules (Chan et al. 2020). Of note, this higher intake of energy that occurs in some cancer types is not necessarily correlated to a higher efficiency of energy production. In fact, given that many cancer cells use aerobic glycolysis as their main mechanism of ATP generation (four ATP molecules generated per one glucose molecule), it can be argued that, in terms of ATP, cancer cells are much less efficient than normal cells that use respiration for obtaining ATP from glucose (approximately 36 ATP molecules generated per one glucose molecule). On the other hand, as has been explained by Vander Heiden et al., this lower efficiency in energy production is related to the fact that cancer cells channel their energy toward biomass production, and their main need is to generate sufficient NADPH in order to produce nucleotides and amino acids to support cell proliferation and not ATP production (Vander Heiden et al. 2009). Thus, cancer metabolism is switched toward anabolism, with the purpose being a higher efficiency in biomass production. In this context, cancer metabolic reprogramming can be viewed as an extension of the maximum power principle described above. *Cancers prioritize energy intake and biomass synthesis over efficiency.* This suggests that cancer cells "hack" existing metabolic pathways, amplifying energy consumption, *not to optimize efficiency* but rather to ensure continued cell growth and survival in the face of selective pressures.

7.8.4.4 Relevance of Functional Information Evolution Theory for Cancer

In their work, Michael Wong et al. outlined three mechanisms through which the rate of evolution in a system can be accelerated: "1) by increasing the number and/or diversity of interacting agents, 2) by increasing the number of different configurations of the system, and/or 3) by enhancing the selective pressure on the system" (Wong et al. 2023). Each of these factors is highly applicable to cancer.

First, cancer treatments frequently introduce selective pressure, which, paradoxically, can accelerate tumor progression rather than inhibit it. This phenomenon, known as therapy-induced cancer progression, is well documented in oncological research. Additionally, certain cancers exhibit spatial and temporal heterogeneity within tumors, a feature that may directly contribute to accelerated evolutionary processes within the tumor population.

Building on Wong et al.'s concept of functional selection, Frédéric Thomas et al. proposed a novel model of tumorigenesis. He and his co-authors suggested that the early stages of tumor formation are driven by Darwinian selection acting both on individual cells *and* on cell clusters, which generate varying phenotypic compositions (GPCs) (Thomas et al. 2024). Subsequently, a second phase of selection occurs, where the functionality of these cellular societies, as defined by Wong et al., becomes the driving force behind tumor growth. This phase sees the tumor's survival and expansion aligned with the functionality induced by the GPC in conjunction with the tumor microenvironment.

Thomas et al. proposed a shift in treatment strategies targeting GPC at the tissue level rather than using conventional targeted agents.

This approach underscores the need for a more holistic view of tumor behavior, where therapeutic interventions are designed to disrupt the functional interactions of the cellular society within the tumor, rather than focusing solely on individual mutations.

The Homeodynamic Theory of Life: An Introduction

8

8.1 Static Maps and Dynamic Territories

In science, the boundary between the map and the territory is often blurred. For instance, quantum physicists typically do not make a distinct separation between theory and fact. In quantum physics, a theory is a mathematical model, consisting of a set of formulas and foundational concepts, which explain how these formulas are used to describe and predict the behavior of nature (Strassler 2024).

By contrast, in biology (Gunawardena 2014), and particularly in medicine, where the distinction between theory and fact can have life or death implications, the accuracy of models is paramount. Incorrect medical theories and models, when applied indiscriminately, can result in fatal outcomes. Medical history is rife with misguided concepts that have led to patient deaths. In medicine, as in other high-stakes professions like aviation, human error can have catastrophic consequences. This highlights why doctors, like commercial pilots, are held to a standard where mistakes are not permissible. This standard partially explains the medical field's inertia toward new ideas and the hesitation of healthcare professionals to be the first to adopt new disease models or treatment strategies.

Our mental representations of reality—our "maps"—arise from everyday experiences and are deeply embedded in the structure of language, which compels us to parse the world into binary categories: nouns and verbs; objects and processes. On top of this foundational dichotomy, we superimpose attributes—properties of objects and actions—as well as instruments of interaction and spatiotemporal coordinates. Core scientific concepts such as space, time, energy, action, and causality are not intrinsic features of the world but abstractions constructed to navigate and interpret it. These constructs belong to our cognitive domain; they reside in the map, not the territory. It is crucial to maintain this epistemological distinction. Maps, while not reality itself, are *realistic approximations*—representations rather than precise replicas. Science, therefore, is an ongoing endeavor to refine these maps, drawing them ever closer to the contours of reality. Our scientific theories and models are provisional frameworks, not ultimate explanations. For instance, Nicholson has questioned the mechanistic view of cells as machines and cautioned against literal interpretations of metaphors such as scaffold proteins functioning like fixed circuit boards that direct cellular information flow, as introduced by Hanahan and Weinberg in their first hallmarks article (Hanahan and Weinberg 2000). Instead, he advocates a paradigm shift toward understanding cells as dynamic, self-organizing systems characterized by the fluidity and adaptability of their constituents and governed by stochastic and nonlinear processes (Nicholson 2019). This perspective urges us to rethink our metaphors and models in favor of ones that better reflect the emergent, processual nature of life.

Our innate model of the external world has been shaped through evolution and is strongly influenced by our visual perception, which tends to segment reality into discrete objects moving against a static background. This object-oriented worldview underpins many foundational scientific frameworks: Newtonian mechanics and classical thermodynamics in physics, the reactant-reaction paradigm in chemistry, Darwinian natural selection in evolutionary theory, and the molecular framework in biology. Oncology is no exception. The prevailing paradigm of precision medicine—celebrated as "The New Paradigm for Healthcare" (Sarwar 2023)—embodies this object-oriented ontology by focusing on the design of therapeutic agents tailored to the structural properties of specific molecules, or the interactions between defined cellular targets. Even time itself has been subject to objectification in modern science: Contemporary quantum physics and Minkowski's space-time formulation, central to Einstein's theory of special relativity, treat time as a quantifiable, geometric entity. Similarly, causality becomes a temporally objectified concept when we define phenomena in terms of discrete causes and effects. This pervasive inclination to conceptualize the world as a collection of bounded entities with linear interactions reflects deep-rooted epistemological commitments that influence not only our

D. Paul, *Cancer: An Integrative Approach*, https://doi.org/10.1007/978-3-032-08808-6_8

understanding of nature but also how we approach complex biological systems such as cancer.

Our concepts often constrain reality within the rigid boundaries of models, which are then subjected to reality checks. When discrepancies arise, we revise or develop new concepts and models to better align with observed phenomena. Concepts digitalize reality, rendering it accessible to logical and mathematical formalism. The reductionistic approach took us to the moon and discovered cures for acute conditions like infections and fractures. However, for designing effective treatments for complex diseases like cancer, a more holistic perspective is required, necessitating the integration of process-oriented concepts in oncology.

8.2 What Is Causality?

Causality is fuzzy and difficult to define. In 1965, Austin Bradford Hill proposed nine criteria for causality without defining it:

- *Strength*—Strong associations are more likely to be causal than weak associations.
- *Consistency* refers to the repeated observation of an association in different case and circumstances.
- *Specificity*—The criterion of specificity refers to the fact that a cause leads to a single effect, not multiple effects.
- *Temporality*—Cause necessarily precedes its effect.
- *Biologic Gradient* refers to the presence of a monotone, unidirectional dose-response curve.
- *Plausibility* refers to the plausibility of a biological mechanism explaining the observed relation.
- *Coherence* implies lack of contradiction between the causal association and what is known of the natural history and biology of the disease.
- *Experimental Evidence* refers to a counterfactual; removal of the cause removes the effect.
- *Analogy* can be drawn with other cause-effect relationships (Hill 1965).

Hill was cautious to state that: "None of my nine viewpoints can bring indisputable evidence for or against the cause and-effect hypothesis, and none can be required as a sine qua non. What they can do, with greater or less strength, is to help us to make up our minds on the fundamental question – is there any other way of explaining the set of facts before us, is there any other answer equally, or more, likely than cause and effect?" (Hill 1965). Importantly, none of Hill's nine criteria is sufficient on its own to establish causality, nor is there a definitive combination or minimum number that guarantees a causal inference. The criteria vary in relevance—that is, strength and consistency of association often carry more weight than specificity—yet how to appropriately weigh them remains uncertain. This challenge is compounded by interdependencies among the criteria; for instance, the presence of experimental evidence can influence the assessment of plausibility (Dekkers 2019).

A first advancement since Hill has been the adoption of the counterfactual framework developed by Judea Pearl, which offers a more rigorous basis for evaluating causal effects of interventions in epidemiology. Pearl has pondered about causality for several decades and wrote extensively on the subject described three levels of inferring causality: The first level referred to as *association* is related to observing some regularities in the environment, the second level, *intervention*, is related to noticing the results of particular actions, and the third level, *counterfactuals*, is associated with imagining and understanding the process (Pearl and Mackenzie 2018). Intervention, the second ladder of causation described by Pearl, is of particular interest as it can help dissect the strength of different causes associated with a given outcome. As an example, using intervention, a team of researchers was able to create the first artificial cells with a minimal genome by progressively reducing the number of genes (Hutchinson et al. 2011) from cells that already had a very small genome (Maniloff 1996; Mushegian and Koonin 1996).

Importantly, neither Austin Bradford Hill nor Judea Pearl offered a definition of causality; both of them rather focused on the criteria of establishing whether the frequency of an event E (effect) will be influenced by a change in another event C (cause). Causality is a slippery concept and their reluctance to pin it down is understandable. The main difficulty with causality is that it belongs to our mental maps, and it is not clear whether and what is (if it exists!) its correspondent in the outside world.

Intuitively, causality is the link between a cause and an effect, but as both Kant and Hume pointed out, it is very difficult to untangle the a priori component of causality embedded in our mental maps and related to psychological temporal mechanisms from the *real-world* causality—whatever we consider *real*—since quantum causality and cosmic causality are different.

Will AI modeling be able to help us untangle this knot of multiple causes and effects and illuminate the foundation of life processes *in general*, or does AI rather offers us only strategies similar to Alexander the Great's sword, which famously cut the Gordian knot, providing us only rough approximations with *restricted* applications?

Arguably, the most important recent AI development has been the introduction of *causal deep learning* (Berrevoets et al. 2024) that has the potential to truly transform the way we solve a large number of real-world problems in the health field. Initially, it was thought that the "ladder of causation" introduced by Judea Pearl (Pearl and Mackenzie 2018) will help define causal structures, but, as shown by the team of Mihaela van der Schaar at Cambridge University, the practical utility of this ladder is limited (Berrevoets et al. 2024). As an alternative to the "ladder of causation," this team proposed a powerful *causal deep learning* AI architecture able to work with "plausible assumptions" and solve a variety of real-world problems by leveraging partial causal knowledge and quantitatively characterizing causal relationships among variables of interest (Weberpals et al. 2025).

8.3 Aristotle's Structure of Causality

Aristotle divided causality in four components using a bronze statue analogy: The *Material cause* refers to the material that is modified through an action, like the bronze that will become the statue, along with the change in shape of the bronze. *Formal cause* is the mold used to create a specific shape. *Efficient cause* refers to the art of making a statue, and *final cause* is the purpose served by the statue (https://plato.stanford.edu/entries/aristotle-causality/).

Much like Euclidean geometry or Newtonian physics, Aristotle's causality schema is an oversimplification. His classification follows a grammatical structure: *A* acts on *B* using instrument *C* to achieve finality in the creation of *D*.

This model, however, is limited since it fails to capture the complexity of biological systems, where the causality structure is very different. In living systems, multiple material causes coexist within intricate causal networks that are challenging to disentangle. Life is teleological, that is, in living systems, the final Aristotelian cause takes precedence, revealing the universal principle that life is at the same time autopoietic and autotelic: self-creating and, at the same time, its own purpose. Also, in biological systems, the final cause is embedded in the cycle of co-dependent material/formal/efficient causes: It does not exist separately. In terms of cancer, research has been mainly focused separately on *material* (metabolic changes), *formal* (structure alterations), and *efficient* (genetic and epigenetics programs) causes and not so much on their co-dependence source of autotelic and teleonomic cyclic causality (*final* cause) (please see Fig. 8.1).

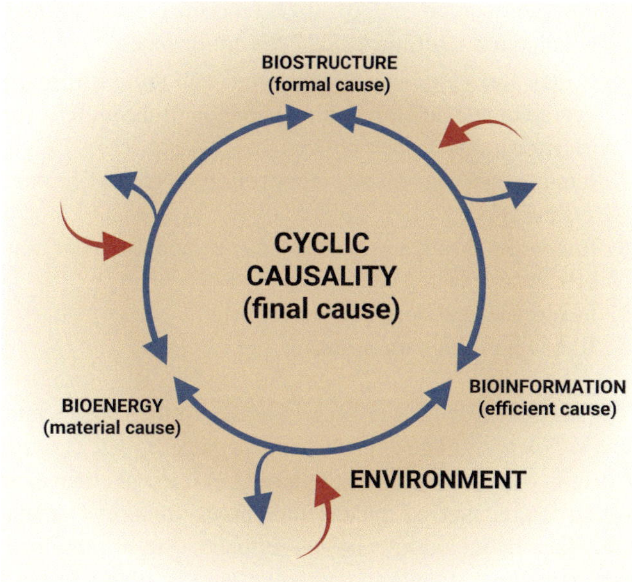

Fig. 8.1 Cyclic causality as its own final cause and effect Created in BioRender. (Courtesy of Laura Pătraş)

8.4 What Can a Cell Do?

A cell belonging to a multicellular organism is capable of a relatively limited set of actions responsible for transiting it from a state to another. This set of actions entails a set of genetic/epigenetic/metabolic programs and, with the exception of meiosis, are present also in unicellular organisms. Both genetic/epigenetic/metabolic components and the environment play a role in transition between cellular states.

In general, a cell's actions repertoire is the following:

1. It may continue existing as it is and perform different activities for the benefit of the whole organism (i.e., secretion of certain hormones by endocrine cells, information communicating and computing by neurons).
2. It may divide into identical or different cells also into one cell identical to the mother cell and one more differentiated (cells with stem cells properties are capable of doing that).
3. It may transform to another type of cell (i.e., epithelial to mesenchymal transformation).
4. It may differentiate or dedifferentiate (stem cells differentiate; some cancer cells may dedifferentiate and acquire a stem cell-like behavior).
5. It may change its chromosomal number, reducing the number in half, through meiosis (only germ cells can do that), or multiplying their number of chromosomes by

polyploidization (polyploidization is the process through which giant multinucleated cells are formed).

6. It may form a new cell (germ cells can unite forming a zygote or egg cell through fecundation or through fusion transform into a polypoid cell).
7. It may attack and destroy other cells (cells of the immune system).
8. It may die through programmed (i.e., apoptosis) or non-programmed (i.e., necrosis) cell death.
9. It may become dormant.
10. It may be stationary or move.

A multitude of factors interact to determine a cell's fate: *Internal factors*: a) genetic alterations; b) germline; c) tissue of origin; d) histone structure changes; e) internal changes in cell structure, that is, nuclear membrane rupture; f) metabolic factors including nucleus-mitochondria communication; g) cellular membrane potential; and h) transcriptional factors; and *External factors:* i) temperature; j) pH; k) stromal pressure; k) cell-cell contact; l) hormonal, neuronal, cytokine, and exosome-mediated signals; m) internal body climate; and n) comorbidities, affecting function of other organs (partial list). All these factors are co-dependent and act in concert through a dynamic "causal network." Not mentioned is the cell's age, which is a pervasive influence on many of the factors cited above. The *causal network* describes how variables (causes and effects) affect each other (Farrokh 2020), and the "weight" of the contribution of different factors can be analyzed and deconvoluted using the counterfactual theory of Judea Pearl (Pearl and Mackenzie 2018).

Depending on the architecture of the causal network that may be represented by different cellular stressors, or signals, internal or external context, etc., a normal cell will follow different trajectories becoming 1. Unchanged, 2. Apoptotic, 3. Senescent, 4. Dormant, and 5. malignant. Using the epigenetic landscape of Waddington as a model, a representation of the causal network of factors influencing a cell's fate can be drawn (please see Fig. 8.2).

Of note, the fates of cells represented above are schematic: *intermediate states exist between a normal and a cancerous cell*. Also, there are multiple different phenotypes of cancer cells and, also, cancer cells can transit between them influenced by similar factors as those depicted in Fig. 8.2. Importantly, some cellular fates are not irreversible; with other words, *some cells can move both uphill or downhill* on the Waddington landscape depicted above, that is, senescent cells may exit senescence, cancer cells may become more or less differentiated, etc. Trans-differentiation by which one cell transforms into another is not a one-way street and, in fact, plasticity and bidirectional programing are key characteristics that support tumor cells growth and metastasis. Notably, as cancer progresses, malignant cells may shed their epithelial characteristics and transition into a hybrid state marked by enhanced plasticity and a predisposition for multi-lineage differentiation. This heightened plasticity, shaped by diverse combinations of transcriptional and epigenetic alterations, along with cues from the surrounding microenvironment, facilitates trans-differentiation into distinct cellular lineages with different functionalities. Cells co-expressing both epithelial and mesenchymal characteristics that exist in a hybrid EMT state—neither completely committed to either epithelial or mesenchymal phenotype—have the greatest malignant and metastatic potential (Pastushenko et al. 2018). The key lineage differentiating factors from epithelial cells to mesenchymal, respectively, neuroendocrine, squamous, and endothelial cells, are partially distinct, that is, Snail, Slug, Twist, Zeb1/2, and GRHL2 for epithelial to mesenchymal trans-differentiation (ubiquitous among cancers); SOX2, BRN2, ASCL1, FOXA2, and N-Myc for epithelial to neuroendocrine trans-differentiation (present in prostate, lung, bladder, and pancreatic cancers); TP63, SOX2, Myc, and c-FOS for epithelial to squamous trans-differentiation (present in lung, bladder, prostate, esophageal, and pancreatic cancers); and Twist, ZEB1, Slug, and VE-cadherin for epithelial to endothelial trans-differentiation (present in lung, breast, prostate, melanoma, and glioblastoma). An important concept is that these various transitions are nonbinary, that is, they tend to be associated with hybrid metastable states (Davies et al. 2023).

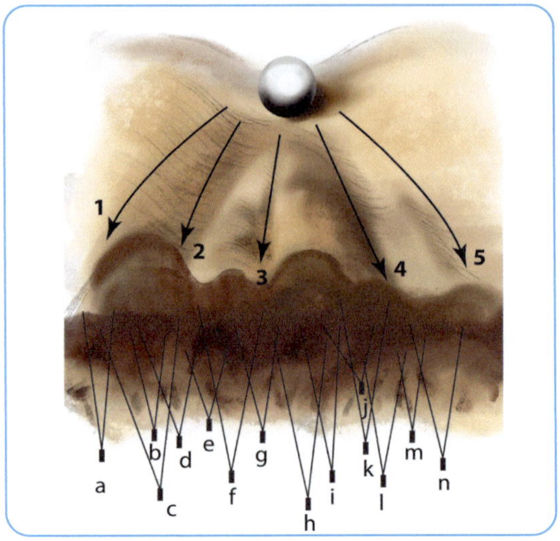

Fig. 8.2 The landscape of the causal network of factors influencing a cell's fate. (Courtesy of Ana Tourian). *Internal factors*: (*a*) genetic alterations; (*b*) germline; (*c*) tissue of origin; (*d*) histone structure changes; (*e*) internal changes in cell structure, that is, nuclear membrane rupture; (*f*) metabolic factors including nucleus-mitochondria communication; (*g*) cellular membrane potential; and (*h*) transcriptional factors. *External factors:* (*i*) temperature; (*j*) pH; (*k*) stromal pressure; (*k*) cell-cell contact; (*l*) hormonal, neuronal, cytokine, and exosome-mediated signals; (*m*) internal body climate; (*n*) comorbidities, affecting function of other organs, etc. *Cellular fates*: *1*. Unchanged. *2*. Apoptotic. *3*. Senescent. *4*. Dormant. *5*. Cancerous. (Adapted from Waddington (1957). The strategy of the genes. London: Geo Allen and Unwin)

A key point regarding the causal networks landscape is that a cell is not a passive bystander pushed in different directions by causal networks. Cells, and organisms, in general, interact continuously with their environment. They behave like rivers incessantly remodeling their banks, and, at the same time, with the banks directing their flow. River and banks, cells and causal networks are co-dependent, a cell's phenotype acting back on its internal and external environment and co-creating its trajectory shaped by homeodynamic forces (please see Fig. 8.4). The co-dependency between a cell or a multicellular organism with the internal and the external environments represents a key characteristic of the living process.

A cancer tumor is heterogeneous and contains both stem-like cancer cells exhibiting "multitasking" capacities able to perform several of the 10 tasks described above and more "specialized" cancer cells that are able to perform limited tasks only.

Both normal stem cells and cancer cells with stem cell behavior appear to have an "open chromatin" that allows them a wider sampling of the transcriptome, while more differentiated cells generally restrict chromatin accessibility and transcriptional diversity as they specialize. As a result, the number of expressed genes per cell appears to be a hallmark of developmental potential, differentiated cells expressing fewer genes per cell than less differentiated cells (Gulati et al. 2020).

Tumors as a whole may exhibit "multitasking" behaviors, as has been demonstrated by studying 250 tumors belonging to 15 cancer types, or a more restricted repertoire or archetype (Hausser et al. 2019). Six tumor types (breast, colon, thyroid, bladder, low-grade glioma, and liver) were found to exhibit one of five types of *archetypes* related to five predominant tasks: cell division, biomass and energy production, lipogenesis, immune interaction, and invasion and tissue remodeling. Of note, these archetypes/tasks do not represent pure cell types (immune cells, stromal cells, malignant cells, etc.) *but rather activities performed by heterogeneous groups of cells.* The clinical relevance of this framework is the prediction that tumors that specialize in a task should be more sensitive to drugs that impair that task. The authors demonstrated that cells coming from tumors with the biomass and energy production archetype are more sensitive to an mTOR inhibitor, for example (Hausser et al. 2019).

More recently, four different archetypes: (a) aerobic energy production, b) protein synthesis, c) immunotolerant tumor microenvironment, and d) "plasmablast-like" differentiation)—with four different set of vulnerabilities—were described in large B-cell lymphoma (DLCBL), pointing to archetypal variability within the same cancer-type tumor and the need for a case-by-case analysis (Blais and Jeukens 2024). Of note, developed in the context of evolutionary theory, the group phenotypic composition (GPC) theory of Thomas et al. presents similar concepts with the archetype framework (Thomas et al. 2024).

8.5 Life as a Bridge Between Quantum and Macroscopic Worlds

Life is a sort of bridge between the quantum and the macroscopic world. Different causality classes are present at different biological scales. For instance, processes at the cellular level are best described through statistical mechanics, where causality operates within a probabilistic framework (Åberg et al. 2016). In contrast to macroscopic systems where cause and effect can be clearly defined, quantum systems often exist in a superposition where the order of events (whether A precedes B or vice versa) is less straightforward, potentially blurring the line between cause and effect. At the cellular scale, causality represents a "network causality," where various combinations of molecular causes exhibit different strengths and types of effects on cell's phenotype depending on the context. By contrast, at higher levels of biological organization, actions are typically digitized, leading to binary outcomes governed by Boolean logic. For example, Schrödinger's cat, on the macroscopic scale, is either alive or dead, with its status easily determinable without relying on complex statistical computations.

Smith and Horowitz defined life as *any process participating in the largest living system on Earth which is the biosphere*. The authors also described life on our planet being at the same time related to the particularities of both the Earth and the cosmic environment (i.e., the Sun as source of energy) and the specific chemical composition of our planet. Almost six decades ago, Morowitz proposed that the phenomenon of *circular causality* of life may have originated from biochemical cycles that spontaneously emerged on Earth (Morowitz 1966).

In order to take into consideration slight variations or homeo-variations, as opposed to *circular causality*, the focus here will be on *cyclic causality*: A visual representation of cyclic causality would be a spiral and not a circle.

8.6 The Shape of Time

Both *directed movements* and *repetitive processes* are fundamental to life. Cells, for example, are powered by electrochemical gradients that drive ATP synthesis and help maintain membrane potentials. These gradients are established and maintained by directed actions controlled by the bioinformation systems of the cell, such as DNA transcription and protein translation, which produce pumps that regulate ion flow across membranes. Within the cell, three interlinked processes—*energy metabolism*, *information communication and processing*, and *structure maintenance and creation*—interact in a continuous, repetitive cycle. The recursivity of these processes creates patterns in both space and time, illustrating the self-organizing nature of cellular life.

A cell is a realm of self-generated directed activity. While the energy that drives cellular processes originates externally, the cell harnesses this energy to organize and direct its internal functions. This mirrors a principle in physics: An open system, continuously infused with energy, has the capacity to spontaneously self-organize. England proposed that merely injecting energy into a far-from-equilibrium system can lead to its organization, highlighting the role of energy flow in structuring biological processes (England 2013).

A cell does not "invent" time; events in a cell unfold regardless of cellular existence. On the other hand, cells are structures that organize the flow of these events, creating their own temporal and spatial order. The structure within the cell imposes direction on events, guiding them toward a particular outcome. In this way, by controlling motion, a cell creates its own temporal structure, which, at its turn, creates the cell, a cell, both being and becoming at the same time the architect and the product of a circular temporal process.

For example, in a voltage-gated ion channel, the flow of current is controlled by the electrical charge it is exposed to—a sort of transistor-like behavior within cells. By controlling the flow of ions and imposing order on molecular agitation, cells structure and maintain the organization of their internal environment in response to external stimuli. This ability to interact with and adapt to the external environment underscores the dynamic and self-regulating nature of living systems.

8.7 Cyclic Causality, The Unifying Concept

Cyclic causality is the key to understand and predict the behavior of organisms and living systems. Over the last 50 years, the cyclic causality present in living systems has been noticed and analyzed by many researchers (i.e., Howard Morowitz, Stuart Kauffman, Robert Rosen, Mae-Wan Ho, etc.). An interesting observation is that biological systems exhibit not only cyclic causality within the same scale (i.e., *horizontal* cyclic causality) but also in between scales (i.e., *vertical* cyclic causality).

Importantly, cyclic causality operates both within living organisms and also between them and their environments, a dynamic that Jessica Baedke and colleagues have described as reciprocal causation (Baedke et al. 2021), with niche formation being a classic example. For the functioning of both unicellular and multicellular organisms, *horizontal* cyclic causality present within the same scale is equally important and is co-dependent with its *vertical* component present between scales. As proposed by Denis Noble, the *vertical* inter-scale cyclic causality is nonsymmetrical: "Downward causation consists in constraints exerted by higher levels on the initial and boundary conditions within which the dynamics of lower-level elements operate. By contrast, upward causation is the way in which those dynamics influence higher level states" (Noble et al. 2019).

Cyclic causality is a *global* property of life processes that ensures their *stability*. When we zoom in, *locally*, biological processes appear as a complex mesh of interconnected causal structures, that are metastable, that is, when perturbed, they can adapt by switching from one dynamic network to another. In the living world, where organisms exist on an intermediate scale between the quanta and the cosmos, rules a different type of causality that is adaptive, context-dependent and emergent, *biocausality*.

Cyclic causality is not just a characteristic of biological feedback or ecological interdependence; it is a cross-scale universal organizational principle. Whether manifested in the entanglement of quantum fields, the self-sustaining loops of metabolism, or the feedback-driven dynamics of galaxies, cyclic causality enables systems to self-regulate, adapt, and endure through time. Cyclic causality represents a universal cosmic property, and thus, it can be used as a unifying framework that links physics, biology, cognition, and cosmology, shedding light on how living processes emerge, persist, and evolve within a recursively structured universe.

For our purpose, we will use the concept of *cyclic causality* to describe the spiral-like causal architecture of biological processes and phenomena present both at microscopic and macroscopic scale. Cyclic causality is the key to understanding and predicting the behavior of organisms and living systems. In contrast to weather, which describes short-term day-to-day transitions and, as Edward Lorenz famously proposed, is fundamentally unpredictable, a flap of a butterfly's wings in Brazil may be setting off a tornado in Texas (Lorenz 1993), blood pressure, for example, and other parameters are kept relatively constant over time in our organisms, and, therefore, human health and disease trajectories are more predictable.

8.8 Biocausality

In order to understand complex biological processes, we must begin by exploring their fundamental causal structure. In general relativity, causality is embedded in the fabric of space-time, while in quantum mechanics, it manifests as probabilistic events and entanglement. Although general relativity is deterministic, quantum mechanics incorporates a substantial random component. Similarly, biological systems, though governed by deterministic physical laws, exhibit stochastic and chaotic behaviors due to the complex interplay of numerous interacting components.

Processes themselves are outcomes of energy transformations. Many natural processes, such as nuclear radioactive decay, are nondeterministic and spontaneous. For instance, Uranium-238 decays to Lead-206 through a series of 14 steps, occurring randomly for individual atoms. These transformations result from the progressive loss of energy as unstable atomic nuclei stabilize. Other quantum processes, like the intrinsic periodicity of elementary particles, are deterministic and predictable.

For over a decade, physicist Donatello Dolce at the University of Camerino in Italy has been developing a novel framework based on cyclic dynamics, a process that loops back on itself (Dolce 2023a, b). Dolce explained that intrinsic periodicity is key to the stability and deterministic behavior of elementary particles. However, it remains unclear why some quantum entities display spontaneous, seemingly acausal behavior while others follow predictable cyclic patterns.

The second principle of thermodynamics rules that in our universe, due to the natural propensity of nonequilibrium mass-energy configurations toward equilibrium, any interaction always involves a spontaneous energy flow from higher- to lower-density potential (Kostic 2020).

As a corollary, every energy exchange involves a portion of the energy that can be used to do work and another portion that is always necessarily lost through heat. As a result, entropy is generated and always increases overall. The spontaneous flow from nonequilibrium toward equilibrium gives time a sort of forced direction and provides events with a *natural causality structure*. In living organisms, energy exchanges are under the control of information processing mechanisms that allows energy to be exchanged only at certain moments and directs it to precise locations. By the intervention of controlled mechanisms, living processes are able to *redirect and maintain* the arrow of time that naturally points to a *remote* universal equilibrium, to *immediate*, finely controlled, far from equilibrium states. This redirected causality is different from the natural causality of events ruled only by the laws of physics and chemistry and therefore can be seen as an example of *functional causality*. The control occurs through feedback mechanisms, present within all living structures both *horizontally* within multiple scales, vertically, *in between* scales, and between the organism and its environment. The living process cycles back on himself. Life is a nonlinear dynamical system that dissipates energy into the environment to maintain its steady state, and as Morowitz noticed, when energy flows through a system at steady state and away from thermodynamic equilibrium, it necessarily creates at least one cyclic flow of matter within the system (Morowitz 1966; Morowitz and Smith 2017). The cyclicity leads to the spontaneous creation of macroscopic structures that are able to dissipate energy more rapidly than thermal relaxation mandated by the second principle of thermodynamic would allow it.

The cyclic causality is a fundamental property of nonlinear dynamical systems and only through nonlinearity "the functions and the goals of the actions of living creatures can be understood. In short, it is the transition from the linear to the nonlinear degree of organization that invests matter with purpose" (Barham 1990). *Cyclicity, though, is necessary but not sufficient for a living system to exist and persist.* In order for a cyclic process to be maintained, it needs to actively distinguish between signals that promote its persistence and signals that don't and adapt accordingly (Barham 1996). In conclusion, it needs a sort of servomechanism, or homeostat. In a cell cycle, for example, the cyclic process can be represented by the Krebs cycle, and the homeostat by the adaptation through transcription factors that can modulate mitochondrial activity. In the human organism, the cyclic process may be represented by the cardiac cycle controlled by the autonomous nervous system, and so on. Besides cyclicity, which represents the generic trajectory of living processes, other noncyclical homeodynamic mechanisms are at work, as described by David Lloyd and his collaborators (Lloyd et al. 2001).

Life is autotelic and autopoietic, its goal is its own persistence; it continuously creates and recreates itself. As opposed to spontaneous cyclic dynamics exhibited in the quantum world by quantum objects (i.e., elementary particles) and at the cosmic level by cosmic objects (i.e., pulsars), life represents a form of *actively preserved* cyclic causality. That is, it needs to consume energy in order to maintain its autopoietic cyclicity. Humberto Maturana and Francisco Varela's concept of autopoiesis describes living organisms as systems that produce and maintain themselves through recursive interactions of their components. These processes are not linearly ordered, but mutually co-constituting, creating organizational closure. "A living system is a network of processes of production such that the components continuously regenerate and realize the network that produces them" (Maturana and Varela 1980).

As opposed to classical causality that follows a reductionist, linear model—*A* causes *B* and *B* causes *C*—, cyclic causality refers to situations where components of a system simultaneously influence and are influenced by one another, creating feedback loops that shape system-level behaviors. These causal cycles are self-sustaining, adaptive, and time-generating, making them essential to understanding both the fabric of the universe and the phenomenology of life. In organisms, cycles occur also in between organizational scales: Molecular interactions (upward causation) give rise to emergent properties of the cells that will ultimately influence molecular function through epigenetic pathways and organism behavior as a whole, which in turn will regulate cells and influence in its turn molecular function (downward causation) via hormonal and neural stimuli (Noble 2012). These feedback loops form causal cycles essential for adaptation and homeostasis.

Life does not merely occupy time; *it is time and generates time* through rhythmic processes like circadian cycles, heartbeats, and metabolic oscillations. These rhythms arise from closed causal loops in genetic and biochemical regulatory networks (Goldbeter 1996). So *the cell itself is a dance that embodies at the same time the immateriality of the rhythm and the materiality of the dancer.*

Biological teleonomy differs from abiotic causality (Shapiro 2023) and should not be conflated with intentionality (Levin 2019, 2023a). Similar to entanglement in quantum physics, cyclic causality is an inherent property of the cosmos, akin to gravitation, and provides a foundational basis for understanding biological systems. As Archimedes famously stated, "Give me a lever long enough and a fulcrum on which to place it, and I shall move the Earth." In my opinion, cyclic causality acts as Archimedes' lever, with the Earth's environment serving as its fulcrum: "*With cyclic causality and the right environment, life on Earth will persist indefinitely.*"

Living organisms can be seen as microcosms governed by cyclicity. From molecular clocks to neural feedback loops and metabolic cycles, life continuously reproduces itself, embodying causal closure as both an autonomous and interdependent principle.

Crucially, Aristotle's framework does not account for the living process that operates upon itself, existing within a self-constructed space and time, and co-dependent with both internal and external environments. Biological systems include delays, and slow processes, and their causal structure is often flexible and involves a chain of distinct steps—it is not rigid and direct like the causality present in the abiotic world governed by physical laws. A different approach is needed to understand biocausality. This approach, which I tentatively call "homeodynamics," will be introduced later.

Cyclic dynamics and biorhythms are not only universal traits of living organisms but of all living systems. For instance, processes within ecosystems or across the biosphere exhibit cyclical patterns. The life criteria proposed by Sarah Walker and Lee Cronin, which focus on identifying multiple copies of identical objects "constructed" through a certain number of steps, implicitly assumes an underlying cyclic dynamic able to repeatedly create copies of the same object. Nanomotions, recently proposed as another universal criterion for life, also hinge on the continuous repetition of metabolic cycles.

While individual living organisms may not exhibit the enduring longevity of quantum structures, such as protons, for example, the life process itself is, theoretically, infinite. The stubborn resilience of the living process is the result of a large variety of control mechanisms. The most obvious control mechanism is the peculiar structural design of life forms. Compartmentalization by membranes, by constraining movements to confined spaces, controls movements. The genetic code also, by specifying the type and amount of proteins in the cell, controls movements inside the cell. Finally, the amount and distribution of ATP produced by a cell controls intracellular movement; as a footnote, ATP can be also exchanged *between* cells.

Less obvious, because they are invisible, is the large variety of cyclic dynamics. Feedback mechanisms, central to cyclic causality, enable a system to self-regulate. Only a system with feedback loops where the outcome of an action can influence the initial conditions, creating repeating cycles of cause and effect, can have cyclic causality. Cyclic causality maintains a dynamic balance between novelty and conservation by controlling both variation and spontaneity and, by the same stroke, ensuring both life's persistence and its quasi-identity. *Cyclic causality is the fundamental law* that underscores life's autopoiesis and autotelism. Two masses are pulled toward each other because of gravity. Life persists because of cyclic causality, which forms the bedrock upon which other biological behaviors can be understood.

Cyclic causality manifests in numerous forms across living systems. *First*, it governs the identity of organisms by maintaining a stable state. All living organisms are autopoietic, maintaining their structure and function for varying durations; for instance, neurons can survive for decades. *Second*, cellular reproduction involves cyclic processes, with cells making copies of themselves through division. *Third*, the reproductive cycles of multicellular organisms illustrate another example of cyclic causality: The human life cycle begins with the fusion of two germ cells to form an egg, leading to the creation of a multicellular organism that generates new germ cells capable of fusing with other germ cells, continuing the cycle.

Cancer itself follows a cyclic process. In addition to uncontrolled division, cancer cells spread distally and divide further, forming metastases and niches, from where it will further spread to other sites forming again other metastases and niches. The formation of niches, from primary to metastatic tumors, mirrors the ecological feedback processes observed in natural ecosystems (Pausas and Bond 2022). These niches support tumor growth and progression, and targeting their formation can interrupt cancer's life cycle.

8.9 Cyclic Causality, A Broader Principle than Evolution

Michael Wong et al. have recently expanded Darwin's theory of evolution to encompass the abiotic realm, proposing a universal evolutionary law that applies to both living and nonliving systems (Wong et al. 2023). They suggest that selection—driven by static persistence, dynamic persistence, and the generation of novelty—constitutes a fundamental process that enhances functional information within systems. This perspective resonates with Alfred Lotka's early proposition that natural selection could represent a fourth law of thermodynamics (Lotka 1922a).

In the model described by Wong et al., evolution with its three orders of selection laws was considered the universal mechanism responsible for the persistence of static, dynamic, or complex systems, both abiotic and biotic (Wong et al. 2023). As explained above, evolution belongs to a larger class of homeopoietic mechanisms, themselves part of the more general homeodynamic processes.

The investigation of cyclic dynamics serves as a conceptual bridge across physics, chemistry, and biology. While evolution exemplifies one form of cyclic dynamics, other essential manifestations include homeostasis and homeorhesis. The broader framework of cyclic causality extends and deepens Wong et al.'s proposal, *positioning evolution as one instance within a universal principle governing the organization and persistence of complex systems.*

Cyclic dynamics are prevalent throughout the universe, directing not only the behavior of living systems but also other abiotic phenomena (Heckman and Best 2023). Many processes (of course, not all) from stellar and galaxy formation to the rotation of the Earth and Moon around the Sun, through the large variety of biological processes, down to the quantum particles that behave like little clocks ticking their own time (Dolce 2023a, b), are cyclical. According to the conformal cyclic cosmology of Penrose, the whole universe may be also cyclical (Penrose 2006, 2012, 2020).

Cyclic causality, the pattern of chains of events causing each other, is the common denominator of the homeodynamic change that ensures stability both for biotic and abiotic processes.

Cyclic causality represents a sort of glue underpinning the recurrent nature of time itself that incessantly flows, repeating itself, with slight variations, moment after moment. The periodicity of time makes possible the existence of clocks as defined by Einstein: "By a clock we understand anything characterized by a phenomenon passing periodically through identical phases so that we must assume, by the principle of sufficient reason, that all that happens in a given period is identical with all that happens in an arbitrary period" (Einstein 1910). If this periodicity (i.e., cyclicity) would not be present, time, as we know it, would not exist, and, of course, we would not be here to talk about it either.

An interesting distinction arises between the cyclicity observed in biotic and abiotic systems. The controlled cyclicity of living processes as opposed to the spontaneous cyclicity of abiotic processes leads, in general, to the presence of a larger number of intermediary steps in biotic cycles versus abiotic cycles. For example, spontaneous inorganic cyclic reactions have fewer individual steps and wear off much faster than organic cyclic reactions that comprise more steps and persist for longer periods of time. Biorhythms have been priorly proposed as *dynamic life signatures* (Yates and Yates 2008; Lloyd 2021). An alternative way from the method proposed by the *assembly theory*, to distinguish and identify biotic processes (Marshall et al. 2017; Marshall et al. 2021; Sharma et al. 2023; Ellis 2023; Walker 2024), is by counting the intermediary steps and the number of identical cycles of a cyclic phenomenon during a certain amount of time. The demarcation line between the abiotic and biotic world may be marked by the number of the maximum number of individual steps present in a spontaneous persistent abiotic cyclic reaction combined with the number of cycles/unit of time. My guess is that cyclic reactions with four or more individual steps that persist for several cycles within a time frame of *milliseconds to seconds* are necessarily organic.

Cyclic causality serves as a bridge between the laws of physics and chemistry that govern the abiotic world and the homeodynamic principles that underlie living systems. The cyclical nature of living processes—both at the molecular level and across larger biological scales—supports the self-organizing, self-sustaining characteristics of life. By considering cyclic causality and homeodynamics introduced further as the foundational principles of life, we can better understand how living systems persist, adapt, and evolve in response to internal and external challenges.

8.10 Klein Bottles and Bio-cyclicity

A two-dimensional representation of a Klein bottle (please see Fig. 8.3) may help visualize the cyclic causality dynamic present in a living organism. This topological curiosity—

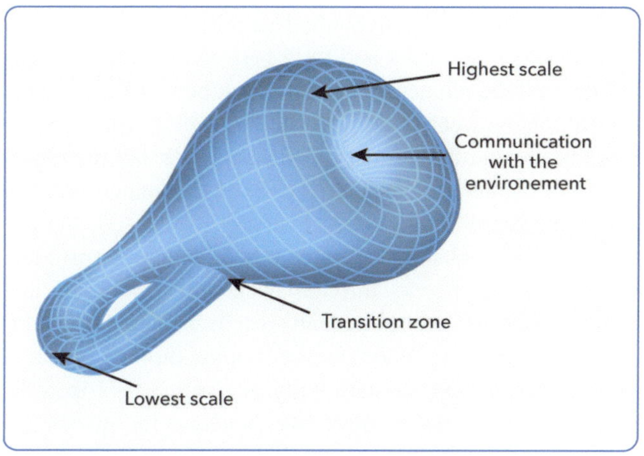

Fig. 8.3 Functional architecture of an organism represented as a Klein bottle. (Courtesy of Ana Tourian)

neither inside nor outside, neither open nor closed—is a suggestive representation of the recursive flows that link molecules, cells, organs, and the whole organism into one dynamic continuum. Please note that this is a *functional representation* and not an anatomical depiction. Cells are sphericals and the human organism has a torus-like (donut) shape. From *Nematodes* to human organism, the digestive system has two separate openings, a mouth and an anus. The earliest animals (i.e., *Cnidarians* like jellyfish, sea anemones, corals), though still present on Earth, have a single opening called a *gastrovascular cavity*, in which digestion takes place, and waste is also expelled through the same opening that resembles Klein bottle's aperture.

We can visualize the hierarchic dynamic present in the organism like an ongoing flow of matter and energy advancing through a set of Klein bottles of different sizes nested one within another like matryoshkas, cycling continuously through this fractal hierarchy in a ceaseless loop: from the subatomic realm of particles and waves, through molecular and cellular architectures, up to the organism's unified experience of self—and back again. The presence of different spatial and temporal scales and their continuous exchange of matter/energy/information may be a necessary requirement for life's existence. *The Klein bottle topology may serve as a universal, functional representation of an organism.* The transition zone depicted in the human organism is represented by the intestinal cells through which the food is absorbed. For a cell, the transition zone depicted in the Klein bottle picture is represented by molecular structures. The precise location of the transition zone in living organisms between molecular structures and elementary particles is still unknown. Quantum events have been demonstrated to take place also at the macroscopic level (Kotler et al. 2021), but the transition zone between the quantum and the macroscopic worlds has not been yet defined.

Feedback mechanisms between the cell and its environment occurs through its cellular membrane represented by the open area of the Klein bottle. In a cell, there are feedback mechanisms going back and forth between the macroscopic cellular and the molecular scale that can be visualized as the connection between the funnel-like larger portion of the Klein bottle and the thinner tubular structure that recoils on itself and reenters the enclosed space of the funnel-like portion of the Klein bottle.

Feedback mechanisms between the human organism and its environment occurs through its natural openings (i.e., mouth, nose, ears, anus, etc.) and its skin. Information reverberates between cells and systems, looping through layers of self-regulation that maintain unity across scales. Likewise, molecular structures themselves—proteins, for instance—engage in recursive dialogues with their surroundings, transmitting and receiving information through conformational folding and through their binding sites such as ATP pockets.

Between the quantum and molecular realms lies an especially intriguing frontier. Although quantum effects are traditionally reserved for the scale of atoms and subatomic particles, they may also manifest at the molecular level—if coherence is preserved. Key processes like photosynthesis occur through quantum coherent energy transfer between photon energy and light harvesting complexes present in plants and bacteria that convert light energy into chemical energy. Tunneling, another quantum process, may be involved in electron transfer during mitochondrial respiration, may enhance the efficacy of enzymatic reactions, and may contribute to DNA mutations (Brookes 2017; McFadden and Al-Khalili 2018; Cao et al. 2020; Mazzoccoli 2022; Feng et al. 2022). In the cosmos, there are similar feedback loops between quantum and macroscopic forces: A star's fate—whether it becomes a white dwarf, black hole, or supernova—hinges on a delicate balance between quantum fusion energy and gravitational collapse.

Biological systems are not stars and feedback dynamics between the quantum level and the molecular components of a cell have not been observed. Biological phenomena are not isolated from the environment, and as Max Tegmark explained, if a quantum system communicates with a macroscopic system or environment, all the quantum properties vanish and everything decoheres (Tegmark 2020). Intriguingly, quantum entanglement of biophotons has been recently described in the myelin sheet surrounding neurons, and quantum entanglement has been proposed as a mechanism for their synchronization (Liu et al. 2024d). If these findings are confirmed, Max Tegmark will need to reconsider his statement.

Inside the cellular cytoplasm, there are liquid-like structured molecular domains—liquid-like droplets in which molecules with specific properties (i.e., disordered regions) aggregate through weak interactions. These condensates formed through liquid separation have been associated with a variety of key molecular processes, that is, enzymatic pathways, cellular signaling, DNA transcription, ribosome biogenesis, and stress responses (Hess and Joseph 2025). Cycles of coherence-decoherence occurring in vivo may explain some of the unusual behaviors of active matter, but they are yet to be found.

As a coda, a protein's three-dimensional shape is essential to its function, and while classical physics accounts for much of this behavior, quantum forces likely influence critical atomic interactions at extremely small scales, where quantum effects like wave-particle duality and entanglement could influence the folding process.

In addition, because they operate in a regime where quantum mechanics governs their structure, behavior, and interactions of the atoms composing them, proteins can be seen as quantum-classical interfaces; the effects of these interactions and the functions resulting from them can be observed in the macroscopic world under the form of biological processes (such as enzyme activity, gene expression, cellular signaling, etc.).

As the smallest functional units of life, proteins translate the probabilistic nature of quantum mechanics into the deterministic, observable outcomes we see in the macroscopic, biological world. Proteins represent a fascinating interface between the quantum world (where the fundamental behaviors of atoms and molecules are governed by quantum mechanics) and the macroscopic biological world (where those behaviors manifest in observable phenomena like metabolism, signal transduction, and muscle movement). They translate the probabilistic grammar of quantum mechanics into the structured language of biology. The quantum properties of molecules within proteins (like electron transfer and tunneling) influence their functions, but these effects are ultimately translated into classical biological processes. This makes proteins a bridge between the microscopic quantum realm and the larger-scale phenomena we can observe and study in biology. Proteins are the alchemical vessels through which matter becomes meaning.

8.11 Life Is Homeodynamic

In biology, since Darwin, evolution has been used as a sort of surrogate TOE explaining both biological structure from the microscopic to the macroscopic world and life behavior. Evolution has been used to explain how molecules turn into organelles, how organelles form cells, and then up, up, higher, and higher on the complexity scale, the behavior of multicellular organisms, like humans, for example. In life sciences, evolution theory has been used to explain almost every single process or event. It has been so successful that recently, a group of authors proposed that evolution is a universal law that rules both the biotic and the biological realms (Wong et al. 2024).

On the other hand, the evolutionary theory itself has evolved. As explained by several authors, in light of new research findings in the first quarter of the twenty-first century, it is becoming more and more clear that, in the context of holobionts or biospheres, other mechanisms such as horizontal transfer of information between same or different organisms' types complement the "struggle for life" and the intergenerational selection mechanisms envisioned by Darwin (Shapiro 2023; Lala et al. 2024).

Despite the multiple revisions of evolutionary theory, there are still fundamental questions that require clarifications, the classic example being the primary level of selection of evolution considered by Darwin to be the individual organism, which subsequently other authors considered to be the genes, the group of genes, the cells, the group of cells, etc.

Over the last century, evolution has been in biology a sort of "emperor of all theories," and other concepts and mechanisms have been less explored. Over the last decade, with the development of the evo-devo field, development is slowly sharing the center stage as many organismal behaviors or features unexplained by evolution become clear when development is taken into consideration.

Besides development, another process equally important for understanding the life is homeostasis. As explained by Genrikh Romanovich Ivanitskii, "Living matter faces two threats: proper determinacy and proper chaoticity. The trajectory of life runs between two extremes, as between Scylla and Charybdis. Were all life processes absolutely deterministic, living matter could not withstand environmental changes. Conversely, were they chaotic, it would lead to permanent conflict between the parts and the whole at both the organismic and biospheric levels; living matter would not have developed in this situation. A deterministic process with a significant chaotic component or a chaotic process with a significant deterministic component forms the mainstay of living matter kinetics" (Ivanitskii 2010).

On August 21, 1865, less than 6 years after the publication of the first edition of Charles Darwin's *Evolution of species*, Claude Bernard presented to the Académie des Sciences in France his newly published *An Introduction to the Study of Experimental Medicine* where he described the importance of "le milieu intérieur" (the internal environment)—an expression coined prior to him by Charles Robin—for an organism existence and survival. The homeostatic mechanisms allow the individual to maintain stability despite fluctuating but predictable environmental conditions, and it is maintained through various regulatory mechanisms (Billman 2020). The internal environment confers the body a certain independence in relationship with the external environment: "This kind of independence which the organism possesses in the external environment derives from the fact that within the living being the tissues are, in fact, removed from direct external influences and are protected by a veritable internal environment, which is constituted particularly by the fluids which circulate in the body [...]. In humans and in warm-blooded animals, the independence of the external environment and the internal environment is such that one could consider these beings as living in their own organic environment" (Holmes 1986).

Subsequently, in 1927, Walter Cannon complemented Bernard's concept of the steady state of the internal environment, by introducing the idea of dynamic control processes needed to maintain the stability of the internal environment. He coined the term "*homeostasis*" to describe the active regulation of different parameters maintained within pre-set limits (homeostatic range) for optimal functioning and survival of the organism despite changes in the environment (Cooper 2008).

Homeostasis can be broadly defined as the process by which organisms maintain a stable internal environment, or steady state, in response to external and internal changes. In an organism, homeostasis relies mostly on negative feedback loops, to maintain the stable internal environment by counteracting external environmental changes and bringing the system back to a set point; essentially, the body reacts to changes by initiating actions that reverse the change and restore balances. Homeostasis is the quasi-stable state of internal chemical and physical conditions of an organism. As explained by Ervin Bauer, homeostasis is needed in order to maintain the living systems far from equilibrium: "The living and only the living systems are never in equilibrium; they permanently invest work on the debit of their free energy budget against that equilibration which should occur for the given the initial conditions of the system on the basis of the physical and chemical laws" (Bauer 1967 cited by Grandpierre et al. 2014).

Homeostasis, robustness of a steady state in biological systems, has gathered much attention as compared to homeorhesis, the stability of the time course against the change of the initial conditions or perturbations, that has been less studied.

Importantly, homeostasis is a property of living systems in general, *and not only of organisms*. Feedback loops operate at multiple levels: cellular, tissue, organismic, ecological, and even planetary. Conrad Waddington expanded the concept of homeostasis to nonstationary systems and called it *homeorhesis*. *Homeorhesis* is the "property of a dynamical system to return to a particular trajectory after an external perturbation or despite the continuous presence of random noise" (Chuang et al. 2019). Various processes like "DNA replication, transcription, translation, protein folding, assembly of macromolecular complexes, cell division, and embryogenesis all exhibit temporal ordering, despite variations in timing and the influence of perturbations at many levels" (Chuang et al. 2019).

Confronted with both internal (i.e., mutations) and external (i.e., environmental pressures) stressors, living systems would quickly perish in the absence of a *third* type of homeostatic process. An appropriate name for this third process is *homeopoiesis*. "Homeo" in Greek means "similar" as opposed to "identical" and "poiesis" is the Greek word for "creation." *Homeopoiesis refers to the control that living systems have on creative processes and novelty*. An example of *homeopoiesis* present in the human organism, for example, is the somatic hypermutation induced during an infection in the antibody variable (V) genes of B lymphocytes by an enzyme called activation-induced cytidine deaminase (AD), allowing for the creation of more specific antibodies capable of attacking the infectious agent. Of note, only the B cells that produce strong affinity antibodies will be selected and survive; the other B cells with lower affinity antibodies will die by apoptosis. I emphasize that homeopoiesis is *controlled* creativity and not spontaneous, emergent creativity. Through homeopoiesis, creativity is governed—it is selection sculpting variation.

Homeostatic control mechanisms comprise three main interdependent components: a receptor, a control center, and an effector that work together to maintain the values of parameters important for the integrity of the living process within a certain range.

Similarly, Darwinian evolution also involves three components: *variation* within a species, that is accompanied by change visible at the organism level, and the *struggle* for existence favoring the survival and reproduction of the winning phenotypes that are selected and transmit the chance of having winning traits to their offspring through the third component, *reproduction.* A key point is that this mechanism favors the species as a whole and not individual organisms, since the *phenotypes are not directly transmitted to* the next generation as Lamarck thought.

As opposed to homeostatic processes that unfold rapidly within a single organism, evolutionary mechanisms span generations and scales. Evolution is not entirely random: While mutations may be stochastic, selection is inherently nonrandom—it is the environment choosing the traits that persist. This makes evolution both open-ended and channeled and both creative and constrained. Living organisms shape their environments (e.g., through niche construction) just as environments shape their evolution. This two-way influence is a form of *homeo*-control—a quasi-equilibrium of reciprocal causality.

Darwinian evolution can be understood as a form of *homeopoiesis*, that is, *controlled creation* or *a strategy to control novelty*—that is, to govern the flow of information within biological systems. In essence, through evolution, random variations are controlled by the environment by selecting the "winning" phenotypes. In evolution, over time, the environment and its pressures essentially feedback information on the species level as it is the species' genetic makeup that is altered over time based on which individuals with certain traits are more likely to survive and reproduce in a given environment, thus contributing to the global genomic pool of the species.

On the other hand, proposed by the theory of adjacent possible (TAP) of Stuart Kauffman (Kauffman 2000, 2019; Björneborn 2020), evolutionary processes may also launch, *indirectly*, *novel functions*, that cannot be predicted in advance. So evolution, besides controlling and channeling novelty, is also *genuinely creative and open-ended.*

The environment is simultaneously an active agent, sculpting the features and functions of the living systems and, at its turn, is shaped by the same living systems through niche formation, for example. This control that goes both ways, and is not perfect, is a *quasi-* or *homeo*-control.

Darwin himself considered evolution a dynamic process akin to a creative process (Beatty 2016, 2019). Importantly, as opposed to the way it is commonly understood, evolution is not about change. Evolution controls change, it channels it. It functions as a sieve. Evolution prunes, selects, channels, and creates through editing. In general, it does not invent, it discovers what is already in place. The fundamental principle underscoring evolution is combinatorics and editing, not de novo creation. It appears that the dice of evolution are loaded; despite the fact that genetic mutations are random, variations and phenotypes are not, as explained by Marc Kirschner and John Gerhart in the *The Plausibility of Life* (Kirschner and Gerhart 2006). For example, in cancer, in addition to genetic driver alterations, additional genetic and epigenetic cofactors may channel the retention of specific mutations in particular cell lineages and tissues. Also, the formation of preprimary tumor and premetastatic niches in cancer point to the presence of other nonrandom, nonevolutionary processes in oncogenesis.

Finally, given the widespread presence of horizontal exchange of genetic material between lineages (Shapiro 2022), the classic vertical tree of life becomes a horizontal rhizome resembling a tangled bush (Soucy et al. 2015). In cancer, horizontal gene transfer may be mediated by exosomes and may occur between tumor cells and the cells of the tumor environment, potentially influencing tumor development and progression (Valcz et al. 2022).

All three, homeostasis, homeorhesis, and homeopoiesis, involve mechanisms where the outcome of a process impacts directly on the continuation of that process, that is, they involve feedback loops. The term *homeodynamic* encompasses all three cyclic control processes that are fundamental to the existence and persistence of living systems.

In contrast to homeostasis, which denotes the relative constancy of an organism's internal environment and the maintenance of stable physiological states, the concept of homeodynamics offers a fundamentally different perspective. As explained by Lloyd et al., in living systems, systemic processes—through dynamic self-organization—can give rise not only to stable (monotonic) states but also to complex behaviors. At certain bifurcation points, qualitative shifts occur wherein the system transitions from monotonic behavior to periodic oscillations or even chaotic dynamics. Ultimately, homeodynamics describes the continuous transformation of a dynamic system into another that is similar—but not identical—at critical junctures (Lloyd et al. 2001). These transitions do not preserve exact identity (*tauto-*) but maintain functional similarity (*homeo-*). The system's evolution follows a cyclic trajectory, akin to the spiraling motion of a hurricane.

All three homeodynamic mechanisms are crucial for life to exist, persist, and evolve. Homeostasis maintains stability, while homeorhesis and homeopoiesis control fluxes of energy and information, respectively. All three homeodynamic processes are examples of *information over mater* processes. The environment acts as the Maxwell's demon, selecting the phenotypes and channeling evolution. In biological evolution, cyclic causality manifests when organisms modify their environment, which then changes the selective pressures acting on them (niche construction), while developmental processes shape phenotypic variation, which in turn influences evolutionary trajectories, and genetic and epigenetic systems co-regulate in recursive feedback cycles. Rather than a single linear causal arrow, evolution is a network of dynamic loops, with causation distributed across levels—from molecules to ecosystems.

Darwin introduced *evolution* theory to explain phenotypic *changes* through history (Darwin 1859). By placing the emphasis on dynamic *persistence*, evolution can be also understood to represent one of the several mechanisms that support persistence: others as important being homeostasis and homeorhesis (i.e., development). Life is a flow, and the diversity of forms supports it. Life needs evolution to persist. The living process cannot go on without the diversity of forms and vice versa. Evolution is one of sources of life's creativity. The living process can be described as dynamic and creative persistence (DCP). Cancer as a form of life itself thrives on the heterogeneity resulting from homeopoietic mechanisms.

For a multicellular organism, the main difference between homeostatic and homeorhetic processes and evolution is that homeostatic and homeorhetic processes are active within *the same* living organism during relatively *short periods of time* versus the components of classic evolution triad that act on *different* scales of living systems and involve both genotypic variation and phenotypic selection that may be separated by *long temporal gaps*. Darwin in his *Origin of the Species* looked at evolution through a very large temporal telescope focusing on gradual transformation of organisms and species over eons. He proposed natural selection as the process responsible for variation control. *Homeostasis* and *homeorhesis* occur much faster than *classical* evolutionary processes.

In essence, *life is homeodynamic*. Without the various homeodynamic control mechanisms, life would not exist. Biological systems are constantly exposed to internal (i.e., genetic mutations) or external (changes in the environment) variations and stressors and homeodynamic mechanisms are needed in order to maintain their quasi-stability.

Homeostasis, homeorhesis, and homeopoiesis are embodiments of cyclic causality, *occurring at different time scales*. Homeostasis maintains the status quo of the organism; homeorhesis, which in essence is an ensemble of feedback and feedforward loops between cells and tissues, ensures the smooth flow of developmental processes; and finally, the homeopoiesis of evolutionary processes is represented by back-and-forth causality cycles, from the genomic level to the phenotypic level and back. The cyclic causality processes underscoring evolution repeated since LUCA billions of times for more than 4 billion years has resulted in the extraordinary diversity of life on Earth today. The deeper meaning of Darwin's evolution theory is that the dynamic and creative persistence of life is trans-organismal. Evolution, as a homeopoietic mechanism, is directed to the living process, not to the individual organisms. Individual organisms are living processes instantiations, meaning through which the life process persists. The outcome of information scrambling at the genetic level during reproduction is information diversification at the species level, and it is beneficial for the species survival, not the individual. The foundation of life is the *global* dynamic persistence of the living process and not of *individual* selfish genes or cells or organisms, as has been proposed by different authors.

Another important homeodynamic mechanism shared by all living organisms is cognition. Both in its elementary forms present in bacteria and, in its more elaborated avatars exhibited all through the Eukarya kingdom, cognitive processes are also examples of cyclic causality. Francis Heylighen and Evo Busseniers define a system capable of specifically responding to perturbations as cognitive and, in their opinion, cognition is a prerequisite for the existence of an autopoietic system (Heylighen and Busseniers 2023). In the context of homeodynamic theory, cognition is one of the forms of cyclic causality that contributes to the living process' persistence.

Ecosystems are maintained through homeostatic mechanisms and their persistence takes priority over species survival. Finally, at the level of the whole biosphere, even bigger forces are at stake, and the maintenance of the global abiotic-biotic dynamic equilibrium supersedes all other aspects.

Life is an all-encompassing planetary process, with its persistence over time intrinsically tied to the biosphere as a whole. The biosphere operates as a homeostatic system, functioning simultaneously as both a hierarchy and a holarchy. Complex feedback and feedforward loops extend in all direction—top-down, bottom-up, and laterally—linking the biosphere with biological scales down to the molecular level.

Originating at the molecular scale, life forms a bridge between the quantum domain and the macroscopic world. Biorhythms serve as key integrative connectors: Nanorhythms synchronize biological functions with quantum phenomena, while circadian rhythms align them with planetary cycles. Life constructs its own temporal domain by introducing delays, segmenting time, and controlling processes through a balance of reversible and irreversible steps, as seen in metabolic cycles. Unlike physical systems governed by Le Chatelier's principle, life does not follow the most efficient path. *Life does not optimize, it compromises*. Life is characterized by balance—by a dynamic equilibrium that favors neither stasis nor unchecked change. Its persistence depends on homeodynamics, a principle that actively modulates the interplay between conservation and innovation.

Biological systems are thus defined not by rigid control mechanisms or linear progression, but by their capacity for adaptive stability. By integrating Cannon's concept of homeostasis (short-term physiological regulation) (Cannon 1932), Waddington's homeorhesis (stability of developmental trajectories) (Waddington 1957), and Darwinian evolution (long-term population-level change), the homeodynamic framework offers a unifying conceptual model. It positions life as a dynamic interplay between constraint and transformation, regulation and emergence, individuality and evolutionary history. This perspective is particularly relevant for interpreting both physiological and pathophysiological phenomena, as the capacity for adaptive regulation lies at the heart of both health and disease.

8.12 Cancer Homeodynamics

Multicellular organisms have evolved intricate stress control mechanisms to maintain integrity under various stressors. These mechanisms function at both cellular and tissue levels, balancing damage and repair. The diversity of programmed cell death modalities—apoptosis, necroptosis, pyroptosis, ferroptosis—contrasts with the relatively limited repertoire of pathways governing cell proliferation and maintenance, emphasizing the biological priority of matter recycling over unrestrained growth.

Both physiological (e.g., homeostasis, allostasis) and pathological processes (e.g., inflammation, fibrosis, thrombosis, cancer) can be viewed as adaptive responses to stress. However, when these responses become dysregulated or maladaptive, pathology emerges.

Cancer, in this framework, represents a failure of the homeostatic and allostatic systems across cellular, tissue, and systemic levels. It is the pathological expression of a survival strategy program, activated when higher-order regulatory mechanisms collapse. As Mark Vincent and others have noted (Vincent 2012), cancer is less a condition "caused" by specific factors than the unleashing of latent possibilities, a default state embedded in the biology of multicellular organisms (Huang et al. 2025a, b).

From this perspective, cancer can be seen as an opportunistic access to basic mechanisms of cellular resilience, at the same time instigating the whole organism to help the cancer survive. Cancer reactivates embryonic, gametogenic, and developmental programs—those active during early life stages and germline formation (Monk and Holding 2001; Asami et al. 2023; Bruggeman et al. 2020). These include genetic programs present during meiosis, zygotic genome activation, and epithelial-to-mesenchymal transition (EMT)—mechanisms used in normal development and repair but hijacked in cancer to facilitate invasion, metastasis, and immune evasion (Youssef et al. 2024).

The imbalance observed in cancer between the unicellular and multicellular gene expression programs (Trigos et al. 2024) is not binary, but fluid. Cancer cells demonstrate remarkable plasticity, shifting between developmental and regenerative EMT trajectories, depending on environmental cues and acquired traits. One trajectory supports tumor progression and metastasis through embryonic-like reprogramming, while the other reflects tissue-repair responses to oncogenic damage. Both are context-dependent and dynamically regulated by cancer-stroma interactions.

Cancer evolution represents a form of *allopoiesis*—the pathological counterpart of *homeopoiesis*—but, to simplify, we will further use the term cancer *homeopoiesis* to refer to the process of controlled creativity of the malignant progress. *Homeopoiesis* is essential for cancer persistence. On one hand, cancers may exhibit a sort of fast-forward evolution (Pressley et al. 2021); on the other hand, they are able to exploit homeostatic processes (Falco et al. 2016) and several embryonal (Lee 2022) and placentation (Costanzo et al. 2018) programs as survival mechanisms. Cancer *homeopoiesis* also explains "the just right" hypothesis of chromosome instability (CIN) in which beyond a threshold, CIN becomes deleterious for cancer cells (Hosea et al. 2024). In recent years, it became apparent that only a subset of cancers evolves through Darwinian evolution and other processes, that is, neutral evolution, development, and Lamarkian epigenetic adaptations are also actively involved; therefore, the notion of *cancer homeopoiesis* better captures cancer's creative mechanisms. Cellular processes need to be well balanced for cancer cells to survive. The telomere crisis, almost universally present during carcinogenesis that induces genome instability is immediately followed by long-term genomic stabilization in order to ensure cancer cells survival.

Ultimately, the cancer process represents the convergence of homeostatic failure, developmental reprogramming, and evolutionary adaptation. It is a cellular survival strategy that transforms a single mutated cell into a multiclonal, tissue-invasive, system-disrupting entity. Cancer does not invent new biology—it co-opts and combines existing mechanisms to persist and expand. Cancer is the ultimate opportunist. It combines developmental, adaptive, and evolutionary forces into a distinct, emergent process that exploits the plasticity of living systems.

In this sense, carcinogenesis is where Haeckel meets Darwin meets Lamarck: Development, evolution, and adaptation interweave into a single pathological trajectory. When challenged by therapeutic pressures, cancer accelerates its progression by forming polyploid giant cells, increasing variability and resistance. Evolutionary mechanisms have no direct impact on a human being's existence. By contrast, evolution is very active at the level of a cancer cell. While it mimics evolution, carcinogenesis is not merely accelerated evolution—it is a distinct life strategy, rooted in the logic of survival and system-level persistence.

The three homeodynamic mechanisms, homeostasis, homeorhesis, and homeopoiesis, shape the trajectory of a cancer cell. The plasticity and resilience of cancer cells and cancerous tumors is explained by the fact they can adapt quickly to stressors (homeostasis), they can change trajectories fast (homeorhesis), and they can exhibit what has been called a fast-forward evolution (homeopoiesis).

These three forces acting at the level of cancer cells are depicted in the cancer homeodynamics cube (please see Fig. 8.4). Of note, the cells depicted are not floating in a vacuum; for simplicity, an environment component was not included. A similar cube can be drawn also for *a group of cells* illustrating the group phenotypic composition (GPC) theory of Frédéric Thomas et al. (2024) where homeostasis, homeorhesis, and homeopoiesis forces apply to a group of cells instead of one. Homeostasis, homeorhesis, and homeopoiesis are all processes that maintain a fine balance between stability and change in living systems ensuring their dynamic and creative persistence. For clarity and conceptual unity, I propose the term *cancer homeodynamics* to encompass all these three forces. They represent life's ability not just to persist, but to persist creatively, with resilience and adaptability. Cancer cells are in continuous transformation—the siblings of a cancer cell are similar but never identical to their parent. This perpetual transformation is illustrated in the homeodynamic cube, below (please see Fig. 8.4).

Cancer is on one hand *caused* by disruptions of homeodynamic mechanisms at the level of the cells, tissues, and organism, and on the other hand, cancer is *causing* a disruption in the homeodynamic mechanisms of other normal cells, tissues, and organism. *Therefore, cancer, both as a life form and as a disease, is essentially a homeodynamic process.*

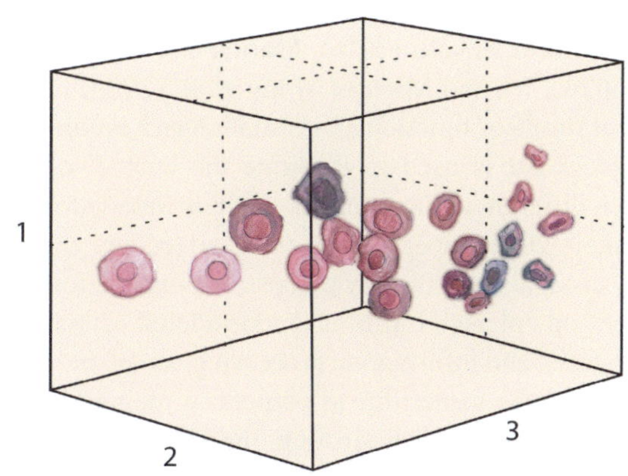

Fig. 8.4 Cancer homeodynamics cube with three axes: *1* = homeostasis; *2* = homeorhesis; *3* = homeopoiesis. (Courtesy of Ana Tourian)

8.13 Bioenergy/Bioinformation/Biostructure

In 1971, Tibor Gánti introduced the chemoton model, proposing that all living organisms have a basic autocatalytic subsystem consisting of metabolism (bioenergy), a replication process (bioinformation), and a membrane enclosing these functions (biostructure) (Gánti 1987, 2003). Recently, Nedilijko Budisa et al. reformulated the chemoton theory and proposed four pillars/cornerstones of cellular life (energy, metabolism, information, and form) in the context of chemoton (Budisa et al. 2020).

As Radu Popa noticed, "*living state is independent of any particular molecules*" (Popa 2004). James Grier Miller found similarities between several layers of life organization, that is, cells, organs, organisms, and communities, and described them all under the large umbrella of "living systems." This sort of functional fractal structure of living systems suggests the existence of general laws or principles that apply equally to the cellular level (both normal and cancer cells) and to higher organizational levels (the human organism and the cancer system).

Life is a cyclical interplay of organizing processes (biostructure), energy transformation (bioenergy), and information processing (bioinformation). Organization/energy/information are codependent. Living systems transform the flow of matter/energy/information present in their inner and outer environments into organized biostructure/bioenergy/bioinformation flows.

Different theories have tried to reduce one of the members of the bioenergy/bioinformation/biostructure triad to the others and to build up a life theory from one member, considering the others as derivative. Erwin Schrödinger, Ilya Prigogine, and, more recently, Douglas Wallace and Nick Lane have used energy as the life global explanandum; Matteo Mossio has used organization; William Miller and John Torday and James Shapiro and Perry Marshall have used cognition and consciousness; Robert Pascal and Addy Pross have used dynamic kinetic stability; and more recently, Carlo Rovelli (2016), Marina Cortês et al. (Cortês et al. 2024a, b), and Michael Wong et al. (2023) have used different forms of functional information and evolution. This incomplete list is used to illustrate the lack of consensus between different authors regarding life's foundation.

In their 2023 paper, Wong et al. advocated that active energy dissipation is the key process responsible for life persistence on planet Earth and considered autocatalysis, homeostasis, and information processing, as secondary functions able at the same time to perpetuate themselves and to propagate the act of dissipation through space and time (Wong et al. 2023). They also noted that natural systems "*evolve by selecting advantageous configurations with respect to systemic persistence*" (Wong et al. 2023). The focus on the prior statement is on "advantageous configurations" and postulate that these configurations are "advantageous."

Prior to Wong et al., Addy Pross proposed a chemical explanation of life and introduced dynamic kinetic stability (DKS) as the guiding principle able to bring life to inanimate matter (Pross 2003, 2011, 2012; Pross and Khodorkovsky 2004; Pascal and Pross 2015, 2019). Using the DKS concept, Robert Pascal and Addy Pross formulated an extension of the second law of thermodynamics that they called *the persistence principle:* "All material entities are driven from less persistent to more persistent forms. […] For replicative systems in the DKS state, the persistence principle enables the direction of change to be specified; that is, from less DKS stable to more DKS stable" (Pascal and Pross 2015, 2019).

8.14 Life's Architecture at Its Minimum

The concept of a minim-gene-set has been defined by Eugene Koonin as "the smallest possible group of genes that would be sufficient to sustain a functioning cellular life form under the most favorable conditions imaginable, that is, in the presence of a full complement of essential nutrients and in the absence of environmental stress" (Koonin 2000).

A theoretical minimum set of 206 genes was proposed by Rosario Gil et al. (2004) and this is close to the smallest natural genome of 182 genes found so far in *Carsonella ruddii*. *Carsonella* is a bacterial symbiont that lives in sap-feeding insects known as psyllids. Importantly, *Carsonella* lives inside a structure called a *bacteriocyte* and her genome lacks genes for cell envelope biogenesis and metabolisms of nucleotides and lipids. These genes are thought to have been transferred from the genome of a *Carsonella* ancestor to the genome of a *psyllid ancestor* and are now expressed under control of the host nucleus, similar to mitochondria in a Eukarya cell.

It is thought that a minimal genome able to sustain life should have approximately 300–500 genes (Koonin 2000). In 2016, JCVI-Syn3.0, the smallest artificial organism capable of basic life functions (i.e., replicating DNA, transcribing RNA, translating proteins, and undergoing cell division), was created by a group led by Craig Venter. JCVI-Syn3.0 has a smaller genome (531,490 bp) than that any other known organism, and it contains a total of 473 genes (438 protein-coding genes and 35 RNA-coding genes) (Hutchison et al. 2016). The smallest known organism capable of independent growth and reproduction, *Mycoplasma genitalium*, a parasitic bacterium with the size of approximately 200 to 300 nm, has the smallest known genome of 450 genes, very close to this figure. Interestingly, when performance was assessed by relative fitness, the minimal cell evolved 39% faster than

Mycoplasma mycoides, the non-minimal cell from which it was derived (Moger-Reischer et al. 2023).

It is important to note that the concept of a minimal gene set has primarily been explored within the context of prokaryotes. Yet the question arises: Are genes absolutely necessary for a cell to be considered alive? Nature itself provides a profound answer. Both red blood cells and platelets (despite lacking nuclei) are considered living entities. Red blood cells lack mitochondria and survive for up to 120 days through anaerobic glycolysis. Although these cells lack DNA, they still maintain a functioning metabolism, cellular membranes, and the ability to process information. Their longevity is limited, with platelets persisting for around 10 days and red blood cells for approximately 120 days, before being cleared by the spleen's macrophages. Conversely, prions, which are misfolded proteins that induce misfolding in other proteins of the same kind, can multiply in the absence of DNA, exhibiting a remarkable ability to persist over extended periods (Eigen 1996; Vázquez-Fernández et al. 2017). Similar to viruses, prions are parasitic entities, incapable of replication without the cellular machinery that produces the proteins to which they attach.

This raises the question: What are the fundamental ingredients necessary for the existence and persistence of unicellular and multicellular life?

The answer appears to encompass three essential requirements: a source of energy capable of being transformed into *bioenergy* through metabolism, a *bioinformation* storage and processing system (including DNA and epigenetic mechanisms), and interfaces both within the organism (such as compartmentalization) and between the organism and its environment (such as the membrane, which forms the biostructure). The *biostructure* incorporates both spatial and temporal dimensions, playing a crucial role in maintaining the dynamic processes of life. Besides these three requirements, a fourth fundamental requirement that should be always taken into consideration is *a supportive environment* present both *inside* and *outside* the life form.

Importantly, living processes are not isolated phenomena. *Life is a network.* Life begets life and thrives only in the presence of life. The external and, in the case of multicellular organisms, the internal environment are swarming multiple other life forms. An often overlook characteristic of life is its codependency on other life forms. Completely autonomous life forms do not exist. Autonomy of living systems is always relative, and their survival depends on their integration into larger networks: cells into colonies or multicellular organisms, organisms into communities, communities into ecosystems, and ecosystems into a biosphere where all these life forms interact.

As emphasized by John Smith and Howard Morowitz, the necessity for integration into the biosphere is an essential requirement for the existence of any life form on Earth, and the codependency of all life forms of the biosphere has been used by the authors as a basic definition of life. Thus, life, in its most fundamental expression, is an intricate, interconnected network of bioenergetic, bioinformatic, and biostructural processes. Each life form, whether cellular or multicellular, is embedded in a dynamic interplay that transcends its individual existence, contributing to and drawing from the broader network of life on Earth.

8.15 Life's Basic Design

Life's dynamic network can be envisioned as a combination of a primary set of relatively conserved and stable components (the core set) that are represented by structures and processes playing a key role in physiology, and a second set of regulatory elements that is more flexible and adaptable and ensure the stability of the core set. Similarly, John Gerhart and Mark Kirschner, as part of their *theory of facilitated variation* (Gerhart and Kirschner 2007), to explain the origin and nature of novelty in evolution, divided an organism's architecture into conserved core processes (unchanging) and into variable regulatory changes (changing). They highlighted the need for the core processes of having special features: "The number and kinds of regulatory changes needed for viable phenotypic variation are determined by the properties of the developmental and physiological processes in which core components serve, in particular by the processes' modularity, robustness, adaptability, capacity to engage in weak regulatory linkage, and exploratory behavior. These properties reduce the number of regulatory changes needed to generate viable selectable phenotypic variation, increase the variety of regulatory targets, reduce the lethality of genetic change, and increase the amount of genetic variation retained by a population" (Gerhart and Kirschner 2007). The core processes referred to biochemical, genetic, cell biological, physiological, and developmental components that were unchanged in all bilateral animals since the Cambrian period. The authors provided a comprehensive list of these conserved processes:

- Components of energy metabolism.
- Biosynthesis of the 60 building blocks.
- DNA replication.
- DNA transcription to RNA.
- Translation of RNA to protein.
- Lipid membrane biosynthesis.
- Transmembrane transport.
- Components of the formation of microfilaments, intermediate filaments, and microtubule cytoskeletons.
- Motor proteins moving materials along the cytoskeletons.
- Contractility processes.

- Movement of the cell by cilia and ruffling membrane action.
- Shuttling of materials between intracellular organelles.
- Phagocytosis.
- Secretion.
- Chromosome dynamics.
- A complex cell cycle driven by protein kinases and protein degradation.
- Sexual reproduction with meiosis and cell fusion.
- Components of 15–20 cell-cell signaling pathways.
- Apical basal polarization of cells.
- Junction formation and cell adhesion processes.
- Epithelium formation.
- Specialization of cells toward physiological ends.
- Some developmental processes of the single-celled egg to the multicellular adult; components of complex developmental patterning, such as anteroposterior axis formation (Wnt/Wnt antagonist gradients) and dorsoventral axis formation (Bmp/antagonist gradients).
- Inductions.
- Complex cell competence.
- Additional specialized cell types.
- Formation of the body plan's map of selector gene compartments (both transcription factors and signaling proteins).
- Carious regulatory processes (Gerhart and Kirschner 2005).

They argued that due to the versatility of these processes, in animals in the post-Cambrian period, there was no need for further functional innovation at the gene product level (protein and functional RNA evolution), and instead further evolution was the result of anatomical and physiological traits that have evolved since the Cambrian period and are the result of regulatory changes in DNA sequences for regulatory elements of DNA, RNAs, and proteins (Gerhart and Kirschner 2007).

Applied to cancer, the *theory of facilitated variation* suggests that cancer cells contain a core of relatively conserved features as well as some variable regulatory elements responsible for cancer progression. In other words, cancer, fundamentally, comprises both of a universal core of relatively conserved elements and, also, some other variable, idiosyncratic components, making every tumor unique. This observation is confirmed by the fact that many of the driver genetic mutations and alterations in cancer are transcription factors, and the presence in cancer of significant epigenetic changes including universal ncRNA involvement. As recently shown, enhancer reprogramming is a key driver of transcriptional dysregulation in cancer, shaping cell fate decisions and promoting the phenotypic plasticity that enables cancer cells to adapt and evolve (Sun et al. 2023a; Yang et al. 2025). The main difference between eukaryotes and prokaryotes is the significant increase in control elements—ncRNAs, for example—in eukaryotes. It is precisely the cellular control elements (i.e., TP 53 mutations, ncRNAs alterations, etc.) that are altered in cancer.

In the context of cancer, Andrea Califano and Mario Alvarez proposed a core of functional master regulator (MR) proteins, theorizing that they are mostly dysregulated in a posttranslational manner, the aberrant activity of which is both necessary and sufficient to maintain a cancer cell state (Califano and Alvarez 2017). Importantly, the MR is conserved within a particular tumor type and represents a sort of bottleneck situated between the variable number and types of genes mutated and the programs that implement and maintain tumor cell homeostasis. In 2011, a Chinese research group described 11 transcription factors playing a significant role in colorectal cancer associated with 18 independent gene modules (Ye et al. 2012).

A basic vehicle-replicator architecture for life was initially suggested by von Neumann (1951). A tetradic structure for a life generator able to perceive and adapt to the environment was subsequently proposed by Barham (1996). This structure contains *two external energy constraints sources* (one high, the other low) that interact with each other, an *epistemon* (i.e., a sensor), able to change its state and induce a biological action once it receives a signal from the low-energy constraints source, and a *biologic oscillator* that originates the biological action. More recently, based on von Neumann's vehicle-replicator concept, Chiara Marletto (2015) also sketched the architecture of a constructor engine composed of three interconnected components: a *replicator* R, the *copier* C, and the *constructor* B (the last two being parts of a *vehicle* V). Predrag Slijepcevic also proposed a tripartite design of a cell comprised of an *"engine"* that explores and exploits thermodynamic imbalances in the environment to carry out work, a *"machine"* that uses the energy obtained by the engine to maintain the structure of the cell, and an *information-processing module*, which coordinates and synchronizes actions of the engine and the machine. The information-processing module consists of the cell sensing system integrated with the cell genome and the cell molecular network that transmit signals from the sensing system to the genome (Slijepcevic 2021). Of note, both the constructor model of Chiara Marletto and the thermodynamic model of Slijepcevic relate to unicellular (i.e., bacteria) and not multicellular life forms. Von Neumann's original aim was not to show that machines can self-replicate—a task he considered trivial—but to develop a logical theory of life.

In the fifth of his Illinois lectures, his focus was on the logical conditions for matter to be evolvable through inherited variation rather than on simple self-replication (von Neumann 1966). As opposed to von Neumann's goal, these modularly models reflect an object-oriented understanding of life.

An alternative design would be represented by a model based on both objects *and* processes instead of static modules. The interplay between static and dynamic components is one of the key characteristics living processes.

8.16 Biointegration

Living systems transform energy and information in ways that differ fundamentally from machines. To capture these distinctions, I propose a new lexicon: *bioenergetics*, *bioinformation*, *biostructure,* and *biointegration*, concepts that transcend traditional computation. *Biostructure* encompasses both organic forms (cytoskeletons, nuclei, organelles) and inorganic substrates (such as water and ions). Living cells function as bio-batteries, with electrolytes and molecular architecture forming the infrastructure, with bioenergetics continuously providing and transforming matter into energy and vice versa, and bioinformation providing control and guidance. *Biointegration* refers to the dynamic, holistic union of all these processes into a functioning, living entity.

In physics, complementarity refers to dualities—wave and particle, space and time. In biology, it applies to *form and process*—anatomy and physiology. A biological system must possess stability through forms and adaptability through processes. These are not dichotomies, but dynamic unities. Consider genetic information flow: DNA's stability ensures continuity, while RNAs shorter lives underscore information exchange and transformation through catalysis (Cech 2024). Proteins, too, embody dynamic potential, shifting conformation, responding to signals, and executing the body's manifold programs.

Genes can be classified based on essentiality: essential (E), nonessential (NE), and quasi-essential (QE), the latter causing impaired but not lethal phenotypes when disrupted (Glass et al. 2017). These distinctions echo the deeper principle: Not all components are equally central, but network topology dictates robustness.

Scale-free networks—marked by a few hyper-connected hubs and many peripheral nodes—are ubiquitous in biological systems, from gene regulation to tissue organization (Wang and Chen 2003). Within a cancer cell, oncogenic hubs dominate protein interaction networks. But a hub may be also represented by a key *process*, such as the shift from localized growth to metastasis. Thus, we must consider both structural and processual hubs—objects *and* transitions.

To encapsulate this duality, I introduce the concept of the *biogenerator*: a coupled network of essential structures (DNA, membranes) and processes (transcription, metabolism) enabling persistence through time. Disruption of any core element may trigger systemic collapse, like a domino cascade. As Wallace noted, "life requires not only anatomy but also the vital force that animates anatomy"—a biointegration of structure, energy, and information (Wallace 2011).

Biogenerators function as central nodes of biological existence, where bioenergy and bioinformation converge within biostructure, forging the dynamic and evolving unity that is the living cell. It mediates the cell's interface with its environment, sustaining coherence and identity. A biogenerator is defined as bioenergy/bioinformation processes together with biostructural components that are essential for life and are conserved within both the life cycle of an organism and within the species to which the organisms belong. The biogenerator of a bacterium, for example, is represented by the minimum processes and structures essential for the bacteria's life. The bacteria biogenerator is preserved during an individual bacteria's existence and during the process of bacteria division and, also, conserved through all bacteria species evolution on Earth. The biogenerator concept is related to the theory of Gerhart and Kirschner's conserved core and, also, to the minimal cell (Glass et al. 2017; Lachance et al. 2019). DNA and the transcription/translation apparatus, essential metabolic reactions, and bacteria's membrane are examples of a bacteria's biogenerator's components, minimal requirements for a bacteria to exist and persist.

The biogenerator, together with other more variable components of the dynamic network of an organism, ensure the organism's persistence and its adaptation to the environment. The biogenerator is essential for the organism's vitality and is relatively conserved not only during the organism life cycle but through long temporal spans transcending sometimes the life of the species, as opposed to the more variable components that are not. The history of life on Earth is also the history of the formation of different types of biogenerators through adding several layers and modules to the primordial electrochemical biogenerator (please see Fig. 8.5).

The biogenerator's integrity is essential for the organism's vitality. The story of life on Earth is also the history of sequential biogenerators.

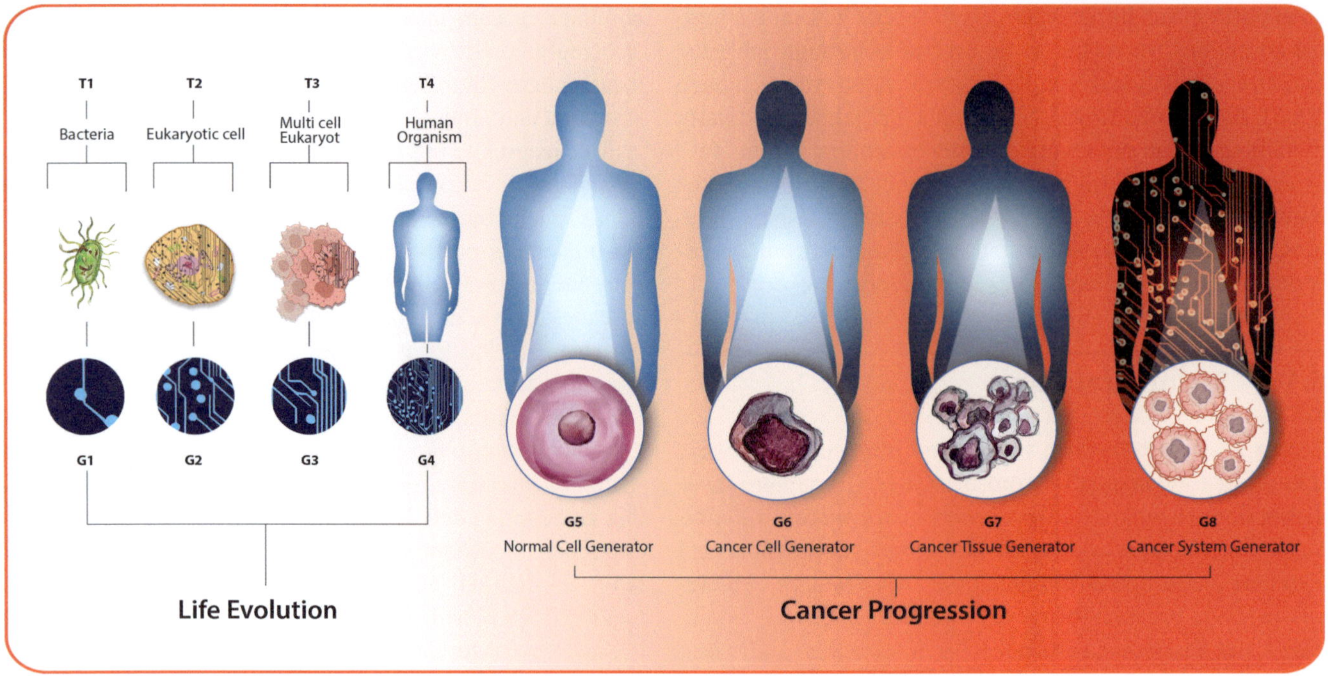

Fig. 8.5 Life evolution and cancer biogenerators. (Courtesy of Ana Tourian)

8.17 Biogenerators

Organisms and both their internal and external environments coevolve together. We are not only part of the nature or part of the universe; nature and universe are also part of us. The geosphere and biosphere are symbiotic. The environment exists both outside and inside an organism. In order to function, organisms also need to be integrated into the external environment, and, at the same time, the internal environment supports the living processes of the organisms. A cell is the living proof (pun intended) of the biotic and abiotic unity. Beyond the static form seen under the microscope, a cell is essentially a stream of coherent electrochemical interactions integrated with biological processes. Life was formed through a symbiosis of abiotic and biotic mechanisms and processes. A cell contains both a colloidal alphabet soup of atoms, ions of molecules of different shapes and sizes in various amounts, and several biologic engines that read, write, and erase several biological codes of DNAs, RNAs, and proteins. *Abiotic electrochemical and biological processes need to be integrated for life to exist.* A cell membrane has thousands of sensors for nutrients, oxygen, and mechanical pressure and is able to adapt within fractions of seconds to environmental challenges without the need to synthetize new proteins—a process that takes much longer. This electrochemical reactivity is a sort of basic life biogenerator universally present in all the cellular living kingdom. There are cells that can live for days (in the case of platelets) or months (in the case of red blood cells) without having a DNA code. A DNA code is an important requirement for life to exist, but it is not the only requirement. As important is the basic electrochemical biogenerator that behaves like a sort of battery, that needs liquid (intracellular water), electrolytes and electricity generated by the system of cellular membranes, in order to function. A cell dies immediately if the basic biogenerator stops functioning. Mitochondria and nuclei are additional layers of control of energy and information fluxes that made possible the development of complex life forms. In eukaryotic cells, mitochondria, with the help of the DNA code, form a sort of supplementary turbo energy generator, controlled and controlling the nucleus, closely connected with the primordial biochemical energy generating system present in the cytoplasm. Multicellular life represents a fundamental transition that is made possible through *communication* and *specialization*.

The *basic biogenerator*, *G0*, may have been represented by chains of biochemical reactions, the ancestors of the cellular metabolism (Smith and Morowitz 2016). Importantly, these biochemical reactions may have been present in systems without membranes. *G0* exists independently of the DNA and mitochondria. In the human organism, for example, red blood cells can live for approximately 120 days and platelets for 8–10 days. Of note, red blood cells lack DNA and mitochondria and platelets lack DNA.

In the ancestors of the first cells, *G0* was integrated with other activities, that is, transcription, translation, and DNA copying leading to the formation of the *G1* generator. The *first biogenerator*, G1, as we know it, present in LUCA and then in bacteria and archea, contained both a metabolic component, the informational component represented by the DNA code, the transcription-translation machinery, and a semipermeable cellular membrane. The *second biogenerator* added power to the system by the symbiosis between and archea and a bacteria, and the *G1* generator was integrated with both the DNA code and the mitochondria forming the *G2* generator. All three, the nucleus, mitochondria, and the cell membrane, are needed for complex multicellular life existence. The *G3* generator appeared first in multicellular eukaryotes by including on the top of *G2* a layer of genes important for cell-cell interaction and development.

Separate from cellular life forms, there are bacteriophages and viruses that represent DNA programs completely dependent on the *G1* generators (for bacteriophage) or dependent on the *G2* generators (for viruses) for their reproduction.

Importantly, besides these microscopic life generators, there are also *G4* macroscopic generators present in multicellular organisms. In the human organism, an important component of *G4* is represented by the network of vital organs. According to James Grier Miller, the design of both uni- and multicellular organisms is similar, that is, the *G1-G4* generators have similar components and architecture (Miller 1978).

Besides the microscopic and macroscopic generators of normal cells and organisms present on Earth, there are also *pathological biogenerators* of conditions like cancer represented by oncogenic networks of genetic, epigenetic, metabolic, biochemical (pH), and biophysical (membrane potential) and other components responsible for several functions, that is, division, maintenance, secretion of different pro-malignant factors, etc., of cancer cells (please see Fig. 8.5). The design of normal cells and cancer cells biogenerators' design is different—the plasticity of the information and metabolic networks of cancer cells allows them to switch back and forth between a multicellular *G3*-like biogenerator system and a unicellular *G2*-like biogenerator. Cancer cells can seamlessly adopt both unicellular and multicellular life forms and also oscillate between dormancy and activity. Cancer cells originate from normal cells but the *G6* generator of cancer cell is very different from the *G5* of normal cells, present in a multicellular organism, from which they originate. Cancer can be considered an adaptation (or maladaptation) to different internal (i.e., DNA mutations) or external (i.e., decreased stromal ph or O2) cellular stressors.

Importantly, the *G0* biogenerator of cancer cells is also modified compared to normal cells, that is, the intracellular ph of cancer cells is more alkaline, the external membrane is relatively depolarized, etc. These differences can be therapeutically exploited. Cancer treatments have been already proposed using deuterium-depleted water (Lu and Chen 2024), cancer alkalinization (Hamaguchi et al. 2022), or ion channels targeting (Capatina et al. 2022). Bioelectric manipulations were also shown to reverse the metastatic phenotype in a Xenopus model (Chernet and Levin 2013; Levin and Martyniuk 2018; Payne et al. 2019; Levin 2021).

Cancer cells are able to create a viable living system despite the significant corruption of the DNA information through mutations and translocations and chromosomal aneuploidies. The extraordinary variety of the DNA alterations and aneuploidies present in cancer genomes is in sharp contrast with the relatively uniform trajectory of cancer cells and tumors. This points to the derepression of a default survival program (Vincent 2012) that may be represented precisely by *G6*.

G6 contains the necessary modules for the function of division, invasion, metastasis, and pro-metastatic niche formation. Cancer progression is related to the transition to the *G7* and *G8* macroscopic cancer biogenerators that contain besides cancer cells other *stromal cells* that support cancer progression, that is, cancer-associated fibroblasts (CAF), macrophages, Tregs, neutrophils (the *G7* generator), and cells present at a distance in other organs such as myeloid cells in the bone marrow that can infiltrate metastatic sites, or lung macrophages that can be instigated by tumor cells to secrete exosomes, contributing to cancer-induced thrombosis—components of the G8 generator (Lucotti et al. 2025). The G8 generator is part of the cancer system presented in Chap. 11.

8.18 What Type of Life Form Is Cancer?

Life on Earth is classically divided into four kingdoms (viruses, bacteria, archea, and eukarya) and three life types: molecular life (viruses), unicellular life (bacteria and archea), and multicellular life (eukarya). Viruses and unicellular and multicellular life forms all have developmental cycles. Of note, viruses can be considered life forms only in combination with viral factories or *viroplasms* (Novoa et al. 2005; Netherton et al. 2007). Similar to the viral cycle, cancers cells may also exhibit a sort of life cycle as they can switch back and forth between an epithelial to a mesenchymal phenotype (the EMT-MET process), they can come in and out of dormancy, and they cocreate cancer niches both within the organ in which they first develop and at distance.

Dictyostelium, a non-metazoan eukaryote belonging to the amoebozoa, exists in two distinct phenotypes: a free-swimming multicellular amoeboid phenotype and a colony-forming unicellular phenotype. Like *Dictyostelium*, cancers can have a multicellular-like organ with a heterogeneous population of cells with different phenotypes that can perform different functions and a unicellular-like phenotype with one or more interacting colonies that may compete or collaborate.

"Under conditions of abundant nutrients, the slime mold remains in its free-swimming amoeboid form; under conditions of low food abundance, the amoeboid free-swimming phenotype forms colonies. Logic would dictate that this organism evolved under high nutrient abundance conditions, and therefore its *unicellular form is the primary phenotype,* the colonial form being derivative" (Torday and Rehan 2017).

Similarly to Dictyostelium, cancer cells may be able to switch between unicellular and multicellular phenotypes depending on environmental nutrients availability.

Cancer may represent a fourth life type, a *dyskariota,* as Mark Vincent suggested (Vincent 2010, 2011) with unique properties compared to the other living processes. Key characteristics of dyskariotas are the ability to switch back and forth between being unicellular and multicellular life forms, and their variable number of chromosomes. Cancer's dyskariota is created by and thrives within the biosphere of the organism that it eventually causes to collapse. The Biosphere 2 experiment comes to mind.

Biosphere 2 is a large-scale scientific experimental and ecological research facility located in Oracle, Arizona, originally designed to explore the viability of closed ecological systems to support human life. Constructed between 1987 and 1991 and spanning 3.14 acres, it was named "Biosphere 2" to distinguish it from Earth's biosphere, considered Biosphere 1.

The most famous experiment that took place in Biosphere 2 began in 1991, when eight individuals were sealed inside the structure for 2 years in an attempt to create a self-sustaining, closed ecological system. The facility included seven biomes such as a rainforest, ocean with coral reef, desert, savannah, mangrove wetlands, agricultural area, and human habitat. The aim was to study the interactions within ecosystems and assess the feasibility of human survival in closed environments—potentially relevant for space colonization. The Biosphere 2 experiment (1991–1993) ultimately did not meet its goal of creating a fully self-sustaining, closed ecological system, and the human participants had to be evacuated after 2 years. Several key factors contributed to its failure, mainly declining oxygen levels from 21% to about 14% due to its consumption by soil microbes and its reaction with exposed concrete, which absorbed CO_2 and trapped oxygen in the form of calcium carbonate, and also due to carbon CO_2 fluctuations mainly related to the absence of proper buffering systems, like the Earth's deep oceans and large forests (Nelson 2021).

As demonstrated by the Biosphere 2 experiment, in an enclosed system, the climate parameters coevolve with the living forms that can significantly influence the whole biosphere (Battistoni 2019).

It took the whole geosphere and eons of time for life as we know it to develop on planet Earth. It takes the human organism several months to develop. Once formed, the biosphere of the organism is closely interrelated with its constituent components. It takes cancer sometimes decades to develop and, through its history, cancer is codependent with the organism's biosphere, which it will eventually pollute and destroy.

8.19 Sing Me the Body Electric

One of the most, if not the most important ingredient for life is electricity. An electric fields-based life origin theories has been recently proposed (Nunn et al. 2022). The universal death criterion in humans is the absence of electrical activity of their hearts (Parish et al. 2018). Thermodynamic disequilibria present in the environment are converted by life into redox disequilibria. Both energy sources used by life, light for phototrophs and chemicals for chemotrophs, are always linked or converted to redox disequilibria which will cause electron exchanges (Hay Mele et al. 2023). The best-known example is the OXPHOS process present in the inner mitochondria membranes. In a mitochondrion's cristae, the overall redox disequilibrium is fragmented into several intermediary steps between an opening electron donor oxidoreductase and an ultimate electron acceptor oxidoreductase (Ahmad et al. 2025). Some of the oxidoreductases participating in the electron transport chain are proton pumps that use the energy of the passing electron to move hydrogen ions from the inner liquid of the mitochondrion into the mitochondrial intermembrane space, creating a chemiosmotic proton gradient that will be used to power the ATP synthase, a transmembrane protein of inner membrane, that will convert ADP to ATP. The decoupling of the electron flow through the electron transport chain from the proton transport across the mitochondrial membrane is able to convert a scalar (directionless) redox chemical reaction into a vectorial (gradient-forming) process, producing chemical and mechanical work (Hay Mele et al. 2023). Most oxidoreductases are metalloproteins that include in their catalytic centers catalytic centers metals such as Fe, Cu, Mo, W, Zn, V, Mn, Ni, Mg, Co, and Se, and nonmetals like S. Metalloproteins are essential ingredients not only for the energy production inside the cells but also for the functioning of our entire planet, since the majority of the earth biogeochemical cycles are controlled by them (Smith and Morowitz 2016; Hay Mele et al. 2023). *Unsurprisingly, trace metals that are essential for life also play important roles in cancer.* The metabolism and homeostasis of metals (metallomics) is correlated to both cancer incidence (Callejón-Leblic et al. 2019; Ge et al. 2022; Zhang et al. 2022d) and progression (Serra et al. 2020; Siddiqui et al. 2023).

Battery-like structures using electricity are fundamental for both cellular and multicellular life. All cells of the human body are batteries-like miniature electrical circuits and the whole human organism can be considered a gigantic battery (Adee 2023). The body uses electrical charges and gradients to power various biological functions, much like a battery stores and releases electrical energy. Adenosine triphosphate (ATP) is created in the mitochondria through a process that involves the production of an electrochemical gradient across the mitochondrial inner membrane, which is used to create ATP production. ATP is then used by the cells to power ion transport, muscle contraction, nerve impulse propagation, substrate phosphorylation, and chemical synthesis.

The focus on electricity is very important in the context of cancer and exploring the distinguishing electrical properties of cancer cells may lead to novel therapeutic options (Di Gregorio et al. 2022). Cancer cells are also batteries-like, with different characteristics from those of normal cells. Recently, changes in membrane potentials both of the cellular membrane and mitochondrial membranes have been proposed as a novel key cancer hallmark (Delisi et al. 2024), unifying the membrane potential studies with the older field of ion channels studies. More than a decade ago, bioelectric manipulations were shown to reverse the metastatic phenotype in a Xenopus model (Chernet and Levin 2013; Levin and Martyniuk 2018; Payne et al. 2019; Levin 2021). Mitochondrial luminoptogenetics (mLumiOpto) is an ingenious mitochondrial-targeted luminoptogenetics gene therapy designed to directly disrupt the inner mitochondrial membrane potential and induce cancer cell death (Chen et al. 2024a). Modulating the activity of ion channels present in cancer cell membranes is another strategy to interfere with both the tumor growth and cancer metastases but, unfortunately, until very recently, the field of channel inhibitors did not advance mainly because of the universal presence of ion channels on normal cells and the toxicity associated with their inhibition (Capatina et al. 2022). More recently, in vivo targeting in mouse models of "targeted osmotic lysis" (TOL) combining pulse electric fields to activate voltage-gated sodium channels (VGSCs) with a simultaneous pharmacological blockade of Na+, K+-ATPase led to impressive results, reducing tumor sizes by 60–100% (Paul et al. 2020b; Gould and Paul 2022). Other cancer treatments that may impact cellular electricity have been also already proposed using deuterium-depleted water (Lu and Chen 2024) and cancer alkalinization (Hamaguchi et al. 2022).

A few years ago, it was shown that individual cristae within a mitochondrion display different membrane potentials and are functionally independent; therefore, mitochondria function more like Tesla batteries composed of multiple small batteries rather than a unique battery (Wolf et al. 2019).

8.20 Life as Dynamic and Creative Persistence

The DNA molecule can be used as a symbol of life's persistence on Earth (Please see Fig. 8.6).

What should be emphasized is the *enduring* presence of the DNA molecule—a thread that has run unbroken through the fabric of life since its very origins. Within the cell, conserved from the Last Universal Common Ancestor (LUCA), are not only the genes that guide the transcription of DNA into RNA and the translation of RNA into proteins but also those genes that orchestrate DNA replication and construct vital molecular machinery, such as ATP synthase.

Alongside DNA, two other ancient pillars have been conserved in all eukaryotic cells: the electro-biochemical engine that transforms energy into life-sustaining bioenergy (i.e., mitochondria) and the structural scaffolding of life: cellular membranes and the intricate cytoskeleton. These core components endure not by chance but because they are indispensable to life's continuation.

Yet life is not defined solely by conserved structures like DNA, mitochondria, membranes, or cytoskeletons. Life is also the conservation of process, of actions that persist through time. The simplest cycle of biochemical reactions was thought to be represented by the citric cycle (Smith and Morowitz 2007). Recently, a phylogenetic reconstitution of LUCA's nature suggested that LUCA appeared approximately 4.2 billion years ago, it had a genome of at least 2.5 Mb encoding approximately 2600 proteins, and importantly, its metabolism already contained a primitive form of the citric acid cycle (Moody et al. 2024).

Classical biology has often focused on static snapshots, inferring meaning from structures frozen in time. Life is an unfolding process, not a fixed state. It is movement, action, and change, and the defining features of this ongoing flow are self-similarity and creativity: Life replicates itself, but the copies are never identical. This inherent variation fuels continual adaptation to environmental challenges. Life is a restless act of becoming, a ceaseless movement of matter, energy, and information. From this perspective, genetic driver alterations in cancer may not represent random accidents but rather active adaptations to various stressors, cellular stress, needed for cell's survival. Mutational signatures of different environmental stressors (smoking, UV light, etc.) have been described (Alexandrov et al. 2020), but regardless of the fact they modify different genes, the malignant cells exhibit similar transitions and trajectories. This suggests the idea that different mutations may trigger similar mechanisms leading to a quasi-universal set of cancer cells phenotypes.

We briefly mentioned above the concept of dynamic and creative persistence (DCP). What exactly persists? The living process itself. *Life is a regulated flow of energy transformation, information processing, and structural organization that persists.* Living systems possess the capacity to continuously harness matter, energy, and information from both their internal and external environments, converting these

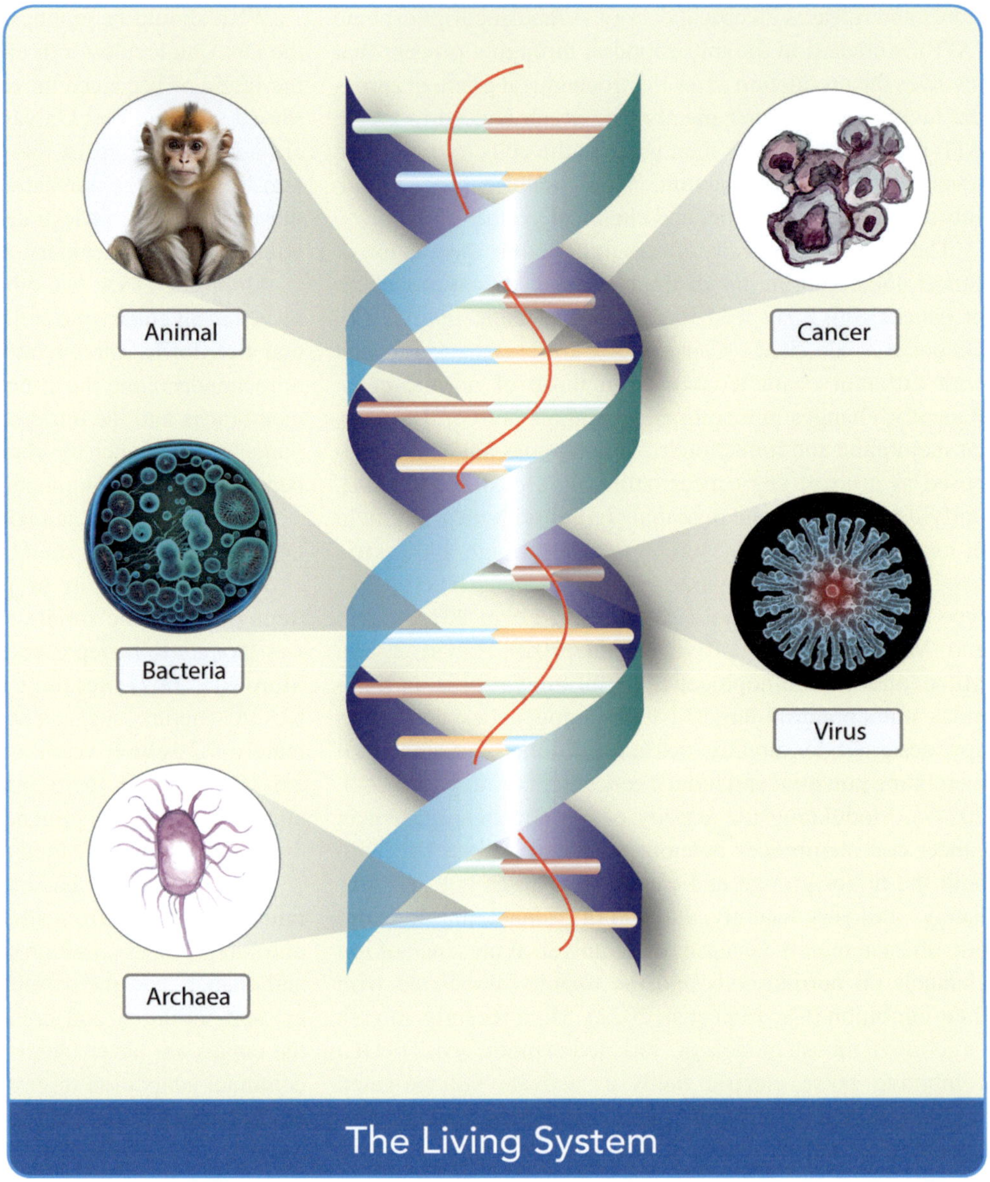

Fig. 8.6 DNA: Life's persistence molecule. (Courtesy of Ana Tourian)

resources into coherent forms of bioenergy, bioinformation, and biological structure (i.e., biostructure).

Life is simultaneously *autopoietic*, self-constructing, and *autotelic*—its purpose is precisely itself. At its core, the living process is homeodynamic: It embodies a continuous balancing act that sustains a dynamic equilibrium between stability and transformation (homeostasis), adjusts this equilibrium in response to environmental changes (homeopoiesis), and fluidly integrates these adaptive processes over time (homeorhesis). Importantly, the environment here refers both to the internal and the external environment.

Expanding the concepts of Robert Pascal and Addy Pross and, more recently, of Michael Wong et al., I am introducing the concept of dynamic and creative persistence (DCP). DCP is not simply an extension of the chemical kinetics and evolutionary principle as proposed by Pascal and Pross, since life *does not* evolve from less persistent processes to more persistent processes. *Fundamentally, the living process is persistent from inception.* Also, it involves not only functional information and evolution, as in Wong's et al. formulation, but also several homeodynamic mechanisms.

The living process—embodied in Dynamic and Creative Persistence (DCP)—is as fundamental to biology as genes, cells, or organisms themselves. Selection and inheritance happen not only at the level of the genes but also at the level of the organisms (Michod et al. 2022). The fact that selection occurs *simultaneously* at multilevels supports the DCP concept.

John Maynard Smith suggested that there are two relevant properties of life: "i) Although the forms of living organisms remain constant, the atoms and molecules of which they are composed are constantly changing; in other words, they have a "metabolism". ii) The parts of the organisms have 'functions"; that is, the parts contribute to the survival and reproduction of the whole" (Smith 1986).

The origin of DCP lies in cyclic causality present at all levels in living systems (Please see also Sect. 8.30 and Fig. 8.8). There are five fundamental cyclic processes:

The first cyclic process refers to the continuous recycling of matter in an organism.

98% of all the atoms in a human body are renewed each year; within 5 years, all the atoms have changed. The fastest-changing component is water. It forms about 70% of the body, and about half the water molecules are replaced every 8 days. Other fleeting elements are carbon, sodium, and potassium.

In our body, 330 billion cells are replaced daily, equivalent to about 1% of all our cells. In 80 to 100 days, 30 trillion will have been replenished—the equivalent of a new organism (Fischetti and Christiansen 2021).

The second cyclic process refers to the *vertical* and *horizontal* feedbacks and feedforwards between the different scales of an organism and, respectively, between the parts and the whole.

The third cyclic process refers to biocognition: the perception-reaction reflex between the environment and the organism.

The fourth cyclic process refers to the process of division of a cell that makes two slightly different copies of itself.

The fifth cyclic process refers to the process of reproduction of multicellular organisms where somatic cells are separated from germ cells. Germ cells from two different organisms are able to fuse together and recreate an entire organism that will contain different types of cells including germ cells themselves.

As Theodosius Dobzhansky famously wrote, "*Nothing in biology makes sense except in the light of evolution.*" Theodosius Dobzhansky was a geneticist and an evolutionary biologist, and his focus was evolution. In biology, over the last century, there has been an artificial split between evolutionary biologists and physiologist. Physiologists like Walter Canon would have disagreed with Dobzhansky, taking the position that, "*Nothing in biology makes sense except in the light of homeostasis.*" More recently, a group of authors considered that development is life's foundation and stated, "*Nothing in biology makes sense except in the light of development*" (Lala et al. 2024).

I see life as an interplay between homeostasis, development (homeorhesis), and evolution (homeopoiesis). Evolution of species on Earth through eons of time and survival of organisms through their life cycles and through daily stressors are all important for life's persistence. A new formulation can be proposed: *Nothing in biology makes sense except in the light of life's dynamic and creative persistence (DCP).*

8.21 A Universal Principle: Dynamic and Creative Persistence

As opposed to Wong et al.'s proposal of selection as being a universal principle, my suggestion is that cyclic causality manifested as DCP is the fundamental principle that links the abiotic to the biotic world: Homeostasis, homeorhesis, and homeopoiesis enable, maintain, and diversify it.

Fundamentally, all three forms of persistence described by Wong et al. (2023) are present in living systems: static persistence governing the world of quanta and atoms described by physics, dynamic persistence governing the world of organic chemistry, and creative persistence governing the world of biology. *Selection and evolution maintain and perpetuate life; they do not improve it.* As pointed out by Mayr, "Natural selection does not optimize organisms, it assures only the survival" (cited by Elek and Müller 2024). DCP underscores both temporal phenomena (the continuity of the living process on planet Earth, and perhaps in the whole Universe, wherever life exists) and spatial/structural phenomena (the persistence of life forms at different scales and for variable durations of time).

Life is a sort of perpetuum mobile, the expression of a self-maintaining process harnessing energy under bioinformation guidance in membrane-bound (or other enclosed) organized spaces, directly dependent on the continuous reciprocal exchange of matter, energy, and information with the environment. As a thermodynamically open process, the living process exchanges matter/energy, information, and structure with the environment and, therefore, is able to exist and persist—in theory at least—*indefinitely*, as long as these exchanges continue. Spore germination, hibernation, and cancer cell dormancy are all living states that require matter/energy, information, and structure, albeit in a lower amount than other more active life forms.

What underscores the persistence of the living process? Physics and chemistry operate using the laws of invariance of inanimate matter. Biology, since Darwin, became the science of variation, and, so far, evolution has been considered the only biological law that has stood the test of time. All organisms change continuously, and evolution has been thought to act like a "blind watchmaker" that, not following any pre-written program or recipes, selects the fittest among a repertoire of phenotypes. In the light of recent developments described above, Darwinian evolution is just one among several other homeodynamic mechanisms. These mechanisms ensure the persistence of the living process; they are the reason for life's extraordinary resilience.

The cyclic causality principle can be stated simply as *Life maintains itself*—of course, only in the context of a permissive and supportive environment.

Postulating that life's goal is its persistence may serve a basis to explain other directed behaviors, for example, cognition, without the need to involve a self-conscious agent as other authors have proposed (Baluška and Levin 2016; Miller and Torday 2017; Levin 2019; Baluška et al. 2021, Reber and Baluška 2021; Miller et al. 2023). Cognition-like properties are embedded in the living process itself that acts as its own supportive (selfless) agent. Metabolism, information processing systems, membranes, internal and external communication, and autotelic organization were Hoffmeyer's five requirements for a system to be alive (Hoffmeyer 1998). I share Godfrey-Smith's opinion that cognition is an embedded property of any living system and *not a precondition for life* to exist (Godfrey-Smith 2021). Cognitive properties have been also attributed to cancer (Miller and Torday 2017).

Living forms appear and disappear, but the living process persists. The dynamic of the living process may vary with time and may be different in different classes of organisms, but its fundamental properties are the same. Life's dynamic is a *homeo* (similar) dynamic and not a *tauto* (identical) dynamic. In order to perpetuate their living state, all organisms need to maintain the union of bioenergy transformation, bioinformation computing, and biostructure recycling and to continuously adjust these processes to environment challenges. Regardless of the particularities of these three processes in different living systems, they are closely interrelated and always present in order for life to exist.

Life's DCP is a characteristic of living systems in general, and not solely of individual life forms. *Importantly, for multicellular organisms or bacterial colonies, the DCP of the higher level of organization supersedes the DCP of individual cellular life forms.* The logic of life appears to be noncommutative in the sense that what promotes the DCP for a superior level in a nested system does not necessarily promotes the DCP of the inferior level and vice versa. In a multicellular organism or in a bacterial colony, some cells may die of programmed cell death (PCD) "sacrificing" their DCP for the benefit of the multicellular organism or the colony (Allocati et al. 2015). At the level of an ecosystem, extinctions of a species may offer opportunity for new species to emerge and diversify. Also, in some tumors, a significant number of cancer cells may also die by apoptosis, sending "ghost messages" to surrounding cancer cells promoting their growth and spread (Zhang et al. 2023c). The mechanism through which dying cancer cells inform the neighbor members of the cancer colony of an imminent danger in glioblastoma appears to be related to exosomal content (Pavlyukov et al. 2018). On the other hand, during cancer progression, the DCP of the cancer cells takes precedence over the DCP of the organism. This is a fundamental observation.

8.22 Persistence Versus Fitness

The DCP theory of life formulated here is an attempt to create a broad, integrated explanatory frame for all living systems, in general. DCP complements and expands and does not contradict evolutionary theories.

Life forms and living processes are not separated. The living process is "embedded" in life forms and at the same time, life forms make the living process conspicuous. Form and process are one. The dichotomy between static and dynamic processes is arbitrary. As Carl Woese put it, "Forms in essence *are* the process. Therefore, fundamental changes in their nature can only mean changes in the underlying evolutionary dynamic" (Woese 2002).

Darwin's focus was changes of life forms through natural selection. As important is the persistence of the living process itself. The outcome of natural selection is diversification of life forms that keep transmitting and multiplying the living process. Natural selection, by creating variety, stabilizes the living process, helping it thrive in many different environments.

Fitness has been defined as the ability of a particular genotype or phenotype to survive and produce offspring in the next generation relative to other genotypes/phenotypes in a population. For asexual reproduction, fitness is assigned directly to the genotype. A higher fitness is related to a higher number of offsprings in the next generation. More offsprings equals more copies of a particular genotype, giving alleles from that particular genotype a higher chance to promote the survival and/or reproduction of other individuals that share that allele, in preference to individuals with a different allele. A key point is that Darwin's selection operates transgenerationally *in a population*. In other words, "fitness is a property not of an individual, but of a class of individuals" (Smith 1989). The same "winning alleles" may be shared by several organisms in a population.

Relative fitness is a numerical way to evaluate fitness and is defined by dividing a particular genotype/phenotype survival and/or reproductive rate by the maximum genotype/survival rate of other genotypes in a population. Intuitively, in a population, if survival rates are equal, a genotype/phenotype is more fit if it leads to more offspring, and if reproductive rates are equal, a genotype/phenotype is more fit if it survives longer.

Natural selection is defined as the process of differential reproduction resulting from differential adaptation (fitness) to the same environment (Brandon 2004). Natural selection acts at the population level, imposing with a variable strength, a specific genotype/phenotype, thus contributing to the population pool of "winning" genotypes/phenotypes.

In order for evolution by natural selection to occur, Darwin proposed three requirements: a) *phenotypic variation,* that is, different individuals in a population have different physiologies, morphologies, and behaviors; b) *differential fitness,* that is, different phenotypes have different rates of survival or reproduction in different environments; and c) *fitness is heritable,* that is, there is a correlation between parents and offspring in terms of contributing of each to future generations. Natural entities (viruses, cells, organisms, species) that have variation, reproduction, and heritability may evolve (Lewontin 1970). Lewontin was among the first to notice that neoplastic growth represents a form of natural selection, in which neoplastic cells reproduce more efficiently than normal cells, competing with them and finally overcrowding them. He also noticed that cancer treatments (radiation or chemotherapy) directed at cell division reduces the fitness advantage of cancer cells "by introducing an environmental factor that selects specifically against dividing cells" (Lewontin 1970).

Besides evolution through natural selection, evolution through drift was proposed as another essential property of life. In 1968, Motoo Kimura, a Japanese biologist, introduced the neutral evolution (or drift) theory. In drift theory, allele frequency in a population may also change by chance in the absence of any Darwinian selection. More recently, the principles of evolution through drift were proposed by Robert Brandon in 2006:

(A) A population at equilibrium will tend to drift from that equilibrium unless acted on by an evolutionary force.
(B) A population on evolutionary trajectory *t*, caused by some net evolutionary force *F*, will tend to depart from the extrapolated path predicted based on *F* alone (in either direction or magnitude or both) even if no other evolutionary force intervenes, unless *F* continues to act.

The accumulation of mutations in many normal tissues may be associated with age and represent a drift phenomenon. Also, progression of some cancerous tumors may be related to drift and not to Darwinian selection. The same tumor may exhibit several types of selection some Darwinian, others non-Darwinian (Davis et al. 2017).

The differential fitness, in Darwin's view, was related to the "struggle for existence" or the "survival of the fittest" (Brandon 2004). "The survival of the fittest" implies, as Robert Brandon explains, a direct link between adaptiveness to an environment to reproductive success: "If *a* is better adapted than *b* in environment E, then (probably) *a* will have greater reproductive success than *b* in E" (Brandon 2004). For bacteria, adaptability, that is, survival and reproductive success, are clearly corelated. In multicellular organisms, adaptability does not necessary imply reproductive success. Mules may be well adapted to an environment but are unable to reproduce. Because fitness relates to both survival *and* reproductive success, in Darwin's theory, such mules would not be considered fit.

An important note: Fitness applies to a particular environment, at a particular moment in time, and this environment is shared by a community of organisms and species. The "struggle for existence" can be also understood as a "complex dynamical rearrangement of the community structure" as a whole "involving the dynamics between the community elements and the interaction with the environment," and not solely as an individual's "red in tooth and claw" fight against the others (Vetsigian et al. 2006). The interaction with the environment is fundamental: Phenotypic selection is separated from genetic response to the selection and *both* genotype and the environment concur to the presence of a certain phenotype (Brandon 2004).

In an ecosystem, the fitness of a particular species will be codependent with the fitness of other species. The survival extends to the whole ecosystem and is not limited to one single organism (or species) and its offspring. If we think of a tumor as an ecosystem, the survival of the tumor will depend not only on the fitness of a single cancer cell or cancer clone or on the group phenotypic composition (GPC) of cancer cells (Thomas et al. 2024) but also on the phenotypic composition of the cancerized stromal cells and properties of the stromal matrix. Finally, the tumor should be considered a functional unity without artificially separating the malignant cells from the stromal environment. Also, the behavior of tumor cells, that is, ability to divide, invade, and metastasize, is closely related to the dynamics of both their internal and external environments, respectively.

As a corollary, cancer cells behavior is not solely the result of competition between individual cancer cells among themselves or between cancer and non-cancer cells, or the result of the pressure imposed by the immune system (i.e., genome editing), or a certain treatment, but also reflects the dynamic of tissular and organismal environments as a whole. Cancer cells with similar genomes behave differently in different environments, which may explain the different growth rates between primary tumors and distant metastases. This is a crucial point: The SMT theory proponents have focused mainly on the role of genes and mutations in cancer, paying less attention to the equally important role of the local and global environments. Mina Bissell and Derek Radisky demonstrated that tissue structure can be dominant over genotype by reversing the phenotype of cancer cells cultured on a reconstituted basement membrane (Bissell and Radisky 2001). More recently, direct compression applied to cancer cells also reversed their malignant phenotype, again demonstrating the importance of the local environment on cancer cells' behavior (Ricca et al. 2018). Systemic environmental factors may play a significant role in cancer's unyielding resistance as demonstrated by the Spontaneous Regression/Complete Resistance SR/CR nude mouse model found to be resistant to sarcoma cells' inoculation mainly due to a very strong anticancer innate immune system (Cui et al. 2003) (see Sect. 11.8).

In order for the Darwinian evolutionary game of *organisms'* fitness to start, self-creating, self-maintaining, and evolving *processes* needed to be present already (Otto 2022). In the absence of entities that lack intrinsic mechanisms that both promote survival and reproduce with fidelity, fitness and natural selection are absent. As explained by Carl Woese, a "Darwinian Threshold" is necessary to start speciation and represents the origin of the evolution tree. Darwin's theory is about the unfolding in time of life forms on Earth after a "solidified, organized cellular design" was established (Woese 2002). Darwin himself in the 1866 fourth revised edition of *The Origin of Species* stated, "If variations useful to any organic being do ever occur, assuredly individuals thus characterized will have the best chance of being preserved in the struggle for life; and from the strong principle of inheritance, they will tend to produce offspring similarly characterized. This principle of *preservation* I have called, for the sake of brevity, Natural Selection" (Darwin 1866).

Darwin's insight was that natural selection helps preserve certain well-adapted life forms. Importantly, life's DCP is not only about preservation of a *well-adapted life form* but also about the persistence of the *process* that underlines the existence of a given life form. The DCP process has been present through all the life's history on Earth and, initially, it was related only incidentally to particular life forms. As mentioned above, it applies both to individual life forms and living systems. As shown by Godfrey-Smith, some organisms are not classical Darwinian individuals and therefore, they are not classical units of selection (Godfrey-Smith 2011, 2013). As he explains (Godfrey-Smith 2013), some organisms are persisters and a high reproduction rate is not an absolute requirement for their existence. The outcome of Darwin's Natural Selection (DNS) is a diversification of forms in which the living process is embedded from the beginning. DNS does not create or improve the living process itself. It diversifies its material embodiments. To clarify, although their metabolism is different, both eukarya and bacteria have bioenergetic mechanisms that are interdependent with an information process (transcription and translation of DNA to proteins through RNA) and their biostructure (e.g., membrane). The basic triad of bioenergy/bioinformation/biostructure can manifest through myriads of life forms. Also, phylogenetic data points to the existence of a unique common ancestor shared by all eukarya including fungi, protists, plants, and animals referred as "The Last Eukaryotic Common Ancestor" (LECA). Lynn Margulis hypothesized that the Last Eukaryotic Common Ancestor (LECA) arose through endosymbiosis, when an archaeal host incorporated an aerobic bacterium that became the progenitor of mitochondria (Margulis 1961). This event drastically changed LECA's metabolism and overall internal cellular design, but the union between bioenergy/bioinformation and biostructure already present before LECA persisted. Life forms and living processes are not separated. As Carl Woese noted, "the forms in essence *are* the process" (Woese 2002). The living process is at the same time "embedded" in the life forms and the life forms are "morphed" into the life process. Forms and processes are the same entity. Darwin's focus was life forms changing through natural selection. Focusing on the persistence of the living process is equally important. The outcome of DNS is diversification of life forms in which the DCP process has been embedded from the beginning.

Brief Note: Nick Lane priorly proposed (Lane 2009) that the vast panorama of life including life itself is due solely to evolution: Besides life, he cites other nine other evolution "inventions": DNA, sex, photosynthesis, complex cells, movement, sight, hot blood, consciousness, and death.

My proposal is that *evolution represents only one process among others* (homeostasis, homeorhesis, etc.) that help the living processes persist, and also contribute to some of the "inventions" mentioned by Nick Lane. Death, for example, is both a homeostatic *and* an evolutionary process. Within the broader framework of homeodynamics, homeostasis, development, and evolution can be understood as similar, and, also interconnected, mechanisms.

8.23 Non-Darwinian Evolution Through Horizontal Genetic Transfer

An important process that supports cancer's DCP is horizontal genetic transfer (HGT). HGT has been postulated by Carl Woese to have been the main mechanism of genetic diversification before the "Darwinian Threshold" and thus active before DNS started to produce genetic novelty (Woese 2002). HGT is still present in many bacteria and fungi and involves the exchange of transposons, integrons, and/or entire gene cassettes between bacteria and fungi of the same or different species. It also plays an important role in the generation of resistance to antibiotic drugs as well as adaptation to new environments (Emamalipour et al. 2020). HGT is an important enabler of local and distal cancer dissemination processes. Cancer cells can exchange genetic material through apoptotic bodies, nanotubes, exosomes, and cell-free DNA (cfDNA) with both cells in the local stroma and cells located in other distal organs (Emamalipour et al. 2020; Valcz et al. 2022).

8.24 Dynamic and Creative Persistence and Evolution

The system of three coreinterdependent processes, bioenergy, bioinformation, and biostructure (Gánti 2003; Otto 2022; Rothschild et al. 2024), together with an environment that keeps the system evolving and far from equilibrium are *life's primary ingredients*. In its essence, life is the persistence of the union of three processes that are codependent with their environment. When a cell divides, it shares with the daughter cells its DNA (bioinformation), its mitochondria (bioenergy), and its membrane (biostructure). The union of these components has been transmitted uninterrupted since the first eukarya appeared on Earth. Today, all multicellular organisms on Earth, including humans, are descendants of that cell. NASA's life definition included Darwinian evolution but, fundamentally, Darwinian evolution by itself with its three components (replication-mutation-selection) is not sufficient for life to come into existence from an abiotic system (Taylor et al. 2016; Duim and Otto 2017) and, furthermore, is not strictly required for life to exist or persist. Evolution is an *auxiliary* process, supporting the living process persistence through time.

The story unfolding on our planet is both the *persistence* of the living process and *evolution* of the living forms. *Evolution can be seen as persistence in action*. A fitted process does not necessarily have to be *the* fittest in order to persist (Cohen and Marron 2020). Life on Earth was the start-up of *a process fitted to persist from the beginning*. Evolution fostered change and pressured it to diversify under various forms. It is estimated that the first cells that already had a metabolism, a DNA code, membranes all united into a functional biosystem are present on Earth for at least 4 billion years. The living process goes on since then and prokaryotic life did not change much since then. It is estimated that prokaryotic cells transitioned into eukaryotic unicellular cells between 2.0 and 1.4 billion years ago.

Multicellularity's offset represented a tradeoff between the benefits of specialized functions and divided labor for the group versus unicellularity's focus on nutrition and reproduction only. Originally, the barrier between unicellular and multicellular was fluid, and there are still species today like Volvox that can make this transition in response to environmental changes. The "invention" of a stable multicellular body was also likely induced by the environment. The persistence of holobionts or ecosystems where different life forms interact, collaborate, and compete is, in the big scheme of things, irrelevant for cells in a multicellular organism or organisms in an ecosystem. Organisms are activities and not static entities. What really matters is the living process that struggles to persist *at any organization scale*. As noted by Craig Holdrege, "In a living being there is activity all the way down." These activities follow goals; in his words, "organisms are creatively expressing beings and not biomachines" (Holdredge 2021).

An organism goal is to survive and/or to reproduce and thus to perpetuate life. Life's goal is to persist, and the reproductive ability and the longevity of a particular organism are means to accomplish this goal. What continues, what is transmitted from generation to generation of living beings is the process itself of transformation of energy, information, and matter that persists trans-generationally. Life is a sort of self-creating and self-perpetuating whirling dance of atoms and molecules that persists over eons of time, apparently changing but fundamentally staying the same. The atoms and molecules continuously come and go; the dance persists. DCP and process ontology advocate the same concept of dynamic persistence through change. As explained by John Dupré, "A storm may also be a very stable element of the world. The Red Spot on Jupiter, for instance, has been observed for several centuries, albeit with gradual changes in shape and size. But unlike the mountain, the Red Spot does not persist because nothing happens to change it, but because a stable pattern is maintained by the very rapid winds that circulate round it. If this activity ceased the Red Spot would dissipate very quickly. Its persistence through time is understood not through unchanging essential properties but through the causal continuity of the processes that maintain the pattern. A process ontology for life starts with the idea that the Red Spot is a more useful paradigm for living systems than the mountain" (Dupré 2017).

Classical Darwinian Natural Selection or other forms of evolution, like drift, for example, prop up life's persistence through a paradoxical Red Queen effect; "it takes all the running you can do, to keep in the same place" (Lewis Carroll *Through the looking glass*). Evolution does not change or improve life's essence; it just helps maintain its dynamic status quo. If we consider life on Earth a single organism, as suggested by Margarida Hermida, evolution supports persistence, as it maintains the life of this organism. On the other hand, the process of evolution itself may evolve, preventing evolution itself to arrive at a halt, through the so-called "evolution of evolution," a term coined initially by Walter Stephen and used subsequently by John Tyler Bonner (2019).

I am using the term "dynamic persistence" in lieu of "fitness" to highlight the primordial role of a process—dynamic persistence—for life's continuance overall versus the fitness of an individual or a group. The focus in Darwin theory of evolution was on the individual and the survival of the fittest. I emphasize here the *persistence of the living process that transcends cells, multicellular organisms, or species*. Through its activity, the living process maintains itself and evolution does not optimize it; it just diversifies the modalities of its existence. Namely, there are no selfish genes, no cheating cells, no champion organisms. It is the process that matters, and its embodiment into forms that produce quasi-identical copies of themselves is preserving it. My accent here is on processes, not forms. Of note, both Herrida's theory of life on Earth as an ontological individual and Torday-Rehan's unicell theory also emphasize life's processual quality and not forms or individual organisms like evolutionary theories.

Persistence is not a "stable non-equilibrium *state*" as defined by Bauer (Müller and Igamberdiev 2024) but rather a "stable nonequilibrium *process*." It is a dynamic nonequilibrium process maintained continuously and made possible by the openness of living systems. For life's nonequilibrium process to exist, it needs an environment both as a continuous source of matter and energy and as a recipient of unused/unusable matter/energy.

What is the best strategy for life's persistence? Rapid reproduction of the living process as in bacteria, or a long survival of an individual life form? At one end of the spectrum, there are pure flow-like life forms that exist very briefly in space and reproduce quickly. The sole purpose of their existence is, so to speak, transmitting the flame of life to future generations. At the other end of the spectrum, there are life forms like a baobab or a tortoise that persists longer in space. Most organisms on Earth have longevities that fall in between these two extremes. Whether or not unicellular or multicellular organisms have a shorter or longer life span, the main purpose of their existence is to propagate and preserve the living process.

Both reproduction and survival strategies are context-dependent so one or the other may prevail in a given environment. Two ecologists, MacArthur and Wilson (1967), introduced the terms "r-selection" and "K-selection" for two opposed types of selection in different animal ecosystems.

r-selection: On one extreme are the species that are highly r-selected, "r" referring to reproduction. In a less competitive, low-quality environment, species tend to be driven by r-selection producing many offspring. The benefit of this strategy is that if resources are limited or unpredictable, some offspring may survive.

K-selection: Species that are highly K-selected invest heavily in each offspring, with longer periods of maturation, parental care, nurturing, and protection of the young. These reproductive strategies are more common in long-lived organisms with overlapping generations, such as whales or elephants, where offspring have a higher probability of survival to maturity. K-selection is associated with populations at or near their carrying capacity, where offspring face intense competition for resources.

Where do we start? As Niels Bohr suggested, life is irreducible to either physical or chemical properties and it should be considered a separate, independent phenomenon (Hoyningen-Huene 1994). To make sense of life's persistence in the first place, a distinct principle or law, encompassing biological phenomena needs to be introduced. DCP theory expands evolutionary theories and proposes an integrated explanatory frame for all living systems, in general.

8.25 Why Do Organisms Reproduce and Struggle to Survive?

If we use life's dynamic and creative persistence (DCP) as a foundation, the behavior and relationships between living systems and their codependence with the environment start making sense. An apple does not choose to fall to Earth; it simply follows the law of gravity. Considering life's conservation principle, the so-called *agency,* that is, directed behavior of organisms, can be understood not as a sign of conscious behavior but as a fundamental necessity. A bacterium does not choose to divide or move in the direction of nutriments. It divides pushed by internal and external forces and moves in the direction of nutriments without choosing to do so. A bacterium does not choose to stay alive. It just does so. For all living forms on earth, there is never a "to be or not to be" question. Organisms simply go on existing without a choice to do so, governed by the *Axiom of Choice* that reads: "*The living process has one single choice: persist.*" Darwin's struggle for existence is thus just one of the corollaries of the living process' auto-conservation principle.

The living process can be understood as a continuous projection of itself from the past into the future—a dynamic phenomenon in which life persistently recreates, regenerates, and sustains itself across time. This continuity evokes Noether's theorem, which posits that for every continuous symmetry in the laws governing a physical system, there exists a corresponding conserved quantity. The physical system itself need not be symmetric; it is the laws of its motion that are symmetric (José and Saletan 1998). Applied to biology, one may ask: What is the conserved quantity that ensures the persistence of life through time and transformation? A compelling candidate is *effective information*—the system's intrinsic "know-how" to generate and maintain the complex processes that constitute living states, that is, directed information, information with a *purpose*.

In a recent framework proposed by Michael Wong and colleagues (Wong et al. 2023), *functional information* was conceptualized as a measure of the information required for a system—capable of existing in numerous configurations—to achieve a specific degree of function. Within this framework, configurations that enhance the system's persistence are more likely to be selected, aligning with principles of both natural selection and self-organizing complexity. This perspective offers a quantifiable link between information, function, and survival. Of note, the *functional information* used by Wong's colleagues is defined in the context of a *spe-*

cific function (i.e., the function of an enzyme to catalyze a specific reaction). The concept of *effective information* introduced here, while closely related to Wong et al.'s notion of *functional information*, is defined more broadly in the context of assembling and sustaining the functionality of the entire living system rather than being limited to a specific function serving that purpose.

I propose that *effective information* in living systems comprises two distinct but complementary components. The first is a stable component, encoding essential life-sustaining functions that are deeply conserved across evolutionary timescales—such as basic metabolic pathways, membrane integrity, and genetic replication machinery. The second is a variable, regulatory component, which embodies the system's capacity for adaptation and flexibility through homeodynamic mechanisms. This component allows for real-time responsiveness to environmental fluctuations, stressors, and internal changes, enabling life to persist not by rigidity but through dynamic modulation and controlled variability.

Moreover, this bifurcation of *effective information*—into a conserved core and an adaptive interface—underpins the theoretical foundation of the biogenerator concept introduced priorly in Sect. 8.16. Biogenerators can be envisioned as organizational hubs where stable core functions intersect with dynamic, context-sensitive regulatory circuits. They act as integrative nodes that coordinate cellular and systemic functions across scales and time frames, contributing to the homeodynamic resilience and evolutionary plasticity of living organisms.

In this view, the persistence of life is not merely a result of replicating molecular structures but of perpetuating *functional architectures*—a choreography of conserved instructions and adaptive improvisations that together sustain the self-constructing, self-maintaining phenomenon we call life.

The living process extends beyond cells, organisms, and species encompassing the whole planet. The fundamental processes of (a) energy generating/transforming, (b) information memorizing/computing, and (c) structuring/organizing/shaping are present not only within cells, organisms, and biosystems but, at a much larger scale, also within Earth's abiotic and biotic environments (Smith and Morowitz 2016).

8.26 Life's Core: Bioenergy/Bioinformation/Biostructure

Life is not solely free energy dissipation and biochemistry. The core of the life process is the codependent union of bioenergy–/bioinformation–/biostructure-related processes that maintains itself through self-copying autocatalytic mechanisms sustained by control processes that are under continuing selective pressure both from inside and outside. For life to exist, metabolism (bioenergy) needs to be tightly coupled with information computing processes (bioinformation), within an organized system (biostructure), not isolated but in continuous codependent interaction with the environment (please see Fig. 8.7), and this, *from the beginning*. This union was present from the get-go. Life forms and environment coevolve together. Therefore, all three, bioinformation-, biostructure-, and bioenergy-related processes, in conjunction with the environment, and other mechanisms maintaining their union (i.e., autopoiesis and homeostasis), are closely linked together and codependent, none of them being primordial in relationship to the other, each of them being a primus inter pares (Otto 2022).

The union of bioenergy–/bioinformation–/biostructure-related processes is present also at the molecular level. Through changes in their shapes (allostery), responding to environmental stimuli such as ions, metabolites, or macromolecules, *proteins* are a sort of logic gates involved in information processing. Besides their bioinformation function, they are also critical components of cell's biostructure, and, sometimes, they can also serve as fuel. The same is true for carbohydrates (mainly used for fuel) and lipids (mainly used for biostructure). In the composition of a cell, water is the most abundant component (as much as 70%). Molecules of water through the hydrogen bonds and their conformational changes are also involved in bioenergetical, bioinformational, and biostructural processes. Water and proteins are the most abundant components of the cell. DNA is the least abundant molecule inside a cell, but it is the main cellular information processor. Proteins and different classes of ncRNAs (miRNA, ltRNA, etc.) and, to some extent, a few metabolites control DNA transcription and translation, genomic replication, and cell division, and other important cytoplasmic activities including metabolism and intracellular transport and communication with the environment.

The model of life as a triad of biostructure/bioenergy/bioinformation processes may serve as working tool to investigate life's principles, in general, and cancer behavior, in particular. The union of the three components of this dynamic and creative core is crucial: Once one of the components is destroyed or separated from the other two, death follows immediately. In cancer, both the individual components of the triad are modified (i.e., aneuploidies, genetic alterations, epigenetic changes, metabolic changes, and structural changes) and the processes that maintain the triad's persistence are transformed, compared to the normal cells. In cancer, homeostasis and autopoiesis are perturbed, the codependence with the local environment transforms into a parasitic relationship, and the unity with the organism is altered.

Bioenergetical/bioinformational/biostructural processes (please see Fig. 8.7) are interdependent and exist both at the cellular and at the organismic level. The cellular level will be discussed below, as an example.

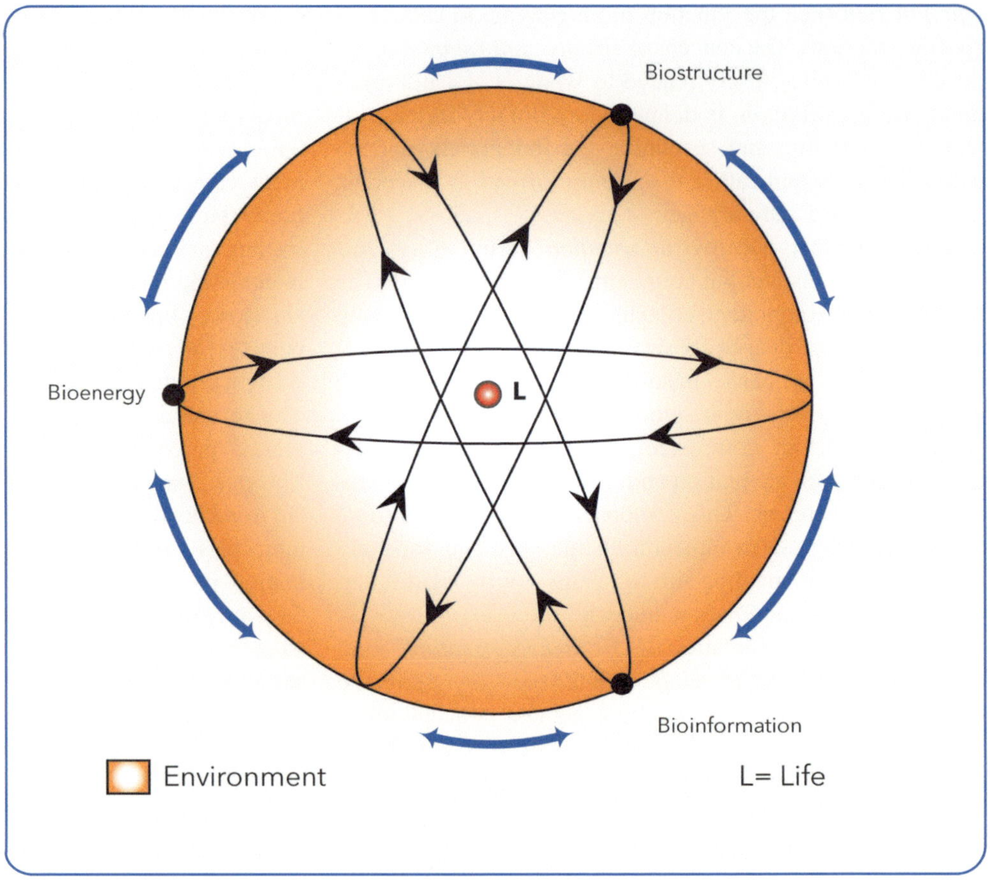

Fig. 8.7 Life's core: biostructure, bioenergy, and bioinformation processes interacting with the environment. (Courtesy of Ana Tourian)

- *Bioenergetic* processes include matter ← → energy transformations that comprise in a cell its metabolism that supports all life processes and mechanisms.
- *Bioinformatic* processes encompass the storage, transmission, and replication of biological information. They include the bidirectional movement of information from DNA to RNA to proteins and back. They also involve epigenetic mechanisms (methylation, ncRNAs, etc.) allowing information to flow back and forth inside the genomic network and between genes and cytoplasm. The information exchange is not restricted to processes occurring within an organism but also, respectively, from the environment to the organism and vice versa. Bioinformation is not restricted only to the transcription/translation and replication processes, part of the so-called "code of life" that involves DNA and RNA. Other communication processes, that is, communication between organelles, between the nucleus and mitochondria, etc., and other biological codes (see below), are equally important.
- *Biostructural* processes represent the ability of living systems to continuously build and rebuild themselves from simpler components forming complex emerging structures in which different organizational levels are codependent and communicate to each other forming Kantian Wholes—a term introduced by Longo et al. (2012) and popularized by Stuart Kauffman (2020a, b). We prefer for its simplicity the term "holon" instead of "Kantian whole." The "holon" was introduced by Arthur Koestler in 1967 in his book *The ghost in the machine* and was presented again at the Alpbach Symposium in 1968 in a paper titled *Beyond atomism and holism: the concept of the holon.* As Koesteler wrote, "The concept of the holon is intended to reconcile the atomistic and the holistic approaches" (Koestler 1969), and he used it in order to describe natural organisms as composed of semiautonomous parts that are linked through a *holarchy* (a term also introduced by Koestler), to form a whole. The biostructure is *holonic*.

8.27 Bioenergy

Bioenergy refers to the transformation and generation of energy in living systems. A eukaryotic cell, for example, transforms chemical energy from large organic molecules into adenosine triphosphate (ATP) which is the cell's universal energy currency. This process consists in a series of metabolic reactions that break down larger molecules, that is, carbohydrates, proteins, and fats, into smaller molecules, that is, carbon dioxide and water, releasing energy, temporarily embedded in the ATP molecule.

In a cell, fluxes of energy, matter, and information are interconnected in a functioning whole that self-generates itself continuously. Metabolism, via metabolic enzymes and metabolites, modulates chromatin—directly or indirectly—through regulation of the activity of chromatin trans-acting proteins, including histone-modifying enzymes, chromatin-remodeling complexes and transcription regulators (Li et al. 2018b), thus controlling gene expression. It has been shown that all essential glycolytic enzymes are translocated to the nucleus (Yu and Li 2017). Within the nucleus, these enzymes play crucial roles in tumor progression, extending beyond their recognized metabolic functions. These nonclassical properties include anti-apoptotic functions, regulation of epigenetic modifications, modulation of transcription factors and cofactors, a role as extracellular cytokines, a protein kinase activity, and involvement in the mTorc1 signaling pathway (Hanselmann and Welter 2016).

In multicellular organisms, cells have evolved sophisticated mechanisms to integrate internal and external signals to maintain homeostasis and adapt to fluctuating environments. Central to this adaptability is the dynamic regulation of metabolism, shifting between anabolic and catabolic states in response to nutrient availability, energy demand, and growth signals. This metabolic flexibility is governed by intricate signaling pathways that integrate intracellular cues with extracellular stimuli. Cellular metabolism is controlled by both genetically encoded enzymes and small-molecule metabolites. Cells have evolved a range of nutrient- and energy-sensing pathways to monitor intracellular nutrient levels such as through mTORC1 and to assess energetic status via indicators like the AMP/ATP and NAD^+/NADH ratios. During periods of nutrient scarcity or energy deficiency, cells suppress nonessential ATP-consuming processes and activate catabolic pathways to mobilize internal energy reserves. Among the most well-characterized regulators of this adaptive response are AMP-activated protein kinase (AMPK) and Silent Information Regulator 1 (SIRT1). AMPK is activated by rising AMP levels, while SIRT1 responds to increased NAD^+ concentrations, both of which signal low-energy states. Upon activation, AMPK and SIRT1 promote ATP-generating catabolic pathways and inhibit energy-intensive anabolic processes, thereby restoring energetic balance. Their complementary and interdependent actions reflect an evolutionarily conserved strategy for maintaining metabolic resilience under stress. Indeed, AMPK and SIRT1 can modulate each other's activity and are often jointly required to initiate and sustain key metabolic pathways. This seamless coordination between signal transduction and metabolic processes is critical for sustaining cellular health and ensuring a rapid and appropriate response to environmental fluctuations (Gomes and Blenis 2015).

8.28 Bioinformation

Causality and information are closely related. Every act of matter/energy exchange is also a process of information exchange. Information needs a material support and a material channel to be transmitted. Matter/energy/information exchanges create causal relationships. "a" is causally related to "b" only if there is a way for "a" to be connected to "b" through an energy/information transmission channel. In a cell, this channel(s) is/are intrinsically linked to its biostructure. The causal structure of biological time informs and is informed by bioinformation. Shannon's information formula ($S = k \log P0$) is similar to Boltzmann's entropy formula ($S = k \log D$) where both P0 and D represent the number of microstates of a system (Shannon 1948). As pointed up by Warren Weaver in the book he cosigned with Claude Shannon, in Shannon's original formulation, Shannon's information is identical to entropy: "Information turns out to be exactly that which is known in thermodynamics as entropy. For in both cases it is a question of number of possibilities and freedom of choice. Entropy, chaos, mixing together, freedom of choice, and information are thus identical as is accepted in information theory and in physics. The higher the number of microstates the higher the information" (Shannon and Weaver 1949). *On the other hand, for Schrödinger, information was the opposite of entropy.* He stated that "It is by avoiding the rapid decay into the inert state of 'equilibrium' that an organism appears so enigmatic [...]. What an organism feeds upon is negative entropy" (Schrödinger 1944). In the same book, Schrödinger mentioned that "negative entropy" really meant free energy, a hypothesis that has been subsequently questioned by Léon Brillouin who considered them related but not identical: "Information can be changed into negentropy and vice versa. If the transformation is reversible, there is no loss, but an irreversible transformation always corresponds to a loss. [...] An isolated system contains negentropy if it reveals a possibility for doing mechanical or electrical work" (Brillouin 1953).

How can these two opposing meanings of entropy can be reconciliated? Obviously, information cannot be both entropy and negentropy at the same time. Going back to the original concepts, it becomes clear that Schrödinger and Shannon were referring to different things. Schrödinger's concept of negentropy relates to order and predictability, while Shannon's information concept relates to disorder and unpredictability. As noticed by Brillouin, *negentropy is not information*. Shannon's information is a general measure of the unpredictability of a configuration of microstates of a system. It represents a *theoretical* maximum potential information that exists in any microstate configurations *before* information exchange. It does not refer to individual message exchanges. As Brillouin made clear, the *actual* information *I* involves an information exchange and has a lower value than S. In an observation act, for example, I = k ln (Po/P1), P0 is the number of equally probable possibilities *before* observation and P1 is the number of equally probable possibilities *after* observation. *I* is always lower than *S,* and it always have a positive value. *Energetically speaking, information is not for free* and Léon Brillouin calculated that one bit of information burns K ln2 = K 0.69 = $0.95264781 \times 10^{-23}$ joules per kelvin. It is a minute amount of course, but as Dennis Gabor, the inventor of holography, famously said: "One cannot get something for nothing, *not even an observation*" (cited by Léon Brillouin 1953).

Bioinformation refers to both the syntactic and the semantic content of communication within and in between the complex network of genes, RNAs, various types of noncoding RNA, and proteins; within and in between the organelles; and between the organelles and the nucleus, as well as between the cell and its environment. It involves RNAs, proteins, signaling molecules, receptors and secondary messengers, transcription factors, and ions perpetually moving inside the cell and between cells.

The most important bioinformation computing processes in a cell are represented by transcription, translation, and genomic replication that together form what has been called the *genetic code*. It was initially believed that the genetic code represents the unique code used across all living organisms (North 2011). Cavalier-Smith was one of the first biologists to point out that the genome does not contain all the information needed to make an organism and that other scripts exist. To offer an example, he introduced the concept of the "membranome": "Membrane heredity, by providing chemically specific two-dimensional surfaces with mutually conserved topological relationships in the three spatial dimensions, plays a key role in the mechanisms that convert the linear information of DNA into the three-dimensional shapes of single cells and multicellular organisms. Animal development creates a complex three-dimensional multicellular organism not by starting from the linear information in DNA […] but always starting from an already highly complex three-dimensional unicellular organism, the fertilized egg, which membrane and DNA heredity together have perpetuated" (Cavalier-Smith 2004).

In a sweeping theoretical expansion, Marcello Barbieri proposed that beyond the genetic code, life is governed by a plurality of other multiple other biological codes, each mediating a distinct class of symbolic correspondences between molecular structures and biological functions. His work draws a foundational distinction between codes that emerge in unicellular life, which sustain intracellular organization, and those that evolved later in multicellular organisms, which enable coordinated tissue, organ, and organismal integration (Barbieri 2015, 2018, 2019).

In unicellular systems, Barbieri identifies what he terms "sequence-based" codes and "structural" codes. These include not only the well-known genetic code (DNA to protein via mRNA) but also the splicing code, which governs how pre-mRNA is edited into mature transcripts, and the signal transduction codes, which translate extracellular stimuli into intracellular actions through cascades of phosphorylation, second messengers, and conformational switches. The ubiquitin code tags proteins for degradation or trafficking, while the histone code—via post-translational modifications like methylation and acetylation—regulates chromatin structure and gene accessibility. The epigenetic code regulates the accessibility of genes to the transcriptional machinery, effectively turning genes on or off in response to environmental cues or developmental stages. This additional layer of information allows for dynamic regulation of gene expression, contributing to cellular differentiation, adaptation, and the inheritance of acquired traits. Other intracellular codes such as the tubulin code influence microtubule function and intracellular transport, while compartment codes organize subcellular domains. More recently, the laminin code has been proposed to orchestrate nuclear shape and transcriptional landscapes, and molecular codes in general highlight how semiotic relations arise even at the level of protein-protein interactions and macromolecular assemblies.

With the evolution of multicellularity, life required new layers of meaning-making, and thus, new codes emerged to ensure spatial, temporal, and functional integration across cells. These include the neural code, which converts electrical and chemical signals into perceptions, behavior, and memory; the Hox code, which governs body plan specification during embryonic development; and the transcriptional codes, which orchestrate combinatorial expression patterns necessary for cell differentiation. The adhesion codes, composed of surface molecules like cadherins and integrins, help cells recognize one another and maintain tissue architecture. The sugar code, involving glycosylation patterns on proteins and lipids, contributes to cell-cell recognition, immune modulation, and developmental signaling. Furthermore, the apoptosis code ensures the programmed removal of cells in a context-dependent manner, crucial for tissue homeostasis. Finally, the bioelectric code, encoded in voltage gradients and ion channel activity, provides a spatial blueprint for tissue patterning and regenerative processes.

What unites all these codes under Barbieri's framework is that they function as conventional correspondences between two independent worlds—material components and symbolic interpretation—requiring a set of rules or molecular "interpreters" that are not dictated purely by physical causality. In this view, *life is not merely a collection of biochemical reactions*, *but also a semiotic system*, layered with multiple levels of code-mediated interpretation and regulation.

Barbieri's pluralistic theory of biological codes reframes the cell and the organism not as linear machines but as multi-coded, multi-agentic systems, where meaning emerges through evolutionarily established correspondences.

It is important to separate the meaning of "bioinformation" from its information theory (IT) connotations and from any link to some "genetic program" or, in general, any "programmed" activity. I use the term "bioinformation" here in order to distinguish it both from Shannon information which refers only to syntactic information and from semantic information that implies the meaning of a message. *Bioinformation is more than syntactic and semantic information.* In the context of living systems, it refers also to *functional and effective information.* The purpose of bioinformation is not only related to feedback or communication in general but also to "in-form" (Kremer-Lecointre and Lecointre, 2023)—to help create forms, structures, and shapes. The cell can be considered a master builder (Martinez Arias 2023). Organisms use and produce *functional and effective information,* that is, information directed toward a desired outcome which is persistence.

As opposed to the simple architecture of signaling pathways in bacteria (Marijuán et al. 2010), eukaryotes have a complex signaling system with at least 21 independent pathways with different functionalities (Marijuán et al. 2013):

- *Early development*: Wnt, hedgehog, Notch, TGF-β, and neurotrophins.
- *Mid development and organogenesis*: integrins, cadherins, nuclear hormones, and Reelin.
- *Tissue physiology*: G-protein coupled receptors, guanylate cyclase, electromolecular transmission (4 main classes: gap junctions, stretch-activated channels, voltage-activated channels, and ligand-gated channels).
- *Stress and criticality*: NFkB, cytokines, autophagy, apoptosis, hippo, and complement cascade.

Cells simultaneously receive many extracellular signals and the actions they take will be contextual depending on the crosstalk between the diverse signaling paths activated. One of the differences between normal cells and cancer cells is the nature of their communication with other cells located either in close proximity, in their microenvironment, or at distance, in the bone marrow, or other organs. Mutations in signaling pathways, such as signal transduction systems, are key triggering mechanisms in different types of cancers (Vogelstein and Kinzler 2004). As a result, communication, which is the essential homeostatic mechanism of living systems, is altered in cancer and, as a result, through a reversal of who hold sway over whom, cancer cells are able to develop parasitic traits exploiting the organism's resources for their own benefit.

During the last two decades, RNAs have been found to play many other significant roles besides transcription and the delivery of genetic information from the nucleus to the cytoplasm and serving as templates for protein production. Within a cell, RNA has important enzymatic activities (Cech, 2024). Many different classes of noncoding RNA molecules have been described (i.e., miRNA, piwi-RNA, snoRNA, long noncoding RNA, tRNA, etc.). The noncoding RNAs are involved in a large number of cell processes, and they can regulate many cellular processes including gene transcription, translation, splicing, and posttranscriptional regulation. It has been demonstrated that they can induce specific gene fusions (Gupta et al. 2018) and as such play a direct role in oncogenesis, regulation of the cell cycle (Bueno and Malumbres 2011), and cell motility (Götte et al. 2010), modulation of cytoskeleton dynamics and endocytosis (Gerasymchuk et al. 2020), and a key role in the control of glucose metabolism (Alamoudi et al. 2018). They have pleiotropic effects not only inside cells but are also actively involved in local and distal intercellular communication (O'Brien et al. 2018). Both miRNA (Van Roosbroeck and Calin 2017) and long noncoding RNA (Gutschner and Diederichs 2012) are involved in all of cancer's hallmarks.

Given their versatile functions both in unicellular and multicellular organisms, RNAs have fundamental roles in maintaining the unity of bioinformation/bioenergy/biostructure, and several RNA-first life origin theories have been proposed (reviewed in Robertson and Joyce 2012).

8.29 Biostructure

Biostructure refers both to the system of membranes delimitating the cells from the outside and to the different intracellular compartments, separating the nucleus and the different organelles from the rest of the cytoplasm, the system of tubules and microtubules of the cytoskeleton inside the cytoplasm and the nuclear matrix inside the nucleus, and also the structured and unstructured water and different ions and organic and inorganic molecules present in finely controlled amount inside the cell. As proposed by Michael Conrad, in order to reconcile stability with evolvability, biostructure incorporates three key features. First, compartmentalization enhances stability by isolating processes and protecting them from disruptive fluctuations. Second, multiple weak interactions among components foster flexibility, allowing systems to explore alternative configurations and adapt, thereby promoting evolvability. Third, redundancy of components provides a safety net: Duplicate elements can compensate for failures, supporting stability while also creating opportunities for divergence and innovation, which enhances evolvability. Together, these features enable living systems to remain robust in the face of perturbations while retaining the capacity to change and evolve (Conrad 1990).

In the context of a multicellular organism, the biostructure refers both to the material components organized at dif-

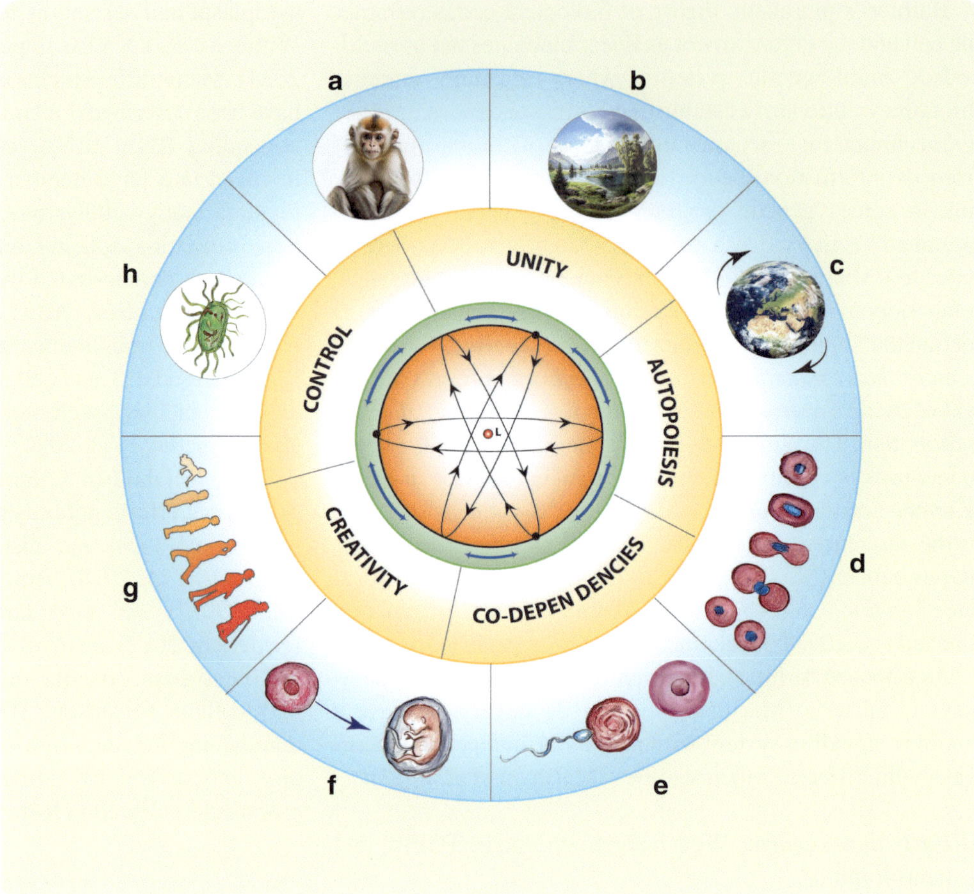

Fig. 8.8 Life's Dynamic and Creative Persistence (DCP) is represented by the bioenergy/bioinformation/biostructure triad (*inner circle*) maintained by unity, control, autopoiesis, codependencies, and creativity mechanisms (*middle circle*). DCP in action is represented by the several life forms and living processes depicted in the *outer circle* (evolution not shown due to lack of space): (*a*) multicellular organisms and (*h*) unicellular organisms; (*b*) ecosystems; (*c*) the Biosphere; (*d*) mitosis and meiosis; (*e*) fertilization and egg formation; (*f*) embryonal development; (*g*) development and aging. (Courtesy of Ana Tourian)

ferent scales (cell, tissue, organism) and to the networks of interactions spanning these different scales facilitated by the circulatory and the nervous systems. The biomechanical properties of the cytoskeleton can influence both transcription, translation, metabolism, and the biostructure itself, as it has been demonstrated by the original work of Ingber on tensegrity (Ingber 2003a, b) and his subsequent work supporting the role of tissue mechanics on cellular growth and cancer inception and progression, respectively (Ingber 2005, 2008; Huang and Ingber 2005; Mammoto and Ingber 2009). In turn, biostructure may be influenced by bioenergy and bioinformation. Epigenetic mechanisms control the cytoskeleton (Walker and Burggren 2020). Also, the histone code (Strahl and Allis 2000; Cosgrove and Wolberger 2005; Rothbart and Strahl 2014) and the tubulin code (Verhey and Gaertig 2007; Gadadhar et al. 2017) may be interconnected. Remarkably, this interconnection may lead to a separate, deterministic evolutionary process ("epigenetic entrainment" is the term proposed) through which "a changing environment could exert directionality by changing chromate-cytoskeletal activity to produce an advantageous phenotype while concomitantly changing the epigenome to drive the emergence of new epimutations that encode that phenotype" (Walker and Burggren 2020).

8.30 Dynamic and Creative Persistence at Work

Dynamic and creative persistence (DCP) is the universal principle underlying both cellular and organismal *processes,* that is, cellular division and death, spore formation and cellular movement, fertilization and egg formation, mitosis and meiosis, embryonal growth, adult development and aging, and the maintenance of unicellular and multicellular individual life *forms*, including individuals, ecosystems, and the entire biosphere (please see Fig. 8.8). Life forms that may persist for a shorter or longer duration of time are embodiments, guardians, and disseminators of the DCP. *The essence of life is the community, and not the individual.* Life is a dynamic network *process* and individual life *forms* are temporal instantiations that support this process. The function of living forms is to perpetuate DCP.

The fact that both cell division *and* cell death promote system stability, even while far from thermodynamic equilibrium, has been previously noted by Frieden and Gatenby (2011). Similarly, other processes like development and aging, adaptation, and the different types of evolution (i.e., classical Darwinian, drift, etc.) also support DCP.

Life is a dynamic network process involving the entire planet. The biosphere of our planet and its climate are tightly connected through nested networks with the different ecosystems and the multitude of life forms that thrive precisely because of this connectivity. There are intricate connections between the macrocosmic climate of planet Earth, the internal body climate of multicellular organisms, and the climates of the individual cells.

8.31 Defining Life Through the Lens of Dynamic and Creative Persistence

Across decades of biological inquiry, theories on the origin of life have postulated a different identity for life's *primum movens*. Some biologists proposed that life began with metabolism; others with genetic translation mechanisms involving DNA, RNA, and proteins; or with the emergence of self-enclosing membranes. Hybrid scenarios, such as Freeman Dyson's, propose that life's birth stemmed from a co-emergence of metabolic and informational systems—a dual genesis of form and function.

Similarly, oncology has mirrored divergent models in its search for cancer's origin: The somatic mutation theory (SMT) foregrounds genetic alterations, while the tissue organization field theory (TOFT) and metabolic hypotheses point instead to emergent properties of structure and function as the seeds of malignancy.

The theoretic framework presented in this book is not concerned with life's creation which is a completely different subject. I analyze here already existing living systems and life forms and the unity among the three primary life processes: transformation through *bioenergy*, computing of *bioinformation*, and maintenance of *biostructure* are taken for granted. These three processes, acting in mutual interdependence, form life's *biocore* and, for simplicity, we can use this term, *biocore*, whenever we refer to the bioenergy/bioinformation/biostructure triad.

Almost a century ago, Ervin Bauer formulated three universal principles that, in his opinion, apply to all living organisms. The second principle was quantitative: "All energy supplied by the environment must necessarily be completely transformed in the living being into forms of energy that work against the onset of the state of equilibrium [...] If we now call all processes in a system that are directed against the onset of equilibrium in a given environment, "*regulatory*", then our second principle in biological terms reads: *All life processes are necessarily regulatory*" (Müller and Igamberdiev 2024). What do all life processes regulate?

What is the outcome of this "*work directed against the onset of state of equilibrium*"? In my opinion, the answer is DCP, the unifying lens through which life can be redefined. Life's mechanisms are supporting DCP. DCP underscores living beings' "*lust for life,*" the essential vitality of the living: its refusal to settle, its drive to persist through flux and adversity (*dynamic persistence*), and its ceaseless invention of novel ways to do so (*creative persistence*). It is the heartbeat of life's adaptability, the signature of its deep time tenacity—from the first stirrings on Earth billions of years ago to the evolution of diverse, resilient forms in ever-changing environments.

Through the concept of DCP, we can formulate two working definitions of life that apply across scales—from cells to ecosystems, from mitosis to speciation. A first definition reads:

> *Life is the dynamic and creative persistence of the union of bioenergy, bioinformation, and biostructure in codependence with the environment.*

The *biocore*, this triadic unity, is not a fixed essence but a process—a movement through time, an ever-adjusting interplay among structural scaffolding, energetic flux, and encoded function.

A second, more abstract definition may be articulated to emphasize life's internal dialectic:

> *The living process is the creative interplay between maintaining a dynamic equilibrium between constancy and change and adjusting this balance within a fluctuating environmental context.*

This complementarity between stability and transformation, homeostasis and novelty, can be illustrated by the chaperone system. Molecular chaperones facilitate proper protein folding, maintaining structural integrity while adapting to physiological and stress-induced challenges. These molecular stewards, poised at the nexus of order and entropy, are rightly described as "standing at the crossroads of life and death," controlling essential cellular functions like genome stability, transcriptional regulation, proteostasis, maintenance of the cellular redox potential, and cytoskeletal dynamics (Quinlan and Ellis 2013). In every one of these functions, they act not merely as guardians of the cell but as agents of dynamic creative persistence (DCP).

Thus, life is not simply what resists death—it is what *persists* by *inventing,* what *endures* by *becoming*. Death is a key survival strategy of the living process. DCP offers a lens through which the vitality of the living can be seen not as a static state but as a sustained act of balancing, improvising, and evolving in concert with the world around it.

9 Homeodynamic Laws and Life Principles

9.1 The Homeodynamic Laws Governing the Living Process

Homeodynamics theory is proposed as a new foundational biology model and, *more geometrico*, a comprehensive formalism consisting of an axiom, three laws, and eight heuristic principles that are introduced as universal rules of the living process.

DCP, amply discussed in the previous chapter, is the axiom underlying the living process. In summary, life is autopoietic and teleonomic and persists by cyclic causation, that is, by being its own goal. This axiom may be formulated in many ways but, in essence, it spells an evidence: Life begets life; the purpose of life is life itself. As introduced earlier, the *Axiom of Choice* reads: For the living process as we know it, there is only *one single choice: persist*. Like *c*, the speed of light in Einstein's special relativity theory, the autotelic nature of life is a foundational empirical observation on which the whole structure of biology may be build up.

For living systems to persist, they must continuously regulate the two tendencies described by the second law of thermodynamics: the drive toward increasing entropy, or greater disorder and a larger number of accessible microstates, and the drive toward decreasing enthalpy, or a shift to more stable, lower-energy states. The balance between entropy, enthalpy, and temperature ultimately determines whether a process is spontaneous. Importantly, living systems do not simply resist entropy since higher entropy can generate diversity and evolvability, nor do they resist enthalpy since lower enthalpy can provide stability and programmability. Instead, life is fundamentally homeodynamic: It actively regulates both thermodynamic parameters and informational parameters. From this perspective, homeodynamic theory serves as a unifying link between thermodynamics and information theory.

Newton formulated the following three laws of motion:

1. An object at rest remains at rest, and an object in motion remains in motion at constant speed and in a straight line unless acted on by an unbalanced force.
2. The acceleration of an object depends on the mass of the object and the amount of force applied.
3. Whenever one object exerts a force on another object, the second object exerts an equal and opposite on the first.

In analogy to Newton's three laws of motion, the homeodynamic laws of living processes may be formulated as follows:

1. *The Conservation Law (Law of Functional Causality).* In the absence of disruptive influences (e.g., a non-supportive environment), the living process persists indefinitely, continuously self-creating and self-maintaining. Its persistence depends on the unity of the bioinformation-biostructure-bioenergy triad, sustained by interdependent homeodynamic mechanisms within a permissive environment.

   In brief, uninterrupted, the living process perpetuates itself through self-copying and evolution. The copies are similar, though never identical. *This law corresponds to homeostasis.*
2. *The Flow Control Law.* When internal or external stressors threaten to alter its trajectory, the biological system responds in an attempt to maintain its course. The living process is inherently homeorhetic—like a choreography governed by rules and constraints, with dynamic adjustments that sustain its flow. *The flow control law corresponds to homeorhesis.*
3. *The Co-dependency law (Law of Reciprocity).* Living processes are inseparable from their internal and external environments, which cocreate their trajectories. In turn,

D. Paul, *Cancer: An Integrative Approach*, https://doi.org/10.1007/978-3-032-08808-6_9

living processes reshape and "sculpt" those environments. An organism is a biosphere within an ecosystem, which, in turn, is part of a larger biosphere. The organism's biosphere ecosystems and the Earth's biosphere are codependent. *The co-dependency law corresponds to homeopoiesis.*

The homeodynamic laws are at the origin of life's autopoietic and teleonomic behavior and ensure life's persistence. Homeodynamic mechanisms contain feedback and feedforward processes that, up to a certain point, can restore a biological system's trajectory.

Darwin's theory of evolution refers to the origin of species and the transition between certain living forms within a species. Darwin focused on organisms, how they appear and why they change forms. Maintenance of form and persistence of process are closely related. Form maintenance comes in different flavors: the static persistence of a crystal form that obeys the laws of physics is different from the dynamic and creative persistence (DCP) of the form of a living organism, a bacterium, a plant, or an animal where both the laws of physics and DCP are at work. The form of an organism is actively maintained and serves the purpose of survival of the organism. Living forms are functional; their purpose is persistence of the living process. There is a never-ending cyclicity: The living process creates the forms that maintain it.

Homeodynamic theory focuses on the persistence of the living process as a whole, with evolution representing only one of its mechanisms. Organisms as such can be seen as guardians and disseminators of the flame of life. Importantly, both life forms and living processes are codependent players in life's game and, in turn, they both depend on the environment.

The living process, present across all life kingdoms and taxa (viruses, archaea, bacteria, eukarya), perpetuating since first life forms appeared on Earth, is a sort of temporal blueprint, a controlled entanglement of bioenergy and bioinformation embedded in a biostructure that continuously interacts and communicates with the environment. This "Temporal Blueprint" can be described by the motto: *E Pluribus Unum* ("out of many, one," Latin). This can be read as: out of many (organisms and processes), one event (life). As explained by Jeffrey Wicken (1980), organisms are far-from-equilibrium systems that "maintain their organizational structures by irreversibly degrading free energy through informed kinetic pathways acquired through evolution. Dissipation through structuring is the strategy of life." Later on, he added that "the ontogenesis of biological organization is powered by the thermodynamic arrow but informed by genetic sequences" (Wicken 1989). In conclusion, both material structures *and* processes are needed for life to exist and function.

Similar to evolution by natural selection that encompasses multiple levels from molecules to biospheres, dynamic and creative persistence (DCP) is present also both at the lower scales of molecules, and regulatory networks and at the higher scales of ecosystems and biospheres. DCP of ecosystems and biospheres shapes the DCP of organisms and vice versa; furthermore, the DCP of organisms shape the DCP of cells and vice versa. In their book *The Origin and Nature of Life on Earth: The Emergence of the Fourth Geosphere,* Eric Smith and Howard Horowitz stated that "the fundamental unit of life is not the individual but the biosphere as a whole" and "the biosphere as a whole rather than any particular kind of individuality or ecology community-structure within it is the locus at which life is identified as a distinctive planetary phenomenon. Life is defined, whether for matter, for events, or for organizational forms, by participation in the biosphere." The two authors also alluded to similar ideas as those presented here: "Conflating life with the capacity for adaptive evolution marginalizes the features that do not change and the mechanisms that maintain them [...] because the nature of life is cast in terms of a dynamic." For them, these "features that do not change" are represented by chemical reactions and metabolism, considered the life core: "Metabolism is not a relic of proto-life but the distributed substrate and the reference for evolutionary coherence. [...] The ordered state of the biosphere itself is distinguished because it is an order that originates in chemistry, and within chemistry it opens distinctive domains of covalently bonded molecules and energy flow through reactions, which would not be accessed by other geospheres in the absence of life. Far from being an incidental substrate to a 'living process' defined by abstractions, we argue that the chemical substrate is the unifying precondition that entails the whole range of other distinctive features of life" (Smith and Morowitz 2016). Microscopic DCP at the molecular level and macroscopic DCP are interconnected in a Klein-like bottle shape.

9.2 Life's Principles

Life as we know it harnesses energy, computes information, and continuously builds, maintains, and destroys biostructure. These three processes are closely interrelated and form an autopoietic and self-copying system in continuous interaction with the environment. This autotelic system is able to persist indefinitely, self-generating itself, through eons of time.

Life's DCP is driven by the triad bioenergy/bioinformation/biostructure maintained through five processes:

- Autopoiesis that includes maintenance and reproduction.
- Control through homeodynamic mechanisms.

- Codependence in which there is interaction with the environment: Adaptation, development, and evolution are codependent phenomena. Also, subsystems of the larger system are codependent. Codependence is fundamental, that is, *the ontological primacy of mutuality represents the foundation of all living systems.*
- Unity allowing a relative autonomy.
- Creativity related to emergence and complexity.

Summing up, eight interconnected life principles pointing to eight fundamental processes of living organisms can be described (please see Fig. 9.1).

1. *Principle of Bioenergy.* **Life as an energy flow. Metabolizing.** Life requires both a continuous supply of free energy and of carbon to drive formation of organic molecules, that is, proteins, DNA, etc. It also requires catalysis and excretion of waste to channel chemical reactions in the right direction (Lane 2015). The principle of energy is generally understood as metabolism which is responsible for energy and matter transformation and generation. The living systems are open systems able to exploit entropy through controlling and organizing actions, structuring space and time, information, and energy flow. Their openness is both horizontal—cells interact with stromal components—and vertical to levels of organization below (the molecular level) and above (the macroscopic level). Living systems are not only able to contain and transform already existing entropy into an organizing principle helping them create and maintain order from disorder but can also actively increase entropy as a homeopoietic mechanism of controlled novelty production.
2. *Principle of Bioinformation.* **Life as an information flow. Communicating.** Life entails communication, codes, information computing, algorithms, memory, learning, meaning creation, and directed behavior.
3. *Principle of Biostructure.* **Life as a structuring flow. Organizing**. Life is structured space and time through hierarchy/holarchy, modules, and compartmentalization. Life forms, circadian rhythm, and embryonal growth and development are all expressions of this structuring flow. Aging and death represent disintegrations of life's organizing flow.
4. *Principle of Control.* **Life as a homeodynamic flow. Controlling.** Life is an interplay between stability/change. Homeostasis, homeorhesis, homeopoiesis, biocognition, the internal climates located at different scales in multicellular organisms, and other control mechanisms, that is, allostasis, are all reflections of this principle.
5. *Principle of Autopoiesis.* **Life as a mirroring flow. Copying.** Autocatalysis, autopoiesis, self-replication, historicity, and reproduction are all manifestations of this principle.
6. *Principle of Unity and Connectivity*. **There is one single flow**. **Integrating**. Life persists through coherence and union. Multicellular organisms are societies of semiautonomous entities, confederations of different species, that is, holobionts, ecosystems, and biospheres. The liv-

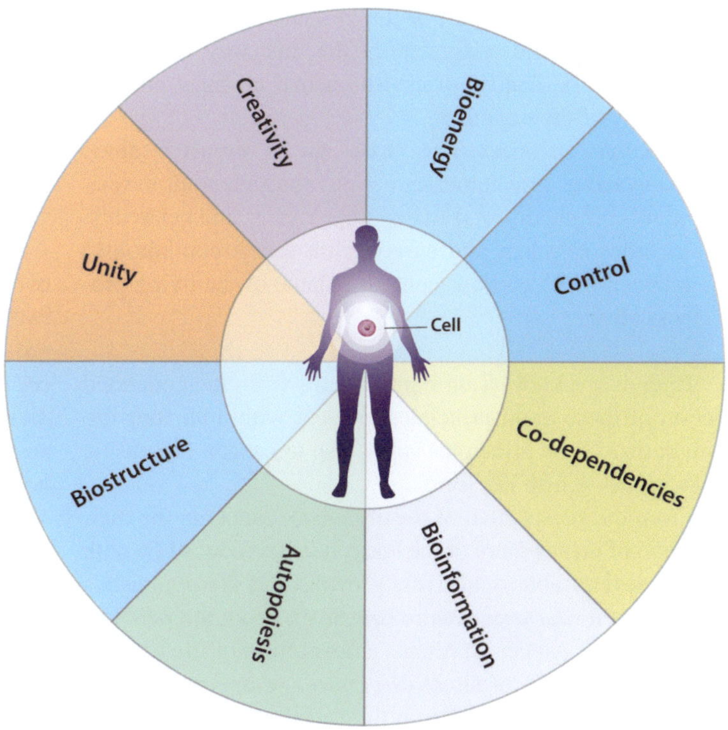

Fig. 9.1 Life principles governing both the human organism and its cells. (Courtesy of Ana Tourian)

ing process exists both in space, and it also extends in time toward the past through ancestry lineage of an organism, toward the immediate future through the developmental life cycle of the organism, and toward the distal future of the species to which the particular organism belongs. Both living processes and life forms are complex nested configurations of fluxes and networks. *Life is a community* of trillions of molecular, cellular, and multicellular organisms interacting and creating coherent and dynamic networks. It does not represent the solitary actions of molecules or cells. *It takes the whole universe for life to exist and persist.*

7. *Principle of Co-dependency.* **Life as a codependent flow. Adjusting.** Living organisms are part of ecosystems and they openly interact between themselves either competing, cooperating, or both. They adjust to the environment through natural selection, mutation, and migration, and, reciprocally, they adjust the environment to their needs through niches formation. The internal environments of organisms are distorted mirrors of external environments. In a multicellular organism, life is a network of inside-outside codependencies with membranes in between: The intracellular ions composition, for example, is very different inside from outside the cells, and actively maintaining this difference makes life possible. During the history of life on Earth, living systems acted as dynamic repositories of their environments, continually internalizing external influences. An organism's evolution is the story of new developmental responses to environmental challenges—responses that, when repeated, become canalized into robust pathways and ultimately become part of the genome through the process of genetic assimilation. In this way, evolution inscribes the history of environmental changes into the biology of the organism.
8. *Principle of Creativity.* **Life as a creative flow. Innovating.** Life harnesses novelty through random variation, complexity, and emergent processes and behaviors. In order to adapt and survive, life forms continuously reinvent themselves, keeping the flame of the living process alive.

Life, as we know it on Earth, results from the concerted action of these eight principles, present within all four life kingdoms. These principles apply also to cancer, a parasitic dyskariota—a fifth life form.

From the perspective of the organism, cancer is the malfunction of one or more of the body systems induced by both one or several abnormal tissue growths and the organism's reaction to it (*the organism perspective*). From the perspective of cancer, cancer represents a novel form of life exploiting the organism's resources (*the cancer perspective*).

Physicians focus mainly on the disease that results from the interaction of the organism and the cancer system (*the organism perspective*) as opposed to the cancer researchers who study the process of carcinogenesis and focus on the cancer cells or the cancer tissues (*cancer perspective*).

In layman's terms, when we talk about "cancer," we usually refer to cancer as a disease, that is, *the organism perspective*.

The organism perspective Energy restriction prolongs human lifespan, while excess calories may lead to obesity, cancer, and shortened lifespan. The organism continuously recreates itself and, besides systemic biostructural, bioenergetic, and bioinformation changes in the organism that may precede cancer for many years, cancer is also a failure of the homeostatic processes, a disorder of organization, of unity disintegration, of arrested differentiation, and an adaptation to a modified intracellular or extracellular environment. Apart from its biological and clinical features, cancer is also a psychosocial disease with a strong existential impact.

The cancer perspective Individual cancer cells and also cancer as a system are functionally and structurally very different from the normal cells and the organism as a whole. Cancer cells and the cancer system as a whole each have different homeostatic parameters from the normal organism. Cancer tissues may contain multiple different cancer clones and, morphologically, they appear very disorganized as compared to normal body tissues or organs but behind this misleading chaotic appearance lurks the implacable logic of a life form struggling to survive at all costs. *The cancer system has a functional unity as opposed to a structural unity.* Cancer cells and the cancer system as a whole adapt, develop, and evolve as an opportunistic life form created and maintained by the organism.

In cancer, homeostasis and autopoiesis are perturbed, the autopoietic function sometimes is on turbo drive, the codependence with the environment becomes a parasitic relationship, and unity with the organism is altered. Cancer cells are extremely adaptable, and their plasticity allows them to survive despite an armamentarium of treatment strategies.

Fundamentally, cancer as a disease is the interaction of two living systems, one cellular, the other multicellular, and two forms of persistence, one mainly temporal (the cancer process), the other mainly spatial (the organism where cancer develops). All cancer processes from excessive cellular division to tissue invasion, metastatic spread, and dormancy are cancer's struggle for life as a life form, displaying the homeodynamic laws and life principles in action.

In a healthy individual, the organism maintains phenotype stability of its trillions of cells both by systemic control through hormones and cytokines and through local interactions between cells themselves and between the cells and the stroma. In cancer, the direction of control going from the organism to the cells is reversed, and cancer cells are able to instigate other normal cells to actively support the malignant process through the formation of niches supporting their growth and also by actively promoting the invasion and the metastatic process. Cancer cells are in a continuous dialogue with the stroma akin to development, and the cancer process is actively supported by the organism during its progression like a pregnancy. The unconstrained division, the local invasion, and the distal metastasis exhibited by cancer cells are all the outcome of *reversed causation*, which represents, in my opinion, the essence of the malignant process.

Metaphorically speaking, malignant cells break the "sustainability" contract with the organism that stipulates that the cells receive a certain amount of nutrients at certain precise times in return for doing their job for the "society of cells" (Sonnenschein and Soto 1998) represented by the organism. Viral life forms persist by forming viral fabrics inside the cell and by using the cellular mechanisms to form virions. Similar to a virus, rogue cells start exploiting the organism's resources, tricking it to contribute actively to their growth and to cocreate cancer-supporting niches. Cancer is the organism's creation, which not only tolerates its presence but it actively enables and supports cancer's progression.

What has been neglected in oncology research, in general, is the role crucial role the entire organism plays in cancer, and the continuous top-down and bottom-up communication between cancer and the organism.

9.2.1 Principle of Bioenergy

Starting with the postulated last universal common ancestor (LUCA), the storyline of life has been relentlessly unfolding on the planet Earth for 3.95 billion years (Weiss et al. 2018) or longer – 4.2 billion years at least according to new data (Moody et al. 2024). This remarkable continuity is driven by far-from-equilibrium nonlinear ordering processes. As explained by Schrödinger, in the presence of free energy transduction mechanisms, the second law of thermodynamics permits the coupling of exergonic processes to endergonic, organized, life structures (Schrödinger 1944). As Glansdorff and Prigogine, cited by Morales and Michaelian, demonstrated, if dissipation of energy increases in a system, the system may become organized: "A non-linear system over which a generalized thermodynamic potential is imposed can 'self-organize' into structures (or more correctly, processes) which can break symmetry in both space and time" (Morales and Michaelian 2020). Basically, "the 'physical purpose of life' is to dissipate energy gradients in both time and space" (Popa 2009).

So two key requirements for the life-ordering process are a continuous external source of energy (plant or animal base nutrients) and an internal "coupling mechanism that channels the energy along specific chemical pathways in order to accomplish a very specific type of work" (Thaxton et al. 2020).

Numerous proposals have been put forward attempting to create a biophysical/biochemical foundation for biology. As shown by the work of Jeremy England, in a thermodynamic system, structural organization is related to energy flow. Processes such as DNA replication or cell division, which can absorb and dissipate large amounts of energy from external sources, are more likely to occur not for mysterious reasons but as a direct consequence of thermodynamic laws (England 2013). The connection between this dissipation-drive adaptive organization and self-replication and evolution has been explicitly stated by England and his collaborators. "When a self-replicator exhibits adaptive traits that match well with its present environment, we can seek an explanation in the hypothesis that the organism's ancestors possessed these traits and found them helpful in the struggle to survive and generate many offspring [...] the link between this picture and the thermodynamic one comes from the fact that, as a result of its statistical irreversibility, *self-replication is necessarily a dissipative process,* so exponential population growth turns out to be one very reliable mechanism of driving external entropy production" (Perunov et al. 2016).

The link between thermodynamic principles and evolution has been also made by Teemu Mäkelä and Arto Annila: "The second law of thermodynamics has been held responsible for the natural patterns for a long time but only recently, when the law was formulated as an evolutionary equation of motion, the association became unambiguous" (Mäkelä and Annila 2010). In an attempt to create a bridge between physics and biology, Arto Annila expanded Maupertuis' principle of the least action to include multiple processes both from the living and from the inanimate world (Annila 2020).

Metabolism is the key process through which living systems transform energy acquired from the environment into a highly ordered state. All metabolic reactions are ruled by the second thermodynamic law. Living systems need a constant supply of energy in order to maintain their life cycles. Organisms harvest energy from their environment to satisfy this constant demand. This energy may be visible light used by plants in photosynthesis or sugars, present, for example, in the chemical bonds of molecules, and used by organisms in their metabolisms to store energy in chemical compounds like ATP, NADH, or FADH and to synthetize biopolymers like DNA, RNA, lipids, and proteins (Weilandt et al. 2021). Recently, Nick Lane's book *Transformer: The Deep Chemistry of Life and Death* placed metabolism at the center of all life processes: "Metabolism is what keeps us alive – it is what being alive is – the sum of the continuous transformations of small molecules on a timescale of nanoseconds, nanosecond after nanosecond" (Lane 2022).

Organism Perspective

Metabolism prevents entropic decay of the cells and organisms by using imported nutrients to generate energy and building blocks, replenishing damaged molecules and organelles (Dang 2015). On the other hand, a "proliferating cell must capture enough energy and mass to replicate, in addition to the energy required to dampen entropy" (Dang 2012).

The essence of organismal metabolism is the coupling of exergonic catabolism through which energy is extracted from breaking chemical compounds with endergonic anabolism that results in growth, reproduction, and maintenance. Metabolic changes at the level of the entire organism may increase the risk of certain cancers. Basal metabolic rate (BMR) is an indicator of overall body metabolism and may be a proxy for the impact of a specific metabolic profile on cancer risk. The European Prospective Investigation into Cancer and Nutrition (EPIC) study included 141,295 men and 317,613 women, who were followed up on average for 14 years. The study demonstrated that a *higher* BMR is associated with the risk of certain cancer types independent of body fatness (Kliemann et al. 2020). Subsequently, the same association between high BMR and cancer risk was found in a Mendelian randomization study (Ng and Schooling. 2021). On the other hand, caloric restriction was found to prolong lifespan of different organisms from yeast to mice and also decreased tumorigenesis in several cancer models (Dang 2015).

An original scheme of the systemic metabolism of the organism was proposed by Ye and Medzhitov (2019). The two authors suggested that the metabolism represents a sort of organismal economy ruled by the laws of supply and demand. "For example, liver and adipose tissue (producers) supply glucose and lipids (metabolic goods) to the circulation (the market), whereas the brain, skeletal muscle and other organs (consumers) take up the glucose and lipids from the circulation according to their metabolic demands. The production and consumption of glucose and lipids is coordinated to balance supply and demand by the action of various metabolic control signals (such as insulin, glucagon, and catecholamines) (Ye and Medzhitov 2019). The whole body's metabolism and the cellular metabolism are intricately connected: "The uptake of glucose by skeletal muscle in response to AMPK activation is an example of demand-driven uptake at the organismal level but a part of homeostatic control to maintain intracellular ATP levels at the cellular level; thus, AMPK is a homeostatic sensor at the cellular level but a demand sensor at the organismal level" (Ye and Medzhitov 2019). In type 2 diabetes, insulin levels are elevated and may increase cancer risk through its mitogenic effect; conversely, antidiabetic medications that lower insulin levels may be beneficial in some types of cancer treatment (Shafiei-Irannejad et al. 2017).

Cancer Perspective

It has been proposed that cancer represents a more stable state of the cell: "Energy loss (e.g., through impaired mitochondria) or disturbance of information (e.g., through mutations or aneuploidy) or changes in the composition or distribution of matter (e.g., through microenvironmental changes or toxic agents) can irreversibly disturb molecular mechanisms, leading to increased local entropy of cellular functions and structures. In terms of physics, changes to these normally highly ordered reaction probabilities lead to a state that is irreversibly biologically imbalanced, but that is thermodynamically more stable" (Hanselmann and Welter 2016).

Hanselmann and Welter suggested that the *primum movens* in oncogenesis is represented by an entropy increase due to a lack of nutrients needed to maintain a low entropic state. Once the cell has entered a high entropic state, replication error repair is impacted leading to an increased rate of mutations and aneuploidy that may be the origin of the cellular cancer state (Hanselmann and Welter 2016).

A similar opinion is shared by Stephen Wolfram: "I believe that cancer is a special kind of adaptation to energetic overload, characterized by multiplication and mutation of genomic DNA (generation of new biomolecules which enhance the probability of survival under harmful conditions), and by chiral alterations (reduction of entropy by entrapping energy) leading to abnormal configurated biomolecules. *In this regard, the genetic alterations are probably secondary changes.* Cancer serves to dissipate energy in a type of developmental process but one in which the results are harmful to the whole organism: an entropic devolution" (Wolfram 2023).

Cancer has a different energetic storyline from the organism and is "in a continuous on state." This, in turn, leads to the establishment of a "perpetual pro-anabolic state of metabolism in affected cells and enables uncontrollable accumulation of transformed cells and tumor expansion" (Pavlova et al. 2022).

More than a century ago, the connection between metabolism and oncogenesis was made by Otto Warburg, who considered cancer to be a mitochondrial structural abnormality, or, alternatively, a mitochondrial malfunction disease, with cancer being a metabolic disorder in which the metabolism has been rerouted to anaerobic glycolysis (Warburg 1956; Seyfried et al. 2014). According to this mitochondrial metabolic theory of cancer, the large genomic heterogeneity seen in tumor cells arises as a consequence, rather than as a cause, of mitochondrial malfunction (Seyfried 2015).

A compelling model linking systemic and cellular metabolism in cancer was proposed by Maurice Israël and Laurent Schwartz (Israël and Schwartz 2011). The two authors advanced the hypothesis that cancer cells may exhibit a hybrid metabolic phenotype that allows them to exploit the catabolic state; they themselves help to induce. Under normal physiological conditions, starvation promotes catabolism, whereas elevated blood glucose levels activate anabolic pathways. These two metabolic states are typically mutually exclusive, regulated by specific enzymatic configurations. In contrast, cancer cells maintain a hybrid enzymatic profile, enabling the simultaneous activation of both catabolic and anabolic processes. Despite the systemic cachectic, starvation-like condition observed in many cancer patients, tumor cells preserve a robust anabolic activity.

To support their hypothesis, the two French authors proposed that cancer cells display a unique metabolic signature in which some enzymes are phosphorylated in a manner characteristic of catabolic hormone signaling via Gs protein-coupled receptors, while others adopt configurations associated with anabolic signaling mediated by tyrosine kinase receptors. This metabolic duality allows for persistent anabolic activity in cancer cells, often driven by constitutively active or overexpressed tyrosine kinase receptors due to oncogenic mutations or gene amplifications.

At the biochemical level, key enzymes such as pyruvate kinase (PK) and pyruvate dehydrogenase (PDH) are inactivated through phosphorylation, mimicking the metabolic configuration seen during gluconeogenesis. Nevertheless, enzymes such as citrate synthase remain active, maintaining glucose flux through glycolysis. This paradoxical regulation supports high glycolytic throughput, supplying biosynthetic precursors necessary for sustained mitotic activity. As a result, cancer cells are able to simultaneously repress mitochondrial oxidative metabolism while enhancing glycolysis, thereby acquiring biosynthetic and energetic advantages that facilitate rapid tumor growth. This metabolic reprogramming also enables tumor cells to utilize host protein and lipid reserves, contributing to systemic cachexia and outcompeting normal cells for nutrients.

Maurice Israël further hypothesized that this hybrid metabolic state originates in pancreatic dysfunction, specifically involving dysregulation of the GABAergic switch that normally modulates the balance between anabolism and catabolism. This dysregulation results in the simultaneous release of glucagon (catabolic) and insulin (anabolic), disrupting systemic metabolic homeostasis and predisposing certain cells—particularly tissue-resident stem cells—to metabolic rewiring (Israël 2012, 2014, 2019).

According to this model, stem cells are the first to adopt a hybrid metabolic phenotype. Over time, the persistence of this altered metabolism, facilitated by somatic mutations or epigenetic modifications, selects for increasingly aggressive clonal populations that ultimately give rise to malignant cells. Initially committed to tissue repair, these metabolically reprogrammed stem cells become neoplastic, leveraging their bioenergetic flexibility to dominate resource competition with surrounding normal cells.

A key translational implication of Israël's model lies in its potential utility for early cancer detection. The model predicts that metabolic alterations precede overt tumorigenesis by several years, offering a window for early diagnosis via metabolomic profiling. Supporting his model, a study by German investigators using data from the European Prospective Investigation into Cancer and Nutrition (EPIC) cohort reported that specific lipidomic signatures—higher levels of phosphatidylcholine PC ae C30:0—were consistently associated with an increased risk of developing breast, prostate, and colorectal cancers (Kühn et al. 2016). These alterations were detectable years prior to clinical diagnosis, suggesting a global metabolic shift, particularly in phosphatidylcholine metabolism, that may serve as an early marker and potential driver of tumorigenesis.

Recently, a similar feedback model of two interconnected "vicious cancer progression cycles" has been proposed by Polina Schwartsburd (2020). In this model, on one hand, cancer-mediated hypoglycemia stimulates glucose sensors that activate the secretion of catabolic hormones—such as adrenaline, glucocorticoids, and glucagon—and stimulates hepatic glucose production. The other complementary cycle redirects the available glucose from host insulin-resistant tissues (muscle and adipose tissue) to cancer cells. This model illustrates how systemic glucose metabolism is reprogrammed by cancer cells via hormonal deregulation.

The essential role of cellular and systemic metabolism in controlling cancer progression has been demonstrated by a group from MD Anderson Cancer Center. Using molecular data of 9125 patient samples from *The Cancer Genome Atlas*, a group of researchers identified distinct metabolic expression subtypes in 27 cancer types based on mRNA expression patterns of seven major metabolic processes (amino acid metabolism, carbohydrate metabolism, integration of energy, lipid metabolism, nucleotide metabolism, tricarboxylic acid cycle, and vitamin and cofactor metabolism) (Peng et al. 2018). The metabolic expression subtypes correlated with clinical outcomes: Subtypes with upregulated carbohydrate, nucleotide, and vitamin/cofactor metabolism most consistently correlated with worse prognoses, whereas subtypes with upregulated lipid metabolism showed the opposite result. The most interesting finding was that these metabolic subtypes were *not* related to specific genetic somatic drivers but were intrinsically coupled with cancer hallmark pathways (i.e., angiogenesis, cell division, etc.) and were modulated by highly recurrent master regulators across cancer types, ultimately leading to consistent survival patterns. As a proof-of-concept in vitro experiment, the authors also demonstrated that knockdown of two master regulators genes of carbohydrate metabolic subtypes (SNAI1 in a lung cancer cell line or RUNX1 in a sarcoma cell line) significantly decreased the concentrations of intracellular glucose. According to this model, the master metabolic regulators identified were key nodes with the greatest influence on systems-level metabolic activities; thus, targeting these metabolic master regulators may inhibit tumor progression. Strikingly, all four master metabolism regulators genes identified in the eight cancer types with significantly worse survival rates due to upregulated carbohydrate metabolism, SNAIL1, RUNX1, RUNX2, and FOSL1 (Peng et al. 2018), also play also a key role in embryonal development and EMT (Cieślik et al. 2013; Hong et al. 2017; Dongre and Weinberg 2019) and might also be master regulators of the metastatic cellular program.

9.2.2 Principle of Bioinformation

Living systems are remarkable in their ability to store, process, and transmit information. Information is a fundamental aspect of life, governing the intricate processes that sustain and propagate biological entities.

While instantiated within material structures, information remains immaterial in nature, a point emphasized by Deacon: "Attempts to attribute a quasi-substantial quality to information or to reduce it to some specific physical property are not only doomed to incompleteness, they ultimately ignore its most fundamental distinctive characteristic" (Deacon 2010). On the other hand, within the constructor theory of Deutsch and Marletto, "a bit is not an abstraction, independent of the physical world: whether or not something has the two counterfactual properties – one, that it is possible for it to be in either of the two physical states ON and OFF; and, two, that the state of ON and OFF can be permuted into one other and also copied into any other physical system that itself has these same two counterfactual properties – entirely depends on the laws of physics" (Marletto 2021).

An organism is an open system, both thermodynamically and informationally. Therefore, the description of the energy flow in biological systems should be complemented with a parallel account of the cellular information flow (Marijuán and Navarro 2022).

The two pathways of information flow and the energy flow are integrated in a cell through common constitutive components. A cell is able to distinguish whether a signal from the environment is a valuable information that needs to be further transmitted or represent a potential energy source that needs to be imported into the cytoplasm in order to extract its free energy. So metaphorically speaking, the cell first "reads" the suitability of a molecule for use as a food before "eating" it (Marijuán and Navarro 2022).

Syntactic information refers to the structure or pattern of data, without regard to its meaning. In biological systems, this corresponds to the sequence of nucleotides in DNA or amino acids in proteins—the "code" itself. Shannon information theory provides a measure of syntactic information, which reflect the amount of statistical correlation between systems.

Semantic information, on the other hand, refers to the meaning or interpretation of that code within a specific context. It is the functional significance of the syntactic information. For living systems, semantic information arises when the syntactic code is interpreted and utilized by cellular machinery to perform vital processes like protein synthesis or gene regulation or by the organism to interpret signals from the environment. Meaning emerges from the cellular capability to self-modify its structures in response to external signals—and to relevant inner changes as well. It is considered that semantic information exists only in the context of living systems (Johannsen 2015). As Weber suggestively put it, "Living cells use DNA to construct the organisms rather than the reverse." Some authors consider evolution the source of semantic information, that is, meaning = information + evolution (Johannsen 2015; Rovelli 2016).

The interplay between these two types of information is crucial for life's existence. The syntactic information encoded in DNA provides the blueprint, but it is the cell's semantic interpretation of this code by the cell that allows for the emergence of functional proteins, regulatory networks, and ultimately the complex behaviors that characterize living systems (Martinez Arias 2023).

Of note, through the book, I am using the term *bioinformation* in a broad sense that covers both information's syntactic and semantic meanings.

As living systems evolve, changes to the syntactic information (mutations) can alter the semantic meanings, potentially producing new traits. In cancer cells, genetic alterations may lead to phenotypic and behavioral changes. Conversely, changes in how the cell interprets and processes the syntactic codes—via regulatory mechanisms evolution—can modify semantic outputs while leaving syntax intact.

This ongoing relationship between syntactic information as code and semantic information as functional meaning enables the incredible complexity, context-dependent behaviors, and evolvability that characterize living systems. Syntactic information provides the "data," while semantic information facilitates its purposeful expression and use for survival. In essence, syntactic information represents the structural patterns, while semantic information captures the contextual, functional significance derived from those patterns by a living system interacting with its environment. Their synergistic relationship is fundamental to biological information processing.

Also, besides the *intracellular* bioinformation flows continuously present between the nucleus and the cytoplasm and vice versa—and between the various organelles within the cytoplasm—there are also signaling systems that connect *extracellular* bioinformation flows with gene expression.

Artemy Kolchinsky and David Wolpert developed a mathematical formalism for semantic information based on the use of counterfactuals and a notion of viability. The authors defined semantic information as "the syntactic information has about its environment which is causally necessary for the environment to maintain its own existence," so, in their opinion, semantic information represents a subset of syntactic information (Kolchinsky and Wolpert 2018). Their formulation uses the state spaces and probability distributions for an agent A and its environment E to characterize the mutual information between the two, while the persistence of A (its ability to maintain a viable) is measured through a *viability function*. The concept of meaning here is thus taken in the most basic sense of being, that is, as related to an agent's continued existence. An important observation is that semantic information is fundamentally *asymmetrical* (unlike some measures of syntactic information, such as mutual information). For instance, as the authors explained, "a chemotactic bacterium swimming around a nutrient solution is presumed to have semantic information about its environment, but the environment is not expected to have semantic information about the bacterium. Our measures of semantic information are fundamentally asymmetrical – even when defined relative to a symmetric syntactic information measure like mutual information – because they are defined in terms of their contribution to the viability of the system, rather than the environment" (Kolchinsky and Wolpert 2018). Recently, Solwinski et al. demonstrated the utility of the semantic information formalism in a resource foraging agent model (Sowinski et al. 2023). Importantly, in Kolchinsky's and Wolpert's opinion, "having high levels of semantic information is a necessary, though perhaps not sufficient, condition for any physical system to be alive."

Meaning for a cell is action; its goal-oriented activities are embedded actions. These goal-oriented activities represent the ability of the cell to adapt its inner state in a context-dependent manner in response both to external signals—and to relevant inner changes. *Agency* is another word for goal-oriented actions and deliberately acting in response to environmental stimuli in ways that are self-maintaining, that is, meaningful. As Stuart Kauffman explained, "Acting in the world is "agency." All living things are agents: self-reproducing, nonequilibrium systems doing thermodynamic work cycles, able to choose and act on their own behalf in their worlds: "yuck" versus "yum" (Kauffman 2000). This is how, in Kaufmann's opinion, we begin to move from "*matter* to *mattering*" (Kauffman 2020). This "mattering" precisely is the ingredient not captured by Kolchinsky's and Wolpert's definition of semantic information.

Semantic information requires a receiver system (e.g., the cell or the human subject) that can interpret and act upon the syntactic code in a contextually appropriate way, extracting functional meaning from the data. This semantic layer is what allows syntactic information to be leveraged for survival and adaptation. Semantic meaning is relevant only in the context of some level of "cognition." As explained by Baluška and Levin, at any level of organization, "cognition" refers to the total set of mechanisms and processes that underlie information acquisition, storage, and processing (Baluška and Levin 2016). Multicellular organisms, such as animals and higher plants, exhibit complex goal-directed behaviors.

In a series of publications, James Shapiro has advanced the hypothesis that cognition represents a fundamental and intrinsic property also of living cells (Shapiro 2021). He posits that cellular behavior is goal-directed, as all cells must continuously adapt to dynamic environmental conditions, a process that he equates with cognitive function. In his paper, he contends that prokaryotic cells are capable of cognition because "they can recognize available nutriments, discriminate between them, utilize them selectively and control their movements towards places where nutrients or energy sources are more available. Prokaryotes engage in chemical communication that allows them to undertake certain complex multicellular activities, such as forming protective biofilms. They can undertake coordinated group migrations, discriminate and exclude unrelated competitors from kin. [...] When subjected to DNA damage, they activate targeted response systems that serve to restore genome activity" (Shapiro 2021). These adaptive, responsive, and coordinated behaviors parallel those observed in cancer cells, suggesting that malignancies may similarly exploit basic cognitive properties for survival and progression.

Building upon Shapiro's framework, Perry Marshall has further explored the concept of cellular cognition, with specific reference to oncogenesis. In line with arguments from his earlier work (Marshall 2015), Marshall proposes that cancer cells demonstrate capacities typically associated with cognitive systems, including self–/non-self-recognition, resistance to therapeutic interventions, and the recruitment of vasculature to sustain tumor growth. He conceptualizes the cell as a "volitional Turing machine," proposing that foundational biological phenomena—including the origin of life, evolution, information processing, cancer, and artificial intelligence—can be unified under the broader paradigm of cognition (Marshall 2021).

Cognition is closely intertwined with learning, the latter defined as an adaptive behavioral modification based on prior experience. This relationship further emphasizes the capacity of individual cells, including cancer cells, to dynamically respond to environmental stimuli in ways that enhance their survival and proliferative potential. Similar to cognition, learning can be defined as an adaptive change in behavior resulting from experience.

A multicellular organism uses multiple channels to communicate between same or different cell types including local paracrine diffusion, electric fields, lymphatic and blood vessels; the properties of the extracellular matrix and the interstitium; gaseous exchanges; functional fluids like urine, sweat, bile, mucus, milk, and tears; and force field conduits (skeletal and muscular systems) (Marijuán et al. 2015).

Organism Perspective

Multicellular life forms are nested hierarchies comprised of complex, embedded, multilayered networks of interactions. All organisms present "top-down causation by information control" (Auletta et al. 2008) through which the "system as a whole exerts causal control over a subsystem (e.g., a gene) via a set of time-dependent constraints" (Walker and Davies 2012). So in biological systems, "information manipulates the matter it is instantiated in" (Walker and Davies 2012), which is another form of hierarchy.

It is important to note that the meaning of information used by Auletta et al. refers to information "selected by the receiver" and not "centered on the signal/noise discrimination" as in Shannon's information theory. The authors explain that "at any biological level, the receiver is, in general, flooded with incoming data and has to separate background data (important but constant) and noise (irrelevant data) from relevant information (data that are needed for some purpose)." Therefore, *information control* consists of information selection—often involving an educated guess—from a *certain* point of view, and this represents the goal of the system (Auletta et al. 2008).

At the cellular level, the communication networks are comprised of genes, metabolic intermediates, miRNAs, and signaling molecules (proteins, lipids, ions). At the tissular level, the communication networks are comprised of interactions between different cell types and between the cells and the supporting stroma. At the organism level, the communication networks are comprised of interorgan communication and homeostatic systems (endocrine, nervous, immune, etc.).

Alongside the nervous and endocrine systems, the immune system may be conceptualized as an information-processing network responsible for both the establishment and maintenance of self-identity and the implementation of adaptive recognition mechanisms through selective processes (Ishida 2004). This dual function enables the immune system to construct a coherent definition of "self" while simultaneously identifying and responding to "nonself" elements. In the context of cancer, tumor cells often express specific antigens—including neoantigens—that, in principle, should elicit immune recognition.

However, in practice, these antigens frequently escape immune detection. This immune evasion can be attributed to multiple factors, including tumor-derived strategies that disrupt antigen presentation pathways, such as deficiencies in human leukocyte antigen (HLA) expression, thereby enabling tumor-associated antigens to remain concealed from immune surveillance (Hazini et al. 2021). Additionally, functional impairments within the immune system, such as T-cell anergy and exhaustion, further compromise antitumor immunity (Crespo et al. 2013).

Effective immune recognition of cancer cells thus requires not only the presence of immunogenic antigens but also the integrity of the informational signals necessary for initiating and sustaining an immune response.

Cancer Perspective

Through the interaction of the host with cancer, new cancer-induced pathologic systemic networks (CISPN) appear and, progressively, the whole organism is "cancerized," or instigated to support cancer development (Paul 2020). Fundamentally, cancer does not invent any system or processes—it "hacks" and uses preexistent modules already existent in normal cells.

Contemporary cancer treatments have historically focused on destruction—killing malignant cells through surgery, chemotherapy, or radiation. Yet, recent research suggests cancer may not merely be a chaotic disease but an *adaptive and information-processing system*, capable of decision-making, environmental sensing, and evolutionary learning. From this perspective, cancer behaves more like a *"cognitive system"* at the cellular and tissue level, responding to microenvironment changes and making dynamic phenotypic choices. This insight opens the door to a novel paradigm: *information therapy*, which aims not to kill but to *"convince" or "persuade"* cancer cells to adopt *less aggressive*, more *cooperative*, or even *dormant* states. This approach parallels communication strategies in immunology, developmental biology, and even social behavior where modifying informational contexts alters behaviors without brute force. The intelligence of cancer has been recently illustrated by an Italian group (Nicolazzo et al. 2023).

Drawing from insights in developmental biology, systems theory, neurobiology, and evolution, we can begin to "talk" to cancer in its own language—rewriting its goals, reshaping its perceptions, and ultimately changing its fate.

The work of Mina Bissel demonstrating the possibility of reversing of the cancer phenotype by normalizing the behavior of breast cancer cells placed in a normal breast environment is also illustrative of the possibility of a "therapeutic dialogue" between the organism and cancer, curbing cancer's aggressivity.

It has been recently shown that carcinomas are electrically excitable and capable of generating nonrandom action potentials (Quicke et al. 2022) that may represent a sort of cellular communication. Mustafa Djamgoz proposed the CELEX (cell electrical excitability) model, which postulates a correlation between ion channel activity and metastatic potential in cancer cells. According to this model, as cells acquire greater metastatic capacity, there is a progressive reduction in outward ionic currents primarily mediated by voltage-gated potassium channels (VGPCs), suggesting a loss of inhibitory electrical signaling. Concurrently, a de novo upregulation of functional voltage-gated sodium channels (VGSCs) is observed specifically in invasive and metastatic cancer cell phenotypes. These changes in ion channel expression and activity contribute to the acquisition of a pro-invasive cellular state. Based on this framework, Djamgoz proposed that a therapeutic strategy combining VGSC inhibitors with VGPC activators may exert a synergistic anti-invasive effect by restoring inhibitory signaling and suppressing pro-metastatic electrical activity (Djamgoz 2024).

Cancer Cell's Programs

If cancer is not a fixed identity but a dynamic cellular condition—capable of shifting between distinct states—then it may not have an irreversible destiny. Rather, it may represent a *reprogrammable cellular mode*, one that can potentially be rerouted away from malignancy and back toward physiological equilibrium. A view focused on the plasticity redefines cancer not solely as a product of immutable genetic lesions but as a malleable biological state—a framework that augments and refines the classical genetic model of oncogenesis. Many traits attributed to cancer cells, particularly those associated with invasion and metastasis, may not arise from autonomous, cell-intrinsic programs alone. Instead, these malignant features often emerge through *adaptive interactions* between cancer cells and their microenvironment—more a reflection of cellular plasticity than of Darwinian selection.

The metastatic journey, in particular, appears to be less about rare, genetically selected "supercells" and more about context-driven reprogramming. It is orchestrated through a molecular dialogue between cancer cells and the surrounding stroma, which secretes a suite of instructive factors, including TGF-β, HGF, TNF-α, Wnt, and PDGF, that modulate cellular behavior. These signals activate in tumor cells a repertoire of master transcriptional regulators, such as Twist, Snail, Slug, Zeb1, and Zeb2, that drive the epithelial-to-mesenchymal transition (EMT), a critical step in metastatic dissemination (Lambert et al. 2017; Riggi et al. 2018; Dongre and Weinberg 2019; Lachat et al. 2021).

This reprogramming is often mediated not by single mutations but by *noncoding drivers* such as microRNAs (miRNAs), which simultaneously modulate entire gene networks. This insight invites a broader conceptual framework in which cancer progression is governed not by a singular "cancer program" but by the interplay of multiple, partially overlapping cellular modules: a *division program* regulating key mitotic genes (Nath et al. 2015); an *invasion program* controlling genes essential for tissue infiltration (Chiaretti and de Curtis 2016; Saini and Courtneidge 2018); and a *metastatic program* driving EMT and systemic dissemination (Dongre and Weinberg 2019) (please see Fig. 9.2). Each of these programs appears to have its own regulatory architecture, yet some elements serve as hubs across them (Smolarz et al. 2022; Sell et al. 2023) For instance, miR-21 and miR-222 promote uncontrolled proliferation, miR-130 and miR-126 support angiogenesis, and miR-373 and miR-155 are implicated in both invasion and metastasis (Ruan et al. 2009).

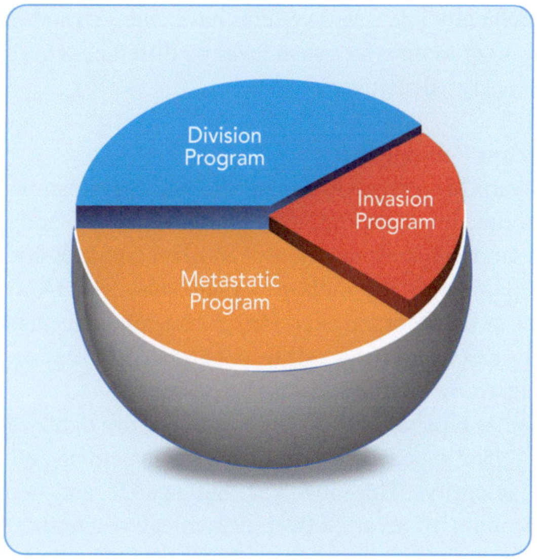

Fig. 9.2 Cancer cell programs. (Courtesy of Ana Tourian)

Such complexity suggests that cancer progression is not the product of a single genetic trajectory but rather a symphonic unfolding of reprogrammable states—each responsive to its internal and external environment. Within this framework, therapeutic strategies might not aim solely to eliminate cancer cells but to *redirect their fate*—rewiring their programs back toward organismal order.

9.2.3 Principle of Biostructure

There are several types of order in biology including synchronic structural (instantaneous) order that may be internal or external to the living system, and diachronic order, namely, order in time. Order can be measured by Shannon's relative information.

Organization refers to both a relatively stable material *structure* and a relatively stable *processual* structure, both underscored by homeodynamic mechanisms that ensure the maintenance of these structures.

Arto Annila and Esa Kuismanen argued that the origin of life's "hierarchical organization of 'systems within systems'" emerges naturally as a consequence of increased rates of energy dispersal ruled by the second law of thermodynamics (Annila and Kuismanen 2009). According to the two authors, "hierarchically integrated systems have emerged and evolved increasingly more effective in leveling differences and gradients in energy densities."

Organism Perspective

For Alvaro Moreno and Matteo Mossio, organization represents a fundamental property that defines all life and underscores an organism's autonomy (Moreno and Mossio 2015). As explained by the authors, organization is ubiquitously present in living organisms. Cellular metabolism must continually reconstruct the cell to counteract internal degradation caused by thermodynamic processes.

First, as biochemical systems, cells require tightly coordinated repair mechanisms to maintain structural and functional integrity. This involves the spatial and temporal orchestration of over 3000 biochemical reactions, whose products not only regenerate cellular components but also sustain the cell's capacity to acquire resources and eliminate waste. Such a self-sustaining, regenerative system is termed autonomy, characterized by internal constraint-based self-regulation—a fundamental prerequisite for life and biological evolution.

Second, multicellularity also requires higher-order organizational structures, such as the body systems (i.e., cardiovascular, lymphatic), to support distributed cellular metabolism across tissues and organs. These systems exemplify complex, hierarchical constraints that integrate cellular and systemic functions. A key observation is that organismal autonomy is shaped not solely by internal metabolism but also by regulated exchanges with the external environment. This dynamic interaction underscores the role of organization not only within organisms but also within their environments.

In summary, using a flight metaphor, the authors explained that "when it comes to the expansion of life on the planet, it may be evolutionary selection that provides the thrust, but it is organization that provides the lift. [...] Without organization, evolution lacks direction; without evolution, organization lacks generative power" (Moreno and Mossio 2015).

In a further analysis, Matteo Mossio et al. defined biological organization as a *closure of constraints.* For them, the organization of a biological system is synonymous with some of the constituents of the system acting as constraints toward the others in a regime of codependence between them that the authors labeled "closure" (Mossio et al. 2016). In plain terms, in order for life to exist, some crucial life parameters acting as constraints toward the others need to be conserved by active homeostatic mechanisms.

"Open systems must be necessarily organized": As inferred from the Morowitz's postulate, any open system (Morowitz 1966, 1968) must have *at least one* and typically will have many, closed-loop processes running through it. Importantly, this postulate applies whether or not the system is *alive*, as earth's climate, as an example, illustrates.

Organisms Are Organized into Modules and Networks

Modularity serves to reduce system complexity by limiting dynamic degrees of freedom, thereby enhancing functional robustness, developmental stability, and maintainability. As described by Cliff Hooker, three primary forms of modularity can be identified (Hooker 2013): (1) spatial or horizontal modularity, which occurs at the same hierarchical level (e.g., social groups, populations); (2) vertical or cross-level modularity, involving nested levels of organization (e.g., social stratifications, organizational hierarchies, or biological scales such as cells, organs, and whole organisms); and (3) process modularity, wherein systems are decomposed into distinct functional mechanisms (e.g., biological processes like respiration or engineered processes such as pulp mill regulation).

True hierarchy is defined by an asymmetric vertical constraint, where control predominantly flows from higher to lower levels. Most biological systems exhibit bidirectional constraints.

Hierarchy, Holarchy, Heterarchy

Hierarchical organization—characterized by systems nested within systems—is a fundamental feature observed across both biotic and abiotic domains of nature. While the structural and functional layering of such hierarchies is well documented in many contexts, the underlying principle driving the evolution of matter toward hierarchical complexity is unclear. Rodenfels et al. suggested that the emergence of increasingly integrated and complex systems can be explained by the universal tendency to reduce free energy in accordance with the second law of thermodynamics. In their opinion, hierarchically organized energy transduction networks arise spontaneously when they facilitate more efficient energy dissipation. Conversely, hierarchical structures are predicted to disassemble into independent subsystems if such decomposition results in a greater overall rate of entropy production. Because energy transfer is inherently interaction dependent, the natural demarcation of nested systems corresponds to the decreasing strength of interactions across increasing spatial scales (Rodenfels et al. 2019).

The proposal of Denis Noble (2006, 2008, 2013) is that the three levels of networks present in multicellular organism—cellular, tissular, and organismic—are codependent and there is no privileged level of causation (please see Fig. 9.3). His opinion was that multicellular organisms have functional *holarchy* architectures.

Organisms may present different types of organization including *heterarchies* (Sasai and Gunji 2008). As opposed to a hierarchy or a holarchy, a heterarchy is a system of organization where the elements of the organization are unranked

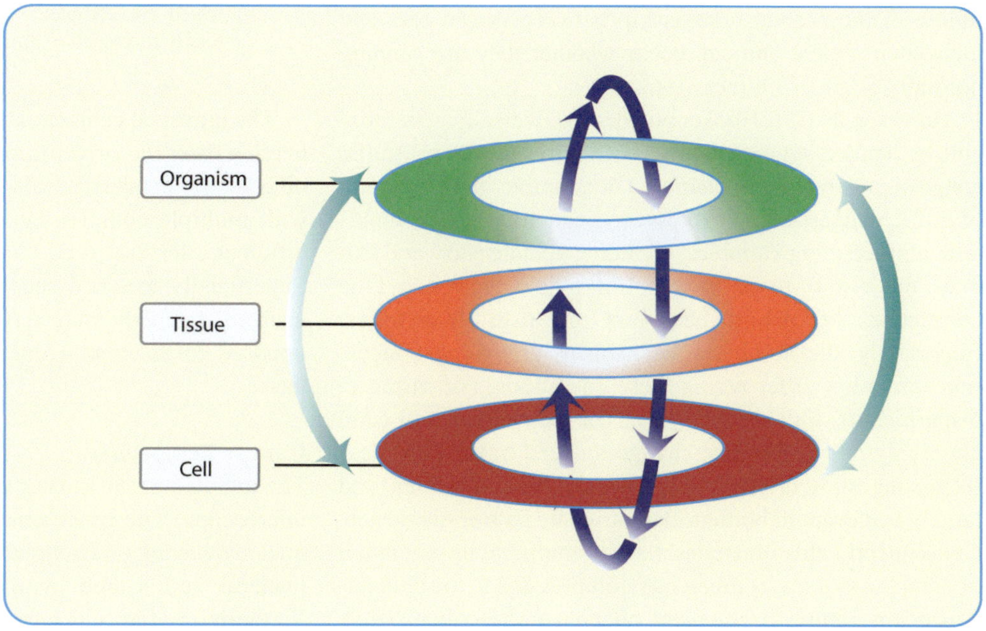

Fig. 9.3 Codependent levels in a multicellular organism. (Courtesy of Ana Tourian)

(nonhierarchical) or where they possess the potential to be ranked a number of different ways.

More recently, Bertolaso and Strauss described cancer as a consequence of interlevel dysregulation due to the regulatory uncoupling of reciprocal processes (differentiation, apoptosis, and proliferation). This view also counters the earlier great emphasis on cell proliferation as the primary causal feature of cancer and holds that "a perturbation of differentiation state is the primary disruption in the chain of cross-level regulation and long-range interactions that maintain normal organismic organization" (Bertolaso and Strauss 2021).

Similar to the two authors, my opinion is that cancer does not represent solely an abnormality of the cell or of the tissue but is related to a systemic derangement of homeostatic mechanisms at both the level of the organism and across organismal scales (see *Homeodynamics* section below).

Although the adult human body comprises 37–40 trillion cells, the majority of these possess negligible potential for neoplastic transformation (Laconi et al. 2020). Most somatic cells are terminally differentiated, specialized for tissue-specific functions, and exhibit limited proliferative capacity. In tissues characterized by high cellular turnover—such as the epidermis, intestinal epithelium, and hematopoietic system—many constituent cells are short-lived. The homeostasis of these tissues is sustained by a relatively small population of tissue-specific stem cells. These stem cells undergo infrequent divisions to generate short-lived, multipotent progenitor cells, which subsequently differentiate into lineage-restricted progenitors and, ultimately, fully differentiated functional cells.

As explained by Laconi et al., this hierarchical organization of cellular differentiation not only ensures efficient tissue regeneration under both homeostatic and stress conditions (e.g., injury or infection) but also serves as an intrinsic tumor-suppressive mechanism. By limiting the proliferative lifespan and genomic plasticity of the majority of cells, and confining long-term self-renewal potential to a small, regulated stem cell compartment, this architecture reduces the probability of oncogenic transformation (Laconi et al. 2020).

More than six decades ago, David Waldron Smithers noted that the life of a complex organism transcends the sum of its individual cells. He argued that organismal organization arises from, but is not reducible to, cellular behavior, much like societal dynamics cannot be fully explained by individual actions. The structure and function of the whole, he emphasized, are emergent properties, shaping—rather than merely resulting from—the behavior of individual cells. He rightly noted that emergent outcomes in complex systems often arise from interactions and also organizational structures rather than from the intrinsic properties of the system's constituent units. In his example, traffic congestion is fundamentally an issue of system-level organization rather than individual vehicle mechanics (Smithers 1962).

Cancer being also a complex system, the same phenotype of unrestrained division, stromal invasion, and metastases may also have very different causes.

In a landmark article, DW Smithers also proposed, famously, the importance of organization for the cancer process: "A breakdown of this orderly process can have many causes. Overgrowth or dedifferentiation are effects of disorganization-repercussions not driving forces. *Cancer is no more a disease of cells than a traffic jam is a disease of cars.* A lifetime of study of the internal-combustion engine would not help anyone to understand our traffic problems. The causes of congestion may be many. A traffic jam is a failure of the normal relationship between driven cars and their environment and can occur whether they are running normally or not" (Smithers 1962).

However, as Cliff Hooker noted, no universally generalizable or detailed dynamical model can fully account for traffic congestion across all scenarios. For example, the formation of traffic congestion on expressways can arise from a multitude of interacting variables, including vehicle entry and exit rates, the ratio of trucks to cars, traffic flow density, and the correlation of driver behaviors or path trajectories. Importantly, the dominant mechanisms governing congestion formation differ across these conditions. For instance, disparities in vehicle speed (e.g., between trucks and cars) are highly relevant in lane-change-induced traffic jams, less so in congestion driven by high-density braking events, and largely irrelevant in bottlenecks caused by path convergence. Consequently, this illustrates that the empirical investigation of complex systems is inherently complex due to the context-dependent nature of emergent phenomena (Hooker 2011). Therefore, generalizing specific causal mechanisms across different contexts may be unfeasible. The same applies to biological complexity, where the manifestation of disease, such as cancer, emerges from a confluence of multilevel interactions that defy reductionist explanations.

In the same article cited above, David Waldron Smithers, in a bold attempt to revise the cancer picture, made 10 statements that, after more than six decades after, sound prophetic:

1. Cancer is a disease of organization.
2. The word cancer, however, merely covers the most disorganized end of a progression in disorganization extending from maldevelopment, malformation, metaplasia, hyperplasia, dedifferentiation, and neoplasia to disintegration.
3. There is no such thing as a cancer cell – only cells behaving in a manner arbitrarily defined as being cancerous.
4. There being no such thing as a cancer cell, there can be no cause of cancer to be found within it.
5. Organization is measured by the amount of information in a system, and the entropy of a system is a measure of its degree of disorganization; gradual increase in entropy occurs, not sudden cellular malignancy.
6. Organizational breakdown commonly leads to progressive loss of growth control with released cell division producing an excess of tissue no longer coordinated with the whole.
7. Organization can become more or less disorganized, and the resulting tumors may progress or regress in behavior patterns.
8. An abnormal cell, particularly a stem cell, may produce a clone of cells reacting abnormally with their environment and so promote disorganization.
9. Environmental stress may disorganize the behavior of many cells within a sphere of influence which may be local or widespread, depending on the character and distribution of the particular cells concerned.
10. There are many causes of organismal disorganization; understanding depends on the explanation of tissue dependencies and relationships which are relationships which are complex, changing, and never likely to be totally foreseeable (Smithers 1962).

The universal characteristic of cancer cells is their *disconnection from the organization plan of the organism.* On the other hand, cancer can also be seen as a nested hierarchy with multiple complex networks of interactions located at different levels (i.e., cellular, tissular, organismic). Superficially, cancer tissues may appear as structurally disorganized, but they may have a *functional organization* as opposed to an *anatomical organization* as normal tissues have.

Cancer Perspective

The architecture of cancer cells appears distorted under the microscope. The cancerous cell is characterized by a large nucleus, having an irregular size and shape, with prominent nucleoli and scarce cytoplasm (Baba and Câţoi 2007). Furthermore, the tumor microenvironment is very different from normal stroma in terms of the composition of both cellular and acellular components (Zheng and Gao 2019).

Alterations in biostructure—ranging from nuclear membrane disruption, cytoskeletal disorganization, and extracellular matrix (ECM) remodeling to systemic microenvironmental changes—contribute to carcinogenesis and metastasis. Structural abnormalities, both intracellular and extracellular, can influence signal transduction, cell polarity, tissue architecture, and systemic homeostasis, facilitating the cancer progression and metastasis.

The cyclic GMP-AMP synthase (cGAS)/stimulator of interferon (IFN) genes (STING) pathway is critical in sensing the presence of cytoplasmic DNA derived from genomic instability events, such as DNA damage and defective cell cycle progression. In cancer, chronic cGAS-STING activation due to the persistence of cytosolic DNA causes inflammation-mediated tumorigenesis (Hoong et al. 2020).

The cytoskeleton maintains cell shape, polarity, and intracellular transport. In cancer cells, the actin cytoskeleton undergoes profound remodeling, promoting motility and invasion. For example, increased expression of actin-binding proteins like cofilin and fascin is associated with enhanced cell migration (Yamaguchi and Condeelis 2007). Microtubule dynamics are also altered in cancer (Legátová et al. 2023). There is an emerging role of microtubule-associated proteins in cancer metastasis (Wattanathamsan and Pongrakhananon 2022).

Tissue structural disruptions also contribute to cancer progression. The loss of apical-basal polarity, commonly seen in epithelial cancers, disrupts cell-cell and cell-matrix interactions, leading to deregulated growth and invasion (Martin-Belmonte and Perez-Moreno 2011). In addition, an active extracellular remodeling is induced by cancer-associated fibroblasts (CAFs) that secrete proteases like matrix metalloproteinases (MMPs) that degrade ECM components, facilitating invasion and metastasis (Lim and Moon. 2016; Kalluri 2016b). Stiffening of the ECM due to increased collagen cross-linking also activates oncogenic signaling pathways and promotes tumor progression (Fischer et al. 2020).

Tumors often reprogram the immune microenvironment, creating immunosuppressive niches. For example, myeloid-derived suppressor cells (MDSCs) and regulatory T cells structurally reorganize immune infiltrates, inhibiting antitumor responses (Gajewski et al. 2013).

Also, the structural characteristics of tissues influence the evolutionary trajectory of tumors. Two phenotypically identical tumors can exhibit markedly different fitness outcomes depending on the structural constraints imposed by their tissue architecture. At the cellular level, a subclone may possess a selective advantage, such as an elevated proliferation rate, but the actual manifestation of this advantage is contingent upon the competitive microenvironment in which the cell resides. Thus, intrinsic phenotypic traits can be modulated or even overridden by the spatial organization of the tissue, ultimately shaping tumor fitness. Initial spatial constraints play a pivotal role in determining the emergent evolutionary dynamics of a tumor, influencing whether evolution proceeds in a neutral or Darwinian manner, independent of changes in mutation rates or intrinsic fitness effects. In ductal tumors, the branching architecture at the point of tumor initiation establishes two key evolutionary accelerators: spatial confinement and cellular mixing. This structural connectivity varies significantly between tumors, contributing to heterogeneity in cellular intermixing across distinct tumor niches. While all tumors ultimately drift toward neutral evolution, those with pronounced spatial constraints may persist longer in a non-neutral evolutionary phase, allowing for the accumulation of a greater number of driver mutations per cell (West et al. 2020).

9.2.4 Principle of Autopoiesis

Autocatalysis is the "ability of a system to exhibit exponential growth of representative measures of size or population in ideal conditions. The property of autocatalysis can appear in different forms—including self-catalysis, cross-catalysis, and network autocatalysis—as long as the effect leads to exponential growth of a suitable metric under ideal conditions" (Bartlett and Wong 2020). Autocatalysis is a necessary (although not sufficient) condition for lifelike behavior (Hordijk and Steel 2018). Stuart Kauffman introduced the concept of an *autocatalytic set* by combining the idea of autocatalysis of life as an organized chemical reaction network. An autocatalytic set is an ensemble of molecules that mutually catalyze each other's formation through "chemical reactions from a basic food source" (Hordijk and Steel 2018).

Maël Montévil and Matteo Mossio expanded the autocatalytic set idea and introduced in several papers the concept of *closure of constrains* as the underlying mechanism of biological organization: "For an organism to be organized means realizing closure of constraints" (Montévil and Mossio 2015; Montévil et al. 2016; Montévil and Mossio 2020). For the two authors, "biological organisms are understood as natural systems realizing a dual causal regime. On the one hand, they are thermodynamically open systems: they are traversed by a flow of energy and matter that enables them to maintain themselves over time in conformity with the second principle of thermodynamic. On the other hand, biological organisms control the thermodynamic flow through the action of structures that, at specific time scales, exert constraints on the ongoing processes and transformations. In particular, organisms are constituted by a set of constraints that (1) are generative—they canalize target processes in such a way to maintain the conditions of existence of other constraints and (2) are dependent—their existence relies on the action of other constraints. The set of constitutive constraints that are both generative and dependent realize mutual dependence, which is usually referred to as *closure*" (Montévil and Mossio 2020).

An organisms's organization is defined as "a closure of constraints" (Mossio et al. 2016). In this model, *closure* and *historicity* (discussed below) both contribute to the functionality of an organism.

Montévil also emphasized the *historicity* of living forms understood as a source of novelty that cannot be formalized and is central for the understanding of life (Montévil 2022a). Organisms have a history that makes them unique, and this is one of the key reasons why, in his opinion, "fixed mathematical structures cannot play a central role" in their description (Montévil 2022b). "The method of physics postulates invariance in order to explain changes. In biology, we postulate instead that there is no underlying invariance behind changes" (Montévil and Mossio 2015; Montévil et al. 2016; Montévil 2022). He introduced together with Mossio a new symbol, χ, pointing to the historical aspects of an organism's identity (Montévil and Mossio 2020).

Autocatalysis and closure of constraints refer to the *autopoietic* (self-creation) properties of organisms: Basically, organisms both copy and continuously build themselves. *Reproduction* exists both at the cellular level as cell division and at the organism level. Reproduction can be seen as the mechanism used by multicellular organisms to make copies of themselves.

Organism Perspective

Autocatalysis and closure of constraints refer to the autopoietic (self-creation) properties of organisms: Basically, organisms copy and make themselves. Because of the dissipative nature, described above, biological systems undergo degradation over time. Therefore, "the whole system preserves its coherence and identity only insofar as it maintains and stabilizes not just some internal states or processes, but the autopoietic system itself as an organized unity" (Mossio et al. 2016). Errors of the copying system at the cellular level may lead to genetic mutations, a first step in the process of cancerization. Historicity is also relevant for the organism. Comorbidities leave their imprint at the organism level and contribute to cancer formation.

Cancer Perspective

As defined by Francisco Varela et al., "an autopoietic organization is defined as a unity by a network of productions of components which (i) participate recursively in the same network of productions of components which produced these components, and (ii) realize the network of productions as a unity in the space in which the components exist" (Varela et al. 1974).

Cancer is not simply the result of unchecked autopoietic processes, but rather a parasitic form of life that strives to persist, with host destruction occurring as collateral damage. Every cancer cell emerges from a normal cell located in a particular tissue. This is important because the behavior of the cancer cell, the development of the primary tumor and metastatic aggressivity will be also dependent on the cellular lineage and the status of differentiation of the cell across the lineage (Sahu et al. 2021). The knowledge that all cancers have idiosyncratic histories and therefore are unique events should not only inform all cancer models but also strengthen the need for personalized treatments tailored for each patient and each cancer type.

Cancer may have a long history before it manifests clinically, and sometimes, it might take one or two decades for cancer to grow from few cells to a tumor mass that can be detected by imaging or palpation. Also, in some cancers, metastasis may be a late event. For example, in pancreatic cancer, metastases occur at least 5 years after the inception of the nonmetastatic founder cell.

The long history of cancerous tumor formations may offer a window of opportunity to detect cancer in its initiation stage before it becomes clinical (see Chap. 12).

9.2.5 Principle of Control

Classically, *homeostasis* is the concept introduced by Walter Cannon to describe the feedback and feedforward mechanisms necessary to maintain the conditions for life persistence. A closely related concept introduced by Conrad Waddington in order to explain the temporal order present in development is the concept of *homeorhesis* (from *rhesis*, flow in Gr.). A third important control present in biological systems is the control of novelty, *homeopoiesis* (from *poiesis*, creation in Gr.), that I have introduced in Chap. 8. Paul Weiss's concept of ordered heterogeneity posited that the overall variance of a complex biological system is significantly lower than the cumulative variances of its individual components. This suggests that a fundamental property of living systems is their capacity to maintain systemic stability despite dynamic fluctuations at the level of constituent elements (Vodovotz et al. 2013). As mentioned before, *homeostasis*, *homeorhesis,* and *homeopoiesis* are all examples of homeodynamic mechanisms. We can think of *homeostasis* as the mechanism that maintains the status quo of a system; *homeopoiesis* as both change control *and* the source of change; and *homeorhesis* as a mechanism to maintain the balance between the two.

Cancer is, on one hand, *caused* by disruptions of homeodynamic mechanisms at the level of the cells, tissues, and organism, and on the other hand, cancer is itself *causing* a disruption in the homeodynamic mechanisms of other normal cells, tissues, and organism. *In essence, cancer represents a homeodynamic disruption.*

As explained by John Torday and Virender Rehan, in early organisms, homeostatic mechanisms developed in order to counterbalance environmental stress (Torday and Rehan 2017). As an example, as a rapidly permeating divalent cation (Ca^{2+}), calcium must be tightly regulated for a cell to function: Intracellular concentrations are kept at $\sim 10^{-7}$ M to prevent deleterious effects such as protein denaturation, chromatin condensation, and activation of degradative enzymes. Calcium is both *vital*—triggering fertilization and regulating development—and *harmful* when unregulated, inducing apoptosis and mitochondrial dysfunction. The evolution of phosphate-based bioenergetics required strict calcium regulation due to the insolubility of calcium phosphate. Mechanisms ensuring low cytoplasmic calcium include limited membrane permeability, intracellular buffering, and active export systems. These systems, conserved from prokaryotes to eukaryotes, reflect calcium's dual role as both life signal and cytotoxin. A complex calcium signaling toolkit, comprised of pumps, exchangers, and channels, allows for stimulus-dependent calcium influx and homeostatic restoration, underscoring its fundamental role in cellular physiology.

A fundamental feature of the cell is its capacity to internalize and compartmentalize environmental factors that would otherwise be cytotoxic—such as oxygen, minerals, heavy metals, microgravity-induced stresses, and even microbial agents. This capability is mediated by the endomembrane system, which enables the sequestration and functional integration of these external elements within distinct intracellular compartments. Comparable in significance to the evolutionary event of endosymbiosis, the controlled internalization of environmental inputs—constituting a cellular "miniclimate"—has been essential for the emergence of complex cellular functions and adaptability (Torday 2015).

Health is the maintenance of homeostasis not only at the macroscopic and microscopic levels but also between the microscopic and macroscopic levels. Disease may appear due to lack of coordination or miscommunication between the macroscopic and the microscopic levels (López-Otín and Kroemer 2021). Cancer does not represent solely an abnormality of the cell or of the tissue but is related to a systemic derangement of homeostatic mechanisms at the level of the organism.

Organism Perspective

A. Several system-wide organism reactions to stress have been described. In July 1936, Hans Selye submitted a letter to the journal *Nature* in which he described the "general adaptation syndrome" (GAS) represented by a three-stage response to different stressors (i.e., cold, surgical injury, spinal shock, muscular exercise, or sublethal doses of a variety of drugs or tissue extracts). The three stages were as follows: a) general alarm reaction: occurred 6–48 hours following exposure to the stimulus, in which the adrenal medulla releases epinephrine and the adrenal cortex produces glucocorticoids, both of which help to restore homeostasis; b) adaptation: from 48 hours until 1–3 months after the beginning of repetitive exposure to the stimulus, during which the fragile new equilibrium is maintained; and finally, c) exhaustion: depending on the severity of the stimulus, animals died, became sick, or died within 3 months from the exposure with symptoms similar to those observed during the general alarm reaction (McCarty 2016). Hans Selye demonstrated the crucial role of the adrenal cortex/hypophysis axis in the stress response (Szabo 1998) while also stressing the nonspecificity of this neuroendocrine response, a concept which was subsequently rejected (McCarty 2016), since it was shown that the hypothalamic-pituitary-adrenal (HPA) axis and the noradrenergic and adrenergic nerves have different patterns of response specifically related to the type of stressor (McEwen 2005).

B. Allostasis is a term introduced initially by Peter Sterling and Joseph Eyer relating to adaptive phenomena that were not explained by homeostatic mechanisms (Sterling and Eyer 1988). Allostasis expanded the homeostasis homework and, as opposed to homeostatic mechanisms responsible for maintenance of quasi-constant states, included flexible physiologic reactions able to anticipate the organism's needs and to promptly meet them (Sterling 2018, 2020).

Subsequently, Bruce McEwen and John Wingfield used the concept of *allostasis* in the context of psychology, in order to explain the mechanisms that, in short run, adjust homeostatic responses to stressful external situations and, in long term, wear down the organism (McEwen and Wingfield 2003). As an example, the body increases blood pressure in response to stress and typically restores it to normal levels once the stressor is removed. However, if the stress becomes chronic, this regulatory mechanism may become impaired, resulting in persistently elevated blood pressure (Golbidi et al. 2015). *Allostasis*, defined as the process of achieving stability through change by anticipating and adapting to external or internal stressors, represents the pathological counterpart of *homeorhesis,* which is related to stability through change in the context of development, which is a physiological process. In the framework of allostasis, Selye's concept of the alarm response was reinterpreted by the two authors as the initial phase of adaptive regulation, wherein glucocorticoids, epinephrine, and other mediators facilitate the organism's adjustment to the stressor. What Selye described as the stage of resistance through the allostasis lenses corresponds to the protective effects achieved through successful allostatic adaptation. However, if the stress response persists and adrenal secretion of glucocorticoids and catecholamines remains chronically elevated, the system may enter a prolonged allostatic state. Over time, this can progress to allostatic overload—a condition that refines Selye's original notion of the exhaustion stage (McEwen 2005).

C. *The Systemic Inflammatory Response State* (SIRS) has been defined by Roger Bone as an abnormal generalized inflammatory reaction in organs remote from the initial insult, and it is most commonly associated with severe sepsis (Bone 1996).

The progression from localized inflammation to systemic immune dysfunction involves five stages. *Stage 1* begins with a *local* injury or infection, triggering the release of pro-inflammatory mediators to initiate wound healing and immune defense, followed by anti-inflammatory signals that restore homeostasis if the insult is minor and the host is healthy. In *Stage 2*, if local control fails, pro-inflammatory mediators enter systemic circulation, recruiting immune cells to the affected site. To dampen inflammation and reestablish equilibrium, a systemic anti-inflammatory response follows. *Stage 3* develops when the systemic inflammatory response becomes excessive or inadequately regulated. This leads to systemic inflammatory response syndrome (SIRS) and early signs of multiple organ dysfunction syndrome (MODS). Alternatively, an overwhelming anti-inflammatory response—either reactive or spontaneous—can lead to *Stage 4* represented by a compensatory anti-inflammatory response syndrome (CARS), marked by profound immunosuppression and heightened infection risk. If homeostasis is reestablished, recovery is possible. Otherwise, progression to *Stage 5,* the terminal phase of multiple organ dysfunction syndrome (MODS), occurs. Here, immune regulation collapses: Some patients experience persistent hyperinflammation, others endure prolonged immunosuppression and secondary infections, while some oscillate between both states, reflecting a final stage of *immune dissonance* (Bone 1996).

Briefly, the five stages can be summarized as following: (1) local reaction at the site of injury or infection, (2) initial systemic response, (3) massive systemic inflammation, (4) excessive immunosuppression, and (5) immunologic dissonance (Bone et al. 1997).

D. Electrolyte abnormalities and volume imbalances are common to both late phases of sepsis and cancer (Lee 2010; Berardi et al. 2019; Turcotte et al. 2022; Amano et al. 2020). Advanced cancer can be considered a *chronic systemic strain syndrome.* In terminal cancer stages, electrolyte abnormalities and fluid imbalances are very common and have been associated with increased morbidity and mortality (Turcotte et al. 2022; Amano et al. 2020) similar to the late phases of sepsis, where cellular damage and ion channel dysfunction have been associated with increased morbidity and mortality (Lee 2010; Zhang et al. 2021a).

9.2.5.1 The Human Organism Biosphere

Cancer cells appear continuously in the human body and they are kept from progressing by the immune system. Therefore, we can think of the human body as a biosphere, where five types of life (viral, bacteria, archea, eukaria, and *oncolife*) are closely linked and codependent, co-operating, and competing. As long as the interactions between them are not interfering with the functioning of the human host, health is maintained. Once imbalances appear due to loss of homeostasis or excessive allostatic load, disease appears.

It took the whole geosphere to develop life on planet Earth. It takes the entire human organism biosphere to develop cancer. Once formed, the human organism biosphere is closely interrelated with its constituent life forms. Once formed, cancer, too, is codependent with the organism's biosphere.

In particular, the interactions between the host organism and cancer are crucial for cancer progression. Cancer behaves toward the host like a parasite deceiving it and hacking its defense systems, exploiting its resources and manipulating it for its own benefit. The cancer behavior that goes from benign to extremely aggressive is influenced directly by the interaction with the organism.

An important component of the human organism biosphere (please see Fig. 9.4) is the internal body climate that will be discussed next.

9.2.5.2 The Internal Body Climate

In an article published in collaboration with Aurora Nedelcu, Biology Professor at University of New Brunswick, we developed the concept of the *internal body climate* that was defined as a "pervasive systemic element that is influenced by, as well as influences, all levels – from cells and tissues to organs and the entire organism." In other words, the organismic climate is a sort of structuring and communicating medium that is influenced by, as well as influences, all levels—from cells and tissues to organs and the entire organism (Paul and Nedelcu 2022). In a multicellular organism, the climate is both the *product* of myriad interactions among cells, tissues, and the organism as a whole (bottom-up dynamics), and the *orchestrator* of their behavior (top-down regulation), sustaining the fragile equilibrium traditionally called homeostasis, expanded here into homeodynamics (please see Fig. 9.5).

Much like the interplay between Earth's biosphere and its climate—where the land, oceans, and organisms cocreate atmospheric patterns that, in turn, shape their fates—the body's internal climate embodies a reciprocal dance of cause and consequence. The body is not merely a passive stage for biological processes but a sentient ecosystem, where each element is both sculptor and a sculpture creation of the whole organism.

Within this ecological framework, cancer can be reimagined as a perturbation of the entire physiological organism's biosphere, seeded and shaped by the internal climate, and not solely a cellular aberration—the rogue *seed*—but rather a disruption of the tissue *microenvironment*, the *soil* in which the seed may sprout and the whole body's *macroenvironment.* The organism itself can modulate the fertility of its soils, altering tissue receptivity through systemic signals. Conversely, tumors—like opportunistic *weeds*—may hijack not only local terrains but also exert far-reaching effects, destabilizing distant organs, reshaping the internal body climate, and undermining the coherence of the whole organism.

In this light, metastasis becomes not a random drift but a climate-mediated reforestation—secondary seeds seeking new soils under shifting internal environments. And cancer, in its deepest essence, is not only a cellular or tissue disease but an *internal climate disorder*—emerging from systemic homeostatic dysregulation and, if uncorrected, leading to systemic collapse.

Thus, cancer must be understood not in isolation but in the full context of the living system—an organism whose internal climate both shapes and is shaped by the disease. Systemic shifts in this body climate—driven by physiological or metabolic perturbations, dietary patterns, aging, or stress—can subtly recalibrate the internal environment, altering the very probabilities of cancer's emergence, its modes of expansion, and its potential for distant colonization. These are the top-down effects—where climate precedes the malignant condition and the internal environment may either sow or suppress malignancy before it takes root.

This reciprocity runs deep. Cancer, once established, does not remain confined to its original location. It instigates the organism to support its progression; it alters and distorts it. It secretes signals, redirects energy flows, and breaks the delicate feedback loops that uphold organismal harmony. These

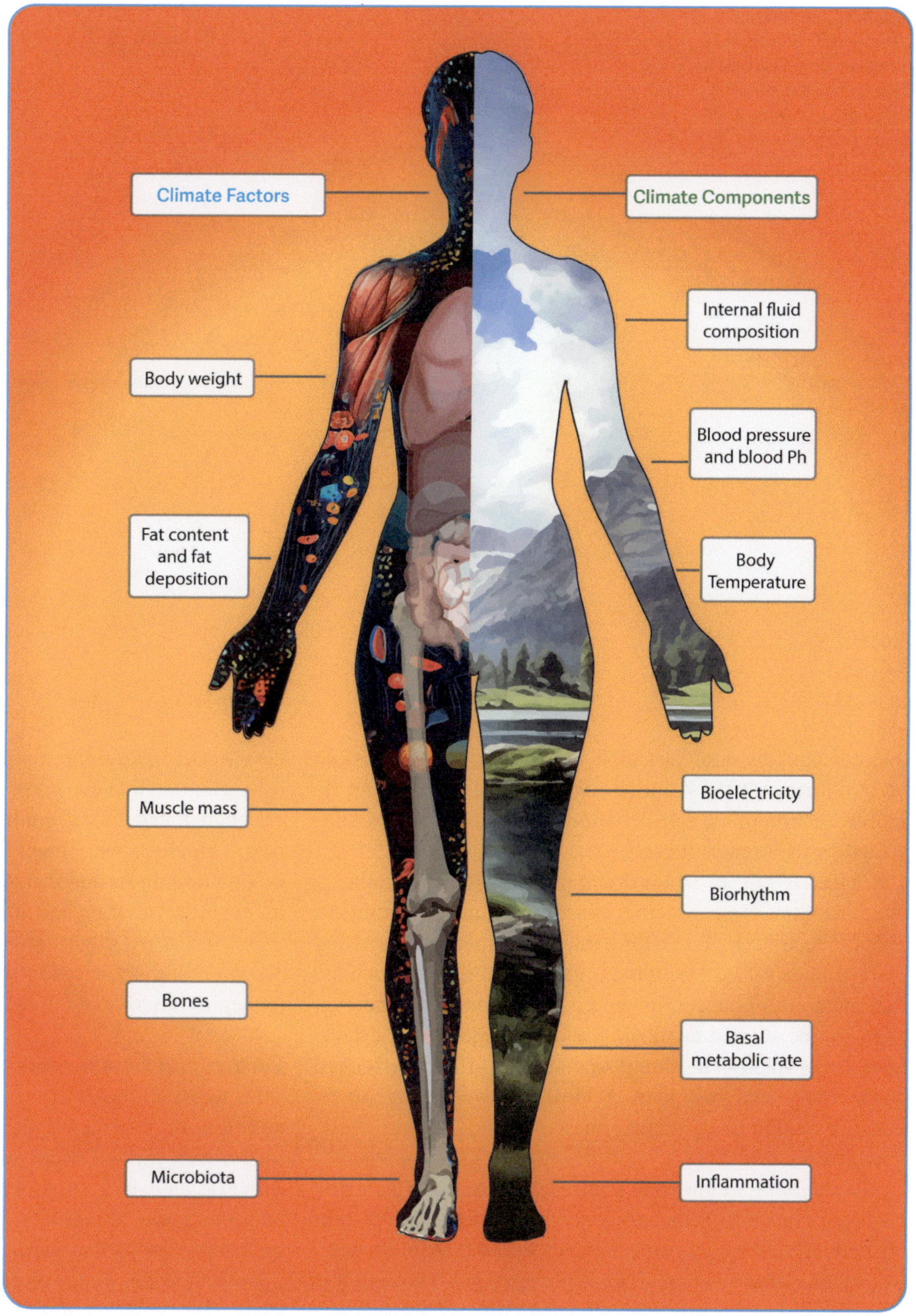

Fig. 9.4 The human organism biosphere and the internal body climate. (Courtesy of Ana Tourian)

bottom-up effects—from cancer cells to tissue and cancer tissues to climate—can unmoor the entire biospheric balance of the organism, gradually transforming its internal climate into one more hospitable to cancer progression, less supportive of resilience. The organism is literally, gradually, *cancerized*.

Through these bidirectional currents, cancer and the internal climate become entangled in positive feedback loops—where tumor-induced changes amplify systemic fragility, and systemic fragility, in turn, nurtures further malignant spread. The metastatic cascade, then, is not merely a mechanical journey of cells but an ecological migration across an altered landscape—fueled by a shifting climate and culminating, at its extreme, in systemic collapse, cachexia, and death.

In this view, cancer is not simply a destructive invader but a reflection of deep imbalances within the organismal biosphere. It is both a symptom and a signal of a dysregulated

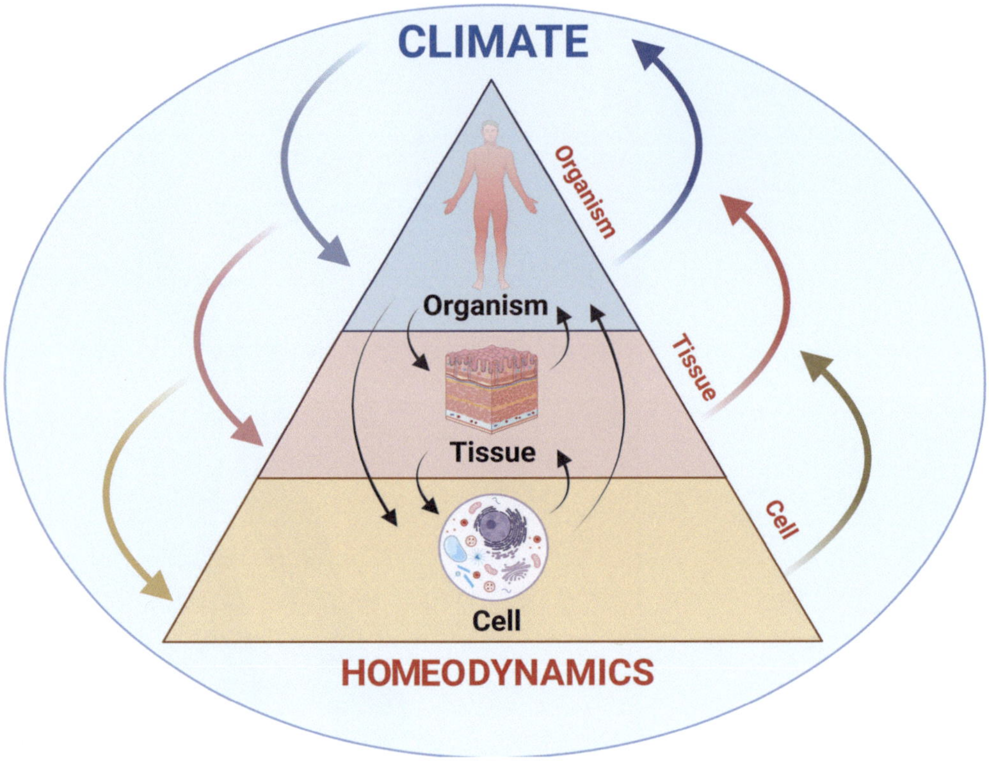

Fig. 9.5 The internal body climate and its reciprocal influence on cells/tissues/organism. Created in BioRender. (Courtesy of Laura Pătraş)

internal climate. And just as it arises from these invisible body climatic shifts, it further distorts this, entrenching its presence through a self-perpetuating storm actively supported by the organism itself.

Cancer is not an isolated cellular rebellion but rather a manifestation of a systemic disequilibrium, an emergent phenomenon born of a shifting internal climate. The internal body climate offers a unifying framework through which the full spectrum of cancer, its emergence, expansion, and evasion, can be studied as internal body climate-driven events. More importantly, it illuminates subtle, often overlooked connections, and can help us design novel interventions through which we might intercept and restore body homeostasis.

By attending to the body's internal climate, we gain new vantage points from which to intervene, not merely to disrupt the cellular machinery of malignancy but to prevent the very environmental shifts that give it fertile ground. Through this lens, therapeutic efforts move beyond correcting mutations or targeting altered phenotypes and microenvironments. Instead, they aim to reengineer the weather system of the organism—to foster conditions in which cancer can no longer arise, progress, or persist.

To implement this vision, several foundational steps are required. First, we must identify the constituents of this internal climate—those physiological, metabolic, hormonal, and molecular forces whose interplay either sustains equilibrium or invites disruption. These are the body climate *factors* and *components*. Then body climate biomarkers must be developed—quantifiable indicators of systemic health or dysregulation that can serve as early harbingers of pathological change capable of guiding both diagnosis and intervention. Finally, we must devise strategies for climate modulation: tools and therapies that prevent unfavorable climate shifts, slow or halt their progression, or even reverse them—what we might term climate reconstitution. In Chap. 12, we will discuss several cancer prevention and treatment strategies based on the internal body climate concept.

Factors and Components of the Internal Body Climate

The internal climate of the human body, a subtle and dynamic equilibrium, emerges from the interplay of a multitude of interwoven *factors*, both intrinsic and extrinsic in nature. *Intrinsic elements*, such as the architecture of the body, its adipose tissue, musculature, skeletal system, and the vast microbial ecosystem within, form the biological substratum upon which this inner climate is built. These elements are shaped by evolution, modulated by genetics, and sculpted by epigenetic and developmental forces. *Extrinsic influences*, on the other hand, flow into the organism from the external world: sunlight and darkness, air and pollution, nourishment and toxin, rest and stress. Diet, exposure to pathogens, and environmental mutagens like tobacco smoke or ultraviolet radiation all act as external climatological agents that tune or disturb the body's internal climate patterns.

Together, these influences regulate a suite of physiological *components* that define the internal milieu: the composition of circulating fluids, osmotic and hydrostatic pressures, thermal gradients, pH levels, bioelectrical fields, circadian rhythms, metabolic rates, and inflammatory tone. These are the shifting tides of the internal ecosystem, subtle yet vital markers of organismal homeostasis or impending disorder.

Cancer, in this context, emerges not merely as a localized aberration but as a profound disturbance of this inner body climate. It can act both as a consequence and a driver of systemic imbalance. The factors and components of the organism's internal climate are interdependent, and in individual patients, their synergistic dysregulation can converge to create a permissive environment that fosters cancer development. In turn, cancer can *perturb internal climate factors*: decrease body weight, deplete muscle and fat reserves, erode bone integrity, and reconfigure the microbiota—directly rewiring the organism's structural identity. At the same time, cancer also reshapes *internal climate components*: altering pH through lactic acid production, generating inflammatory cascades, and disrupting electrical potential through tumor-host ionic interactions. Reciprocally, these very climate components can influence tumor behavior—shaping its capacity to grow, invade, and resist. An acidic microenvironment, for instance, favors invasion and immune evasion; chronic low-grade inflammation primes tissues for malignancy; and altered biorhythms disrupt hormonal regulation and immune surveillance. In this complex and reciprocal dialogue, cancer and the body's climate are not separate—they are coauthors of a shared pathophysiological environment.

Internal Factors Modulating the Body's Internal Climate

Adiposity and Inflammation

Excess body weight and fat distribution significantly contribute to a pro-inflammatory internal environment and increased cancer risk. Between 2011 and 2015 in the United States, approximately 4.7% of cancer cases in men and 9.6% in women were attributed to excess body weight (Islami et al. 2019). Adiposity is associated with 12 cancer types, with central (visceral) obesity posing a greater risk than peripheral fat (Barberio et al. 2019). Fat distribution, such as high waist-to-hip ratio, has been linked to increased risk of colorectal and pancreatic cancers, while subcutaneous versus visceral fat exhibits differential inflammatory potential (Luo et al. 2008; Tchkonia et al. 2013; Fu et al. 2015).

Skeletal Muscle as an Endocrine Organ

Muscle mass exerts protective systemic effects through its endocrine function, secreting myokines that influence metabolism and tumor progression. The "obesity paradox," where overweight cancer patients show improved survival, may partly be explained by preserved muscle reserves (Caan et al. 218). Elevated muscle mass is associated with better outcomes in cancers such as prostate cancer (Lopez et al. 2022).

Bone as a Dynamic Endocrine Organ

Bone actively regulates systemic physiology via hormones like osteocalcin, which is implicated in prostate cancer. Bone also influences metastasis: Osteoblasts can maintain cancer cell dormancy, while osteoclasts may facilitate reactivation (Atkinson and Delgado-Calle 2019). Tumor-secreted osteopontin can mobilize bone marrow-derived cells, promoting the outgrowth of distant tumors (Mc Allister et al. 2008).

Gut Microbiota and Systemic Modulation

The gut microbiome exerts wide-ranging effects on host metabolism, immune function, and cancer progression. Influenced strongly by diet, it modulates host gene expression and energy homeostasis. The concept of "commensal homeostasis" highlights the dynamic host-microbiota balance, which can be monitored via host-derived biomarkers (Zhang et al. 2018). Furthermore, gut microbiota can affect therapy outcomes, for example, by synthesizing androgens during androgen deprivation therapy, potentially promoting prostate cancer progression (Pernigoni et al. 2021).

Age

The vast majority of cancers appear after age 50. Age is an important internal factor as it directly impacts the body climate (please see Fig. 9.6).

Internal Body Climate Components Relevant to Cancer

The internal body climate—shaped by interactions among cells, tissues, organs, microbiota, and external influences—plays a dual role in cancer, acting both as a modifiable target for prevention/treatment and as a dynamic reflection of disease progression.

Internal Fluid Composition

The composition of internal fluids, such as blood, lymph, and interstitial fluid, serves as a prime example of an internal climate component, arising from the interplay of numerous physiological processes and their interactions with the external environment. These fluids not only mirror the functional state of various internal climate factors but also actively participate in cancer progression, either facilitating or impeding it through direct or indirect mechanisms. Certain blood parameters function as general indicators of systemic health (body climate markers), while others serve

Fig. 9.6 Internal body climate changes with age (Courtesy of Ana Tourian)

as specific prognostic tools or markers of cancer progression. Similarly, the interstitial fluid, shaped by cellular and tissue activity as well as the composition of blood and lymph, has been increasingly recognized for its role in influencing cancer development and therapeutic response (Munson and Shieh 2014).

- *Cellular Components*: Blood cells may serve as health indicators. Neutrophilia, anemia, and reduced NK cell activity are adverse prognostic factors (Ocana et al. 2017; Shaul and Fridlender 2018; Gvirtzman et al. 2021). CTCs and ctDNA can be used as biomarkers; a high platelet count may be associated with cancer progression as platelets aid immune evasion and metastasis (Menter et al. 2014; Palacios-Acedo et al. 2019).
- *Macromolecules*: Altered levels of lipids, carbohydrates, and proteins reflect both dietary inputs and tumor metabolic reprogramming. Elevated branched-chain amino acids are linked to cancer risk (Gumpper-Fedus et al. 2022), while cachexia alters macronutrient profiles (Naumann et al. 2019).
- *Hormones*: Hormonal imbalances (endogenous or tumor-derived) are implicated in multiple cancers. Tumors may secrete hormones that enhance progression (e.g., ACTH, BBS-like peptides) (Castellone et al. 2015).
- *Metabolites*: Specific metabolite changes are associated with cancer risk and progression across tissue types, which has been useful in identifying potential diagnostic and therapeutic targets (e.g., MMA, acylcarnitines) (His et al. 2019; Gomes et al. 2022).
- *Exosomes*: Tumor-derived exosomes facilitate metastasis and immune evasion (Zhou et al. 2014; Tominaga et al. 2015; Tai et al. 2018), while their cargo (e.g., miRNAs) may serve as diagnostic/prognostic markers (Hoshino et al. 2020).
- *Circulating PD-L1*: Elevated soluble PD-L1 in various cancers impairs immune function; thus, its therapeutic removal (e.g., plasmapheresis) shows promise (Ali et al. 2017; Oh et al. 2021).
- *Oxygen*: Local hypoxia fosters tumor growth, but systemic oxygen's role remains uncertain. Gut oxygen gradients affect microbiota and potentially cancer (Igarashi et al. 2020).
- *Microelements*: Trace elements (e.g., potassium) influence immune responses and are dysregulated in cancer (Eil et al. 2016). Zinc, copper, and phosphorus were found to be increased in cancer tissues as opposed to iron and calcium that were found to be decreased (Saeed et al. 2024).

Blood Pressure

Tumor interstitial fluid pressure (IFP) impairs drug delivery and correlates with poor prognosis (Heldin et al. 2004). Systemic hypertension and heart failure are risk factors for several cancers, potentially due to shared inflammatory and metabolic pathways (Seretis et al. 2019; de Boer et al. 2019; Meijers and de Boer 2019; Han et al. 2017; Sobczuk et al. 2017; Bertero et al. 2018; Meijers et al. 2018; Kim et al. 2020; Xuan et al. 2021).

Temperature

Body temperature reflects systemic metabolic activity. Experimental data suggest thermoneutral housing enhances antitumor immunity, implying a role for ambient and body temperature in cancer progression (Zhao et al. 2017).

pH

Internal fluids, such as blood and interstitial fluid, maintain a relatively stable pH of approximately 7.4, with fluctuations tightly regulated by buffering systems—particularly in the bloodstream. Although systemic blood pH in cancer patients typically remains unaffected, localized acidosis is frequently observed within advanced tumors (Ward et al. 2020). This acidic tumor microenvironment primarily arises from cancer-associated metabolic reprogramming, notably the accumulation of lactic acid produced via anaerobic glycolysis (Ali et al. 2016). Such acidity has been shown to facilitate both tumor growth and metastasis (Rohani et al. 2019; Boedtkjer and Pedersen 2020). Beyond tumors themselves, the pH of the surrounding tissue microenvironment can also modulate cancer progression. For instance, an acidic environment has been implicated in promoting the development of certain cancers, including oral and bladder cancers (Alguacil et al. 2011; Ramya et al. 2015), whereas alkaline conditions may predispose to others, such as cervical cancer (Clarke et al. 2012). These findings underscore the significance of pH dynamics not only within tumors but also in the broader internal climate that supports or constrains oncogenesis.

Bioelectricity

Resting membrane potential reflects cell differentiation (Yang and Brackenbury 2013; Prevarskaya et al. 2018). Depolarization is associated with stemness and tumorigenesis; bioelectric signals may initiate or support malignancy (Lobikin et al. 2012; Chernet and Levin 2013; Sheth and Esfandiari 2022).

Biorhythms

Circadian disruption (e.g., shift work) contributes to carcinogenesis. Tumors can decouple from systemic circadian regulation, gaining growth advantages (Blask et al. 2002; Kinouchi and Sassone-Corsi 2020; Patel and Kondratov 2021).

Basal Metabolic Rate

Elevated BMR correlates with increased cancer risk in specific subtypes, suggesting a role for metabolic demand in tumor initiation (Kliemann et al. 2020).

Inflammation

Both local and systemic inflammation play pivotal roles in cancer progression. In a bidirectional relationship, inflammation can promote tumor development (cancer-promoting climate change), while advancing malignancy can amplify systemic inflammation (cancer-induced climate change). Among key contributors, obesity has emerged as a significant factor. Beyond its general health effects, excess adipose tissue secretes pro-inflammatory adipocytokines, linking metabolic dysfunction to cancer—a relationship still under active investigation (Howard et al. 2010). This chronic low-grade inflammation is marked by aberrant cytokine production, which may contribute to both tumor initiation and progression. Notably, adipocyte-conditioned media have been shown to enhance tumor cell migration (Hoda et al. 2007; Lang and Ratke 2009). Conversely, tumor-derived cytokines circulating in the blood can serve dual roles: as biomarkers of systemic climate dysregulation and as potential therapeutic targets (Chehrazi-Raffle et al. 2021; Kartikasari et al. 2021). Elevated levels of C-reactive protein (CRP), a marker of systemic inflammation, have been associated with poor outcomes in patients with solid tumors. In the general population, high CRP levels are also linked to an increased future risk of developing cancer (Allin and Nordestgaard 2011).

The generalized inflammation seen in cancer progression may represent a physiological response that has spiraled into global dysregulation—a form of chronic systemic inflammatory state. In late-stage disease, this culminates in cachexia, driven by a fatal "cytokine storm" that mirrors the dysregulated immune response seen in other critical conditions. The energy demands of prolonged inflammation, rather than being adaptive, may become a maladaptive metabolic burden, accelerating decline.

A full list of references regarding the internal body factors and components can be found free online at the end of our article: https://pmc.ncbi.nlm.nih.gov/articles/PMC9815514/ (Paul and Nedelcu 2022).

9.2.5.3 Organism's Homeostasis

Optimal organismal function is sustained by a hierarchy of homeostatic control systems (Billman 2020). At the *first level*, intrinsic feedback loops regulate cellular physicochemical parameters and organ-specific processes. The *second level* comprises autonomous regulatory circuits—such as the baroreceptor reflex—that monitor deviations in first-level variables and initiate corrective responses. The *third level* involves central nervous system (CNS) networks that integrate second-level signals with environmental inputs to coordinate systemic physiological and behavioral responses. A *fourth layer* encompasses social and communal influences, which further modulate homeostatic regulation. Central to this integrative framework is the hypothalamus, which processes inputs across all levels to orchestrate adaptive responses and maintain systemic equilibrium.

Within this multilevel model, specialized homeostatic subsystems regulate energy balance, body weight, and composition. Leptin, an adipose-derived hormone, plays a pivotal role in energy homeostasis; genetic disruptions of leptin or its receptor result in severe early-onset obesity in both humans and rodents (Münzberg et al. 2020). Complementing this, a leptin-independent mechanism, the "gravitostat," has been described, whereby mechanical loading of the lower limb bones generates feedback signals that modulate body weight and fat mass (Jansson et al. 2018). Body composition is also tightly regulated: Minor reductions in hepatic glycogen content trigger hyperphagia to restore glycogen stores, whereas protein and lipid reserves are subject to less precise control (Müller et al. 2010).

In addition, circadian rhythms act as master regulators of homeostatic processes across taxa. The molecular basis of circadian regulation—recognized by the 2017 Nobel Prize in Physiology or Medicine awarded to Jeffrey C. Hall, Michael Roshbash, and Michel W. Young—governs numerous physiological functions, including metabolism, sleep-wake cycles, and hormonal secretion (Schwartz et al. 2022).

Cancer Perspective

In their original paper published in *PNAS* describing the *theory of facilitated variation*, Mark Gerhart and John Kirschner considered that regulatory changes, rather than structural gene alterations, are the predominant drivers of genetic innovation and that evolution appears to favor mechanisms that minimize the number and complexity of regulatory modifications, often achieving new phenotypic outcomes by redeploying conserved molecular components and pathways. In their opinion, regulatory changes operate on a core repertoire of ancient, conserved processes, selecting specific subsets that are then subject to natural selection and phenotypic plasticity, whether environmentally induced or developmentally embedded. These ancient processes serve as a rich and evolutionarily favored reservoir of variation, often requiring minimal regulatory changes. Importantly, this perspective suggests that the capacity for *facilitated phenotypic variation* likely emerged as a by-product of natural selection acting on the functional robustness, adaptability, modularity, exploratory capacity, and flexible regulatory connectivity of core processes themselves. These properties were selected at the level of the individual organism for their effectiveness in maintaining physiological and developmental stability under variable internal and external conditions. Thus, similar molecular features that enable short-term adaptability also underpin long-term evolutionary change as regulatory innovations become genetically stabilized (Gerhart and Kirschner 2007).

From this perspective, cancer may be conceptualized as a homeodynamic disease, characterized by disruptions in the dynamic regulatory processes that maintain cellular and systemic stability. This view is supported by the observation that many cancer-driving mutations affect transcription factors, which play pivotal roles in orchestrating gene expression programs, and by the recognition that epigenetic alterations, particularly within noncoding regions of the genome, represent fundamental hallmarks of malignancy.

Importantly, the homeodynamic disturbances in cancer are not confined to intracellular pathways but extend to the tumor microenvironment, notably involving the stroma. These alterations are linked to phenomena such as the loss of contact inhibition, enabling uncontrolled proliferation, and extend further to the organismal level, where deficiencies in immune surveillance contribute to tumor escape and progression. This multiscale disruption of regulatory equilibrium underscores the systemic nature of cancer as a disease of compromised biological homeodynamics.

Luo et al. noted that in addition to the 11 cancer cell hallmarks described by Hanahan and Weinberg, mostly related to genetic alterations (i.e., gain-of-function mutation, amplification and/or overexpression of key oncogenes together with the loss-of-function mutation, deletion, and/or epigenetic silencing of key tumor suppressors), five other intracellular stress hallmarks (DNA damage/replication stress, proteotoxic stress, mitotic stress, metabolic stress, and oxidative stress) and two extracellular one (hypoxia and nutritional stress) play a significant role (Luo et al. 2009) in cancer. A key point idea introduced by the Luo et al. is that genetic instability due to genomic stress is only one of several stressors present in cancer, some of the stressors being located outside the cell. Also, oncogenic mutations in proto-oncogenes can be selected for in cancer cells subjected to metabolic stress (Ward and Thompson 2012).

The article of Luo et al. raises the possibility that controlled evolvability may represent a cellular stress response. The team of Susan Rosenberg demonstrated more than two decades ago that in bacteria, both genetic mutations and amplification are inducible genetic instabilities allowing rapid evolution under stress and escape from growth inhibition (Hastings et al. 2004). Susan Rosenberg suggested that the occurrence of random genetic mutations may represent a tightly controlled and, in a sense, "purposeful" response to stressful environments, and her team presented evidence of stress-induced mutagenesis programs (Galhardo et al. 2007; Rosenberg and Queitsch 2014). Her hypothesis was that stress-inducible mutagenesis mechanisms can "accelerate adaptive evolution in populations specifically when organisms are maladapted to their environments, i.e., when they are stressed, and then return genomes to low mutation rates in rare, adapted mutants that thrive in the new environment and so are stressed no longer" (Galhardo et al. 2007).

In cancer, the stress pathways are hyperactive and represent a potential therapeutic target. Recently, long noncoding RNA have been recognized as major players in helping cancer cells regulate and overcome genotoxic, oxidative, and metabolic stresses (Connerty et al. 2020).

The major sources of ROS in cancer are the mitochondria and the nicotinamide adenine dinucleotide phosphate (NADPH) oxidases (Kalyanaraman et al. 2018). ROS are increased in cancer and are involved in all steps of oncogenesis from cancer initiation to metastasis (Tafani et al. 2016) and act as signaling transducers during cell proliferation, migration, invasion, angiogenesis, and drug resistance (Nakamura and Takada 2021). The role of ROS in tumor initiation is mainly related to protein damage bearing a thiol group. ROS may inactivate key tumor suppressors like PTEN, for example, leading to hyperactivation of the PI3K/Akt pathway (Silva et al. 2008).

ROS is a collective term for a number of related molecules with very different reactivity. ROS, once viewed merely as by-products of oxygen metabolism, have been found to play many roles in normal cellular physiology and are now recognized to the induction of cellular defense genes, the activation of transcription factors, and for playing a significant role in phosphorylation of kinases and mobilization of ion transport systems (Touyz and Briones 2011). The key redox signaling agents hydrogen peroxide (H2O2) and the superoxide anion radical (O2-) are generated under the control of growth factors and cytokines by more than 40 enzymes, prominently including NADPH oxidases and the mitochondrial electron transport chain (Sies and Jones 2020). Hydrogen peroxide (H2O2) is recognized as the principal ROS in redox regulation of biological activities (Stone and Yang 2006). The major mechanism by which ROS mediate their effect is through thiol-based modification of target proteins (Zeida et al. 2019). Similarly to Ca2+, the intracellular concentration of H2O2 is much lower inside than outside of the cell and is maintained in the low nanomolar range (approximately 1–100 nM), under tight homeostatic control (Sies and Jones 2020).

Oxidative stress is defined as an imbalance between production of oxidants and the ability to detoxify them with antioxidants, favoring the oxidants. Excess ROS causes physical DNA and protein damage, apoptosis induction, or lipid peroxidation. Various stressor such as growth factors, chemokines, or physical stressors (exercise) may increase the production of H2O2, and its removal is achieved by efficient reducing systems. A high exposure leads to inflammatory response, growth arrest, and cell death by various mechanisms.

The concept of bioenergetic health index has been introduced as a measure of oxidative stress that can inform preventive measures (Chacko et al. 2014).

ROS production is increased in cancer cells due to high levels of aerobic glycolysis that generates oxidative stress, and ROS are involved in all steps of oncogenesis from cancer initiation to metastasis (Tafani et al. 2016), acting as signaling transducers during cell proliferation, migration, invasion, angiogenesis, and drug resistance (Nakamura and Takada 2021). Importantly, environmental factors present in the tissue environment, like hypoxia for example, act on the electron transport chain (Tafani et al. 2016) and may generate reactive oxygen species (ROS) known to damage DNA, RNA, and proteins (Torday 2015). Given the fast grow of some tumors, due to lack of proper vascularization, hypoxia occurs and ROS is generated, leading to increased oxidative stress.

Faced with cellular or extracellular stress, normal cells activate death pathways like apoptosis, for example. Cancer cells have the apoptotic pathways blocked and, instead, activate their stress survival pathways. RAS/MAPK and PI3K, two key signaling pathways altered in cancer, also play a key role in organism development (Arya and White 2015).

Cancer frequently hijacks host homeostatic systems to redirect energy and nutrient substrates in support of tumor growth and progression. Within this regulatory framework, the hypothalamus plays a central role in integrating physiological and environmental cues, the so-called "internal climate," to maintain systemic stability. When perturbations exceed the hypothalamus's regulatory capacity, chronic dysregulation may ensue, facilitating disease states including carcinogenesis. In advanced stages, cancer itself may impair hypothalamic function or impose irreversible systemic shifts, effectively disrupting homeostasis.

Peripheral cytokines such as IL-1β and TNF-α, commonly elevated in cancer, can promote neuroinflammation in hypothalamic nuclei, altering neuronal circuits that regulate energy balance and muscle mass. This inflammatory activation can stimulate the hypothalamic-pituitary-adrenal (HPA) axis, leading to glucocorticoid-mediated catabolism and muscle wasting (Braun et al. 2011; Burfeind et al. 2016). Concurrently, autonomic dysregulation affecting skeletal muscle, liver, adipose tissue, and endocrine organs exacerbates metabolic imbalance and promotes systemic inflammation. These mechanisms collectively position the hypothalamus as a central node in the pathophysiology of cancer-associated cachexia (Molfino et al. 2017).

Homeostasis and homeorhesis, dynamic processes that govern biological equilibrium and developmental trajectories, respectively, are fundamental not only to organismal physiology but also to tumor biology. Cancer cells frequently exhibit intrinsic homeostatic disruptions, such as altered metabolic programming (e.g., the Warburg effect), and secrete bioactive factors in excess, thereby modifying their own internal regulatory milieu. At the cellular level, transitions through the cell cycle (G1, S, G2, G0) are tightly regulated by homeorhetic mechanisms. Mutations in genes encoding key cell cycle regulators, such as cyclin-dependent kinases, can impair these checkpoints, enabling unrestrained proliferation (Ghafouri-Fard et al. 2022). Similarly, loss of tumor suppressor TP53 impairs apoptosis in genomically unstable cells, further illustrating how cancer disrupts intrinsic cellular control systems.

Conversely, tumors can exert systemic effects by dysregulating host homeostatic and homeorhetic processes. As demonstrated by Slominski et al., cancer cells secrete a broad array of signaling molecules, including cytokines, neurotransmitters, pituitary and hypothalamic hormones, biogenic amines, melatonin, and glucocorticoids, that can modulate central regulatory axes (Slominski et al. 2023). These tumor-derived factors can influence neuroendocrine centers such as the hypothalamus, pituitary, adrenal glands, and thyroid, thereby altering systemic physiology.

In hematologic malignancies, such as leukemia, tumors may induce a state of adaptive homeostasis to subvert host metabolic pathways and gain a proliferative advantage (Ye et al. 2018). Tumor-driven systemic changes can resemble diabetic-like metabolic derangements, further impairing host resilience. Notably, maintenance of normal glucose metabolism appears to exert protective effects, underscoring a mechanistic link between metabolic syndromes and cancer vulnerability.

Emerging evidence also supports a role for biological rhythms in cancer dynamics. Circadian regulation affects cell cycle, DNA repair, and metabolism, all of which are critical to tumor growth and treatment response. Chronotherapy, the alignment of treatment timing with circadian rhythms, has shown promising results in optimizing therapeutic efficacy and minimizing toxicity (Lévi 2001; Mormont and Lévi 2003). Recent studies suggest that malignant cells may exploit or disrupt host circadian regulation to enhance invasion and metastasis (Cortés-Hernández et al. 2020), highlighting the potential of chronobiology-informed interventions in cancer management.

9.2.6 Principle of Unity and Connectivity

Unity is what glues together bioenergy/bioinformation/biostructure within themselves and with the control processes that ensure life's persistence.

An organism can be seen from different perspectives as being simultaneously an autonomous individual, a network of relationships, and multiple codependent systems with different functions. Regardless of the perspective considered, unity is fundamental.

Unity is achieved through communication that is ubiquitous in living systems and is present inside a cell (signaling pathways between organelles); between cells through direct contact (gap junctions, etc.), nanotubes, or through secreted molecules and exosomes; and at the organism level through nerves, circulating cells through blood and lymphatic channels, exosomes, and various molecules (metabolites, cytokines, hormones).

In the human body, although each body system performs a different role, all the systems work together to keep the entire body functional and healthy. Due to this high interconnectivity, a disorder in one system can cause other systems to break down.

A body system is defined as an integrated set of organs and tissues that function cooperatively to perform specific physiological tasks. For instance, the immune system comprises a complex network of organs, tissues, and immune cells, such as leukocytes, that identify and neutralize pathogenic threats, thereby maintaining internal homeostasis.

Physiological processes often require the interaction of multiple systems. For example, the respiratory and circulatory systems collaborate to ensure efficient gas exchange: The lungs facilitate the diffusion of oxygen into the bloodstream and the removal of carbon dioxide, which is then expelled from the body. Some organs contribute to multiple systems. The pancreas, for example, functions as both an exocrine organ in the digestive system, by secreting enzymes that facilitate the breakdown of fats, carbohydrates, and proteins in the small intestine, and an endocrine organ in the hormonal system, by producing insulin and glucagon to regulate blood glucose levels.

Glucose serves as a primary energy substrate for cellular metabolism. Insulin facilitates glucose uptake by cells, reducing blood glucose concentrations, whereas glucagon promotes glycogenolysis and gluconeogenesis in the liver, increasing blood glucose availability. These hormones work in a feedback loop to maintain glucose homeostasis. Although body systems are classified separately for analytical purposes, they operate in a highly integrated manner. Dysregulation in one system can propagate pathophysiological effects in others.

The human body functions as a complex, interconnected network of systems, each contributing to overall physiological stability. Understanding the interdependence of these systems offers critical insights into how lifestyle factors such as nutrition, physical activity, and disease influence systemic health.

Moreover, biological time exhibits a multifaceted structure that transcends purely stochastic or deterministic models. It operates simultaneously across multiple scales and temporal rates, serving as a unifying principle for coordinating system-wide physiological processes.

As argued by Cliff Hooker, autonomy is a key property that can be used to differentiate a living from a nonliving system and is also a particular expression of global functional coherence (Hooker 2011). In Hooker's formulation, "an organism's adaptiveness is its capacity to alter its specific traits in mutually coordinated ways so as to adapt to, that is, satisfy autonomy in, different life-environments" (Hooker 2011).

As shown by Richard Michod et al., the passage from the unicellular to the multicellular level always introduces potential conflicts between the levels of selection (Michod et al. 2003). Cancer is the passage from commensalism to parasitism. Its behavior can be considered an adaptation to a hostile environment that includes the immune system attack. As part of this behavior, cancer cells are able to go back and forth between unicellular and multicellular modality. A phylostratigraphic analysis of tumor and developmental transcriptomes suggested recently that "not only does oncogenesis recapitulate ontogenesis, and ontogenesis recapitulates phylogenesis, but also oncogenesis recapitulates phylogenesis" (Zhou et al. 2018). Some organisms, such as slime molds and certain algae, can alternate between unicellular and multicellular states, typically in response to environmental cues like nutrient availability. Under scarcity, individual cells aggregate into a mobile multicellular structure to search for food, later reverting to the unicellular form for reproduction. In a similar way, cancer resists classification as strictly unicellular or multicellular; its hallmark plasticity enables context-dependent transitions between organizational states, supporting its adaptability and survival.

The Body's Wireless Internet System

In multicellular organisms, the functional unity is ensured by a complex web of communication that coordinates function across the entire system. Over the past decade, the significance of interorgan communication has gained increasing recognition across a broad spectrum of diseases, both acute and chronic. This includes conditions such as sepsis (Ilaiwy et al. 2019), diabetes (Shirakawa et al. 2017), cardiovascular disease (Shalhoub et al. 2013), hepatic and renal dysfunction (Zhang et al. 2018; Tatsumi et al. 2016), as well as immune and neurodegenerative disorders (Armutcu 2019). The emerging focus on organ network interactions at the organismal level underscores the systemic complexity of disease states (Oishi and Manabe 2020), reinforcing earlier foundational work on organ-to-organ signaling (Droujinine and Perrimon 2013; Ivanov and Bartsch 2014).

This interorgan communication, mediated through hormones, cytokines, metabolites, neural signals, and extracellular vesicles, forms the physiological infrastructure for systemic homeostasis (Tokizane and Imai 2025). Disruption of this network underlies a wide range of chronic diseases, including metabolic syndrome, cancer, cardiovascular dysfunction, and neurodegenerative disorders.

Each organ functions within a tightly regulated system of feedback and feedforward loops. Signals from one organ often influence the behavior of others, contributing to the overall homeodynamic stability of the organism (Kotas and Medzhitov 2015). The hypothalamus integrates inputs from the periphery, including nutritional, inflammatory, and circadian signals, to coordinate endocrine responses via the hypothalamic-pituitary axis, influencing distant tissues such as the adrenal glands, liver, and gonads. Similarly, adipose tissue, long thought to be a passive energy reservoir, is now recognized as an active endocrine organ. It secretes adipokines like leptin, which signals satiety to the brain, and adiponectin, which modulates insulin sensitivity in the liver and muscle (Ahima and Flier 2000). These endocrine loops illustrate how tissue-derived signals participate in the dynamic allocation of energy and resources across the body.

In the context of cancer, the notion that tumors behave as functional organs interacting with the rest of the body has been proposed for over a decade (Egeblad et al. 2010). This perspective has been further supported by the development of model systems—such as the Drosophila platform established by Perrimon's group—to explore interorgan signaling in vivo (Droujinine and Perrimon 2016, 2019). Increasingly, cancer is being framed as a systemic disease that disrupts homeostasis through widespread molecular and metabolic crosstalk (Egeblad et al. 2010; McAllister and Weinberg 2010, 2014; Al-Zoughbi et al. 2014; Paul 2015, 2020; Borniger, 2019).

A particularly exciting dimension of interorgan communication is the role of extracellular vesicles (EVs), such as exosomes, which transport proteins, lipids, and RNAs—including microRNAs—between distant tissues (Guay and Regazzi 2017).

These vesicles allow for nonhormonal, highly specific messaging and have been implicated in processes ranging from muscle regeneration to tumor metastasis (Kalluri 2016a). For example, tumor-derived EVs can prime the premetastatic niche in distant organs by altering vascular permeability and immune cell recruitment (Hoshino et al. 2015).

Exosomes are nano-sized vesicles (up to approximately 150–200 nm diameter) secreted by most cells. They are enclosed by a lipid bilayer and carry various biomolecules, including proteins, glycans, lipids, metabolites, RNA, and DNA. When exosomes are taken up by other cells, these cargoes are transferred and influence the phenotype of recipient cells. As such, exosomes are essential mediators of cell-cell communication, importantly, not only locally in the microenvironment but, importantly, also at distance (Wortzel et al. 2019).

Organism Perspective

The human organism is a complex, multicellular system composed of specialized tissues and organs that must operate in concert to maintain homeostasis and ensure survival. This systemic unity is achieved through a variety of tightly regulated communication mechanisms that operate at molecular, cellular, and systemic levels. Among these, hormonal signaling, neuronal networks, immune surveillance, and through secreted factors, such as cytokines, chemokines, metabolites, hormones, as well as EVs, particularly exosomes, play pivotal roles in integrating physiological responses and maintaining intercellular coordination. Beyond chemical signaling, mechanical and bioelectrical cues also contribute to the synchronization of cellular and tissue functions. Mechanical forces, for example, transmitted through the extracellular matrix, influence cell behavior, differentiation, and organ development (Discher et al. 2005). These non-chemical forms of communication are critical for maintaining structural and functional coherence.

Cancer Perspective

From a macroscopic perspective, cancer exhibits characteristics akin to those of a parasitic organism nested within the host body. It operates not merely as a localized mass of transformed cells but as a complex, spatially distributed system composed of multiple interlinked pathological sites—including the primary tumor, regional extensions, and distant metastatic colonies. These anatomically dispersed lesions form an integrated *cancer system*, which functions as a cohesive pathological entity within the broader host organism.

This cancer system is not autonomous; rather, it exists in a state of dynamic codependence with the physiological systems of the host. Through continuous biochemical, cellular, and extracellular exchange—mediated by factors such as cytokines, hormones, metabolites, and extracellular vesicles—the cancer system establishes a bidirectional interaction with the host's regulatory networks. These interactions lead to the emergence of *Cancer-Induced Systemic Pathologic Networks (CISPN)*: maladaptive, cross-organ communication pathways that progressively reconfigure the host's internal environment.

The formation of CISPNs marks a profound systemic shift in the organism. Homeostatic processes, once governed by organism-centric regulation, become subverted or co-opted to favor tumor growth, survival, immune evasion, and metastatic dissemination.

As a result, the host body undergoes a pathological reorganization—a process of *organismal cancerization*—in which normal physiological functions are repurposed to meet the demands of the cancer system. In this cancerized state, the boundaries between normal and malignant biology blur, and the organism becomes an active, though involuntary, participant in its own pathological transformation.

Cancer, whether confined to a single anatomical site during early stages or dispersed across multiple loci—including the primary tumor, regional extensions, and distant metastatic outposts—can be conceptualized as a dynamic, self-organizing system embedded within and interacting with the host organism (Paul 2020). In this systemic framework, cancer does not remain a localized pathology but progressively integrates with and reshapes the physiological networks of the host, leading to the emergence of a "cancerized" organismal state (Paul 2020).

This transformation is mediated through complex and multi-scalar signaling networks involving cellular components such as tumor-associated macrophages and bone marrow-derived myeloid cells; molecular mediators including pro-inflammatory cytokines, chemokines, and metabolic by-products; as well as extracellular vesicles, particularly exosomes (McAllister and Weinberg 2014; Hoshino et al. 2020; Kalluri and LeBleu 2020). These signals establish a pathological form of interorgan communication that subverts homeostatic mechanisms, reprogramming distant tissues to create permissive niches that facilitate tumor growth, immune evasion, and metastatic colonization (Peinado et al. 2017; Wishart 2019).

Moreover, the cancer system competes with and disrupts the interorgan communication process. This hijacking of systemic signaling not only alters the behavior of individual organs but also destabilizes the broader regulatory networks responsible for maintaining physiological integrity. Through these interactions, cancer progression becomes a process of progressive systemic infiltration—whereby the malignancy co-opts the body's own regulatory circuits to propagate a malignant agenda across space and time. The "circulome" databases are still in their infancy (Wu et al. 2020).

The metabolic dialogue between tumors and distant organs is also a critical component of cancer-induced interorgan communication. Tumors impose a high metabolic demand on the host, often leading to systemic metabolic reprogramming. For instance, studies by Petruzzelli and Wagner describe how tumor-induced cachexia involves the mobilization of lipids and proteins from adipose tissue and skeletal muscle, mediated by tumor-derived factors such as PTHrP and IL-6 (Petruzzelli and Wagner 2016). These signals act on distant organs to degrade host tissue, providing metabolic substrates to fuel tumor growth.

In addition, recent work highlights how the nervous system is implicated in tumor-host communication. Magnon et al. demonstrated that neural infiltration and sympathetic nervous signaling contribute to tumor initiation and progression (Magnon et al. 2013), while other studies suggest that tumors can influence brain function, inducing cognitive and behavioral alterations in cancer patients (Pendergrass et al. 2018).

Collectively, these findings converge on a new understanding of cancer as an active participant in a pathological reconfiguration of the body's internal communication networks. By manipulating interorgan signaling through molecular, cellular, and vesicular mechanisms, tumors effectively transform the host organism into a pro-tumorigenic environment.

9.2.7 Principle of Codependency

The environment has a key role as it is actively shaping and is being shaped by the living organisms in a *codependent relationship.*

The "holobiont" is a term initially coined by Adolf Meyer-Abich in 1943 and subsequently used by Lynn Margulis. A holobiont is a host and the many other species living in or around it, which together form a discrete ecological unit. The whole organism and cancer are two interacting living systems that form together a holobiont. Of note, the "holobiont" is close to the original meaning coined by Meyer-Abich that defined the "holobiont" as a "specific biont within a strongly integrated whole"—in his words, a "holobiome created through holobiosis, rather than as the whole itself." Living systems, that is, holobionts, ecosystems, biospheres, formed by interactions between individual life forms, can be seen as strategies of life persistence. Carbon-14 is produced in the stratosphere continuously by nuclear reactions of atmospheric nitrogen with thermic neutrons produced naturally by cosmic rays. Subsequently, Carbon-14 atoms interact with oxygen leading to the creation of carbon dioxide. The carbon dioxide (CO2) will subsequently cycle between the atmosphere, plants, sols, and the ocean. The oxygen of the atmosphere is produced by plants who incorporate the CO2 that is cycled between the atmosphere, plants, sols, and the ocean. In light of this, the whole Earth biosphere is codependent with its abiotic environment. Ecosystems are also codependent with the whole biosphere and with the life forms that compose them. Organisms are codependent with their biotic and abiotic environments that shape their evolution. Organism themselves can be considered holobionts or biospheres (see above) formed by codependencies between different types of life forms. Both adaptation and different forms of evolution including Darwinian evolution can be seen as codependencies between life forms and the environment. Adaptation responds to changes in the environment during an organism's lifetime and evolution copes with environmental changes during generational timescales (Bonner 2019). So adaptation and evolution go hand in hand. The first is intragenerational; the other one is intergenerational. All this happens in the context of a particular environment. Besides adaptation and evolution, development also plays a key role. Evolution brings novelty; development uses old tricks. Developmental programs are life's most elaborate dynamic persistence mechanisms.

Bauer defined "life as relationship of a system to its given environment" (Müller and Igamberdiev 2024). Yury Zhuravlev and Vladik Avetisov noted, "Agents of life are transmuted in such a way to produce a diversity of genovariants among which the more fitted will be selected during future changes in the environment. However, nobody knows whether the environmental changes are accidental or whether they follow a certain scenario, predetermining a direction in the development of life" (Zhuravlev and Avetisov 2006). We can similarly define the disease labeled "cancer," as the *relationship of a cancer system to the organism environment.* Defining cancer in relativistic terms is important. Essentially, there is no separation of cancer from the organism where it develops, and what we call cancer can be only understood in a specific context. Cancer's system relationship with the organism determines the clinical behavior of the disease we call cancer.

Organism Perspective

The human body exists in a state of dynamic symbiosis with the gut microbiota—a complex community of microorganisms that collectively outnumber human somatic and germline cells by approximately tenfold. This microbial ecosystem exerts profound influence over host immunity, metabolism, and systemic homeostasis. Increasing evidence underscores the role of gut microbiota in modulating inflammation, both locally within the tumor microenvironment and at the level of systemic circulation. The gut microbiota, in particular, constitutes a densely populated microbial ecosystem with essential roles in metabolism, immune development, barrier function, and neural signaling (Sommer and Bäckhed 2013; Belkaid and Hand 2014). This reciprocal relationship has led to the concept of the host and its microbiota as a "holobiont"—a single ecological unit where codependence is key to maintaining homeostasis (Bosch and McFall-Ngai 2011).

The immune system is deeply influenced by signals from the microbiota. Early-life colonization is critical for immune maturation, and germ-free animal models show defective immune responses that can be partially restored upon microbial colonization (Round and Mazmanian 2009). Microbial metabolites such as short-chain fatty acids (SCFAs), particularly butyrate and propionate, modulate regulatory T-cell differentiation and anti-inflammatory pathways (Arpaia et al. 2013; Furusawa et al. 2013), reinforcing the role of the microbiota in maintaining immune tolerance. The microbiota provides metabolic capabilities that the host lacks. For instance, microbial fermentation of indigestible dietary polysaccharides produces SCFAs, which serve as energy sources and signaling molecules for the host (Koh et al. 2016). Furthermore, gut microbes contribute to bile acid metabolism, vitamin synthesis (e.g., vitamin K and B12), and xenobiotic detoxification, underlining their metabolic indispensability (Nicholson et al. 2012).

Disruption of the host-microbiota codependence, termed dysbiosis, has been implicated in a broad range of diseases, including cancer (Lynch and Pedersen 2016; Woting and Blaut 2016). The mutual dependence is evident in how both microbial imbalances and host-derived inflammation can exacerbate each other, contributing to chronic disease states.

Two pivotal studies published in 2013 in *Science* demonstrated that commensal microbes shape antitumor immune responses and can influence the efficacy of cancer therapies through modulation of inflammatory pathways and immune tone (Iida et al. 2013; Viaud et al. 2013).

Beyond immunomodulation, the gut microbiota is integral to hematopoietic homeostasis. It regulates steady-state myelopoiesis and supports neutrophil development and function under non-pathological conditions (Gorjifard and Goldszmid 2016). Perturbations in this microbial-immune crosstalk can have pathophysiological consequences. In murine models, specific microbial compositions have been shown to facilitate mammary tumorigenesis through neutrophil-dependent mechanisms (Kostic et al. 2013). Furthermore, microbiota-induced expansion and mobilization of myeloid-derived suppressor cells (MDSCs) promote systemic tumor-supportive inflammation, thereby accelerating malignant progression (Rutkowski and Conejo-Garcia 2015). These findings collectively position the gut microbiome as a critical regulator of cancer-associated inflammation, influencing both the initiation and progression of malignancy via immune-mediated and hematopoietic pathways.

Cancer Perspective

In advanced stages, cancer develops inside the organism, a distinct *onco-holobiont* which comprises besides cancer cells, stromal cells, infiltrating immune populations, vascular and lymphatic networks, and potentially resident or co-opted microbial communities to grow within the tumor microenvironment. This integrated unit forms together with the local primary tumor stroma, and the distal metastatic sites, an *oncosphere*, that functions as an autonomous pathophysiological system that interacts dynamically with the host organism, shaping the trajectory of cancer progression.

The interaction between cancer and its host shapes the behavior and evolution of cancer as a disease. The more benign course or the aggressivity of cancer is influenced directly by this interaction. Communication is crucial for the development and the functioning of the cancer system. Cancer-Induced Systemic Pathologic Networks (CISPN) are formed through communication; in cancer, not only the stromal cells including fibroblasts and immune cells, and the local microbiota, but the entire organism is prompted to support cancer growth, invasion, and metastasis (Paul 2015, 2020; Lieff 2020) through an "instigation" process previously described by Robert Weinberg and his team (McAllister et al. 2008).

Cancer's behavior—whether indolent or aggressive—is determined by the continuous bidirectional interactions between tumor and host, including metabolic exchange, immune modulation, extracellular matrix remodeling, and neuroendocrine signaling. These host-tumor communications influence critical hallmarks of cancer such as immune evasion, angiogenesis, and metastatic dissemination (Hanahan 2022).

Crucial to the functioning and progression of the onco-holobiont is the establishment of intricate intercellular communication networks. Cancer cells exploit and reprogram surrounding stromal cells—such as cancer-associated fibroblasts, tumor-associated macrophages, myeloid-derived suppressor cells, and endothelial cells—to create a supportive niche. In addition, local microbiota may contribute to shaping immune responses and modulating inflammatory pathways, further embedding the tumor within a permissive microenvironment (Garrett 2015; Dzutsev et al. 2017).

At a systemic level, malignancies can induce widespread physiological remodeling through the formation of Cancer-Induced Systemic Pathologic Networks (CISPNs). These networks are constructed via a constellation of signals—circulating cytokines, hormones, exosomes, and metabolic intermediates—that reprogram distant organs and tissues to support tumor growth, cachexia, immune suppression, and metastasis. This systemic reorientation echoes the concept of tumor instigation, first introduced by Weinberg and colleagues, where tumor-secreted factors prime distant tissues to create a pro-metastatic niche, facilitating the outgrowth of otherwise indolent tumor cells (McAllister et al. 2008; McAllister and Weinberg 2014).

Thus, cancer must be understood not merely as a localized pathology but as an organism-scale, dynamically regulated system that co-opts host physiology to secure its persistence and expansion.

Cancer cells collaborate between themselves and with other cells in the stroma (Cheung and Ewald 2016). Recently, Aurora Nedelcu and her team demonstrated that when a less aggressive breast cancer line was cocultured with a highly metastatic breast line, the invasive potential of both lines was increased. The mechanism for the synergy was an autocrine-paracrine TGF-β1-dependent instigating of the weakly metastatic clone into expressing an enhanced malignant phenotype that benefited both clones with a "help me help you" strategy, as described by the authors (Carneiro et al. 2023). Cancer cells also collaborate with other cells from the stroma such as cancer-associated fibroblasts via exosomes (Naito et al. 2022).

The cancer process is also codependent with both local and distal organismal factors. In terms of local factors, for example, it has been recently described a neural addiction of cancer cells (Magnon and Hondermarck 2023). The groundbreaking experiments of Mina Bissell's group demonstrated the crucial influence of the stromal factors on cancer development and progression (Bissell et al. 1982; Dolberg and Bissell; 1984, Roskelley et al. 1994; Roskelley et al. 1995, Weaver et al. 1997). Furthermore, there is a long history of seminal observations highlighting the importance in cancer of non-cell autonomous factors (reviewed in Feunteun et al. 2022) starting with Paget's "seed and soil" hypothesis pointing to the importance of the distal organ environment for metastases.

9.2.8 Principle of Creativity

Complexity and Emergence
Creativity represents the ability to create new knowledge and new behaviors. Darwin considered evolution a dynamic process akin to a creative process (Beatty 2016, 2019). As noted by Mark Kirschner and John Gerhart, the authors of the *theory of facilitated variation*, introduced above, creativity in evolution is not related to selection alone; rather, the phenotype itself, through its established molecular and physiological architecture, constrains and channels the spectrum of viable phenotypic variations that can emerge in response to regulatory shifts (Kirschner and Gerhart 2005).

The accent is on *regulatory changes*, rather than structural gene alterations, and the predominant drivers of genetic innovation and evolution appear to favor mechanisms that minimize the number and complexity of regulatory modifications, often achieving new phenotypic outcomes by redeploying conserved molecular components and pathways. Regulatory changes operate on a core repertoire of ancient, conserved processes, selecting specific subsets that are then subject to natural selection. Phenotypic plasticity, whether environmentally induced or developmentally embedded, serves as a rich and evolutionarily favored reservoir of variation, often requiring minimal regulatory changes. Importantly, this perspective does not align with Lamarckian inheritance, nor does it imply selection for future benefit. Rather, the capacity for *facilitated phenotypic variation* likely emerged as a by-product of natural selection, acting on the *functional robustness, adaptability, modularity, exploratory capacity, and flexible regulatory connectivity* of core processes. These properties were selected at the level of the individual organism for their effectiveness in maintaining physiological and developmental stability under variable internal and external conditions. Thus, similar molecular features that enable short-term adaptability also underpin long-term evolutionary change, as regulatory innovations become genetically stabilized (Gerhart and Kirschner 2007).

Variation and novelty are intrinsically related to entropy, and evolution, as described in Chap. 7, can be understood as mechanisms of entropy control. Variation is a source of randomness contrasting to the deterministic functionality of evolution. Entropy may also lead to the phenomenon of emergence, which may be more or less controlled. Outside evolution, genuine creativity may be the by-product of any homeopoietic process since random noise can never be completely controlled (Kamb and Garguli 2025). David Chalmers distinguished between two types of emergence: *weak* and *strong*. A phenomenon at a high scale is *weakly emergent* with respect to a lower scale if, given the laws governing the latter, the patterns observed in the former are unexpected, but they can be deducible in principle from advanced calculations and/or computation. If such a deduction is not possible even in principle, then the phenomenon is *strongly emergent*. An example of a strong emergent phenomenon is consciousness (Chalmers 2006).

As noted by Giuseppe Longo et al., evolutionary processes inherently give rise to novel "adjacent possible" unoccupied niches, even in the absence of direct selective pressures (Longo et al. 2012). These emergent opportunities facilitate new evolutionary trajectories, exemplifying a form of radical emergence wherein novel life forms evolve from preexisting biological contexts. In the authors' opinion, the emergence of organisms capable of exploiting these new niches frequently originates from singular or sequential quantum-level molecular events that are intrinsically acausal. Consequently, the niche itself does not serve as a causal determinant but functions as an enabler of radical evolutionary novelty. Life, therefore, is not solely embedded within deterministic causal networks but also participates in a dynamic web of mutual enablement and emergent complexity. Stuart Kauffman's "adjacent possible" considered an unanticipated collateral effect of evolution is another example of strong emergence (Kauffman 2016, 2019).

Besides TAP, which relates to traits selected initially for a function that may serve another different function (Kauffman's example is a screwdriver that can have many other uses besides screwing in a screw; it can also be used to open a can of paint, etc.), genuine novel functions may raise also indirectly through a sort of collateral creation process (CCP) where traits serving no functions, in particular, that were not prior selected (like the color of the handle of the screwdriver, using Kauffman's example) achieve a function in an unprecedented context. In the context of cancer, both TAP, that is, the activation of embryonal programs, and CCP, that is, cancer passenger mutations (Castro-Giner et al. 2015; Wodarz et al. 2018; Kumar et al. 2020), may be sources of creativity.

Hoping to find smooth ways of integrating biology within physics and chemistry without adding any additional laws or special behaviors, scientists have been modelling complex systems that reside at the border between order and disorder using chaos theory, power laws, fractals, transition points, and criticality. Fundamentally, though, the emergent behavior of complex system containing large numbers of homogeneous units is very different from the complexity of the living systems that have large number of heterogeneous and composite elements.

The principle of complexity is closely related to organic life. As stated by Mayr, "every organic system is so rich in feedbacks, homeostatic devices, and potential multiple pathways that a complete description is quite impossible" (Mayr 1961).

Complexity of a system has been described by Daniel Mc Shea as an "increasing function of the number of different types of parts or interactions it has" (McShea 1996).

More recently, complex systems have been described as "systems whose states change as a result of interactions and whose interactions change concurrently as a result of states. Due to this chicken–egg-type problem, complex systems show an extremely rich spectrum of behavior: they are adaptive and co-evolutionary; they show path-dependence, emergence, power laws; they have rich phase diagrams; they produce and destroy diversity; they are inherently prone to collapse; they are resilient, and so on" (Thurner et al. 2018).

Complexity has been considered a prerequisite of persistence: "For a system to persist in time in a complex environment, it should possess a certain degree of complexity that is greater than that of its surrounding (this is also known as the law of requisite variety). In order for a system to maintain its structure and keep increasing its complexity in order to stay in competition with the ever-changing complex environment, the system needs to self-organize by increasing the amount of free-energy needed to do work, thereby decreasing its internal entropy, and creating more action efficient flow channels at every instant in time" (Chatterjee et al. 2017).

Different kinds of individuals have different degrees of complexity as measured by their number of parts, degree of integration, or hierarchical nestedness.

Complexity has been also described as a feature of living systems (Sharma et al. 2023; Shapiro 2022) but order, complexity, and heterogeneity are not exclusively restricted to living entities (Popa 2004).

Physical complexity in biological systems has been defined as information about the environment that can be used to make predictions about it. Physical complexity is relevant for the fitness of a living system because making predictions about an environment allows it to exploit it more efficiently. More precisely, we can think of physical complexity as the amount of information that is coded in the genomes of an adapting population, about the environment to which it is adapting (Adami 2002).

It has been suggested that macroevolution of life is driven toward increased complexity (Yaeger 2009). More recently, Carlo Rovelli and his collaborators published an article in which they formulated a similar opinion: "Evolution is traversal of an extremely complex energetic landscape, progressing from less-complex to more-complex states as each increase in complexity offers an increased number of new channels for further entropy growth" (Jeffery et al. 2019).

On the other hand, there is also evidence that some dynamic systems may evolve toward lower entropy states, not by increasing complexity but by increasing organization. Organization refers to both order and complexity and it is important to differentiate between the two. It is possible for a dynamic system to experience changes in organization in ways that do not elicit changes in complexity (Popa 2009).

Organism Perspective

Life is creative. For life to exist, a controlled balance between randomness and order is needed. This controlled balance that ensures dynamic persistence is itself continuously evolving and represents a source of creativity. Living systems are autopoietic, but the autopoiesis is not perfect. In unicellular organisms, mutations induced by the internal processes or by the external environment are a source of continuous novelty. This novelty is controlled by evolutionary mechanisms. Similarly, in multicellular organisms, variation happens during meiosis when recombination ensures that the species continues to evolve. As pointed up by Paul Nurse (Nurse 2021), any entity that has the ability to reproduce and transmit variable information from one generation to the next *can and will* evolve. Evolution encompasses four broad processes: the process of natural selection described by Darwin which is adaptative in the sense that it depends on the fitness of individuals and three others that are nonadaptative and responsible for variation (genetic mutation, chromosomal recombination that can occur within and among chromosomes, and genetic drift through which gene frequencies deviate from generation to generation independent of other forces) (Lynch 2007). Darwinian natural selection is what we have in mind when we talk about evolution, and it derives from three basic principles that are all required: Intergenerational transmission of information, the variability of information transmitted, and the competition between living forms. As opposed to natural selection, the nonadaptive processes of evolution are stochastic in nature.

As noted by Marijuán and Navarro, classical Darwinian evolution is only one of the plethora of mechanisms able to generate variability. They mention many others that occur at the cellular, organismic, or population level: "mutations (neutral, significant, short-scale, large-scale); sex (meiosis, recombination) and populations (allele distribution, genetic drift); gene modification (horizontal transmission, frameshift change, intron/exon swapping, domain recombination, duplications); chromosome rearrangements (translocations, inversions, deletions, crossover, duplications); mobile, transposable, retrotransposable, and viral elements (with additional effects via ncRNAs, siRNAs, miRNAs, piwiRNA, gene silencing); developmental (biased gene expression, repetitive DNA, ncRNAs, enhancers and regulators, neoteny); epigenetic (histone code, DNA methylation, hormonal imprinting, stress-metabolic-behavioral axis); whole genome duplications; symbiosis; plus a variety of other behavioral and environmental (and niche) effects which impinge on organisms" (Marijuán and Navarro 2022).

A recent study demonstrated that somatic mutations are ubiquitous across species, with the annual somatic mutation rate varying substantially among them. Notably, this rate exhibited a strong inverse correlation with species lifespan, a relationship not paralleled by any other life history trait examined (Cagan et al. 2022).

Importantly, a multicellular organism's evolution, like the human organism, occurs across multiple successive generations and not during its lifetime. This is radically different from cancer cells which evolve continuously. The plasticity and continuous evolution of cancer cells allows them to progressively gain control over the organism's metabolism and redirect its resources toward their own growth and development.

Cancer Perspective

Cancer plasticity, the ability of cancer cells to adapt and change their characteristics—an expression of its creativity—is significantly influenced by genomic instability. It has been recently shown in a cohort of patients with refractory metastatic cancers characterized by a short survival time that several types of genomic instability were strongly enriched, particularly the mutation rate, the frequency of whole genome doubling (WGD), and the fraction of the genome covered by focal copy-number alterations.

Centrosome Aberrations

Another non-cell autonomous mechanism capable of inducing changes in the biomechanical properties of epithelia has been uncovered in tissues harboring cells with centrosome aberrations (CA). Several lines of evidence support the progressive involvement of CA in the transition from early to advanced stages of carcinogenesis, being also found in preneoplastic lesions and sometimes even in normal tissue away from the tumor. CA represents the major mechanism leading to chromosomal instability and aneuploidy, via the formation of multipolar spindles and chromosomal missegregations. CA are associated with aneuploidy, defective cell division symmetry, and altered ciliogenesis. Although these manifestations a priori hamper the ability of cells to divide successfully, they do generate chromosomal instabilities necessary for the few surviving cells to acquire malignant properties, thus suggesting that they are not bystanders in the development of the tumor but true drivers of tumorigenesis. Indeed, although it usually affects only subsets of cells within the tumors, the amplification of centrosomes is a hallmark of most human cancers. CA have been observed during the early stage of tumorigenesis in different types of preneoplastic lesions such as in situ breast ductal carcinoma and prostate intraepithelial neoplasia, suggesting that they could promote the early stage of tumorigenesis (Chan 2011).

Retrotransposons

Transposable elements (TEs) are discrete pieces of DNA that can move from site to site within (and sometimes between) genomes. TE are the most prevalent selfish genetic elements found in humans; approximately 45% of the human genome is made of TE (Cordaux and Batzer 2009). The scale of the contribution of TEs to the human genome is all the more remarkable when considering that protein-coding regions account for just 1.5% of the human genome. TEs can be separated into two major classes: DNA transposons and retrotransposons. DNA transposons, which make up ~3% of the human genome, are currently not mobilizing in the human genome. However, a correlation between HERVs and human cancer has been described and many tumors, such as melanoma, breast cancer, germ cell tumors, renal cancer, or ovarian cancer, express HERV proteins, mainly HERV-K (HML6) and HERV-K (HML2) (Gonzalez-Cao et al. 2016).

Retrotransposons can be subdivided into two groups, distinguished by the presence or absence of Long Terminal Repeats (LTRs). Retrotransposons duplicate via RNA intermediates that are reverse-transcribed and inserted at new genomic locations. Human LTR elements are endogenous retroviruses which account for ~8% of the genome and are probably not active. On the other hand, the vast majority of human TEs result from the present and past activity of non-LTR retrotransposons, including long interspersed elements (LINEs) (~21%) and short interspersed elements (SINEs) (~11%) that collectively account for about one-third of the human genome. Of these, the most active elements able of retrotransposing in the human genome include autonomous LINE-1 (L1) from the LINEs and nonautonomous *Alu* from the SINEs (Kazazian and Moran 2017). Retrotransposons, which may be remnants of ancient viruses, are normally repressed by epigenetic mechanisms, but this repression can be lost in cancer, leading to their transcriptional activation that can cause harmful mutations and inflammation.

Approximately 50% of human tumors contain somatic retrotranspositions of L1 elements (Tubio et al. 2014; Helman et al. 2014).

The patterns and mechanisms of somatic retrotransposition was analyzed in 2954 cancer genomes from 38 histological cancer subtypes within the framework of the Pan-Cancer Analysis of Whole Genomes (PCAWG) project. The authors identified 19,166 somatically acquired retrotransposition events, which affected 35% of samples and spanned a range of event types. LINE1 insertions emerged as the first most frequent type of somatic structural variation in esophageal adenocarcinoma and the second most frequent in head-and-neck and colorectal cancers (Rodriguez-Martin et al. 2020). Aberrant L1 integrations can delete large regions of a chromosome, which sometimes leads to the removal of tumor-suppressor genes and can induce complex translocations and large-scale duplications. Somatic retrotranspositions can also initiate breakage-fusion-bridge cycles, leading to high-level amplification of oncogenes (Rodriguez-Martin et al. 2020).

Recently, an integrative genomic analysis of 214 multiregional rapid autopsy samples from pancreatic ductal adenocarcinoma (PDAC) patients demonstrated for the first time the coevolution of mutant TP53, inflammation, retrotransposition, and the display of viral mimicry in PDAC. The findings were validated in an additional large autopsy cohort of 35 PDAC tumors and several PDAC cell lines. Immunostimulatory SINE expression was found to be negatively regulated by L1 retrotransposon activity in the setting of TP53 mutation while, in the setting of wildtype TP53 status, tumors were using ADAR1 as a response mechanism to SINEs. Both mechanisms highlight the opportunistic behavior of cancer cells that exploit multiple pathways to promote tumor growth via immune manipulation (Sun et al. 2023a).

Also, hypomethylation of retrotransposable elements corresponds with genomic instability in squamous cell lung cancer (Zhang et al. 2019).

Interestingly, emerging evidence suggests that retrotransposons may behave also as tumor suppressors. Treatment of cancer cells with demethylating agents, which removes silencing of retrotransposons, triggered cell death via activation of ERVs and dsRNA sensing pathway (Chiappinelli et al. 2015). Remarkably, transposon-triggered innate immune response also confers cancer resistance to the blind mole rat. These findings suggest that in some cancers, forced reactivation of retrotransposons can have an antitumor effect (Zhao et al. 2021).

Meiosis

Meiosis is a special type of cell division that occurs only in the reproductive cells of sexually reproducing organisms. Unlike somatic cell division which only has one round of division, it comprises of two rounds of cell division called meiosis I and meiosis II. Before meiosis begins, the cell's DNA is replicated once during an S-phase. However, this single replication is followed by two consecutive divisions. As a result, there are four daughter cells produced at the end of meiosis, each containing only half the number of chromosomes as the original parent cell. The reduction in chromosome number is crucial for sexual reproduction. When the egg and sperm cells fuse during fertilization, the resulting zygote regains the full complement of chromosomes specific to that species (Comaills and Castellano-Pozo 2023).

Importantly, in healthy somatic cells, meiotic genes are permanently silenced; however, in cancer, failure of meiotic gene silencing is a common phenomenon. Cancer testis antigens or genes (CTA, CTG) are predominantly expressed in adult testes while silenced in the rest somatic tissues. They are involved in meiotic chromosome metabolism and specifically in meiotic recombination. Only a few CTGs (typically between 0 and 2) were found to be highly expressed in a given tumor (Wang et al. 2016). More recently, the repertoire of CTGs has been significantly expanded (Bruggeman et al. 2018, 2020) and a "germline program" hypothesis promoting cancer cell malignancy has been proposed (Bruggeman et al. 2023). Reexpression of meiotic genes may lead to changes in chromatin structure. Chromatin structure is essential both for maintaining differentiation and cellular identity and is highly remodeled during the malignant transformation process. A meiosis-like architecture structure may promote the malignant state (Sou et al. 2023). As many CTGs are highly specific to cancer, and absent in adult germ cells and somatic tissues, targeting their gene products may be an interesting therapeutic avenue.

10 Biological Time

10.1 Life as Verb and Pattern

One of the difficulties to capture life into a comprehensive definition is understanding and agreeing what are we talking about when we talk about life. Is life a cell? An organism? An ecosystem? Or is it the entire processes and transformations that began some 4.2 billion years ago with LUCA and created the biosphere itself? Michael Russell provocatively suggested that life is "a verb, not a noun"—a ceaseless becoming rather than a static thing (Russell 2017). Through this lens, life is not confined to form but flows as a process-generating forms—cells, organisms, ecosystems—across both time and scale. It is a dynamic continuum: cellular cycles, organismal rhythms, ecological networks, and evolutionary spirals all woven into a self-perpetuating choreography of change. As Edmund Ambrose once remarked, "The matter in life has no permanence; only the pattern according to which it is arranged or organized has permanence. Life is basically a pattern of organized activity" (Ambrose 1982). In this sense, life becomes a unique form of time—biological time—distinct from the cosmic or quantum chronologies, a temporal architecture, all its own.

While Newtonian physics views time as an external, universal parameter, in biology, time is constitutive, being the very essence of being alive. Conrad Waddington captured this beautifully: "Perhaps the main respect in which the biological picture is more complex than the physical one, is the way in which time is involved in it. In the Newtonian system, time was one of the elements in the physical world, quite separate from any of the others; a material body given mass just existed, unchanging and, indeed, quite indifferent to the passage of time. But time and change are part of the essence of life. Not only so; to provide anything like an adequate picture of a living thing, one has to consider it as affected by at least three different types of temporal change, all going on simultaneously and continuously. These three time-elements in the biologic picture differ in scale. On the largest scale is evolution; any living thing must be thought of as the product of a long line of ancestors and itself the potential ancestor of a line of descendants. On the medium scale, an animal or plant must be thought of as something which has a life history. It is not enough to see the horse pulling a cart past the window as the good working horse it is today; the picture must also include the minute fertilized egg, the embryo in its mother's womb, and the broken-down old nag it will eventually become. Finally, on the shortest timescale, a living thing keeps itself going only by a rapid turnover of energy or chemical change; it takes in and digests food, it breathes, and so on" (Waddington 1957). The living process is defined by three layered temporal tapestries of different durations: evolution stretching across generations, individual development from fertilized egg to the aging adult, and the cycles of metabolic activity, sustained through constant energetic turnover. Far from being static, life *is* precisely this nested set of unfolding stories. *This temporal scaling is one of the key conditions for life to exist.* The combination of slow and fast processes is essential for life. Living organisms are able to *process and integrate information across temporal scales*. They continuously sense both their internal and external environments, and they act in the direction of maximizing their survival chances.

Cancer, like all phenomena of life, is both a noun *and* a verb. Rather than focusing solely on the static identity of cancer cells or tissues, cancers as tumors, we should also consider the dynamic systems and feedback loops that sustain their existence, cancers as processes. A new paradigm in oncology is emerging—one that seeks not merely the molecular object but the shifting network of codependencies. This demands on a theoretical framework shaped by evolving networks, relational ontologies, and a blend of object-oriented and process-informed models. Curing cancer may reside not in targeting frozen 3D molecular shapes but in reshaping its 4D fluid temporal narrative.

D. Paul, *Cancer: An Integrative Approach*, https://doi.org/10.1007/978-3-032-08808-6_10

10.2 Energy Transfer and Directed Movements

A cell is a realm of self-generated directed activity. The origin of activity does not come from the cell; it comes from outside. Jeremy England considered that the simple fact of continuously infusing energy into a far from equilibrium, open system will determine it to spontaneously self-organize (England 2011). A cell organizes itself by creating its own spatial and temporal structures and its own directed movements, imposing order on the molecular agitation and the herding of ions. *Cells are organized flows of events that are channeled to self-create themselves anew.* Cells are powered by electrochemical gradients responsible for ATP synthesis and the maintenance of the membrane potential. The electrochemical gradient is maintained by directed actions that are under control of the bioinformation systems of the cell (i.e., DNA is transcribed and then translated into proteins that can function as pumps that maintain the gradient of protons or ions across various membranes). Inside a cell, there are three interconnected, continuous, and repetitive processes: energy metabolism, information communication and processing, and structure maintenance and creation. Their recursivity creates patterns in space and time.

10.3 Life Uninterrupted

Over half a century ago, John Tyler Bonner posited that the true unit of biological meaning is not the adult form but the life cycle itself: "*Not only the adult, but the whole life cycle will be considered the organism*" (Bonner 1965). More recently, Margarida Hermida, a researcher from Portugal, described cellular life, from LUCA to the present, *as a singular ontological organism*: "The zygote is not only genetically related to the parents, but it is spatio-temporally continuous with the mother's oocyte and the father's sperm cell, which are spatio-temporally continuous with other cells in their bodies, which in turn are continuous with their own parent's reproductive cells, and so on. Every living cell today is spatio-temporally continuous with previously existing living cells, up until the unicellular ancestor of all subsequent life on Earth. In these long sequences of reproductive events, the life processes have 'kept going', as living things do. [...] A life is a spatio-temporally restricted event, which continues while the life processes are occurring in a particular chunk of matter (or, arguably, when they are temporally suspended, but can be restarted at any moment), even though there is continuous replacement of parts. Life is organized in discrete packages, particular cells and multicellular organisms with differing degrees of individuality. [...] Assuming that all life on Earth has a common origin, all living organisms, cells, and tissues descending from this origin exhibit continuity of the life processes at the cellular level, as well as many of the features that define the individual character of species: spatio-temporal localization and restriction, continuity, historicity, and cohesiveness. *Therefore, life on Earth is an ontological individual*" (Hermida 2016). Of note, prior to Margarida Hermida, Olivier Rieppel proposed also that species should be considered biological individuals (Rieppel 2013).

In parallel, John Torday developed the model of a zygote-centered evolution, called the "unicell" (Torday 2015), in which he suggested that the germ cells are being selected for and "the adults are disposable." In his view, "multicellularity is an effective mechanism of further maintaining the integrity of the unicellular state by extending its ability to encounter and cope with environmental stimuli and stresses. [...] Embryogenesis and organic development must be regarded then as an act of cellular choreography elaborating developmental stages as the means of enabling the acquisition of pertinent epigenetic experiences. The sum total of these will be recapitulated again in the unicell (zygote state). This explains the primacy of meiosis, which is mathematically the best means of averaging the communal epigenetic experiences of populations of like organisms (which now is the same as saying 'like unicells')" (Torday 2014).

Subsequently, John Torday, further developed this concept in a book written in collaboration with Virender Rehan: "The multicellular form may merely be a derivative of the unicellular state, acting as a matrix for it to monitor the oncoming environment so that the gene pool knows what epigenetic marks acquired during the multicellular phase of the life cycle to include or exclude in the next generation. For example, Dictyostelium exists in two forms, a free-swimming amoeboid form and a colonial fruiting body. Under conditions of abundant nutrients, the slime mold remains in its free-swimming amoeboid form, under conditions of low food abundance, the amoeboid free-swimming phenotype forms colonies. Logic would dictate that this organism evolved under high nutrient abundance conditions, and therefore its *unicellular form is the primary phenotype,* the colonial form being derivative" (Torday and Rehan 2017).

In the two authors' opinion, the cell is considered the basic structural and functional unit of all life forms. Yet, while conceptually compelling, the unicell model obscures the distinct identity and autonomy of multicellular life. Organisms have a life on their own, *distinct* from cellular life. If we make the distinction between living systems and life, that Michael Lachmann and Sara Walker did: "*a dead cat is not alive, but it is life,*" viruses may be also considered life forms on their own (Lachmann and Walker 2019). In my opinion, the fact that for some time after a cat's death its cells are still alive does not make a dead cat a form of life. Similarly, a virus cannot be considered to be alive, or a life form, in the absence of a cellular viral fabric.

10.4 Schrödinger's Aperiodic Solid and Time, the Mysterious Vital Force

The idiosyncrasies of the living systems have been noted for a long time, and at the beginning of the nineteenth century, Jöns Jakob Berzelius stated that organic molecules can be formed only by living cells, and the synthesis of organic molecules requires a "vital force" present only in living organisms. The theory was discarded subsequently by Friedrich Wohler who demonstrated that organic compounds (i.e., urea) can also be created in the laboratory. In his book, *The Vital Question*, Nick Lane brought arguments to the hypothesis that life originated in deep sea hypothermal vents and proposed chemiosmosis as a universal cellular life mechanism. In his opinion, chemiosmosis, present in both deep hypothermal vents, and all life forms, and, in general, the shuttling of electrons and protons, or, more specifically the process of mitochondrial respiration, represents a true universal vital force. For Nick Lane, "the flame" that keeps all organisms "growing, dividing, conquering" is the flux of bioenergy (Lane 2015). The focus on bioenergy is not new and, historically, the first half of the twentieth century until the discovery of DNA and genes has been dominated by biochemical and biophysical life theories (Morange 2021).

Several authors have proposed that evolution is a sort of information computing mechanism so, in this model, biological computation is also seen like a sort of "vital force" (Bäck et al. 2002; Crutchfield and van Nimwegen 2002; Freeland 2002; Landweber and Kàri 2002; Valiant 2013; Mayfield 2013; Cohen and Marron 2023).

More recently, models of evolution as cognition have been put forward, and cognition has been proposed as a universal mechanism unifying all living domains, from the proto cognition of bacteria and other unicellular organisms to multicellular eukariota, like human beings (Miller and Torday 2018; Miller 2023; Torday 2025; Corning 2023; Baluška et al. 2023; Kull 2023; Uller 2023). Four orders of biological cognition were introduced by František Baluška and his collaborators: first order of cellular cognition (FiOCC) present in bacteria and archea, "likely based on their limiting excitable plasma membranes that support their cognitive cellular clocks based on reactive oxygen species (ROS) and excited free electrons[...], a second order of cellular cognition (SeOCC) which is supported by an initial aggregation of host and symbiotic cellular (acting as organelles) that are integrated via intracellular organismal synapses, [...] a third order of cellular cognition (ThOCC) is specific to eukaryotic multicellular organisms, as an aggregation and integration via extracellular cell-cell synapses, including immunological, neuronal, plant, and symbiotic synapses [...] and a possible fourth order of cellular cognition (FoOCC), as a supracellular consciousness, permitting the effective integration of interlinked ThOCCs to better ensure their survival through coordinated responses to environmental stresses" (Baluška et al. 2023). As it directs behavior, cognition is another candidate of the *vital force*.

Temporal and spatial structures of an organism are codependent. Schrödinger imagined the source of order in the organism to be an "aperiodic solid," a sort of blueprint that would exercise control over the organism's activities (Kauffman 2020a, b). When the structure and function of DNA were discovered, it was thought that DNA is life's "code script"; however, maybe, the "aperiodic solid" envisioned by Schrödinger is not DNA; it is not an aperiodic molecule, but rather similar to a molecular dance or a song that follows the music of biological time. As George Ellett Coghill put it, "organisms are organized patterns in time" (cited by Russell 1945). Genes are static patterns on a molecule; they are like notes of a musical score. Arguably, the greatest life's invention was not DNA or some other biological code, it was biological time, a temporal complex resembling a musical score that follows different rules than those of the oscillatory dynamics that may be sometimes present also in abiotic systems (e.g., a pulsar). Reversely, the possibility of life itself lies in the deep structures of time and, once the right environment is present, life emerges. Clearly, regardless of the direction of this relationship, life and time are inseparable. Within the cell, ultradian rhythmicity plays a central coordinating role, integrating feedback loops that connect the cell's energetic and redox states with the dynamic development of its ultrastructure. This model serves as a foundational framework and reflects a universal principle of cellular organization, wherein ultradian rhythms—oscillations with periods shorter than 24 hours—act as synchronizing signals that maintain the coherence and integrity of the living state (Jenkins et al. 2025).

We do not need vital forces to explain life. In living organisms, bioenergetical, bioinformational, and biostructural processes are unified into a coherent process, and biological time, at the same time a key facilitator and the outcome of this process, acts like a mysterious vital force.

10.5 Life in Fifteen Steps and Many Copies

In 2017, assembly theory (AT) was introduced (Marshall et al. 2017; Marshall et al. 2021; Sharma et al. 2023; Ellis 2023; Walker 2024) originally as a method to identify life signs on other planets. An assembly is described by two values: the number of copies of the observed objects and the objects' assembly indices (an assembly index is the number of steps on a minimal path producing the object). The main claims of the authors are that complex *objects with a high assembly index are being produced by selection* and objects with higher assembly indexes have longer histories, representing a sort of time capsule. In support of their model, living and nonliving samples were analyzed by high-definition mass spectrometry and infrared and nuclear magnetic resonance (NMR) spectroscopy. The conclusion of their analyses was that no molecules created by random processes 15 or more steps and only living systems had complex molecules with an assembly index above an experimentally determined value of 15 steps. In AT, time is considered fundamental as the number of steps on a minimal path to create the object is what defines the complexity of an object measured by the assembly index. In order for a complex living molecule to be produced, a "memory," that is, a recording of actions or processes that are needed to build the molecule is needed (Walker and Cronin 2023). Living molecules are considered to be a sort of life recorders and the complexity of an object depends on its history. In the authors' opinion, the explanation for the striking difference between abiotic and biotic objects is due to two fundamentally different types of physics: classical physics and "the physics of evolution and selection" that "must take over from other forms of physics to explain how a molecule is formed" (Walker and Cronin 2023).

Inferring from a specific object's complexity and quantity, its organicity, and making assumptions about the presence of selection, is not, however, straightforward. "The difference between an inorganic and an organic system lies not in the greater complexity of the latter, but in the *orderly directiveness* of its activities towards the ends of living, developing and reproducing" (Russell 1945). This "orderly directiveness" is only partially captured by the AT's assembly index.

Another interesting proposal for an alien life detection strategy, based on processes and not objects, relates to *ultradian rhythms* (Yates and Yates 2008), and, in general, biological rhythms, oscillations, and synchronized temporal platforms that can be considered *dynamic signatures of life* (Lloyd 2021).

10.6 An Integrative Perspective

Recently, a processual-ontological view of evolution was called "it is the song not the singer has been proposed" (ITSNTS) by Ford Doolitle and Andrew Inkpen (Doolittle and Inkpen 2018). The authors propose that the *processes* implemented by genes, cells, species, or communities are "units of selection" in their own right. ITSNTS theory implies that the main beneficiary of evolution is not the individual or the collective that implements a process but the process itself. The two authors of ITSNTS see "taxa and the communities they form" as "adaptations of the processes they implement, not the other way around" (Doolittle and Inkpen 2018).

Songs and singers cannot be separated (Doolittle and Inkpen 2018). In 2021, a theory integrating both song and singer, ITSATS ("It's the song, *and* the singer"), both processes and "things" like genes, cells, and organisms, has been put forth (Bapteste and Papale 2021).

10.7 The Music of Biological Time: A Primer

Fundamentally, intertwined with motion and transformation, time ceases to exist in the absence of either. In Buddhist philosophy, time is devoid of intrinsic, independent properties and is thus approached through the concept of *śūnyatā*—emptiness or codependent origination. Movement may be stochastic, and in nonliving macroscopic systems, it is typically governed by entropic forces. Here, time adheres to the second law of thermodynamics, unfolding irreversibly in a single direction dictated by the anisotropy of energy. *Every energetic transaction, every motion, entails simultaneously an exchange of information.* Actions in nature transmit not just energy but also information, generating causal linkages that bind past, present, and future. *Causality is precisely the link created through energy and information transfer.* The past informs and energizes the present, and the present informs and energizes what is yet to come.

Living systems harvest external energy—the sun for the plants and organic nutrients for many organisms—and transform it into organized matter. In living systems, movement is under tight control, and given the open nonlinearity of all living systems, life is not constrained to the inexorable dissolution imposed by entropy and is able to exploit creatively, to its own benefit, the irreversibility associated with it. In living systems, time is cyclical, has a rhythm, and advances in spiral-like loops called biorhythms, repeating themselves after minutes, hours, days, or even years. In biology, time, like the mythical Phoenix bird, turns upon itself again and again, recreating itself anew in an autotelic gesture that blends simultaneity and succession in a polyphony of interwoven patterns, continuously adapted to the environment with which they coevolve. The astonishing continuity of life on Earth is closely related to the musical qualities of biological time, evolution diversifying, and creating nested polyphonies of biorhythms adapted to their environments.

In the inanimate world, physical time arises spontaneously and remains largely unstructured. By contrast, living organisms are territories of self-generated temporal structures. The origin of time in the living world is closely related to energy provided, in general, by the environment. Although Brownian movement is also present in all cells, the energy provided by it is minimal.

Energy exchange processes in living systems are under strict regulation, with time serving as a central organizing principle. What underlies this temporal control? Just as blood pressure or glucose levels are stabilized by homeostatic mechanisms, developmental and physiological processes are guided by *homeorhesis*—a form of dynamic equilibrium that preserves developmental trajectories while allowing adaptive coevolution with the environment. Through *homeorhesis*, biological development proceeds in a temporally ordered, algorithmic sequence, with events unfolding in precise succession, like beads strung along a thread. Complementing this directional flow, living systems also embody *cyclicity*, a second fundamental temporal dimension. *Cyclicity* imparts rhythm to life, structuring it into recurring loops or biorhythms, where processes repeat at quasi-regular intervals and together create the temporal architecture of biology. *Cyclicity* is not unique to living systems. Numerous physical and abiotic processes generate highly regular oscillations that function as natural timekeepers—for example, the vibrations of Cesium atoms, the periodic orbit of the Earth around the Sun, or other astronomical cycles. At the quantum level, all particles display wave-particle duality, and waves by their very nature embody periodic oscillations.

Another special temporal aspect in biological systems is the embodied *present*. Unlike classical physics, where the present has no defined role and practically does not exist, in biology, the present is real and of utter importance. A stone or a cyclone don't sense their inner or outer environment. They are completely unaware of themselves and their surroundings. Living organisms sense their environments and are acutely aware of it. They react to stimuli through context-sensitive mechanisms informed by past experiences. Memory is fundamental to survival, and *the present* is deeply tied to *simultaneity*—the concurrent unfolding of multiple processes. In a cell, for example, many processes occur at the same time and an action is never simple, related to a cluster of other actions that concur in producing a given outcome. *The present* in biological systems functions as an organizing principle, like the wave collapse in the quantum world, structuring simultaneous events. Living systems sense their environment and respond to stimuli through context-appropriate reactions based on recorded prior exposures and memory/memories vital for their survival. *The biological present is probably the most important mechanism for integrity preservation of a living organism.* Organisms use the present's platform in order to integrate internal and external information in real time (*real time <u>is</u> the present*) and respond to challenges and stressors from the environment. The experience of the *present moment*, sometimes termed the *specious present*, is central to both perception and biological action. Recent hypotheses suggest that cell membranes may function as active components in cellular biocognitive processing (Baluška et al. 2021), contributing to the organism's capacity to perceive and act in real time. The contextual behavior of organisms, a form of *proto-presence*, or of *proto-awareness*, is related to the existence of the biological present. In living systems, information reverberates through the whole system, and, sometimes, responses are slightly delayed, allowing more elaborate reactions to develop. Different levels of awareness present from unicellular to multicellular organisms are evolutionary strategy to survive in an ever-changing environment.

Importantly, the present may not be uniquely related to the biological world, and over the last decade, Lee Smolin developed a model where the present may also play a role in the abiotic world of quantum physics (Smolin 2023).

To resume, biological time manifests in three interrelated forms (please see Fig. 10.1):

1. *Linear Biological Time*—Analogous to *melody* in music, linear time in biology often follows algorithms shaped by genetic or epigenetic blueprints and experiential memory. A cell embodies a dual legacy: It is both a living archive of evolutionary processes and a predictive agent acting on anticipated future states. The tempo of biological processes is modulated by random events, which introduces the novelty essential for persistence. Crucially, *evolvability*—the capacity for adaptation—should not be conflated with evolution itself. The variation of the pace is not in contradiction with the observation of the duration conservation of some living processes. In cancer, for example, the duration of mitosis is relatively well conserved across different cells and cancer types, as opposed to the duration of the migration process of cancer cells that is variable—cancer cells may be present in the circulation for a variable period of time. Both internal (i.e., energy levels) and external factors (signals from the environment) influence the duration of a given living process. Evolvability may sometime lead to an acceleration of the oncogenic process, since it depends both on the presence of certain genetic alterations (internal factors) and the stromal rigidity (external factor).
2. *Cyclical Biological Time*—Similar to *rhythm* in music, cyclical time defines the repetitive processes essential to life. In a cell, cyclicity is ubiquitous: a) It is present in metabolic reactions of the Krebs cycle which runs sequentially, in eight consecutive steps, starting with the condensation of acetyl CoA to form citrate, and ending with the malate which is dehydrogenated to form oxaloacetate, which combines with another molecule of acetyl CoA and starts the new cycle again; b) it is present in the DNA being transcribed into mRNA and the translated into proteins; noncoding RNAs (ncRNAs) and proteins acting back on the DNA in feedback loops, c) it is present in the system of membranes that connects through a labyrinth of tubules the outside to the inside of the cell. Cells are always busy constructing and deconstructing their selves growing, renewing themselves, copying, and dividing, like a river cascading back on itself in an unending flow, following patterns of cyclical regeneration, maintaining coherence and continuity. In yeast, oscillatory patterns

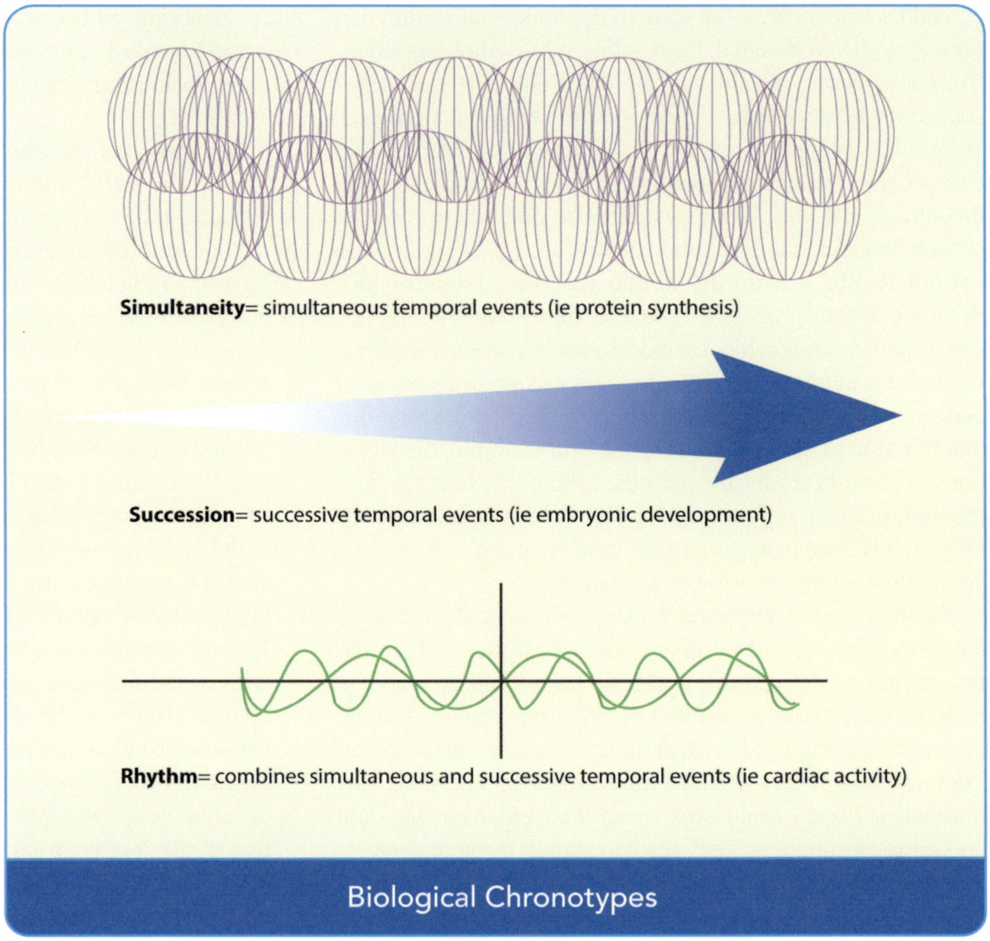

Fig. 10.1 Biological time: simultaneity-succession-rhythm. (Courtesy of Ana Tourian)

have been identified at multiple distinct timescales—specifically at 4 minutes, 40 minutes, and 12 hours—revealing a self-similar temporal architecture (Jenkins et al. 2025). These nested rhythms exhibit fractal correlations across time domains, suggesting that dynamic processes at shorter intervals are structurally and functionally related to those occurring over longer periods. This hierarchical yet nonlinear interdependence of temporal scales supports the notion of what Yates called "heterarchical organization," wherein regulatory interactions are distributed across multiple levels without a fixed top-down or bottom-up control, reflecting the complex, multilayered temporal orchestration characteristic of living systems (Yates 1992).

If temporal flow did not have *directionality*—the so-called arrow of time—there would be no agency, and, as a corollary, no life. If *cyclicity* would be absent, life, as we know, would not persist, either.

By dissipating energy, life maintains organization not only in space but also in time. As Alfred Goldbeter pointed out, rhythms are themselves dissipative structures, corresponding to self-organizing temporal processes. Rhythms are made possible by both time's directionality and circularity (Fig. 10.1). The mechanisms underlying self-organization are a combination of interrelated feedback and feed-forward loops. Rhythms appear precisely because of these underlying homeostatic mechanisms. Several examples of rhythms are also presented by Alfred Goldbeter in his book (Goldbeter 2010)—ionic channel control leads to cardiac and neuronal rhythms, enzymatic control leads to glucose oscillations, transport control leads to intracellular calcium oscillations, and, finally, gene expression control underlies circadian rhythms. Life cycles are also rhythmical, and in living systems, time turns upon itself again and again, recreating itself anew in an autotelic gesture, that blends simultaneity and succession in a polyphony of interwoven patterns.

3. *Biological Simultaneity*—Similar to *harmony* in music, *an organism is an organized ensemble of coherent events occurring at different temporal scales united by biorhythms.* Simultaneity in living systems is a distinctive quality that plays important roles. Simultaneity refers both to events occurring at a certain biological level—the biochemical reactions in a cell, for example—and in-between levels. The issue of simultaneity occurring

between levels is fundamental and is responsible for the causal circularity present in living systems. This cyclic causality resembling the Ouroboros serpent of Greek mythology is nonsymmetrical. As explained by Noble et al., "downward causation consists in constraints exerted by higher levels on the initial and boundary conditions within which the dynamics of lower-level elements operate. By contrast, upward causation is the way in which those dynamics influence higher level states" (Noble et al. 2019).

In an organism, synchronization is of utter importance. In order to function, nervous, cardiac, and intestinal systems all have pacemaker cells pulsating and creating a functional whole. Biological simultaneity is also at the origin of the present, a sort of *superstructured* simultaneity.

Actions in biological systems are both *simultaneous* and *coherent*; with other words, there is an intrinsic order in their simultaneous occurrence. This order is considered to be mainly related to biorhythms. In humans, the circadian rhythm orchestrates systemic coherence over 24-hour cycles, regulating both cellular and physiological activities. In addition, *ultradian* rhythms—oscillating at 12-, 8-, and 4-hour intervals—have been identified. These harmonics, particularly those occurring every 4 or 8 hours, appear vital for protein and amino acid turnover, facilitating rapid cellular adaptations to environmental stressors (Zhu et al. 2018; Asher and Zhu 2023).

Biological time spans a vast spectrum of scales (Lloyd et al. 2008). For example, the isomerization of the 11-cis retinal chromophore in rhodopsin occurs within 200 femtoseconds, initiating the phototransduction cascade leading to vision (Gruhl et al. 2023). Quantum effects likely underlie the coherence of such ultrarapid events.

Remarkably, across species from bacteria to humans, 95% of the reaction time lies within only one order of magnitude (100-fold range) of the overall geometric mean of about 25 ms, compared to the much larger variation in body mass differing by up to 20 orders of magnitude (10 power 20-fold range). This suggests that a fundamental biochemical constraint governs response speed, unrelated to the structure or the scale of the organism (Rospars and Meyer-Vernet 2021).

In humans, the conscious formation of a sensory experience takes approximately 100–200 milliseconds (Tononi 2004), a duration echoed in single-cell organism's response lag times. At the molecular scale, ribosomes polymerize proteins at a rate of ~5 amino acids per second, yielding a typical 400-residue protein in about 80 seconds (Baboo et al. 2014). Glycolysis, the Krebs cycle, followed by oxidative phosphorylation, unfolds over several seconds to a couple of minutes. These overlapping timescales, from femtoseconds to minutes, suggest that biological processes are scaffolded across vast temporal layers. The millisecond range, in particular, may serve as a unifying temporal platform linking cellular and cognitive functions—a shared temporal "thickness" of the present.

Different cancer treatments target distinct temporal vulnerabilities. Chemotherapy, radiation, targeted therapies, and electromagnetic frequencies all act on specific timescales. Interventions operating within the millisecond window may hold special therapeutic promise, as seen in the use of low-frequency alternating current to disrupt cancer cell behavior. Understanding biological time's layered architecture offers new insights into both disease mechanisms and therapeutic strategies.

10.8 Biological Chronotypes

Biological time has the following three fundamental properties:

- *Structure* (**S**), unfolding in (quasi) precise, partially predefined steps that may be reversible. Biological time has both non-repetitive, algorithmic and cyclical, repetitive properties. It is both simultaneous and successive, stochastic and ordered.
- *Presence* (**P**) is probably the most significant invention of the living world. In order to survive, organisms need to respond to stimuli in real time. These responses are at the same time determined by past experiences encoded in different cellular programs, and stochastic due to unpredictable nature of new challenges. Therefore, biological time is a continuous source of novelty.
- *Historicity* (**H**)—Some processes at work in living organisms are the result of billions of years of evolution. Some of the survival strategies acquired over these eons of time have been conserved. A cell is on one hand a living *memory* of all the processes that lead to its existence, and also, a continuous instantiation of these processes that keep it alive.

In a cell, directed by information, matter becomes energy, and energy becomes structured matter in never-ending repeated cycles. An organism is an embodiment of both temporal creation and creativity, and the SPH structure of biological time functions as a glue unifying both intracellular processes between themselves and, also cells and organs into coordinated activity, assuring coherence between the nested levels of cells/tissues/organs integrating them into the functioning whole of the organism. Importantly, biological time helps also integrate organisms in their environments and is a key component of their continuous evolution and adaptation. When we look through a microscope at a cell, we see structured space, but this is just a snapshot of a movie running in multiple dimensions simultaneously, mirroring a complex temporal organization.

This temporal organization refers to:

(a) *The order of steps* in which processes unfold in a cell—like a melody.
(b) *The duration of processes*—that may belong to different timescales.
(c) *Coordination of simultaneous events*—like the harmonies in a song.
(d) *The flow of time* that can be moving faster or slower—The tempo of living processes is under tight control.
(e) *Jazzy like improvisations* in the music score related to stochastic processes.
(f) *The history of the species* imposing, for example, a certain duration of life cycles.
(g) *The "thickness" of the present* that may be very different from an organism to another.

In biology, as opposed to physics, like Bach's fugues, time has a complex and recursive structure. Underlying temporal creativity, there are fundamental rhythms responsible for the adaptivity and the resilience of organisms.

Pushed by the spiraling wind of biological time, biological forms, like clouds, are ephemeral. They are born, last for a while, and then dissolve into thin air as if they have never existed. On the contrary, life's perpetuum mobile-like process persists, advancing through eons like an immortal beating heart.

10.9 Biological Time and Cancer

Biological time is present across life forms including cancer, and remodeling of biological time is one of cancer's key characteristics. Besides being a disease of spatial disarray, cancer is also a dyschronia, a disease of time. In cancer, alterations of spatial organization and remodeling of temporal structures always come together. Although cancer cells are more or less disconnected from the temporal structures of the multicellular organism that created them, they continue to need temporal structures and sometimes, as it has been shown in brain cancer by the team of Frank Winkler, behave like pacemakers and follow the tempo of their own creation (Hausmann et al. 2023).

Synchronization between various processes is present *within* the organisms, at all levels, molecular, microscopic, and macroscopic, and, importantly, it is also present *between* these scales. At the organism level, the circadian clock regulates numerous physiological activities through tightly coordinated modulation of gene expression and biochemical functions all glued together through daily rhythms. The circadian system is hierarchically organized into two major components: a central pacemaker located in the hypothalamic suprachiasmatic nucleus (SCN) and peripheral oscillators within cells. When the retina receives photons from the environment, it signals the SCN to synchronize peripheral clocks via neural and humoral pathways (Lin et al. 2020). The core molecular circadian clock is built on transcription-translation feedback loop machinery that involves the BMAL1 and CLOCK transcription factors that form heterodimers that bind and activate rhythmic expression of clock-controlled genes, including the transcriptional repressors PER1–PER3, CRY1, and CRY2 (Wang et al. 2024d).

The cell cycle and circadian clock are intricately connected. The circadian clock regulates or "gates" key phases of the cell cycle (Masri et al. 2013; Gaucher et al. 2018). Uncontrolled growth of cancer cells may sometimes relate to desynchronization from the body's rhythms. Abnormalities of the circadian clock have been demonstrated in several cancers, particularly breast cancer (Hammarlund et al. 2023).

Four mechanisms linking circadian clock and cancer have been described (Sulli et al. 2019):

(a) Circadian clock components regulate the expression of numerous genes involved in cellular processes like DNA repair, protein folding redox regulation, autophagy, nutrient metabolism, and cellular secretion. These processes are disrupted by alterations of circadian rhythms, creating a cancer-prone microenvironment through chronic inflammation and metabolic reprogramming, among others.
(b) Circadian clock proteins may physically interact with oncogene protein products. Modulating circadian clock function or expression of clock proteins can therefore promote or inhibit cancer.
(c) Circadian clock proteins and interacting partners sense the cellular environment. Oncogenic changes can alter clock protein stability, localization, or function. For example, the cell's redox state can affect CLOCK/BMAL1 binding to DNA. Also, posttranslational modifications (acetylation, phosphorylation) of clock proteins are modulated by the cell's nutrient status or signaling events.
(d) The circadian clock regulates secreted factors like cytokines, hormones, and neurotransmitters that can affect clock function in different tissues through paracrine/endocrine signaling, serving as biomarkers. Some tumors may overproduce these factors, disrupting distant tissue clocks.

In addition to the above mechanisms, oncogenic programs may directly disrupt circadian regulation, which in turn further fuels tumor growth. This bidirectional relationship between the circadian clock and cancer highlights the importance of maintaining proper circadian rhythms for cellular homeostasis and cancer prevention.

Recently, it has been also shown that circadian disruption is associated with enhanced metastasis via dysregulation of key biological processes and modulation of cancer stem cells (CSCs) and their specialized microenvironment (reviewed in Wang et al. 2024e). Circadian disruption promotes development of the epithelial to mesenchymal transformation and increase in the cancer stem cells populations within tumor masses. Also, a recent study documented that extracellular vesicles content involved in cell-cell crosstalk are controlled by the circadian clock, with their cargo reaching a peak at the onset of night (Yeung et al. 2022). Another study has documented the presence of breast cancer circulating tumor cells (CTSs) at night during sleep (Diamantopoulou et al. 2022), but it is not clear whether sleep per se or circadian rhythm influences the dynamics of CTCs and whether this phenomenon is unique to breast cancer (Dauvilliers et al. 2022).

It is yet unclear whether direct targeting of clock proteins may affect the cancer phenotypes (Lin et al. 2020). On the other hand, the emerging role of melatonin as a circadian clock gatekeeper has informed several ongoing clinical studies treating a range of human cancers through modulation of the TME, EMT, immunosuppression, and angiogenesis. Also, there is renewed interest in investigating the benefit of delivering systemic cancer therapy at a given time of day or during the night (Wang et al. 2024d).

10.10 Heterochronies

In evolutionary developmental biology, heterochronies refer to differences in genetically controlled algorithmic temporal programs, including a) shifts in initiation, duration, or rate of a process compared to ancestors/other species; b) initiation/duration changes of a process related to regulatory sequence variation or expression dynamics; or c) tempo changes (allochrony) that involve the same genetic program operating at different speeds, not explained by regulatory sequence variations (Rayon 2023).

There are several mechanisms involved in normal developmental timing including epigenetic mechanisms (histone 3D structure, epigenetic regulators like EZH2, EHMT1/2, or DOT1L), changes in protein stability, miRNA regulation, and mitochondrial rate and mitochondrial activity (Rayon 2023).

More than three decades ago, a genetic regulatory pathway of temporal events—called the heterochronic pathway—has been identified in Caenorhabditis (C) elegans. This pathway controls the order of both tempo and events of the four larval stages present in C. elegans after embryogenesis. The genes controlling developmental temporal events are called heterochronic genes and encode both microRNAs and factors controlling RNA expression. Interactions between heterochronic pathway members act as switches governing transitions between larval stages.

MicroRNAs (miRNAs) are a family of small noncoding RNAs that regulate a wide array of biological processes. Lin-4, the first miRNA identified in 1993 by Lee et al. (1993), is encoded by a heterochronic gene (Peng and Croce 2016). Strikingly, the first name used for miRNAs was ***small temporal RNAs*** (stRNAs), precisely because of the miRNAs direct influence on temporal developmental processes.

It comes then as no surprise to discover that miRNAs, in general, and stRNAs, in particular, are involved in all steps of oncogenesis from initiation to metastasis. For example, as determined by analyzing 46 studies on 20 different types of human cancers, a decreased expression of the tumor suppressor let-7 microRNAs and an increased expression of oncogenic miR-21 were very frequently associated with poor prognosis (Balzeau et al. 2017). Both the genes coding for the let-7 miRNA and the genes coding for the LIN28 A and LIN28B, RNA-binding proteins, that directly inhibit the biogenesis of mammalian let-7 miRNA (Piskounova et al. 2008), are heterochronic genes.

The immediate future of cancer therapeutics may be represented by ncRNA, which are gatekeepers, enablers, and enhancers of biological time.

10.11 The Importance of Time in Oncology

Cancerogenesis is different from embryonic development and from the human life cycle. Also, the time of a cancer cell is different from the time of a normal cell, in fundamental ways. Under the microscope, cancer cells look disorganized; however, the main difference in cancer cells is in their biological time: Cancer cells may divide continuously ignoring inhibitory signals from the environment and they are "immortal" in theory due to the presence of telomerase. A better understanding of the cancerogenesis and the temporal makeup of cancer cells will hopefully lead to more efficient novel treatments.

Beyond chronobiology-informed cancer treatments, there are other important temporal aspects pertinent for oncology. The cancer process has different timelines and follows different trajectories related both to its molecular makeup and its environment in a particular individual.

The first fundamental point is that most often, cancer takes years or decades to develop from a cell with mutations to a full-blown cancerous tumor. Intercepting cancer before it becomes clinically manifest is arguably the most important oncology project in the near future.

Since inception, there are multiple factors that can influence its trajectory and the progression is nonlinear from *precancerous* lesions to *clinical detectable* cancer to *clinical symptomatic* cancer and, sometimes further, to *metastatic* cancer and *cachexia* and *terminal cachexia*. In as many as 10% of cases, metastasis can develop before the primary tumor that sometimes will never be detected. Some cancers remain local and never metastasize either to lymph nodes or distal organs, while other cancers do metastasize, but the metastatic lesions never progress. A significant number of cancerous tumors that never manifested clinically are incidentally discovered at autopsy.

When a patient is diagnosed with cancer, it is important to use all the available data in order to establish where, on the nonlinear trajectory described above, the cancer process is situated in relation to the patient's lifeline. Treatments and prognostics will vary according to cancer's stage, patient's age, and comorbidities. Overall, 55%–60% of cancers are currently cured using the present therapeutic armamentarium. The majority of these did not spread either locally to the lymph nodes or distally to other organs. There are cancers that have metastasized to the lymph nodes (regional disease) or to only a few distal places from the primary tumor (oligometastatic disease) that can still be cured, and there are cancers that are local but incurable because of the proximity of vital organs that are incurable. A minority of metastatic cancers can be cured at present, and the cure rate significantly varies according to the cancer type.

Despite their nonlinearity, cancer's trajectories are relatively deterministic in the sense that particular points on its progression trajectory are related to specific survival probabilities. In the metastatic stage, some treatments are able to prolong lives by several months but very few treatments have been associated with long-term survival. This is why systemic treatments administered in the metastatic stages are considered palliative as opposed to treatments administered in early stages considered potentially curative.

Life is fundamentally a regulated flow, and organisms are dynamic streams – ephemeral figurations unfolding in time (Zajicek 1995). Cancer is in many respects a temporal disease. Although tumorigenesis can be simplistically interpreted as a regression along the evolutionary trajectory (Nagahata and Kawamoto 2023), the intrinsic plasticity of cancer cells enables them to dynamically reposition along the differentiation axis and, depending on the context, exhibit features characteristic of either unicellular or multicellular organisms. In metastatic stages, the majority of current treatments are a sort of temporal decelerators; they delay but are unable to change an inexorable outcome.

In advanced cancer, the art of oncology is literally often about "buying some more time." The most frequent question patients with cancer ask their doctors is "How long I am going to live?" When patients are terminally ill and present certain pathognomonic clinical findings, experienced doctors can predict with high certainty their imminent demise and their families may benefit from having this information. However, in all other cases, the honest answer to this question is always, "I don't know." Survival data of large intervention studies cannot be completely extrapolated to a particular patient, and therefore, an additional explanation can be provided to patients and their families: "And, it *cannot* be known. We have only statistical data that may or may not apply to you. Nothing in medicine is 100%."

The Temporal Landmarks of Cancer

11

11.1 What Is Needed for a Cancer Cell to Form?

Cancer cells contain many genetic errors and many abnormal proteins. In order for a cancer cell to come into existence, the correction proof mechanism (i.e., TP53, for example) must be missing. Of note though, TP53 mutations may also be present in non-cancerous conditions, and only a fraction of patients with the Li-Fraumeni syndrome develop cancer. In order for a cancer cell to divide uncontrollably, mutations or epigenetic alterations of either a division pathway or an apoptotic pathway, or both, are needed. In order to avoid chromosome shortage leading to recombinations that may be potentially lethal, cancers need telomerase activation. They also need their metabolism to become adapted to the growth needs. They need to be able to change their phenotype and adapt to different environments, i.e., to exhibit phenotypic plasticity. The need for specific genetic alterations in a normal cell to become cancerous has been demonstrated experimentally by the team of Robert Weinberg, and was presented in the classical hallmark series of articles written in collaboration with Douglas Hanahan.

In order for cancer to develop, genetic mutations do not suffice. A key step in cancerization is the cooptation of both the local tissue and also of the whole organism, instigated in order to support cancer development. The instigation process occurs through activation of various developmental programs. Very few genetic alterations are involved in this process. Driver mutations are mainly involved during the initiation phases of carcinogenesis when they confer a growth advantage to the transformed cell. Subsequent steps are choreographed through epigenetic and extracellular cues.

The main differences between a malignant and a benign tumor are their invasive and metastatic features, not its growing abilities. By definition, invasive cancer disrupts the tissular basal membrane. For the malignant phenotype to develop, interaction with the environment plays a fundamental role. For epithelial cancers, epithelial to mesenchymal transition (EMT) is the key step that differentiates a benign from a malignant tumor, and it may be induced by both intracellular factors (i.e., oxidative stress) and extracellular factors (i.e., hypoxia and inflammation) (López-Novoa and Nieto 2009). Many signals received from the tumor microenvironment can initiate EMT including TGFβ, HIF-1α, EGF, WNTs, and Notch (Ribatti et al. 2020).

The similarity between cancer and development has been noticed for more than a century. Similar developmental signals (WNT, HH, and Notch) are carried by exosomes during development and during cancer local invasion and metastases (McGough and Vincent 2016). Melanoma-derived exosomes transfer the receptor tyrosine kinase MET to bone marrow progenitors, programming them to support tumor angiogenesis, invasion, and metastasis (Peinado et al. 2012). Recently, pancreatic ductal adenocarcinomas have been found to produce exosomes that induce liver cells to secrete transforming growth factor β (TGFβ), which in turn contributes to pre-metastatic niche formation (Costa-Silva et al. 2015). The ability of tumors to prime other tissues for metastasis is an emerging theme in cancer biology, and exosomes appear to be important mediators of this phenomenon by transferring signaling molecules or modulators of signaling (Rak 2015). For example, exosomes derived from mesenchymal stem cells (MSCs) were shown to promote angiogenesis by activating the extracellular signal-regulated kinase 1/2 (ERK1/2) and p38 MAPK pathways (Zhu et al. 2012). Exosomes derived from colon cancer cells enhance growth through the transfer of oncogenic KRAS to recipient cells (Demory Beckler et al. 2013).

Immune cell dysfunction is also an important contributor to cancer development (Philip and Schietinger 2022). It has been demonstrated that starting in the early stages of tumorigenesis, tumor-specific CD-8 cells become non-responsive (Schietinger et al. 2016), and the signal-independent dysfunctional state is epigenetically imprinted (Philip et al. 2017). It appears that an increase in Tox expression through NFAT signaling is responsible for the widespread epigenetic and gene-expression changes associated with the exhausted differentiation state (Scott et al. 2019). A potential therapeutic intervention would be to reduce Tox expression in combina-

D. Paul, *Cancer: An Integrative Approach*, https://doi.org/10.1007/978-3-032-08808-6_11

tion with simultaneous treatment with drugs that alter CD8+ T cell epigenetic states and anti-PD-1 checkpoint blockade (Mann and Kaech 2019). It has been recently suggested that the spatial positioning within the tumors of CD4-CD8-APC "immune triads" is the critical determinant of effective antitumor immunity (Espinosa-Carrasco et al. 2024).

11.2 Cancer as a Process

In 1991, Michael Sporn noted in relation to the cancer process: "the actual disease is an evolving molecular and cellular process, not a static circumstance, the onset of which can be dated to the time when a pathologist can finally see cells invading through a basement membrane on a slide. Classical terminology, which originated more than a century ago, ignores the importance of the critical dimension of time in the development of epithelial carcinogenesis; by now, we know that there is a latency period, often 20 years or more, between the initiation of carcinogenesis and the onset of the terminal, invasive, and metastatic phase of the disease. Classical pathology and clinical oncology have dealt with cancer in the three dimensions of space (the size and location of lesions); to understand the process of carcinogenesis, which is the reality of the disease, we must now add the dimension of time" (Sporn 1991).

So, cancer is a process, not a state. In order to understand it, we should move from a static picture of tissues and cells seen under a microscope to the process of cancer—a movie of a system evolving in time. It is important to expand our picture of cancer by looking at it not as a bunch of abnormal genes or proteins, but as a choreography of cells progressing through time. The metastatic process is also a meta-dynamic process, not only a change (meta) in the place (stasis) of cancer cells, but also a change in their dynamic. The metabolism of a cancer cell has to adapt to the new location, and the metastatic cell will adapt a new metabolic phenotype.

More than 60 years ago, Leslie Foulds noted that cancer "a dynamic process advancing through stages that are *qualitatively different,*" leading the way to the contemporary understanding of the oncogenic process (Foulds 1958; McMahon et al. 2017).

There are several key temporal milestones spanning sometimes decades that mark the progression from a normal cell to a cancer cell. In some cancers, for example, pancreatic cancer, the development of the primary tumor typically takes around 12 years, but progression from cancer cells present in the primary tumor to metastatic seeding is relatively linear, taking an average of 6.8 years (Yachida et al. 2010). In other cancer types, like colorectal, lung, and breast, the primary tumor growth and metastases appear to occur in parallel since it has been demonstrated that systemic spread typically begins 2–4 years before the diagnosis of the primary tumor (Hu et al. 2020). For esophageal cancer, on the other hand, metastases may appear very quickly, and a distinct "diaspora" model with multiple subclones, each seeding multiple metastatic sites, has been proposed (Noorani et al. 2020). As described by Turajlic et al., both early/rapid and late/slow metastatic spread occur in renal cell carcinoma, and different evolutionary processes may be responsible for the difference in behavior. The authors analyzed in detail the patterns of mutations of the renal cancers and concluded that tumors exhibiting rapid progression may have represented cases of punctuated evolution driven by multiple clonal drivers as opposed to tumors with an "attenuated progression" that may have represented cases of branched or linear evolution (Turajlic et al. 2018).

11.3 Computing Cancer Cells' Trajectories

Several software programs have been developed able to infer the future of a cell using RNA data (La Manno et al. 2018; Street et al. 2018; Bergen et al. 2020; Pandey and Zafar 2022). RNA velocity represents the change in mRNA abundance and is calculated by distinguishing between unspliced (nascent) and spliced (mature) mRNAs, allowing estimation of the time derivative of the gene expression state (La Manno et al. 2018). Cells that show a higher abundance of unspliced mRNA than expected in a steady state for a particular gene have a positive RNA velocity, which represents an indication that a gene is upregulated. Conversely, cells that show a lower abundance of unspliced mRNA than expected in a steady state for a particular gene have a negative RNA velocity, which represents an indication that a gene is downregulated. La Manno et al. used single-cell RNA sequencing (scRNA-seq) data to compute their individual RNA velocity and estimated the future state of individual cells by predicting the direction and speed of transcriptional changes.

Subsequently, this model has been refined, and RNA velocity was estimated using scVelo, a novel AI software that dropped the assumption of either the presence of steady states or a common splicing rate across genes (Bergen et al. 2022). *scVelo* is able to identify transition states and stages of cell fate commitment, and, importantly, putative driver genes of these transcriptional changes. As Bergen et al. explained, a similar method can be used to identify the activated pathways in a cell which is crucial for evaluating the potential benefit of targeting agents.

Another software, *Sligshot*, used for inferring cell lineages from single-cell gene expression data, is particularly useful for understanding cell progression and differentiation. This method begins by clustering cells based on their gene expression profiles and then identifies *pseudotime*, i.e., a one-dimensional variable representing each cell's transcriptional progression toward the terminal state and infers the most likely developmental paths (Street et al. 2018).

More recently, *MARGARET* has been used to reconstruct complex trajectory topologies using a deep unsupervised metric learning and a graph-partitioning approach based on a novel connectivity measure, automatically detecting terminal cell states (Pandey and Zafar 2022). This AI was able to map the sequence of events of hematopoiesis and identified all major lineages and associated gene expression trends. It was also able to identify the ectodermal, endodermal and mesodermal lineages and also novel transitional populations in embryogenesis. (Pandey and Zafar 2022).

No cancer cell is an island and "cancers are more akin to interconnected biomes of a single continent" (Rolong et al. 2021). The tumor microenvironment (TME) has multiple cellular components, including immune, vascular, neuronal, adipose, and stromal cells that can influence tumor initiation and progression.

Ecotyper is a software able to transform bulk transcriptomics data into cell populational data. Luca et al. analyzed 12 cell types representing the majority of immunological and structural cells found in human epithelial tumors: B cells, plasma cells, CD8 T cells, CD4 T cells, NK cells, monocytes/macrophages, dendritic cells, mast cells, neutrophils, fibroblasts, endothelial cells, and epithelial cells in 16 types of carcinoma. Sixty-nine cell states were found (each cell type being associated with several cell states) (Luca et al. 2021). Most states were ubiquitous across carcinomas, highlighting common features of the oncogenic process independent of the tumor site. The authors were also able to delineate 10 ecotypes defined as co-associated transcriptional programs across multicellular communities—a community being defined as a collection of cell states that cooccur across independent tissue samples, independent of spatial neighborhoods. The clinical utility of this method was both the ability of Ecotypes to classify tumor subtypes and patient groups, and to predict survival in several tumor types including colon, ovarian, and gastric cancers.

The same team also used *Ecotyper* to analyze samples of Diffuse B Cell Lymphoma (DLBCL). Among the 12 different cell types analyzed, they identified 39 cell states (including 5 states belonging to DLBCL tumor cells themselves) grouped in 9 ecotypes. The clinical utility was demonstrated by finding a subset of patients with DLBCL that can benefit from adding Bortezomib based on a unique ecotype. Interestingly, treatment responses were related to their unique stromal pattern and to the prevalence of a distinct CD8 T cell state, and not to the malignant B cell phenotype signatures (Steen et al. 2021).

A Chinese group applied the same method to characterize colon cancer, identifying 42 states and seven ecotypes. Strikingly, FOLFOX and FOLFIRI improved survival only within particular Monocytes/Macrophages, and CD8 T cells states, and had no significant effect on patients within other cell states (Li et al. 2023b).

Cancer is heterogeneous and its behavior is nonlinear. Individual cancer cells trajectories and tumor trajectories overlap only partially. The micro- and macro-environments have a significant impact on cancer progression.

11.4 Twelve Steps Leading to the Malignant Phenotype

Three temporal phenomena are tightly controlled in multicellular organisms: cell growth, cell division, and apoptosis. Furthermore, four classes of extracellular signal proteins contribute to this control, mirrored by three intracellular pathways controlled by three classes of molecules (Alberts et al. 2002):

- *Growth factors* promote an increase in cell mass by stimulating the synthesis and inhibiting the degradation of molecules. In the cell, there are growth factor pathways (EGFR, KRAS, etc.).
- *Mitogens* that stimulate the rate of cell division by removing intracellular brakes that restrain cell cycle progression in G1.
- *Survival factors* that increase cell numbers by inhibiting apoptosis.
- A fourth class of extracellular signals *inhibits cell growth or division, and induces cells to undergo apoptosis* contributing to size control.

In cancer, all three pathways: cell growth, cell division, and apoptosis are significantly altered.

Titia de Lange and collaborators (Maciejowski and de Lange 2017) proposed a scenario consisting of five distinct genetic processes needed for the cancer phenotype to develop:

1. The *mutator phenotype* developed through loss of TP53/RB may represent a form of adaptation to these stresses. Loss of TP53/RB is one example of a mutator phenotype.
2. The *mutator phenotype* leads to *increased replication,* followed by telomere shortening, leading to a telomere crisis.
3. The telomere crisis leads to *diploidization* and offers additional genes to compensate for the loss of function through gene loss. Duplications and chromotripsis occur in a search for solutions in order for the cell to survive.
4. The accumulation of mutations leads to telomerase activation.
5. Once telomerase is activated, it leads to chromosome healing and genetic stabilization.

Activation of telomerase provides a path out of telomere crisis, ultimately leading to the formation of a cancer clone with a heavily rearranged genome.

For cancer to continue to grow below 1–2 mm^3, three additional steps are required:

1. An activation of a growth pathway (i.e., by KRAS mutations).
2. Activation of microenvironmental programs related to local invasion and.
3. Stimulation of angiogenesis.

Cancer's unimpeded growth is only half of the story, and this is where things get really interesting. The local and the distal environment actively support cancer growth; they are not passive bystanders. In addition to unchecked cellular division, cancer, in order to progress, also needs to incite the organism to assist its growth. Neither the classical theory of cancer (SMT) nor TOFT discusses the importance of the stromal instigation induced by cancer. The authors of TOFT consider variation and motility the default state of all cells. The capacity to proliferate is considered a constitutive quality of living cells, and so would require no stimulation or explanation. For them, cancer is caused by the release of the brakes placed by stroma on cells, and as a result, cancer cells regain their natural state of dividing and migrating. Basically, according to Carlo Sonnenschein and Anna Soto, the two proponents of TOFT, cancer represents development gone out of control due to tissue factors (Sonnenschein and Soto 2016).

The blind spot of both the SMT and TOFT scenarios is the lack of inclusion of the reciprocal effects of cancer itself on both the micro- and the macroenvironment of the organism. Cancer is neither an independent growth nor a simple release of brakes on cellular division and migration. A simple comparison of cancer with embryonal development falls short, since it does not include the opposite impact of cancer on the local stroma and the whole organism. Factors that instigate the organism to support cancer growth have been highlighted for more than a decade (McAllister and Weinberg 2014) and are essential to understand cancer progression. Factors secreted by the tumor (i.e., exosomes) are as important. The malignant condition is a two-way street in which both cancer and the organism influence each other and play equally important roles in cancer's progression and prognosis (McAllister and Weinberg 2010; Alečković et al. 2019).

The normal organism is not a passive bystander. It actively contributes to the cancer process, and *the whole organism's involvement* did not receive enough attention in the hallmarks articles series.

For metastatic cancer to occur, further conditions are required:

1. Activation of a trans-differentiation cellular program, i.e., EMT transition.
2. Activation of organismal programs related to niche formations.
3. A program of maintenance of the stemness of the state.

Finally, a 12th step is needed for cancer to become clinical.

Of note, these steps are not necessarily sequential. Cancer can become symptomatic without being metastatic. Also, in carcinoma of unknown origin, metastases may develop before the primary cancer becomes clinical.

In different cancers, the twelve steps unfold at a different pace. There are cancers taking decades to become clinical, with others developing much faster. There are some tumors in which most cells are dividing, other tumors in which most cells are in a quiescent state.

For these steps to unfold, there is no need for twelve separate genetic or epigenetic events to occur. There are protein hubs like TP53 that, once mutated, can contribute to all these steps, including stemness (Ghatak et al. 2021).

The mechanisms of transitions between different cellular states are only partially known. Cancer treatments may also influence these transitions: it has been shown for example that different treatment modalities used in cancer treatment may induce metastases (Martin and Anderson 2018 including: surgery (Tohme et al. 2017; Grewal et al. 2021; Jiao and Lv 2023) radiotherapy (Blyth et al. 2018; Zheng et al. 2020; Piffkó et al. 2025), chemotherapy (Karagiannis et al. 2018; D'Alterio et al. 2020), and even immunotherapy that may induce stem-cell like states of cancer cells and drive metastatic progression (Beziaud et al. 2023).

Recently, as a proof of principle, it has been demonstrated that the malignant state of a cell can be *reversed* by inhibiting only three master regulator genes: MYB, HDAC2, FOXA2 (Gong et al. 2025). It is yet unclear whether a similar genetic engineering can be applied to human patients with cancer and whether it will have the same impact on cancer cells in vivo.

11.5 Preclinical Cancer Initiation Stage: Group Events vs Individual Events

Zhaolian Lu et al. demonstrated that initiation in colorectal cancer may involve recruitment of non-mutated cancer clones by mutated cancer clones (Lu et al. 2024). *Cooperation and not competition between clones may be at the origin of cancer inception.* Lu's team's genomic and clinical data support a model of polyclonal-to-monoclonal transition, with monoclonal lesions representing a more advanced stage. As proven by the scRNA-seq data, early inflammation-driven polyclonal lesions lead to increased intercellular communication that may represent novel cell–cell and extracellular matrix interactions. Some oncogenes, like KRAS, for example, in colorectal cancer, may confer a selective advantage for the clonal outgrowth and might be responsible for polyclonal-to-monoclonal transition. A proof-of-principle model using *BRAF*^{V600E}/TP53 transgenic zebra fish suggested that the presence of oncogenic driver mutations in the initial phases of cancer transformation may represent a sort of cancerized field effect. Using live imaging, Kaufman et al. were able to see in real time how a single mutated cell becomes malignant through activation of embryonal programs (Kaufman et al. 2016).

11.6 Cancer's Big-Bang

In the majority of cases, only one single cell out of dozens of trillions of cells is responsible for the malignant disease. Mutations are not a sine qua non for the malignant phenotype, as there are cancers without any driver mutation identified. It has been shown that chromatin acts like a filter for activating the competence of oncogenic driver mutations, and they are only activated in a specific cellular/tissular context. As shown by Baggiolini et al., the melanoma phenotype can be readily induced in melanoblasts as opposed to melanocytes (Baggiolini et al. 2021). Molecular profiling of the melanoblasts showed that they had a higher expression than melanocytes of chromatin-modifying enzymes, allowing for expression of downstream oncogenic and neural crest programs. PTEN mutations were required for malignization of melanocytes. As shown by Köhler et al. a study in a mouse melanoma model also suggested that malignization may not be linked to a discrete cell type or differentiations stage, and different cells along the development from neural crest- and melanoblast-like cells to melanocytes may be able to produce cancerous tumors (Köhler et al. 2017). Clearly, mutations do not suffice—besides mutations, malignant transformation also requires cell-type specific transcription factors and lineage-specific developmental chromatin modifiers allowing for a permissive chromatin landscape (Baggiolini et al. 2021). Other authors added microenvironmental factors, organized in a tumor niche (Murgai et al. 2015), to the list of factors that are essential for cancer initiation (Köhler et al. 2017). In conclusion, internal and external factors concur in cancer inception, and a causal network rather than a unique cause is responsible for cellular malignization. Cancer is defined not only by a growth rate increase, but also by the ability to invade locally and to migrate at a distance. A phenotype that includes these functions is already present in the first cancer cell from which the malignant tumor(s) develop. This first cancer cell has similar properties to the fertilized zygote, and both the malignant phenotype and cancer progression are closely related to the activation of embryonic programs and embryonic development. Cancer instigates the normal organism to support its progression by hacking normal developmental programs and, in this way, is similar to a pregnancy.

The integrative framework introduced in this book advances the idea that *cancer from inception is a systemic disease* caused by disturbances present at multiple levels of the organism, and the cancerization process is the result of the *interaction* of a score of factors present at the cellular, tissular and organism level.

11.7 Cancer: A Tale of Two Life Forms

There is an ongoing dialogue between cancer cells, tissues, and the organism. Initially, according to Macfarlane Burnet and Lewis Thomas (Burnet 1957, 1970; Thomas 1959) the organism keeps cancer cells' growth under control through immune surveillance (Thomas 1982; Dunn et al. 2002). Simplistically, we can think of this dialogue as a two-way influence: one that initially goes from the organism (**O**) towards the cancer cell (**Cc**), and then, the tumor **(Tu)**: **O → Cc,** and **O → Tu**; and, subsequently, another that goes in the opposite direction: **Tu→O**. (→signifies control).

Initially, cancer cells and tissues are controlled by the organism: **O → Cc, and O → Tu.** At a certain point, tumors start impacting the organism and, progressively, take control over it: **Tu→O.**

Based on the above, we can classify the cancer cells/tissues–organism interactions into four types:

Type I In the Type I interaction, cancer cells/tissues have a minimal effect on the organism, and their growth and spread are kept under control (preclinical and early clinical stages of cancer, some late stages and clinical cases of cancer remission). The general climate of the organism initially does not allow cancer to develop and, early in cancer development, cancer effect on the climate is minimal.

Type II The organism "ignores" cancer. The climate is permissive for cancer cells to grow and invade locally and sometimes even to spread distally, but even so the influence of cancer on the climate is undetectable. Also, there are cases where, as part of type II interaction, a sort of balance is established between cancer and the organism. In these cases, the climate has a neutral influence on cancer growth and cancer is kept under control with or without treatment, and stable disease occurs (SD) for variable periods of time.

Type III Cancer starts to influence the internal body climate and its effect on the body systems are detectable (for example, advanced clinical stages of cancer when cancer does not respond to treatment leading to progressive disease).

Type IV Cancer has a significant systemic effect on the organism, and it attunes its internal environment in order to support its own growth—the cachexia stage.

It is apparent from the above scheme that the organism can impact cancer in two ways: it can keep cancer under control (type I interaction); or it can promote its growth (type IV interaction). The cancer surveillance theory of Burnet and Thomas consider cancer as evolving through loss of homeostatic control from a Type I interaction to a type III or IV interaction (Burnet 1957; Thomas, 1982). Initially, according to this theory, the organism keeps cancer under control through immune surveillance (type I interaction). At a certain point, the tumors escape immune surveillance, the organism seems to "ignore" cancer and cancer starts to grow and spread (type II interaction). Subsequently, cancer starts to influence body's systems (type III and IV interactions). In the early phases, in general, the cancer effect on the organism is minimal and patients are asymptomatic. When cancer becomes more advanced, symptoms may appear. In many instances, in the terminal phases, it is not the cancer per se that kills the patient, but the cancer-induced generalized inflammation, a process similar to the systemic inflammatory response syndrome (SIRS) (Bone et al. 1997; Kaukonen et al. 2015) that is followed, in its final stage, by the multiple organ dysfunction syndrome (MODS) that often is a harbinger of imminent death. From the scheme above, it appears that an important focus of cancer treatment should be not only to stop cancer progression by inhibiting its local growth, but also controlling the systemic effects of cancer on the normal body systems by inhibiting its spread and modulating its impact on the organism.

11.8 The Temporal Landmarks of Cancer

We can now formulate the key temporal landmarks of cancer, at least as important, in my opinion, as the classical cancer hallmarks (Please see Fig. 11.1).

1st landmark Transformation of a normal cell into a cell with oncogenic mutations. We learned recently that driver oncogenic mutations are also present in noncancerous tissues. On the other hand, given the high frequency of driver mutations in cancer, although not always sufficient, with few exceptions, in general, mutations are an integral part of the malignant transformation process.

2d landmark Transformation of a cell with oncogenic mutations into a neoplastic cell. *A neoplastic cell may be either benign or malignant.* The possibility of metastatic transformation appears even at this stage.

3d landmark Transformation of a neoplastic cell into a microscopic cancerous tumor. The possibility of metastatic transformation continues to be present.

4th landmark Transformation of a microscopic cancerous tumor into a macroscopic cancerous tumor. The possibility of metastatic transformation continues to be present.

5th landmark Once a malignant tumor is formed, metastases may occur anytime.

6th landmark Transformation of a macroscopic tumor, either primary or metastatic, into a clinically significant tumor.

Elucidating the factors involved in these transformations is key to understanding oncogenesis.

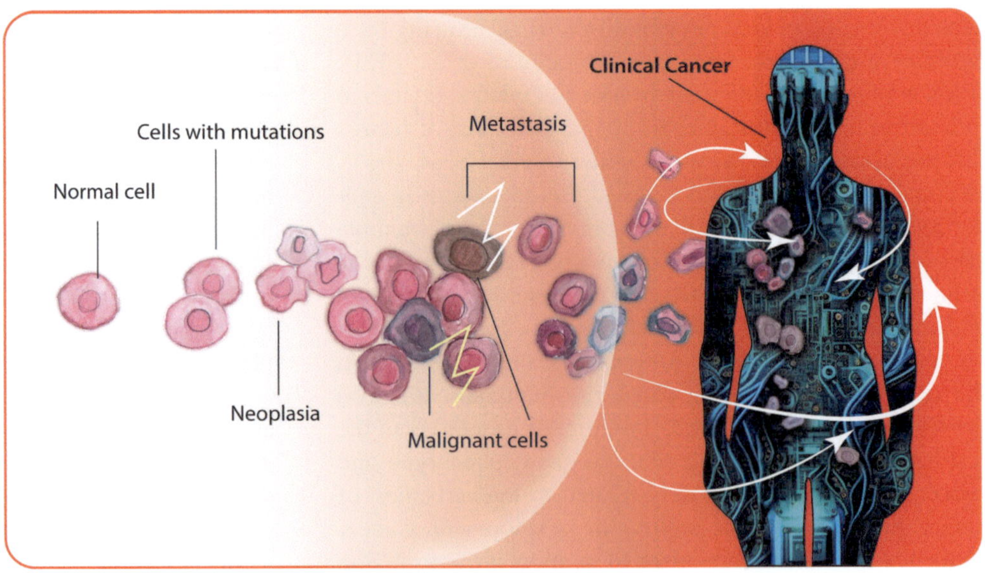

Fig. 11.1 Simplified depiction of the temporal landmarks of cancer. (Courtesy of Ana Tourian)

In my opinion, the reason why we are not able to prevent, predict, and cure cancer is related to our imperfect knowledge regarding the nature of the factors responsible for the transitions between cancer's temporal landmarks.

The importance of cell-intrinsic processes during premalignant evolution has been demonstrated in an organoid system of gastric cancer in which TP53 deficiency was able to recapitulate multiple hallmarks of CIN gastric cancer including aneuploidy and transcriptional programmes (Karlsson et al. 2023).

1st landmark

It is thought that imperfect copying of DNA during division may represent the main culprit for the transformation of a normal cell into a cancer cell. Recent data suggests that besides genetic factors, the transformation from a normal to a cancer cell within different tissues also involves epigenetic events, tissue factors, exposure to different carcinogens, micro- and macro-environmental factors, and aging. Importantly, single-celled organisms divide primarily in response to nutritional signals, whereas the division of metazoan cells is contingent on cell–cell contacts, cell–matrix contacts, nutrition, numerous growth factors, and hormonal signals from distant tissues (Gerhart and Kirschner 1997).

11.9 Do Mutations Play a Physiologic Role?

There is a fitness advantage conferred by driver mutations as shown by Rozhok and DeGregori (2019). Recently, experiments done in yeast confirmed the classical "less is more" hypothesis that loss-of-function mutations play a basic role in rapid adaptation to ecological conditions. The loss-of-function mutations in cancer that may arise as the oncogenic process unfolds, (i.e., TP53 or RB genes) may be nonrandom and play an adaptive role to different stressors and thus confer a fitness advantage, similar to driver mutations in oncogenes.

By applying an integrative-omics approach to the analysis of gene expression, copy-number alteration, and mutation data for in situ ductal carcinoma (DCIS), invasive ductal carcinoma (IDC), and metastatic (Met) site samples (Sarhadi et al. 2020), it has been suggested that the malignant traits are mostly acquired during DCIS. Cell motility pathways, for example (e.g., Focal adhesion, Gap junction, and calcium signaling pathways) as the basis of cancer cell invasion were among the most perturbed pathways shared in DCIS, IDC, and Met. The authors found also three notable differences: the number of genes in the same pathway was higher in later stages; the same pathways were increasingly heterogeneous in terms of direction of expression and regulation of their targets in the later stages and the perturbation of enriched pathways significantly increased in the later stages, indicating fine-tuning evolution of the earlier-developed pathways by further genetic alteration in late-stage tumors.

In the majority of cases, the reason DCIS does not lead to cancer may be related to the age of the patient. In young tissues, the microenvironment is able to control the clonal development of cancer cells with driver mutations as opposed to older or injured tissues, where cancer cells may outcompete normal cells (De Gregori 2018).

2d Landmark

The somatic mutation theory (SMT) considers cancer a genetic disease. Iñigo Marticorena found in normal cells the same type of genetic mutation drivers present in cancer cells and cancer tumors. Remarkably, the number of mutations in breast carcinoma in situ may be higher than the number of mutations in invasive breast cancer, so the malignant state does not appear to be related either to a special *kind* or to an increased *number* of mutations, as previously thought. Only a small proportion of carcinomas in situ became cancer. Benign neoplastic transformation appears to be related to the selection of some clones with growth advantage (Martincorena and Campbell 2015).

3d Landmark

This landmark is very important. For a fertilized egg to start developing, it needs a molecular trigger inside the egg that jump-starts the process of fertilization. The protein responsible for turning a dormant egg into an embryo is called phospholipase C zeta (PLCζ), and it is considered to be the physiological stimulus responsible for generating calcium (Ca2+) oscillations that will induce egg activation (Saleh et al. 2020). Stem cell theory postulated that modified stem cells are at the origin of cancerous tumors, but what causes a cancer stem cell to start dividing and creating a tumor? It is unclear whether a process similar to fertilization is also occurring in cancer stem cells.

The link between embryogenesis and tumorigenesis has been hypothesized for more than a century, and there are striking similarities between the two processes. They differ, though, in two fundamental ways: normal stem cells lead to the formation of differentiated normal cells, in contrast to the progeny of cancer stem cells that are cells that never fully mature. Furthermore, the process of normal development is highly regulated both spatially and temporally, while the oncogenic process is much less predictable due to genomic and chromosomal instability (Ma et al. 2010).

Interestingly, deregulated Ca2 + signaling has been linked to all of the "cancer hallmarks" (Reviewed in Stewart et al. 2015). The importance of calcium for cellular life is reflected by the highly complex regulation of calcium levels inside the cell. On the plasma membrane, there are no less than five different Ca2+ channels: voltage-gated Ca2+ channels (VGCC), store-operated Ca2+ channels (SOCC), transient receptor potential (TRP) channels, Na+/Ca2+ exchanger (NCX), and plasma membrane Ca2+ ATPase (PMCA). On the ER membrane, there are four types of receptors present: sarco/endoplasmic reticulum Ca2+-ATPase (SERCA), STIM, inositol-1,4,5-trisphosphate receptor (IP3R), and ryanodine receptor (RyR) and present in mitochondria are the mitochondrial Ca2+ uniporter (MCU) complex and Na+/Ca2+/Li + exchanger (NCLX). Calcium signaling is required in the early G1 phase and in the transition from G1 to S, G2 to M, and metaphase to anaphase (Kumari et al. 2024).

Also, spontaneous Ca2+ oscillations were observed in highly metastatic human prostate and breast cancer cells in contrast to the absence of such oscillations in weakly/non-metastatic cancer cells (Rizaner et al. 2016). Various parameters of the oscillations depended on extracellular Ca2+ and voltage-gated Na + channel activity. Treatment with either tetrodotoxin (a general blocker of voltage-gated Na + channels) or ranolazine (a blocker of the persistent component of the channel current) suppressed the Ca2+ oscillations, suggesting a functional voltage-gated Na + channel expression in the generation of oscillatory intracellular Ca2+ activity in highly metastatic cells.

The ubiquitous and varying role of Ca2+ in normal and cancer cells complicates targeting approaches and raises safety concerns, mainly cardiac (Kumari et al. 2024). As a proof of principle, a phase I study by Holdhoff et al. that enrolled glioblastoma patients demonstrated the safety of a sequential combination of mibefradil, a calcium channel blocker, and temozolomide, with some patients demonstrating radiographic responses (Holdhoff et al. 2017).

11.10 What Determines the Transition from Cells with Mutations to Full-Blown Malignancy?

In my opinion, the most important step in cancerization is the co-optation of not only of the local tissue, but also the whole organism to support the cancer process. How does it happen?

Classically, cancer has been considered a disorder of individual cancer cells with a behavior mainly determined by internal processes, with progress in oncology being mainly driven by discoveries in molecular biology. The Swanton et al. hallmarks article (Swanton et al. 2024) pointed to the complexity of tumorigenesis and proposed a more nuanced view in which the local and distal environment also play significant roles in the cancer process (Swanton et al. 2024).

Recently, David Krakauer and his collaborators proposed that individuality can be non-binary and continuous, with some processes having more individuality than others. Their broad definition of individuals as aggregates that propagate information from the past to the future and have temporal integrity allowed them to define three classes of individuality:

- colonial individuality in which there is a combination of self-determination by internal processes and environmental determination by external processes.
- organismal individuality in which the system is primarily self-determined by internal processes.
- environmental-driven individuality in which the internal processes are strongly influenced by the environment (Krakauer et al. 2020).

In the light of the Krakauer article, we can see cancer as a process where individuality is fluid and where cells can switch during the oncogenic process between different classes of individualities, i.e., organism-like, clonal-like, or environmentally determined.

As suggested by Thomas Deisboeck and Iain Couzin, individual vs. collective behavior or, more generally, the emergence of cooperation within cancer cells, may be related to environmental conditions. A tumor may have heterogeneous patterns both in time and in space, sometimes exhibiting collective behaviors, and at other times not; and some locations within the tumor may show collective behavior, while others do not (Deisboeck and Couzin 2009).

The key cancer cells' behaviors (uncontrolled division, invasion, metastasis) may represent stress adaptations related to both internal and external stressors, followed by a fundamental change in the relationship between cells and their micro- and macro-environments. Cancer cells switch to a survival mode in which they literally hack normal cellular programs using the organism's resources for their own development. Cancer cells are not independent intruders; to survive, they form alliances between themselves and with other normal cell that they corrupt for their own benefit. The phenotype of cancer cells is fluid, and the survival mechanism of their plasticity allows them to adapt into more or less differentiated cellular states, switching between division and quiescence, rest and movement.

Cancer inception is not related solely to genetic alterations or metabolic changes within a cell or a group of cells. The key change is the inauguration of a novel relationship between the organism and the novel cancer life form that starts developing with its support, similar to a pregnancy. It is still unclear whether this relationship develops very fast within a single cell's life, like fertilization, or whether it represents a gradual process that may take several years. The timing is likely variable, some cancer cells developing slowly from pre-malignant cells, others emerging quickly within days.

4th Landmark

Angiogenic switch" is considered the main factor that transforms a few millimeter tumor into a macroscopic tumor. The "angiogenic switch" described by Judah Folkman and collaborators refers to new blood vessel formation and is followed by rapid growth of malignant cells (Hanahan and Folkman 1996). Recently, co-option has been described as another mechanism through which tumors gain access to the blood supply (Kuczynski et al. 2019).

A parallel progression of primary tumors and metastases has been proposed by Christoph Klein in 2009 (Klein 2009). Recently, analyzing genomic data from paired primary colorectal cancers and distant metastases to the liver and brain from five different patient cohorts, Christina Curtis and her team, demonstrated that metastatic seeding often occurs early, years before diagnosis and surgery, when the carcinoma is clinically undetectable (0.0001–1 cm^3). This finding is clinically relevant, indicating that in some cases, patients with early-stage colorectal cancers may benefit from adjuvant chemotherapy (Hu et al. 2019). Subsequently, similar findings were described by the same group in lung and breast cancer (Hu et al. 2020).

5th Landmark: What Causes Metastasis?

Importantly, although metastatic programs are distinct from cancer initiation programs as suggested recently (Ganesh 2023), a large study involving more than 25,000 primary tumor–metastases pairs (Nguyen et al. 2022) demonstrated the hypothesis of René Bernards and Robert Weinberg (Bernards and Weinberg 2002) that, in general, primary tumors and metastases have very similar oncogenic mutations. The recurring finding increased in metastatic versus primary tumors is whole genome duplications (Priestley et al. 2019; Nguyen et al. 2022). Sometimes, enhanced overexpression of the same amplified oncogenic pathways found in primary tumors has been described in metastases (Patel et al. 2021), but the metastatic behavior could not be inferred from genomic pattern analysis between primary tumors and metastases (Nguyen et al. 2022). The metastatic process appears to be mainly epigenetic (Patel and Vanharanta 2017; Patel et al. 2021).

Epigenetic factors may be both *intracellular*: i.e., DNA methylation (Esteller 2008; McMahon et al. 2017), chromatin accessibility (Corces et al. 2018), histone modifications (Markouli et al. 2021) and 3D chromatin conformation (Peng et al. 2022) and *extracellular*: inflammation, acidosis, and hypoxia. Cancer's most striking trademark is the reactivation of physiological programs repressed in terminally differentiated cells, i.e., developmental, regenerative, and stem cell programs, and not genetic alterations or genomic instabilities. *Intracellular* stressors including genetic alterations, epigenetic changes and metabolic shifts, and *extracellular* cues like hypoxia or inflammation may trigger the reactivation of these programs that involve the cooptation of both local factors and the whole organismal support for cancer to thrive.

A sodium leak channel non-selective protein (NALCN) has been recently identified as a key regulator of both cancer metastasis *and* nonmalignant cell dissemination. Deletion of NALCN from gastric, intestinal, or pancreatic adenocarcinomas in mice did not alter tumor incidence, but markedly increased the number of circulating tumor cells (CTCs) and metastases, thus suggesting that tumorigenesis and metastasis may be regulated by different mechanisms (Rahrmann et al. 2022).

Classically, for a tumor cell to form a metastatic lesion, it must undergo several steps, including local tissue invasion and penetration through the basal membrane, intravasation, survival and arrest in the circulation, extravasation, dormancy (sometimes), and proliferation at the metastatic site (Fidler 2003; Giancotti 2013; Welch and Hurst 2019). In order to metastasize, cancer cells must orchestrate diverse cellular functions to overcome the difficulties of the metastatic cascade. These functions are not only limited to cell-autonomous traits but are also highly dependent on the interaction of the metastatic cell with the tumor and host stroma. The stromal elements, including activated fibroblasts and immune and vascular cells, foster tumor progression through paracrine mechanisms. Primary tumors are complex ecosystems consisting of many cell types, with different involvement in the invasion and metastatic process. (Joyce and Pollard 2009; Obenauf and Massagué 2015).

Cancer metastasis also demonstrates organotropism, in which multiple factors are involved including cancer cell-intrinsic factors (i.e., specific genes, epigenetic factors, specific exosomes release etc.), circulation patterns, organ-specific pre-metastatic niches, and the interaction between tumor cells and the host microenvironment (Gao et al. 2019).

Cell–cell interactions are central to metastasis, coordinating tumor progression through the tumor microenvironment, premetastatic niche (PMN), and secondary sites. Primary tumors precondition distant organs via secreted factors and extracellular vesicles (EVs), driving PMN formation and promoting immune evasion and metastatic spread. This process involves dynamic crosstalk among tumor, stromal, and immune cells, leading to extracellular matrix (ECM) remodeling at future metastatic sites. These findings underscore the therapeutic potential of targeting intercellular communication in metastatic disease.

11.11 Cancer Life Cycle and Niches

11.11.1 Cancer's Life Cycle

The concept of specialized tumor-supportive microenvironments, such as *primary tumor niches* (Murgai et al. 2015) and *pre-metastatic niches* (Kaplan et al. 2006), is well established in oncology. Building upon this framework, I propose that cancer, like all living systems, follows a life cycle—

beginning not with the primary tumor itself, but with the formation of a *pre-primary tumor niche*. This early niche is shaped by both local stromal components and systemic environmental cues, which collaborate to create a permissive milieu for tumor initiation. For instance, in gastric cancer, it has been demonstrated that cancer stem cells can originate from the bone marrow, highlighting the distal contributions to early tumor development (Houghton et al. 2004).

As malignant clones emerge and proliferate, they organize the primary tumor niche as described by Meera Murgai and colleagues (Murgai et al. 2015). This niche serves not only as a physical aggregation of tumor cells but also as a highly dynamic and interactive ecosystem comprising stromal cells, extracellular matrix (ECM) components, immune cells, and signaling molecules. These elements cooperate to sustain tumor growth, promote immune evasion, and facilitate local invasion. In some cases, the cancer life cycle may pause at this stage or enter a state of dormancy, characterized by quiescent cancer cells that can remain clinically silent for extended periods (Phan and Croucher 2020).

Sometimes, the life cycle of cancer further progresses from the primary tumor niche to the formation of *pre-metastatic niches*—specialized microenvironments in distant organs that are conditioned by primary tumor-derived factors to receive and support disseminated tumor cells. These cells may then colonize these permissive sites, establishing *metastatic niches*. From these secondary locations, cancer cells can continue the cycle either by further metastasizing to additional organs or by engaging in *tumor self-seeding*—migrating back to the primary tumor site and contributing to its growth and heterogeneity (Kim et al. 2009). This recursive, multistage process illustrates that cancer behaves not as a static mass but as a dynamic, evolving system with a life cycle that mirrors, in many ways, the developmental trajectories of other living entities.

11.11.2 The Premetastatic Niches

The concept of pre-metastatic niches are important add-ons to the classical metastatic steps (Psaila and Lyden 2009). The "pre-metastatic niche" concept borrowed from stem cell biology refers to the specialized microenvironment that supports and actively regulates the function and proliferation of cancer cells growing in a distal site. This concept has an important clinical impact, since it suggests that anti-metastatic systemic therapies targeting the pre-metastatic niches should be employed earlier rather than later, perhaps in combination with the adjuvant treatments directed to the primary tumor. Certain cancer types presenting parallel progression (breast, lung, colorectal) may be particularly suitable for this approach.

The experimental demonstration of the pre-metastatic niche suggested that cancer is a systemic disease with the tumor affecting multiple systems in the host that can result in the spread of cancer cells and enhancement of metastasis (Joyce and Pollard 2009). The ability of tumors to prime other tissues for metastasis is an emerging theme in cancer biology, and exosomes appear to be important mediators of this phenomenon by transferring signaling molecules or modulators of signaling. The work of David Lyden's team has shown the role of melanoma-derived exosomes in supporting tumor growth and metastasis through the education of bone marrow progenitors by horizontally transferring the MET oncoprotein from melanoma to BM progenitor cells (Peinado et al. 2012). Subsequently, the same team demonstrated an important role for exosomes, in general, in dictating organ-specific metastasis. A specific repertoire of integrins expressed on tumor-derived exosomes, distinct from tumor cells, was identified, dictating exosome adhesion to specific cell types and extracellular molecules in particular organs. Exosomes expressing ITG$\alpha_v\beta_5$ integrins were found to specifically bind to Kupffer cells, mediating liver tropism, whereas exosomal ITG$\alpha_6\beta_4$ and ITG$\alpha_6\beta_1$ integrins were found to specifically bind to lung-resident fibroblasts and epithelial cells, governing lung tropism (Hoshino et al. 2015). A distinct mechanism of non-integrin and extracellular matrix was identified in the case of bone metastases. A hyaluronan-binding protein (CEMIP, KIAA1199) was found to promote brain metastasis by generating a pro-metastatic environment (Rodrigues et al. 2019), highly enriched in brain metastatic breast and lung tumor-derived exosomes. Besides, pre-metastatic niche formation, exosomes were found to cause a plethora of systemic effects including pro-thrombotic changes, immune dysregulation, cachexia, and organ failure (Lucotti et al. 2022; Zhang et al. 2024). Recently, tumor-derived exosomes were found to dysregulate hepatic function and induce fatty liver formation (Wang et al. 2023a).

Carefully dissecting the metastatic process into distinct steps may lead to tailored treatments specific to each step:

1. *Pre-metastatic niche (PMN) formation*. To date, tailored therapies targeting PMN formation are scarce (Pătras et al. 2023b). Epigenetic therapy that uses low-dose DNA methyltransferase and histone deacetylase inhibitors, 5-azacytidine and entinostat, has been shown to disrupt the premetastatic niche by inhibiting the trafficking of myeloid-derived stem cells (MDSCs) through the downregulation of CCR2 and CXCR2, and by promoting MDSC differentiation into a more-interstitial macrophage-like phenotype (Lu et al. 2020). It has been demonstrated that neutrophils facilitate ovarian cancer premetastatic niche formation in the omentum (Lee et al. 2019). Inhibiting neutrophil-extracellular traps (NET) using

indirect NET inhibitors (i.e., Methotrexate, Hydroxychloroquine) or enzymes involved in the NET formation, such as PAD4 and MPO is currently under investigation (Chamardani and Amiritavassoli 2022).

The discovery of the metastatic tumor niches highlights the importance of the organism for both primary and metastatic tumor formation. The presence of mutations inside a cell may be due to random phenomena, but the persistence of these mutations and the further transformation into a full—blown cancer cell needs the cooperation of the entire body. To develop from one cell to a tumor that may kill the body that created it, cancer needs to be nurtured by the organism itself. The fact that we can cure local cancer has obscured the fact that cancer is a disease of the whole organism. Even when it is confined to an organ or to lymph nodes, cancer represents the local manifestation of a global disorder of the organism, and therefore, it might recur, or a different second malignancy may arise, caused by the same organismal disorder.

2. *Epithelial to mesenchymal transition (EMT).* The ability of epithelial cells to shift between epithelial and mesenchymal identities is critical to fundamental biological processes, such as embryonic development, differentiation, regeneration, carcinogenesis, and metastasis (Thiery et al. 2009).

Recent studies suggest that the epithelial plasticity of tumor cells cannot be classified simply as either epithelial or mesenchymal. Instead, cancer cells may possess a spectrum of intermediate states (Pastushenko et al. 2018). This hybrid epithelial/mesenchymal state has been termed "partial EMT" (P-EMT) and may be regulated by epithelial protein internalization, in contrast to the transcriptional repression of epithelial genes in complete EMT (Aiello et al. 2018). P-EMT PDAC cells preferably traveled as clusters of CTCs , whereas PDAC cells that underwent a complete EMT invaded as single CTCs (Reichert et al. 2018). The ability of cells to switch from epithelial to mesenchymal phenotype may be of clinical importance as demonstrated by a study in which patients with adenocarcinoma of the pancreas (PDAC) developed liver metastasis with cancer cells capable of mesenchymal-to-epithelial transition (MET), as opposed to lung metastasis colonized by cells that were not MET-capable (Reichert et al. 2018). Importantly, PDAC patients with recurrent pulmonary metastases demonstrate significantly longer overall survival compared to patients with recurrence at other sites (40.3 vs. 20.9 months).

Epithelial-mesenchymal transition (EMT), a process by which epithelial cells acquire mesenchymal characteristics, plays essential roles in various physiological and pathological contexts. EMT is broadly categorized into three distinct types: Type I occurs during embryogenesis, contributing to processes such as gastrulation, renal morphogenesis, and neural crest development; Type II is involved in tissue repair and regeneration in adult organisms; and Type III is associated with cancer progression and metastasis. While developmental EMT is tightly regulated, the EMT observed during tumor progression is markedly dysregulated and is characterized by aberrant activation of signaling pathways. This process is governed by a complex regulatory network involving microRNAs, epigenetic modifications, post-translational modifications, and alternative splicing mechanisms. Several factors derived from the tumor microenvironment have been identified as key inducers of EMT, including transforming growth factor-β (TGF-β), oxidative stress, inflammation, hypoxia-inducible factor 1-alpha (HIF-1α), epidermal growth factor (EGF), WNT ligands, and Notch signaling pathways (Ribatti et al. 2020).

The genes involved in EMT are rarely mutated (Xue et al. 2020). Numerous studies have demonstrated that exosomes encapsulate various components, such as microRNAs and proteins, and transfer information between tumor cells or between tumor cells and the tumor microenvironment, thereby regulating the EMT process (Jiang et al. 2022).

The mechanism of cancer cell metastasis is similar to embryonic development, where epithelial cells migrate and differentiate into all tissues of the embryo. In this period, epithelial cells differentiate into mesenchymal cells, migrate, and return to the epithelial cells when necessary. The important cell signaling pathways in embryonic development, such as growth factor receptor tyrosine kinases (FGF, HGF, TGF-α), TGF-β, Wnt, NOTCH, MAPK, JAK/STAT, and inflammatory cytokines such as NF-κB, TNF-α, IL6, cathepsin, and HIF-1α. All play crucial roles in EMT during cancer cell invasion and metastasis. In cancer progression, similar cell signaling pathways and molecular activation occur in embryonic development; therefore, it is important to understand the molecular mechanisms of cell signaling pathways related to cancer metastasis and improve new diagnostic and therapeutic approaches for resistance to drugs.

Zhe Jiang et al. analyzed in detail the oncogenic drivers promoting either primary mammary tumors, lung metastases, or both. The latter, termed shared oncogenic drivers, were shown—via transposon insertion site sequencing—to be clonally related, excluding the possibility of independent primary tumors. Rapid growth is enabled by glycolysis, whereas OXPHOS dominates in circulating tumor cells. Successful metastasis requires plasticity, allowing shifts between metabolic states, EMT/MET transitions, and cycling between proliferation and dormancy. The authors concluded that this plasticity may have been driven by shared oncogenic drivers, while metastasis-enriched drivers may reflect genetic or epigenetic changes triggered by the tumor microenvironment (TME). Interactions with the TME can activate specific pathways at metastatic sites, making these interactions the key factors for successful metastasis (Jiang et al. 2023).

3. *Local tissue invasion and penetration through the basal membrane.* The concept of *migrastatics* has been proposed as specific anti-invasion/anti-metastatic drugs (Gandalovičová et al. 2017), and several invasion mechanisms have been described (Nürnberg et al. 2011; Baro et al. 2024; Nair et al. 2024); but no migrastatic drug is currently approved for clinical use.
4. *Intravasation.* Cancer cells' intravasation has been recently inhibited in vitro by targeting the fluid shear stress sensor TRPM7 (Yankaskas et al. 2021), and has been inhibited in vivo by targeting tetraspanin CD151 (Zijlstra et al. 2008); but no agents inhibiting cancer cells' intravasation are currently approved for clinical use.

   Cesar Lopez-Camarillo et al. introduced the concept of metastamiR (metastasis-associated miRNAs) that are involved in the regulation of key steps in the metastatic program and processes, such as EMT, apoptosis, and angiogenesis, but do not influence the primary tumor either in development or initiation steps of tumorigenesis (Lopez-Camarillo et al. 2012).

   Tumor cell migration and invasion take place at the tumor–host interface and are accompanied by a desmoplastic stroma reaction. This desmoplasia ifibroinflammatorynflammatory process and an established feature of solid cancers such as pancreatic, lung, and breast cancer. The so-called "cancer-associated fibroblasts" are involved in the desmoplastic reaction. Cancer-associated fibroblasts represent activated stromal cells that not only surround, supply, and protect the tumor but also provide a soiland-seed platform for tumor metastases (Hoeppner and Bronsert 2021).
5. *Survival and arrest in the circulation.* The survival of circulating tumor cells (CTSs) in the blood circulation is a complex process involving cancer cell clusters, platelet aggregation, cytokines, exosomes, immune cells, hypoxia, and autophagy (Wang et al. 2018). Surgery, chemotherapy and immunotherapy have all been demonstrated to decrease the number of CTCs (Gostomczyk et al. 2024). Targeting tumor-associated macrophages and inhibiting their role as tumor promoters and/or aiming at their activation and reeducation are attractive therapeutic options (Mantovani et al. 2017). There are several clinical trials currently investigating macrophage-centered strategies for CTCs inhibition. (Gostomczyk et al. 2024).
6. *Extravasation, survival,* (sometimes) *dormancy,* and *proliferation* at the metastatic site leading to micrometastases. The crucial role played by integrins during intravasation and extravasation within the metastatic cascade has been recognized for many years, but despite sustained efforts, none of the integrin-targeted approaches were able to demonstrate a significant clinical benefit (Sökeland and Schumacher 2019; Liu et al. 2023b). Disintegrins can target specific integrins, and it is conceivable that they could target CTCs (Arruda Macêdo et al. 2015). In vitro and animal studies using disintegrins have been promising (Minea et al. 2012; Schönthal et al. 2020).
7. *Growth of clinically significant macro-metastases.* Currently, there are no specific systemic treatments, i.e., chemotherapies, targeted agents, or immunotherapies *specifically* targeting the growth of metastases versus primary tumors.

Arguably, the most important temporal landmark of the cancer process is the development of clinical symptoms that impact the functioning of the organism and the well-being of the patient. The majority of cancer models stop short and do not include the most significant step: clinical cancer. The clinical significance of cancer can go from localized symptoms and interaction with body systems to a global effect on the body's metabolism, like cachexia.

It has been demonstrated recently that a large number of cancers, although macroscopically visible, may go undetected because of a lack of symptoms. As many as 59% male patients over the age of 79 are found to have asymptomatic prostate cancers (Bell et al. 2015). Incidental invasive breast cancer was detected in 1.5% of women, carcinoma-in-situ in 8.5% of women, and atypical hyperplasia in 9.8% of women. Thyroid cancer was found in 11.2% of patients (Furuya-Kanamori et al. 2016). Incidental lung cancers havealso been reported (Manser et al. 2005), but this is much rarer. A systematic review of 13 studies of 16,730 autopsies across seven countries and seven decades demonstrated a pooled prevalence of 0.4% of subclinical lung cancer diagnosed post-mortem (Bonney et al. 2023).

A recent post-mortem histopathological analysis of 797 cases performed over 2 years in a hospital from India reported 46 tumors, of which 28 (50.9%) were malignant.

6th Landmark: What Makes Cancer Symptomatic?

The seminal frameworks proposed by Hanahan and Weinberg established a set of core cellular and tissue-level hallmarks of cancer, including sustained proliferative signaling, evasion of growth suppressors, resistance to cell death, and activation of invasion and metastasis, among others. These hallmarks have since served as foundational concepts for understanding the intrinsic biology of cancer cells and their immediate microenvironment (Hanahan and Weinberg 2000, 2011).

Building on this foundation, Danny Welch and Douglas Hurst proposed a refined set of metastasis-specific hallmarks, which encompass: (1) cellular motility and invasion, (2) colonization at secondary sites, (3) cellular plasticity, and (4) modulation of the metastatic niche (Welch and Hurst 2019). While these features represent essential components of the metastatic cascade, they align closely with the multi-step model of metastasis—comprising dissemination, intravasation, survival in circulation, extravasation, and colonization earlier described by Massagué and colleagues (Nguyen and Massagué 2007; Chiang and Massagué 2008; Ganesh and Massagué 2021). Thus, the proposed metastatic hallmarks largely recapitulate mechanistic stages previously delineated in metastasis biology.

11.12 The Systemic Hallmarks of Cancer

In 2001, I proposed a complementary set of six novel systemic cancer hallmarks, which emerge from the complex, bidirectional interactions between the tumor and the host organism. These hallmarks reflect macroscopic alterations that extend beyond the tumor microenvironment, affecting distant organ systems and reprogramming the whole-body physiology (Please see Fig. 11.2). The historical analogy of cancer to a crab, showing cancer impacting all body systems, is illustrated in Fig. 11.2.

The first hallmark is the cancer system as a functional unit—a systemic architecture formed through dynamic communication between the primary tumor, bone marrow-derived cells, and distant metastatic sites. This interconnected network functions as an integrated pathophysiological entity, modulating hematopoiesis, immune responses, and organ function to support cancer progression. The other five systemic hallmarks are as following.

Systemic inflammation—characterized by elevated levels of circulating pro-inflammatory cytokines (e.g., IL-6, TNF-α), acute-phase proteins, and chronic immune cell activation, contributing to tumor progression, angiogenesis, and paraneoplastic syndromes.

Immune suppression—involving both local and systemic mechanisms of immune evasion, including expansion of myeloid-derived suppressor cells (MDSCs), regulatory T cells (Tregs), and the expression of checkpoint ligands (e.g., PD-L1), which blunt antitumor immunity and foster immune escape.

Metabolic reprogramming and cancer-associated cachexia—systemic alterations in energy balance, muscle catabolism, adipose tissue browning, and hepatic metabolism lead to progressive wasting and functional decline. These changes are orchestrated via neuroendocrine and inflammatory mediators such as Growth Differentiation Factor 15 (GDF15), Interleukin-1 beta (IL-1β), and Parathyroid Hormone-related Protein (PTHrP) (Xiao et al. 2025).

Thrombo-inflammatory state—a hypercoagulable condition often observed in cancer patients, driven by tumor-derived tissue factor, platelet activation, neutrophil extracellular traps (NETs), and inflammatory cytokines, contributing to increased risk of venous thromboembolism (VTE) and tumor dissemination.

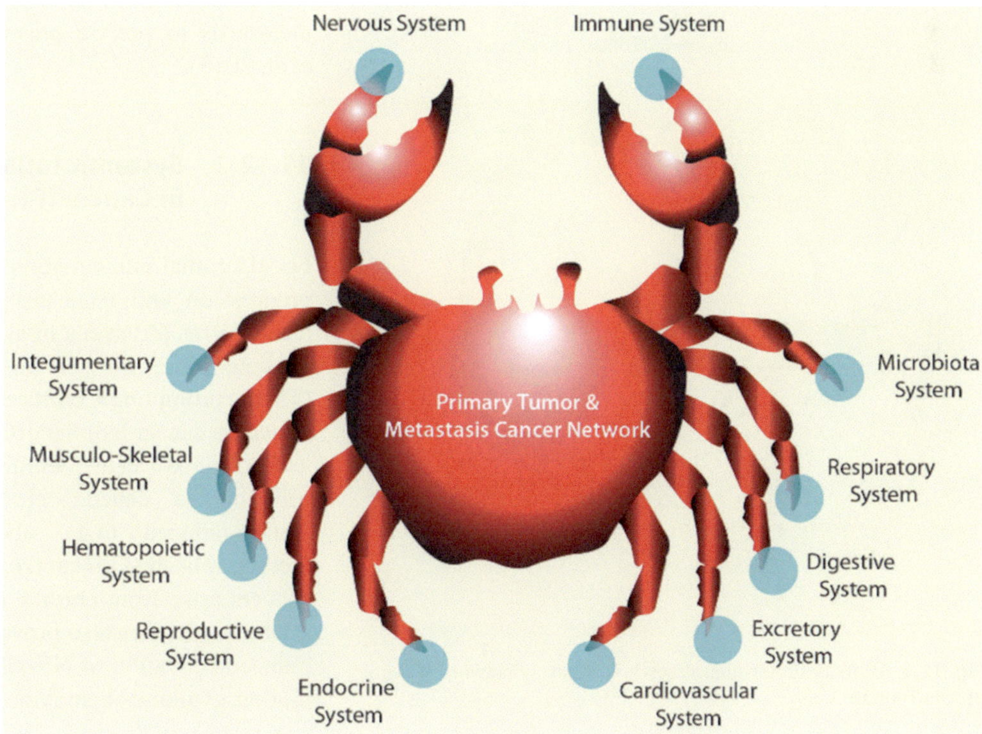

Fig. 11.2 Cancer impacting all body systems (Courtesy of Ana Tourian)

Neuroendocrine dysregulation—encompassing stress axis activation (e.g., HPA axis hyperactivity), autonomic nervous system involvement, and disruption of circadian rhythm, all of which influence tumor growth, immune function, metabolism, and host resilience.

Together, these systemic hallmarks reflect the capacity of cancer to rewire the organism's internal networks, transforming host physiology to sustain its growth and dissemination. Recognizing these macroscopic features of cancer may yield novel therapeutic opportunities aimed not only at tumor eradication but also at restoring systemic balance and organismal integrity.

11.12.1 Primary Tumor-Metastasis-Bone Marrow Axis (Please See Fig. 11.3)

Clinical and experimental evidence supports a systemic interaction between primary tumors, metastatic sites, and the bone marrow. In renal cancer, primary tumor resection may induce metastasis regression, while in some models, it accelerates metastasis. This bidirectional influence forms a trialogue: tumor-metastasis-bone marrow.

Bone marrow-derived cells (BMDCs) contribute to metastasis by forming pre-metastatic niches and potentially seeding malignant cells. BMDCs expressing VEGFR-1 localize to future metastatic sites, promoting angiogenesis and niche formation. Cytokines and tumor-derived vesicles also facilitate this process. Six key features define pre-metastatic niches: angiogenesis, lymphangiogenesis, inflammation, immunosuppression, organotropism, and reprogramming. The best described tumor secreted soluble molecules responsible for the niche formation and development are: VEGF-A, placental growth factor, and versican. Besides the cancer cells, there are also factors secreted by the tumor stroma that play a similar role: TGF-β, TNF-α, hypoxia-inducible factor-1 (HIF-1), and granulocyte colony-stimulating factor (G-CSF). Some of these factors are secreted by the BDMCs themselves: VLA-4 (integrin α4β1), matrix metalloproteinase (MMP)9, and ID3 protein) (Liu and Cao 2016).

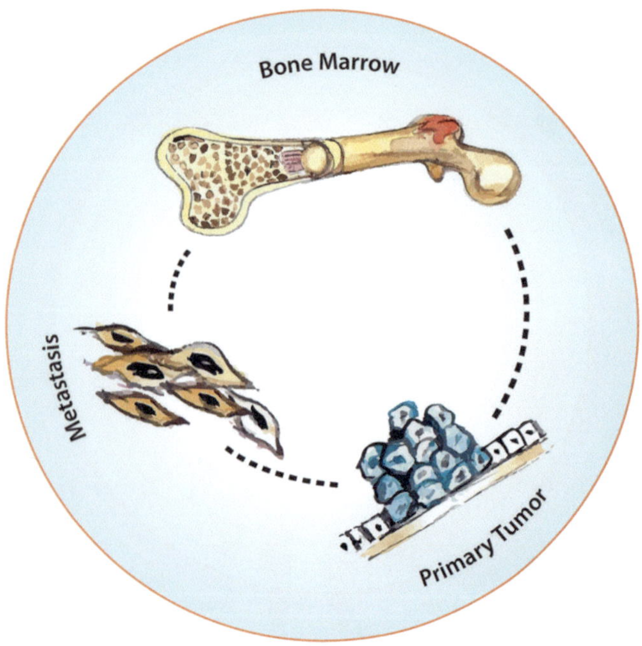

Fig. 11.3 Primary tumor-metastases-bone marrow network (Courtesy of Ana Tourian)

Exosomes are pivotal in mediating systemic communication. They transfer bioactive molecules (proteins, mRNAs, miRNAs, DNA) that influence tumor progression and drug resistance (Guo and Guo 2015; Anfossi et al. 2018). Syndecan-dependent mechanisms regulate exosome biogenesis; tumor exosomes modify ECM via CD44 and hyaluronan interactions (Karampoga et al. 2022). Certain miRNAs are enriched in exosomes coming from the cancer cells, indicating that the exosomes' composition seems to be controlled (Hessvik and Llorente 2018). Exosomes from KRAS mutant colorectal cancer cells, for example, show a distinct miRNA profile compared to wild type cells (Cha et al. 2015). By releasing soluble factors, apparently, tumor cells are able to specifically direct bone marrow-derived cells to the sites where they were supposed to go (Costa-Silva et al. 2015). An "integrin code" present on the surface of the exosomes seems to be responsible for the homing of future metastasis to precise prespecified distal organs (Hoshino et al. 2015).

11.12.2 Systemic Inflammation Networks in Cancer (Please See Fig. 11.4)

Local stromal inflammation is a recognized driver of cancer progression and metastasis, with well-defined molecular mechanisms (Allavena et al. 2015). Inflammatory cells and mediators are integral to the tumor microenvironment (TME), prompting Colotta et al. to classify inflammation as a seventh cancer hallmark (Colotta et al. 2009).

Grivennikov et al. identified four inflammation types contributing to cancer: chronic (infection/autoimmunity), irritant-induced (e.g., obesity), tumor-associated, and therapy-related (Grivennikov et al. 2010). Roughly 20% of cancers arise from chronic inflammation (Kundu and Surh 2008), yet tumors also provoke inflammation via oncogene-induced activation of NF-κB and STAT3, and HIF-1α, promoting cytokine/chemokine production (e.g., IL-6, TNF-α,

IL-17, TGF-β), which may be either pro- or anti-tumor (e.g., IL-12, IFN-γ) (Grivennikov et al. 2010).

Systemic inflammation is common in advanced cancers (Terlizzi et al. 2014). In colorectal cancer, stromal cells produce key cytokines (IL-6, CXCL8, CCL2, GM-CSF) at higher levels than tumor cells (Tuomisto et al. 2019).

Tumor-derived exosomes further propagate inflammation by delivering IL-8, CCL2–5, CCL20, and TGF-β (Becker et al. 2016; Manning and Danielson 2018). The liver plays a central role via acute-phase protein synthesis (e.g., CRP, serum amyloid A) and Kupffer cell cytokine secretion, with macrophage depletion reducing inflammatory mortality in mice (Bankey 1996; Jain et al. 2011).

Gut microbiota also modulates local/systemic inflammation and influences myelopoiesis and neutrophil function (Iida et al. 2013; Viaud et al. 2013). It promotes carcinogenesis via neutrophil recruitment and MDSC mobilization (Zhang and Frenette 1999; Benešová et al. 2023; Rutkowski and Conejo-Garcia 2015).

Systemic inflammatory indices—such as NLR, PLR, Glasgow Prognostic Score, and the systemic immune-inflammation index (SII = P × N/L)—correlate with poor prognosis across malignancies (Hui 2015; Wang et al. 2020a). In advanced cancers, inflammation reflects tumor aggressiveness (Dupré and Malik 2018).

Despite promising preclinical data, cytokine-targeting therapies (e.g., TNF-α, IL-6, IL-8 inhibitors) have shown limited clinical efficacy due to the context-dependent roles of cytokines (the "yin-yang" paradigm) (Dinarello 2010; Nakamura and Smyth 2017).

11.12.3 Immunity Inhibition Systemic Networks in Cancer (Please See Fig. 11.5)

Tumor evolution is shaped by the interplay of pro-tumor inflammation and antitumor immunity (Chen et al. 2018). While tumors elicit systemic inflammation, they concurrently suppress immune responses locally and systemically. Immune checkpoint inhibition (e.g., PD-1/PD-L1 blockade) has demonstrated clinical utility; yet tumors also exploit extracellular vesicles (EVs) to evade immunity. For instance, metastatic melanoma-derived exosomes express PD-L1, with IFN-γ enhancing PD-L1 levels, suppressing CD8+ T cells and promoting tumor growth (Chen et al. 2018). Tumor exosomes may also impair immunity via miR-203 (TLR4/IL-12 downregulation) (Zhou et al. 2014), Treg expansion (Yin et al. 2014), NK cell inhibition (Berchem et al. 2015), and suppression of general immune cell function (Liu et al. 2015; Whiteside 2016).

The gut microbiota co-regulates immunity, shaping immune development and function (Honda and Littman 2016; Belkaid and Harrison 2017). Microbial signals modulate myeloid cell responses, influencing both innate and adaptive immunity (Gorjifard and Goldszmid 2016). In murine models, antibiotic-treated or germ-free animals failed to respond to chemotherapy, highlighting the microbiota's role in antitumor immunity (Iida et al. 2013; Viaud et al. 2013). In humans, specific gut microbes (e.g., *Bifidobacterium longum*, *Collinsella aerofaciens*, *Enterococcus faecium*) correlate with better immunotherapy responses in melanoma

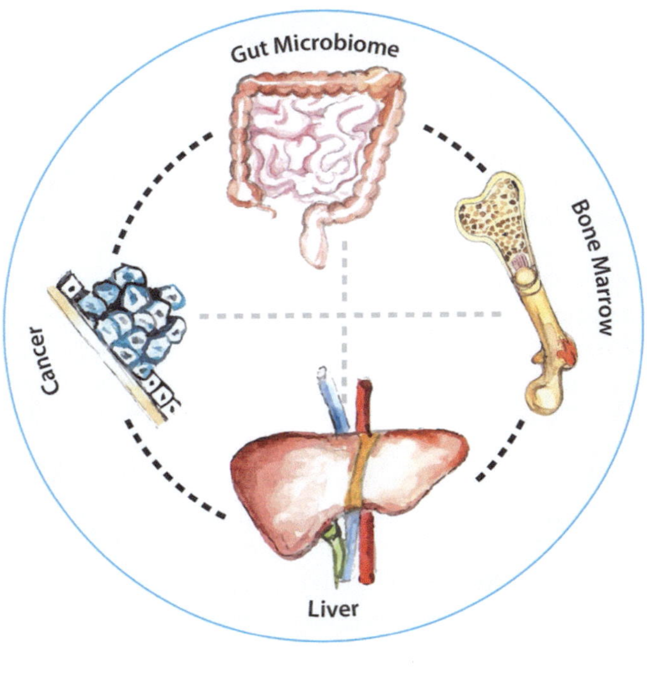

Fig. 11.4 The systemic inflammation network. (Courtesy of Ana Tourian)

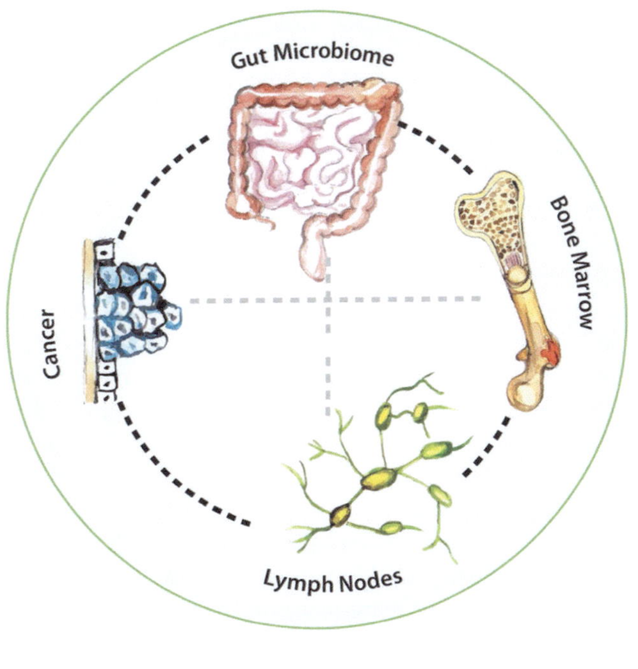

Fig. 11.5 The immunity inhibition systemic network. (Courtesy of Ana Tourian)

(Matson et al. 2018), while *Akkermansia muciniphila* is enriched in responders across several cancer types (Routy et al. 2018). Antibiotic use before immune checkpoint therapy reduces efficacy and survival.

Immune cells display dual roles in cancer (Tamura et al. 2018; Shitara and Nishikawa 2018). While M1 macrophages and CD8+ T cells are typically antitumor, M2 macrophages and Tregs often promote tumor growth. However, this dichotomy is context-dependent. For instance, Tregs correlate with poor prognosis in most cancers, but favorable outcomes in colon and breast carcinomas (Whiteside 2015). Tumor-associated macrophages (TAMs) and neutrophils also exhibit pro- and antitumoral activities (Zamarron and Chen 2011; Shalapour and Karin 2015). Neutrophils, recruited by platelets, contribute to metastatic niche formation (Labelle et al. 2014). Elevated neutrophil-to-lymphocyte ratios are linked to poor prognosis in several cancers, and G-CSF has been implicated in metastasis promotion (Ocana et al. 2017; Mouchemore et al. 2018). Furthermore, neutrophil extracellular traps (NETs), induced by tumor cells, facilitate metastasis, and NET-degrading agents have shown efficacy in preclinical models (Cools-Lartigue et al. 2014; Park et al. 2016).

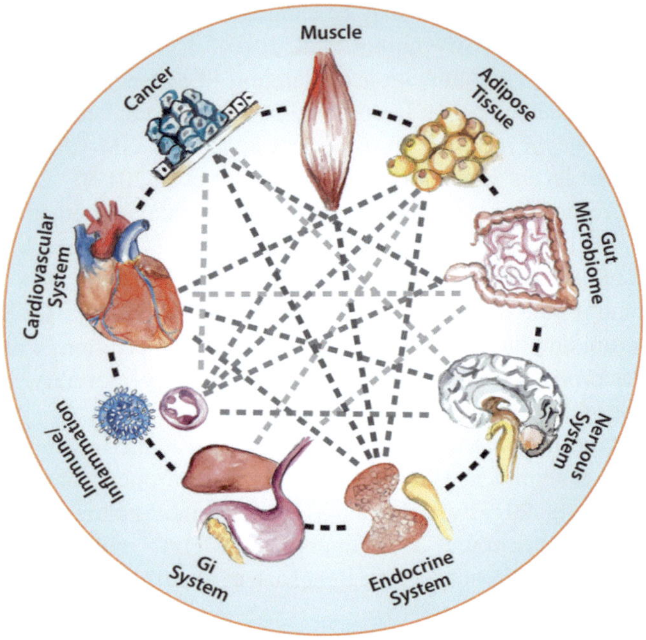

Fig. 11.6 The global metabolism/cachexia network. (Courtesy of Ana Tourian)

11.12.4 Metabolic Reprogramming and Cancer-Associated Cachexia (Please See Fig. 11.6)

Cancer cells underscore a profound metabolic reprogramming, a hallmark of cancer first observed by Otto Warburg (1956). Characteristic metabolic changes include aerobic glycolysis, glutaminolysis, and enhanced amino acid and lipid metabolism. Tumors often secrete lactate despite sufficient oxygen, a feature known as the Warburg effect, though mitochondrial respiration remains active and necessary for tumor growth (Warburg 1956; Vaupel and Mayer 2012; Davidson et al. 2016). In some cases, this lactate re-enters hepatic metabolism through the oncogenic Cori cycle, an energetically costly process (Fearon et al. 2012; Payen et al. 2016; Porporato 2016) with a significant impact on the whole-body metabolism. Clinically, 70–80% of patients with gastric and pancreatic cancer develop cachexia that is associated with an increase in the resting energy rates (RER) (Argilés et al. 2018; Kim et al. 2013). The changes in RER increase slowly, though during cancer progression, and several studies described similar RER in normal vs cancer patients (Reeves et al. 2006; Johnson et al. 2008; Cao et al. 2010). In the last 3 months of life of patients with colorectal cancer and terminal cachexia though, due to an increase in the tumor masses and in the metabolic rates, especially in the liver and tumor tissues, the RER was found to be sometimes even higher than 17,000 kcal/day (Lieffers et al. 2009).

Beyond glucose, tumors depend on glutamine, glycine, aspartate, serine, branched-chain amino acids (BCAAs), acetate, and lipids to sustain nucleotide, protein, and membrane biosynthesis (Vander Heiden and DeBerardinis 2017). In some tumors, glycolysis and oxidative phosphorylation operate concurrently (Alam et al. 2016). Notably, the "reverse Warburg effect", proposed by Pavlides and Lisanti, describes a form of tissue-level metabolic parasitism where cancer cells induce oxidative stress in stromal cells, forcing them into glycolysis and utilizing the byproducts (lactate, ketones) as fuel for tumor growth (Pavlides et al. 2010; Martinez-Outschoorn et al. 2011).

At the systemic level, tumors secrete inflammatory cytokines (i.e., TNF-α, IL-1, IL-6) and pro-cachectic factors (e.g., proteolysis-inducing factor), triggering widespread catabolism and energy loss in the host (Tsoli and Robertson 2013; Flint et al. 2016). This host-wide metabolic reprogramming, termed a "metabolic dictatorship" (Lee et al. 2016) culminates in cachexia.

In cachexia inter-organ communication forms pathologic metabolic networks, driven by tumor-derived cytokines and host responses (Wang et al. 2021).

Cachexia involves metabolic and inflammatory disruptions across multiple organs, muscle, adipose tissue, liver, pancreas, brain, gut, and heart (George et al. 2007, Argilés et al. 2014). Al-Zoughbi proposed the concepts of "metabolic cancer syndrome" and "macroenvironment" to describe this coordinated hijacking of host energy regulation (Al-Zoughbi et al. 2014).

Muscle wasting results from inflammation-induced disruptions in protein metabolism, apoptosis, and regenerative capacity (Argilés et al. 2014). Simultaneously, adipose tissue undergoes lipolysis, marked by elevated circulating fatty acids, glycerol, and triglycerides (Mantovani et al. 2000; Das et al. 2011) Pro-inflammatory cytokines such as IL-6, IL-1β, TNF-α, and TGF-β promote a white-to-brown fat transition, elevating energy expenditure—a process enhanced by tumors secreting parathyroid hormone-related protein (Kir et al. 2014; Petruzzelli et al. 2014).

Tumor-derived exosomes and microvesicles also contribute to cachexia by transmitting inflammatory signals and miRNAs. For example, it has been shown that adipocytes secrete exosomes in abundance, which are then taken up by tumor cells, leading to increased migration and invasion (Lazar et al. 2016).

Cancer-induced cachexia is not a passive byproduct of tumor progression, but a systemic metabolic reprogramming driven by tumor–host interactions. Through inflammatory and metabolic crosstalk, tumors exploit multiple organ systems to meet their energy demands—transforming the body into a metabolic host that sustains, rather than resists, cancer growth. Understanding this multi-organ pathology is critical to developing interventions that not only target tumors but also restore systemic metabolic balance.

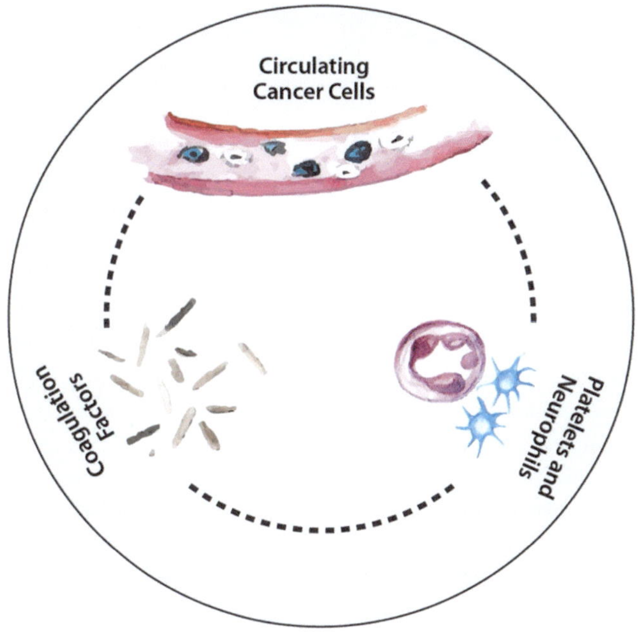

Fig. 11.7 The thrombosis network. (Courtesy of Ana Tourian)

11.12.5 Thrombosis Networks in Cancer Progression (Please See Fig. 11.7)

Thrombosis and the lymphatic system are increasingly recognized as key players in metastasis, with ~20% of cancer patients experiencing thrombotic events (Khorana et al. 2007). A bidirectional relationship exists between cancer and thrombosis: tumors promote coagulation, while thrombotic elements enhance tumor progression (Falanga et al. 2017). Cancer induces a hypercoagulable state via procoagulant release (e.g., tissue factor, TF) and activation of endothelial cells, platelets, and leukocytes. Cancer cells also directly engage host cells, amplifying thrombogenic responses.

TF is central to cancer-associated thrombosis, expressed by tumor cells or induced in vascular tissues through soluble mediators and cell contact. Its regulation is linked to oncogenic pathways (EMT, TGF—β, EGFR, PTEN, Src, hypoxia) (van den Berg et al. 2012). TF is abundant in many epithelial cancers and may influence tumor dormancy by recruiting stromal elements (Magnus et al. 2014).

Platelet and neutrophil interactions facilitate tumor cell adhesion, extravasation, and niche formation (Labelle and Hynes 2012). Platelet-derived signals are required for the rapid intravascular recruitment of neutrophils to circulating tumor cells (CTCs) thrombi, contributing to early metastatic niches (Labelle et al. 2014). Platelet-derived TGF-β activates NF-κB in circulating tumor cells (CTCs), promoting EMT (Labelle et al. 2011). There are three members of the selectin family: P-selectin expressed on activated platelets and endothelial cells, L-selectin present on leukocytes, and E-selectin expressed on activated endothelial cells (Läubli and Borsig 2010a). Selectins (P-, L-, E-) mediate tumor cell aggregation and vascular adhesion; P-selectin notably enhances lung metastasis, as shown by its absence reducing tumor burden in mice (Becker et al. 2016). Integrin blockade on platelets reduces pulmonary colonization, indicating anti-metastatic potential (Lavergne et al. 2017).

Recently, it has been found that pro-thrombotic small extracellular vesicles (sEVs) are released by CXCL13-reprogrammed interstitial macrophages within the non-metastatic lung microenvironment in various cancers, forming a specialized pro-thrombotic niche (PTN) (Lucotti et al. 2025). These vesicles are enriched in clustered integrin β2, which forms dimers with integrin αX and binds to platelet glycoprotein Ib (GPIb), thereby promoting platelet aggregation. Inhibition of integrin β2 reduces both sEV-induced thrombosis and lung metastasis. Notably, elevated levels of sEV-associated β2 integrin are detectable in the plasma of pancreatic ductal adenocarcinoma (PDAC) patients before thrombotic events, distinguishing them from those without a thrombosis history. These findings indicate that formation of the lung PTN is a systemic manifestation of cancer progression and identify sEV-β2 as both a prognostic biomarker for thrombotic risk and a potential therapeutic target to prevent cancer-associated thrombosis and metastasis (Lucotti et al. 2025).

11.12.6 Neuroendocrine Networks in Cancer (Please See Fig. 11.8)

The local and systemic interactions between cancer cells and cancer niches and key components of the nervous system, including neurons, astrocytes, oligodendrocytes, microglia, Schwann cells, and peripheral nerves, are emerging as important cancer drivers. Neuro-cancer dynamics can influence tumor initiation, progression, immune microenvironment modulation, and metastatic dissemination (Mancusi and Monje 2023). The CNS and autonomic nervous system critically influence tumorigenesis (Jobling et al. 2015). Stress-related activation of the sympathetic-adrenal-medullary axis elevates catecholamines (epinephrine, norepinephrine), which suppress cellular immunity and promote tumor growth. Immune cells express receptors for stress hormones (i.e., CRH, ACTH, cortisol), enabling direct neuroimmune modulation. Sympathetic innervation of lymphoid organs further links stress and immune dysregulation. One of the most studied CNS–cancer links is psychological stress, which activates the sympathetic-adrenal-medullary axis, leading to systemic release of catecholamines (epinephrine, norepinephrine). These neurotransmitters suppress cellular immunity and promote tumor growth, inflammation, angiogenesis, epithelial-mesenchymal transition (EMT), and metastasis (Elenkov et al. 2000; Magnon 2015; Cole et al. 2015). Immune cells express receptors for key stress mediators—cortisol, CRH, ACTH, catecholamines—allowing neuroendocrine signals to modulate immune responses and disease progression (Li et al. 2010).

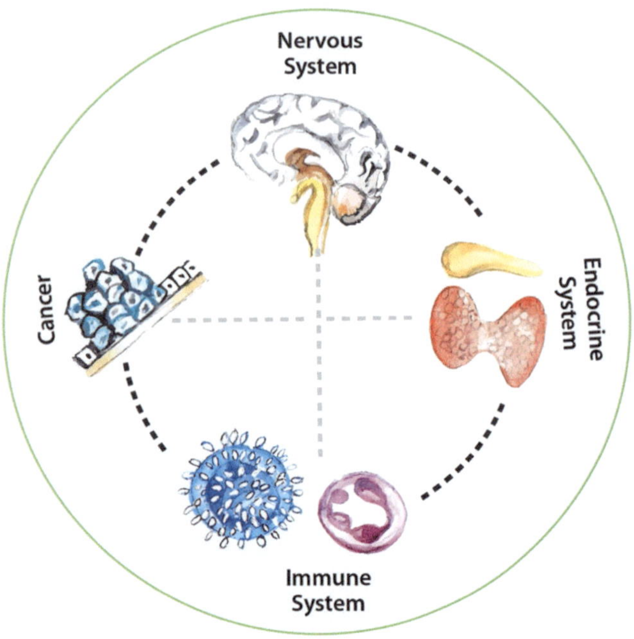

Fig. 11.8 The neuroendocrine network

For example, sympathetic innervation of the bone marrow enhances myeloid cell production, which may infiltrate tumors and facilitate metastasis (Katayama et al. 2006; Elefteriou 2016). In prostate cancer, β-adrenergic signaling via stromal receptors promotes tumor progression (Magnon et al. 2013).

Experimental models show that environmental enrichment (EE) reduces tumor burden via hypothalamic brain-derived neurotrophic factor (BDNF), which downregulates leptin via β-adrenergic sympathetic pathways. These effects are independent of physical activity and point to a central neuroendocrine mechanism influencing cancer outcomes (Cao et al. 2010).

The hypothalamus acts as a key integrator of systemic inflammatory signals in cancer. Cytokines such as IL-1β and TNF-α induce hypothalamic inflammation, disrupting autonomic and metabolic homeostasis, and activating the hypothalamic-pituitary-adrenal (HPA) axis, contributing to cachexia and muscle wasting (Braun et al. 2011; Burfeind et al. 2016, 2018; Mravec et al. 2019). Functional MRI studies have confirmed reduced hypothalamic activity in cancer-associated cachexia (Molfino et al. 2017).

A striking study in a prostate cancer model demonstrated that neural progenitor cells from the brain migrated via the circulation to tumors, where they differentiated into adrenergic neurons that supported early tumor development, suggesting tumors can recruit neural resources from the CNS (Mauffrey et al. 2019).

While the sympathetic system's role in cancer is well-established, parasympathetic influences, particularly via the vagus nerve, are now gaining attention. Cholinergic signaling regulates immune responses through the inflammatory reflex, modulating cytokine release (e.g., TNF-α, IL-2) by T and B cells (Pavlov and Tracey 2012; Fujii et al. 2017). However, parasympathetic input may also promote cancer: vagotomy in mice suppressed gastric tumorigenesis, and blocking muscarinic receptors reduced prostate tumor invasion and improved survival (Magnon et al. 2013; Zhao et al. 2014; Yu et al. 2017).

There is a complex interplay between cholinergic and adrenergic signaling, and indiscriminate vagal inhibition may enhance sympathetic-driven cancer progression (Cole et al. 2015).

Approximately 8% of cancers are linked to paraneoplastic endocrine syndromes (Pelosof and Gerber 2010). Specific tumor types are associated with distinct hormone secretions—i.e., parathyroid hormone-related protein by squamous cell carcinomas, ACTH or bombesin-like peptides by small cell lung cancer, many of which serve as autocrine growth factors (Sherbet 2005; Castellone et al. 2015).

Beyond classical hormones like estrogens and androgens (Sherbet 2005), thyroid hormones have been shown to promote metastasis (Mousa et al. 2018), while melatonin has proven to be protective in a mouse model (Karadas et al. 2021). Low melatonin levels, linked to poor sleep in cancer patients, correlate with depression and diminished tumor control, particularly in breast cancer (Su et al. 2017; Zaki et al. 2020).

Cancer is not only a cellular or molecular disorder but a disease of systemic neuroimmune and neuroendocrine dysregulation. The nervous system, both central and autonomic, plays an active role in shaping the tumor microenvironment, systemic immunity, metabolism, and progression. Understanding and therapeutically targeting neurobiological circuits, through stress modulation, neuromodulators, or nerve–tumor interactions, may unlock novel strategies for integrated cancer therapy.

11.12.7 Cancer-Induced Systemic Pathologic Networks (CISPN)

Advancements in liquid biopsies and multi-omics technologies have revolutionized precision oncology, enabling real-time, minimally invasive profiling of tumors. These tools provide insight into tumor heterogeneity, facilitate early diagnosis, and refine prognostic and therapeutic strategies. Beyond tumor-centric analysis, there is growing recognition that systemic cancer hallmarks—features reflecting cancer's impact on the whole organism—also warrant characterization, particularly through circulating biomarkers.

A critical future direction is the identification of circulating intersystemic program nodes (CISPN)—key molecular mediators linking tumors with systemic responses. Among the most promising CISPN candidates are exosomal microRNAs (miRNAs) and proteins, which not only reflect tumor presence but may function as master regulators of metastasis and systemic dysfunction (Anfossi et al. 2018; Hoshino et al. 2020). In parallel, metabolomics, which captures global metabolic shifts driven by cancer, offers a powerful lens into systemic reprogramming.

The Consortium of Metabolomics Studies (COMETS), established in 2014, has advanced large-scale research connecting the human metabolome to disease progression (Yu et al. 2018).

Indeed, systemic metabolic changes have been observed in various advanced cancers (Xiao and Zhou 2017; Dean et al. 2018), underscoring metabolism as a core systemic hallmark of malignancy. An analysis of 9125 patient samples across 27 cancer types revealed metabolic expression subtypes based on transcriptional activity of seven metabolic processes—including amino acid, carbohydrate, lipid, and nucleotide metabolism (Peng et al. 2018).

Interestingly, these metabolic subtypes were not related to specific genetic somatic drivers but were intrinsically coupled with cancer hallmark pathways (i.e., angiogenesis, cell division, etc.) and were modulated by highly recurrent master regulators across cancer types, ultimately leading to consistent survival patterns. Subtypes enriched in carbohydrate, nucleotide, and vitamin/cofactor metabolism were associated with poor prognosis, whereas lipid metabolism-enriched tumors tended to fare better. As a proof of concept, the authors conducted in vitro experiments showing that silencing two key regulatory genes associated with carbohydrate metabolic subtypes, SNAI1 in a lung cancer cell line and RUNX1 in a sarcoma cell line, led to a significant reduction in intracellular glucose levels. These findings support the model in which master metabolic regulators function as central nodes, exerting broad control over systems-level metabolic processes. Targeting these regulators could therefore offer a strategic avenue for inhibiting tumor progression. Notably, the four master regulators identified in eight cancer types with the poorest survival outcomes (SNAI1, RUNX1, RUNX2, and FOSL1) are also pivotal in embryonic development and epithelial-mesenchymal transition (EMT) (Cieślik et al. 2013; Hong et al. 2017; Dongre and Weinberg 2019), processes that are deeply intertwined with metastatic transformation.

In my opinion, the most important step in cancerization is the co-optation not only of the local tissue, but also of the whole organism in the assistance of cancer progression. The key to curing metastatic cancer lies in a detailed understanding of the cancer-organism dialogue. In multicellular organisms, cells have programs related to intracellular activities, and also to local (tissular) and distal (organism) interactions. Once activated, these cellular programs normally used during development, transform the organism–cancer relationship into a sort of mother–embryo dynamic, the entire organism being coaxed to support cancer progression. This parasitic dynamic is the essential characteristic of cancer, an opportunistic life form, with the sole purpose of persistence. Cancer activation of embryonal programs are key survival strategies in cancer. Other cancer strategies include adaptation, clonal heterogeneity, phenotypic variation and modulation of evolvability.

The Path to Cancer Cure

12

12.1 Cancer's Homeodynamics Laws

We started this book with the premise that a better understanding on cancer's nature may lead us to a change of paradigm that will get us closer to cancer cure. So, how can we cure cancer?

Cancer is a form of life and, as a corollary, it is immediately apparent that both the life conservation law and life's principles apply to cancer. So, cancer follows the same rules as all other living processes:

1. *In the absence of factors (lack of nutrients, environmental changes) that will alter its basic conditions of existence, cancer will persist. In the right environment, cancer is able to go on indefinitely (see the HeLa cells, for example). In brief: Uninterrupted, under* in vitro *conditions, the cancer will go on indefinitely.* The persistence of the human organism is based solely on homeorhetic (growth and development) and homeostatic mechanisms, and needs reproduction to keep on transmitting the living process from generation to generation. Cancer is different. In addition to homeostatic and homeorhetic mechanisms, cancer's persistence is also due to homeopoietic mechanisms (i.e., cancer may exhibit fast-forward evolution), allowing the living process of cancer cells to go on, in theory, indefinitely. By the same token, as a unicellular living process, cancer cell's persistence mechanisms are very robust as they developed over four billion years or so, as opposed to the human organism's persistence mechanisms which are less robust; as a multicellular life form, the human organism is a much more recent event in the history of life on the Planet Earth and the multicellular living process had less time to build-up resilience mechanisms.
2. *Whenever an internal or external stressor is acting on the cancer system threatening to change its trajectory, the cancer system reacts in an attempt to maintain its trajectory.* ***The cancer system conserves its flow.*** **Conservation of flow** refers to the cancer system as whole and not to individual cancer cells. Individual cancer cells may accelerate their division rate, may enter dormancy, or may reach a relative balance between the two; flow maintenance strategies for the cancer system are diverse. These different mechanisms may be combined and form cancer cycles (Please see Fig. 12.1).
3. *Cancer and its environment co-create each other.* Initially, local and global environmental elements create a pre-primary tumor niche, launching cancer's development. In its turn, cancer contributes to the creation of local and metastatic niches that further support its own growth and spread.

There are several concepts that can be derived from the above that may get us closer to curing cancer:

First. Targeting cancer's homeodynamics.
Second. Targeting the key components of the cancer biogenerator and disrupting cancer's homeodynamics.
Third. The codependence of cancer and the environment makes interception, early detection and cancer prevention feasible. The best way to cure cancer is to prevent it from occurring or to catch it in an early curable phase. Metastatic prevention is another important strategy. The same cancer-organism codependence makes interventions directed against the micro or macro cancer environment possible treatment options. Modulating the cancer–organism interactions by curbing inflammation, inhibiting cancer exosomes secretions, targeting the cancer-induced systemic pathologic networks (CISPN), and preventing pre-metastatic niche formations are other treatment ideas. Treatments can also be directed against the internal body climate components that are modified in cancer.

D. Paul, *Cancer: An Integrative Approach*, https://doi.org/10.1007/978-3-032-08808-6_12

12.2 Cancer's Life Cycles and Treatment Intersections

Viruses, unicellular organisms, and multicellular life forms all follow distinct developmental cycles. Similarly, the human organism undergoes a series of well-defined developmental stages—from embryonic and fetal phases to childhood, adolescence, adulthood, and finally, senescence (Please see Fig. 9.6). Most cancers emerge during the adult and aging phases of life.

As outlined in the previous chapter, cancer itself appears to follow distinct developmental trajectories, notably the metastatic life cycle and the dormancy life cycle (Please see Fig. 12.1). Additionally, recent evidence suggests that circulating tumor cells (CTCs) exhibit circadian dynamics, with elevated levels observed during nighttime hours (Diamantopoulou et al. 2022). This phenomenon may reflect interactions with immune rhythms or the relative hypoxic state characteristic of sleep.

Cancers differ markedly in their temporal and spatial dynamics. Some, such as Burkitt lymphoma or aggressive leukemias, are typified by rapid cellular proliferation and short-lived persistence—what may be described as temporal dominance. Others, like well-differentiated thyroid cancers, are characterized by slow cellular turnover and long-term spatial persistence. These differences have profound therapeutic implications: rapidly dividing tumors are generally sensitive to chemotherapy, while slowly proliferating cancers often exhibit resistance to cytotoxic agents and may, in certain metastatic contexts (e.g., low-grade prostate or thyroid cancers), be safely monitored without immediate intervention.

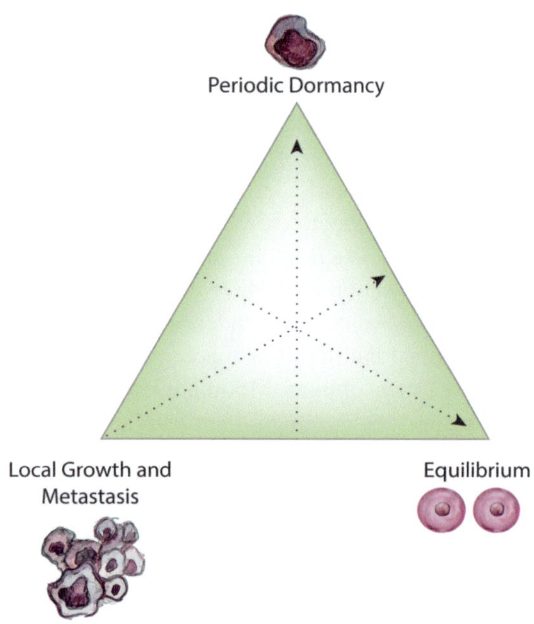

Fig. 12.1 Cancer cycle: local growth and metastasis, periodic dormancy and stasis (Courtesy of Ana Tourian)

The most therapeutically challenging tumors—glioblastoma, anaplastic thyroid carcinoma, sarcomatoid carcinomas, and certain gastrointestinal adenocarcinomas—occupy an intermediate state, exhibiting both high temporal and spatial persistence. These cancers frequently present with robust stromal components that serve as protective niches, limiting the effectiveness of conventional therapies including chemotherapy, targeted agents, and immunotherapy.

Cancer can thus be conceptualized as a complex adaptive system composed of diverse cellular populations—ranging from generalists to specialists—and spatially distributed networks involving primary tumors, metastatic sites, and supportive bone marrow niches. These components form non-random, functionally integrated networks geared not toward the survival of individual cells, but toward the persistence of the cancer system as a whole. This systemic persistence transcends the survival of any single cell population, underscoring the need to complement cell- or tissue-directed therapies with system-level strategies that disrupt the cancer's organizational architecture. This is a key point, because at present the cancer treatment strategies do not include cancer-system targeting.

In planning cancer treatment, both trajectories—the patient's and the cancer's—must be considered. Gaining insight into the cancer's evolutionary path is essential for choosing the appropriate intervention point. Different patients and cancer types—defined by stage, tempo, and biological behavior—require tailored strategies. Rapidly progressing tumors demand urgent, aggressive treatment, while indolent tumors may not benefit from intensive cytotoxic regimens.

Cancer evolution reflects a trade-off between proliferation and survival, akin to the *r/K—selection* theory in ecological systems.

Tumor environments are not uniform: the center of tumors is crowded, hypoxic, and nutrient-poor, and fluctuates in nutrients, space, and interaction between the components in the microenvironment at the edge of the tumors (Li et al. 2021c). As cell density rises, K-selected phenotypes—optimized for survival in crowded niches—tend to dominate. Although intra-tumoral competition may reduce short-term fitness, it drives long-term niche specialization, allowing distinct cell subpopulations to optimize resource use and resist external stressors, including drug therapy. Ultimately, tumor expansion is driven by the peripheral zone, governed by r-selection, while the central core serves as a crucible for the emergence of more aggressive, treatment-resistant phenotypes. These phenotypes, once disseminated to the tumor edge, fuel invasion, and further expansion. In simple terms, given the increased competition between cancer cells, the core of the cancer tumor may be responsible for the inception of more aggressive and resistant phenotypes, while at the periphery these phenotypes will be responsible for the tumor invasion and expansion. The coexistence of both r- and K-selected cell populations within tumors highlights the necessity of designing therapeutic strategies that simultaneously target the distinct vulnerabilities associated with each evolutionary mode.

12.3 The Significance of Dynamic and Creative Persistence Theory in Oncology

The theory of Life's Dynamic and Creative Persistence (DCP) expands the proposal of the Endogenous Molecular-Cellular Network Cancer concept proposed by Gaowei Wang and his collaborators, which described molecular and cellular networks centered around genes and proteins as critical hubs for the normal functioning of cells and organs (Wang et al. 2018). DCP extends this model into a more integrative and comprehensive framework, encompassing not only proteomic components, but also non-protein-coding entities such as non-coding RNAs (ncRNAs), along with metabolic and structural networks that sustain the viability and adaptability of living systems within specific environmental contexts. Importantly, DCP emphasizes both temporal continuity—the unbroken progression of life across evolutionary time—and spatial or structural persistence—the maintenance of organized life forms for varying durations. In the context of clinical oncology, two manifestations of DCP confront one another: the DCP of the host organism and the DCP of the evolving cancer process.

DCP offers a unifying paradigm that reconciles various theories of cancer, each of which traditionally foregrounds a specific domain—genetic, metabolic, or structural. Recognizing that bioenergy, bioinformation, and biostructure are in constant dynamic interplay within living systems, it becomes clear that these models are not mutually exclusive, but rather emphasize different facets of the same integrated system. For example, metabolic alterations can drive genetic and epigenetic changes as well as structural remodeling, while cytoskeletal and membrane perturbations can retroactively affect genomic stability and gene expression. Indeed, recent findings by Wang et al. have identified alterations in the cytoskeletal protein network as a novel driver of carcinogenesis, with mechanotransduction pathways linking microenvironmental properties to tumor cell behavior (Wang et al. 2022).

Within this broader systemic framework, Antonio Mazzocca's Systemic-Evolutionary Theory of Cancer (SETOC) posits that cancer arises through an uncoupling of nuclear and mitochondrial communication, leading to the "de-emergence" of the integrated eukaryotic cell system and the "re-emergence" of more primitive, archaea-like or prokaryotic subsystems characterized by dysregulated and uncoordinated activity (Mazzocca 2019). While such mitochondrial–nuclear disconnection has been documented in hepatocellular carcinoma (Mello et al. 2019) and other malignancies (Xia et al. 2019), the long-term success of cancer as a pathological system hinges on the re-establishment—albeit pathological—of a coordinated relationship among bioinformation, bioenergy, and biostructure. Cancer cells often circumvent mitochondrial dysfunction by activating alternative energy production pathways. As noted by Nick Lane, the uncoupling of mitochondria from the nucleus generally results in diminished cellular fitness—yet cancer cells defy this trend, maintaining high adaptability and proliferative capacity (Lane 2015).

In malignancy, all three elements of the DCP triad—bioinformation, bioenergy, and biostructure—undergo profound transformation. This includes genomic instability, epigenetic reprogramming, metabolic rewiring, and architectural reorganization. Moreover, the processes that sustain their integrated persistence—homeostasis, autopoiesis, and dynamic adaptability—are disrupted. Cancer severs its co-regulatory ties with the host organism and reshapes its relationship with the environment into a parasitic one. It no longer contributes to the unity of the organism, but rather subverts it. Notably, cancer cells exhibit complex, adaptive, and even creative behaviors—potentially driven by their open chromatin architecture—which allow them to innovate survival strategies, resist therapies, and evolve as a quasi-autonomous biological system.

12.4 Can We Intercept Cancer *Before* It Starts?

In biological systems, wherever structural or temporal order is preserved, it is underpinned by active control mechanisms. The organism sustains its internal coherence—homeostasis—through tight regulation of vital parameters such as oxygen saturation, pH, glucose concentration, blood pressure, and electrolyte levels. The internal equilibrium of the internal body macroclimate is intimately connected to the microclimate of tissues and the microclimate within individual cells. Consequently, maintaining a stable internal body climate directly supports both tissue and cellular homeodynamics. In this context, assessing health becomes synonymous with assessing the organism's homeostatic capacity.

It is generally accepted that a malignant mass becomes clinically detectable (on physical exam or imaging) once it reaches a volume of approximately 1 cm3 (1 cm diameter), at which point it contains upwards of 1 billion cells and weighs approximately 1 gram (Del Monte 2009). Using these doubling times, corresponding latency times of 4–25 years can be calculated for various human cancers, which are consistent with the observed clinical timeline of cancer development in humans.

As an example, let's consider the natural history of breast cancer shaped by a complex interplay of tumor biology, growth kinetics, and detection timing. Breast cancer is often perceived as a rapidly progressing disease, yet empirical data from serial imaging studies suggest that most primary tumors exhibit remarkably slow growth rates. Assuming exponential growth, a tumor originating from a single cell and doubling every 200 days would require nearly 20 years to reach a clinically detectable size of 1–2 cm. Even tumors with a more aggressive doubling time of 100 days would need approximately 10 years to reach that threshold, and only those with extremely rapid kinetics—doubling every 20 days—might become palpable within 2 years of their initiation. However, real tumor growth is rarely constant. Contemporary modeling favors Gompertzian kinetics and power laws, where growth decelerates as the tumor enlarges due to factors like hypoxia, limited angiogenesis, and immune interactions (Benzekry et al. 2014), but it has been shown that some models represent a better fit to certain cancer types than others (Sarapata and de Pillis 2014). Accordingly, preclinical latency may be shorter than linear models suggest, and early exponential growth could be followed by plateaus or spurts. Metastatic lesions appear to grow significantly faster than their primary counterparts (Kobets et al. 2000).

In a classic study, tumor growth varied considerably between subjects, with 5% of tumors taking less than 1.2 months to grow from 10 mm to 20 mm in diameter, and another 5% taking more than 6.3 years. The mean time a tumor needed to grow from 10 mm to 20 mm in diameter was estimated as 1.7 years, and the mean time increases with age (Weedon-Fekjaer et al. 2008). These estimates underscore the considerable heterogeneity in tumor biology, with some lesions growing indolently over many years and others more aggressively.

More recently, in a population-based retrospective study using the Munich Cancer Registry and analyzing 66.800 patient datasets over four time periods since 1978 and including patients enrolled in prevention and adjuvant treatment clinical trials, several principles regarding primary tumors and metastases growth were derived (Engel et al. 2022). One of the most important findings of this study was that although primary tumors may take more than a decade to grow, metastases can already be initiated by very small, 1 mm tumors; the vast majority of metastases appear to be initiated at early stages and not later on during tumor development (Lambert et al. 2017).

Recent advances in liquid biopsy technologies have spurred significant interest in the use of circulating tumor DNA (ctDNA)—fragments of tumor-derived genetic material released into the bloodstream—as a minimally invasive tool for cancer detection and monitoring. However, while ctDNA has shown great promise in detecting residual disease and monitoring treatment response in advanced cancers, its application in the early detection of asymptomatic tumors remains challenging due to fundamental biological and technical constraints.

One key limitation lies in the extremely low abundance of ctDNA in the blood during the earliest stages of tumor development. In many cases, early-stage tumors—particularly those less than 1 cm in diameter—shed insufficient quantities of DNA into circulation to be reliably detected in a standard 10 mL blood sample (Phallen et al. 2017; Wan et al. 2017; Fiala and Diamandis, 2018). The shedding of ctDNA appears to correlate with tumor volume, cellular turnover, and vascularity, all of which are typically minimal in small, indolent, or early lesions (Diehl et al. 2007; Dang and Park 2022; Panet et al. 2024).

As a result, current ctDNA assays struggle to achieve the necessary sensitivity and specificity thresholds required for population-wide screening, especially for cancers with low mutational burdens or those arising in tissues that naturally shed little DNA into circulation. These limitations have led to concerns about false negatives in early-stage cancers, where a negative ctDNA result may provide unwarranted reassurance. While technological advances—such as improved assay sensitivity, error suppression, and integration of multi-omics features (e.g., methylation, fragmentomics)—are being explored to overcome these challenges, the clinical utility of ctDNA for early cancer detection remains an area of active investigation rather than current clinical practice.

Therefore, the best strategy to curb the cancer epidemics would be the early detection of *the propensity of the organism to engender* cancer and to act system-wide, on the organism, making it less permissive for cancer development. The propensity for cancer development is closely related to systemic homeostatic imbalances.

Human plasma is estimated to contain ~4600 circulating metabolites (Psychogios et al. 2011), and these metabolites have systemic effects as they can function as signals (Wang et al. 2020b).

Systemic metabolomics—the spectrum of metabolites present in plasma at a given time—is a kind of a fingerprint of the state of the organism integrating information about different variables including nutrition, genetic and epigenetic factors, age, lifestyles, medication use, and substance abuse, environment, and disease. The main methods for detecting metabolites in the plasma include liquid chromatography (LC-MS), gas chromatography (GC-MS), and NMR spectroscopy. In the last few years, metabolome of exosomes became an intense field of research in oncology (Altadill et al. 2016; Zebrowska et al. 2019) as it has been found that exosomes are instrumental in transporting cargoes such as proteins, nucleic acids or metabolites between the tumor, stromal and immune cells in the local tumor microenvironment and systemically, thus facilitating cancer progression, angiogenesis, metastasis, drug resistance, and immunosuppression by reprogramming the metabolism of cancer cells and their surrounding stromal cells (Yang et al. 2020a; Yang et al. 2022).

The potential for systemic metabolomics for detecting patients at risk for developing pancreatic cancer was demonstrated by Mayers et al. who reported that the increase of leucine, isoleucine, and valine, collectively known as branched-chain amino acids (BCAA), is associated with a twofold increase in pancreatic ductal adenocarcinoma (PDAC) and may precede the clinical appearance of PDAC by 2–5 years (Mayers et al. 2014). These findings were substantiated by another large study done in Japan that found an association between increased levels of BCAA and increased risk of PDAC for more than 10 years before diagnosis (Katagiri et al. 2018). BCAA dysmetabolism may represent a unifying component of several mutually dependent pathological conditions, i.e., obesity, insulin resistance, and PDAC (Rossmeislová et al. 2021).

Metformin and sulphonylurea treatment results in lower BCAA levels (Walford et al. 2013) and BCAA levels decrease as HbA1c levels improve, indicating that improper glucose metabolism may contribute to elevated BCAA levels. As a result, serum BCAA levels could be also a new indicator for assessing metabolic disorders and glycemic management (Iwasa et al. 2015).

Other epidemiological studies supported the interception concept. For instance, a nested case-control study of 2248 women from the European Prospective Investigation cohort found that concentrations of arginine, asparagine, and phosphatidylcholines were inversely associated with breast cancer risk, while the acylcarnitine C2 was positively associated with disease risk (His et al. 2019). More recently, variations in levels of glycine, serine, sphingomyelin, and free carnitine were linked to endometrial cancer development (Dossus et al. 2021). Cancer-specific circulating metabolites have also been described in other types of cancer, for example, lung—(Wan et al. 2017); kidney—(Guida et al. 2021), ovarian—(Zeleznik et al. 2020); colorectal—(McCullough et al. 2021); pancreas—(Mayerle et al. 2018); prostate—(Adams et al. 2019); pointing to potentially novel biomarkers and treatments. Similarly, methylmalonic acid (MMA), a by-product of propionate metabolism, was found to be upregulated in the serum of older people and functions as a mediator of tumor progression (Gomes et al. 2020). This suggests that MMA targeting may be a novel therapeutic approach for advanced carcinomas (Gomes et al. 2020).

Recently, associations between 1463 plasma proteins and the incidence of 19 cancers were investigated in nine cancer subsites of UK Biobank participants (average 12 years follow-up). Protein–cancer associations were most prevalent for blood cancers, or in tissues with a role in the maintenance of blood composition, or with a high throughput of blood such as the liver, kidneys, and lungs. A smaller number of protein associations was found for cancers that involve organs that are not directly involved in blood composition (breast, prostate); and limited evidence for protein associations was observed in cancers of the pancreas, thyroid, lip, oral cavity, and melanoma. Importantly, significant protein-cancer associations were found in some cancer cases (mainly cancers of the blood, liver, lung, and stomach) diagnosed seven years after the blood draw. Four proteins, in particular, had strong associations and early prognostic significance for the development of three types of cancer: CD74 and TNFRSF1B for Non-Hodgkin Lymphoma (NHL), and ADAM8 and SFTPA2 for leukemia and lung cancer, respectively (Papier et al. 2024).

Besides identifying multiple associations between blood proteins and cancer risk more than seven years before cancer diagnosis, many proteins had concordant evidence from genetic analyses, suggesting they may play an active role in cancer development. Also, proteins related to early cancer processes among carriers of established cancer risk variants were also identified and may be used in the future as potential biomarkers for risk stratification and early diagnosis. A subsequent proteomic study using the UK databank identified proteins significantly related to multiple myeloma and NHL (Carrasco-Zanini et al. 2024). In August 2025, the UK Biobank Consortium released a dataset comprising whole-genome sequencing from 490,640 participants (UK Biobank Consortium 2025), offering a foundation for future AI-driven studies of genetic cancer susceptibility.

A recent study looked at the commitment point of a tumor by identifying the most recent common ancestor (MRCA). For most cancers, the MRCA was found to arise some three to six months before diagnosis, but there was a spectrum of latency, culminating in several cancer types where the MRCA was present between two and twelve years before diagnosis. Importantly, in 30% of cases, genome duplication was present between five and ten years before diagnosis, a very long time before commitment. For ovarian cancers, this timescale extended to a median of 14 years, with some cases showing lag times of multiple decades. In the immediate future, analyzing and integrating both omics (genomics, metabolomics, epigenomics, transcriptomics) data and components of the exposome data may allow specific cancer risk identification, leading to individual disease prevention and mitigation (Barouki et al. 2018; Motsinger-Reif et al. 2024).

12.5 Early Cancer Detection

Early detection of cancer offers the highest potential for curative intervention and significantly improved patient survival. Traditional diagnostic modalities, including tissue biopsies and radiologic imaging, often detect malignancies at later stages when the disease burden is higher and therapeutic efficacy diminished, and we will not discuss them here. In recent years, liquid biopsy has emerged as a minimally invasive, repeatable, and potentially cost-effective tool for early cancer detection, risk stratification, and real-time monitoring. Liquid biopsy encompasses a suite of technologies aimed at analyzing circulating tumor-derived analytes in body fluids—primarily blood—including circulating tumor DNA (ctDNA), circulating tumor cells (CTCs), extracellular vesicles (e.g., exosomes), proteins, metabolites, and nucleic acid fragment patterns (Yu et al. 2021; Connal et al. 2023; Ma et al. 2024) and even microbial biomarkers (Luo et al. 2024). Integrating multi-omics platforms with advanced imaging modalities holds promise to refine early cancer detection and tailor individualized screening paradigms.

1. *Circulating Tumor DNA (ctDNA)*

Circulating tumor DNA refers to fragments of DNA shed by tumor cells into the bloodstream, typically less than 200 base pairs in size. Detection of ctDNA provides a noninvasive avenue to monitor tumor dynamics, detect mutations, and assess minimal residual disease (MRD). For early detection, however, the low abundance of ctDNA poses a significant analytical challenge. Ultrasensitive methods such as digital droplet PCR and next-generation sequencing (NGS) platforms using deep sequencing methods (e.g., CAPP-Seq, Safe-SeqS, and TEC-Seq) have enabled the detection of ctDNA at variant allele frequencies as low as 0.01% (Newman et al. 2016; Zviran et al. 2020). Recently, it was shown that Deep Learning may improve the accuracy of ctDNA detection (Widman et al. 2024). Lung, ovarian, liver, and gastric tumors tend to shed more cell-free DNA (cfDNA) than do thyroid, breast, and prostate cancers, and early-stage and premalignant growths also release less cfDNA, so some type of early cancers may be easier to detect than others (Eisenstein 2025).

Large-scale initiatives like the Circulating Cell-free Genome Atlas (CCGA) have demonstrated that whole-genome and targeted methylation sequencing of ctDNA can detect over 50 cancer types with a single blood draw, many at early stages (Klein et al. 2021).

The CancerSeek platform used in the DETECT-A study combined ctDNA mutation detection with eight protein biomarkers, to achieve higher sensitivities for cancer detection at early stages (Cohen et al. 2018).

2. *Circulating Tumor Cells (CTCs)*

CTCs are exceedingly rare—often as few as 1–10 cells per 10 mL of whole blood—but carry critical information reflecting the molecular, transcriptomic, and phenotypic landscape of the tumor. Their presence, even in early-stage cancers, has been increasingly documented, suggesting that dissemination may precede overt clinical or radiographic evidence of disease (Pantel and Alix-Panabières 2019). In particular, epithelial-mesenchymal transition (EMT) enables early tumor cells to invade surrounding tissues and enter circulation, where they may contribute to eventual metastasis. Isolation platforms include epithelial cell adhesion molecule (EpCAM)-based immunoaffinity capture (e.g., CellSearch®), size-based filtration, and microfluidic technologies (Alix-Panabières and Pantel 2014). Several studies have provided evidence for the utility of CTCs in early-stage cancers.

The benefit of using CTCs for early cancer detection is currently being explored, since in early disease, their number is very limited and due to the low abundance, ongoing refinements in enrichment and single-cell analysis will be needed (Lawrence et al. 2023).

3. *Proteomics*

Protein biomarkers in blood have long been used in oncology (e.g., PSA, CA-125, CEA), but their specificity and sensitivity for early detection are often suboptimal. Mass spectrometry-based proteomics and aptamer-based assays (e.g., SOMAscan) have revolutionized the identification of low-abundance proteins with greater precision.

A recent study using ten proteins showed high accuracy for early-stage cancer detection both in men (area under the curve (AUC): 0.98, 95% CI 0.96, 1) and in women (AUC: 0.983, 95% CI 0.95, 1.00). This panel was able to identify 93% of stage 1 cancers among men and 84% of stage 1 cancers among women with a specificity of 99% (Budnik et al. 2024). Eighteen cancer types were analyzed as follows: bladder, breast, brain, cervical, colorectal, kidney, liver, non-small cell lung cancer, small cell lung cancer, esophagus, osteosarcoma, ovarian, pancreas, prostate, stomach, testis, thyroid, and uterus.

4. *Fragmentomics*

Fragmentomics refers to the analysis of physical characteristics of circulating DNA fragments, such as size distribution, end motifs, and nucleosome positioning. These features reflect tissue-of-origin and degradation patterns specific to tumor-derived DNA.

Aberrant fragmentation patterns have been shown to differentiate between healthy individuals and those with early cancers. Integrating machine learning with fragmentomic signatures enhances discrimination power, offering high specificity in multicancer detection (Cristiano et al. 2017). Fragmentomics motifs have emerged as orthogonal biomarkers to imaging in lung cancer, improving diagnostic accuracy when combined (Mazzone et al. 2024).

5. *Methylomics*

DNA methylation, particularly at CpG islands, is a critical epigenetic modification regulating gene expression. Tumorigenesis is associated with widespread methylation aberrations—both hypermethylation of tumor suppressor genes and global hypomethylation (Liu et al. 2023a).

ctDNA methylation profiling offers tissue-of-origin specificity, and methylation changes often precede genomic mutations during oncogenesis. Assays like GRAIL's targeted methylation sequencing use bisulfite conversion and NGS to detect minute methylomic changes associated with early tumorigenesis. Methylomics has been particularly effective for cancers like pancreatic and ovarian, which often lack specific early symptoms or reliable protein biomarkers.

6. *Metabolomics*

Metabolomics explores global metabolic signatures derived from tumor–host interactions. Alterations in amino acid metabolism, lipid profiles, and energy metabolism are hallmarks of early tumorigenesis.

Techniques such as liquid chromatography-mass spectrometry (LC-MS) and nuclear magnetic resonance (NMR) spectroscopy have identified diagnostic metabolic profiles in blood, urine, and saliva (Boumehira et al. 2022; Xu et al. 2024). A highly sensitive urine-based test to identify patients with colonic adenomatous polyps (Wang et al. 2014) was developed more than a decade ago and, after being tested on a large population (Deng et al. 2017), is currently approved for use in Canada. Several serum metabolomics platforms have been described that may help the early diagnosis of ovarian cancer (Saorin et al. 2020). Approximately 25% of pancreatic cancers are CA 19–9 negative, and early pancreatic cancer diagnostic via plasma metabolomics is also an active area of research (Perazzoli et al. 2023).

The main challenge of metabolomics is on the one hand, the contribution of multiple factors, i.e., dietary intake, microbiome, liver, and other organs to serum metabolomics and on the other hand, the chemical complexity and vast number of metabolites that exist. The best use of metabolomics is to combine it with other -omics platforms. (Schmidt et al. 2021).

7. *Exosomes* and Extracellular Vesicles

Exosomes are 30–150 nm extracellular vesicles secreted by cells, including tumor cells, and contain molecular cargo (RNA, DNA, proteins, lipids) reflective of their origin. They play critical roles in intercellular communication and can be isolated from various body fluids.

Tumor-derived exosomes carry oncogenic miRNAs, long non-coding RNAs (lncRNAs), and proteins that serve as diagnostic and prognostic biomarkers. Exosomal miRNA panels have shown promise for early detection of multiple cancer types (Kalluri and LeBleu 2020; Liu et al. 2024a; Semeradtova et al. 2025). Moreover, their stability in circulation and abundance compared to CTCs or ctDNA make them attractive candidates for scalable screening platforms.

8. *Microbe Biomarkers*

The most robust evidence linking microbial biomarkers to cancer comes from colorectal cancer (CRC). Several studies have identified an enrichment of *Fusobacterium nucleatum*, *Peptostreptococcus anaerobius*, and *Parvimonas micra* in CRC tissues and fecal samples compared to healthy individuals (Wang et al. 2016; Yu et al. 2017). These bacteria may contribute to tumorigenesis through pro-inflammatory pathways, modulation of the tumor immune microenvironment, and direct genotoxic effects. Fecal microbiota-based classifiers combining multiple bacterial taxa have achieved sensitivities and specificities comparable to or better than traditional screening tests like fecal occult blood tests (Wirbel et al. 2019).

Microbial signals have also been detected in non-gastrointestinal cancers. For example, distinct oral microbiome signatures have been associated with pancreatic cancer (Wei et al. 2020), with increased abundance of *Porphyromonas gingivalis* and *Aggregatibacter actinomycetemcomitans* linked to increased risk (Fan et al. 2018). Similarly, alterations in the lung microbiome have been correlated with lung cancer, potentially reflecting immune modulation and chronic inflammation (Greathouse et al. 2018). Intra-lung microbiota has also been found to be linked to malignancy, EGFR mutation, first-line treatment outcome, and survival of NSCLC patients (Zhang et al. 2022c).

Increasing evidence suggests that interactions between the intratumoral microbiota and the immune system can modulate host tumor immune responses, thereby influencing the effectiveness of immunotherapy (Liu et al. 2024c).

9. *Integrated Multi-Omics and Imaging Approaches*

Due to tumor heterogeneity and interindividual variability, no single biomarker or platform is likely sufficient for pan-cancer early detection. Integrated approaches combining multiple liquid biopsy modalities, such as ctDNA with exosomal miRNAs or proteomics with metabolomics, may improve diagnostic performance through complementary data (Di Sario et al. 2023).

Moreover, the combination of liquid biopsy with imaging—termed "liquid biopsy-guided imaging"—is gaining traction. For instance, a positive blood-based multi-omics test can be followed by high-resolution imaging (e.g., PET-CT, MRI, low-dose CT) focused on specific anatomical regions, thereby improving diagnostic yield and reducing false positives. Artificial intelligence (AI)-driven integration of liquid biopsy data with radiomics further augments sensitivity and localization of occult malignancies (Wei et al. 2023; Acharya and Mukhopadhyay 2024).

Liquid biopsy technologies represent a transformative paradigm for the early detection of cancer. Advances in ctDNA sequencing, CTC enrichment, proteomics, methylomics, and other omics platforms have brought us closer to accurate, minimally invasive screening strategies. When used in concert with modern imaging modalities and AI-guided analytics, these tools offer the potential to detect cancer before clinical symptoms emerge, enabling earlier intervention and improved outcomes. Ongoing clinical trials and real-world validation studies will determine how these platforms are best integrated into population-wide cancer screening programs.

10. *Multi-Cancer Early Detection Testing*

Currently, there are several cancer detection tests, but all of them are still associated with a very high rate of false positive test results. The DETECT-A study that combined a blood test with whole-body imaging had a positive predictive value (PPV) of 28%, meaning that approximately 3 in 4 positive tests were false positives (Lennon et al. 2020). The Detect-A study enrolled only women. The Pathfinder study, which used a blood-based cfDNA only, resulted in a PPV of 38%, meaning that approximately 2 in 3 tests were false positives (Schrag et al. 2023). The Pathfinder study enrolled 73,5% women. Importantly, the Cancerguard test used in the Detect-A study does not reveal the cancer location, while the Galleri test used in the Pathfinder study reveals tumor location with high accuracy (Eisenstein 2025).

As reviewed by Ros Wade et al., the diagnostic performance of currently available multi-cancer early detection (MCED) tests shows substantial variability across platforms (Wade et al. 2025). Galleri® (GRAIL, Menlo Park, CA, USA) demonstrated a sensitivity ranging from 20.8% to 66.3% and specificity between 98.4% and 99.5%, based on data from three studies. CancerSEEK (Exact Sciences, Madison, WI, USA) reported a sensitivity of 27.1% to 62.3% and specificity of 98.9% to 99.1% across two studies. SPOT-MAS™ (Gene Solutions, Ho Chi Minh City, Vietnam) exhibited a notably higher sensitivity of 72.4% to 100% and specificity between 97.0% and 99.9% in two reported studies. In a single study, Trucheck™ (Datar Cancer Genetics, Bayreuth, Germany) showed 90.0% sensitivity and 96.4% specificity, while Cancer Differentiation Analysis (AnPac Bio, Shanghai, China) reported a sensitivity of 40.0% and specificity of 97.6%. In contrast, AICS® (AminoIndex Cancer Screening; Ajinomoto, Tokyo, Japan) screens for individual cancers rather than offering a pan-cancer test, and thus, aggregate test performance metrics are not available. Across platforms, detection sensitivity tends to be lower for early-stage cancers (stages I–II) compared to later stages (III–IV), underscoring an ongoing challenge in early cancer detection. Investigating patient-relevant outcomes and the service impacts resulting from performing these tests is currently ongoing (Wade et al. 2025).

12.6 Cancer Prevention

12.6.1 Cancer Prevention: Insights from the Nordic Twin Studies

Understanding the relative contributions of genetic and environmental factors to cancer risk has long been a central question in oncology and epidemiology. The Nordic Twin Studies, particularly those conducted by researchers in Denmark, Sweden, and Finland, provide one of the most robust datasets for disentangling these influences. By comparing cancer concordance rates between monozygotic (MZ, genetically identical) and dizygotic (DZ, genetically nonidentical) twins, these studies offer empirical insights into cancer heritability across different organ systems.

A landmark study by Lichtenstein et al., which analyzed over 44,000 twin pairs from the Swedish, Danish, and Finnish twin registries, concluded that the heritability of cancer varies significantly by cancer type (Lichtenstein et al. 2000). The overall estimate suggested that approximately 26–28% of the risk of developing cancer is attributable to inherited genetic factors, while the remaining 70–75% is due to environmental and lifestyle-related influences. Notably, the study found the highest heritability in cancers such as prostate, colorectal, and breast, while other cancers, like lung cancer, for example, showed much stronger environmental associations. More recently, it was found that the lifetime risk of any hematologic malignancy was increased to 4.5% (95% CI 3.1–6.5%), conditional on hematologic malignancy in a dizygotic co-twin, and to 7.6% (95% CI 4.8–11.8%), if a monozygotic co-twin had a hematologic malignancy as compared to an incidence of hematological malignancies of 2.5% (95% CI 2.4–2.6%) in the background population (Clemmensen et al. 2021).

These findings highlight the organ-specific nature of cancer heritability. Cancers of hormonally regulated tissues and those with well-known inherited syndromes (e.g., BRCA1/2 in breast and ovarian cancer) tend to show higher genetic contribution. Conversely, malignancies closely tied to environmental exposures—such as smoking (lung cancer), or HPV infection (cervical cancer)—are predominantly driven by nongenetic factors.

Subsequent twin studies have generally confirmed these results (Mucci et al. 2016; Harris et al. 2019). For instance, Mucci et al. revisited and expanded the Nordic dataset, incorporating newer cancer diagnoses and follow-up. They affirmed the overall heritability estimates and reinforced the conclusion that most cancers arise from the interplay of modest genetic predisposition and modifiable environmental exposures. Also, as Mucci et al. have shown, familial cancer risk is across different locations, with a minority of pairs where both twins were diagnosed with cancer having cancer at the same site.

Importantly, while only a minority of cancers are caused directly by high-penetrance inherited mutations (typically 5–10% of all cancers), low-penetrance variants identified by genome-wide association studies (GWAS) may collectively increase risk but rarely do so in isolation. Therefore, the majority of sporadic cancers likely arise from a complex mosaic of stochastic events, somatic mutations, lifestyle factors, and environmental exposures, acting on a background of polygenic susceptibility (Sud et al. 2017).

12.6.2 Cancer Prevention: Insights from GLOBOCAN and the Interplay of Genetics, Lifestyle, and Infection

The GLOBOCAN database, maintained by the International Agency for Research on Cancer (IARC), offers a comprehensive snapshot of global cancer incidence and mortality. Its latest estimates reveal striking geographic variations in cancer types and burdens, reflecting complex interactions among genetic predisposition, environmental exposures, lifestyle behaviors, and infectious agents (Sung et al. 2021).

A clear example of lifestyle-driven cancer incidence is lung cancer in China, where it remains the leading cause of cancer-related death. This high burden is largely attributable to tobacco smoking, with over 300 million smokers in the country, combined with indoor air pollution from biomass fuel use and industrial emissions (Yang et al. 2015). Despite some progress in tobacco control, smoking prevalence remains high, particularly among men, fueling the lung cancer epidemic.

Infectious agents play a major role in cancer etiology in many low- and middle-income countries. In sub-Saharan Africa, for instance, hepatocellular carcinoma (HCC) ranks among the top cancers, driven predominantly by chronic hepatitis B virus (HBV) and hepatitis C virus (HCV) infections acquired perinatally or in early childhood (Samant et al. 2021). The coexistence of aflatoxin exposure from contaminated grains further exacerbates liver cancer risk in this region.

Similarly, in Eastern European countries such as Romania, Bulgaria, and Ukraine, the incidence of cervical cancer remains significantly higher than in Western Europe, a pattern strongly linked to persistent infection with high-risk human papillomavirus (HPV) strains and historically low screening coverage (Bruni et al. 2021). Despite the availability of effective HPV vaccines, implementation challenges and cultural resistance have delayed widespread immunization in some of these regions. As an example, in Romania, cervical cancer is the third most common cancer in women, its incidence rate being three times higher than the EU average after breast and colorectal cancer. According to the WHO, Romania occupies the first place in the EU regarding the mortality rates due to cervical cancer (11.23/100,000) (Todor et al. 2021).

Contrastingly, some unexpected findings in GLOBOCAN data raise interesting epidemiological questions. For example, Bolivia reports among the lowest rates of colorectal cancer worldwide, a pattern observed across several Andean countries (Sung et al. 2021). This may reflect a combination of protective dietary factors, such as high fiber intake and traditional plant-based foods, limited red and processed meat consumption, and potentially underdiagnosis due to limited screening infrastructure. Nonetheless, the relative rarity of colorectal cancer in some low-income settings invites further study into diet–microbiome–cancer interactions and underscores the need to distinguish between low incidence and under-reporting.

Genetic predisposition also shapes cancer patterns, though often in synergy with environmental influences. In Japan and Korea, the incidence of gastric cancer remains high despite declining trends, likely due to a historical burden of *Helicobacter pylori* infection, dietary salt intake, and host genetic susceptibility, including polymorphisms in inflammatory pathway genes (Tsugane and Sasazuki 2007; Fang et al. 2015).

Altogether, the GLOBOCAN data underscore the diverse etiological landscape of cancer across the globe, illustrating how local context—be it viral prevalence, lifestyle, diet, or healthcare access—modulates risk. These insights not only inform cancer prevention strategies tailored to national needs but also highlight the global interconnectedness of oncogenic factors—from tobacco industry marketing to vaccine implementations and nutrition transitions.

12.6.3 Cancer Prevention: Insights from Canine and Feline Malignancies

Cancer is a leading cause of death in companion animals, particularly in aging dogs and cats. Among the various types of neoplasms, sarcomas—malignancies arising from connective tissues such as bone, muscle, and fat—are disproportionately common in both species, particularly in certain breeds and anatomical sites. This observation provides valuable insight not only into veterinary oncology but also into the broader interplay between environmental exposure and cancer susceptibility across species.

Cancer is the most common cause of death in adult dogs, and the overall cancer incidence has been estimated at over 1 in 4, with rates increasing dramatically with age (Schwartz et al. 2022; Rafalko et al. 2023; Fonti et al. 2024). Among large-breed dogs, osteosarcoma and soft tissue sarcomas are especially prevalent. As an example, ~50% of Irish Water Spaniels and Flat-Coated Retrievers die of cancer, whereas cancer-related mortality is significantly lower in breeds such as Shih Tzus and Dachshunds (Rafalko et al. 2023).

Interestingly, in dogs, lymphomas, lymphoid leukemias, and sarcomas are much more common than solid tumors and in a large American study, represented 63% of the total cases (Rafalko et al. 2023). Also, similar to humans, leukemias, lymphomas, and sarcomas tend to occur at an earlier age (i.e., <8 years), whereas malignant melanoma and cancers of the mammary gland, lung, and urinary bladder or urethra had median ages at diagnosis of ≥11 years. Intact females showed a significantly younger age at malignant tumor diagnosis compared to their neutered counterparts (9.0 vs. 10.0; $p < 0.001$). The same result is observed in males (9.6 vs. 10.0; $p < 0.001$), pointing to the importance of the hormonal environment in cancer incidence (Fonti et al. 2024).

In cats, while cancer is somewhat less common overall, injection-site sarcomas (ISSs)—often fibrosarcomas—typically occurring at vaccine or injection sites have been described for more than three decades (Hendrick and Goldschmidt 1991). The incidence of Feline Injection-Site Sarcoma (FISS) is very high. It has been estimated to be one to four in every 10,000 vaccinated cats in the United States (Hartmann et al. 2023).

The high incidence of sarcomas in dogs and cats highlights a key principle: the interaction between genetic predisposition and environmental or iatrogenic triggers in cancer development. For example, in cats, the occurrence of fibrosarcomas following vaccination or depot injections reflects how chronic localized inflammation can drive a malignant transformation. Similarly, in dogs, the elevated risk of bone sarcomas in large breeds may be exacerbated by selective breeding, rapid growth rates, and possibly exposure to environmental carcinogens in urban or household settings (Hartmann et al. 2023).

Importantly, these cancer patterns are rarely seen in wild animals, including feral populations of the same species. The discrepancy suggests that domesticated animals, like humans, are subject to unique environmental conditions—including processed food, air and soil pollutants, synthetic chemicals, and medical interventions—that may contribute to the elevated cancer burden. Studies comparing tumor types in domestic pets and wild mammals have shown that wild animals tend to develop fewer epithelial and connective tissue cancers, and when tumors are present, they often arise from infectious etiologies rather than environmental toxins or chronic low-grade inflammation (Pesavento et al. 2018).

This contrast underscores the notion that the anthropogenic environment—urban pollution, indoor air contaminants, household chemicals, and chronic exposure to synthetic compounds—may be a shared carcinogenic milieu for both humans and their companion animals, but not for their wild counterparts. Thus, cancer in pets may serve as an early warning system, or "sentinel,"—a canary in the mine—for environmental risks that are otherwise difficult to quantify in humans due to longer latency periods and complex exposure histories (Backer et al. 2001).

12.6.4 Cancer Prevention Strategies

Given the contribution of the whole organism to cancer inception and progression and the relatively long duration of cancer development—several years and even decades—the best way to address the current cancer epidemic is cancer prevention. It has been demonstrated that metastatic clones may occur early in tumor development, often in preclinical stages, so prevention efforts should be directed to tumorigenesis per se since, once cancer becomes clinical, it may be too late to address metastatic spread (Zhao et al. 2016).

Estimation of the relative proportion of genetic and environmental influences for each specific cancer showed that for all cancer types except thyroid cancers, environmental factors (including lifestyle factors) predominate (Czene et al. 2002; Irigaray et al. 2007). The original estimate of the role of the hereditary genetic contribution to cancer incidence appears to be 5–10% (Vogelstein and Kinzler 1998; Anand et al. 2008). A genome-wide complex trait analysis for common cancers at nine anatomical sites in Chinese population (14,629 cancer cases vs. 17,554 controls) estimated heritability based on the common SNPs and described higher figures for heritability for nine cancer sites: gastric cancer (20.26%), esophageal squamous cell carcinoma (19.86%), colorectal cancer (16.30%), lung cancer (LC) (15.17%), and epithelial ovarian cancer (13.31%); and a similar heritability around 10% for hepatitis B virus-related hepatocellular carcinoma, prostate cancer, breast cancer, and nasopharyngeal carcinoma (Dai et al. 2017).

Almost half a century ago, epidemiologic studies demonstrated that environment and lifestyle are responsible for 40–80% of cancers (Higginson and Muir 1977), and the differences in number appear to be due mainly to how "environment" is defined and to the variation in population taken into consideration (McGuinn et al. 2012). The term "exposome" was introduced to clarify what exactly "environment" means to define an individual's environmental exposure (Wild 2005).

In terms of the environmental factors, the most important cited factor is diet (30–35%), followed by smoking (25–30%), and infections (15–20%), with remaining factors such as radiation, environmental pollutants, sedentarism, and stress (Anand et al. 2008). Prevention programs of the most common cancer types (i.e., breast exams and mammograms in women, PSA testing, colonoscopies, low resolution CT scans in prior and active smokers, and PAP smears) are very important for detecting cancer in its earlier, treatable stages, but, probably more impactful, is a lifestyle without smoking, alcohol abuse, obesity, excessive exposure to sunlight, unhealthy stress and poor diet. Cancer prevention through vaccines and treatment of viral and bacterial infections, and prevention of carcinogen exposure are also key interventions that may help reduce cancer incidence.

Over the last decade, a group from Harvard led by George Church has been tinkering with the genome of prokaryotes and eukaryotes, trying to modify them in such a way that they become resistant to viral infections (Church and Regis 2012). In the same vein, but a lot safer, there are general lifestyle interventions able to change the body's internal climate and make the organism less prone to cancer.

Here are some examples of lifestyle interventions that can be implemented for cancer prevention:

1. *Paying attention to what we eat.* In the United States, it is estimated that obesity accounts for about 4.7% of new cases of cancer in men and 9.6% new cases of cancer in women (Islami et al. 2019). The International Agency for Research on Cancer (IARC) 2020 report found strong evidence for obesity and the risk of 13 different cancers: post-menopausal breast, colorectal, endometrial, esophageal, pancreatic, renal, liver, stomach, gallbladder, ovarian, thyroid, multiple myeloma, and meningioma (Wild et al. 2020). There is a clear link between cancer and inflammation in which the gut microbiota plays a key role. Avoiding high quantities of animal protein and introducing vegetables and plants containing natural anti-inflammatory ingredients in the diet is important both for cancer prevention and treatment. Tea polyphenols, and other compounds like sulforaphane, genistein, resveratrol, and curcumin, found in foods like green tea, cruciferous vegetables, soybeans, grapes, and turmeric, respectively, are some of the epigenetic diet compounds that have demonstrated their effectiveness against cancer and should be part of a health-promoting diet (Aslam et al. 2024). *Food used properly may become medicine.*

   A range of lifestyle interventions have been proposed to reduce the risk of obesity and, by extension, cancer. Four primary strategies are discussed below:

1.1 Regular Meal Timing and Circadian Synchronization

Maintaining consistent eating schedules enhances circadian rhythm alignment. Reinforcing the host's circadian timing system through fixed meal times was shown to induce 24-hour rhythmic gene expression even in tumors with clock gene deficiencies, resulting in inhibited cancer growth (Li et al. 2010). A decade later, these findings were corroborated by subsequent studies (Das et al. 2021). Circadian regulation of immune cells has also been well documented, with studies demonstrating that circulating CD4+ and CD8+ T cell counts are nearly twice as high in the evening compared to the morning (Lévi et al. 1998; Druzd et al. 2017; Scheiermann et al. 2018; Downton et al. 2020). Moreover, it has recently been shown that administering immunotherapy in the morning or early afternoon nearly doubled overall survival compared to administration later in the day (Qian et al. 2021; Karaboué et al. 2024). Additionally, key circadian hormones—such as melatonin, testosterone, and glucocorticoids—have been found to modulate circulating tumor cell (CTC) dynamics, with most spontaneous CTC intravasation occurring during sleep. This supports a rationale for circadian-timed therapies, particularly in cancers prone to metastasis (Diamantopoulou et al. 2022).

1.2 Intermittent Fasting as an Immunomodulatory Strategy

Interestingly, caloric restriction—by decreasing the food intake by 30–40% has been shown across multiple studies to significantly inhibit the growth of diverse tumor types, including breast, lung, prostate, brain, pancreatic, and colorectal cancer (reviewed in Levine et al. 2014).

Intermittent fasting (IF) is defined as a dietary strategy that cycles between periods of voluntary fasting (no or minimal caloric intake) and eating over a defined period of time. Unlike traditional diets that focus on what to eat, intermittent fasting emphasizes when to eat.

IF typically includes the following regimens:

- Time-restricted feeding (TRF): Eating is confined to a daily window of 4–12 hours (e.g., 16:8 fasting:eating hours). The remaining hours constitute the fasting period (e.g., 16-hour overnight fast). Aligned with circadian rhythms, this is the most widely studied and practiced IF pattern.
- Alternate-day fasting (ADF): Fasting every other day, either completely (zero calories) or with very low caloric intake (~25% of daily needs).

- 5:2 diet: Five days of normal eating per week, combined with two non-consecutive days of caloric restriction (~500–600 kcal/day).
- Periodic prolonged fasting: Involves fasting for 24–72 hours or more, performed less frequently (e.g., monthly), and typically under medical supervision.

Preclinical models demonstrated that IF can enhance immune-mediated tumor surveillance and restructure metabolic and inflammatory pathways unfavorable to cancer (Longo and Panda 2016; Delconte et al. 2024). Intermittent fasting seems to confer health benefits to a greater extent than can be attributed just to a reduction in caloric intake (de Cabo and Mattson 2019). In theory, IF interventions may play a role in cancer prevention and cancer recurrence, but currently randomized human studies are lacking (Patterson et al. 2015; Chaix et al. 2019; Tiwari et al. 2022; Faris et al. 2025).

1.3 Adherence to a Balanced Diet

Nutriments are energy, and the quantity and quality of diet impact cancer development and progression. Universal cancer prevention diet recommendations are difficult to formulate because there is a high variability between different cancer types and, sometimes, between primary tumors and their metastases. The China study investigated the impact of nutrition on health and disease (Campbell and Campbell 2004). Dr. Campbell estimated that protein makes up about 15% of American diets, with 80% of these proteins coming from animal sources; whereas the diets of rural Chinese people contain only 10% of protein, of which only 10% was from animal sources. Interestingly, Chinese people consumed more calories on average than Americans, but had lower weights. The study suggested that low-protein diets (5% or less) prevent dangerous carcinogens from forming and may decrease the cancer risk. The authors' conclusions were that a diet higher in *animal protein* may be related to cancer and other chronic conditions, and plant-based diets may prevent some cancer types and other chronic conditions.

The main criticism of the study was related to the correlation fallacy, i.e., correlation and causation are distinct and should not be confused. A strong statistical correlation of two events should not be used as proof of their causal relationship (Minger 2010).

Another study helped settle the question. The European Prospective Investigation into Cancer and Nutrition (EPIC) was a multicenter prospective study carried out in 23 centers from 10 European Western countries including 519,978 subjects (366,521 women and 153,457 men), most aged 35–70 years, specifically designed to investigate the relationship between diet and cancer. The five main findings were:

(a) Consumption of fruit is negatively associated with cancer of the lung, but probably not with prostate cancer and breast cancer.
(b) Consumption of vegetables, mainly onion and garlic, probably reduces the risk of intestinal stomach cancer, but probably is not associated with cancer of the lung, prostate, and breast cancer.
(c) Consumption of red and processed meat is positively associated with colorectal cancer and with non-cardia stomach cancer in those infected by *Helicobacter pylori.*
(d) Fish intake is negatively associated with colorectal cancer risk.
(e) High alcohol intake increases the risk of breast cancer (Gonzalez and Riboli 2006).

We see from the EPIC study that, except for some protective effect of fruit consumption on lung cancer and a detrimental effect of alcohol abuse on breast cancer, nutrition habits mainly affect gastrointestinal (GI) malignancies. GI malignancies represent more than a quarter of all cancers, and approximately 35% of cancer-related deaths are from GI malignancies. Given their high rate of mortality, the preventive effect of nutrition is very important in this cancer subset.

Diet composition also significantly influences cancer risk. A large study involving over 6000 participants found that individuals aged 50–65 who consumed high-protein diets had a fourfold increased cancer risk compared to those on low-protein diets. Interestingly, above age 66, subjects on a low-protein diet had an increased cancer risk. In this age group, subjects with high protein consumption had a 60% reduction in cancer mortality. These effects appear to be mediated by IGF-1: for every 10 ng/mL increase in IGF-1, cancer mortality risk increased by 9% among 50–65 year-olds on high-protein diets, while no such association was observed in older participants. Individuals with growth hormone receptor deficiency (GHRD), characterized by low serum IGF-1 and insulin levels, exhibited significantly reduced cancer and diabetes mortality. In order to optimize health-span and longevity, the authors recommended a low protein intake during middle age, followed by a moderate to high protein consumption in older adults (Levine et al. 2014).

More recently, a systematic review of 32 prospective cohort studies and a meta-analysis of 31 prospective studies concluded that intake of total and animal protein was not significantly associated with risk of cancer mortality. Then, the same study reported a significant inverse dose-response association between intake of plant protein and all-cause mortality but found no association between intake of plant protein and cancer mortality (Naghshi et al. 2020).

In contrast, evidence linking carbohydrate intake with cancer is limited. While dietary fiber intake is consistently associated with a reduced risk of colorectal cancer and improved clinical outcomes, the influence of dietary sugars remains less well defined. Strikingly, although vilified by the media, the evidence of an association between carbohydrate consumption and cancer risk is very scant. A common assumption is that cancer cells are dependent on their access to glucose for growth and reproduction, and controlling blood glucose at an optimal level in cancer patients may be a practical way to avoid increasing the size of a tumor (Gatenby and Gillies 2004), but solid data proving this assumption is currently missing. Limited data suggest an increased risk of colorectal cancer associated with high intakes of sucrose and an increased risk of ovarian cancer associated with high intakes of lactose (Key and Spencer 2007). Also, several studies described a relationship between hyperglycemia and cancer at the cellular, tissular, and organismic levels, but at the same time, recent Mendelian randomization studies proved a significant causal relationship only between hyperglycemia and breast cancer (Stan and Paul 2024). There is some data that increased fiber consumption is associated with reduced risk of cancer development and mortality after diagnosis (Maino et al. 2019).

Fatty acids support membrane biosynthesis during rapid proliferation and are an important energy source for cancer cells during conditions of metabolic stress (Koundouros and Poulogiannis 2020). Omega-6 (n-6) polyunsaturated fatty acids (PUFA) (e.g., arachidonic acid (AA)) and omega-3 (n-3) PUFA (e.g., eicosapentaenoic acid (EPA)) are precursors to potent lipid mediator signaling molecules, termed "eicosanoids," which have important roles in the regulation of inflammation. In general, eicosanoids derived from n-6 PUFA are proinflammatory, while eicosanoids derived from n-3 PUFA are anti-inflammatory (Patterson et al. 2012). An extensive systematic review assessing the relationship between PUFAs consumption and cancer risk found that increasing total PUFA may very slightly increase cancer risk, offset by small protective effects on cardiovascular diseases (Hanson et al. 2020). On the other hand, it has been previously shown that a low dietary ratio of omega-6 to omega-3 fatty acids may delay progression of prostate cancer (Apte et al. 2013). Also, dietary lipids can promote metastasis through fatty acids interacting with the fatty-acid receptor CD36 on metastasis-initiating cells (Pascual et al. 2017). Furthermore, it has been recently shown that dietary palmitic acid (PA), but not oleic acid or linoleic acid, promotes metastasis in oral carcinomas and melanoma in mice (Pascual et al. 2021).

Cholesterol, on the other hand, has an ambiguous role in oncogenesis. A major meta-analysis found no clear preventive effect of cholesterol-lowering strategies (Emberson et al. 2012). However, a case-control study involving 295,925 cancer patients suggested a link between statin use and a slight reduction in cancer-related mortality for 13 different cancer types, including prostate, colon, lung, and urinary (Nielsen et al. 2012). These conflicting results may be due to the heterogeneity of lipid metabolism in different cancer types—in some cancer cases, cholesterol-synthesizing enzymes may be upregulated, in others downregulated (Yang et al. 2020b). For instance, ATP-binding cassette transporter A-1 (ABCA1) is upregulated in melanoma but downregulated in colon cancer (Yang et al. 2020). In some cancers, like pancreatic cancer, combining inhibitors of low-density lipoprotein receptor (LDLR) with chemotherapy may lead to synergistic antitumor effects (Guillaumond et al. 2015).

These associations may be mediated by the gut microbiota, whose composition and function are shaped by dietary inputs and play a pivotal role in modulating cancer susceptibility. Diet, as an extrinsic climate factor, modulates the activity of the host and gut microbiota synchronously and influences their interaction. For example, a high-fat diet enhanced gene expressions of both the host and microbiota (Abu-Toamih Atamni et al. 2019; Yao et al. 2020). The gut microbiota could induce or modulate the signal transmission directly or indirectly to affect energy homeostasis (Rosenbaum et al. 2015).

The microbiota is governed by a distinct regulatory system referred to as "commensal homeostasis" (Zhang et al. 2018; Chen et al. 2024b), a finely tuned balance maintained through continuous interaction with host physiology and environmental factors. Intriguingly, the host appears capable of governing microbiome stability, using physiological parameters, such as oxygen and nitrate gradients along the intestinal axis, or shifts in bile acid profiles, as sensors of microbial balance (Lee et al. 2022). In prostate cancer, microbiota can also directly affect cancer progression and therapy by producing androgens that may promote prostate tumor growth (Pernigoni et al. 2021).

Diet acts as a master regulator within this axis, coordinately modulating the expression of host and microbial genes. For example, high-fat diets have been shown to upregulate gene activity in both microbial and host compartments (Abu-Toamih Atamni et al. 2019; Yao et al. 2020). In return, the microbiome exerts regulatory control over host energy metabolism via both direct molecular interactions and secondary signaling pathways (Rosenbaum et al. 2015). Importantly, variations in dietary patterns give rise to distinct microbial ecologies, which have been linked to differing levels of colorectal cancer risk (Hoang et al. 2023).

It is important to note that these dietary approaches are primarily validated for cancer prevention. In the context of established cancer, dietary personalization is more complex, given the heterogeneity of cancer metabolism across individuals, tumor sites, and metastatic lesions.

1.4 Reducing Systemic Inflammation through Diet

Inflammation is a central mediator linking carcinogenic and anticarcinogenic factors (Anand et al. 2008). NF-κB acts as a transcriptional master regulator of inflammatory pathways, integrating pro-inflammatory signals (e.g., TNF, IL-1, IL-6, COX-2, 5-LOX, matrix metalloproteinase-9, intercellular adhesion molecule 1), endothelium leukocyte adhesion molecule 1, and vascular cell adhesion molecule 1 and controlling the expression of other gene products linked with tumorigenesis (c-myc, cyclin D1, Bcl-2, Bcl-xL, survivin, cFLIP, and TRAF-1, vascular endothelial growth factor, etc). A dietary inflammatory index has been developed based on the effect (or lack of effect) of six inflammatory biomarkers: IL-1β, IL-4, IL-6, IL-10, TNF-α and C-reactive protein (Shivappa et al. 2014). Anti-inflammatory food recipes can be designed using this information.

2. Regular Physical Activity

The link between physical activity and cancer prevention has long been proposed. Siversten and Dahlstrom first observed a lower cancer incidence among physically active farmers over a century ago (Siversten and Dahlstrom 1921). Independent of physical activity levels, sedentary behavior has been associated with increased risks for colon, endometrial, and lung cancers (Katzmarzyk et al. 2019). Physical activity lowers cancer risk through multiple mechanisms, including immune-modulatory system modulation, reduction in circulating estrogen levels, decreased oxidative stress, insulin resistance, inflammation, and epigenetic alterations, decreased telomere attrition, and enhances DNA repair mechanisms (Amraiz et al. 2024). Multiple studies confirm the preventive role of exercise for several cancers, including bladder, breast, colon, endometrial, esophageal adenocarcinoma, renal, and gastric cancers (McTiernan et al. 2019; Patel et al. 2019). Evidence for colon and breast cancer is particularly robust (Rezende et al. 2018). A recent study reported that in breast cancer survivors, engaging in 2.5–5 hours per week of moderate-intensity exercise or 1.25–2.5 hours per week of vigorous-intensity aerobic exercise weekly reduced recurrence and cancer-specific mortality by more than 40% (Cannioto et al. 2021).

3. Optimizing Sleep

Disruption of sleep patterns may promote tumorigenesis through effects on energy regulation, immune function, and stress hormone release (Berisha et al. 2022). Epidemiological data have linked night shift work with increased breast cancer risk (IARC Monographs, Vol. 124, 2019), an association further supported by a Finnish co-twin study (Schernhammer et al. 2023). In a Japanese prospective study of nearly 24,000 women, those who slept ≤6 hours had a significantly higher breast cancer risk (HR 1.62), while longer sleep duration (≥9 hours) was associated with a lower risk (HR 0.72) (Kakizaki et al. 2008). The authors hypothesized that reduced melatonin secretion underlies this association. However, a Chinese meta-analysis found no consistent link between sleep duration and cancer risk (Chen et al. 2018). More recent data suggest that chronotype—being a "night owl" versus a "morning lark"—rather than sleep duration itself, may influence breast cancer risk (Hurley et al. 2019; Von Behren et al. 2024).

4. Stress Reduction

A major contributing factor to rising cancer incidence in industrialized societies may be the chronic stress associated with prolonged work hours and psychosocial burden. It has been recently shown in a mouse model that chronic stress increases metastasis via neutrophil-mediated changes to the microenvironment (He et al. 2024). In humans, there is also evidence that stress may promote cancer progression, partly via neuroimmune pathways (Lempesis et al. 2023). Neurotransmitters and neuropeptides released during stress can directly modulate immune cell function and inflammatory signaling, fostering a tumor-permissive environment. While stress reduction is beneficial for overall health and may indirectly influence cancer risk, the evidence for a direct causal link between stress reduction and cancer prevention is still inconclusive (Gosain et al. 2020). Adopting healthy lifestyle habits, managing stress effectively, and engaging in regular exercise are recommended for overall well-being and may play a role in cancer prevention.

12.7 Cancer Prevention Through Climate Stewardship of the Human Body

Cancer prevention is traditionally framed around lifestyle interventions aimed at minimizing mutagenic exposure-avoiding ultraviolet radiation, refraining from tobacco use, and limiting contact with chemical carcinogens. While these measures are critical, the framework developed in this book advocates for a broader and more systemic perspective: the cultivation and preservation of a *healthy internal climate* within the human organism. This approach emphasizes two central strategies: (A) fostering internal conditions that actively buffer against mutagenic stress by reducing inflammation, oxidative stress, and metabolic dysregulation, and (B) preventing internal climate disruptions that create fertile ground for tumorigenesis and cancer progression.

Except for a small fraction of hereditary cancers, the vast majority—potentially more than 90% (the other 10% being related to hereditary factors)—of cancers can be traced back to *disruptions in the internal body climate*. This includes malignancies clearly initiated by external factors—such as small-cell lung cancer driven by tobacco exposure or cervical adenocarcinoma induced by human papillomavirus—as well as cancers whose genesis and growth are deeply entangled with internal shifts in hormonal, immune, and metabolic equilibrium, such as breast cancer in women and prostate cancer in men. Both *external exposures* and *internal physiological* states shape the genomic landscape of the organism and its vulnerability to malignant transformation.

Mutations arising from environmental insults, radiation, or endogenous reactive species may act as seeds of disease, but whether those seeds take root depends largely on the internal terrain. The organism's climate determines whether a mutation remains neutral, becomes oncogenic, or is silenced by intrinsic repair and immune mechanisms. Most individuals exposed to carcinogens do not develop cancer, a fact that underscores the critical role of modulation by the internal body climate.

Take oxidative stress, for instance. Free radicals generated by smoking, UV radiation, or industrial chemicals can inflict DNA damage that may lead to mutations. Yet, the body is not a passive recipient, it produces endogenous antioxidants and extracts others from the diet to neutralize oxidative threats. This capacity to maintain redox balance acts as a line of defense against carcinogenesis.

However, once a tumor is established, the equation shifts. Antioxidants may now serve the enemy, shielding cancer cells from oxidative damage and supporting their survival. Notably, the *source* of antioxidants matters. While dietary intake of beta-carotene has shown a mild protective effect against lung cancer, supplementation in smokers was paradoxically associated with increased cancer risk. Similarly, natural dietary vitamin E appears to lower prostate cancer risk, whereas synthetic supplements have shown the opposite effect. Context is key: in one study, a specific antioxidant regimen, including coenzyme Q10, vitamins A, C, and E, selenium, folic acid, and beta-carotene, extended survival in terminal cancer patients by several months, suggesting that antioxidant therapy may have time- and condition-dependent benefits.

Diet as a whole exerts a profound influence on the body's internal climate. Specific macronutrients also modulate cancer risk and progression in nuanced ways. Several studies have demonstrated that caloric restriction, also, involving a 30–40% decrease in food intake, significantly inhibits the growth of various tumor types, including breast, lung, prostate, brain, pancreatic, and colorectal cancers. Additionally, specific dietary components can, directly or indirectly, affect cancer development (Lien and Vander Heiden 2019).

In a landmark study, a machine-learning model trained on an array of host metrics—blood biomarkers, dietary habits, body composition, physical activity, and microbiota profiles—was able to predict individual glycemic responses to real-world meals. Personalized interventions guided by this model not only reduced postprandial glucose spikes but also remodeled the gut microbiota (Rein et al. 2022).

A parallel approach holds promise in oncology. Nutritional strategies, tailored through biomarkers such as circulating tumor DNA, methylated DNA, glycoproteins, or exosomes detected via liquid biopsies, could reshape the internal terrain—recalibrating host–microbe symbiosis and contributing to climate-oriented cancer care.

Beyond nutrition, other lifestyle elements shape the body's internal ecosystem. Anti-inflammatory diets, restorative sleep, emotional regulation, and regular physical activity all support immune competence and metabolic homeostasis. Exercise, in particular, has a well-documented role in cancer prevention, reducing both incidence and recurrence across multiple cancer types. In breast cancer, physical activity enhances immune function, modulates inflammatory mediators, and lowers insulin levels, all factors implicated in tumor initiation and progression (Miyamoto et al. 2022). Similar effects have been reported for other cancers. Mechanistically, the benefits of exercise extend beyond immune support. It counters obesity, modulates insulin and hormone levels, and tempers chronic inflammation, all key components of the body's inner climate.

Taken together, this evidence suggests that cancer prevention is not only a matter of avoiding harm, but of cultivating *biological resilience*—by tending to the internal terrain in which cells live, communicate, and evolve. In this view, health is not simply the absence of disease, but the presence of a stable, adaptive, and resilient internal climate (Pruimboom and Muskiet 2018). In such a climate, cancer finds no fertile soil.

In order to persist, the multicellular organism tightly controls certain cellular functions: division, local invasion, and migration. In a certain sense, like cells protect themselves from viruses through anti-viral programs, organisms protect themselves from cells through tightly controlled growth and developmental programs. Anti-viral programs fail, and anti-cellular programs fail. The lack of protection of the organism allows cancer, a cellular life form, to develop. Similar to viruses that construct a viral fabric inside the cell, tumor cells construct both pro-primary tumor niches and pro-metastatic niches to help their development. A logical cancer prevention would be to prevent the pro-primary tumors and pro-metastatic niches from developing. It has been shown that in order for niches to develop, many contributing cells need to be involved. The main way cancer cells communicate between themselves and with neighbor or with cells located at a distance, in other organs, is through exosomes (Kaplan et al. 2005; Peinado et al. 2012; Becker et al. 2016).

12.7.1 Improving Mitochondrial Function

Targeting Mitochondrial Function in Cancer: Therapeutic Strategies and Emerging Opportunities.

Mitochondria are central to cellular metabolism, redox balance, and apoptotic signaling, making them both a vital resource and a vulnerability in cancer cells. While the Warburg effect—describing cancer cells' preference for aerobic glycolysis—has historically overshadowed the role of mitochondria, it is now widely recognized that mitochondrial metabolism remains essential for tumor progression, metastatic competence, and therapeutic resistance (Vyas et al. 2016; Martínez-Reyes and Chandel 2021; Chen et al. 2024d). Mitochondria represent a dual-edged axis in cancer biology—a target for disruption in malignant cells and a potential node of support in host defenses (Dickerson et al. 2017). Consequently, strategies aimed at restoring or disrupting mitochondrial function represent a promising frontier in cancer therapy.

On one hand, improving mitochondrial function in non-transformed host cells—such as immune or stromal cells—may enhance systemic antitumor immunity and resilience. Restoring the function of exhausted T cells to improve the response to immunotherapy through mitochondria-targeted strategies, such as metabolic remodeling, enhanced biogenesis, and regulation of antioxidant and reactive oxygen species, and activating mitochondrial mitophagy are all strategies that are actively investigated (Yang et al. 2024; Chen et al. 2024c). Dichloroacetate (DCA) has been proposed as an agent able to specifically block the aerobic glycolytic pathway and restore steady-state OXPHOS, and have been shown to be effective against various cancers in vitro and in vivo (Bhat et al. 2015). Some trials have reported reductions or stabilization in tumor size, but not consistently across all patients or cancer types, and, at present, DCA is not FDA-approved and will likely never become a standalone chemotherapeutic compound because of its associated toxicity at therapeutic doses (Koltai and Fliegel 2024).

Conversely, many cancers exhibit aberrant mitochondrial dependencies that can be selectively targeted. Mutations in isocitrate dehydrogenase (IDH1/2), for instance, confer *neomorphic* enzymatic activity that leads to the production of the oncometabolite 2-hydroxyglutarate (2-HG), which disrupts mitochondrial and epigenetic regulation.

Beyond IDH mutations, cancer cells often exhibit increased reliance on mitochondrial oxidative phosphorylation (OXPHOS), particularly in therapy-resistant and stem-like cancer populations. Inhibitors targeting components of the electron transport chain (ETC)—such as IACS-010759 (Complex I inhibitor) or metformin—have been explored for their ability to impair ATP production and induce metabolic collapse in OXPHOS-dependent tumors (Molina et al. 2018). Similarly, BCL-2 family inhibitors (e.g., venetoclax) exploit mitochondrial apoptotic priming and have shown notable activity in hematologic malignancies by restoring mitochondrial-mediated cell death. Besides venetoclax, other strategies to inhibit BCL-2 family proteins are currently being investigated including BH3-mimetics as well as proteolysis targeting chimeras (PROTACs), antibody–drug conjugates (ADCs), and tools targeting the BH4 domain of BCL2 (Vogler et al. 2025).

12.7.2 Improving Hypothalamic Function

In the context of systemic inflammation, the hypothalamus acts as a central integrative hub, receiving and processing afferent signals from peripheral tissues and translating them into neuroendocrine responses, altered neuronal signaling, and broad metabolic adjustments (Burfeind et al. 2018). This central node of homeostatic regulation becomes critically relevant in cancer, which constitutes a profound homeostatic disruption to the organism (Francis and Borniger 2021). In fact, cancer not only challenges systemic homeostasis but also actively disturbs neural circuitry, including key hypothalamic pathways responsible for maintaining physiological balance. As cancer progresses, it both reflects and exacerbates a breakdown of homeostatic mechanisms; this relationship is bidirectional. The inability of the hypothalamus to sustain organismal equilibrium in advanced disease stages is particularly evident in the onset of cancer cachexia—a multifactorial syndrome marked by systemic inflammation, neuroendocrine imbalance, and profound metabolic dysregulation (Braun et al. 2011; Burfeind et al. 2016, 2018).

This breakdown in hypothalamic regulation suggests a potential role for hypothalamus-centered therapeutic strategies aimed at monitoring or restoring systemic balance. Systemic biomarkers that reflect internal "climate" shifts—encompassing inflammatory, metabolic, hormonal, and neuroendocrine changes—can serve not only as early indicators of disease onset but also as targets for intervention designed to recalibrate hypothalamic control.

Experimental data from environmental enrichment (EE) models provide compelling support for this hypothesis. EE, which includes enhanced physical, social, and cognitive stimuli, has been shown to activate the hypothalamic-sympathoneural-adipocyte (HSA) axis (Foglesong et al. 2019). The activation of the HSA axis leads to increased expression of brain-derived neurotrophic factor (BDNF) in the hypothalamus, stimulating sympathetic nervous system activity that promotes white adipose tissue browning and reduces leptin levels, as shown by the work of Lei Cao and his collaborators (Cao et al. 2010b; Cao and During 2012; Xiao et al. 2016; Mansour et al. 2021).

Mice housed in EE displayed significantly reduced serum leptin levels and elevated hypothalamic expression of brain-derived neurotrophic factor (BDNF), a key neuropeptide implicated in energy homeostasis and neuroplasticity. Overexpression of hypothalamic BDNF in vivo recapitulated the antitumor effects of EE, while genetic knockdown of

BDNF abrogated them. Mechanistically, BDNF exerted its anti-cancer effects through the suppression of leptin synthesis in adipocytes, mediated via sympathetic β-adrenergic signaling from the hypothalamus (Cao et al. 2010b). These findings support a neuroendocrine axis in which hypothalamic outputs modulate peripheral metabolic signals with direct relevance to tumor biology (Borniger 2019; Francis and Borniger 2021).

Pro-inflammatory cytokines like IL-1β and TNF-α can alter hypothalamic function, disrupting appetite regulation and energy expenditure. Cancer and its treatments can impair hypothalamic-pituitary function, leading to hormonal deficiencies that affect metabolism and immune responses (Slominski et al. 2023; Kanter et al. 2024). Therapeutic approaches that reduce systemic inflammation may help restore hypothalamic control over metabolism, potentially mitigating cachexia and improving patient outcomes. Hormone replacement therapies targeting deficiencies in growth hormone, thyroid hormones, or cortisol may help restore systemic homeostasis. Careful assessment and management of these hormonal axes are essential in the comprehensive care of cancer patients.

Beyond its role in cancer, the hypothalamus has also been implicated in the regulation of aging. A landmark study identified hypothalamic inflammation and loss of homeostatic signaling as key drivers of age-related decline (McCarthy 1994). This observation is particularly intriguing in light of the well-established association between aging and increased cancer incidence. It is unclear whether age-associated changes in the hypothalamus may also be cancer-promoting (Hajdarovic et al. 2022). Strategies aimed at rejuvenating hypothalamic function—such as micronutrient interventions using agents like chromium picolinate—may have broader implications not only for aging-related disorders, but also for maintaining systemic resilience in cancer patients (McCarthy 1994).

Taken together, these insights support the conceptualization of an internal "body climate"—a dynamic equilibrium shaped by the interplay of internal and external forces, including stress, inflammation, metabolic flux, neuroendocrine signals, and cancer itself. When the body's internal climate deviates beyond the regulatory bandwidth maintained by the hypothalamus, systemic perturbations can arise that foster tumor initiation and progression. From this perspective, cancer prevention can be reframed as the effort to maintain homeodynamic climate stability—ensuring that internal fluctuations remain within the adaptive capacity of hypothalamic governance.

In the therapeutic realm, interventions aimed at halting or reversing this internal climate disruption hold promise. Such approaches might involve either recalibrating individual climate components—such as inflammatory mediators, energy balance, or endocrine outputs—or enhancing the function of the central "rheostat," the hypothalamus itself. Ultimately, the goal is to restore systemic resilience by reinforcing homeostatic networks, thereby impeding disease progression and potentially improving patient outcomes through both preventive and therapeutic strategies.

Therapeutic strategies that focus on restoring hypothalamic function, through environmental, pharmacological, or hormonal interventions, hold promise in re-establishing systemic homeostasis disrupted by cancer.

12.7.3 Cancer Vaccines for Prevention

Cancer vaccines represent a promising avenue for cancer prevention and immune surveillance enhancement. Traditionally, prophylactic cancer vaccines have aimed to prevent cancers caused by infectious agents. The most successful examples include the hepatitis B virus (HBV) vaccine, which reduces the risk of hepatocellular carcinoma (Chang et al. 2000), and the human papillomavirus (HPV) vaccines, such as Gardasil and Cervarix, which effectively prevent cervical, anal, and oropharyngeal cancers associated with HPV strains 16 and 18 (Cheng et al. 2020). These vaccines work by eliciting strong humoral and cellular immune responses against viral antigens, thereby halting oncogenic viral persistence and transformation.

However, the scope of cancer vaccines is now expanding beyond infection-related cancers. Based on official reported COVID-19 deaths, it has been estimated that vaccinations prevented 14.4 to 19.8 million deaths from COVID-19 in 185 countries and territories between Dec 8, 2020, and Dec 8, 2021 (Watson et al. 2022).

The success of COVID-19 vaccines has propelled renewed interest in using RNA platforms for cancer prevention in high-risk populations and therapeutic immunization in early cancer stages. RNA vaccines, particularly mRNA vaccines, have shown promise in inducing robust immune responses against various tumors. This innovative approach leverages the body's own mechanisms to produce antigens that stimulate an immune response, making it a compelling strategy for cancer prevention and treatment (Dailey et al. 2023).

12.8 Cancer Treatments

12.8.1 How Is Cancer Treated? The Way Ahead

Current cancer treatments belong to three broad categories:

- inhibition of activity of molecules present in cancer (i.e., proteins encoded by mutant cancer driver genes) either inside the cancer cells (small molecules mainly tyrosine kinase inhibitors) or on their surface (antibodies, antibody drug conjugates, etc);

- targeting the cancer-normal organism interaction i.e., stimulation of the immune system using checkpoint inhibitors.
- use of living cells (CAR-T, CAR-NK, adoptive T cell treatments) to attack the cancer cells.

These approaches are examples of *precision oncology* treatments.

All three strategies can be improved:

- new targets located inside or outside the cancer cells can be described and used for treatment,
- other cancer-normal organism interactions can be exploited. For example, finding effective agents to target cancer-induced and cancer-promoting inflammation (both local and systemic) would literally revolutionize the oncology field. Other important cancer treatment strategies include inhibiting cancer exosome secretion and cancer niche formation.
- using other modalities to improve the efficacy of immune cells targeting cancer cells. An example is the important work of Vogelstein and his collaborators that developed dual antibodies able to target HLA peptide complexes specific for cancer cells harboring TP53 and KRAS gene alterations (Hsiue et al. 2021; Douglass et al. 2021).

Differentiation therapy represents a paradigm shift in cancer treatment, aiming not to destroy cancer cells through cytotoxic means but to coax them into terminally differentiated, non-proliferative states. This approach is particularly effective in malignancies where a block in cellular differentiation is a key driver of tumorigenesis, as seen in certain leukemias. Two agents central to this strategy are all-trans retinoic acid (ATRA) and arsenic trioxide (ATO), which have transformed the treatment of acute promyelocytic leukemia (APL) (Wang and Chen 2008; Lo-Coco et al. 2013). While the success in APL has been striking, applying differentiation therapy to other cancers has proven more challenging. *Solid tumors often lack a single dominant differentiation block and possess more complex mutational landscapes*. Combination strategies, including epigenetic modulators and immunotherapy, may expand the therapeutic reach of differentiation approaches to some solid tumors (de Thé 2018).

An alternative therapeutic strategy involves leveraging the inherent plasticity of cancer cells to steer their behavior toward less aggressive phenotypes. Grounded in the framework of homeodynamic theory, such interventions—i.e., homeodynamic treatments—aim not to eradicate cancer cells, but to restore systemic balance and stability within the tumor-host ecosystem.

12.8.2 From Precision Targeting to Network Medicine and Process Oncology

The field of personalized oncology, which later evolved into precision oncology, originated from the premise that unique intracellular targets could be identified in individual tumors and therapeutically exploited. This approach was grounded in the concept of oncogene addiction, wherein cancer cells depend on specific signaling pathways driven by protein mutations or amplifications. The classification of genetic alterations into "drivers" and "passengers" reflected this logic—drivers enhance cellular fitness and promote tumorigenesis, while passengers are either neutral or deleterious.

However, it soon became evident that many driver oncogenes are highly interconnected within complex protein interaction networks. For example, TP53 and EGFR have interaction networks comprising over 300 and 250 proteins, respectively (Collavin et al. 2010; Li et al. 2013). This extensive connectivity, combined with functional redundancy within intracellular signaling pathways, has been strongly linked to the emergence of treatment resistance—particularly in the context of targeted therapies for solid tumors (Sabnis and Bivona 2019).

The precision targeting paradigm is inherently aligned with the continuous treatment model, based on the premise that cancer cells are dependent—or "addicted"—to specific molecular pathways, which must be persistently inhibited. Consequently, in metastatic settings, targeted therapies are often administered indefinitely, provided there is ongoing clinical response. In select cases, such as the use of osimertinib in EGFR-mutated metastatic lung adenocarcinoma, this approach has demonstrated significant clinical success.

However, this strategy is not universally applicable. Extrapolating the benefits observed with continuous EGFR inhibition in this molecularly defined subset to broader oncologic contexts is scientifically unsound. Cancer is highly heterogeneous, and the profound pathway dependence seen in EGFR-mutated lung adenocarcinoma represents an exception rather than the norm. Most cancers do not exhibit such a singular molecular vulnerability; instead, they display adaptive plasticity and evolving resistance mechanisms that undermine the effectiveness of sustained targeted inhibition. Functional targeting approaches using 2D cultures, 3D Organoids, and patient-derived xenografts used to predict drug responses did not yet demonstrate their clinical efficacy, and, currently, are all considered experimental (Letai et al. 2021).

Recently, the concepts of network pharmacology and network medicine have emerged, promoting a "network target–multi-component" paradigm that expands upon the traditional focus on single molecular targets by addressing the complexity of interconnected biological systems (Gan et al. 2023). Using the network medicine paradigm as a start-up, this book introduces an alternative framework—*process modulation therapy*—which will be elaborated upon in the sections that follow. Unlike static pathway targeting, this strategy emphasizes dynamic interventions, aiming to modulate simultaneously multiple evolving biological processes and systemic interactions that underlie cancer progression.

12.8.3 Targeting Cancer

What are we targeting with our current arsenal of systemic cancer therapies?

Chemotherapy agents primarily act on proliferating cells, targeting their division and exploiting their high mitotic activity. However, metastatic cancer cells often prioritize motility over division, particularly as they adopt mesenchymal-like phenotypes, rendering these cells less susceptible to agents that disrupt mitosis in epithelial cell populations. Tyrosine kinase inhibitors (TKIs) act by competitively inhibiting the ATP-binding site of tyrosine kinases, disrupting intracellular signaling cascades. Monoclonal antibodies are designed to recognize and bind specific surface antigens expressed on tumor cells, while immune checkpoint inhibitors interfere with regulatory interactions between cancer cells and immune effector cells, thereby reactivating antitumor immunity. Anti-angiogenic agents such as bevacizumab inhibit vascular endothelial growth factor (VEGF), while VEGF receptor inhibitors block receptor-mediated signaling essential for neovascularization. Additionally, fibroblast growth factor receptor (FGFR) tyrosine kinase inhibitors aim to disrupt fibroblast-mediated support within the tumor microenvironment.

Significantly, targeting specific molecular nodes within these networks has led to durable responses—and in select cases, even cures—underscoring both the intricate codependencies within molecular networks and the surprising reliance of some tumors on a limited number of molecular drivers. This reveals a form of bottom-up vulnerability, where disrupting a single molecular axis can destabilize the broader tumor system.

The immune environment may have facilitating or inhibitory effects on cancer growth and spread, depending on the genetic, epigenetic, and metabolic make-up of the cancer cells. A fascinating example is the case of a "super mouse," discovered by serendipity by Dr. Zheng Cui and his team at Wake Forest University in the late 1990s: a single mouse that, against all odds, resisted aggressive cancer cell injections (Cui et al. 2003). Rather than dismissing the anomaly, Cui thoughtfully bred this exceptional mouse, establishing a lineage now known as "spontaneous regression/complete resistance" (SR/CR) mice. Remarkably, approximately 40% of descended mice inherited this potent, natural immunity against a broad range of transplanted tumors. As Cui and colleagues demonstrated, the mechanism behind this resistance is related to an exceptional response of the innate immune system—specifically neutrophils, macrophages, and natural killer cells—that mounts a swift and powerful attack against a specific mouse sarcoma cancer cell line (S180) (Hicks et al. 2006; Koch et al. 2013).

The further work of Cui and collaborators suggested that the cancer resistance in SR/CR mice is due to at least two separate processes, one related to the host, i.e., leukocyte migration/infiltration to the site of cancer cells, and another related to the cancer cells, i.e., the presence of specific antigens on the surface of the mouse sarcoma cancer cells. The infiltration of SR/CR leukocytes appears to be based on both the innate ability of SR/CR mouse leukocytes to respond to chemotactic signals produced by mouse sarcoma cancer cells and on whether mouse sarcoma cancer cells produced these chemotactic signals. Importantly, SR/CR mice are not resistant to all cancer types. Some non-mouse sarcoma cancer cells can escape from SR/CR resistance because they do not induce infiltration of SR/CR leukocytes (Riedlinger et al. 2010).

The discovery of SR/CR mice has profound implications. First, it reveals that innate immune systems can eradicate some forms of cancer, through rapid, immediate responses reminiscent of those mounted against microbial invaders, different from the slower adaptive immunity mechanisms. White blood cells alone were sufficient to reproduce the dramatic antitumor effect also in normal recipients inoculated with the S180 mouse sarcoma cell line, offering therapeutic remission in mice burdened with large tumors.

The CR/SR mouse is a striking example of nature's successful experiment in making a whole organism impervious to cancer. Cui's team showed that this cancer resistance trait to mouse sarcoma was transferable to further generations. The fact that resistance is inheritable suggests a transferable factor—possibly a genetic mutation. Unfortunately, decades later, identifying the causal mutation transferred between the SR/CR mice generations remains a challenge.

Unfortunately, the SR/CR mouse resistance/remission paradigm does not extend to the majority of malignancies, which are governed by a multiplicity of molecular pathways and redundancies and are not controlled solely by the innate immunity. Ideally, precision oncology would enable the identification and exploitation of specific vulnerabilities across all tumor types. Yet, as exemplified by the limited outcomes of the NCI-MATCH trial, the promise of universally effective, mutation-directed therapies remains far from realized. Most tumors resist simple categorization, and their behavior reflects complex, dynamic interactions rather than single-point dependencies.

In clinical practice, oncologists face time constraints and the necessity of making rapid treatment decisions, often favoring standardized approaches over highly personalized ones. As a result, precision oncology remains largely confined to a narrow spectrum of actionable targets for which effective agents have been developed. Moreover, the field still lacks reliable predictive biomarkers for guiding the use of chemotherapy or immunotherapy, and most current treatments for metastatic disease achieve only modest benefits.

An alternative strategy gaining momentum is phenotypic targeting, particularly through the emerging field of theranostics—a fusion of diagnostics and therapy using radiolabeled agents. This approach has shown promise in targeting tumors based on phenotypic traits rather than specific intracellular mutations. For example, PSMA-targeted radioligand therapy is now used in advanced prostate cancer, and lutetium-177 dotatate therapy is approved for treating neuroendocrine tumors expressing somatostatin receptors. Both therapies act on surface-expressed phenotypic markers—prostate-specific membrane antigen (PSMA) and somatostatin receptors, respectively—allowing for simultaneous tumor visualization and treatment. These modalities represent a shift from targeting internal molecular abnormalities toward exploiting stable, tumor-specific surface phenotypes, and exemplify how phenotypic targeting can complement molecular approaches in systemic cancer therapy.

12.8.4 Targeting Biogenerators Inside and Outside the Cancer Cell

The presence of biogenerators as introduced in Chap. 8, may represent cancer's Achilles' heel. Biogenerators are hubs within the network that sustain the malignant phenotype and offer potential leverage points for intervention. Identifying the components (molecular, functional, or both) in a particular cancer would represent an important advancement in the oncology field.

Evidence for the existence of such cancer biogenerators is supported by the recurrence of *common functional genetic/epigenetic alterations* across diverse tumor types. These include genomic instability driven by TP53 mutations, telomerase reactivation, and somatic mutations in at least one of ten canonical pathways: cell cycle, Hippo, Myc, Notch, Nrf2, PI3K/Akt, RTK-RAS, TGF-β, p53, and Wnt/β-catenin (Sanchez-Vega et al. 2018). Also, besides their key role in cellular division, some of the driver mutation gene products have often pleotropic functions, pointing towards a biogenerator function.

Supporting the biogenerator concept, 50% of all early clonal driver mutations are located in only nine driver genes, and subclonal mutations occur in 35 different genes, pointing to the relatively small number of genes playing a significant role in cancer (Gerstung et al. 2020; Ostroverkhova et al. 2023). TP53 (51.4%), LRP1B (13.4%), PIK3CA (11.6%), KRAS (11.1%), EGFR (10.6%), and APC (10.5%) are identified as the top mutated cancer genes in China (Ma et al. 2022), and these genes are similar to the genes mutated in the United States (Mendiratta et al. 2021) and also display similar mutation frequencies (Mendiratta et al. 2022).

The architecture of genetic networks involved in cancer *is not universal*: it may exhibit either modular or non-modular organization, but not both simultaneously (Iranzo et al. 2018). Therefore, each cancer is likely driven by a unique set of biogenerators whose specific configuration, while potentially adhering to general organizational principles, must be individually characterized, case by case.

A key step forward in oncology would be to identify the core biogenerator components: proteins, non-coding RNAs (ncRNAs), or other biomolecules. Several candidates have already been described. Both products of driver mutation genes and ncRNAs exhibit pleiotropic functions. TP53 is a classic example: beyond its role in genomic stability and cell cycle regulation, the p53 protein also orchestrates metabolic control, including glycolytic flux and biosynthetic pathways (Valente et al. 2020; Yun et al. 2009). Likewise, oncogenes such as KRAS (Pupo et al. 2019), EGFR (Makinoshima et al. 2014), and PIK3CA (Hao et al. 2016) not only regulate signal transduction but also profoundly influence metabolic processes. These multifunctional oncoproteins link bioinformation (genetic control), bioenergy (metabolism), and biostructure (cellular architecture), forming a triadic biocore.

The oncogenic "addiction" of a cancer cell to a specific driver mutation may relate to the degree to which its protein product integrates and governs the bioenergy-bioinformation-biostructure axis. In support of this hypothesis, the "strongest" driver genes whose protein products have been targeted with significant clinical benefit, i.e., BCR-ABL, BRAF, EGFR, are related to all three components of the biocore, i.e., bioenergy, bioinformation, and biostructure. As an example, in chronic myeloid leukemia, the BCR-ABL fusion protein function is not related only to cellular survival (Neshat et al. 2000). BCR-ABL plays a key role also in controlling both the cellular metabolism (Kato et al. 2018; Feng et al. 2023) and cellular structure (Salgia et al. 1997). Blocking the tyrosine kinase activity of the BCR-ABL protein with imatinib leads to more than 80% of the patients being alive at ten years (Hochhaus et al. 2017). It can be argued that the CML is an idiosyncratic cancer being driven by only one "strong" driver. Even though most solid tumor cancers, in general, present multiple driver mutations, "strong" drivers that can be targeted with significant clinical results have also been identified in solid tumors. Two prominent examples are V600E BRAF mutations in melanoma and several genetic alterations of the EGFR gene in lung adenocarcinoma. BRAF mutations induce both metabolic (Haq et al. 2013) and structural (Klein et al. 2008) changes in melanoma cancer cells. Also, EGFR is a cytoskeleton-associated protein (Song et al. 2008), and disrupting EGFR interactions with the cytoskeletal components inhibited the growth of an EGFR T790M-mutant lung cancer cell (Chiu et al. 2012). EGFR mutations are able to rewire the metabolism of adenocarcinoma cancer cells (Mendes et al. 2023).

Besides protein-coding genes, non-coding RNAs also act as potential biogenerators. ncRNA with pleiotropic functions have also been described (Pasut et al. 2016; Montico et al. 2022; Amicone et al. 2023; Yang et al. 2023). For example, miR-142 is a master regulator of several functions of hematological malignant cells, directly or indirectly controlling division, metabolism, and cytoskeleton remodeling (Zareifar et al. 2024; Huang et al. 2024). Using it as a therapeutic target, though, is not straightforward, as miR-142 can behave both as an oncogene and as a tumor repressor in different cellular contexts, and its inhibition may lead to rapid cancer progression (Zhang et al. 2023a). Another double-edged sword is the miRNA-452 (Karimi Dermani et al. 2023) which may also either promote or inhibit the oncogenic process depending on the cancer type. Although attractive, before being introduced to the oncology clinic, the targeting of ncRNAs needs rigorous testing.

To reconcile the limited number of recurrent mutations with the shared malignant phenotypes—uncontrolled proliferation, invasion, and metastasis—Andrea Califano and Mariano Alvarez proposed a model of "oncotecture": a tumor's regulatory architecture governed by a small set of master regulator (MR) proteins. These MRs function within tightly autoregulated tumor checkpoints that integrate upstream mutational signals and canalize them into tumor-sustaining transcriptional programs (Califano and Alvarez 2017). Notably, the MR-based architecture is tumor-agnostic, suggesting a universal regulatory logic. A related view has been proposed by Douglas Brash and Junmei Cairns (Brash and Cairns 2009) and Armando Aranda-Anzaldo and Myrna Dent (Aranda-Anzaldo and Dent 2021) who proposed that cancer is a disease set up by cellular stress responses that are universally present in every cell. As a corollary to this, defects in these stress response mechanisms (although evolutionary deleterious) would protect against cancer, a hypothesis that could be experimentally tested (Brash and Cairns 2009). Stress response regulators such as HSF1 (heat shock factor 1) are emerging as potent modulators of this biocore. Overexpression of HSF1 in several cancers correlates with malignancy and poor prognosis (Gumilar et al. 2023), further reinforcing its role as a system-level driver.

Some of the biogenerators essential for tumor maintenance may be found within housekeeping gene networks. Joshi et al. demonstrated that the expression patterns of housekeeping genes differ between cancer and normal cells and could serve as disease classifiers or therapeutic targets (Joshi et al. 2022). Similarly, Joan Smith and Jason Sheltzer showed that many top-scoring prognostic markers in TCGA analyses are housekeeping genes involved in cell cycle progression (Smith and Sheltzer 2022). These two authors analyzed the results of the clinical trials targeting the top-scoring unfavorable prognostic genes and found that the majority of them failed, leading to significant side effects when their proteins were inhibited. Evolution would not exist without differences in fitness, and it is still being argued whether mutations that produce variability and drive evolution happen inside the genes (Hoekstra and Coyne 2007) or outside the genes (in the cis-regulatory regions) (Stern and Orgogozo 2008). In cancer, recurrent driver mutations overwhelmingly reside in coding regions. Though certain non-coding mutations (e.g., TERT promoter) contribute to tumorigenesis, most appear infrequently and lack consistent selection (Reiter et al. 2019).

The original concept of a fixed intracellular protein network, as proposed by Hanahan and Weinberg in their foundational "Hallmarks of Cancer" paper, envisioned protein interactions as components of a stable, solid-state molecular circuit—akin to an electronic wiring diagram. However, this static view is increasingly challenged by contemporary evidence. Alexei Kurakin has argued that such a rigid framework fails to capture the dynamic, context-dependent nature of intracellular processes (Kurakin 2005). In a more recent publication, Daniel Nicholson emphasizes that the cell is not a machine or a factory, and that proteins in vivo rarely adopt a single, well-defined conformation (Nicholson 2019). Instead, many proteins exist in intrinsically disordered or highly flexible states, forming transient, pleomorphic complexes composed of weakly interacting and continuously fluctuating subunits. These assemblies vary in size and composition, display functional promiscuity, and exhibit behaviors that are highly sensitive to cellular context. Unlike classical macromolecular machines, these dynamic ensembles resist crystallographic analysis and operate through transient, opportunistic interactions. In light of this evolving understanding of protein behavior, the architecture of cancer-related protein networks—the "oncotecture"—is unlikely to be universal. While general organizing principles may exist, the specific configuration of bioregulatory generators within these networks likely varies between tumor types. Therefore, each cancer requires a distinct characterization of its underlying biogenerator architecture.

To date, several key genetic driver mutations have been identified, enabling the development of targeted therapies directed against them. However, beyond these genetic drivers, a growing body of research has uncovered additional potent nongenetic drivers of malignancy. These include molecular mechanisms such as epigenetic regulators (e.g., non-coding RNAs), metabolic components (specific metabolites and enzymes), and dynamic cellular processes (e.g., elevated glucose uptake, increased autophagy). These non-genetic drivers represent promising targets for therapeutic intervention.

Among these, the biochemical (e.g., metabolism, pH, mechanical stress) and biophysical (e.g., membrane potential) components of cancer cell regulation are among the most evolutionarily conserved. Targeting these ancient regulatory elements may induce profound changes in cancer cell behavior and potentially disrupt the malignant phenotype at its roots.

The concept of the *biogenerator* provides a useful framework for understanding how various cancer therapies exert their effects. Some chemotherapies and targeted agents are **tumoricidal**, meaning they completely disrupt the activity of a specific intracellular biogenerator (e.g., G6), leading to cell death. Others are **tumoristatic**, selectively inhibiting only a specific module *within* the G6 generator- such as blocking the proliferation modules—forcing the cell into alternative states like maintenance or quiescence.

Immunotherapies, by contrast, primarily act on the G7 macroscopic biogenerator, which encompasses tumor–host interactions at the tissue level. These therapies can also be tumoricidal or tumoristatic, but their actions are directed not solely at individual tumor cells, but at the tumor as an integrated system. For example, in small cell lung cancer, where tumor cells often lack PD-L1 expression, checkpoint inhibitor efficacy is mediated through indirect mechanisms, reflecting systemic or microenvironmental modulation rather than direct cytotoxicity.

The cancer biogenerators are not solely represented by genetic, epigenetic or metabolic alterations and they allow us to broaden the concept of cancer drivers outside the cell and include *extracellular drivers of malignancy*—whether located in the tumor microenvironment (part of the G7 biogenerator)—or embedded within the organism's internal milieu or distal organs (part of the G8 biogenerator) (Please see Fig. 14), offering new avenues for therapeutic development. Targeting, for example, cell-cell exosomes communication between cancer cells themselves or with other cells located within the tumor stroma or at a distance may disrupt both the G7 and the G8. This integrative perspective paves the way for systemic interventions that target cancer not only as a cellular disease, but as a dynamic and adaptive pathological system embedded within the host organism.

12.8.5 Targeting Biochemical and Biophysical Cancer Biogenerators

Acetyl-CoA plays a pivotal role as a central metabolic hub, both in modern cellular physiology and in prebiotic chemistry preceding the evolution of the Krebs cycle (Smith and Morowitz 2016). As a key energetic and biosynthetic intermediate, acetyl-CoA integrates multiple metabolic pathways. Under anabolic conditions, it is primarily generated from glucose, while during catabolic states, it derives from a variety of substrates including medium- and long-chain fatty acids, acetate, amino acids, and β-hydroxybutyrate (Guertin and Wellen 2023). Within mitochondria, acetyl-CoA feeds into the citric acid cycle to drive ATP production, and it also serves as a precursor for the biosynthesis of lipids, cholesterol, and ketone bodies. Beyond its metabolic roles, acetyl-CoA is essential for epigenetic regulation: it donates acetyl groups in post-translational modifications that influence protein activity and gene expression. Thus, acetyl-CoA is not merely a carbon carrier but a master regulator orchestrating the interplay between cellular metabolism and gene regulation. Remarkably, this molecule emerged in Earth's early geochemical environment, potentially over 4 billion years ago—before the last universal common ancestor (LUCA) (Martin 2020)—and continues to represent a crucial intersection within the diverse array of physiological and pathological biogenerators that sustain or disrupt life.

Acetyl-CoA recently came into focus (Guertin and Wellen 2023) as its key metabolic enzymes were found to be extensively dysregulated in cancer. ATP citrate lyase (ACLY) is the predominant enzyme synthesizing cytoplasmic and nuclear acetyl-CoA. Highest alteration frequency was observed in esophagogastric adenocarcinoma (16.13% with amplification) but they are also present in neuroblastoma, skin cancer, invasive breast carcinoma, and esophagogastric cancer, High ACLY expression is associated with poor differentiation and prognosis in various malignancies, such as glioblastoma, lung cancer, gastric cancer, breast cancer, liver cancer, cervical cancer, prostate cancer, osteosarcoma, and acute myelogenous leukemia. The earliest and most significant ACLY inhibitor is hydroxycitrate acid (HCA), a substance used to treat obesity.

Another important enzyme for the production of cytosolic acetyl-CoA is acyl-CoA synthetase short-chain family member 2 (ACSS2). ACSS2 alteration is frequent in lung cancer (12.82%), and uterine endometrioid carcinoma (12.43%), Hodgkin's lymphoma, and esophagogastric cancer. ACSS2 is highly expressed in triple-negative breast cancer, GBM, liver cancer, cervical cancer, or renal cell carcinoma. Both ACLY and ACSS2 inhibitors are attractive targets for cancer therapy and are actively studied (Chen et al. 2023a).

Finally, some tumors depend on ketolysis and on the specific ketolytic enzyme: succinyl-CoA: 3-oxoacid-CoA transferase (SCOT) for making acetylCoA in their mitochondria (Israel and Schwartz 2020). Pharmacological and natural inhibitors of SCOT have been identified (Schwartz et al.2020; Israël et al. 2024), and in some cancers, SCOT inhibition may lead to therapeutic results.

The acetylCoA example underscores the importance of investigating the G0 electrochemical layer present in all cellular life forms. Exploiting the dependency of cancer cells on G0 and targeting specifically this electro-chemical generator is an attractive therapeutic opportunity that needs to be further explored.

Cellular and multicellular organisms are literally "living batteries" and to function properly they need *electricity*, the right amount of *fluid,* and the right amount of *electrolytes*. All three components of this "battery" are modified in cancer cells as compared to normal cells. Compared to normal cells, cancer cells typically exhibit a more alkaline intracellular pH and reduced membrane potential. These electrochemical distinctions offer exploitable therapeutic windows. Bioelectric interventions—ranging from deuterium-depleted water (Lu and Chen 2024), alkalinization therapy (Hamaguchi et al. 2022), and ion channel modulation (Capatina et al. 2022), to electric field-based approaches—have demonstrated promising results, including reversal of metastatic phenotypes in *Xenopus* models (Chernet and Levin 2013; Levin and Martyniuk 2018; Payne et al. 2019; Levin 2021).

12.8.6 Timing Is the Key

Targeting the cancer's ciclicality at the cellular level, i.e., targeting cellular division, with different chemotherapy agents was one of the first systemic strategies that demonstrated clinical benefit.

Besides the cellular ciclicality, there are cycles, rhythms, and biorhythms that can be targeted at the tissular and at the organismic level. Importantly, these cycles are expressed both by the cancerous cells and the normal organism. Reinforcing the body's natural immune defenses and weakening cancer's virulence are both valid therapeutic strategies (Su et al. 2025).

The integrity of the circadian time clock can be assessed using BodyTime, a blood NanoString-based test assessing the expression of 12 gene expression profiles (Wittenbrink 2018).

As a proof of concept, a retrospective study of 713 patients (Cohort 1 from France, n = 165; Cohort 2 from China, n = 548) that received who received first-line immuno-chemotherapy for stage IIIC or IV NSCLC demonstrated that patients receiving immuno-chemotherapy before 11:30 am showed statistically significant ($p < 0.0001$) OS improvement (33.0 months [95% CI, 27.5–41.0] vs patients receiving the treatment later in the day (19.5 months [18.0–22.5]) (Huang et al. 2025). Similar findings were described in patients with kidney, bladder, esophageal cancer, or melanoma (Landré et al. 2024; Karaboué et al. 2024).

Further supporting the notion that timing is the key, a study led by Colinda Scheele from VIB-KU Leuven, in Belgium Center for Cancer Biology demonstrated that chemotherapy treatments administered in women with breast cancer in the follicular phase when progesterone levels are low are associated with higher median survival than chemotherapy administered in the second part of the menstrual cycle when progesterone levels are relatively high (Bornes et al. 2025).

Also, the following chronotherapeutic interventions can enhance or restore circadian rhythms to reduce cancer pathogenesis and improve the response to anticancer treatments: morning bright light, melatonin, glucocorticoids, caloric restriction, and intermittent fasting (Lee 2021).

12.8.7 Limitations of the Targeted Therapy Paradigm and Emerging Avenues

In chronic myeloid leukemia (CML), the BCR-ABL fusion gene resulting from a chromosomal translocation is a well-characterized pathophysiologic driver. This aberration underpins the malignant phenotype, rendering the leukemic cells dependent on its signaling—a vulnerability effectively exploited by Imatinib, which has demonstrated remarkable therapeutic success. Similarly, other malignancies such as melanoma (BRAF mutations) and non-small cell lung cancer (EGFR, ALK, ROS1, NTRK alterations) harbor key oncogenic drivers that are critical for tumor cell survival. However, the paradigm of targeting single molecular drivers is increasingly challenged by clinical and biological complexities.

Several limitations of the driver-centric model have emerged:

Heterogeneity of Oncogenic Drivers: Not all tumors harbor a dominant, targetable driver mutation. Co-occurring genetic alterations can modulate the oncogenic dependency and render single-agent targeting insufficient or ineffective.

Multiplicity of Oncogenic Signals: Most cancers, particularly solid tumors, do not rely on a single actionable mutation. Instead, they often display a mosaic of oncogenic pathways, where inhibition of one driver does not result in sustained therapeutic benefit due to compensatory mechanisms.

Emergence of Resistance: Resistance to targeted therapies is nearly universal over time, often due to secondary mutations, pathway reactivation, or phenotypic plasticity, limiting the long-term efficacy of these agents.

Checkpoint inhibitors, while revolutionary, also exhibit substantial limitations. The therapeutic response in metastatic solid tumors is highly variable and difficult to predict, largely due to the complexity and immunosuppressive nature of the tumor microenvironment. While according to the BRAF mutational status, with Ipilimumab/Nivolumab combinations, melanoma has shown 46–72% survival at 6.5 years (Wolchok et al. 2022), outcomes are less favorable in other settings—for instance, only ~15% five-year survival in metastatic lung cancer (Garon et al. 2019; Gettinger et al. 2018).

Overall, only about 20% of patients with advanced solid tumors experience durable benefit from immunotherapy, and in many cases, these responses do not translate into meaningful long-term survival gains.

Crucially, malignancy is not solely driven by genetic mutations leading to abnormal protein structures. Nongenetic mechanisms also play pivotal roles. Epigenetic alterations, including changes in chromatin architecture and regulation by non-coding RNAs, are emerging as central drivers. In addition, cancer cells often exhibit profound metabolic reprogramming. Beyond molecular and metabolic factors, biophysical properties such as altered membrane potential and dysregulated intracellular pH represent underexplored therapeutic targets. These nongenetic determinants offer promising avenues for the development of next-generation, integrative therapeutic strategies.

Historically, targeting the ciclicity of the cancer at the cellular level, i.e., targeting cell division, with different chemotherapy agents demonstrated impressive clinical benefit. Besides the cellular ciclicality, there are cycles, rhythms, and biorhythms that can be targeted at the tissular and at the organismic level.

12.8.8 War and Peace Strategies

We now understand that, in most cases, the war against metastatic cancer cannot be decisively won through sheer force—whether by chemotherapy's carpet bombing, the sniper-like precision of targeted agents, or the search-and-destroy tactics of immunotherapy combinations. While these approaches can yield remarkable results in select situations, they often provoke resistance and collateral damage. Increasingly, we are learning that in many forms of advanced cancer, direct confrontation may not be the wisest path. Instead, therapeutic strategies that seek balance—delicate negotiations rather than outright assault—may offer greater long-term benefit. In fact, for the majority of metastatic cases, perhaps we should shift our perspective entirely away from warfare and toward peaceful coexistence, a manageable truce between the tumor and the host.

Malignant cells are far from uniform in their behavior. The cancer stem cell model posits that a subset of tumor cells, those with stem-like properties, are responsible for initiating and sustaining the disease, while the majority of cells within the tumor lack this capacity. Only some can invade, metastasize, or give rise to new colonies. Moreover, not all tumors arise from a single origin; many are polyclonal, with up to 30% of cancers exhibiting multiple founding cell lineages. The spatial and temporal heterogeneity of cancer, across genetic, epigenetic, metabolic, phenotypic, and microenvironmental layers, compounds the challenge of designing effective treatments.

If we continue to use the metaphor of war, then each cancer demands a tailored campaign strategy. But the goal of this "war" is not to annihilate the enemy, but to prolong life and preserve quality of life. With this shift in intention, our tactics should also change. Rather than seeking total eradication, we may aim to manage cancer as a chronic condition, much like diabetes, containing it, adapting to it, and evolving alongside it.

Cancer does not follow a fixed trajectory. Its goal, like all life forms, is to survive no matter what. In order to achieve that, it might either press on the accelerated division button, or, on the contrary, it can go slower, or even dormant—it might also invade locally, or metastasize. All these are different survival strategies. Our therapies must be dynamic, responsive, and attuned to these transitions and moving targets. Tumor growth, invasion, and metastasis are not acts of aggression in the classical sense; they are, at their core, stress-induced survival responses. If we fail to recognize this, we risk provoking the very behaviors we aim to suppress. In some cases, aggressive treatments may paradoxically accelerate disease progression by activating these intrinsic programs. Understanding this nuanced biology calls for a reimagining of our roles: less as warriors, and more as careful pathfinders navigating a complex, evolving ecosystem.

This represents a major paradigm switch: focusing on balancing and guiding cancer through micro and macroenvironment manipulations into less aggressive phenotypes as opposed to killing it. Secondary plant metabolites like Paclitaxel, extracted from the bark of the Pacific yew tree (Taxus brevifolia), or irinotecan, extracted from the happy tree (Camptotheca acuminata), have already demonstrated their efficacy in several cancer types. Carefully designed plant extracts containing combinations of secondary metabolites can be used as cancer treatments by acting synergistically, by targeting multiple pathways, and potentially, by restoring the normal balance (homeostasis) within cells and tissues (Seca and Pinto 2018; Basak et al. 2020; Jamal et al. 2025).

12.8.9 Patterns of Disease

In the trajectory of disease, from its earliest stirrings to its resolution or culmination, one can discern a structural similarity with Aristotle's tripartite narrative of beginning, climax, and end. This classical framework—originally conceived for tragedy and storytelling—finds a parallel in the natural history of an illness, offering a conceptual lens through which the dynamic unfolding of pathology can be understood. The existence of disease patterns is related to the compensatory nature of homeodynamic mechanisms. Illness is not a random sequence of biological events; it has the structure of a story unfolding within and with the assistance of the body. Timing in clinical medicine is essential, and the existence of disease patterns allows timely temporal interventions, allowing physicians to intervene in specific phases of the malignant trajectory.

The ***beginning*** of disease often unfolds imperceptibly, marked by subtle perturbations in the organism's homeodynamic network. These disturbances—molecular, psychosocial, environmental, and/or systemic—initiate a shift away from physiological harmony. Biomedically, this corresponds to the **prodromal phase**, characterized by early molecular or metabolic shifts, preclinical dysfunction, and incipient dysregulation across interconnected systems. During this phase, the body attempts to compensate through tightly regulated homeostatic and homeorhetic feedback mechanisms aimed at restoring equilibrium.

In the ***middle phase***, the failure of these compensatory responses becomes evident. This is the stage of clinical disease, when homeodynamic systems are overwhelmed or maladaptively engaged. The *middle* represents the full manifestation of the disease state, where symptoms are clinically evident and pathology reaches its crescendo. This is the phase of diagnostic clarity and therapeutic engagement. It is here that the organism is locked in a struggle to re-establish coherence-mobilizing immune responses, metabolic adaptations, and behavioral changes—while medical interventions aim to reverse, stabilize, or contain the pathological process.

The ***end*** may take divergent forms: resolution, chronic adaptation, remission, or death. In some cases, successful treatment restores homeodynamic balance and leads to recovery. In others, the organism adapts to a new state of altered function, as seen in chronic diseases. Alternatively, disease may culminate in systemic failure, marking the end not only of the pathological process but of the organism's coherence as a living system. Importantly, this final phase also carries profound symbolic and emotional weight, often reframing the meaning of the illness experience for patients, families, and caregivers.

The progression from health to disease—whether acute or chronic—can be understood through a unified physiological and patho-physio-psychological pattern that links Hans Selye's General Adaptation Syndrome (GAS), Roger Bone's concept of the Systemic Inflammatory Response Syndrome (SIRS), and the model of allostasis.

Selye's GAS describes the body's nonspecific response to stress in three phases: *alarm*, *resistance*, and *exhaustion*. Initially, the organism activates neuroendocrine and immune mechanisms to restore internal balance. If the stressor persists, the body enters a resistance phase, mobilizing resources to maintain homeostasis. But prolonged stress leads to exhaustion—a state in which adaptive capacity is depleted and vulnerability to disease increases. Selye's model, though developed in the context of acute stress, presciently maps onto the dynamics of chronic illness and systemic failure.

Similarly, Roger Bone's SIRS framework describes how systemic inflammation—whether triggered by infection, trauma, or malignancy—can escalate into a *self-propagating state of dysregulation*, independent of the original trigger. SIRS represents a tipping point where normal homeodynamic feedback loops, designed to resolve local injury or infection, become *amplified and maladaptive*, driving tissue damage, metabolic imbalance, and organ dysfunction. This shift from localized to systemic inflammation mirrors the transition from physiological to pathological adaptation.

Allostasis describes how the body anticipates needs and adjusts set points through regulatory mediators like cortisol, cytokines, and neurotransmitters. However, repeated or prolonged activation of these mediators results in *allostatic load*—a cumulative physiological burden that erodes resilience. In cancer, chronic inflammation, oxidative stress, circadian disruption, and metabolic dysregulation reflect such a load, fostering a permissive environment for tumor initiation, progression, and immune escape.

At the heart of this pathophysiologic narrative lies the concept of *homeodynamics*: the capacity of living systems to maintain adaptive stability through dynamic, non-linear feedback loops. When these feedback loops designed to resolve stress falter or become dysregulated, a predictable cascade unfolds, self-amplifying, creating a runaway inflammatory cascade that damages tissues and erodes physiological resilience. This pattern underlies the pathogenesis of many complex human diseases, including cancer.

12.8.10 Cancer: The Chronic Systemic Strain Syndrome

Similar to the general adaptation syndrome, the systemic inflammatory response and allostasis, the maligant process also presents a three-stage dynamic:

- a) an initial local reactive stage,
- b) a systemic adaptation stage which, in some cases, is followed by.
- c) an exhaustion stage that may represent the harbinger of multiple organ failure and death.

Advanced and metastatic cancer represents a *chronic systemic strain syndrome* (CSSS)—a pathologic state that is often accompanied by metabolic disturbances, anorexia and fatigue that involves a pre-cachectic state that may subsequently progresses to advanced/terminal cachexia.

Cachexia is a complex, multi-organ pathophysiological state characterized by progressive weight loss, skeletal muscle wasting, adipose tissue depletion, systemic inflammation, and metabolic dysregulation (Petruzzelli and Wagner 2016; Setiawan et al. 2023). It involves inter-organ communication (Argilés et al. 2018) and it represents a chronic form of organism-wide metabolic and inflammatory collapse. According to the consensus definition proposed by Fearon et al. cancer cachexia develops through several stages: pre-cachexia, cachexia, and refractory (terminal) cachexia (Fearon et al. 2011). Cachexia affects approximately 70–80% of pancreatic cancer patients, with muscle loss reported in up to 63% of those with advanced disease (Baracos et al. 2018; Kays et al. 2018). During cachexia, persistent inflammatory and neuroendocrine stressors override homeostatic mechanisms across multiple organ systems. Tumor-derived catabolic factors (e.g., proteolysis-inducing factor, lipid-mobilizing factor) and pro-inflammatory cytokines such as TNF-α, IL-6, and IFN-γ drive a hypercatabolic state, triggering proteolysis, lipolysis, and mitochondrial dysfunction (Argilés et al. 2014; Zhang et al. 2020c). Simultaneously, the neuroendocrine axis is disrupted, with aberrant cortisol signaling, hypothalamic inflammation, and impaired appetite regulation exacerbating tissue breakdown and anorexia (Tsoli and Robertson 2013). The liver becomes a key metabolic hub, reprioritizing amino acid and lipid metabolism to support acute-phase protein synthesis at the expense of peripheral tissue integrity. Beyond the skeletal muscle, cachexia affects adipose tissue, liver, gut, brain, and even the tumor itself, creating a network of maladaptive metabolic interactions. The loss of white and brown adipose tissue alters lipid metabolism and thermogenesis. Cross-talk between muscle, fat, and liver is mediated by cytokines, stress hormones, and metabolites, leading to a disruption of key anabolic pathways and systemic energy wasting (Tsoli and Robertson 2013). Cognitive changes, fatigue, and altered neurochemical signaling reflect brain involvement, while gut dysbiosis and impaired barrier function further amplify systemic inflammation.

From both the perspective of the organism and that of the tumor, cancer progression emerges as a dysregulated homeodynamic response, an escalating cascade driven by unbalanced feed-forward loops. These loops, meant to restore equilibrium, instead overshoot their mark, pushing the system further from stability. Thus, cancer's course can be read as the patterned overreaction of a living system out of sync with itself, a dynamic unraveling that echoes the maladaptive rhythms of prolonged physiological strain.

Cancer, therefore, is not simply a collection of rogue selfish cells. It is a systemic disease arising within a host environment that *loses feedback control*. Tumor growth both reflects and reinforces breakdowns in homeodynamic regulation—co-opting hormonal, immune, metabolic, and neural circuits to sustain its evolution. The tumor-host interface exemplifies a closed-loop failure, where adaptive responses become chronic drivers of pathology.

Viewed through this integrative lens, cancer represents different manifestations of *homeodynamic collapse*—a failure of the organism to recalibrate in the face of prolonged stress or injury. Whether manifesting as sepsis, metabolic syndrome, neurodegeneration, or cancer, the underlying pattern is one of dysregulated feedback, maladaptive allostasis, and exhaustion of systemic resilience. Restoring health, then, requires more than treating symptoms or isolated pathways; it calls for *rebalancing the entire adaptive network.*

Healing, in this framework, is not the suppression of symptoms, but the restoration of the organism's capacity for dynamic creative persistence (DCP)—the very principle that sustains life.

In critical care, the systemic inflammatory response syndrome (SIRS) is managed as a life-threatening perturbation of homeostasis, characterized by widespread inflammation, fluid imbalance, metabolic dysregulation, and often, loss of vascular and bioelectrical integrity. The aim is to stabilize the internal milieu by restoring key physiological parameters such as electrolyte composition, acid–base balance, and adequate perfusion and hydration.

The CSSS of cancer, by contrast, is not an acute failure of systemic function, but rather a chronic, evolving disturbance—a sustained homeodynamic imbalance driven by tumor-host interactions, metabolic reprogramming, immune modulation, and systemic inflammation. Despite these differences in tempo and trajectory, both conditions share a common thread: the progressive breakdown of organismal regulation.

Conceptualizing the human body as a *living battery,* or more precisely, an *electrolytic system*, offers a unifying strategy for both Intensive Care Units (ICU) management and oncology interventions. This framework emphasizes the role of *ionic gradients, bioelectric fields, and fluid composition* in maintaining physiological function. In this analogy, shock represents an *acute battery failure*, where voltage collapses due to rapid loss of fluid or systemic conductivity, while the CSSS of cancer can be seen as a *chronic battery dysfunction*, in which long-term metabolic and electrochemical imbalances undermine systemic resilience and cellular coordination.

Drawing from this model, it may be therapeutically advantageous in oncology to explore controlled replenishment of the body's "electrolytic bath"—that is, restoring fluid balance, electrolyte concentrations, and pH stability—as part of a broader effort to reestablish homeodynamic integrity. Beyond passive fluid resuscitation, *active modulation of bioelectric signaling* through ion channel manipulation, targeted electrochemical therapies, or metabolic interventions aimed at restoring mitochondrial and membrane potential could help reset dysfunctional cellular states.

Incorporating these strategies into the oncology paradigm would require a shift from purely tumor-centered approaches to a systems-level view that recognizes the cancer-bearing organism as an integrated, *electrically active system*. Such an approach could open new therapeutic avenues aimed at *restoring systemic coherence* and *breaking the pattern of cancer progression*, not merely eliminating tumor cells. This cross-disciplinary synthesis—linking critical care physiology, bioelectricity, and oncology—may hold promise for developing future cancer treatments grounded in the restoration of systemic homeodynamics.

Ultimately, cancer is not merely a genetic, cellular, or tissue disease, but a systems-level phenomenon driven by dynamic *biogenerators* that integrate informational, energetic, structural, and electrochemical processes. Mapping and targeting these layered biogenerators may offer a host of novel therapeutic opportunities.

12.9 Homeodynamic Interventions in Cancer

Homeodynamic interventions diverge fundamentally from conventional precision medicine approaches. *First*, they may rely on *functional* selectivity—modulating specific intracellular signaling pathways downstream of a single receptor—rather than receptor subtype selectivity. *Second*, they can act simultaneously on agonistic and antagonistic axes, balancing opposing forces within a regulatory system. Third, unlike most precision oncology treatments, which target tumor cells selectively, homeodynamic therapies aim to *restore balance at the organismal level*, influencing both malignant and nonmalignant cells.

Organism homeostasis is impacted by cancer. This is how cancer kills. Cancer homeodynamics at its turn may be modulated by the organism and systemic treatments. By improving the organism's homeostasis, cancer's virulence can also be curbed.

For example, deuterium-depleted water (DDW) and systemic alkalinization have shown anti-cancer effects by globally modulating metabolic and bioelectric states (Lu and Chen 2024; Hamaguchi et al. 2022). Similarly, growth hormone-releasing hormone (GHRH) agonists—used in breast and prostate cancers—act not on peripheral tumors directly, but through central neuroendocrine regulatory pathways (Kach et al. 2015).

A particularly intriguing distinction lies in receptor versus functional selectivity. Receptor selectivity targets specific receptor subtypes, while functional selectivity modulates discrete downstream pathways linked to the same receptor (Berg and Clarke 2018). In oncology, where pathway specificity is crucial, tuning drugs for functional selectivity may enhance therapeutic precision.

Another therapeutic paradigm—rooted in the Yin-Yang duality of Traditional Chinese Medicine (TCM)—involves the simultaneous or sequential use of agonists, antagonists, and harmonizers, a strategy long used in psychiatric treatments (Rose and Levin 1992; Wang et al. 2024f). This dual modulation strategy reflects the homeodynamic principle of adaptive tension between opposing regulatory forces.

Modern combinatorial approaches, particularly in immunotherapy, use similar principles. For instance, in hepatocellular carcinoma, combining bevacizumab (an anti-angiogenic agent) with atezolizumab (an immune checkpoint inhibitor) significantly improved survival (Finn et al. 2020), exemplifying systemic rebalancing rather than single-target intervention.

An innovative strategy involves computationally designed polypharmacological agents. In a recent study, Avchaciov et al. extended the lifespan of *C. elegans* by over 40% using a cocktail of 22 geroprotective compounds—each individually associated with longevity—highlighting the potential of intelligently designed multi-target interventions (Avchaciov et al. 2024).

Thus, understanding cancer through the lens of *homeodynamics*, and integrating developmental, evolutionary, and systemic stress responses offers a new framework for therapeutic innovation, one that aims not only to kill cancer cells but to *restore systemic integrity* across the scales of the organism.

Over the past decade, precision oncology has emerged as the prevailing paradigm in cancer treatment, driving research efforts toward the identification of molecular targets, some of which have yielded substantial clinical benefits. However, cancer cannot be fully explained as a mere consequence of the progressive accumulation of genetic mutations or immune evasion. Its origins lie in complex, multi-scale homeodynamic disturbances—subtle but persistent dysregulations at the cellular, tissue, and organismal levels that often precede the appearance of clinically detectable tumors. As cancer progresses, it further perturbs the homeostatic balance of the host organism, initiating a pathological feedback loop between the disease and its biological context.

Living systems maintain stability and adaptability through an intricate web of feedback and feedforward regulatory loops. The delicate tuning achieved via antagonistic yet complementary forces—such as stimulation and inhibition—is central to survival. This dynamic equilibrium, referred to as homeostasis in Western science, has long been paralleled in Chinese philosophy by the concepts of Yin and Yang, which represent opposing but interdependent forces that govern change and stability in the universe. These principles, foundational to Traditional Chinese Medicine (Wang et al. 2024g), find echoes in modern biology: kinases and phosphatases regulate protein activation states; autophagy maintains protein quality through degradation and recycling; the sympathetic and parasympathetic branches of the autonomic nervous system counterbalance each other; and the immune system's activating and suppressive arms work in tandem to preserve physiological integrity. Such Yin-Yang dynamics are embedded across all levels of biological organization.

Cancer, too, displays its own form of homeodynamic regulation. Its evolution and expansion depend on finely tuned opposing processes—akin to Yin and Yang—that govern proliferation, dormancy, invasion, and immune evasion. Therapeutic approaches rooted in homeodynamic principles have already demonstrated clinical utility. A notable example is the adaptive therapy strategy pioneered by Dr. Robert Gatenby, which leverages evolutionary dynamics to manage prostate cancer more effectively (Zhang et al. 2022a).

Precision medicine and homeodynamic interventions are not mutually exclusive, but inherently interconnected. Targeted therapies that inhibit or activate specific molecules inevitably alter cellular homeostasis. Moreover, top-down interventions—those aimed at restoring systemic balance—can reshape lower-level biological networks. Unlike unidirectional interventions, which may provoke compensatory rebounds, homeodynamic approaches aim to restore equilibrium across both arms of the regulatory dyad. Thus, therapeutic strategies that may seem paradoxical from a reductionist viewpoint—such as the stimulation of oncogenic pathways—become intelligible when framed within a homeodynamic systems model (Dias 2024).

Cancer should not be viewed solely as a cellular defect but rather as a multi-level failure of homeostatic regulation spanning molecular, cellular, tissue, and systemic domains. Disruptions in cell cycle regulation, proteostasis, immune surveillance, and endocrine signaling collectively establish a permissive environment for tumor initiation, progression, and metastasis. A key example is the immune system, which—although tasked with eliminating abnormal cells—often becomes dysregulated in cancer. Tumors disrupt immune homeostasis through a convergence of molecular, metabolic, and systemic mechanisms.

Chronic inflammation fosters the expansion of immunosuppressive cell populations, including regulatory T cells (Tregs) and myeloid-derived suppressor cells (MDSCs), which inhibit cytotoxic T cell function (Gajewski et al. 2013). Nutrient competition between cancer and immune cells—especially for glucose and glutamine—often favors tumor cells, resulting in metabolic exhaustion and impaired immune function (Chang et al. 2015). Additionally, hypoxia and lactic acidosis suppress dendritic cell maturation and dampen cytotoxic responses (Brand et al. 2016). While some patients experience temporary benefit from immunotherapy, relapses are common. Thus, the goal of treatment should extend beyond immune activation to include restoration of a resilient, adaptive homeodynamic balance that can sustain anticancer immunity and prevent recurrence.

A central challenge in immunotherapy is immune exhaustion- a state of dysfunctional immune responsiveness. Overcoming this requires therapeutic strategies that go beyond transient activation to establish durable immune memory, metabolic and epigenetic resilience, and system-level homeostatic restoration. For example, IL-15 superagonists such as ALT-803 (N-803) enhance the expansion of memory-like NK and CD8+ T cells without triggering exhaustion (Romee et al. 2018). TIGIT blockade, often combined with PD-1 inhibitors, has shown synergistic effects by reversing dysfunction in tumor-infiltrating lymphocytes (Johnston et al. 2014). Epigenetic modulators such as histone deacetylase (HDAC) and DNA methyltransferase (DNMT) inhibitors can also rejuvenate exhausted T cells and upregulate antigen presentation machinery in tumor cells, thereby enhancing immune recognition and reducing relapse risk (Topper et al. 2017).

An innovative therapeutic concept involves introducing engineered, "tamed" tumor cells into the tumor microenvironment. These cells would be both therapeutically controllable and competitively dominant over more aggressive clones—a strategy that, in essence, uses cancer to fight cancer—"let's give cancer, cancer"—by leveraging the dynamics of the tumor ecosystem (Bin Masroni et al. 2025).

Beyond the cellular level, interventions can be developed to target cancer-specific homeodynamic mechanisms at higher organizational scales and to modulate the reciprocal interactions between tumor and host. A defining feature of cancer is its plasticity, both at the cellular and system levels. Reversible transitions—between glycolysis and oxidative phosphorylation, between stem-like and differentiated states, between dormancy and proliferation, between unicellular and multicellular behaviors, between epithelial and mesenchymal phenotypes, and between polyploidy and euploidy—are essential to cancer's adaptability. As shown in various biological systems, such transitional states often represent points of vulnerability (Di et al. 2021). Exploiting these transitional junctures may offer new therapeutic opportunities to disrupt cancer progression at its most adaptable and unstable moments.

12.9.1 Manipulating the Internal Body Climate for Therapeutic Purposes

One way of exploiting cancer cells' plasticity is to design interventions that change the micro and macroenvironments of the organism for therapeutic purposes. In Chap. 9, we introduced the concept of an internal body climate. While most current oncology treatments target cellular (e.g., chemotherapy, targeted agents) and tissue-level (e.g., immunotherapy, anti-angiogenics) processes, emerging strategies seek to modulate the organism's internal climate to both inhibit cancer progression and restore systemic homeostasis. Cancer not only thrives in, but also reshapes, this internal environment. Thus, body climate-targeted interventions may complement conventional treatments by influencing tumor behavior, improving host resilience, and enhancing therapy efficacy. The components of the organism's internal climate are interdependent, and in individual patients, their synergistic dysregulation can converge to create a permissive environment that fosters cancer development.

The depigmentation exhibited by the Himalayan rabbit is an example of the impact of the environment on an organism's phenotype. A Himalayan rabbit is completely white at birth, but within weeks, the fur on the rabbit's tail, feet, nose, and ears darkens. The rabbit is homozygous for a gene that codes a heat-sensitive enzyme. The enzyme is involved in the synthesis of melanin. On the body of the rabbit, heat inactivates the enzyme, and melanin production is blocked. Since the tail, feet, nose, and ears of the Himalayan rabbit are colder, they turn black (Utzeri et al. 2021; Demars et al. 2022) (Please see Fig. 12.2). The effect of body temperature changes on fur color is related to the presence of the heat-sensitive enzyme in the Himalayan rabbit's cell. Similarly, in cancer cells, specific genetic alterations may be associated with specific vulnerabilities to environmental factors that can be exploited for therapeutic purposes.

In a seminal study, Cao et al. investigated the impact of environmental enrichment (EE) on tumor progression in mouse models, revealing that a stimulating physical and social environment could significantly improve survival in cancer-bearing animals (Cao et al. 2010a, b; Cao and During 2012). EE consisting of complex housing with sensory, cognitive, motor, and social stimulation through increased physical activity, cognitive stimulation, and social interaction, has long been associated with neuroplasticity and improved stress resilience. Cao and colleagues extended this paradigm into oncology by demonstrating that EE induces a robust systemic effect that alters the host's internal milieu, thereby influencing tumor biology.

Fig. 12.2 When raised in low temperatures, Himalayan rabbits may exhibit black fur—when raised in higher temperatures, only their ears, tail, nose, and feet are black. (Courtesy of Ana Tourian)

Using mouse models of melanoma and colon cancer, the team housed mice either in standard conditions or in enriched environments featuring toys, running wheels, and enhanced social housing. Strikingly, mice in the enriched environments exhibited a significant reduction in tumors compared to controls. Mechanistic analysis revealed that EE activates the hypothalamic brain-derived neurotrophic factor (BDNF) signaling pathway, which in turn upregulates sympathetic nervous system activity and induces lipolysis in white adipose tissue. This cascade culminated in reduced leptin levels—a hormone known to promote cancer progression—thus creating an internal environment less supportive of tumor growth.

The findings by Cao et al. highlight a powerful neuro-immune-metabolic axis through which the brain, influenced by the environment, can regulate cancer progression.

Meditation and stress reduction techniques may also modulate immune surveillance, with recent data supporting immune activation following intensive retreats (Chandran et al. 2021). A mindfulness-based intervention of 48 randomized controlled studies totaling 4683 subjects found that compared to the control group, individuals practicing mindfulness meditation had decreased levels of C-reactive protein and interleukin-6, systemic inflammation markers (Dunn and Dimolareva 2022).

Age is the number one carcinogen as Ron De Pinho famously stated, and aging significantly impacts the internal body climate (De Pinho 2000). There are four hallmarks common to aging and cancer: genomic instability, epigenetic alterations, chronic inflammation, and dysbiosis, and thus, they can be considered meta-hallmarks of both (López-Otín et al. 2023).

Age-related biological differences have been observed across five common cancer types (Van Herck et al. 2021). In breast cancer, tumors in older individuals tend to exhibit a more indolent phenotype, characterized by lower histological grade, increased prevalence of luminal subtypes, and higher estrogen receptor (ER) positivity. In lung cancer, histological subtype distribution shifts with age, with a greater proportion of squamous cell carcinomas (SCCs), which are associated with poorer prognosis, compared to adenocarcinomas in older patients. Similarly, prostate cancer and melanoma demonstrate increased aggressiveness with advancing age. This is reflected in prostate cancer by higher Gleason or ISUP grades and elevated D'Amico risk classifications, and in melanoma by increased Breslow thickness, ulceration, and mitotic activity. Moreover, melanoma shows distinct age-associated patterns in tumor location and histological subtype. In colorectal cancer, tumor localization also varies with age: the incidence of right-sided colon cancers—which generally carry a less favorable prognosis—increases in older patients, whereas the occurrence of rectal cancers declines. In colorectal cancer, whereas incidence increases with age, metastatic spread declines (Pretzsch et al. 2022). These findings highlight the significance of age in shaping the biological characteristics and clinical behavior of various cancer types.

Senolytic agents targeting pro-tumorigenic senescent cells are being explored to rejuvenate the aged internal environment, particularly relevant given the increased cancer burden with age (Zhu et al. 2015; Saccon et al. 2021; Novais et al. 2021). On the other hand, the relationship between aging in cancer is very complex, and some hallmarks of aging have a tumor suppressive action (telomer attrition, stem cell exhaustion); thus, carefully designed and conducted rejuvenation trials are needed before senolytic agents can be recommended in the oncology setting.

Intermittent fasting has shown promise in reducing tumor growth, metastasis, and chemotherapy toxicity in preclinical models, though human data remain limited and context-dependent (Bauersfeld et al. 2018; de Groot et al. 2019; Clifton et al. 2021; Morita et al. 2021; Kalam et al. 2023). Similarly, deuterium-depleted water has demonstrated antiproliferative effects in vitro and in several clinical study reports (Somlyai et al. 1993; Krempels et al. 2008; Mirica 2010; Kovács et al. 2011; Syroeshkin et al. 2018; Zhang et al. 2020a; Yaglova et al. 2020; Boros et al. 2021; Somlyai et al. 2023, 2025).

Microbiota modulation via prebiotics, probiotics, or fecal transplantation can influence responses to immunotherapy, though translational consistency is still lacking (Gopalakrishnan et al. 2018; Goubet et al. 2018; Gong et al. 2019; Bouferraa et al. 2021). Circadian reinforcement (e.g., fixed meal timing) has also been shown to restore rhythmic gene expression and enhance immune function, with chrono-immunotherapy has demonstrated improved survival in animal and early human studies (Lévi et al. 1998; Li et al. 2010; Druzd et al. 2017; Scheiermann et al. 2018; Das et al. 2021; Downton et al. 2020; Qian et al. 2021; Diamantopoulou et al. 2022).

Bioelectric modulation, via Tumor Treating Fields (TTFields) (Mun et al. 2018; Rivera et al. 2019; Ceresoli et al. 2019; Ghiaseddin et al. 2020) and amplitude-modulated radiofrequency electromagnetic fields (Jimenez et al. 2019; Sharma et al. 2019; Blackstock et al. 2021), has gained traction, with devices now FDA-approved for glioblastoma, mesothelioma, non-small cell lung cancer, liver cell cancer, and soon to be approved pancreatic cancer (Rivera et al. 2019; Babiker et al. 2025).

Systemic physiological parameters like core body temperature offer additional therapeutic levers. *Hyperthermia*, long recognized for its synergistic effects with radiation and chemotherapy, also shows promise in immunomodulation (Repasky et al. 2013; Datta et al. 2020; Li et al. 2020c).

pH modulation, though limited by physiological buffering capacity, may help prevent metastasis and can be enhanced through targeting acid-base transporters or alkaline diets (Wike-Hooley et al. 1984; Hamaguchi et al. 2022; Bogdanov et al. 2024).

Using the laws and principles of processual biology and homeodynamics, integrative oncology therapies, defined as targeted interventions directed towards both the systemic and local malignant processes and functions, will, I hope, soon supplement precision oncology modalities and lead to more effective cancer prevention and cancer treatment programs.

12.9.2 Exploiting Cancer Cells' Plasticity for Treatment

Cancer cell plasticity—the ability of malignant cells to reversibly shift their phenotype in response to internal or external cues—is increasingly recognized as a central hallmark of malignancy (Hanahan 2022). This dynamic trait underlies cancer's remarkable adaptability, allowing tumor cells to survive in hostile microenvironments, resist therapy, evade immune surveillance, and metastasize. Traditionally considered a barrier to durable treatment responses, plasticity is often associated with the emergence of drug-resistant clones, cancer stem cell phenotypes, epithelial–mesenchymal transition (EMT), and lineage infidelity. However, this same plasticity that empowers cancer's resilience may also be harnessed as a therapeutic lever.

Rather than targeting fixed molecular features, a new therapeutic paradigm proposes to exploit the malleable nature of cancer by manipulating the surrounding environment in ways that influence phenotypic trajectories. Cellular behavior is not hardwired; instead, it is shaped by a network of microenvironmental cues—including cytokines, exosomes, extracellular matrix components, metabolic substrates, mechanical forces, and neuroendocrine signals. By altering this signaling landscape, it may be possible to steer plastic cancer cells away from aggressive, dedifferentiated states and toward more stable, quiescent, or even terminally differentiated phenotypes.

This plasticity-oriented approach aligns with the broader vision of *systems oncology*, advocated in this book, where the tumor is considered a complex adaptive system embedded in and responsive to its host environment. Interventions such as metabolic modulation, chronotherapy, immunomodulation, mechanical reconditioning, and even behavioral or psychosocial interventions may, when properly timed and combined, constitute powerful tools to "tame" cancer—not by eradicating every malignant cell, but by altering the evolutionary fitness landscape such that aggressive phenotypes are disfavored.

In this light, cancer cell plasticity is no longer viewed solely as a threat but as a therapeutic vulnerability. By embracing the dynamic and context-dependent nature of malignancy, oncology may move beyond static targets and enter an era of cancer modulation—a strategy not of blunt elimination, but of reprogramming, adaptation, and intelligent containment.

12.9.3 Taming Cancer Through Cellular Behavioral Therapies

Cancer cells are cognitive agents, and their adaptability and innate intelligence are untapped features that can be therapeutically exploited. Building on recent progress in predictive AIs, current IT tools now make it feasible to develop sophisticated models of a cancer cell that link pharmacological signals and internal/external environmental states to targeted modulation of metabolic and receptor pathways and can predict what component of these processes needs to be modulated for therapeutic effects, and also test those predictions. Through a novel approach that has been termed "*somatic psychiatry*" positive and negative reinforcement-based strategies borrowed from cognitive and behavioral sciences may exploit the plasticity of tumor cells and induce long-term adaptive changes (Mathews et al. 2023). Michael Levin and his collaborators have been advocating the use of bioelectricity to induce behavioral switches of cancer cells (Levin 2011; Mathews and Levin 2018; Davies and Levin 2023; Lagasse and Levin 2023). The use of "electroceuticals," devices able to manipulate neuronal signaling using electric pulses—like subcutaneous implantable vagal nerve stimulators championed by Kevin Tracey (Olofsson and Tracey 2017; Pavlov and Tracey 2022; Tracey 2025), the neurosurgeon that first described the inflammatory reflex (Tracey 2002, 2025; Pavlov and Tracey 2012), demonstrated some efficacy in inflammatory disorders (Tracey 2025) but its ability to impact cancer progression is still unproven. Noninvasive electroceuticals as, for example, ion channel targeting drugs (Pio-Lopez and Levin 2023; Balasubramanian et al. 2024) can also, potentially alter cancer behavior.

12.9.4 Hormesis and Its Therapeutic Potential in Cancer Treatment

Hormesis is defined by a dose-response relationship phenomenon characterized by low-dose stimulation and high-dose inhibition. The word "hormesis" was introduced in 1943, when Chester Southam and John Ehrlich discovered that at low concentrations, an extract of cedar heartworm stimulated growth in wood-decaying fungi; whereas at high doses, the same substance inhibited growth (Southam and Ehrlich 1943). Subsequently, it was found out that hormesis represents a universal phenomenon independent of chemical/physical agent, biological model, or endpoint measured (Calabrese and Baldwin 2003). The hormetic phenomenon demonstrated by Edward Calabrese and Linda Baldwin is a paradigm shift in toxicology, challenging the long-held belief in linear or threshold dose-response models. Mounting evidence supports the prevalence of U-shaped (hormetic) dose-response curves, where a low dose of a potentially harmful agent induces adaptive beneficial effects, while higher doses exert toxic effects. Historically observed in toxicology and radiation biology, hormesis has increasingly attracted interest in oncology, both for its implications in cancer prevention and its potential for therapeutic modulation. Key processes involved in hormesis include upregulation of antioxidant systems (i.e, the nuclear factor erythroid 2-related factor 2-Kelch-like ECH-associated protein 1 (Nrf2-Keap1) and sirtuin-forkhead box O (SIRT-FOXO) pathways), heat shock proteins, DNA repair enzymes, and anti-inflammatory mediators (Calabrese 2013). These adaptive responses are mediated by signaling cascades such as MAPK, PI3K/Akt, and AMPK, which help restore homeostasis and confer increased resistance to subsequent stressors. In both aging and cancer, hormetic responses represent double-edged swords. In aging, mild stress may induce stress tolerance and extend the lifespan, whereas excessive stress may accentuate the aging process (Schumacher 2009). This is known as the hormesis paradox in aging research (Rattan 2008; Salminen and Kaarniranta 2010). In cancer, it has been proposed that subclinical cancer may stimulate the immune system to fight cancer (Bordonaro and Lazarova 2024) as low levels of reactive oxygen species (ROS) may activate Nrf2, which can enhance cellular defense mechanisms, potentially preventing cancer initiation. However, high levels of ROS can overwhelm these defenses, leading to oxidative damage and promoting cancer progression (Bhakta-Guha and Efferth 2015).

Inducing hormetic responses to improve therapeutic outcomes and stimulate resilience of the organism tissues may open new avenues in cancer care (Wan et al. 2024).

Various hormetic interventions have been proposed: radio-hormesis using low doses of radiation (Scott, 2014); chemo-hormesis using low doses of chemotherapy (Yoshimasu et al. 2015), phyto-hormesis using phytochemicals for cancer prevention (Surh, 2003), neuro-hormesis using dietary and lifestyle interventions (Mattson 2008; Longo and Panda 2016; Pruimboom and Muskiet 2018). Combination regimens using immunotherapy in combination with low-dose cyclophosphamide have been shown to deplete regulatory T cells (Tregs), enhancing the efficacy of cancer vaccines and immune checkpoint inhibitors (Noordam et al. 2018). Also, combinations of low doses of chemotherapy and radiotherapy may be synergistic (Tang et al. 2023).

Leo Pruimboom and Frits Muskiet recommended the use of multiple "low dose" hormetic triggers as part of an "intermittent living protocol" of life-style challenges, combined in seven to ten-day interventions consisting of intermittent fasting, intermittent eating and food variety, intermittent water drinking, intermittent heat and cold, intermittent hypoxia, and intermittent hypercapnia. The outcome of this non-randomized protocol applied to 55 volunteers for ten days was an increase in transaminases and highly sensitive C-reactive protein (hsCRP). All study participants except one described a subjectively better feeling of well-being and health immediately after the ten-day intervention. The hypothesis that should be further investigated in a randomized study is that intermittent living protocol interventions may have beneficial effects on immunological, metabolic, and anthropometric parameters (Pruimboom and Muskiet 2018).

Despite promising preclinical data, translating hormetic principles into clinical practice is not easy. Cancer heterogeneity and evolutionary dynamics may render some tumors resistant to hormetic modulation. Moreover, the nonlinear nature of hormetic dose-responses complicates dose optimization and requires precise titration and timing. Recently, an in vitro study using a drug-resistant ovarian cancer cell line demonstrated that cisplatin (CDDP) and paclitaxel (PTX), paradoxically stimulate rather than inhibit cancer cell proliferation (Chang et al. 2025a). Also, it has been shown that, in general, high, rather than low, doses of SBRT have been associated with a hormetic-like abscopal (local irradiation leading to distal responses) effect (Endo et al. 2023; Swamy 2022; Wang et al. 2024). On the other hand, it was shown recently that high doses of radiation may increase metastasis (Piffkó et al. 2025); thus, clearly, the therapeutic effect of low versus high doses of interventions is context dependent.

In conclusion, in some contexts, hormesis-based interventions may be useful and serve as cancer treatments by themselves or adjuncts to standard treatments. As an example, in the next section, I will describe the design and the results of a Pilot study that investigated the therapeutic benefit of a radioactive glucose derivative (Paul et al. 2020a).

12.9.5 Using Low Doses of Radioactive ^{18}F-FDG as a Hormetic Cancer Treatment

The widespread observation of elevated glucose uptake across numerous human malignancies forms the foundation for the extensive clinical use of ^{18}F-fluorodeoxyglucose (^{18}F-FDG), a radiolabeled glucose analogue in which the hydroxyl group at the 2-position is replaced with a fluorine atom. Chemically known as 2-fluoro-2-deoxy-D-glucose, ^{18}F-FDG is a cornerstone of modern oncologic imaging. Its application in positron emission tomography (PET) has become a well-established diagnostic and monitoring modality for a broad spectrum of cancers—including non-small cell lung cancer, gastrointestinal malignancies, head and neck cancers, breast cancer, melanoma, and both Hodgkin and non-Hodgkin lymphomas—which consistently exhibit high FDG avidity.

Upon systemic administration, ^{18}F-FDG is transported into cells via glucose transporters and subsequently phosphorylated by hexokinase to form ^{18}F-FDG-6-phosphate. However, because this radiotracer is not a substrate for further glycolytic processing, it becomes metabolically trapped within cells (Gallagher et al. 1978). This property provides a favorable mechanism for delivering localized radioactivity to metabolically active tumors. Importantly, while 2-deoxyglucose (2-DG), a non-radiolabeled glycolytic inhibitor, has been explored as a therapeutic agent, high-dose administration has resulted in unacceptable toxicity, as demonstrated in clinical trials (Stein et al. 2010).

The therapeutic concept of using ^{18}F-FDG was inspired in part by the clinical success of radioactive iodine (^{131}I) in treating thyroid cancer. During my second year of fellowship, in 2000, I experienced a pivotal "aha" moment when I noted that the dosimetric properties of 100 mCi of ^{131}I-tositumomab compared favorably with those of just 2 mCi of ^{18}F-FDG. This observation sparked the idea of repurposing ^{18}F-FDG from a diagnostic to a therapeutic tool in cancers that are highly FDG-avid. The rationale was supported by the favorable physical properties of fluorine-18: a relatively short half-life of approximately 110 minutes, decay via high-yield (96%) β^+ emission, and the release of positrons with an average energy of 0.250 MeV and a tissue path length of approximately 0.1–0.2 cm.

As these positrons traverse tissue, they lose kinetic energy through interactions akin to those of electrons, producing localized ionization and tissue damage. This process culminates in annihilation events that emit two 511 keV photons in opposite directions—a principle central to PET imaging, but also indicative of potential therapeutic impact. The combined attributes of short-range, high-energy positron emission and selective metabolic trapping suggest that ^{18}F-FDG may serve not only as an imaging agent, but also as a novel, systemically deliverable radiotherapeutic—particularly for cancers with high glycolytic activity. Given these physicochemical characteristics, ^{18}F-FDG merits further exploration as a candidate for targeted metabolic radiotherapy.

The concept of using ^{18}F-fluorodeoxyglucose (^{18}F-FDG) as a therapeutic agent in cancer has been proposed for nearly three decades, and while I am neither the first nor the only proponent of this idea, I was able to demonstrate that relatively high systemic doses of ^{18}F-FDG can be safely administered to humans. Furthermore, preliminary evidence suggests this *systemic positron-based treatment* may offer clinical benefit. Early work by Meyer and colleagues in 1996 and 2006 showed tumor regression in glioma xenograft models following direct intratumoral FDG injection. Subsequently, researchers at the Albert Einstein College of Medicine reported that systemic administration of ^{18}F-FDG in a murine breast cancer model had cytotoxic effects on tumor cells and extended survival in mice bearing aggressive tumors (Moadel et al. 2003, 2005; Jaini and Dadachova 2012). Based on organ-level dosimetry data (Emami et al. 1991) and additional studies (Cristy and Eckerman 1987; Mejia et al. 1991; Deloar et al. 1998; Hays et al. 2002), the normal tissue uptake of FDG appears predictable and tolerable, with bone marrow toxicity being the dose-limiting factor. With an established marrow dose tolerance of 200 cGy, theoretical calculations suggest that systemic administration of up to 4.3 Ci of ^{18}F-FDG could be delivered safely to cancer patients.

To assess the safety, tolerability, and potential therapeutic utility of ^{18}F-FDG in humans, I designed a Phase I pilot study involving incremental dosing in patients with advanced, treatment-refractory cancers. After obtaining an Investigational New Drug (IND) approval from the FDA (IND #103704), we initiated a clinical study administering 100–200 mCi/m^2 of ^{18}F-FDG over a 1–5-day interval. Eligible patients included those with advanced solid tumors and lymphomas with high FDG-avidity and no remaining standard-of-care treatment options.

To date, four patients have been treated with a single systemic dose of 3.7–7.4 GBq/m^2 (100–200 mCi/m^2) of ^{18}F-FDG. All patients were monitored closely for toxicity and treatment response. Encouragingly, no treatment-related adverse events were observed, and no increased radiation exposure to research personnel occurred. One month post-treatment, all patients exhibited stable disease in the primary index lesion. Notably, two patients (#1 and #3) showed marked reductions in lesion standardized uptake values (SUVs), with declines of 17–33% on Day 1 and 25–31% on Day 30. Patient #3, with rapidly progressive diffuse large B-cell lymphoma, demonstrated a 31% SUV reduction at 24 hours and one month, accompanied by disease stabilization on CT per RECIST criteria and increased central necrosis of a renal lesion. Additionally, two lung nodules in this patient decreased in both size and metabolic activity. Post-treatment survival durations were as follows: Patient #1, 6 months; Patient #2, 14 months; Patient #3, 6.7 months; and Patient #4, 6.5 months. These outcomes are notable, given that approximately 70% of patients with advanced malignancies undergoing conventional chemotherapy have a median survival of less than 6 months (Haslem et al. 2018). For instance, patient #2 with brain-metastatic small cell lung cancer survived 14 months, exceeding the expected median of 6–9 months (Lekic et al. 2012; Bernhardt et al. 2017). Importantly, no dose-limiting toxicities (DLTs) or grade ≥ 1 adverse events were observed, supporting the safety profile and feasibility of dose escalation or repeat dosing in combination with immune checkpoint blockade.

Although the primary aim of this Phase I study was safety assessment, preliminary efficacy signals were observed in a heavily pretreated cohort. The disease stabilization seen across all patients following a single low dose, the observed tumor necrosis, and the prolonged survival in metastatic small cell lung cancer patients collectively provide a clinical correlate to preclinical data demonstrating ^{18}F-FDG-induced tumor apoptosis (Wang et al. 2016). These effects are particularly striking given that the estimated absorbed radiation doses to target lesions (0.63–0.8 cGy) were approximately 100-fold lower than the doses typically required for radiotherapeutic efficacy (Xu et al. 2008).

A potential explanation for these findings may lie in a hormetic effect—low-dose radiation-induced immunomodulation—which has been proposed in several publications (Duport 2003; Jolly and Meyer 2009; Calabrese 2014; Stone et al. 2018; Pruimboom and Muskiet 2018). This effect may have primed the immune system in a way that potentiated antitumor responses.

From a translational perspective, low-dose ^{18}F-FDG may function as a systemic, non-toxic priming agent that enhances the efficacy of immunotherapies such as PD-1 and PD-L1 inhibitors. Unlike focal radiation strategies that rely on the abscopal effect, or molecularly targeted agents that become ineffective with antigen downregulation, ^{18}F-FDG theoretically affects all metabolically active tumor sites concurrently (Weichselbaum et al. 2017; Siva et al. 2015; Bartelink et al. 2019).

Recent studies have also provided insight into metabolic competition within the tumor microenvironment. At the whole-tumor level, cancer cells account for roughly two-thirds of glucose uptake, while myeloid cells consume the remaining third. Uptake of other nutrients, such as glutamine and lipids, is largely driven by cancer cells, both in per-cell and total uptake metrics (Reinfeld et al. 2021). Given that glutamine is the most abundant amino acid in plasma and many tumors are glutamine-dependent (Li et al. 2024), dual inhibition of glucose and glutamine metabolism may be therapeutically advantageous.

In this context, the development of radiolabeled glutamine analogs such as ^{18}F-(2S,4R)-4-fluoroglutamine has enabled noninvasive glutamine imaging via PET (Zhu et al. 2017). A therapeutic strategy combining ^{18}F-FDG and ^{18}F-fluoroglutamine—or another ^{18}F-glutamine-based tracer—may enhance the effects observed in this pilot study, offering a dual metabolic disruption approach with broad systemic reach and minimal toxicity.

12.9.6 The "Scottish Shower" Strategy in Cancer Therapy: Alternating Receptor Modulation as a Novel Strategy

Biological systems are homeodynamic and their existence depends on maintaining key parameters within a certain range. This balance is achieved through complex complementary yin-yang mechanisms. For example, in cancer, apoptosis acts as a double-edged sword (Morana et al. 2022). While traditionally viewed as a tumor-suppressive process, particularly in its role in removing genetically unstable cells, its function in cancer is paradoxical. High-grade tumors often display elevated basal levels of apoptosis, reflecting their complex role. On one hand, apoptosis functions as a cell-autonomous tumor suppressor and is a target of many conventional therapies. On the other hand, limited or localized apoptosis within tumors can paradoxically promote survival and therapeutic resistance by shaping the tumor microenvironment—modulating phagocytic responses and influencing surrounding viable tumor cells—thereby contributing to tumor progression. A deeper understanding is needed to distinguish between deletional apoptosis, which irreversibly removes cells (as seen in developmental tissue sculpting), and regenerative apoptosis, which may stimulate compensatory proliferation and tissue remodeling. The dual role played by apoptosis in cancer raises the question of whether a combination of treatments that both inhibit and promote apoptosis may be beneficial in some clinical scenarios.

Classical approaches to cancer therapy typically involve sustained inhibition of oncogenic pathways and continuous administration of targeted agents. Other administration strategies are also effective. As an example, sometimes, high doses of tyrosine kinase inhibitors (TKIs) can overcome the resistance associated with lower doses of TKIs. For instance, pulsatile administration of high-dose EGFR inhibitors has shown efficacy in EGFR-mutant non-small cell lung cancer (NSCLC) in a cohort of patients that developed resistance to regular doses (Zhu et al. 2016). A similar strategy was used in a patient with brain and leptomeningeal metastases that progressed on a standard dosage of icotinib (125 mg three times daily). The patient had a near-complete response and prolonged survival using high doses of pulsatile icotinib (1125 mg every 3 days) (Li et al. 2020a). Also, in the context of prostate cancer, intermittent androgen deprivation therapy (IADT) has been shown to delay resistance and reduce toxicity (Crook et al. 2012).

Another strategy is a dynamic modulation of receptor signaling, through concomitant or sequential activation and inhibition. This principle mirrors the "Scottish shower" method in physiology, wherein alternating exposure to hot and cold stimuli enhances vascular and systemic tone. Applied to oncology, this approach involves the use of partial agonist-antagonist agents or alternating stimulation and inhibition of key receptors, aiming to destabilize tumor cell homeostasis and induce therapeutic stress. The "Scottish shower" analogy creates a state of *controlled instability* that impairs cancer cell adaptation and enhances immune or therapeutic engagement. This strategy could be implemented through:

- *Partial antagonists-agonists* drug development.
- *Pulsatile ligand delivery* followed by receptor inhibition.
- *Drug cycling* of agonists and antagonists.
- *On-off kinase inhibition regimens*, leveraging kinetic windows of cell cycle vulnerability.

Such dynamics may be especially relevant in tumors with *high plasticity,* where signaling fluctuations influence differentiation, migration, and resistance.

Many cancer-associated receptors, including hormone receptors (e.g., estrogen, androgen), immune checkpoints (e.g., PD-1, CTLA-4), and growth factor receptors (e.g., EGFR, HER2), are part of tightly regulated signaling networks that exhibit feedback loops, adaptive resistance, and context-dependent outcomes. Continuous inhibition of these pathways often leads to adaptive rewiring, drug resistance, or activation of compensatory circuits (Holohan et al. 2013). This resistance is partially avoided by the use of partial agonists-antagonists. Classical agents like Tamoxifen, for example, widely used in estrogen receptor (ER) positive breast cancer, are partial ER receptor agonists-antagonists (Jameera Begam et al. 2017).

Using agents like partial agonist-antagonist agents like Tamoxifen and alternating activation and inhibition may prevent adaptation of cancer cells by disrupting feedback regulation, expose transient vulnerabilities during signaling transitions, induce cancer cells exhaustion, metabolic stress, or apoptotic priming, also possibly improve immune visibility through intermittent activation-induced antigen presentation.

12.10 Using Life's Principles for Cancer Treatment

12.10.1 Cancer Treatments Using the Bioenergy Principle

Using Thermodynamics for Cancer Treatment

Can entropy differences between cancer tissues and normal tissues be exploited for cancer treatment? The cancer entropy generation can change in time and in space as a consequence of tumor growth (Lucia 2013). In the case where no external energy is applied, Luo calculated that for different sizes of cells in the human body, the rate of entropy production of cancerous cells is always higher than that of healthy cells (Luo 2009).

An Italian group led by Umberto Lucia tested the hypothesis that the extremely low-frequency electromagnetic field (ELF-EM) induces an increase in entropy coupled to a decrease in cell growth (Lucia et al. 2017, 2020). The negative effects of electromagnetic fields on cancer proliferation has been demonstrated in vitro on six cancer cell lines (Lucia et al. 2017).

Interestingly, the same electromagnetic wave frequency had distinct effects on different cell types, and in every cell type, different parameters of the ELF-EM impacted differently the energy utilization and proliferation.

In recent years, the use of electricity to treat cancer has led to promising results. Tumor-treating Fields (TTFields) are low intensity, intermediate frequency, alternating electric fields delivered externally through noninvasive transducers placed locoregionally around the anatomic region of the tumor (Mun et al. 2018). TTF has been FDA approved for the treatment of brain cancer (Ghiaseddin et al. 2020) and mesothelioma (Ceresoli et al. 2019). Low-energy amplitude-modulated radiofrequency electromagnetic fields also appear to be a promising modality for systemic cancer treatment (Tuszynski and Costa 2022). Since 2008, a mouthpiece using tumor-specific amplitude-modulated radiofrequency electromagnetic fields (AM RF EMF) has been also approved in Europe for the treatment of liver cancer (Jimenez et al. 2019).

The mechanism of TTF and AM RF EMF treatments is thought to be the activation of specific calcium channels, resulting in increased levels of cytosolic Ca2+ (Jimenez et al. 2019). A similar mechanism has been described for ELF-EM in different cell types (Wei et al. 2015).

In another application of thermodynamic concepts to living cells considered dynamic non-equilibrium systems, it has also been suggested that cancer can be seen as a first-order dynamic phase transition akin to water freezing or boiling (Davies et al. 2011). As a corollary of this model the authors suggested that given the "long-range correlation effect, which is a characteristic property of systems undergoing phase transitions [...] a truly successful therapy would require a global change of conditions disfavoring the cancer phenotype and not simply a local excision or destruction of cancer cells in their micro-environment." As examples of global interventions Paul Davies and his co-authors recommended "metabolic interventions including caloric restrictions, nutritional supplements and vitamins (as antioxidants), physical exercise, metabolic modulators that activate mitochondrial enzymes and/or inhibit glycolytic pathways, and [...] anti-inflammatory strategies" (Davies et al. 2011).

Targeting metabolic vulnerabilities of the cancer cells is an attractive alternative to molecular targeting, but, in general, metabolic networks are currently constructed based on group-level metabolic covariance rather than individual-level connectivity, and therefore, cannot be directly used for individualized disease prediction. Only by developing tools able to analyze individualized metabolic networks on a case-by-case basis can personalized successful metabolic targeting approaches be designed in the future (Meng et al. 2024).

Mitochondria Transfer

Mitochondrial transfer has recently garnered attention due to its potential role in cancer (Artusa et al. 2025). The cancer process needs a lot of energy, and cancer can deplete T cells of their mitochondria. Horizontal MT transfer between mammalian cells is mediated by tunneling nanotubes (TNTs), gap junctions, macrovesicles, or direct extrusion and internalization, and it was found that TNT formation is the most common mechanism for intercellular mitochondrial transport (Zhang et al. 2023a). Disrupting mitochondria transfer through actin filament network regulators, inhibitors, or through targeting the cell adhesion molecule PVRL2 (encoding Nectin-2) are some of the strategies currently evaluated.

An opposing strategy, powering the T cells by transferring mitochondria from normal cells to T cells, is also pursued by the team of Luca Gattinoni, who has developed an innovative method to transfer healthy mitochondria from donor cells into T cells (Baldwin et al. 2024). This process involves using bone marrow stromal cells (BMSCs) to form tiny, nanotubular connections with T cells. Through these connections, the BMSCs can deliver their mitochondria directly into the T cells. This transfer is not just a simple exchange; it supercharges the T cells, which gives them more energy and enhances their ability to fight cancer.

Nutrition Interventions

Intermittent Fasting (IF) induces a metabolic switch from glucose-based to ketone-based energy utilization, with implications for insulin sensitivity, inflammation, autophagy, and cellular stress resistance. IF has been proposed as a means to decrease tumor burden as well as the negative effect of chemotherapy (Nencioni et al. 2018; Omar et al. 2022). For instance, in several mouse models, intermittent fasting reduced tumor growth, as measured by weight or volume, decreased the rate of metastasis, and prolonged survival (de Groot et al. 2019). Some of these studies are controversial and suggest potential detrimental effects in certain oncologic conditions. Also, the human data is currently scarce (Bauersfeld et al. 2018). Preliminary studies suggest that prolonged fasting in some patients who have cancer is safe and capable of decreasing chemotherapy-related toxicity and tumor growth (Clifton et al. 2021).

Cancer dependency on certain nutrients may also be utilized for therapeutic purposes. Importantly,

cancer cells have different nutritional preferences based on genetic alterations, tissue of origin, and degree of interaction with systemic hormones and metabolites, so clearly dietary interventions and a combination of systemic pharmacologic partners with dietary interventions need to be personalized for cancer treatment (Taylor et al. 2022). Unfortunately, the strong dependency of cancer cells on glucose and glutamine, the preferred nutrients by most cancer cells, is not easily translated into a targetable vulnerability (Kanarek et al. 2020). There are several nutriments other that can impact cancer development and can be targeted (Morita et al. 2021). The most significant impact of nutritional manipulation in humans was reported in a randomized study of women who used a low-fat diet (LFD). Compared with the control arm in which women consumed a usual diet, the women enrolled in the LFD group had a statistically significant lower incidence of deaths after breast cancer (Chlebowski et al. 2006).

In a mouse model of breast cancer, asparagine restriction reduced the metastasis of breast cancer without affecting the growth of the primary tumor (Knott et al. 2018). However, asparagine restriction is difficult to implement in human patients. Other amino acid-depleted diet approaches may be more feasible for patients (Kanarek et al. 2020). Serine depletion, for example, either by removing it from the diet or by enzymatic inhibition, may have a therapeutic benefit, especially in tumors with TP53 mutations (Maddocks et al. 2013). Glycine removal from the diet reduced tumor growth in xenograft and allograft models (Maddocks et al. 2017). Also, adenosine depletion by preventing the conversion of adenosine monophosphate to adenosine, using Oleclumab, a human IgG1 mAb that selectively inhibits CD73, is safe and open to use in clinical trials (Segal et al. 2024). Time-restricted feeding was shown to reduce the rates of tumor metastasis in mice (Li et al. 2010; Yan et al. 2019; Das et al. 2021).

A modified lipid diet may curb the growth of certain cancer types (Carracedo et al. 2013). As an example, depriving ovarian cancer cells of unsaturated fatty acids by inhibiting the stearoyl Co-A desaturase (SCD) enzyme involved in their production, and simultaneously introducing a saturated fatty acid-enriched diet, may block both ovarian cancer cells' growth and their dissemination (Zhao et al. 2022).

Another nutrition intervention strategy is time-restricted feeding. Early time-restricted feeding limiting daily food intake to a period of 10 hours or less, followed by a daily fast of at least 14 hours, was demonstrated to improve insulin sensitivity, blood pressure, and oxidative stress, without weight loss in men with prediabetes (Sutton et al. 2018) and represents an attractive anti-cancer intervention that should be tested in clinical studies.

Other Approaches Using Cancer Cells' Addiction to Certain Nutrients

The most commonly used nutrients of cancer cells are glucose and glutamine. It has been demonstrated that cancer cells need both a durable supply of energy—usually obtained through aerobic glycolysis using glucose as fuel to produce ATP, and intermediate metabolites, especially citrate, which can be further converted to acetyl-CoA as a building block for the synthesis of fatty acids. Both glucose through the pentose phosphate cycle and glutamine through the TCA cycle can provide these intermediate metabolites (Vander Heiden et al. 2009; Wang et al. 2018). Glucose and glutamine pathways are interdependent, since depletion of glucose causes cell growth inhibition and cell death in most cell types, even when glutamine supply is present, and glutamine shortage also inhibits glucose uptake in pancreatic cancer cells, for example (Kaadige et al. 2009). Cells that convert glucose and glutamine into biomass most efficiently will proliferate the fastest (Vander Heiden et al. 2009).

2-deoxyglucose (2-DG) has long been recognized as an effective inhibitor of glucose metabolism in vitro and in different animal models. A stage I study using 2-DG using was able to identify the therapeutic dose of 2-DG, but the clinical efficacy was minimal due to a parallel increase in autophagy. The dose-limiting toxicity of 2-DG was QT prolongation (Stein et al. 2010), and further studies were no longer pursued. Also, a Phase I study of telaglenastat, a glutaminase inhibitor, demonstrated a disease control rate (DCR = PR + SD) of 50% in renal cell carcinoma (Harding et al. 2021) that prompted subsequent clinical studies using combinations of telaglenastat and everolimus or, telaglenastat and cabozantib. The Phase I study of telaglenastat, in combination with either everolimus or cabozantinib, showed for both combinations encouraging clinical activity and tolerability in heavily pretreated patients with metastatic RCC for both combinations (Meric-Bernstam et al. 2022). These results were confirmed in subsequent Phase II studies just for the combination of telaglenastat and everolimus, and also showed a modest non-statistically significant improvement in PFS from 3.8 to 1.9 months (Lee et al. 2022; Tannir et al. 2022). Interestingly, glutamine supplementation, although not currently FDA approved, has been demonstrated to mitigate toxicities from chemotherapy or radiotherapy and is generally considered safe (Muranaka et al. 2024). The fine balance between glutamine, on one hand, providing a substrate for normal cellular functions, and, on the other hand, potentially supporting malignant cells, underscores the importance of careful consideration and individualized approaches in the clinical use of both glutamine inhibition and supplementation. Cancer cells can use other amino acids (i.e., asparagine, aspartate, glutamate) to compensate for glutamine limitation or inhibition of glutamine catabolism. Perturbing their acquisition and metabolism should be considered as a complementary approach when we target glutamine metabolism in cancer (Jiang et al. 2019).

The Danger of Ketogenic Diets

A ketogenic diet has been advocated for several years now as a beneficial diet for cancer patients, but caution is needed as ketone body utilization may drive tumor growth and metastasis (Israël and Schwartz 2020).

It has been shown, for example, that catabolic fibroblasts present in the tumor stroma, with mitochondrial dysfunction, produce ketone bodies that are further used by cancer cells. Cancer cells are able to use ketone bodies as mitochondrial fuels for energy oxidative phosphorylation (OXPHOS), through the activity of ACAT1/2 and OXCT1/2 mitochondrial genes, thus driving anabolic tumor growth (Martinez-Outschoorn et al. 2012). The authors noted: "To prevent this form of "two-compartment tumor metabolism," ketone inhibitors should be designed to halt ketone body production in cancer-associated fibroblasts and ketone body re-utilization in epithelial cancer cells. This simple strategy could effectively starve cancer cells to death by "cutting off their fuel supply" (Martinez-Outschoorn et al. 2012). Bela Ozsvari et al. proposed a new class of drugs labeled "mitoketoscins," designed specifically to target ketone re-utilization and mitochondrial function by inhibiting ACAT1 and OXCT1 genes (Ozsvari et al. 2017).

The key enzyme in ketolysis is Succinyl-CoA: oxoacid—CoA transferase (SCOT), product of the OXCT1 gene. In cancer metabolism, SCOT, becomes a vulnerable point, since cancer cells are unable to use glucose to synthesize acetyl-CoA, as normal cells would do. Maurice Israël et al. have recently published a comprehensive description of several drugs able and natural compounds able to inhibit SCOT (Israël et al. 2024).

It has been demonstrated in several large longitudinal cohort studies that specific metabolome changes can be identified several years before clinical cancer inception. Currently, metabolic networks are generally constructed based on group-level metabolic covariance rather than individual-level connectivity, and therefore, cannot be directly used for individualized disease prediction (Meng et al. 2024). The characterization of personalized metabolic networks represents an important future research direction in the field.

Given the importance of metabolism on multiple essential cancer cell characteristics (i.e., division, invasion, metastasis) and the need shown by aggressive cancers for increased energy supplies (Lieffers et al. 2009), metabolism is an attractive target in cancer treatment.

Due to the plasticity of cancer cells that can quickly shift the metabolism in different directions and use different metabolic pathways to survive, targeting cancer metabolism is very challenging (Läsche et al. 2020). This metabolic flexibility is responsible for the largely inefficient use of targeted nutrition interventions.

12.10.2 Cancer Treatments Using the Bioinformation Principle

The strategy of targeting abnormal cancer pathways has been demonstrated by the development of small molecules and targeted antibodies. In the United States, there are currently more than 100 targeted approved agents (NCI website: https://www.cancer.gov/about-cancer/treatment/types/targeted-therapies/approved-drug-list). Epigenetic-directed treatments are currently limited, but soon, given the significant involvement of ncRNA in cancer, RNA-based or directed therapeutic strategies will be developed. Synthetic lethality targets specific gene interactions. Novel bioinformation-informed cancer treatments can also be designed using other cellular codes besides the genetic, epigenetic, or the second messenger codes already exploited for therapeutic purposes (see Marcello Barbieri's list of codes Sect. 8.2). As an example, exosomes—messengers of an intercellular communication code—can be targeted for treatment.

Epigenetic Directed Treatments

Epigenetic targeted therapies are currently limited to the treatment of hematological malignancies (i.e., DNA methyltransferase inhibitors, Azacitabine and Decitabine used in myeloid neoplasms; histone deacetylase inhibitors, Vorinostat and Belinostat used in T-cell lymphomas; Panobinostat used in multiple myeloma, etc.). Alone or in combination, Bromodomain and extraterminal domain (BET) inhibitors, lysine demethylase (KDM) inhibitors, and lysine methyltransferase (KMT) inhibitors are tested in several solid tumor clinical trials (Yu et al. 2024).

Synthetic Lethality

The signaling pathways altered in a particular cancer can be compensatory or mutually exclusive. Examples of mutual exclusivity include KRAS and EGFR mutations in human lung adenocarcinomas, NRAS and BRAF mutations in melanoma, and PTEN and PIK3CA in sarcoma. Importantly, these mutual exclusivities are tissue-specific. The same mutations that are mutually exclusive in a particular tissue can coexist in a different tissue (Peters et al. 2024). In some cases, genetic alterations of signaling pathways may present a "conditional dependence" on other partner genes or compensatory pathways, and if we therapeutically or genetically inactivate the partner, it will selectively affect cancer cells with the inactivated gene, sparing normal cells and leading to synthetic lethality (Cereda et al. 2016). Synthetic lethality/sickness (SL) is a well-described type of genetic interaction, defined as cell death or reduced cell viability that occurs under the combined inactivation of two genes, but not under the inactivation of either gene alone (Cheng et al. 2021). One application of synthetic lethality in oncology currently introduced in clinical practice is the use of poly (adenosine diphosphate (ADP)-ribose) polymerase (PARP) inhibitors in patients with germline-loss-of-function mutations in either BRCA1 or BRCA2 genes. Three BRCA mutations can impair DNA-repairing mechanisms and improve cancer cells' sensitivity to DNA-damaging chemotherapeutics and PARP inhibitors. Olaparib was the first PARP inhibitor (PARPi) approved in 2014 to treat women with advanced ovarian cancer associated with defective BRCA genes (Falchi et al. 2017). PARPi treatments are currently approved in ovarian, primary peritoneum, fallopian tube, breast, pancreatic, and prostate cancers that have BRCA1/2 pathogenic variants or other markers of homologous recombination repair deficiency (HRD).

Several other gene-based synthetic lethal combinations have been described in in vitro systems (Ashworth and Lord 2018). Genetic and epigenetic synthetic lethality pairs have been also recently described (Ngoi et al. 2025).

Exosomes-Based Cancer Treatments

Inter-cellular and inter-organ communication is fundamental for maintaining physiological homeostasis and becomes dysregulated in the context of disease, particularly during the initiation and progression of systemic pathologies such as cancer. In oncogenesis, the host organism plays an active, rather than passive, role by contributing to tumor development from its earliest stages.

Once established, the tumor exerts widespread systemic effects by co-opting host signaling networks to facilitate its own growth and progression. A key mechanism by which tumors mediate long-range communication is through the release of extracellular vesicles, including exosomes. These vesicles have been implicated in the crosstalk between primary tumors and distant metastatic sites (Wortzel et al. 2019). Notably, tumor-derived exosomes are enriched with microRNAs (miRNAs) that have been shown to modulate multiple systemic hallmarks of cancer, influencing processes such as immune evasion, angiogenesis, and metastasis (Schwarzenbach et al. 2019).

Exosomes are essential players in biological communication. A potential role of exosomes is using them in early cancer diagnostics as biomarkers (Hoshino et al. 2020; Yu et al. 2022). Exosome-directed therapies are currently in development (Lucotti et al. 2022).

It has been recently demonstrated, for example, that prothrombotic small extracellular vesicles (sEVs), secreted by interstitial macrophages in the non-metastatic lung microenvironment—termed the pro-thrombotic niche (PTN)—promote platelet aggregation via β2 integrin. Blocking β2 integrin reduces both sEV-induced thrombosis *and* lung metastasis, and potentially represents a novel treatment strategy (Lucotti et al. 2025). Exosomes can also be removed through plasmapheresis (Orme et al. 2020), they can be targeted through genetic manipulations or using new classes of drugs (Li et al. 2022b), or they can be used as drug delivery carriers (Koh et al. 2023). Exosome-based drug delivery is emerging (Akbari et al. 2022; Koh et al. 2023; Basyoni et al. 2025), and exosome-based treatments have the potential to revolutionize therapeutic interventions in cancer.

12.10.3 Cancer Treatments Using the Biostructure Principle

Mina Bissell's research has significantly contributed to our understanding of how the physical properties of the tumor stroma, particularly the extracellular matrix (ECM), can regulate cancer cell behavior. Her studies emphasize the concept of "dynamic reciprocity," where cells and their surrounding ECM engage in continuous bidirectional communication, affecting gene expression and cellular phenotype (Bissell et al. 2005). Mina Bissell introduced the theory of dynamic reciprocity, highlighting that the ECM not only provides structural support but also transmits biochemical and mechanical signals that influence cellular function. This interaction is crucial for maintaining tissue architecture and function. Disruption of this balance can lead to loss of tissue organization and promote malignant transformation. For instance, degradation of laminin-111 by matrix metalloproteinase-9 (MMP9) disrupts tissue architecture, leading to increased cell proliferation and tumor growth (Spencer et al. 2011). Inhibiting the formation and composition of primary (Murgai et al. 2015), or pre-metastatic tumor niches (Hiratsuka et al. 2002; Kaplan et al. 2005; Hiratsuka et al. 2008; Peinado et al. 2008; Psaila and Lyden 2009) may interrupt cancer's life cycle (Pătraș and Paul 2024).

The biostructure of an organism can be conceptualized as complex, hierarchically embedded networks of interactions spanning multiple layers. At the cellular level, networks consist of genes, various RNA species (e.g., microRNAs, long non-coding RNAs), signaling molecules (including proteins, lipids, and ions), and metabolic intermediates. At the tissue level, networks encompass interactions among diverse cell types as well as between cells and the surrounding stromal environment. At the organismal level, networks involve communication and coordination among organs (such as the heart and lungs) and physiological systems (including the endocrine, nervous, and immune systems). These networks are not isolated; they interact in two principal ways: horizontal interactions, which occur between networks operating at the same level (e.g., between metabolic and genetic networks), and vertical interactions, which connect networks across different hierarchical levels (e.g., linking cellular, tissue, and organismal dynamics). Overall, network interactions are multidirectional, with both *horizontal* and *vertical* influences shaping biological function and regulation (Paul and Komarova 2021). Effectively dismantling the complex, multi-layered network that characterizes cancer requires innovative, multi-scale therapeutic strategies that concurrently target critical components of cancer networks at the cellular, tissue, and systemic levels. Certain pharmacological agents, such as beta-blockers (Kokolus et al. 2017; Switzer et al. 2024) and calcium channel blockers (Jacquemet et al. 2016; Tajada and Villalobos 2020; Liu et al. 2024c), can act *vertically* across different scales, simultaneously influencing organismal, tissue, and cellular pathways involved in tumor progression, and, in some cancer types, have demonstrated clinical benefit. However, the antitumor effect of beta-blockers appears to be context-dependent, with emerging evidence highlighting potential pro-tumorigenic effects associated with β-adrenergic blockade in some cancer types. A recent study, for example, described worse relapse-free survival associated with beta-blockers used in patients with head and neck cancers (Chen et al. 2023b). These observations indicate that β-blockade may exert dual, and at times opposing, effects on both cancer cell behavior and immune modulation. This duality has not been observed for calcium channel blockers, and their use is probably safer across multiple malignant tumors (Liu et al. 2025).

12.10.4 Cancer Treatments Using the Autopoiesis Principle

Can Cancer Be Cured Using Chemotherapy?

Chemotherapy agents or tyrosine kinase inhibitors are widely used to curb cancer division and, in some settings, can lead to cancer cures. In a 2014 paper, I described the case of a 72-year-old woman diagnosed in 2000 with a stage IV locally metastatic colon cancer with large liver metastases treated with surgery for her primary tumor, with the liver metastases left in place due to their sizes (Paul et al. 2014). After surgery, the patient received chemotherapy for one year, using the Roswell Park regimen consisting of weekly 5-Fluorouracil (5-FU) (500 mg/m^2) and leucovorin (LV) (20 mg/m^2) for 6 weeks, followed by a 2-week drug-free period that led to a complete remission. 14 years later, at the time of the publication of the case report, the patient, then 86 years old, was still alive without any sign of recurrent disease. Despite our efforts to understand the reason for her clinical evolution that included a whole exome analysis performed by Bert Vogelstein at Johns Hopkins University, her cure remained a mystery. Interestingly, one month following the initial diagnosis of colon cancer, the patient was readmitted to the hospital for an acute saddle pulmonary embolism. Anticoagulation with coumadin was initiated, and the patient was maintained on coumadin for the full period of her follow-up. I wonder whether the coumadin anticoagulation given in combination with the 5FU/LV chemotherapy contributed to this patient's long-term survival. More than four decades ago, a clinical study demonstrated a greater overall median survival for patients who were treated with both Warfarin and 5-FU, but these earlier results were never reproduced (Chlebowski et al. 1982).

Autophagy Dysregulation

Autophagy, a cellular degradation process, plays a dual role in cancer by, depending on the context, either promoting or inhibiting tumor growth. It can support cancer cell survival under stress by recycling cellular components, but excessive autophagy may lead to apoptosis. Autophagy helps maintain cellular homeostasis by degrading and recycling cellular components, which can support cancer cell survival under stress conditions like nutrient deprivation (Chen et al. 2021; Vitto et al. 2022). It acts as a double-edged sword: promoting tumor growth in advanced stages while inhibiting tumor development in early stages by enhancing tumor suppressor activity (Chavez-Dominguez et al. 2020; Vitto et al. 2022). Autophagy can lead to treatment resistance by providing energy for cancer cell proliferation and evading apoptosis induced by therapies (Carretero-Fernández et al. 2025). Therapeutic strategies aim to modulate autophagy, either by inhibiting it to induce cancer cell death or stimulating it to overcome resistance in certain contexts (Tonkin-Reeves et al. 2023; Ganzleben et al. 2021). The dual role of autophagy necessitates a nuanced understanding of its mechanisms to effectively harness it for therapeutic purposes.

Splicing Alterations

The spliceosome is a complex molecular machine responsible for removing non-coding introns from precursor messenger RNA (pre-mRNA) transcripts, a process known as splicing. Dysregulation of splicing is a common hallmark of cancer cells and is associated with altered expression of splicing factors and spliceosome components. It has been discovered that splicing is a key pathway deranged in cancer (Agrawal et al. 2018), and alterations in the splicing of primary transcripts of protein-coding genes also contribute to carcinogenesis (Zhang et al. 2021b). Changes in the splicing pattern of proto-oncogenes may generate active oncoproteins, and abnormal splicing of tumor suppressor genes may lead to inactive forms of tumor suppressor proteins (Dvinge et al. 2016). Epithelial splicing regulatory proteins (ESRPs), including ESRP1 and ESRP2, and RBFOX2 as have been identified as important metastasis regulators (Lu et al. 2015). Several approaches have been explored to target the spliceosome in cancer:

- *Small molecule inhibitors*: Compounds like H3B-8800, Indisulam, and Tetrocarcin A have been identified as direct spliceosome inhibitors. These molecules interfere with different stages of spliceosome assembly, leading to the accumulation of unspliced or mis-spliced transcripts, which can trigger cell death pathways in cancer cells (Bashari et al. 2023).
- *Targeting splicing factors*: Specific splicing factors, such as SRSF6, RBM39, and PRMT5, have been implicated in cancer progression. Inhibiting these factors can modulate alternative splicing events and potentially reverse the protumorigenic effects of aberrant splicing (Ivanova et al. 2023).
- *Oligonucleotide-based therapies*: Antisense oligonucleotides and small interfering RNAs (siRNAs) can be designed to target and degrade specific splicing factor mRNAs or modulate alternative splicing patterns of oncogenes or tumor suppressors (Bowling et al. 2021; Bonner and Lee 2023). The inhibition of the spliceosome can lead to various cellular consequences in cancer cells, including: unproductive splicing and nonsense-mediated decay of transcripts involved in DNA repair (e.g., CHEK2), sensitizing cells to DNA damage; generation of pro-apoptotic protein isoforms (e.g., Mcl-1S), promoting cell death; accumulation of double-stranded RNA in the cytoplasm, triggering antiviral signaling and apoptosis; modulation of alternative splicing events that regulate tumor cell proliferation, invasion, metastasis, and chemoresistance.

Currently, the direct targeting of the spliceosome is not used for the treatment of cancer due to increased toxicity, but targeting splicing regulatory proteins is an active area of investigation (Marzese et al. 2018; Stanley and Abdel-Wahab 2022; Bonner and Lee 2023). Ongoing research aims to improve the specificity, potency, and delivery of spliceosome-targeting agents, as well as exploring combination therapies to enhance their efficacy.

Proteostasis Targeting

Most cancers require a fine regulation of protein homeostasis (proteostasis) to maintain growth; thus, proteins comprise the most important class of drug targets (Miklos and Maleszka 2001).

Disruption of proteostasis networks contributes to the development and progression of cancer (Voutsadakis 2017), and the levels of proteasome in cancer cells are, in general, higher compared to their normal counterparts (Chen et al. 2017). The proteasome is involved in a multitude of cellular processes including cell cycle, cell growth, apoptosis, immune system, cancer stemness, oxidative stress, gene transcription, signal transduction, and differentiation (Li and Zhan 2021). Many cancers need an active proteasome for degrading tumor suppressors and cell cycle checkpoint inhibitors necessary for rapid cell division (Thibaudeau and Smith 2019).

PROTACs (Proteolysis-Targeting Chimeras) are a novel class of therapeutic agents that induce targeted protein degradation by hijacking the cell's natural protein degradation machinery. They consist of two ligands connected by a linker—one ligand binds to the target protein of interest, and the other recruits an E3 ubiquitin ligase. This ternary complex formation leads to ubiquitination and subsequent proteasomal degradation of the protein of interest. Conceptually, PROTAC aims to target the "undruggable" majority of proteins in the human proteome (Li et al. 2023a). PROTACs offer several advantages over traditional small molecule inhibitors (Li et al. 2022a; Huo et al. 2022, Liu et al. 2022; Kelm et al. 2023; Sincere et al. 2023):

- They eliminate all functions of the target protein, not just inhibiting one activity.
- They can overcome drug resistance mechanisms that arise from target mutations.
- They can simultaneously degrade multiple targets by incorporating ligands for different POIs.

However, PROTACs face challenges like poor oral bioavailability due to their large size, limited repertoire of E3 ligases utilized, cell permeability, in vivo stability, and potential off-target toxicity from complete target degradation (Sincere et al. 2023; Li et al. 2023). Ongoing research aims to improve PROTAC design principles, explore new E3 ligase recruiters, and develop novel PROTAC modalities like Antibody-based PROTACs (AbTACs) that are able to degrade PD-L1 via internalization and lysosome-mediated degradation and Ribonuclease-targeting chimeras (RIBOTACs) that re able to target miRNA involved in cancer (Li et al. 2023).

Targeting Cancer Stem Cells

Cancer stem cells (CSC) play an essential role in cancer progression and metastasis. Various strategies have emerged to target CSCs effectively, focusing on their unique properties and microenvironmental interactions. The main strategies that are currently investigated in clinical trials are: targeting surface molecules of CSCs, targeting signal pathways of CSCs, targeting metabolic pathways of CSCs, targeting CSC niches, targeting differentiation mechanisms of CSCs, and targeting CSC-immune cells crosstalk (Huang et al. 2021). The main problem with most anti-CSCs therapeutic targeting strategies relates to their similarity with normal stem cells leading to off-target side effects (Osum and Kalkan 2023).

12.10.5 Cancer Treatments Using the CoDependency Principle

Cancer cells communicate with immune and other stromal cells (i.e., fibroblasts). Some of these cells support the development of cancer cells (macrophages M2, neutrophils, etc.), others inhibit or destroy them. As proof of principle, for more than a decade now, the immune check inhibitors have demonstrated the benefit of targeting the PD1-PDL-1 interaction between cancer cells and the immune cells. Other key interactions between cancer cells and other cells (i.e., neurones) are currently investigated.

One of the most pressing unmet needs in oncology is the development of effective anti-inflammatory therapies for cancer. Inflammation represents a dynamic interface between the host organism and the tumor, actively shaping the trajectory of cancer from its earliest preclinical manifestations to advanced disease and cancer-associated cachexia. Rather than being a mere byproduct, systemic inflammation functions as both a hallmark and a driver of malignancy, orchestrating key processes such as tumor initiation, immune evasion, and metastatic dissemination.

Chronic inflammatory signaling contributes to oncogenesis by promoting angiogenesis, genomic instability, immunosuppression, and tissue remodeling. Key pro-inflammatory cytokines- including interleukin-1β (IL-1β), interleukin-6 (IL-6), and tumor necrosis factor-alpha (TNF-α)—play central roles in these processes. While several agents targeting these cytokines have been repurposed or developed for potential oncologic use, none have yet gained regulatory approval specifically for cancer treatment. The CANTOS trial, which evaluated canakinumab, a monoclonal antibody targeting IL-1β, demonstrated a significant reduction in lung cancer mortality among high-risk individuals with atherosclerosis (Ridker et al. 2017). However, these findings were not replicated in subsequent lung cancer-specific trials (Lythgoe and Prasad 2022). To enhance antitumor immunity and to mitigate immune-related adverse events, ongoing studies are investigating the therapeutic potential of IL-6 receptor blockade (tocilizumab) and TNF-α inhibition (etanercept), particularly in combination with chemotherapy and immune checkpoint inhibitors.

In parallel to these studies, the intricate interplay between cancer and the microbiota has gained attention as a modifiable axis of inflammation and disease modulation. Specific microbial communities have been shown to influence tumor behavior, and microbiota-targeted strategies are being developed both for prevention and treatment. Personalized microbiome-based interventions are emerging across multiple tumor types, each with distinct microbial signatures that modulate immune tone and treatment responsiveness (Almonte et al. 2025; Fernandes et al. 2022; Mishra and Mishra 2024). Recently, TOPOSCORE, a 21-bacteria-based qPCR has been proposed by the team of Laurence Zitvogel as a tool to assess dysbiosis in a given individual at any point in time and also as a guide for microbiota-targeted interventions (Derosa et al. 2024).

Over the past decade, the neuroimmune axis has emerged as a critical domain in cancer biology, revealing complex interactions between the nervous system, immune cells, and tumor microenvironment. Understanding the molecular underpinnings of this tripartite communication may unlock new therapeutic opportunities that improve both survival outcomes and patient well-being. Neural signals—nociceptive, adrenergic, and cholinergic—have been shown to regulate immune cell trafficking, influence tumor progression, and modulate treatment efficacy (Khanmammadova et al. 2023).

Among neural pathways, the vagus nerve stands out as a potential master regulator. It transmits a multitude of signals—immune, metabolic, mechanical, and chemical—between tumors and the central nervous system (CNS), influencing both local tumor behavior and systemic host responses. Bidirectional communication between the tumor and CNS via vagal afferents and efferents is increasingly recognized, although the exact mechanisms remain incompletely understood. The vagus nerve also interfaces with key regulatory systems such as the gut microbiota and the renin-angiotensin system (RAS), reinforcing its role as a central node in cancer-related signaling.

Therapeutically, novel strategies are being explored to modulate vagal activity in ways that disrupt pro-tumor neuroimmune circuits. For instance, pharmacologic blockade of vagal afferents through inhibition of metabolic receptors such as GLP-1 or 5-HT_3 may attenuate cancer-promoting feedback loops. Other approaches target central autonomic structures—such as the paraventricular nucleus (PVN) of the hypothalamus—using optogenetics or angiotensin receptor type 1 (AT1R) antagonists to suppress cancer-associated neural signaling. Notably, vagal denervation has been shown to inhibit gastric tumorigenesis (Zhao et al. 2014), and more recent studies have identified peptidergic sensory neurons within vagal axons that release calcium into the cytoplasm of tumor cells, thereby enhancing growth and metastasis (Zhi et al. 2025).

Taken together, these findings position the vagus nerve as a promising therapeutic target in the emerging field of cancer neuromodulation. Modulating vagal tone may provide a novel avenue to simultaneously impact tumor progression, systemic inflammation, and host resilience (Martinez and Sabatier 2025).

12.10.6 Cancer Treatments Using the Unity Principle

Considering the non-scale-free architecture of cancer-induced pathological networks at the organismal level (Paul and Komarova 2021), therapeutic strategies aimed at dismantling cancer unity by targeting communication mediators like extracellular vesicles (EVs) secreted by malignant tissues may be less susceptible to the resistance commonly observed with interventions targeting intracellular or tissue-specific networks. Expanding therapeutic approaches to encompass the organismal level, and, integrating EV-targeted strategies with existing treatment modalities, may substantially enhance the efficacy of cancer therapies in the near future. The implementation of trans-scale targeting, using agents active at the same time at the cellular, tissue, and organismic level, is arguably more effective than molecular scale targeting, which is now the norm (Paul 2020).

Another concept is the disruption of the cancer system by removal the primary tumor. The primary tumor is not merely a passive reservoir of malignant cells; it actively orchestrates metastatic progression through circulating tumor cells (CTCs), pro-metastatic secretomes, immune suppression, and exosome-mediated organotropism. Its presence can theoretically maintain systemic immunosuppression and contribute to the pre-metastatic niche, fostering further dissemination. In some cases, removing the primary lesion may disrupt these signaling networks, reduce tumor-induced immunosuppression, and restore antitumor immune surveillance (Fidler 2003; Danna et al. 2004). This strategy has been proven effective in some cases of breast cancers (Pérez-Fidalgo et al. 2011; Rashid et al. 2013; Lopez-Tarruella et al. 2019) and renal cancer (Ferriero et al. 2023); however, this strategy needs to be carefully analyzed on a case-by-case basis, since sometimes, removing the primary tumor may accelerate progression and induce metastases (Jiao and Lv 2023).

A strategy that may be potentially curative in some patients is the treatment of oligometastatic disease (OMD) by using stereotactic radiotherapy (Kinj et al. 2022). Recent research has demonstrated that oligometastatic behavior is associated with distinct biological and molecular features, including specific microRNA expression profiles, chromosomal alterations, driver gene mutations such as ERBB2, PBRM1, SETD2, KRAS, PIK3CA, and SMAD4, as well as polymorphisms like TCF7L2. Additionally, the extent of immune cell infiltration into metastatic lesions, varying by tumor type, has been linked to this phenotype. These findings support the notion that OMD may represent a unique clinical and molecular entity (Ottaiano et al. 2023). On the other hand, a single genetic snapshot of the primary tumor is unlikely to reflect the genetic diversity and evolution of metastatic lesions. Despite their potential, genetic assessments in oligometastatic disease (OMD) face notable limitations, including intratumoral heterogeneity, spatial and temporal variation in tumor evolution, technical difficulties in obtaining biopsies, challenges in interpreting complex genomic data, and important ethical concerns (Santorsola et al. 2023). Rather than relying purely on absolute numbers or genetic profiles, the use of clinical criteria, i.e., a low probability of occult metastases, a low tumor growth rate, and a low tumor burden, may help select patients who may benefit from this approach (Szturz and Vermorken 2022).

12.10.7 Cancer Treatments Using the Creativity Principle

Copying Strategies from Cancer to Treat Cancer

Cancer, paradoxically, may hold the keys to its own undoing. Among its most formidable qualities are its adaptability and resilience—traits that have long confounded clinicians but can now be turned into a therapeutic opportunity. A growing body of research has begun to explore how mimicking cancer's own survival strategies, particularly in immune cells like T lymphocytes, can inform and dramatically enhance T cell-based cancer immunotherapies, especially chimeric antigen receptor (CAR)-T cell therapy.

Any cell type in the human body is susceptible to oncogenic transformation, including T cells themselves, as observed in T-cell lymphomas. These malignancies often arise in the immunosuppressive and hostile tumor microenvironment (TME), a complex landscape populated by inhibitory immune cells, suppressive cytokines, hypoxia, and metabolic stress (Joyce and Fearon 2015). Remarkably, malignant T cells manage to not only survive but thrive under such harsh conditions. This resilience is rooted in specific genetic alterations that enhance their functionality and persistence, adaptations that natural selection within the TME appears to have favored over time.

From a therapeutic perspective, this presents a striking insight: if cancerous T cells have evolved to withstand the same suppressive forces that thwart therapeutic T cells, then perhaps their molecular blueprints can be repurposed to fortify engineered T cells used in immunotherapy.

Building on this concept, Garcia and colleagues conducted a comprehensive study of genomic alterations in malignant T cells, identifying 71 candidate genetic modifications that appeared to enhance survival and function in the TME. When select alterations were introduced into CAR T cells in preclinical models, the engineered cells demonstrated enhanced cytotoxicity, improved persistence, and greater proliferative capacity—all while maintaining safety and specificity (Garcia et al. 2024).

The most striking discovery emerged from a fusion gene between CARD11 and PIK3R3. CARD11 is known to regulate T cell receptor (TCR) signaling and NF-κB activation, while PIK3R3 is a regulatory subunit of phosphoinositide 3-kinase, involved in cell growth and survival pathways. The fusion construct appeared to activate multiple beneficial signaling cascades, mimicking the robust survival machinery of cancerous T cells. In mouse models, CAR T cells expressing this fusion not only persisted longer, but also maintained heightened functionality specifically in the presence of their target antigen—ensuring tumor-restricted activation and reducing the risk of off-target effects.

This work reflects a broader principle in modern immunotherapy: adaptive bioengineering, inspired by evolutionary biology. Rather than viewing cancer solely as an adversary, researchers are beginning to study it as a repository of biological innovations, forged under intense selective pressures. Tumors, in effect, have already conducted a vast natural experiment in optimizing cell survival under extreme conditions. These lessons can be reappropriated to engineer more robust and context-sensitive immune responses.

Importantly, this strategy emphasizes contextual activation, an approach that enhances safety by ensuring that supercharged T cells only unleash their full potency within the tumor milieu. This represents a significant advance in addressing one of the major limitations of current CAR-T therapies: on-target, off-tumor toxicity (June and Sadelain 2018).

The notion of borrowing survival tactics from cancer to combat cancer may seem counterintuitive, but it reflects a shift in thinking from destruction to reprogramming. By identifying and harnessing the very features that enable malignant cells to evade immune control, scientists are now able to arm immune cells with comparable resilience and adaptability.

This approach exemplifies a new generation of immunotherapy: precision-guided, biologically inspired, and dynamically regulated. As our understanding deepens, the boundary between cancer biology and therapeutic immunology continues to blur, revealing not just how cancer survives, but how we might finally learn to turn its logic against itself.

Targeted Therapy Paradigm and Emerging Avenues Targeting Cancer's Stress-Driven Plasticity

Cancer is increasingly understood not merely as a mass of mutated cells, but as an adaptive system—a biological entity shaped and driven by stress. From its earliest origins, cancer engages in what may be described as a stress-driven exploration of different cellular adaptations, wherein instability and disruption serve not only as challenges to overcome but also as creative forces that enable survival, plasticity, and evolution. However, this same stress-adaptive machinery, if pushed beyond its limits or appropriately sabotaged, may offer a critical point of therapeutic vulnerability.

Recent studies propose that cancer may arise from an initial stress response, due to, for example, hypoxia, inflammation, or DNA damage, that becomes permanently engaged, transforming what should be a transient adaptation into a self-sustaining pathological state (Baron et al. 2020). At the cellular level, stress induces a highly plastic state, reminiscent of developmental programs such as stemness and epithelial-mesenchymal transition (EMT) (Shomar et al. 2022). This plasticity allows cancer cells to escape death, detach from tissue structures, resist drugs, and seed metastases.

Importantly, this adaptive state is not isolated to single cells—it propagates across the cancer system as a whole, allowing for coordinated changes in cell fate, metabolism, immune evasion, and niche remodeling (Gyurkó et al. 2013). Every metastatic seeding event may represent a fresh iteration of this process: stress triggers openness, openness enables exploration, and exploration generates survival solutions.

Two key molecular drivers of this adaptive potential are genomic instability (GIN) and chromosomal instability (CIN). While GIN refers to the accumulation of point mutations and DNA damage, CIN involves large-scale missegregation of chromosomes during mitosis. Both mechanisms can result in genetic novelty—aneuploidy, gene amplifications, deletions—that provide raw material for selection, especially under therapeutic stress (Ferguson et al. 2015).

Yet, this instability is a double-edged sword. There is a critical threshold beyond which genetic chaos becomes lethal, either by overwhelming a cancer cell's repair capacity or by increasing its visibility to the immune system. This precarious balance between adaptive instability and self-destruction offers a unique therapeutic window.

Emerging strategies now aim to exploit this fragile equilibrium by targeting cancer's stress-adaptive responses directly. One avenue involves enhancing CIN: essentially pushing cancer over the edge. CIN-promoting drugs, such as microtubule poisons or inhibitors of mitotic checkpoint proteins, force mitotic errors beyond tolerable limits, inducing mitotic catastrophe and immune recognition (Hosea et al. 2024). Another promising strategy is to inhibit stress-response pathways that enable plasticity, such as the DNA damage response (DDR) and oxidative stress signaling. Disabling these repair and buffering systems deprives cancer cells of their capacity to adapt, especially under the selective pressure of chemotherapy or immune attack (Strauss and Pursell 2023).

Additionally, therapeutic approaches that limit plasticity itself, by locking cells into differentiated states or preventing EMT, may reduce cancer's ability to explore alternate evolutionary trajectories. This could be particularly powerful when applied during metastatic transition, where re-engagement of plastic programs is critical for survival.

In this framework, cancer is no longer regarded as a genetic aberration, but rather as a maladaptive state of survival programs exploration. Cancer does not evolve despite stress, but because of it. Targeting stress-survival adaptations use by cancer cells is a promising research avenue.

Transposable Elements in Cancer: A Hidden Vulnerability with Therapeutic Promise

The human genome, once viewed as a stable repository of genetic information, is now understood to be richly dynamic, shaped in large part by the activity of transposable elements

(TEs)—mobile sequences of DNA capable of changing their position within the genome. Among these, the LINE-1 (Long Interspersed Nuclear Element-1) family stands out both for its abundance, comprising nearly 20% of the human genome, and for being the only active autonomous transposon still capable of mobilization in humans (Lander et al. 2001). While normally silenced in adult tissues, LINE-1 reactivation has emerged as a striking feature of many cancers, representing both a potential driver of malignancy and a novel target for therapy.

LINE-1 elements are transiently active during early embryogenesis, where they contribute to chromatin remodeling and the maintenance of epigenetic plasticity, functions critical to early development (Jachowicz et al. 2017). However, in cancer, LINE-1 reactivation is often pathological. Loss of epigenetic repression, through DNA hypomethylation or defective chromatin remodeling, can lead to LINE-1 expression and retrotransposition, creating genomic instability, disrupting gene regulation, and possibly fostering metastatic behavior (Schneider et al. 2021).

Moreover, the RNA and proteins encoded by LINE-1, particularly its reverse transcriptase (RT) and endonuclease, are capable of inducing double-stranded DNA breaks, activating inflammatory responses, and generating neoantigens, all of which contribute to the cancer phenotype. The dual role of LINE-1 as both an instigator of genomic chaos and a visible target to the immune system makes it an especially intriguing molecular lever for therapeutic intervention.

A groundbreaking proof-of-concept study by Schneider et al. demonstrated that LINE-1 activity in colorectal cancer can be pharmacologically inhibited using antiretroviral agents initially developed for HIV/AIDS. The combination of Zidovudine, a nucleoside reverse transcriptase inhibitor (NRTI), and Efavirenz, a non-nucleoside reverse transcriptase inhibitor (NNRTI), significantly reduced tumor cell viability in vitro and curtailed tumor growth in vivo in xenografted mice. These results suggest that reverse transcriptase inhibitors, by blocking the catalytic machinery of LINE-1, may suppress its oncogenic activity (Schneider et al. 2021).

This represents a compelling example of drug repurposing, where safe, well-characterized agents are redirected toward novel clinical indications. Given the wide availability and established pharmacokinetics of antiretroviral drugs, their application in oncology could accelerate the timeline from bench to bedside, particularly for malignancies characterized by LINE-1 reactivation, such as colorectal, prostate, and ovarian cancers (Rodić et al. 2014).

Targeting LINE-1 is not only about halting transposition—it may also impact downstream oncogenic pathways that are activated by LINE-1-associated transcripts and proteins. Furthermore, LINE-1 activity may serve as a biomarker for patient stratification, helping to identify tumors with high transposon burden and likely responsiveness to RT inhibitors.

In addition, LINE-1 inhibition may synergize with immune checkpoint inhibitors by reducing immunosuppressive signaling and enhancing antigen presentation—especially given the capacity of LINE-1 to generate endogenous retroviral antigens that are otherwise absent in normal tissue (Chiappinelli et al. 2015).

12.11 The Importance of Treating Co-morbidities

Cancer, though localized in its origin, ultimately engages the entire organism through complex systemic interactions involving inflammation, metabolism, immunity, and neuroendocrine regulation. As such, survival in cancer patients is not solely determined by tumor-targeted therapies, but also by the integrity and resilience of other organ systems. Comorbidities negatively impact cancer prevention strategies and treatment outcomes, contributing to avoidable excess cancer mortality. Managing co-morbidities, such as diabetes, hypertension, and COPD, may also improve survival and reduce systemic climate disruptions (Fowler et al. 2020; George et al. 2021; Abravan et al. 2024).

Co-morbidities modulate tumor biology by reshaping the tumor microenvironment, altering cancer cell metabolism, disrupting the gut microbiome, and modifying the tumor's responsiveness to therapy. Mechanistically, comorbidities may accelerate cancer progression through pathways involving local and systemic inflammation, remodeling of the tumor immune landscape, induction of mesenchymal phenotypes in cancer cells, and the establishment of pro-metastatic niches in distant organs (Panigrahi and Ambs 2021). Patients with comorbidity have poorer survival, poorer quality of life, and higher health care costs (Sarfati et al. 2016). Improving systemic health and addressing comorbidities may significantly enhance outcomes by restoring organismal homeostasis and optimizing the body's internal environment. Homeostatic balance is thus restored not only through tumor control, but also by improving the total physiological milieu. AI techniques can help integrate and manage heterogeneous data through advanced machine learning and bioinformatics algorithms and help define individualized treatments for each patient with multimorbidities and cancer (Cesario et al. 2021).

12.12 Can Cancer's Trajectory Be Anticipated?

Although we can deconstruct life into its fundamental molecular constituents, examine cells under the microscope, and anatomically describe human organs, the creation of even the simplest living system from scratch remains beyond our reach. A single cell contains approximately 42 million molecules, whose interactions are far from random. The core challenge lies in the fact that these molecules are in constant motion, each following distinct, dynamic trajectories. This movement is characterized as *non-ergodic*—that is, the molecules do not explore all spatial configurations freely—and resembles behavior observed in "glassy" systems, where elements remain largely immobilized for long periods before occasionally shifting significant distances (Åberg and Poolman 2021). This stands in stark contrast to the random, ergodic Brownian motion typical of non-living systems.

Life orchestrates the interplay of information, energy, and structure with extraordinary precision to generate organized complexity—a feat still far beyond what current laboratory techniques can replicate. Consider, for comparison, the three-body problem in astronomy: it lacks a general, closed-form solution and resists prediction using a finite set of standard mathematical operations. If we cannot fully predict the interaction of just three bodies, how could we hope to precisely map the choreography of millions of interacting molecules within a living cell and predict its behavior? Living systems, more intricate than any physical model, exhibit intrinsic unpredictability. Biology, then, is the science of moving targets—systems in flux that defy static analysis.

Molecular biology has made significant strides in cataloging oncogenic alterations—driver mutations, signaling disruptions, and epigenetic modifications—that can, in principle, be therapeutically targeted. However, the clinical efficacy of targeted therapies often diverges from preclinical expectations. Drugs that inhibit specific molecular targets with high precision in vitro may behave unpredictably in the complex in vivo tumor microenvironment. Hence, despite the aspirations of precision oncology, many of its approaches remain fundamentally imprecise.

Predicting the exact survival time of an individual cancer patient remains elusive, though population-level statistical models offer general survival estimates. These models, while informative, cannot account for the individual biological variability that shapes clinical outcomes.

Histopathology continues to serve as a cornerstone of cancer diagnosis and classification, offering detailed insights into tumor architecture, cellular morphology, and the tumor microenvironment. Beyond diagnosis, advances in computational pathology now enable image-based prognostication using digital analysis of histological slides. By leveraging deep learning models and image-derived biomarkers, these tools can extract clinically relevant information that exceeds the perceptual limits of the human eye.

Traditional grading systems—such as the Gleason score in prostate cancer and the Nottingham grade in breast cancer—link morphological features such as glandular architecture, nuclear pleomorphism, and mitotic activity to clinical outcomes (Epstein et al. 2005; Rakha et al. 2008). However, these manual systems are subject to variability among observers and can suffer from subjectivity (van der Laak et al. 2021).

The digitization of whole-slide images (WSIs), combined with machine learning—particularly convolutional neural networks (CNNs)—has enabled the automated extraction of prognostic features. For instance, Coudray et al. showed that CNNs trained on WSIs of lung adenocarcinoma could classify histological subtypes and predict overall survival based on learned morphological patterns (Coudray et al. 2018). Similar successes have been reported in breast and colorectal cancers, where AI-driven models have identified features predictive of recurrence and progression (Bychkov et al. 2021; Wulczyn et al. 2021).

One of the key strengths of digital histopathology lies in its capacity to quantify and analyze spatial heterogeneity within tumors—an important driver of therapeutic resistance and disease progression. Graph-based modeling of tumor microarchitecture has demonstrated that the spatial distribution of tumor-infiltrating lymphocytes and stromal elements carries independent prognostic significance (Saltz et al. 2018). More advanced approaches are now integrating histological data with multi-omic profiles to enhance precision in risk stratification (Kather et al. 2020).

In a large-scale study of 47,625 cancer patients, Nunez et al. demonstrated that survival outcomes at 6, 36, and 60 months could be predicted with an area under the curve (AUC) exceeding 0.9 using both traditional models and natural language neural networks, based solely on the initial oncology consultation (Nunez et al. 2023).

Altogether, the integration of digital pathology with molecular profiling and clinical metadata heralds a new era in oncology—one that promises to deliver truly personalized, data-driven care.

12.13 A Brief Note on Clinical Trials

Monkey Wrench Treatments and the Benefit/Neutral/Harmful Axiom in Oncology

The evolution of oncology from empiricism to molecular precision has been transformative—shifting from serendipitously discovered chemotherapies to targeted therapies, immune checkpoint inhibitors, CAR-T cells, and bispecific antibodies. Yet, despite this therapeutic arsenal, only a fraction of these agents are truly personalized, guided by patient-specific biomarkers confirmed in tumor tissue. This underscores a central challenge in modern oncology: the unpredictable complexity of therapeutic intervention in dynamic biological systems.

To illustrate this complexity, consider the metaphor of throwing a monkey wrench into a runaway engine. Three outcomes are possible: the engine stalls or slows (a *beneficial* effect), accelerates chaotically (a *harmful* effect), or continues unaffected (a *neutral* effect). This analogy is apt for many contemporary molecular therapies, which—like the monkey wrench—exert multifaceted effects. Most agents act on multiple intracellular targets and influence diverse cellular and systemic pathways. A drug may trigger proliferation, apoptosis, senescence, or exert no effect at all, depending on the cell's internal state. Similarly, at the tissue level, the same agent may induce tumor regression, stability, or progression. At the organismal level, the treatment may affect non-malignant tissues and organs in ways that indirectly promote or inhibit tumor growth—or leave it unchanged altogether.

This leads to the formulation of the *Benefit–Neutral–Harm (BNH) axiom*: any treatment administered to a living system can be classified as beneficial (B), neutral (N), or harmful (H). In an organism with cancer, in general, a beneficial effect is typically associated with tumor shrinkage—either partial response (PR), complete response (CR), or, occasionally, with durable stable disease (SD). Neutral or harmful effects, on the other hand, usually manifest as disease progression (PD), defined clinically as a $\geq 20\%$ increase in the sum of diameters of target lesions from the nadir, with an absolute increase of at least 5 mm (Eisenhauer et al. 2009).

The phases of clinical trials mirror the BNH framework:

Phase I trials are designed primarily to assess harm (toxicity). If H is excessive, the trial is halted.
Phase II trials aim to evaluate both harm and benefit. They proceed if a sufficient signal of B is detected.
Phase III trials focus on confirming B by comparing the new therapy to standard-of-care treatments.

Due to the prolonged timelines required to demonstrate overall survival (OS) benefits, surrogate endpoints such as overall response rate (ORR), PR, CR, and progression-free survival (PFS) are often used in trial design. ORR is defined as the proportion of patients who achieve a PR or CR within a defined time frame. However, the field remains divided on the validity of these surrogates. One prevailing argument is that only responses classified as PR or CR correlate with meaningful clinical outcomes, such as improvements in PFS and OS, whereas SD does not (Vidaurre et al. 2009). Another concern is that PFS itself often correlates poorly with OS, and therapies that merely prolong SD without improving OS are viewed as offering limited clinical value (Belin et al. 2020; Huang et al. 2024). Of note, PFS refers both to cancer and the organism with an accent on cancer, as opposed to OS that refers mainly to the organism. Importantly, assessing the impact of treatment on ECOG performance status (PS) and quality of life (QOL) should be also considered during a clinical trials, as many studies indicate that higher quality of life, particularly in areas like physical functioning and reduced pain, is associated with longer survival times in cancer patients (Quinten et al. 2009; Kypriotakis et al. 2016; Lim et al. 2025). This relationship is observed across various cancer types and treatment settings, though its strength can vary.

In practice, however, the landscape is far more nuanced. A recent retrospective analysis found that 57% of cancer drug approvals (128 out of 223) were based on studies that did not use OS as a primary endpoint. Among the 95 trials that did include OS, fewer than one-third demonstrated a statistically significant OS benefit at the time of approval, often due to immature data (Naci et al. 2024). As Mittal et al. have noted, the majority of modern cancer therapies are approved based on ORR or PFS, with little evidence of OS extension or improvements in quality of life (QoL) (Mittal et al. 2024).

Randomized controlled trials (RCTs) remain the cornerstone of evidence-based medicine, and randomization plays a pivotal role in preserving the validity and integrity of clinical research. Its primary functions are to eliminate bias, control confounding variables, and enable statistically sound inferences.

Eliminating bias: Random assignment ensures that each participant has an equal chance of receiving any given treatment, minimizing conscious or unconscious selection bias.

Controlling confounders: Randomization helps distribute both known and unknown confounding variables (e.g., age, genetic predisposition) evenly across treatment groups, allowing any observed treatment effect to be attributed more confidently to the intervention.

Statistical rigor: Randomization provides the foundation for valid statistical analyses—enabling the use of p-values, confidence intervals, and other inferential tools necessary for evaluating causality and generalizability.

Nevertheless, despite decades of rigorously conducted RCTs, global cancer mortality remains high. Approximately half of all patients diagnosed with cancer continue to die from the disease. This sobering reality reflects a critical failure in metastatic cancer management. While localized cancers are often curable, metastatic disease remains recalcitrant to current interventions. This suggests that many strategies employed in advanced disease are ineffective—or worse, potentially harmful. Ultimately, the purpose of randomized trials is to maximize benefit while minimizing harm, yet the high mortality of metastatic cancer reveals the urgent need to reevaluate how success is measured and how therapies are developed.

A crucial point that oncology practitioners should always remember is that, despite randomization, individual-level effects of a drug investigated in a randomized Phase 3 clinical trial (RCT) can not be automatically extrapolated from the hazard ratio (HR) of the population-level benefits (Fay and Li 2024). Relying on data from RCTs to guide treatment decisions in real-world practice has become the prevailing paradigm, yet it rests on a problematic assumption—that average treatment effects observed in trial populations can be readily extrapolated to patients whose personal or disease characteristics differ from those studied. This assumption is deeply flawed, as it overlooks the marked heterogeneity of treatment responses both across individual patients and between highly selected trial populations and the broader, more complex real-world cohorts. As a result, the application of RCT-derived averages to clinical decision-making may misrepresent therapeutic benefit, obscure risks, and ultimately compromise the personalization of care (Curth et al. 2024).

12.14 The Way Ahead

Individualization of Treatment Goals

Cancers differ profoundly not only between tumor types but also between patients. No two cancers are identical, and no two individuals experience cancer in the same way. Therefore, treatment goals must be tailored to both the biological features of the tumor and the physiological, psychological, and personal context of the patient. Precision in oncology must be both molecular and human.

Biomarker-Driven Precision and the Promise of Third-Generation Therapies

Third-generation targeted therapies have achieved unprecedented response rates in selected cancers. Yet, the central challenge of oncology remains: transforming short-term tumor responses into durable survival, particularly in metastatic disease. The use of predictive biomarkers—those causally linked to tumorigenesis or to the mechanism of action of a therapy—can dramatically enhance the statistical power, biological relevance, and interpretability of clinical trials. These biomarkers allow for rational patient stratification, reduce inter-patient heterogeneity, and improve causal inference. In most cancers, different tumors grow for different reasons; thus, selecting patients whose cancers are driven by the specific molecular target of a drug ensures biological homogeneity and improves signal detection. For example, Trastuzumab (Herceptin) is effective only in HER2-positive breast cancer, and enrolling patients based on HER2 overexpression optimizes therapeutic effect and minimizes noise. Similarly, EGFR mutations predict response to EGFR inhibitors in lung cancer. When biomarkers illuminate a drug's biological rationale, trials become faster, smarter, and more predictive—marking biomarkers as a cornerstone of modern precision oncology.

12.15 Randomization and the Limits of Predictability

Both biomarker stratification and RCTs are human strategies to impose order on biological variability. Randomization, by design, separates therapeutic effect from statistical noise. Yet in practice, RCTs may be most effective at quantifying toxicity rather than predicting benefit. Even in successful Phase III trials, the control arm benefits a subset of patients—highlighting the complexity of treatment effects. It is not uncommon for an individual patient to derive greater benefit from a control arm regimen than from a newly approved agent. The details of trial design, inclusion criteria, and endpoints are therefore critical for interpretation and application. The recent advances in causal machine learning (Curth et al. 2024; Feuerriegel et al. 2024; Weberpals et al. 2025) and, also, the digital-twins applications, may offer pioneering solutions to personalization of cancer clinical trials (van der Schaar et al. 2025).

12.16 Reconsidering Dose Assumptions: The Maximum Tolerated Dose Fallacy

A less recognized limitation in trial design is the traditional reliance on the Maximum Tolerated Dose (MTD) determined in Phase I studies. This approach assumes a linear dose-efficacy relationship: more drug equals more benefit. Increasingly, this assumption is being questioned. Different doses may engage different molecular targets or yield distinct clinical effects. For example, low-dose paclitaxel exerts anti-metastatic activity through vascular remodeling—an effect lost at higher, cytotoxic doses. Lower doses are typically better tolerated and may, paradoxically, be more effective in certain settings. This principle underlies metronomic chemotherapy strategies, which utilize frequent low-dose administration to sustain therapeutic pressure with minimal toxicity. In clinical reality, the same dose can be beneficial in one context, harmful in another, or altogether inert. Furthermore, increasing the dose within the same context may invert the therapeutic effect, leading to progression. These complexities necessitate a shift toward adaptive, context-aware dosing strategies.

12.17 Redefining Therapeutic Goals in Metastatic Disease

The overarching aim of clinical trials is to improve both the quality and duration of life. While we have not cured diabetes, we have transformed it into a manageable chronic illness. A similar reframing may be appropriate in metastatic cancer, where the realistic goal may be to stabilize disease and preserve quality of life (QOL), rather than pursue short-lived tumor shrinkage. Stable disease (SD), maintained over time with good QOL, may be more meaningful than achieving partial or complete responses (PR or CR) that do not translate into survival benefits. This paradigm aligns with homeodynamic treatment approaches, which seek to support systemic balance rather than eradicate all tumor cells.

We are now entering the era of AI-augmented oncology, where artificial intelligence can synergize biomarker-driven personalization with the statistical rigor of randomization. Yet, the Benefit-Neutral-Harm (BNH) axiom remains a guiding principle: every treatment can help, do nothing, or cause harm. Clinical trials, however rigorous, cannot fully capture the complexity of human physiology. In reality, each patient treated for cancer is a de facto Phase I trial—a one-patient experimental treatment requiring thoughtful titration, continuous monitoring, and adaptive decision-making. Rather than imposing a maximal dose upfront, in metastatic cancer where the cure using standard agents is rare, a more sophisticated approach involves gradual escalation guided by real-time feedback. The convergence of AI, precision biomarkers, and adaptive trial design marks a promising path forward. Nevertheless, the human organism remains inherently unpredictable—and the BNH axiom will always apply.

12.18 What to Do When No Treatment Option Is Available

In the fast-paced world of clinical oncology, the day to day scenario often involves selecting an evidence-based treatment directly from established guidelines and administering it to the patient. While clinical guidelines are invaluable tools that help standardize care and ensure quality, it is essential to remember that they are not mandates, but recommendations. Every patient is different and a "one fits all" strategy should be avoided. The practice of medicine, particularly in oncology, resides at the intersection of science and art. Science demands data, reproducibility, and rigor. Art, on the other hand, draws on clinical experience, intuition, and a holistic perspective that encompasses not only tumor biology, but also the patient's comorbidities, age, performance status, psychological resilience, and broader social and existential contexts.

In practice, many oncologists encounter challenging scenarios when symptom control is difficult to achieve, significant side effects associated with prior treatments are lingering, or disease progresses despite multiple lines of guideline-based therapy. When standard options have been exhausted, clinical trials are unavailable or unsuitable (e.g., due to prior malignancies, etc.), and the patient retains adequate performance status to continue systemic therapy, treatment decisions become increasingly complex.

A randomized study in patients with newly diagnosed metastatic non-small-cell lung cancer demonstrated that compared with patients receiving standard care, patients receiving early palliative care had longer survival, despite receiving less aggressive care at the end-of-life (Temel et al. 2010).

At the end of my three-year Hematology-Oncology fellowship, I was still not feeling comfortable having end-of-life discussions with my patients and their families and so I decided to pursue training in Hospice and Palliative Care. Achieving palliative care skills and, especially, pain treatment proficiency, was extremely rewarding as it enabled me to accompany and offer comfort to my patients through all their difficult cancer odyssey.

13 Real-Time Adaptive Cancer Therapies Guided by Longitudinal Biomarkers, Multiomics Integration, and Artificial Intelligence-Assisted Strategies

As mentioned in Chap. 1, the triumph of artificial intelligence, such as DeepMind's AlphaGo, over world champions in complex strategy games such as chess and Go, marked a paradigm change in computational cognition. These victories did not merely showcase brute-force processing power but also demonstrated the ability of AI systems to learn, adapt, and outmaneuver human intuition in dynamic, rule-bound environments. In oncology, particularly in the context of metastatic cancer, a condition marked by profound heterogeneity, evolutionary plasticity, and therapeutic resistance, a similar paradigm shift is emerging. The question is no longer whether AI can contribute to cancer care, but how it can revolutionize our approach to treating what has long been considered an incurable disease.

Conventional treatment strategies often follow predefined, static protocols, which may fail to account for the continuous molecular evolution of tumors. Recent advances in deep learning, liquid biopsy, and multiomics integration offer the potential to transform cancer therapy into a real-time, adaptive process—guided by artificial intelligence (AI) and optimized through game theory principles.

By integrating continuous inputs from liquid biopsies, multi-omics data, imaging, and clinical parameters, AI has the potential to transform cancer treatment from reactive to proactive; from protocol-driven to dynamically individualized. Similarly to how diabetes is managed today through continuous monitoring, lifestyle adaptation, and personalized pharmacological regimens, oncology-focused AI could reframe metastatic cancer as a chronic but controllable disease.

The architecture of DeepMind's AlphaZero's—an updated version of AlphaGo—is a continuous feedback loop that combines a Deep Neural Network DNN's "intuition" that guides the Monte Carlo Tree Search (MCTS) which in turn generates better training data to further improve the DNN's predictions. Given its ability to learn new strategies from scratch, an AI with abilities could be used to guide individual cancer treatments, simulating the evolutionary trajectories of tumors, anticipating cancer progression and resistance mechanisms, and recommending adaptive treatment interventions in *real time*. Basically, once capture cancer's evolving rules are captured, it can be beaten at its own game.

Biomarkers are foundational tools in precision oncology, guiding prognosis, therapeutic selection, and disease monitoring. Taking into consideration both cancer and the host, an integrative, biomarker-oriented treatment can be designed. While cancer-specific biomarkers —such as circulating tumor DNA (ctDNA), exosomes, and fragmentomic signatures—have been extensively studied, they provide only a partial picture. Increasing evidence supports the integration of host-derived, systemic biomarkers including immune profiling, metabolic markers, and physiological indicators. Tumors evolve in a dynamic interplay with the host, involving local tissue environments, immune surveillance, metabolism, and systemic physiology. Importantly, each cancer/host characteristic is different, and, in my opinion, personalizing treatment strategies is a must. A composite *Global Biomarker Vulnerability Score* (GBVS) that integrates tumor-derived and host-derived indicators across the organism's scales could represent a key step toward comprehensive, adaptive cancer management.

Liquid biopsies, which detect circulating tumor cells (CTCs), cell-free tumor DNA (ctDNA), exosomes, and other biomarkers in body fluids, provide a minimally invasive window into the tumor's current genetic and epigenetic landscape (Cha et al. 2023; Nicholas et al. 2024; Siravegna et al. 2017). Using multiomics data, including genomics, transcriptomics, proteomics, and metabolomics (Menyhárt and Győrffy 2021; Shakyawar et al. 2024), we can sequentially update the profile of cancer's evolving phenotype and its interaction with the host microenvironment.

In the context of DCP and the homeodynamic theory introduced priorly, cancer's progress follows certain rules and principles, and its evolution can be interfered with and redirected. Given the relatively stable homeodynamic cancer trajectories, drawing parallels from strategic gameplay, a powerful AI system could assist oncology clinicians to

D. Paul, *Cancer: An Integrative Approach*, https://doi.org/10.1007/978-3-032-08808-6_13

transform cancer into a benign condition. In this envisioned model, the oncologist and the AI co-develop a continuously evolving therapeutic strategy that stays one step ahead of tumor adaptation, leveraging predictive analytics, real-time feedback, and systemic modeling.

Deep learning approaches—including machine learning algorithms, convolutional neural networks (CNNs), recurrent neural networks (RNNs), and the emerging paradigm of digital twins—can be trained on large-scale, multimodal datasets to detect complex biological patterns, forecast cancer progression, and recommend individualized treatment strategies. AI models can integrate heterogeneous inputs to compute patient-specific indices, such as the GBVS score, thereby supporting precision oncology through predictive and adaptive decision-making. When continuously fed with real-time data from liquid biopsies, an AI can help switch cancer treatment strategies from reactive to proactive, anticipate cancer's "next move," identifying emergent driver mutations, and adjust therapeutic regimens accordingly—shifting basically from a static to a feedback-informed adaptive strategy (LeCun et al. 2015; Yoon et al. 2017; Topol 2019; Esteva et al. 2019; West et al. 2020; Hernandez-Boussard et al. 2021).

Machine learning (ML), reinforcement learning (RL), and game-theoretic AI frameworks provide both conceptual and practical innovations for oncology. By framing cancer as an adaptive, strategic opponent within an evolutionary game, AI can model tumor–host dynamics and learn to anticipate cancer's tactics, such as drug resistance or immune evasion (Kaznatcheev et al. 2017; Stanková et al. 2019; Bin Masroni et al. 2025). For instance, multi-agent RL systems can simulate heterogeneous tumor cell populations—such as therapy-sensitive versus resistant clones—and design treatment policies that maintain long-term tumor control while minimizing selective pressures that accelerate resistance. Also, it has been shown that causal machine learning (ML) is able to predict how outcomes change due to treatment, enabling personalized predictions and comparisons across different treatment options (Curth et al. 2024; Feuerriegel et al. 2024; Weberpals et al. 2025). This game-theoretic perspective suggests that adaptive therapy, guided by AI, may prolong treatment efficacy and better align with the homeodynamic logic of cancer progression (Please see Fig. 8.4). An example of this approach, known as *Adaptive Therapy*, was developed by Robert Gatenby at Moffitt Cancer Center and has already demonstrated its efficacy in the treatment of prostate cancer (Chahoud et al. 2023).

An AI system could recommend when to intensify, de-escalate, or switch therapies based on a probabilistic forecast of tumor behavior, aiming not necessarily to eradicate every cancer cell, but to manage the ecosystem in a homeostatic, controlled equilibrium—thereby transforming metastatic disease into a manageable condition. A similar strategy using continuous glycemic monitoring and personalized nutrition interventions has been successfully used by Eran Segal and Eran Elinav in diabetic patients (Zeevi et al. 2015; Segal et al. 2024).

In sum, by integrating real-time multiomic insights with deep learning and strategic game theory, Process Oncology protocols can be applied to each individual patient and to the vulnerabilities of each cancer type. The key is to integrate data both from different sources and biologic levels: from cancer and the host, from cell/tissue/organism levels, and to include proactive actions as opposed to waiting for cancer to progress. This paradigm shift holds the promise of curbing cancer progression with unprecedented precision and turning metastatic cancer from a terminal illness into a chronic, controllable process.

A Global Biomarker Vulnerability Score-Based AI-Assisted Cancer Treatment Architecture

As an example, I propose a five-step toy model. At present, this model represents at present an idealization, and some of the tests proposed are still under investigation.

Step 1: Systems Interrogation—Building the Biomarker Atlas

Before intervention, a comprehensive profiling phase is conducted to characterize the tumor and the organism as a system in flux. This involves multi-dimensional interrogation:

Tumor-Centric Biomarker Profiling

Genetic and epigenetic landscape:

Routine Biopsy:

- Image analysis of biopsy by AI;
- Tumor-infiltrating lymphocytes (TILs), PD-L1 expression, immune score;
- Spatial single cell multiomics (Nam et al. 2021);
- NGS to assess actionable mutations, tumor mutational burden (TMB), neoantigen load including HLA profiling;
- Tumor transcriptome; soon we will also be able to visualize the whole human transcriptome within a cell (Khafizov et al. 2024);
- Phosphoproteomics profiles of the tumor cells (Schneider et al. 2024; Lee et al. 2024).

Liquid biopsy:

- Fragmentomics and ctDNA for spatial-temporal tumor evolution;
- Proteomic Signatures for prediction and targeting;
- Exosomes for dynamic tumor communication;

- Circulating tumor cells (CTCs) for metastatic potential and pharmacodynamics;
- Metabolomics signatures for prediction and targeting.

Imaging:

Whole-body anatomical (i.e., CTs) and functional (i.e., PET scans) imaging to define lesion spread, treatment response zones, and oligometastatic potential.

Host-System Biomarker Profiling

Global Homeostatic Biomarkers:

- Inflammatory (CRP, IL-6), neuroendocrine (cortisol, HRV), and metabolic (glucose variability, ketones);
- Immune Competence Markers: NLR, LMR, TCR clonality, CD8+ exhaustion profiles;
- Epigenetic Clocks, Biological Age, Patient's Circadian Clock: Methylation age, sarcopenia, frailty scores, the BodyTime Nanostring clock gene expression assay to determine circadian clock abnormalities (Wittenbrink et al. 2018);
- Microbiome-Associated Modulators: Diversity indices, gut-derived metabolites, fecal microbial profiling using the TOPOSCORE (Derosa et al. 2024);
- Heart Rate Variability (HRV);
- Organ Functional Resilience: hepatic, renal, cardiac reserve;
- Charlson Comorbidity Index;
- Perceived Stress Scale;
- Performance status (ECOG), symptom burden.

Step 2: AI-Guided Treatment Selection and Sequencing
2A. GBVS Generation

Using machine learning models trained on multimodal datasets, AI integrates tumor and host biomarkers to:

- Prioritize therapeutic vulnerabilities;
- Forecast immune responsiveness;
- Predict systemic tolerance and long-term treatment benefit.

A Global Biomarker Score (GBS) is generated, quantifying the dynamic balance between tumor aggression and organismal resistance. GBS guides not only treatment choice but also the timing and sequence of interventions.

2B. Treatment Architecture
Tumor-Directed Agents:

Targeted therapies (e.g., EGFR, ALK, BRAF inhibitors) and antibody drug conugates in the presence of actionable mutations or targetable antigens present on tumor cells, respectively;
Immune checkpoint inhibitors if TMB/PD-L1 high, TCR diversity adequate, and inflammation low;
Chemotherapy as a bridging or debulking strategy if no molecular vulnerabilities are found;
Bispecific antibodies or CAR-T in high antigen-load, immune-active environments;
Repurposing drugs for the specific patient.

Host-Tuned Modulations ("Internal Climate Control")

- Nutritional regimens (e.g., anti-inflammatory diets to modulate inflammation)—design a personalized nutrition based on *foodomics* data (Menichetti et al. 2025);
- Anti-inflammatory protocols (e.g., low-dose naltrexone, NSAIDs);
  - pH and oxygen manipulation, promoting normoxia to reduce hypoxia-driven resistance;
- Microbiome editing, via pre/pro/synbiotics or fecal transplants;
- Chrono-therapeutics, aligning drug delivery with circadian cycles;

2C. Sequencing Principles:

Begin with combination therapy across three planes:

- Intracellular (i.e., small molecules, targeted agents);
- Microenvironmental (i.e., anti-stroma, anti-angiogenic therapies);
- Systemic (e.g., metabolism, immune priming approaches).

Adapt treatment based on evolving biomarker landscape; flexible, non-linear paths allow for simultaneous therapy and recalibration. Pro-actively induce changes in cancer's trajectory pushing it into dormancy/equilibrium (Fig. 12.1). Use trans-scale and multi-scale targeting.

Step 3: Monitoring—Efficacy, Resistance, and Tolerance

- Real-time tracking using different omics (i.e., ctDNA, fragmentomics, metabolomics, exosomics, imaging);

- Systemic biomarkers (e.g., CRP, ketone levels, HRV) related to host resilience;
- AI-assistance in modeling resistance emergence and suggesting alternate strategies.

Step 4: Re-Interrogation and Adaptive Interventions

At each disease milestone or sign of progression, re-profile the tumor and host:

- Update molecular profiles, re-scan with liquid biopsies;
- Assess systemic state via homeostatic indices;
- AI recalculates the GBS to re-sequence therapeutic interventions.

Step 5: Maintenance or Intervention Transition

If cancer control is reached: Switch to homeostatic interventions (e.g., microbiome stabilization, inflammation control, circadian alignment) to prevent progression and reignite immune surveillance.

If cancer progresses: Return to Step 2 with a new treatment axis informed by updated biomarkers.

This model acknowledges cancer not as an isolated cellular aberration, but as a systemic process influenced by the host's internal climate. By integrating tumor-driven vulnerabilities and organism-based biomarkers, by using AI to navigate this complexity, we establish a flexible, adaptive, treatment framework that transcends static treatment lines.

This represents a paradigm shift repositioning cancer care as a co-dependent system, where both malignancy and host are continuously interacting and changing, offering a route toward more durable, and patient-centric, therapeutic outcomes.

14 The Essential Integration: Patient-Oriented Oncology

14.1 Psycho-Neuro-Endocrine-Immuno-Oncology

Psycho-neuro-endocrine-immuno-oncology (PNEIO) is an interdisciplinary field that investigates the dynamic interactions between psychological processes, the nervous system, the endocrine system, the immune system, and cancer (Vollhardt 1991; Bauer 1994; Mészáros Crow et al. 2023; Bottaccioli and Bottaccioli 2020). At its core, PNEIO explores the profound influence of mental and emotional states on physiological responses, with a specific focus on understanding how these interactions impact cancer progression.

The intricate connections between psychology, the nervous system, and the immune system underscore the complexity of the mind–body relationship. Psychological factors, such as stress, emotions, and coping mechanisms, have been shown to exert tangible effects on the functioning of the nervous and immune systems. The interplay between them involves a network of signaling pathways, neurotransmitters, and immune responses, shaping the body's ability to defend against external threats, including the development and progression of diseases such as cancer. Studies indicate that stress-induced changes in immune function may create permissive conditions for cancer progression.

Feedback loops within the neuroendocrine-immune axis regulate responses to internal and external stimuli. Psychological factors, including stress and emotions, can influence these feedback loops, modulating hormonal release and immune responses. Hans Selye's General Adaptation Syndrome included the shrinkage of the thymus gland as a part of the body's stress response (Tachè 2014). The central nervous system (CNS) communicates with the immune system via the hypothalamic-pituitary-adrenal (HPA) axis and the sympathetic-adrenal-medullary (SAM) axis. Stress activates the HPA axis, leading to the release of cortisol, which modulates immune function by suppressing pro-inflammatory cytokines and altering leukocyte distribution. Simultaneously, the SAM axis releases catecholamines like norepinephrine, influencing immune cell activity and cytokine production (Miller et al. 2007). Conversely, immune cells release signaling molecules that affect the nervous and endocrine systems. Understanding this bidirectional communication provides a holistic perspective on the integration of psychological factors in cancer biology.

Recognizing the bidirectional communication between different body systems emphasizes the need for integrated psychosocial interventions in cancer treatment. Targeting both psychological well-being and physiological reactions to disease may enhance treatment efficacy and overall patient outcomes.

14.2 Emotions and Coping Mechanisms

Emotional states, including depression and anxiety, have been implicated in cancer susceptibility and progression. Epidemiological evidence suggests a correlation between prolonged states of emotional distress and increased cancer risk. The neurobiological underpinnings of the connection between emotions and the immune system involve intricate neurotransmitter and hormonal pathways. Dysregulation in these pathways may impact immune cell function, creating a permissive environment for cancer development and progression.

Coping mechanisms play a pivotal role in individuals' responses to stressors, thereby influencing health outcomes, including cancer. Various coping strategies, such as problem-focused coping, emotion-focused coping, and social support seeking, can exert distinct effects on the immune system. The dichotomy between adaptive and maladaptive coping mechanisms shapes the psychological and physiological responses to stress. Adaptive coping strategies, like positive reframing or acceptance, may bolster immune function, while maladaptive strategies, including avoidance or substance use, may compromise it. Examining the implications of coping strategies on cancer outcomes provides a nuanced understanding of the role of psychological resilience in cancer resilience.

D. Paul, *Cancer: An Integrative Approach*, https://doi.org/10.1007/978-3-032-08808-6_14

Addressing emotional states in cancer patients is integral to comprehensive care. Psychosocial interventions, such as cognitive-behavioral therapy, mindfulness-based stress reduction, and pharmacological approaches, have demonstrated efficacy in managing emotional distress (see below).

14.3 The Role of Family Support and the Impact of Loneliness on Cancer Prognosis

Cancer, while fundamentally a biological disease, presents itself as an intricate psychosocial phenomenon that deeply reshapes the emotional, social, and psychological landscape of patients' lives. Increasingly, research in psycho-oncology has illuminated the extent to which psychosocial factors, particularly family support and loneliness, influence not only the experience of cancer but also its clinical trajectory. As science moves beyond reductionist models focused solely on tumors and molecular pathways, there is growing recognition that the social environment of the patient significantly modulates outcomes, from treatment adherence to immune competence and survival.

Family support encompasses a wide spectrum of emotional, informational, and practical resources provided by close relations throughout the patient's cancer journey. It plays a fundamental role in buffering against psychological distress, enhancing treatment adherence, and even modulating biological mechanisms linked to cancer progression. Emotionally, family support can stabilize the patient's mental state, mitigating the effects of anxiety, depression, and fear. A meta-analysis demonstrated that patients with strong perceived social support had a better quality of life (Ding et al. 2024). This psychological stability is not merely comforting; it translates into physiological resilience. Supportive relationships attenuate stress-induced activation of the hypothalamic-pituitary-adrenal (HPA) axis and sympathetic nervous system, both of which are implicated in cancer progression. Thus, the emotional grounding provided by family members may contribute to reducing systemic inflammation and slowing tumor growth.

Furthermore, family support plays a pivotal role in facilitating treatment adherence. The cancer treatment pathway is often long, complex, and physically taxing. Family members frequently act as informal caregivers, managing medication schedules, providing transportation to medical appointments, and encouraging continued engagement with healthcare providers. In a comprehensive analysis, DiMatteo, found that patients with robust social support systems were 1.74 times more likely to adhere to prescribed treatment regimens (DiMatteo 2004). Adherence, in turn, is directly associated with improved clinical outcomes, particularly in contexts such as chemotherapy, radiation therapy, and long-term hormonal treatments. In essence, the psychological support provided by loved ones sometimes can determine whether a patient completes life-prolonging therapies.

Biologically, the impact of social support extends into the immune system. Studies have shown the close interrelationship between factors secreted by neurons, the immune system and cancer progression (Reznik and Torjani 2024; Wang et al. 2025a). Similarly, studies have reported that breast cancer patients with high levels of family support demonstrated significantly longer disease-free survival, underscoring the immune-enhancing and potentially tumor-suppressing effects of psychosocial well-being.

In stark contrast to these protective effects, loneliness and social isolation have emerged as independent risk factors that may accelerate disease progression and reduce survival in cancer patients. Loneliness is not merely a subjective feeling; it triggers a cascade of biological stress responses that can impair immune function and promote oncogenesis. It has been shown that socially isolated individuals exhibited elevated cortisol levels and increased expression of pro-inflammatory genes such as IL-6 (Cole et al. 2007). These inflammatory states are known to support tumor angiogenesis, invasion, and metastasis. Similarly, Susan Lutgendorf and her team demonstrated that ovarian cancer patients experiencing greater social isolation had tumors with higher expression of genes associated with angiogenesis, a critical process for tumor growth (Lutgendorf et al. 2008).

From a clinical perspective, lonely patients also report a higher symptom burden and lower quality of life. Social isolation was associated with pain, fatigue, and depression in elderly adults (Powell et al. 2022). Loneliness is strongly correlated with depression, which in itself is an established predictor of decreased cancer survival. The factors associated with social isolation include age, gender, comorbidity burden, education level, residence, medical insurance, occupation status, personality, race, smoking status, having children, not living alone, household income level, marital status, the role of primary caregiver, physical health status, mental health status, social health status, coping styles, and the level of social support and social interaction (Wang et al. 2024h). A lack of social connectedness may impair a patient's ability to process medical information, make informed decisions, and navigate the healthcare system, factors that cumulatively result in poorer outcomes.

Perhaps most strikingly, social isolation has been linked to increased mortality. It is known for almost two decades now that women with breast cancer have fewer social ties had a 66% higher risk of death compared to their more socially integrated counterparts (Kroenke et al. 2006). A broader study that included almost 6000 subjects concluded that loneliness independently predicts the development of the symptom cluster of pain, fatigue, and depression and social isolation and is a mortality risk factor similar to smoking and obesity, placing it squarely within the realm of modifiable determinants of health (Holt-Lunstad et al. 2015).

Recognizing the influence of these social determinants, several clinical strategies can be employed to mitigate the negative effects of isolation while reinforcing the protective role of support. First, involving family members in the care process, from diagnosis to treatment planning, should be a routine practice. Family-inclusive care improves communication, promotes shared decision-making, and ensures that patients are not facing their illness in isolation. Additionally, training programs for caregivers can empower them with the skills necessary to support the patient effectively while also addressing their own mental health needs.

Second, healthcare systems should incorporate structured psychosocial support programs into oncology care. Group therapy, peer mentoring, and psychological counseling offer valuable emotional resources, particularly for patients who lack family support. Advances in telemedicine have made it increasingly feasible to extend these services to patients in rural or underserved areas. Virtual support groups and online therapy platforms can bridge the social gap for those facing geographic or mobility barriers.

Third, routine psychosocial screening should be integrated into standard oncology assessments. Instruments such as the UCLA Loneliness Scale (Russell et al. 1978) and the eight-item modified Medical Outcomes Study (MOS) Social Support Survey (Moser et al. 2012) can help clinicians identify at-risk patients early in the disease course, allowing for timely interventions. These assessments can be administered during routine clinic visits and used to tailor individualized care plans that address not only the biological aspects of cancer but also its psychosocial dimensions.

Family support and social connectedness are not ancillary elements of cancer care, they are essential. Mounting empirical evidence affirms that robust family support enhances prognosis by strengthening psychological resilience, promoting adherence to therapy, and modulating key immune and endocrine pathways. In contrast, loneliness exerts a silent yet powerful influence, fostering biological dysregulation, increasing symptom burden, and elevating mortality risk. As such, a paradigm shift is needed in oncology, one that recognizes the patient not merely as a host of cancer cells, but as a socially embedded, emotionally complex human being. By integrating psychosocial assessments and interventions into routine care, we can move closer to delivering truly comprehensive, person-centered cancer treatment.

14.4 Psychosocial Interventions

Psychosocial interventions play a crucial role in cancer care, aiming to address the psychological and emotional challenges faced by individuals affected by cancer. These interventions encompass a broad spectrum of approaches, including counseling, psychoeducation, support groups, and cognitive-behavioral therapy.

Several evidence-based practices have demonstrated efficacy in improving the psychological well-being of cancer patients and survivors. Cognitive-Behavioral Therapy (CBT) (Breitbart et al. 2018) and Mindfulness-Based Stress Reduction (MBSR) (Carlson et al. 2013; Victorson et al. 2020) are among the interventions with already proven effectiveness. These practices not only help individuals cope with the emotional impact of cancer but also equip them with valuable skills to navigate the challenges associated with diagnosis, treatment, and survivorship.

A recent meta-analysis included eight psychosocial interventions (behavior therapy, cognitive therapy, cognitive behavior therapy (CBT), CBT plus additive treatment or mode of delivery that augmented the CBT, bereavement or supportive therapy, multiple or combined interventions, other psychotherapy, and psychoeducation), seven immune outcomes (proinflammatory cytokine or marker levels, anti-inflammatory cytokine levels, antibody levels, immune cell counts, natural killer cell activity, viral load, and other immune outcomes), and nine moderating factors (intervention type, intervention format, intervention length, immune marker type, basal vs stimulated markers, immune marker measurement timing, disease state or reason for treatment, age, and sex). Across 56 randomized clinical studies and 4060 participants, psychosocial interventions were found to be statistically significantly ($P < 0.001$) associated with enhanced immune system function (Shields et al. 2020).

Overall, the study found an average of 14.7% (95% CI, 5.7%–23.8%) improvement in beneficial immune system function over control and an 18.0% (95% CI, 7.2%–28.8%) decrease in harmful immune system function over time over controls. These associations persisted for at least 6 months following treatment and were robust across age, sex, and intervention duration. They were more pronounced for cognitive behavior therapy.

Developed by Dr. William Breitbart and colleagues at Memorial Sloan Kettering Cancer Center, meaning-centered psychotherapy (MCP) is grounded in Viktor Frankl's logotherapy, emphasizing the human drive to find meaning even amid suffering. MCP has been shown to enhance spiritual well-being, increase hope, and improve quality of life in patients with advanced cancer. A randomized controlled trial demonstrated that MCP significantly reduced psychological distress and improved spiritual well-being among patients with advanced cancer (Breitbart et al. 2018). Recently, the benefits of CBT and MCP were found to be equivalent in a randomized controlled study (Marco et al. 2024).

Created by Dr. Harvey Max Chochinov, Dignity Therapy (DT) involves guided conversations that help patients reflect on meaningful aspects of their lives, culminating in a legacy document for loved ones. This process has been shown to alleviate psychological distress and enhance spiritual well-being in terminally ill patients. DT not only benefits the patient but often leaves a valuable emotional and narrative legacy for the family. In a randomized controlled trial, patients who received DT reported higher satisfaction and a heightened sense of dignity compared to those receiving standard palliative care (Chochinov et al. 2011).

Mounting evidence suggests that psychological interventions can exert a positive impact on immune function and, consequently, cancer outcomes. Reductions in stress and anxiety through psychosocial interventions have been associated with improved immune responses, potentially influencing the body's ability to mount an effective defense against cancer. Mind–body practices, such as meditation, yoga, and Tai Chi, impact gene expression (Saatcioglu 2013; Huston and McFarlane 2016; Buric et al. 2017; Venditti et al. 2020; Dutcher et al. 2022) and have shown promise in influencing psycho-neuro-immunological pathways.

These practices emphasize the connection between mental, emotional, and physical well-being, fostering a holistic approach to health. Research suggests that mind–body interventions may contribute to stress reduction, improved emotional states, and modulation of immune responses, thereby potentially impacting the progression and management of cancer.

While not substitutes for conventional treatments, these approaches aim to address aspects of well-being that may be overlooked in traditional medical care, offering patients additional tools to manage symptoms and improve their overall quality of life. While psychosocial interventions may help improve the quality of life of patients with breast cancer, their effect on cancer prognostic and actual survival has not been consistently demonstrated (Falagas et al. 2007; Anghel et al. 2025).

The integration of complementary medicine in cancer care raises ethical considerations regarding safety, efficacy, and patient autonomy. Collaboration between healthcare providers, including oncologists and integrative medicine specialists, is essential to ensure coordinated ethical care. Open communication, informed consent, and respect for patient preferences are critical elements in navigating the integration of these approaches with conventional cancer treatments.

14.5 Educating and Empowering Cancer Patients

At the moment of diagnosis, many patients feel helpless seeing that their bodies have become unrecognizable battlegrounds and their futures have been overtaken by uncertainty, many patients feel helpless. Yet, within this clinical and existential upheaval lies an often-underappreciated therapeutic force: patient empowerment. Emerging evidence supports the view that educating and activating patients to take an engaged role in their cancer care, physically, psychologically, and behaviorally, can improve not only their quality of life but also enhance clinical outcomes and survival.

Knowledge is power, and in oncology, it can also be medicine. Patient education improves understanding of disease processes, treatment options, and side effect management, enabling individuals to make informed decisions and adhere more effectively to complex treatment regimens. In randomized trials, better-informed patients show higher treatment compliance, fewer emergency interventions, and reduced symptom burden (McCorkle et al. 2011).

For example, in breast cancer, women who received structured education and decision-support tools were significantly more likely to complete adjuvant therapies and report higher satisfaction with care, factors associated with improved long-term survival (Ter Stege et al. 2024).

Empowerment extends beyond knowledge to encompass psychological resilience and self-efficacy; the belief in one's capacity to affect outcomes. Multiple studies have shown that cancer-related stress, helplessness, and depression are associated with increased inflammation, weakened immune surveillance, and poorer survival (Sephton et al. 2009; Antoni et al. 2006). Conversely, psychological interventions that foster a sense of control, such as CBT, MBSR or DT, mindfulness, and meaning-centered therapies, can restore psychoneuroimmunological balance, potentially influencing tumor biology itself (Lutgendorf and Andersen 2015).

In a landmark study, psychosocial intervention in patients with early breast cancer not only reduced distress but also led to improved survival rates after 11 years, suggesting that mind–body factors modulate disease trajectory in meaningful ways (Andersen et al. 2008).

Educated and empowered patients are more likely to adopt protective health behaviors, including improved nutrition, physical activity, smoking cessation, and adherence to follow-up care, all of which have independent effects on survival (Rock et al. 2020). Education amplifies the uptake of such behaviors, while empowerment ensures their sustainability. For instance, colorectal cancer survivors who engaged in moderate physical activity post-treatment had significantly lower recurrence and mortality rates, especially those who participated in structured lifestyle education programs (Meyerhardt et al. 2006).

The shift from patient as passive recipient to proactive partner reframes the clinical encounter. Empowered patients ask more questions, clarify treatment goals, and seek second opinions when needed. This leads to higher satisfaction, fewer medical errors, and more personalized care. Importantly, patients who feel involved in their care process are more likely to follow through with complex or long-duration treatments, which is especially crucial in multimodal cancer therapies.

Educating and empowering cancer patients is more than a humanistic ideal: it is a clinically impactful strategy. It modulates not only mind and behavior, but also immune function, stress response, and survival itself. The growing body of evidence makes clear that empowering patients is not a luxury, but a cornerstone of contemporary, evidence-based oncology. In a disease that robs individuals of certainty, restoring a sense of agency may be one of the most powerful therapies we have both for cancer treatment, and for preventing cancer from recurring.

14.6 Beyond the Tumor: Toward an Integrative Model of Cancer and Care

The need for integration arises from a quiet crisis: many patients feel unseen in the very institutions designed to help them live. They emerge from consultations with little understanding and less peace. The suffering of patients is almost always medicalized; their voices are eclipsed by survival data. At the end of life, quality of life becomes essential. Integrating a disease-oriented model with a patient-oriented model restores balance. It ensures that treatment decisions, whether for curative intent or palliation, reflect not only what medicine can offer, but what the person can bear, and what they ultimately desire.

Cancer, in all its biological intricacy and emotional gravity, has long been approached as a problem to be solved, an invader to be staged, measured, and subdued. Within this paradigm, the cancer-centered model of care has offered great strength. It is precise, evidence-bound, and therapeutically assertive.

The Cancer-Centered Model is built on the precision of biology and the power of science. It is structured around evidence-based protocols, staging systems, and therapeutic regimens. Oncologists operate as navigators of a complex terrain—tumor genomics, imaging results, trial data—working to extend life and, when possible, achieve a cure. The current oncology model thrives in the language of protocols, in tumor boards and staging systems, in targeted agents and survival curves. Yet in its clarity, it sometimes forgets the complexity of its subject—not the tumor, but the person who hosts it. In contrast, the patient-centered model of care does not see disease in isolation, but as a lived experience. It listens before it prescribes and pauses to ask not only "What can we do?" but "What matters to you?"

The Patient-Centered Model, by contrast, places the person—rather than the pathology—at the center of care. It begins not with the tumor board, but with the patient's story. This model champions respect, communication, and partnership. As the Institute of Medicine framed it, it is "care that is respectful of and responsive to individual patient preferences, needs, and values." It is a model of empathy, flexibility, and shared decision-making.

These two models, one anchored in the molecular choreography of malignancy, the other in the delicate architecture of the self, need not exist in opposition (Tinetti et al. 2016). Rather, their integration promises a fuller, more humane practice of oncology. Such integration acknowledges that while cancer is a biological process requiring methodical and often aggressive intervention, healing cannot be extricated from the person's story, context, and inner life. It is a weaving together of cure and care, of cytotoxic precision and empathic presence.

To understand the fullness of this integration, we must first see the contours of each model clearly. The cancer-centered model is steeped in the legacy of biomedical science. It thrives in academic centers, is shaped by consensus guidelines, and measures success in tumor shrinkage, recurrence-free intervals, and median overall survival. It is the realm of PET scans, pathology reports, and clinical trials. Yet, for all its discipline, it often silences the subjective experience of illness, the fatigue that lingers long after remission, the spiritual reckoning that accompanies metastatic diagnosis, the desire to be seen not only as a patient, but also as a parent, a partner, or simply as a suffering human being.

This integration begins in the structure of care itself. True multidisciplinary teams are not just aggregates of specialties; they are communities of attention, where oncologists, nurses, social workers, psychologists, and palliative care specialists work not in parallel, but in concert, and always in dialogue with the patient. Treatment becomes a negotiation between survival and meaning, between risk and relevance. It becomes truly personalized, not in the reductionist genomic sense alone, but in the fullest humanistic sense.

Integrated oncology also requires the reanimation of language. Communication becomes more than the transfer of information; it is the creation of connections. Models like the ASCO VALUE framework (Schnipper et al. 2016) guide clinicians in these conversations, reminding them to listen, to acknowledge emotion, and to elicit what lies beneath the question: "What are my chances?" Through these exchanges, patients are not only informed, but empowered—and importantly, also comforted.

14.7 Why Integration Matters

In oncology, *cure and care* must be complementary, not competing goals. Life-prolonging therapies may achieve clinical success, yet leave patients diminished in spirit or function. The challenge is not merely to extend life, but to enrich it. When cancer-centered and patient-centered approaches unite, treatment becomes both effective and humane.

Perhaps nowhere is integration more impactful than in the early inclusion of palliative care. Too often miscast as a final chapter, palliative care, when introduced early, becomes a parallel narrative—one that focuses on symptom relief, psychosocial support, and the alignment of treatment with life goals. Evidence continues to affirm what intuition already suggests: that patients who receive early palliative care live not only better, but sometimes longer.

In addition to structured psychotherapies, spiritual care plays a crucial role in alleviating end-of-life anxiety. Spirituality—defined broadly as the search for meaning, connection, and transcendence—can be a powerful coping resource. Patients who experience a sense of spiritual peace or connection with something larger than themselves often report less fear of death and greater acceptance. Chaplaincy support, spiritual counseling, and non-denominational practices such as mindfulness and meditation can offer patients a framework for making peace with mortality. These approaches also help them cultivate a sense of presence and inner stability, even in the face of physical decline.

Spiritual interventions may improve the quality of life of cancer patients as has been proven by several studies including a randomized study I initiated in my second year of Hematology-Oncology residency that took more than a decade to complete enrollment due to the difficulty of patient enrollment in end-of-life studies (Moadel et al. 2005). During our encounters with patients living their last days, we glimpsed a profound truth: spiritual care, when fully realized, becomes its own form of medicine.

Challenges, of course, remain. Health systems are frequently built for throughput, not depth. Time is rationed. Outcome measures prioritize the visible tumor over the invisible soul. Yet these are not insurmountable barriers—they are design flaws awaiting revision. Solutions are at arm's length. Metrics of care must evolve to include patient-reported outcomes, experiences, and values. Time spent in listening and presence must be recognized as central, not supplemental. And new roles—nurse navigators, communication coaches, survivorship counselors—can redistribute the burden of care without losing its intimacy.

At its core, integrated oncology is not about merging two systems, but about returning to the essence of medicine: to heal. It reclaims the ancient understanding that healing is not only about eradicating disease, but about restoring coherence in the face of chaos. Cancer challenges not only the body, but also—essentially—the narrative of one's life. To integrate cancer-centered and patient-centered care is to pay attention to both. It is to treat not only the malignancy that divides the body, but also the mental suffering that fractures the self.

In this integration, oncology expands beyond biology. The microscope images remain sharp, the scans detailed, the protocols precise, but the perspective widens. And in that widening, a deeper kind of medicine begins.

Cancer care has long been a disease-centered paradigm: rigorous, protocol-driven, and focused on the measurable: tumor size, survival curves, therapeutic response. This cancer-centered model, forged in the tradition of biomedical excellence, has transformed once-some lethal diagnoses into chronic or curable conditions. Yet, despite these achievements, many patients journey through their treatments feeling unseen, processed, and reduced to their disease. In response, an integrated patient-centered model emerged, one that honors not just the biology of cancer but the biography of the person who bears it. The synthesis of these two models into an integrative approach, one that treats the tumor, the other that supports the patient, restores the human face of medicine.

Shared decision-making offers one such bridge. It allows patients to engage meaningfully with their care choices, aligning clinical options with personal values. In parallel, psychosocial support, addressing distress, depression, and existential fear, has been shown to not only improve mood and function but also prolong survival. Integration humanizes science.

14.8 Cancer: Major Existential Stressor

As cancer progresses toward advanced stages, patients are increasingly confronted not only with physical symptoms and functional decline, but also with a range of profound existential concerns. These challenges, such as the fear of death, loss of meaning, disruption of identity, and anxieties about leaving loved ones behind, can generate an emotional distress that often surpasses physical suffering. The recognition and management of these existential issues have become a central concern in palliative oncology and integrative cancer care. Understanding the psychological and philosophical dimensions of dying is essential to providing compassionate and holistic care for patients nearing the end of life.

The awareness of impending death tends to awaken a heightened sense of vulnerability and existential uncertainty. Cancer patients often experience a confrontation with their finitude, which may provoke intense fear, not only of physical death, but of the process of dying itself, with its associations of suffering, dependency, and loss of control. This fear of death is not monolithic; it encompasses multifaceted anxieties such as fear of pain, fear of non-existence, fear of the unknown, fear of leaving unfinished business, and concern over the well-being of family members after one's departure. These fears may coexist with spiritual distress and questions about the meaning or value of one's life and suffering.

From a clinical perspective, the existential dimension of cancer care demands approaches that address the emotional, spiritual, and psychosocial aspects of the dying process. Psychological interventions, spiritual counseling, meaning-centered therapies, and end-of-life philosophical counselling have all been developed and validated to reduce existential distress and improve emotional well-being in patients nearing life's end.

Importantly, the relational aspect of care should not be overlooked. The Charaka Samhita, a foundational text in Indian Ayurveda, emphasizes four key pillars for successful treatment: the physician (*vaidya*), the patient (*rogi*), the person who helps the patients, who may be a family member (*upastha*), and the medicine (*dravya*). Each of these four active components, along with their respective qualities, is crucial for a comprehensive and effective healing process (Valiathan 2009). The presence of empathic clinicians, loved ones, and palliative care teams, who can bear witness to the patient's fears without judging them, provides an essential anchor through the turbulent times of dying. The therapeutic alliance itself, built on compassion, respect, and authenticity, can be a container for existential anxiety. Simply allowing space for patients to speak openly about death, without euphemism or avoidance, can reduce their sense of isolation and promote emotional relief. Pharmacological interventions may also be considered, particularly when existential distress overlaps with clinical depression or severe anxiety. Low-dose anxiolytics or antidepressants can help alleviate the physiological symptoms of emotional suffering, but these should never replace the psychosocial and spiritual dimensions of end-of-life care.

Ultimately, addressing the fear of death in cancer patients requires a multidimensional, individualized approach that honors the patient's subjective experience. It involves acknowledging death not merely as a medical endpoint, but as an existential reality with profound emotional impact. In the final stages of life, when curative treatment is no longer possible, care must shift toward supporting patients' psychological integrity, affirming their inherent dignity, and helping them find meaning within the last days of their lives. Such care not only alleviates suffering, but also transforms the end-of-life experience into one of relational depth, personal clarity, and, whenever possible, inner peace.

15 Conclusions

15.1 The Time-Bound Paradox of Cancer: Harnessing Its Processual Nature to Cure It

Cancer is not a static thing; it is a process, a moving form, a continuous unfolding narrative. Rooted in the logic of processual ontology, cancer does not simply "exist" in the traditional sense; it becomes, it evolves, it adapts. It is a dynamic condition, shaped by cellular plasticity, environmental cues, and recursive interactions with the host organism. This temporal fluidity, which makes cancer so formidable, may also be its greatest vulnerability.

Unlike diseases fixed in form or confined to anatomical boundaries, cancer is a shape shifter, capable of transforming its molecular identity, evading immune detection, resisting therapies, and re-emerging in new territories. Its swift progression is often perceived as a biological curse: a race against time. Yet, embedded within this very fickleness of form and function lies an unexploited potential: if cancer can change quickly, then it is not rigid. And if it is not rigid, then its trajectory can be modulated and steered toward more benign behaviors.

Indeed, the same plasticity that allows cancer cells to adopt aggressive phenotypes, undergo epithelial-mesenchymal transitions, metabolic rewiring, and immunological escape, suggests that these phenotypes are not final. They are reversible states, governed not solely by permanent genetic lesions, but by epigenetic landscapes, microenvironmental conditions, and temporal rhythms. In this view, cancer becomes less of an immutable enemy and more of a reprogrammable system, one that can be rerouted away from malignancy and back toward physiological coherence.

Time, then, is not only cancer's accomplice; it may be its undoing. If we learn to understand and entrain the rhythms of cancer's becoming, its oscillations between quiescence and proliferation, its molecular switches of suppression and expression, we may discover how to interrupt its trajectory. Therapies that are temporally intelligent, that adapt in synchrony with cancer's internal dynamics, may prove more effective than static, one-size-fits-all interventions. *Dynamic problems require dynamic solutions.*

Just as the tides can be predicted and the seasons anticipated, so too might cancer's fluctuations be mapped, modulated, and ultimately mastered. Its capacity to change fast means it lives on the edge of instability, a fact that could be used to push it beyond the brink of viability. What if we could confuse it, outpace it, or restore it to order *before* it progresses into lethality?

To tame cancer, we must not fight against its fluidity, but exploit it, *recognizing that in its processual quality lies the very possibility of transformation.* The key to curing cancer may not be in more and more precise molecular bombs and cellular weapons, but in guiding its evolution toward benign behaviors and extinction.

15.2 An Integrative Approach in Cancer Research and Treatment

A fundamental paradigm shift is necessary in oncology: the human organism, *not* merely its genes, cells, tissues, or organs, must be returned to the center of focus. The central thesis of this book is that effective cancer treatment, especially of advanced and metastatic disease, requires reintegration of the organism as a complex, dynamic whole.

Cancer is not simply a cellular anomaly; it represents a systemic conflict between two competing life forms, the host organism and the evolving cancer system. For decades, molecular biology has guided cancer research with tremendous success in understanding pathways and mutations. However, this reductionist approach has often overlooked the integrative physiology of the human body.

To advance beyond current therapeutic limitations, particularly in metastatic cancer, we must conceptualize the human organism as a biosphere, a complex network comprising a *global internal climate*, regional *microclimates,*

D. Paul, *Cancer: An Integrative Approach*, https://doi.org/10.1007/978-3-032-08808-6_15

and countless local *miniclimates* maintained by somatic cells, immune cells, microbiota, and tumor subclones. A better characterization of biological climates from nanoscales (i.e., molecular nanoclimates) up to the macroclimate of the whole organism will turn cancer into a manageable condition.

Reducing the future burden of cancer depends critically on the development of both cancer prevention and early detection and early interception strategies, proven to be the most effective tools for decreasing cancer incidence. At present, we lack treatments that prevent or treat metastatic disease, and developing innovative therapeutic approaches capable of limiting cancer aggressiveness and achieving durable cures in metastatic disease is fundamental. *Precision targeting,* must expand beyond the cellular and tissue level to encompass the full complexity of the organism. Cancer is a disease of time and *precision targeting* must be complemented by *process oncology*, a systems-level framework focused on organismal dynamics, and by *homeodynamic treatment strategies* that reinforce physiological resilience and restore systemic balance.

Technological advancements, particularly in artificial intelligence (AI) and emerging AI architectures, will play a vital role in this evolution. Equally essential is the rise of *transdisciplinary collaborations,* bringing together diverse teams of experts across medicine, biology, data science, and engineering to co-create the next generation of cancer care.

Crucially, a deeper understanding of *biocausality* and the reciprocal interactions between the organism and the cancer system is indispensable. As Walter Cannon, a pioneer of integrative physiology, emphasized, the body possesses an inherent *wisdom*—a self-regulating intrinsic intelligence that guides its pursuit of internal balance. Long-lasting cures for metastatic cancer may only emerge when we learn to recognize this wisdom, to respect it, and to support the organism in restoring its own integrity. Healing, in this light, is not imposed—it is enabled.

As the ancients rightly observed: *"Medicus curat, natura sanat"*—the physician treats, but nature heals. In this sense, the nature itself of the organism remains our most powerful therapeutic ally.

Correction to: Cancer: An Integrative Approach

Correction to:
D. Paul, *Cancer: An Integrative Approach*,
https://doi.org/10.1007/978-3-032-08808-6

The original version of the book was inadvertently published with errors on pages x, 2, 3, 21, 25, 31, 33, 49, 55-56, 61, 63, 85, 86, 100, 109, 110, 118, 119, 121, 143, 156, 160, 162, 167, 173, 174, 176, 185, 188, 190, 196, 199, 226, 241, 242, 243, 244, 245, 247, 249, 264. The corrections have been updated in the respective chapters.

The updated version of this book can be found at
https://doi.org/10.1007/978-3-032-08808-6

D. Paul, *Cancer: An Integrative Approach*, https://doi.org/10.1007/978-3-032-08808-6_16

References

Abdel-Haleem AM, Lewis NE, Jamshidi N et al (2017) The emerging facets of non-cancerous Warburg effect. Front Endocrinol (Lausanne) 8:279. https://doi.org/10.3389/fendo.2017.00279

Åberg C, Poolman B (2021) Glass-like characteristics of intracellular motion in human cells. Biophys J 120(11):2355–2366. https://doi.org/10.1016/j.bpj.2021.04.011

Åberg C, Varela JA, Fitzpatrick LW, Dawson KA (2016) Spatial and structural metrics for living cells inspired by statistical mechanics. Sci Rep 6:34457. https://doi.org/10.1038/srep34457

Abou-Alfa GK, Macarulla T, Javle M et al (2020) Ivosidenib in IDH1-mutant, chemotherapy-refractory cholangiocarcinoma (ClarIDHy): a multicentre, randomised, double-blind, placebo-controlled, phase 3 study. Lancet Oncol 21(6):796–807. https://doi.org/10.1016/S1470-2045(20)30157-1. Epub 2020 May 13. Erratum in: Lancet Oncol. 2020 Oct;21(10):e462. https://doi.org/10.1016/S1470-2045(20)30547-7. Erratum in: Lancet Oncol. 2024 Feb;25(2):e61. https://doi.org/10.1016/S1470-2045(24)00013-5

Abravan A, Faivre-Finn C, Gomes F, van Herk M, Price G (2024) Comorbidity in patients with cancer treated at The Christie. Br J Cancer 131(8):1279–1289. https://doi.org/10.1038/s41416-024-02838-w

Abu-Toamih Atamni HJ, Kontogianni G, Binenbaum I et al (2019) Hepatic gene expression variations in response to high-fat diet-induced impaired glucose tolerance using RNAseq analysis in collaborative cross mouse population. Mamm Genome 30(9–10):260–275. https://doi.org/10.1007/s00335-019-09816-1

Acanda de la Rocha AM, Berlow NE, Fader M et al (2024) Feasibility of functional precision medicine for guiding treatment of relapsed or refractory pediatric cancers. Nat Med 30(4):990–1000. https://doi.org/10.1038/s41591-024-02848-4

Acanda, de la Rocha AM, Berlow NE, Azzam DJ (2025) Functional precision medicine: the future of cancer care. Trends Mol Med 31(5):404–408. https://doi.org/10.1016/j.molmed.2024.10.015

Acharya D, Mukhopadhyay A (2024) A comprehensive review of machine learning techniques for multi-omics data integration: challenges and applications in precision oncology. Brief Funct Genomics 23(5):549–560. https://doi.org/10.1093/bfgp/elae013

ACS (American Cancer Society) (n.d.) website: https://www.cancer.org/

Adamatzky A, Akl S, Burgin M et al (2017) East-West paths to unconventional computing. Prog Biophys Mol Biol 131:469–493. https://doi.org/10.1016/j.pbiomolbio.2017.08.004

Adami C (2002) What is complexity? Bioessays 24(12):1085–1094. https://doi.org/10.1002/bies.10192

Adami C (2004) Information theory in molecular biology. Phys Life Rev 1:3–22. https://doi.org/10.1016/j.plrev.2004.01.002

Adami C (2016) What is information? Philos Trans A Math Phys Eng Sci 374(2063):20150230. https://doi.org/10.1098/rsta.2015.0230

Adams CD, Richmond R, Ferreira DLS et al (2019) Circulating metabolic biomarkers of screen-detected prostate cancer in the Protect study. Cancer Epidemiol Biomarkers Prev 28(1):208–216. https://doi.org/10.1158/1055-9965.EPI-18-0079

Adee S (2023) We are electric. Grand Central Publishing, New York

Agrawal AA, Yu L, Smith PG, Buonamici S (2018) Targeting splicing abnormalities in cancer. Curr Opin Genet Dev 48:67–74. https://doi.org/10.1016/j.gde.2017.10.010

Ahima RS, Flier JS (2000) Adipose tissue as an endocrine organ. Trends Endocrinol Metab 11(8):327–332. https://doi.org/10.1016/s1043-2760(00)00301-5

Ahmad AS, Ormiston-Smith N, Sasieni PD (2015) Trends in the lifetime risk of developing cancer in Great Britain: comparison of risk for those born from 1930 to 1960. Br J Cancer 112:943–947. https://doi.org/10.1038/bjc.2014.606

Ahmad TA, Gopal DP, Chelala C et al (2023) Multimorbidity in people living with and beyond cancer: a scoping review. Am J Cancer Res 13(9):4346–4365

Ahmad M, Wolberg A, Kahwaji CI (2025) Biochemistry, electron transport chain. In: StatPearls [Internet]. StatPearls Publishing, Treasure Island (FL). Available from: https://www.ncbi.nlm.nih.gov/books/NBK526105/

Aichel O (1911) Vorträge und Aufsätze über Entwickelungsmechanik der Organismen: Über Zellverschmelzung mit qualitativ abnormer Chromosomenverteilung als Ursache der Geschwulstbildung (in German) [Lectures and essays on the developmental mechanics of organisms: on cell fusion with qualitatively abnormal chromosome distribution as the cause of tumor formation]. Verlag Von Wilhelm Engelmann, Leipzig

Aichel O (1911a) In: Roux W (ed) Vorträge und Aufsätze über Entwickelungsmechanik der Organismen. Wilhelm Engelmann, Leipzig, pp 1–115

Aichel O (1911b) Vorträge und Aufsätze über Entwickelungsmechanik der Organismen: Über Zellverschmelzung mat qualitativ abnormer Chromosomenverteilung als Ursache der Geschwulstbildung. Wilhelm Engelmann, Leipzig

Aiello NM, Maddipati R, Norgard RJ et al (2018) EMT subtype influences epithelial plasticity and mode of cell migration. Dev Cell 45(6):681–695.e4. https://doi.org/10.1016/j.devcel.2018.05.027

Akbari A, Nazari-Khanamiri F, Ahmadi M et al (2022) Engineered exosomes for tumor-targeted drug delivery: a focus on genetic and chemical functionalization. Pharmaceutics 15(1):66. https://doi.org/10.3390/pharmaceutics15010066

Akinyemiju T, Moore JX, Pisu M et al (2019) Association of baseline inflammatory biomarkers with cancer mortality in the REGARDS cohort. Oncotarget 10(47):4857–4867. https://doi.org/10.18632/oncotarget.27108

Aktipis A (2020) The cheating cell: how evolution helps us understand and treat cancer. Princeton University Press, Princeton

Aktipis CA, Nesse RM (2013) Evolutionary foundations for cancer biology. Evol Appl 6(1):144–159. https://doi.org/10.1111/eva.12034

Aktipis CA, Boddy AM, Jansen G et al (2015) Cancer across the tree of life: cooperation and cheating in multicellularity. Philos Trans R Soc Lond B Biol Sci 370(1673):20140219. https://doi.org/10.1098/rstb.2014.0219

Alam MM, Lal S, FitzGerald KE, Zhang L (2016) A holistic view of cancer bioenergetics: mitochondrial function and respiration play fundamental roles in the development and progression of

D. Paul, *Cancer: An Integrative Approach*, https://doi.org/10.1007/978-3-032-08808-6

diverse tumors. Clin Transl Med 5(1):3. https://doi.org/10.1186/s40169-016-0082-9

Alamoudi AA, Alnoury A, Gad H (2018) miRNA in tumour metabolism and why could it be the preferred pathway for energy reprograming. Brief Funct Genomics 17(3):157–169. https://doi.org/10.1093/bfgp/elx023

Alberts B, Johnson A, Lewis J (2002) Molecular biology of the cell, 4th edn. Garland Science, New York

Albuquerque TAF, Drummond do Val L, Doherty A, de Magalhães JP (2018) From humans to hydra: patterns of cancer across the tree of life. Biol Rev Camb Philos Soc 93(3):1715–1734. https://doi.org/10.1111/brv.12415

Alcolea MP, Alonso-Curbelo D, Ambrogio C et al (2024) Cancer hallmarks: piecing the puzzle together. Cancer Discov 14(4):674–682. https://doi.org/10.1158/2159-8290.CD-24-0097

Aldana M, Robeva R (2021) New challenges in systems biology: understanding the holobiont. Front Physiol 12:662878. https://doi.org/10.3389/fphys.2021.662878

Alečković M, McAllister SS, Polyak K (2019) Metastasis as a systemic disease: molecular insights and clinical implications. Biochim Biophys Acta Rev Cancer 1872(1):89–102. https://doi.org/10.1016/j.bbcan.2019.06.002

Alexandrov LB, Nik-Zainal S, Wedge DC et al (2013) Signatures of mutational processes in human cancer. Nature 500(7463):415–421. https://doi.org/10.1038/nature12477

Alexandrov LB, Kim J, Haradhvala NJ et al (2020) The repertoire of mutational signatures in human cancer. Nature 578(7793):94–101. https://doi.org/10.1038/s41586-020-1943-3

Alguacil J, Kogevinas M, Silverman DT et al (2011) Urinary pH, cigarette smoking and bladder cancer risk. Carcinogenesis 32(6):843–847. https://doi.org/10.1093/carcin/bgr048

Ali M, Alam SP, Kumar S et al (2016) Does blood pH change in cancer patients? Int J Curr Res 8(4):29543–29544

Ali A, Krecko L, Leitzel K et al (2017) Circulating PD-L1 (programmed death-ligand 1) and outcomes in a HER2-positive metastatic breast cancer cohort treated with first-line trastuzumab. J Clin Oncol 35(Suppl 15):1024–1024. https://doi.org/10.1200/JCO.2017.35.15_suppl.1024

Alix-Panabières C, Pantel K (2014) Challenges in circulating tumour cell research. Nat Rev Cancer 14(9):623–631. https://doi.org/10.1038/nrc3820

Allavena P, Germano G, Mantovani A (2015) Molecular links between inflammation and cancer. In: Thiagalingam S (ed) Systems biology of cancer. Cambridge University Press

Alles J, Fehlmann T, Fischer U et al (2019) An estimate of the total number of true human miRNAs. Nucleic Acids Res 47(7):3353–3364. https://doi.org/10.1093/nar/gkz097

Allin KH, Nordestgaard BG (2011) Elevated C-reactive protein in the diagnosis, prognosis, and cause of cancer. Crit Rev Clin Lab Sci 48(4):155–170. https://doi.org/10.3109/10408363.2011.599831

Allocati N, Masulli M, Di Ilio C, De Laurenzi V (2015) Die for the community: an overview of programmed cell death in bacteria. Cell Death Dis 6(1):e1609. https://doi.org/10.1038/cddis.2014.570

Almonte AA, Thomas S, Zitvogel L (2025) Microbiota-centered interventions to boost immune checkpoint blockade therapies. J Exp Med 222(7):e20250378. https://doi.org/10.1084/jem.20250378

Altadill T, Campoy I, Lanau L et al (2016) Enabling metabolomics-based biomarker discovery studies using molecular phenotyping of exosome-like vesicles. PLoS One 11(3):e0151339. https://doi.org/10.1371/journal.pone.0151339

Altea-Manzano P, Decker-Farrell A, Janowitz T et al (2025) Metabolic interplays between the tumour and the host shape the tumour macroenvironment. Nat Rev Cancer. 25(4):274–292. https://doi.org/10.1038/s41568-024-00786-4

Alvarez MJ, Shen Y, Giorgi FM et al (2016) Functional characterization of somatic mutations in cancer using network-based inference of protein activity. Nat Genet 48(8):838–847. https://doi.org/10.1038/ng.3593

Al-Zoughbi W, Huang J, Paramasivan GS et al (2014) Tumor macroenvironment and metabolism. Semin Oncol 41(2):281–295. https://doi.org/10.1053/j.seminoncol.2014.02.005

Alzubi MA, Turner TH, Olex AL et al (2019) Separation of breast cancer and organ microenvironment transcriptomes in metastases. Breast Cancer Res 21(1):36. https://doi.org/10.1186/s13058-019-1123-2

Amano K, Liu D, Bruera E, Hui D (2020) Collapse of fluid balance and association with survival in patients with advanced cancer admitted to a palliative care unit: preliminary findings. J Palliat Med 23(4):459–465. https://doi.org/10.1089/jpm.2019.0392

Ambrose EJ (1982) The nature and origin of the biological world. Ellis Horwood, Chichester

Amend SR, Pienta KJ (2015) Ecology meets cancer biology: the cancer swamp promotes the lethal cancer phenotype. Oncotarget 6(12):9669–9678. https://doi.org/10.18632/oncotarget.3430

Amend SR, Torga G, Lin KC et al (2019) Polyploid giant cancer cells: unrecognized actuators of tumorigenesis, metastasis, and resistance. Prostate 79(13):1489–1497. https://doi.org/10.1002/pros.23877

Amicone L, Marchetti A, Cicchini C (2023) The lncRNA HOTAIR: a pleiotropic regulator of epithelial cell plasticity. J Exp Clin Cancer Res 42(1):147. https://doi.org/10.1186/s13046-023-02725-x

Amraiz D, Kiani AK, Awan UA, et al (2024) Cancer prevention and treatment based on lifestyles. In: Saeed RF, ul Shaheed S (eds) Nutrition and dietary interventions in cancer. Springer, Berlin

Anand P, Kunnumakkara AB, Sundaram C et al (2008) Cancer is a preventable disease that requires major lifestyle changes. Pharm Res 25(9):2097–2116. https://doi.org/10.1007/s11095-008-9661-9

Anari F, Ramamurthy C, Zibelman M (2018) Impact of tumor microenvironment composition on therapeutic responses and clinical outcomes in cancer. Future Oncol 14(14):1409–1421. https://doi.org/10.2217/fon-2017-0585

Anastasiadou E, Jacob LS, Slack FJ (2018) Non-coding RNA networks in cancer. Nat Rev Cancer 18(1):5–18. https://doi.org/10.1038/nrc.2017.99

Andersen BL, Yang HC, Farrar WB et al (2008) Psychologic intervention improves survival for breast cancer patients: a randomized clinical trial. Cancer 113(12):3450–3458. https://doi.org/10.1002/cncr.23969

Anfossi S, Fu X, Nagvekar R, Calin GA (2018) MicroRNAs, regulatory messengers inside and outside cancer cells. Adv Exp Med Biol 1056:87–108. https://doi.org/10.1007/978-3-319-74470-4_6

Ang DC, Warrick AL, Shilling A et al (2014) Frequent phosphatidylinositol-3-kinase mutations in proliferative breast lesions. Mod Pathol 27(5):740–750. https://doi.org/10.1038/modpathol.2013.197

Anghel T, Melania BL, Costea I et al (2025) Review of psychological interventions in oncology: current trends and future directions. Medicina (Kaunas) 61(2):279. https://doi.org/10.3390/medicina61020279

Annila A (2020) Back to reality. Privus Press, Helsinki

Annila A (2024) On the origin of cognition. Biol Theory. https://doi.org/10.1007/s13752-024-00472-6

Annila A, Kuismanen E (2009) Natural hierarchy emerges from energy dispersal. Biosystems 95(3):227–233. https://doi.org/10.1016/j.biosystems.2008.10.008

Antoni MH, Lutgendorf SK, Cole SW et al (2006) The influence of biobehavioural factors on tumour biology: pathways and mechanisms. Nat Rev Cancer 6(3):240–248. https://doi.org/10.1038/nrc1820

Antony J, Thiery JP, Huang RY (2019) Epithelial-to-mesenchymal transition: lessons from development, insights into cancer and the potential of EMT-subtype based therapeutic intervention. Phys Biol 16(4):041004. https://doi.org/10.1088/1478-3975/ab157a

Apiz Saab JJ, Muir A (2023) Tumor interstitial fluid analysis enables the study of microenvironment–cell interactions in cancers.

Curr Opin Biotechnol 83:102970. https://doi.org/10.1016/j.copbio.2023.102970
Apte SA, Cavazos DA, Whelan KA, Degraffenried LA (2013) A low dietary ratio of omega-6 to omega-3 fatty acids may delay progression of prostate cancer. Nutr Cancer 65:556–562. https://doi.org/10.1080/01635581.2013.775316
Aractingi S, Kanitakis J, Euvrard S et al (2005) Skin carcinoma arising from donor cells in a kidney transplant recipient. Cancer Res 65(5):1755–1760. https://doi.org/10.1158/0008-5472.CAN-04-2783
Aranda-Anzaldo A, Dent MAR (2021) Is cancer a disease set up by cellular stress responses? Cell Stress Chaperones 26(4):597–609. https://doi.org/10.1007/s12192-021-01214-4
Arcamone F, Cassinelli G, Fantini G et al (1969) Adriamycin, 14-hydroxydaunomycin, a new antitumor antibiotic from S. peucetius var. caesius. Biotechnol Bioeng 11(6):1101–1110. https://doi.org/10.1002/bit.260110607
Argilés JM, Busquets S, Stemmler B, López-Soriano FJ (2014) Cancer cachexia: understanding the molecular basis. Nat Rev Cancer 14(11):754–762. https://doi.org/10.1038/nrc3829
Argilés JM, Stemmler B, López-Soriano FJ, Busquets S (2018) Inter-tissue communication in cancer cachexia. Nat Rev Endocrinol 15(1):9–20. https://doi.org/10.1038/s41574-018-0123-0
Aristotle (1933, 1989) Metaphysics, Book 5, section 1013a. In: Aristotle in 23 Volumes, vols. 17, 18 (trans. Tredennick H). Harvard University Press/Heinemann, Cambridge, MA/London
Aristotle (1997) Poetics. (trans. Heath M). Penguin Classics, London
Armitage P, Doll R (1954) The age distribution of cancer and a multi-stage theory of carcinogenesis. Br J Cancer 8(1):1–12. https://doi.org/10.1038/bjc.1954.1
Armutcu F (2019) Organ crosstalk: the potent roles of inflammation and fibrotic changes in the course of organ interactions. Inflamm Res 68(10):825–839. https://doi.org/10.1007/s00011-019-01271-7
Arney K (2020) Rebel cell: cancer evolution and the new science of life's oldest betrayal. Benbella Books, Texas
Arpaia N, Campbell C, Fan X et al (2013) Metabolites produced by commensal bacteria promote peripheral regulatory T-cell generation. Nature 504(7480):451–455. https://doi.org/10.1038/nature12726
Arruda, Macêdo JK, Fox JW, de Souza CM (2015) Disintegrins from snake venoms and their applications in cancer research and therapy. Curr Protein Pept Sci 16(6):532–548. https://doi.org/10.2174/1389203716666150515125002
Artusa V, De Luca L, Clerici M, Trabattoni D (2025) Connecting the dots: mitochondrial transfer in immunity, inflammation, and cancer. Immunol Lett 274:106992. https://doi.org/10.1016/j.imlet.2025.106992
Arya R, White K (2015) Cell death in development: signaling pathways and core mechanisms. Semin Cell Dev Biol 39:12–19. https://doi.org/10.1016/j.semcdb.2015.02.001
Asami M, Lam BYH, Hoffmann M et al (2023) A program of successive gene expression in mouse one-cell embryos. Cell Rep 42(2):112023. https://doi.org/10.1016/j.celrep.2023.112023
Asher G, Zhu B (2023) Beyond circadian rhythms: emerging roles of ultradian rhythms in control of liver functions. Hepatology 77(3):1022–1035. https://doi.org/10.1002/hep.32580
Ashworth A, Lord CJ (2018) Synthetic lethal therapies for cancer: what's next after PARP inhibitors? Nat Rev Clin Oncol 15:564–576. https://doi.org/10.1038/s41571-018-0055-6
Aslam S, Iqbal R, Saeed RF, Akram N, Ijaz F, Liaqat I, Aslam AS (2024) Nutritional Genomics and Cancer Prevention. In: Saeed RF, ul Shaheed S (eds) Cancer prevention and treatment based on lifestyles, nutrition and dietary interventions in cancer. Springer
Atkinson EG, Delgado-Calle J (2019) The emerging role of osteocytes in cancer in bone. JBMR Plus 3(3):e10186. https://doi.org/10.1002/jbm4.10186
Attal R, Bakkar A, Bouillaud F et al (2023) From electrons to cancer: redox shift as a driving force of tumorigenesis. Adv Redox Res 10(1):100087. https://doi.org/10.1016/j.arres.2023.100087
Auletta G, Ellis GFR, Jaeger L (2008) Top-down causation by information control: from a philosophical problem to a scientific research programme. J R Soc Interface 5:1159–1172
Avchaciov K, Clay KJ, Denisov K et al (2024) Multiple targets, one goal: compounding life-extending effects through polypharmacology. bioRxiv. https://doi.org/10.1101/2024.06.23.600269
Aytes A, Giacobbe A, Mitrofanova A et al (2018) NSD2 is a conserved driver of metastatic prostate cancer progression. Nat Commun 9(1):5201. https://doi.org/10.1038/s41467-018-07511-4
Baba AI, Câţoi C (2007) Comparative oncology. The Publishing House of the Romanian Academy, Bucharest
Babiker HM, Picozzi V, Chandana SR et al (2025) Tumor treating fields with gemcitabine and nab-paclitaxel for locally advanced pancreatic adenocarcinoma: randomized, open-label, pivotal phase III PANOVA-3 study. J Clin Oncol. https://doi.org/10.1200/JCO-25-00746
Baboo S, Bhushan B, Jiang H et al (2014) Most human proteins made in both nucleus and cytoplasm turn over within minutes. PLoS One 9(6):e99346. https://doi.org/10.1371/journal.pone.0099346
Bäck T, Cock JN, Rozenberg G (2002) Evolutionary computation as a paradigm for DNA-based computing. In: Landweber LF, Winfree E (eds) Evolution as computation. Springer, Berlin
Backer LC, Grindem CB, Corbett WT et al (2001) Pet dogs as sentinels for environmental contamination. Sci Total Environ 274(1-3):161–169. https://doi.org/10.1016/s0048-9697(01)00740-9
Bae T, Tomasini L, Mariani J et al (2018) Different mutational rates and mechanisms in human cells at pregastrulation and neurogenesis. Science 359(6375):550–555. https://doi.org/10.1126/science.aan8690
Baedke J (2019) O Organism, Where Art Thou? Old and new challenges for organism-centered biology. J Hist Biol 52(2):293–324. https://doi.org/10.1007/s10739-018-9549-4
Baedke J, McManus SF (2018) From seconds to eons: Time scales, hierarchies, and processes in evo-devo. Stud Hist Philos Biol Biomed Sci 72:38–48. https://doi.org/10.1016/j.shpsc.2018.10.006
Baedke J, Fábregas-Tejeda A, Prieto GI (2021) Unknotting reciprocal causation between organism and environment. Biol Philos 36(5):1–29. https://doi.org/10.1007/s10539-021-09815-0
Baer JM, Zuo C, Kang LI et al (2023) Fibrosis induced by resident macrophages has divergent roles in pancreas inflammatory injury and PDAC. Nat Immunol 24(9):1443–1457. https://doi.org/10.1038/s41590-023-01579-x
Baggiolini A, Callahan SJ, Montal E et al (2021) Developmental chromatin programs determine oncogenic competence in melanoma. Science 373(6559):eabc1048. https://doi.org/10.1126/science.abc1048
Bailey KD (2006) Living systems theory and social entropy theory. Syst Res Behav Sci 23(3):291–300. https://doi.org/10.1002/sres.728
Bailey MH, Tokheim C, Porta-Pardo E et al (2018) Comprehensive characterization of cancer driver genes and mutations. Cell 173(2):371–385.e18. https://doi.org/10.1016/j.cell.2018.02.060
Baillie KE, Stirling PC (2021) Beyond kinases: targeting replication stress proteins in cancer therapy. Trends Cancer 7(5):430–446. https://doi.org/10.1016/j.trecan.2020.10.010
Baish JW, Jain RK (2000) Fractals and cancer. Cancer Res 60(14):3683–3688
Baixauli F, López-Otín C, Mittelbrunn M (2014) Exosomes and autophagy: coordinated mechanisms for the maintenance of cellular fitness. Front Immunol 5:403. https://doi.org/10.3389/fimmu.2014.00403
Baker SG (2018) The detached pericyte hypothesis: A novel explanation for many puzzling aspects of tumorigenesis. Organisms J Biol Sci 2:25–41

Bakhoum SF, Compton DA (2012) Chromosomal instability and cancer: a complex relationship with therapeutic potential. J Clin Invest 122(4):1138–1143. https://doi.org/10.1172/JCI59954

Bakhoum SF, Landau DA (2017) Cancer evolution: no room for negative selection. Cell 171(5):987–989. https://doi.org/10.1016/j.cell.2017.10.039

Balachandran S, Narendran A (2023) The developmental origins of cancer: a review of the genes expressed in embryonic cells with implications for tumorigenesis. Genes (Basel) 14(3):604. https://doi.org/10.3390/genes14030604

Balasubramanian S, Weston DA, Levin M, Davidian DCC (2024) Electroceuticals: emerging applications beyond the nervous system and excitable tissues. Trends Pharmacol Sci 45(5):391–394. https://doi.org/10.1016/j.tips.2024.03.001

Baldwin JG, Heuser-Loy C, Saha T et al (2024) Intercellular nanotube-mediated mitochondrial transfer enhances T cell metabolic fitness and antitumor efficacy. Cell 187(23):6614–6630.e21. https://doi.org/10.1016/j.cell.2024.08.029

Balkwill FR, Mantovani A (2012) Cancer-related inflammation: common themes and therapeutic opportunities. Semin Cancer Biol 22(1):33–40. https://doi.org/10.1016/j.semcancer.2011.12.005

Balmain A (2020) The critical roles of somatic mutations and environmental tumor-promoting agents in cancer risk. Nat Genet 52(11):1139–1143. https://doi.org/10.1038/s41588-020-00727-5

Baluška F, Levin M (2016) On having no head: cognition throughout biological systems. Front Psychol 7:902. https://doi.org/10.3389/fpsyg.2016.00902

Baluška F, Witzany G (2014) Life is more than a computer running DNA software. World J Biol Chem 5(3):275–278. https://doi.org/10.4331/wjbc.v5.i3.275

Baluška F, Miller WB Jr, Reber AS (2021) Biomolecular basis of cellular consciousness via subcellular nanobrains. Int J Mol Sci 22(5):2545. https://doi.org/10.3390/ijms22052545

Baluška F, Miller WB Jr, Reber AS (2023) Cellular basis of cognition and evolution: from protists and fungi up to animals, plants, and root-fungal networks. In: Corning PA, Kauffman SA, Noble D, Shapiro JA, Vane-Wright RI, Pross A (eds) Evolution "on purpose": teleonomy in living systems. MIT Press, Cambridge, MA

Balzeau J, Menezes MR, Cao S, Hagan JP (2017) The LIN28/let-7 pathway in cancer. Front Genet 8:31. https://doi.org/10.3389/fgene.2017.00031

Bamford DH, Grimes JM, Stuart DI (2005) What does structure tell us about virus evolution? Curr Opin Struct Biol 15(6):655–663. https://doi.org/10.1016/j.sbi.2005.10.012

Bandea C (2009a) A unifying scenario on the origin and evolution of cellular and viral domains. Nat Prec. https://doi.org/10.1038/npre.2009.3888.1

Bandea C (2009b) The origin and evolution of viruses as molecular organisms. Nat Prec. https://doi.org/10.1038/npre.2009.3886.1

Bankey PE (1996) Hepatic regulation of systemic inflammation following acute injury. Curr Opin Crit Care 2:280–286

Bányai L, Trexler M, Kerekes K et al (2021) Use of signals of positive and negative selection to distinguish cancer genes and passenger genes. Elife 10:e59629. https://doi.org/10.7554/eLife.59629

Bapteste E, Papale F (2021) Modeling the evolution of interconnected processes: it is the song and the singers: tracking units of selection with interaction networks. Bioessays 43(1):e2000077. https://doi.org/10.1002/bies.202000077

Barabási AL, Albert R (1999) Emergence of scaling in random networks. Science 286(5439):509–512

Baracos VE, Martin L, Korc M et al (2018) Cancer-associated cachexia. Nat Rev Dis Primers 4:17105. https://doi.org/10.1038/nrdp.2017.105

Barbazán J, Matic Vignjevic D (2019) Cancer associated fibroblasts: is the force the path to the dark side? Curr Opin Cell Biol 56:71–79. https://doi.org/10.1016/j.ceb.2018.09.002

Barberio AM, Alareeki A, Viner B et al (2019) Central body fatness is a stronger predictor of cancer risk than overall body size. Nat Commun 10(1):383. https://doi.org/10.1038/s41467-018-08159-w

Barbieri M (2015) Code biology a new science of life. Springer

Barbieri M (2018) What is code biology? Biosystems 164:1–10. https://doi.org/10.1016/j.biosystems.2017.10.005

Barbieri M (2019) A general model on the origin of biological codes. Biosystems 181:11–19. https://doi.org/10.1016/j.biosystems

Barbour J (1999) the end of time: the next revolution in physics. Oxford University Press, Oxford

Barham J (1990) A Poincarean approach to evolutionary epistemology. J Soc Biol Struct 13:193–258

Barham J (1996) A dynamical model of the meaning of information. Biosystems 38(2–3):235–241. https://doi.org/10.1016/0303-2647(95)01596-5

Baro L, Almhassneh RA, Islam A, Juanes MA (2024) Tumor invasiveness is regulated by the concerted function of APC, formins, and Arp2/3 complex. iScience 27(5):109687. https://doi.org/10.1016/j.isci.2024.109687

Baron M, Tagore M, Hunter MV et al (2020) The stress-like cancer cell state is a consistent component of tumorigenesis. Cell Syst 11(5):536–546.e7. https://doi.org/10.1016/j.cels.2020.08.018

Baronzio G, Parmar G, Baronzio M, Kiselevsky M (2014) Tumor interstitial fluid: proteomic determination as a possible source of biomarkers. Cancer Genom Proteom 11(5):225–237

Barouki R, Audouze K, Coumoul X et al (2018) Integration of the human exposome with the human genome to advance medicine. Biochimie 152:155–158. https://doi.org/10.1016/j.biochi.2018.06.023

Barresi V, Cinnirella G, Valenti G et al (2018) Gene expression profiles in genome instability-based classes of colorectal cancer. BMC Cancer 18(1):1265. https://doi.org/10.1186/s12885-018-5174-z

Bartelink IH, Jones EF, Shahidi-Latham SK et al (2019) Tumor drug penetration measurements could be the neglected piece of the personalized cancer treatment puzzle. Clin Pharmacol Ther 106(1):148–163. https://doi.org/10.1002/cpt.1211

Bartlett S, Wong ML (2020) Defining Lyfe in the universe: from three privileged functions to four pillars. Life (Basel) 10(4):42. https://doi.org/10.3390/life10040042

Bartman CR, Weilandt DR, Shen Y et al (2023) Slow TCA flux and ATP production in primary solid tumours but not metastases. Nature 614(7947):349–357. https://doi.org/10.1038/s41586-022-05661-6

Basak SK, Bera A, Yoon AJ et al (2020) A randomized, phase 1, placebo-controlled trial of APG-157 in oral cancer demonstrates systemic absorption and an inhibitory effect on cytokines and tumor-associated microbes. Cancer. 126(8):1668–1682. https://doi.org/10.1002/cncr.32644

Basanta D, Anderson AR (2013) Exploiting ecological principles to better understand cancer progression and treatment. Interface Focus 3(4):20130020. https://doi.org/10.1098/rsfs.2013.0020

Basanta D, Anderson ARA (2017) Homeostasis back and forth: an eco-evolutionary perspective of cancer. Cold Spring Harb Perspect Med 7(9):a028332. https://doi.org/10.1101/cshperspect.a028332

Bashari A, Siegfried Z, Karni R (2023) Targeting splicing factors for cancer therapy. RNA 29(4):506–515. https://doi.org/10.1261/rna.079585.123

Basyoni AE, Atta A, Salem MM, Mohamed TM (2025) Harnessing exosomes for targeted drug delivery systems to combat brain cancer. Cancer Cell Int 25(1):150. https://doi.org/10.1186/s12935-025-03731-z

Battistoni A (2019) A repair manual for spaceship earth. Logic Magazine: Nature 9:219–237. https://www.ryestrategy.com/blog/biosphere-2-learning-from-failure

Bauer SM (1994) Psychoneuroimmunology and cancer: an integrated review. J Adv Nurs 19(6):1114–1120. https://doi.org/10.1111/j.1365-2648.1994.tb01195.x

Bauersfeld SP, Kessler CS, Wischnewsky M et al (2018) The effects of short-term fasting on quality of life and tolerance to chemotherapy in patients with breast and ovarian cancer: a randomized crossover pilot study. BMC Cancer 18:476. https://doi.org/10.1186/s12885-018-4353-2

Baylin SB, Jones PA (2016) Epigenetic Determinants of Cancer. Cold Spring Harb Perspect Biol 8(9):a019505. https://doi.org/10.1101/cshperspect.a019505

Beatty J (2016) The creativity of natural selection? Part I: Darwin, Darwinism, and the mutationists. J Hist Biol 49(4):659–684. https://doi.org/10.1007/s10739-016-9456-5

Beatty J (2019) The creativity of natural selection? Part II: The synthesis and since. J Hist Biol 52(4):705–731. https://doi.org/10.1007/s10739-019-09583-4

Becker A, Thakur BK, Weiss JM et al (2016) Extracellular vesicles in cancer: cell-to-cell mediators of metastasis. Cancer Cell 30(6):836–848. https://doi.org/10.1016/j.ccell.2016.10.009

Becker KA, Beckmann N, Adams C et al (2017) Melanoma cell metastasis via P-selectin-mediated activation of acid sphingomyelinase in platelets. Clin Exp Metastasis 34(1):25–35. https://doi.org/10.1007/s10585-016-9826-6

Bedessem B, Ruphy S (2015) SMT or TOFT? How the two main theories of carcinogenesis are made (artificially) incompatible. Acta Biotheor 63(3):257–267. https://doi.org/10.1007/s10441-015-9252-1

Bejan A (2000) Shape and structure, from engineering to nature. Cambridge University Press, Cambridge

Bejan A, Lorente S (2010) The constructal law of design and evolution in nature. Philos Trans R Soc Lond B Biol Sci 365(1545):1335–1347. https://doi.org/10.1098/rstb.2009.0302

Belikov AV, Leonov SV (2022) Cancer types with high numbers of driver events are largely preventable. PeerJ 10:e12672. https://doi.org/10.7717/peerj.12672

Belin L, Tan A, De Rycke Y et al (2020) Progression-free survival as a surrogate for overall survival in oncology trials: a methodological systematic review. Br J Cancer 122(11):1707–1714. https://doi.org/10.1038/s41416-020-0805-y

Belkaid Y, Hand TW (2014) Role of the microbiota in immunity and inflammation. Cell 157(1):121–141. https://doi.org/10.1016/j.cell.2014.03.011

Belkaid Y, Harrison OJ (2017) Homeostatic immunity and the microbiota. Immunity 46(4):562–576. https://doi.org/10.1016/j.immuni.2017.04.008

Bell KJ, Del Mar C, Wright G et al (2015) Prevalence of incidental prostate cancer: a systematic review of autopsy studies. Int J Cancer 137(7):1749–1757. https://doi.org/10.1002/ijc.29538

Bell DC, Leanza L, Gentile S, Sauter DR (2023) News and views on ion channels in cancer: is cancer a channelopathy? Front Pharmacol 14:1258933. https://doi.org/10.3389/fphar.2023.1258933

Benešová I, Křížová Ľ, Kverka M (2023) Microbiota as the unifying factor behind the hallmarks of cancer. J Cancer Res Clin Oncol 149:14429–14450. https://doi.org/10.1007/s00432-023-05244-6

Benjamin DJ, Haslam A, Prasad V (2024) Cardiovascular/anti-inflammatory drugs repurposed for treating or preventing cancer: a systematic review and meta-analysis of randomized trials. Cancer Med 13(5):e7049. https://doi.org/10.1002/cam4.7049

Bensellam M, Jonas JC, Laybutt DR (2018) Mechanisms of β-cell dedifferentiation in diabetes: recent findings and future research directions. J Endocrinol 236(2):R109–R143. https://doi.org/10.1530/JOE-17-0516

Benzekry S, Lamont C, Beheshti A et al (2014) Classical mathematical models for description and prediction of experimental tumor growth. PLoS Comput Biol 10(8):e1003800. https://doi.org/10.1371/journal.pcbi.1003800

Berardi R, Torniai M, Lenci E et al (2019) Electrolyte disorders in cancer patients: a systematic review. J Cancer Metastasis Treat 5:79. https://doi.org/10.20517/2394-4722.2019.008

Berchem G, Noman MZ, Bosseler M (2015) Hypoxic tumor-derived microvesicles negatively regulate NK cell function by a mechanism involving TGF-β and miR23a transfer. Oncoimmunology 5(4):e1062968. https://doi.org/10.1080/2162402X.2015.1062968

Berg KA, Clarke WP (2018) Making sense of pharmacology: inverse agonism and functional selectivity. Int J Neuropsychopharmacol 21(10):962–977. https://doi.org/10.1093/ijnp/pyy071

Bergen V, Lange M, Peidli S et al (2020) Generalizing RNA velocity to transient cell states through dynamical modeling. Nat Biotechnol 38(12):1408–1414. https://doi.org/10.1038/s41587-020-0591-3

Berger MF, Lawrence MS, Demichelis F et al (2009a) The genomic complexity of primary human prostate cancer. Genes Dev 23(7):781–791. https://doi.org/10.1101/gad.1781509

Berger SL, Kouzarides T, Shiekhattar R, Shilatifard A (2009b) An operational definition of epigenetics. Genes Dev 23(7):781–783. https://doi.org/10.1101/gad.1787609

Bergers G, Fendt SM (2021) The metabolism of cancer cells during metastasis. Nat Rev Cancer 21(3):162–180. https://doi.org/10.1038/s41568-020-00320-2

Berisha A, Shutkind K, Borniger JC (2022) Sleep disruption and cancer: chicken or the egg? Front Neurosci 16:856235. https://doi.org/10.3389/fnins.2022.856235

Bernard MJ, Goldstein AS (2024) A metabolic-epigenetic mechanism directs cell fate and therapeutic sensitivity in breast cancer. Cancer Res 84(9):1382–1383. https://doi.org/10.1158/0008-5472.CAN-24-0460

Bernards R, Weinberg RA (2002) A progression puzzle. Nature 418(6900):823. https://doi.org/10.1038/418823a

Bernhardt D, Adeberg S, Bozorgmehr F et al (2017) Outcome and prognostic factors in patients with brain metastases from small-cell lung cancer treated with whole brain radiotherapy. J Neurooncol 134(1):205–212. https://doi.org/10.1007/s11060-017-2510-0

Berrevoets J, Kacprzyk K, Qian Z, van der Schaar M (2024) Causal deep learning: encouraging impact on real-world problems through causality. Found Trends Signal Process 18(3):200–309. https://doi.org/10.1561/2000000123

Bertero E, Canepa M, Maack C, Ameri P (2018) Linking heart failure to cancer: background evidence and research perspectives. Circulation 138(7):735–742. https://doi.org/10.1161/CIRCULATIONAHA.118.033603

Bertolaso M, Dupré J (2018) A processual perspective on cancer. In: Daniel J, Nicholson D, Dupré J (eds) Everything flows: towards a processual philosophy of biology. Oxford University Press. https://doi.org/10.1093/oso/9780198779636.001.0001

Bertolaso M, Strauss B (2021) The search for progress and a new theory framework in cancer research. In: Bertolaso M (ed) Rethinking cancer: a new paradigm for the postgenomics era. MIT Press, Cambridge

Beumer J, Clevers H (2024) Hallmarks of stemness in mammalian tissues. Cell Stem Cell 31(1):7–24. https://doi.org/10.1016/j.stem.2023.12.006

Beziaud L, Young CM, Alonso AM et al (2023) IFNγ-induced stem-like state of cancer cells as a driver of metastatic progression following immunotherapy. Cell Stem Cell 30(6):818–831.e6. https://doi.org/10.1016/j.stem.2023.05.007

Bhakta-Guha D, Efferth T (2015) Hormesis: decoding two sides of the same coin. Pharmaceuticals (Basel) 8(4):865–883. https://doi.org/10.3390/ph8040865

Bhat TA, Kumar S, Chaudhary AK et al (2015) Restoration of mitochondria function as a target for cancer therapy. Drug Discov Today. 20(5):635–643. https://doi.org/10.1016/j.drudis

Bianchi G, Martella R, Ravera S et al (2015) Fasting induces anti-Warburg effect that increases respiration but reduces ATP-synthesis to promote apoptosis in colon cancer models. Oncotarget 6(14):11806–11819. https://doi.org/10.18632/oncotarget.3688

Bianchi JJ, Zhao X, Mays JC, Davoli T (2020) Not all cancers are created equal: tissue specificity in cancer genes and pathways. Curr Opin Cell Biol 63:135–143. https://doi.org/10.1016/j.ceb.2020.01.005

Bich L, Green S (2018) Is defining life pointless? Operational definitions at the frontiers of biology. Synthese 195:3919–3946. https://doi.org/10.1007/s11229-017-1397-9

Billman GE (2020) Homeostasis: the underappreciated and far too often ignored central organizing principle of physiology. Front Physiol 11:200. https://doi.org/10.3389/fphys.2020.00200

Bin Masroni MS, Koay ES, Lee VKM et al (2025) Sociobiology meets oncology: unraveling altruistic cooperation in cancer cells and its implications. Exp Mol Med 57(1):30–40. https://doi.org/10.1038/s12276-024-01387-9

Binnewies M, Roberts EW, Kersten K et al (2018) Understanding the tumor immune microenvironment (TIME) for effective therapy. Nat Med 24(5):541–550. https://doi.org/10.1038/s41591-018-0014-x

Birsoy K, Sabatini DM, Possemato R (2012) Untuning the tumor metabolic machine: targeting cancer metabolism—a bedside lesson. Nat Med 18(7):1022–1023. https://doi.org/10.1038/nm.2870

Bissell MJ, Radisky D (2001) Putting tumours in context. Nat Rev Cancer 1(1):46–54. https://doi.org/10.1038/35094059

Bissell MJ, Hall HG, Parry G (1982) How does the extracellular matrix direct gene expression? J Theor Biol 99(1):31–68. https://doi.org/10.1016/0022-5193(82)90388-5

Bissell MJ, Kenny PA, Radisky DC (2005) Microenvironmental regulators of tissue structure and function also regulate tumor induction and progression: the role of extracellular matrix and its degrading enzymes. Cold Spring Harb Symp Quant Biol 70:343–356. https://doi.org/10.1101/sqb.2005.70.013

Biswas S, Rao CM (2018) Epigenetic tools (the writers, the readers and the erasers) and their implications in cancer therapy. Eur J Pharmacol 837:8–24. https://doi.org/10.1016/j.ejphar.2018.08.021

Bizzarri M, Cucina A (2014) Tumor and the microenvironment: a chance to reframe the paradigm of carcinogenesis? Biomed Res Int 2014:934038. https://doi.org/10.1155/2014/934038

Bjerkvig R, Tysnes BB, Aboody KS, Najbauer J, Terzis AJ (2005) Opinion: the origin of the cancer stem cell: current controversies and new insights. Nat Rev Cancer 5(11):899–904. https://doi.org/10.1038/nrc1740

Björneborn L (2020) Adjacent possible. In: Glaveanu V (ed) The Palgrave encyclopedia of the possible. Palgrave Macmillan, Cham. https://doi.org/10.1007/978-3-319-98390-5_100-1

Blackiston DJ, McLaughlin KA, Levin M (2009) Bioelectric controls of cell proliferation: ion channels, membrane voltage and the cell cycle. Cell Cycle 8(21):3527–3536. https://doi.org/10.4161/cc.8.21.9888

Blackstock AW, Benson AB, Kudo M, et al. (2021) Safety and efficacy of amplitude-modulated radiofrequency electromagnetic fields in advanced hepatocellular carcinoma. 4open 4:3. https://doi.org/10.1051/fopen/2021003

Blagosklonny MV (2012) Answering the ultimate question "what is the proximal cause of aging?". Aging (Albany NY) 4(12):861–877. https://doi.org/10.18632/aging.100525

Blagosklonny MV (2013) Aging is not programmed: genetic pseudo-program is a shadow of developmental growth. Cell Cycle 12(24):3736–3742. https://doi.org/10.4161/cc.27188

Blagosklonny MV (2021) The hyperfunction theory of aging: three common misconceptions. Oncoscience 8:103–107. https://doi.org/10.18632/oncoscience.545

Blais J, Jeukens J (2024) Pareto task inference analysis reveals cellular trade-offs in diffuse large B-cell lymphoma transcriptomic data. Front Syst Biol 4:1346076. https://doi.org/10.3389/fsysb.2024.1346076

Blask DE, Dauchy RT, Sauer LA, Krause JA, Brainard GC (2002) Light during darkness, melatonin suppression and cancer progression. Neuro Endocrinol Lett 23(Suppl 2):52–56

Blokzijl F, de Ligt J, Jager M et al (2016) Tissue-specific mutation accumulation in human adult stem cells during life. Nature 538(7624):260–264. https://doi.org/10.1038/nature19768

Bloomfield M, Duesberg P (2016) Inherent variability of cancer-specific aneuploidy generates metastases. Mol Cytogenet 9:90. https://doi.org/10.1186/s13039-016-0297-x

Blyth BJ, Cole AJ, MacManus MP, Martin OA (2018) Radiation therapy-induced metastasis: radiobiology and clinical implications. Clin Exp Metastasis 35(4):223–236. https://doi.org/10.1007/s10585-017-9867-5

Boedtkjer E, Pedersen SF (2020) The acidic tumor microenvironment as a driver of cancer. Annu Rev Physiol 82:103–126. https://doi.org/10.1146/annurev-physiol-021119-034627

Bogdanov A, Verlov N, Bogdanov A et al (2024) Tumor alkalization therapy: misconception or good therapeutics perspective? – the case of malignant ascites. Front Oncol 14:1342802. https://doi.org/10.3389/fonc.2024.1342802

Boire A, Burke K, Cox TR et al (2024) Why do patients with cancer die? Nat Rev Cancer 24(8):578–589. https://doi.org/10.1038/s41568-024-00708-4

Bolhaqueiro ACF, Ponsioen B, Bakker B et al (2019) Ongoing chromosomal instability and karyotype evolution in human colorectal cancer organoids. Nat Genet 51(5):824–834. https://doi.org/10.1038/s41588-019-0399-6

Bonafè M, Barbi C, Storci G et al (2002) What studies on human longevity tell us about the risk for cancer in the oldest old: data and hypotheses on the genetics and immunology of centenarians. Exp Gerontol 37(10–11):1263–1271. https://doi.org/10.1016/s0531-5565(02)00137-7

Bone RC (1996) Immunologic dissonance: a continuing evolution in our understanding of the systemic inflammatory response syndrome (SIRS) and the multiple organ dysfunction syndrome (MODS). Ann Intern Med 125(8):680–687. https://doi.org/10.7326/0003-4819-125-8-199610150-00009

Bone RC, Grodzin CJ, Balk RA (1997) Sepsis: a new hypothesis for pathogenesis of the disease process. Chest 112(1):235–243. https://doi.org/10.1378/chest.112.1.235

Bonner JT (1965) Size and cycle: an essay on the structure of biology. Princeton University Press

Bonner JT (2019) The evolution of evolution. J Exp Zool B Mol Dev Evol 332(8):301–306. https://doi.org/10.1002/jez.b.22859

Bonner EA, Lee SC (2023) Therapeutic targeting of RNA splicing in cancer. Genes (Basel) 14(7):1378. https://doi.org/10.3390/genes14071378

Bonney A, Togawa K, Ng M et al (2023) Prevalence of subclinical lung cancer detected at autopsy: a systematic review. BMC Cancer. 23:794. https://doi.org/10.1186/s12885-023-11224-3

Bordenstein SR, The Holobiont Biology Network (2024) The disciplinary matrix of holobiont biology. Science 386(6723):731–732. https://doi.org/10.1126/science.ado2152

Bordonaro M, Lazarova D (2024) Hypothesis: Cancer hormesis and its potential for cancer therapeutics. Life (Basel) 14(3):401. https://doi.org/10.3390/life14030401. PMID: 38541725; PMCID: PMC10972146

Borges JL (2000) Borges: selected poems. Penguin Books

Bornes L, van Winden LJ, Geurts VCM et al (2025) The oestrous cycle stage affects mammary tumour sensitivity to chemotherapy. Nature 637(8044):195–204. https://doi.org/10.1038/s41586-024-08276-1. PMID: 39633046; PMCID: PMC11666466

Borniger JC (2019) Central regulation of breast cancer growth and metastasis. J Cancer Metastasis Treat 5:23. https://doi.org/10.20517/2394-4722.2018.107. PMID: 31773065; PMCID: PMC6879058

Borniger JC, Walker WH II, Surbhi EKM, Zhang N, Zalenski AA et al (2018) A role for hypocretin/orexin in metabolic and sleep abnormalities in a mouse model of non-metastatic breast cancer. Cell Metab 28:118–129.e5. https://doi.org/10.1016/j.cmet.2018.05.004

Boros LG, Somlyai I, Kovács BZ et al (2021) Deuterium depletion inhibits cell proliferation, RNA and nuclear membrane turnover to enhance survival in pancreatic cancer. Cancer Control 28:1073274821999655. https://doi.org/10.1177/1073274821999655

Bosch TCG, McFall-Ngai MJ (2011) Metaorganisms as the new frontier. Zoology 114(4):185–190. https://doi.org/10.1016/j.zool.2011.04.001

Bottaccioli F, Bottaccioli AG (2020) PsychoNeuroEndocrine Immunology and Science of the Integrated Care: The Manual. Editorial Edra

Boucherat O, Vitry G, Trinh I et al (2017) The cancer theory of pulmonary arterial hypertension. Pulm Circ 7(2):285–299. https://doi.org/10.1177/2045893217701438. PMID: 28597757; PMCID: PMC5467931

Bouferraa Y, Chedid A, Amhaz G et al (2021) The role of gut microbiota in overcoming resistance to checkpoint inhibitors in cancer patients: mechanisms and challenges. Int J Mol Sci 22(15):8036. https://doi.org/10.3390/ijms22158036. PMID: 34360802; PMCID: PMC8347208

Bougeard G, Renaux-Petel M, Flaman JM et al (2015) Revisiting Li-Fraumeni syndrome from TP53 mutation carriers. J Clin Oncol 33(21):2345–2352. https://doi.org/10.1200/JCO.2014.59.5728

Boumehira AZ, Akchiche YF, Cherfi W et al (2022) Metabolomics approaches for early cancer diagnosis: a review. Asian J Agric Biol 202012566. https://doi.org/10.35495/ajab.2020.12.566

Bowling EA, Wang JH, Gong F et al (2021) Spliceosome-targeted therapies trigger an antiviral immune response in triple-negative breast cancer. Cell 184(2):384–403.e21. https://doi.org/10.1016/j.cell.2020.12.031. PMID: 33450205; PMCID: PMC863524

Boyle EA, Li YI, Pritchard JK (2017) An expanded view of complex traits: From polygenic to omnigenic. Cell 169(7):1177–1186. https://doi.org/10.1016/j.cell.2017.05.038. PMID: 28622505; PMCID: PMC5536862

Bozic I, Antal T, Ohtsuki H et al (2010) Accumulation of driver and passenger mutations during tumor progression. Proc Natl Acad Sci USA 107(43):18545–18550. https://doi.org/10.1073/pnas.1010978107

Brand A, Singer K, Koehl GE et al (2016) LDHA-associated lactic acid production blunts tumor immunosurveillance by T and NK cells. Cell Metab. 24(5):657–671. https://doi.org/10.1016/j.cmet.2016.08.011. PMID: 27641098

Brandon RN (2004) Adaptation and environment. Princeton University Press, Princeton

Brandon RN (2006) The principle of drift: biology's first law. J Philos 103(7):319–335

Brash D, Cairns J (2009) The mysterious steps in carcinogenesis. Br J Cancer 101(3):379–380. https://doi.org/10.1038/sj.bjc.6605171

Braun TP, Zhu X, Szumowski M et al (2011) Central nervous system inflammation induces muscle atrophy via activation of the hypothalamic-pituitary-adrenal axis. J Exp Med 208(12):2449–2463. https://doi.org/10.1084/jem.20111020. PMID: 22084407; PMCID: PMC3256966

Breitbart W, Pessin H, Rosenfeld B et al (2018) Individual meaning-centered psychotherapy for the treatment of psychological and existential distress: A randomized controlled trial in patients with advanced cancer. Cancer 124(15):3231–3239. https://doi.org/10.1002/cncr.31539. PMID: 29757459; PMCID: PMC6097940

Brennan P (2002) Gene-environment interaction and aetiology of cancer: what does it mean and how can we measure it? Carcinogenesis 23(3):381–387. https://doi.org/10.1093/carcin/23.3.381

Brenner S (2012) Turing centenary: life's code script. Nature. 482(7386):461. https://doi.org/10.1038/482461a

Brier S, Joslyn C (2013) What does it take to produce interpretation? Informational, Peircean and code-semiotic views on biosemiotics. Biosemiotics 6:143–159

Brillouin L (1953) The negentropy principle of information. J Appl Phys 24:1152. https://doi.org/10.1063/1.1721463

Bronchud MH (2018) Are aggressive epithelial cancers 'a disease' of Eutherian mammals? Ecancermedicalscience 12:840. https://doi.org/10.3332/ecancer.2018.840. PMID: 30034517; PMCID: PMC6027979

Bronchud MH, Tresserra F, Xu W, et al. (2016) Placental immune editing switch (PIES): learning about immunomodulatory pathways from a unique case report. Oncotarget 7(50):83817–83827. https://doi.org/10.18632/oncotarget.13306. PMID: 27852037; PMCID: PMC5347808

Brookes JC (2017) Quantum effects in biology: golden rule in enzymes, olfaction, photosynthesis and magnetodetection. Proc Math Phys Eng Sci 473(2201):20160822. https://doi.org/10.1098/rspa.2016.0822. PMID: 28588400; PMCID: PMC5454345

Brosius J (2003) The contribution of RNAs and retroposition to evolutionary novelties. Genetica 118(2–3):99–116. PMID: 12868601

Brown JS, Amend SR, Austin RH, et al. (2023) Updating the definition of cancer. Mol Cancer Res 21(11):1142–1147. https://doi.org/10.1158/1541-7786.MCR-23-0411. PMID: 37409952; PMCID: PMC10618731

Broz MT, Ko EY, Ishaya K et al (2024) Metabolic targeting of cancer-associated fibroblasts overcomes T-cell exclusion and chemoresistance in soft-tissue sarcomas. Nat Commun. 15(1):2498. https://doi.org/10.1038/s41467-024-46504-4. PMID: 38509063; PMCID: PMC10954767

Bruce JIE, James AD (2020) Targeting the calcium signalling machinery in cancer. Cancers (Basel) 12(9):2351. https://doi.org/10.3390/cancers12092351. PMID: 32825277; PMCID: PMC7565467

Brücher BL, Jamall IS (2014) Epistemology of the origin of cancer: a new paradigm. BMC Cancer 14:331. https://doi.org/10.1186/1471-2407-14-331. PMID: 24885752; PMCID: PMC4026115

Brücher BL, Jamall IS (2016) Somatic Mutation Theory—Why it's wrong for most cancers. Cell Physiol Biochem 38(5):1663–1680. https://doi.org/10.1159/000443106. PMID: 27160408

Bruggeman JW, Koster J, Lodder P et al (2018) Massive expression of germ cell-specific genes is a hallmark of cancer and a potential target for novel treatment development. Oncogene 37(42):5694–5700. https://doi.org/10.1038/s41388-018-0357-2. PMID: 29907769; PMCID: PMC6193945

Bruggeman JW, Irie N, Lodder P et al (2020) Tumors widely express hundreds of embryonic germline genes. Cancers (Basel) 12(12):3812. https://doi.org/10.3390/cancers12123812. PMID: 33348709; PMCID: PMC7766889

Bruggeman JW, Koster J, van Pelt AMM et al (2023) How germline genes promote malignancy in cancer cells. Bioessays 45(1):e2200112. https://doi.org/10.1002/bies.202200112

Bruni L, Saura-Lázaro A, Montoliu A et al (2021) HPV vaccination introduction worldwide and WHO and UNICEF estimates of national HPV immunization coverage 2010–2019. Prev Med 144:106399. https://doi.org/10.1016/j.ypmed.2020.106399. Erratum in: Prev Med. 2022 Feb;155:106925. doi: 10.1016/j.ypmed.2021

Budisa N, Kubyshkin V, Schmidt M (2020) Xenobiology: a journey towards parallel life forms. Chem Bio Chem 21(16):2228–2231. https://doi.org/10.1002/cbic.202000141. PMID: 32323410

Budnik B, Amirkhani H, Forouzanfar MH, Afshin A (2024) Novel proteomics-based plasma test for early detection of multiple cancers in the general population. BMJ Oncol 3(1):e000073. https://doi.org/10.1136/bmjonc-2023-000073. PMID: 39886137; PMCID: PMC11235013

Bueno MJ, Malumbres M (2011) MicroRNAs and the cell cycle. Biochim Biophys Acta 1812(5):592–601. https://doi.org/10.1016/j.bbadis.2011.02.002. PMID: 21315819

Buicu AL, Cernea S, Benedek I, Buicu CF, Benedek T (2021) Systemic inflammation and COVID-19 mortality in patients with major noncommunicable diseases: chronic coronary syndromes, diabetes and obesity. J Clin Med 10(8):1545. https://doi.org/10.3390/jcm10081545. PMID: 33916917; PMCID: PMC8067631

Burfeind KG, Michaelis KA, Marks DL (2016) The central role of hypothalamic inflammation in the acute illness response and cachexia. Semin Cell Dev Biol 54:42–52. https://doi.org/10.1016/j.semcdb.2015.10.038. PMID: 26541482; PMCID: PMC4872506

Burfeind KG, Zhu X, Levasseur PR et al (2018) TRIF is a key inflammatory mediator of acute sickness behavior and cancer cachexia. Brain Behav Immun 73:364–374. https://doi.org/10.1016/j.bbi.2018.05.021. PMID: 29852290; PMCID: PMC6129432

Buric I, Farias M, Jong J, Mee C, Brazil IA (2017) What is the molecular signature of mind-body interventions? A systematic review of gene expression changes induced by meditation and related practices. Front Immunol 8:670. https://doi.org/10.3389/fimmu.2017.00670. PMID: 28670311; PMCID: PMC5472657

Burnet M (1957) Cancer: a biological approach. I. The processes of control. Br Med J 1(5022):779–786. https://doi.org/10.1136/bmj.1.5022.779. PMID: 13404306; PMCID: PMC1973174

Burnet M (1957d) Cancer: a biological approach. III. Viruses associated with neoplastic conditions. IV. Practical applications. Br Med J 1(5023):841–847. https://doi.org/10.1136/bmj.1.5023.841. PMID: 13413231; PMCID: PMC1973618

Burnet FM (1970) The concept of immunological surveillance. Prog Exp Tumor Res 13:1–27. https://doi.org/10.1159/000386035

Bychkov D, Linder N, Turkki R et al (2021) Deep learning-based tissue analysis predicts outcome in colorectal cancer. Sci Rep 11(1):3390. https://doi.org/10.1038/s41598-021-82618-

Byrne HM (2010) Dissecting cancer through mathematics: from the cell to the animal model. Nat Rev Cancer 10(3):221–230. https://doi.org/10.1038/nrc2808. PMID: 20179714

Caan BJ, Cespedes Feliciano EM, Kroenke CH (2018) The importance of body composition in explaining the overweight paradox in cancer—counterpoint. Cancer Res 78(8):1906–1912. https://doi.org/10.1158/0008-5472.CAN-17-3287. PMID: 29654153; PMCID: PMC5901895

CabAnnilas ME, Dadu R, Ferrarotto R et al (2024) Anti-programmed death ligand 1 plus targeted therapy in anaplastic thyroid carcinoma: a nonrandomized clinical trial. JAMA Oncol 10(12):1672–1680. https://doi.org/10.1001/jamaoncol.2024.4729. PMID: 39446377; PMCID: PMC11581602

Cagan A, Baez-Ortega A, Brzozowska N et al (2022) Somatic mutation rates scale with lifespan across mammals. Nature 604(7906):517–524. https://doi.org/10.1038/s41586-022-04618-z. PMID: 35418684; PMCID: PMC9021023

Cairns J (1975) Mutation selection and the natural history of cancer. Nature 255(5505):197–200. https://doi.org/10.1038/255197a0. PMID: 1143315

Calabrese EJ (2013) Hormetic mechanisms. Crit Rev Toxicol 43(7):580–606. https://doi.org/10.3109/10408444.2013.808172

Calabrese EJ (2014) Hormesis: from mainstream to therapy. J Cell Commun Signal 8(4):289–291. https://doi.org/10.1007/s12079-014-0255-5. PMID: 25366126; PMCID: PMC4390802

Calabrese EJ, Baldwin LA (2003) Hormesis: the dose-response revolution. Annu Rev Pharmacol Toxicol 43:175–197. https://doi.org/10.1146/annurev.pharmtox.43.100901.140223. PMID: 12195028

Califano A, Alvarez MJ (2017) The recurrent architecture of tumour initiation, progression and drug sensitivity. Nat Rev Cancer 17(2):116–130. https://doi.org/10.1038/nrc.2016.124. PMID: 27977008; PMCID: PMC5541669

Calin GA, Vasilescu C, Negrini M, Barbanti-Brodano G (2003) Genetic chaos and anti-chaos in human cancers. Med Hypotheses 60(2):258–262. https://doi.org/10.1016/s0306-9877(02)00383-3

Callejón-Leblic B, Arias-Borrego A, Pereira-Vega A et al (2019) The metallome of lung cancer and its potential use as biomarker. Int J Mol Sci 20(3):778. https://doi.org/10.3390/ijms20030778. PMID: 30759767; PMCID: PMC6387380

Campbell TC, Campbell TM II (2004) The China study: startling implications for diet, weight loss, and long-term health. BenBella Books, Dallas

Cannioto RA, Hutson A, Dighe S et al (2021) Physical activity before, during, and after chemotherapy for high-risk breast cancer: relationships with survival. JNCI J Natl Cancer Inst 113(1):54–63. https://doi.org/10.1093/jnci/djaa046

Cannon WB (1932) The wisdom of the body. W. W. Norton, New York

Cao L, During MJ (2012) What is the brain-cancer connection? Annu Rev Neurosci 35:331–345. https://doi.org/10.1146/annurev-neuro-062111-150546. PMID: 22462541

Cao DX, Wu GH, Zhang B et al (2010a) Resting energy expenditure and body composition in patients with newly detected cancer. Clin Nutr 29(1):72–77. https://doi.org/10.1016/j.clnu.2009.07.001. PMID: 19647909

Cao L, Liu X, Lin EJ et al (2010b) Environmental and genetic activation of a brain-adipocyte BDNF/leptin axis causes cancer remission and inhibition. Cell 142(1):52–64. https://doi.org/10.1016/j.cell.2010.05.029. PMID: 20603014; PMCID: PMC3784009

Cao J, Cogdell RJ, Coker DF et al (2020) Quantum biology revisited. Sci Adv 6(14):eaaz4888. https://doi.org/10.1126/sciadv.aaz4888. PMID: 32284982; PMCID: PMC7124948

Capatina AL, Lagos D, Brackenbury WJ (2022) Targeting ion channels for cancer treatment: current progress and future challenges. Rev Physiol Biochem Pharmacol 183:1–43. https://doi.org/10.1007/112_2020_46. PMID: 32865696

Cappell KM, Sherry RM, Yang JC et al (2020) Long-term follow-up of anti-CD19 chimeric antigen receptor T-cell therapy. J Clin Oncol 38(32):3805–3815. https://doi.org/10.1200/JCO.20.01467. PMID: 33021872; PMCID: PMC7655016

Carlson LE, Doll R, Stephen J et al (2013) Randomized controlled trial of mindfulness-based cancer recovery versus supportive expressive group therapy for distressed survivors of breast cancer. J Clin Oncol 31(25):3119–3126. https://doi.org/10.1200/JCO.2012.47.5210. PMID: 23918953

Carneiro CS, Hapeman JD, Nedelcu AM (2023) Synergistic inter-clonal cooperation involving crosstalk, co-option and co-dependency can enhance the invasiveness of genetically distant cancer clones. BMC Ecol Evol 23(1):20. https://doi.org/10.1186/s12862-023-02129-7. PMID: 37226092; PMCID: PMC10207807

Carracedo A, Cantley LC, Pandolfi PP (2013) Cancer metabolism: fatty acid oxidation in the limelight. Nat Rev Cancer 13(4):227–232. https://doi.org/10.1038/nrc3483. PMID: 23446547; PMCID: PMC3766957

Carrasco-Zanini J, Pietzner M, Davitte J et al (2024) Proteomic signatures improve risk prediction for common and rare diseases. Nat Med 30(9):2489–2498. https://doi.org/10.1038/s41591-024-03142-z. PMID: 39039249; PMCID: PMC11405273

Carretero-Fernández M, Cabrera-Serrano AJ, Sánchez-Maldonado JM et al (2025) Autophagy and oxidative stress in solid tumors: mechanisms and therapeutic opportunities. Crit Rev Oncol Hematol 212:104820. https://doi.org/10.1016/j.critrevonc.2025

Cartwright J (2013) Physicists discover a whopping 13 new solutions to three-body problem. Science Now. March 8

Carvalho J (2020) Cell reversal from a differentiated to a stem-like state at cancer initiation. Front Oncol 10:541. https://doi.org/10.3389/fonc.2020.00541. PMID: 32351900; PMCID: PMC7174973

Carvalho J (2021) A bioelectric model of carcinogenesis, including propagation of cell membrane depolarization and reversal therapies. Sci Rep. 11(1):13607. https://doi.org/10.1038/s41598-021-92951-0. PMID: 34193902; PMCID: PMC8245601

Casanova A, Wevers A, Navarro-Ledesma S, Pruimboom L (2023) Mitochondria: It is all about energy. Front Physiol 14:1114231.

https://doi.org/10.3389/fphys.2023.1114231. PMID: 37179826; PMCID: PMC10167337

Castaño Z, Tracy K, McAllister SS (2011) The tumor macroenvironment and systemic regulation of breast cancer progression. Int J Dev Biol 55(7–9):889–897. https://doi.org/10.1387/ijdb.113366zc. PMID: 22161844

Castellone MD, Laukkanen MO, Teramoto H et al (2015) Cross talk between the bombesin neuropeptide receptor and Sonic hedgehog pathways in small cell lung carcinoma. Oncogene 34(13):1679–1687. https://doi.org/10.1038/onc.2014.104. PMID: 24747971; PMCID: PMC4539617

Castro-Giner F, Ratcliffe P, Tomlinson I (2015) The mini-driver model of polygenic cancer evolution. Nat Rev Cancer 15(11):680–685. https://doi.org/10.1038/nrc3999. PMID: 26456849

Cavalier-Smith T (2004) The membranome and membrane heredity in development and evolution. In: Hirt RP, Horner DS (eds) Organelles, genomes and eukaryote phylogeny. CRC Press, Boca Raton, pp 335–351

Cavallo F, De Giovanni C, Nanni P et al (2011) 2011: the immune hallmarks of cancer. Cancer Immunol Immunother 60(3):319–326. https://doi.org/10.1007/s00262-010-0968-0. PMID: 21267721; PMCID: PMC3042096

Cavenee WK, Dryja TP, Phillips RA et al (1983) Expression of recessive alleles by chromosomal mechanisms in retinoblastoma. Nature 305:779–784

Cech TR (2024) The catalyst: RNA and the quest to unlock life's deepest secrets. W.W. Norton & Company, New York

Çengel YA (2023) Eighteen distinctive characteristics of life. Heliyon 9(3):e13603. https://doi.org/10.1016/j.heliyon.2023.e13603. PMID: 37101483; PMCID: PMC10123176

Cereda M, Mourikis TP, Ciccarelli FD (2016) Genetic redundancy, functional compensation, and cancer vulnerability. Trends Cancer 2(4):160–162. https://doi.org/10.1016/j.trecan.2016.03.003. PMID: 28741568

Ceresoli GL, Aerts JG, Dziadziuszko R et al (2019) Tumour treating fields in combination with pemetrexed and cisplatin or carboplatin as first-line treatment for unresectable malignant pleural mesothelioma (STELLAR): a multicentre, single-arm phase 2 trial. Lancet Oncol 20(12):1702–1709. https://doi.org/10.1016/S1470-2045(19)30532-7

Cesario A, D'Oria M, Calvani R et al (2021) The role of artificial intelligence in managing multimorbidity and cancer. J Pers Med 11(4):314. https://doi.org/10.3390/jpm11040314

Cha DJ, Franklin JL, Dou Y et al (2015) KRAS-dependent sorting of miRNA to exosomes. Elife 4:e07197. https://doi.org/10.7554/eLife.07197

Cha Y, Kim S, Han SW (2023) Utilizing plasma circulating tumor DNA sequencing for precision medicine in the management of solid cancers. Cancer Res Treat 55(2):367–384. https://doi.org/10.4143/crt.2023.446. PMID: 36915241; PMCID: PMC10101775

Chacko BK, Kramer PA, Ravi S et al (2014) The bioenergetic health index: a new concept in mitochondrial translational research. Clin Sci (Lond) 127(6):367–373. https://doi.org/10.1042/CS20140101. PMID: 24895057; PMCID: PMC4202728

Chaffer CL, Weinberg RA (2015) How does multistep tumorigenesis really proceed? Cancer Discov 5(1):22–24. https://doi.org/10.1158/2159-8290.CD-14-0788. PMID: 25583800; PMCID: PMC4295623

Chahoud J, Anderson ARA, Zhang J et al (2023) Evolutionary dynamics and intermittent therapy for metastatic cancers. J Clin Oncol 41(28):4469–4471. https://doi.org/10.1200/JCO.23.00647. PMID: 37418680; PMCID: PMC10553063

Chaix A, Manoogian ENC, Melkani GC, Panda S (2019) Time-restricted eating to prevent and manage chronic metabolic diseases. Annu Rev Nutr 39:291–315. https://doi.org/10.1146/annurev-nutr-082018-124320. PMID: 31180809; PMCID: PMC6703924

Chalmers D (2006) Strong and weak emergence. In: Davies P, Clayton P (eds) The re-emergence of emergence: the emergentist hypothesis from science to religion. Oxford University Press, Oxford

Chamardani TM, Amiritavassoli S (2022) Inhibition of NETosis for treatment purposes: friend or foe? Mol Cell Biochem 477(3):673–688. https://doi.org/10.1007/s11010-021-04315-x. PMID: 34993747; PMCID: PMC873633

Chan JY (2011) A clinical overview of centrosome amplification in human cancers. Int J Biol Sci 7(8):1122–1144. https://doi.org/10.7150/ijbs.7.1122. PMID: 22043171; PMCID: PMC320440

Chan HP, Liu WS, Liou WS, Hu C, Chiu YL, Peng NJ (2020) Comparison of FDG-PET/CT for cancer detection in populations with different risks of underlying malignancy. In Vivo 34(1):469–478. https://doi.org/10.21873/invivo.11797. PMID: 31882515; PMCID: PMC6984113

Chandra D, Singh KK (2011) Genetic insights into OXPHOS defect and its role in cancer. Biochim Biophys Acta 1807(6):620–625. https://doi.org/10.1016/j.bbabio.2010.10.023. PMID: 21074512; PMCID: PMC4681500

Chandran V, Bermúdez ML, Koka M et al (2021) Large-scale genomic study reveals robust activation of the immune system following advanced inner engineering meditation retreat. Proc Natl Acad Sci U S A 118(51):e2110455118. https://doi.org/10.1073/pnas.2110455118

Chang MH, Shau WY, Chen CJ et al (2000) Hepatitis B vaccination and hepatocellular carcinoma rates in boys and girls. JAMA 284(23):3040–3042. https://doi.org/10.1001/jama.284.23.3040. PMID: 11122592

Chang CH, Qiu J, O'Sullivan D et al (2015) Metabolic competition in the tumor microenvironment is a driver of cancer progression. Cell 162(6):1229–1241. https://doi.org/10.1016/j.cell.2015.08.016. PMID: 26321679; PMCID: PMC4864363

Chang X, Tamauchi S, Nakagawa A et al (2025a) Chemotherapeutic hormesis induced by the tumor microenvironment in refractory ovarian cancer. Sci Rep 15(1):596. https://doi.org/10.1038/s41598-024-84290-7. PMID: 39747971; PMCID: PMC11696277

Chang L, Liu J, Zhu J (2025b) Advancing precision medicine: the transformative role of artificial intelligence in immunogenomics, radiomics, and pathomics for biomarker discovery and immunotherapy optimization. Cancer Biol Med 22(1):33–47. https://doi.org/10.20892/j.issn.2095-3941.2024.0376

Chao BN, Carrick DM, Filipski KK, Nelson SA (2022) Overview of research on germline genetic variation in immune genes and cancer outcomes. Cancer Epidemiol Biomarkers Prev 31(3):495–506. https://doi.org/10.1158/1055-9965.EPI-21-0583

Charlson ME, Pompei P, Ales KL, MacKenzie CR (1987) A new method of classifying prognostic comorbidity in longitudinal studies: development and validation. J Chronic Dis 40(5):373–383. https://doi.org/10.1016/0021-9681(87)90171-8. PMID: 3558716

Chatrath A, Przanowska R, Kiran S et al (2020) The pan-cancer landscape of prognostic germline variants in 10,582 patients. Genome Med 12(1):15. https://doi.org/10.1186/s13073-020-0718-7

Chatterjee A, Georgiev G, Iannacchione G (2017) Aging and efficiency in living systems: complexity, adaptation and self-organization. Mech Ageing Dev 163:2–7. https://doi.org/10.1016/j.mad.2017.02.009. PMID: 28267566

Chatterjee A, Rodger EJ, Eccles MR (2018) Epigenetic drivers of tumourigenesis and cancer metastasis. Semin Cancer Biol 51:149–159. https://doi.org/10.1016/j.semcancer.2017.08.004. PMID: 28807546

Chavez-Dominguez R, Perez-Medina M, Lopez-Gonzalez JS et al (2020) The double-edge sword of autophagy in cancer: from tumor suppression to pro-tumor activity. Front Oncol 10:578418. https://doi.org/10.3389/fonc.2020

Chehrazi-Raffle A, Meza L, Alcantara M et al (2021) Circulating cytokines associated with clinical response to systemic therapy in metastatic renal cell carcinoma. J Immunother Cancer 9(3):e002009. https://doi.org/10.1136/jitc-2020-002009. PMID: 33688021; PMCID: PMC7944971

Chen JL, Lucas JE, Schroeder T et al (2008) The genomic analysis of lactic acidosis and acidosis response in human cancers. PLoS Genet 4(12):e1000293. https://doi.org/10.1371/journal.pgen.1000293. PMID: 19057672; PMCID: PMC2585811

Chen H, Lin F, Xing K, He X (2015) The reverse evolution from multicellularity to unicellularity during carcinogenesis. Nat Commun 6:6367. https://doi.org/10.1038/ncomms7367

Chen Y, Zhang Y, Guo X (2017) Proteasome dysregulation in human cancer: implications for clinical therapies. Cancer Metastasis Rev 36(4):703–716. https://doi.org/10.1007/s10555-017-9704-y. PMID: 29039081

Chen H, Li C, Peng X et al (2018) A pan-cancer analysis of enhancer expression in nearly 9000 patient samples. Cell 173(2):386–399.e12. https://doi.org/10.1016/j.cell.2018.03.027. PMID: 29625054; PMCID: PMC5890960

Chen C, Gao H, Su X (2021) Autophagy-related signaling pathways are involved in cancer (Review). Exp Ther Med 22(1):710. https://doi.org/10.3892/etm.2021.10142

Chen J, Facchinetti F, Braye F et al (2022) Single-cell DNA-seq depicts clonal evolution of multiple driver alterations in osimertinib-resistant patients. Ann Oncol 33(4):434–444. https://doi.org/10.1016/j.annonc.2022.01.004. PMID: 35066105

Chen G, Bao B, Cheng Y et al (2023a) Acetyl-CoA metabolism as a therapeutic target for cancer. Biomed Pharmacother 168:115741. https://doi.org/10.1016/j.biopha.2023.115741. PMID: 37864899

Chen HY, Zhao W, Na'ara S et al (2023b) Beta-blocker use is associated with worse relapse-free survival in patients with head and neck cancer. JCO Precis Oncol 7:e2200490. https://doi.org/10.1200/PO.22.00490. PMID: 37285560; PMCID: PMC10309540

Chen DW, Carr G, Worden FP (2024) Survival prognostication in patients with differentiated thyroid cancer and distant metastases: a SEER population-based study. Thyroid 34(7):837–845. https://doi.org/10.1089/thy.2023.0709

Chen C, Xiang A, Lin X et al (2024a) Mitophagy: insights into its signaling molecules, biological functions, and therapeutic potential in breast cancer. Cell Death Discov 10(1):457. https://doi.org/10.1038/s41420-024-02226-6

Chen F, Xue Y, Zhang W et al (2024b) The role of mitochondria in tumor metastasis and advances in mitochondria-targeted cancer therapy. Cancer Metastasis Rev 43(4):1419–1443. https://doi.org/10.1007/s10555-024-10211-9

Chen K, Ernst P, Sarkar A et al (2024c) mLumiOpto is a mitochondrial-targeted gene therapy for treating cancer. Cancer Res 84(23):4049–4065. https://doi.org/10.1158/0008-5472.CAN-24-0984. PMID: 39288077; PMCID: PMC11609628

Chen Y, Xiao L, Zhou M, Zhang H (2024d) The microbiota: a crucial mediator in gut homeostasis and colonization resistance. Front Microbiol 15:1417864. https://doi.org/10.3389/fmicb.2024.1417864. PMID: 39165572; PMCID: PMC1133323

Cheng J, Demeulemeester J, Wedge DC, et al. (2017) Pan-cancer analysis of homozygous deletions in primary tumours uncovers rare tumour suppressors. Nat Commun 8(1):1221. doi: 10.1038/s41467-017-01355-0. Errata in: Nat Commun (2018) 9(1):5397; Nat Commun (2019) 10(1):525. PMID: 29089486; PMCID: PMC5663922

Cheng L, Wang Y, Du J (2020) Human papillomavirus vaccines: an updated review. Vaccines (Basel) 8(3):391. https://doi.org/10.3390/vaccines8030391. PMID: 32708759; PMCID: PMC7565290

Cheng K, Nair NU, Lee JS, Ruppin E (2021) Synthetic lethality across normal tissues is strongly associated with cancer risk, onset, and tumor suppressor specificity. Sci Adv 7(1):eabc2100. https://doi.org/10.1126/sciadv.abc2100. PMID: 33523837; PMCID: PMC7775773

Cheng E, Lee DH, Tamimi RM et al (2022) Long-term survival and causes of death after diagnoses of common cancers in 3 cohorts of US health professionals. JNCI Cancer Spectr 6(2):pkac021. https://doi.org/10.1093/jncics/pkac021

Cheong JH, Park ES, Liang J et al (2011) Dual inhibition of tumor energy pathway by 2-deoxyglucose and metformin is effective against a broad spectrum of preclinical cancer models. Mol Cancer Ther 10(12):2350–2362. https://doi.org/10.1158/1535-7163.MCT-11-0497. PMID: 21992792; PMCID: PMC3237863

Chernet B, Levin M (2013) Endogenous voltage potentials and the microenvironment: bioelectric signals that reveal, induce and normalize cancer. J Clin Exp Oncol Suppl 1:S1-002. https://doi.org/10.4172/2324-9110.S1-002. PMID: 25525610; PMCID: PMC4267524

Chesney J, Lewis KD, Kluger H et al (2022) Efficacy and safety of lifileucel, a one-time autologous tumor-infiltrating lymphocyte (TIL) cell therapy, in patients with advanced melanoma after progression on immune checkpoint inhibitors and targeted therapies: Pooled analysis of consecutive cohorts of the C-144-01 study. J Immunother Cancer 10(12):e005755. https://doi.org/10.1136/jitc-2022-005755. PMID: 36600653; PMCID: PMC9748991

Cheung KJ, Ewald AJ (2016) A collective route to metastasis: Seeding by tumor cell clusters. Science 352(6282):167–169. https://doi.org/10.1126/science.aaf6546. PMID: 27124449; PMCID: PMC8183671

Cheung KJ, Padmanaban V, Silvestri V et al (2016) Polyclonal breast cancer metastases arise from collective dissemination of keratin 14-expressing tumor cell clusters. Proc Natl Acad Sci U S A 113(7):E854–E863. https://doi.org/10.1073/pnas.1508541113. PMID: 26831077; PMCID: PMC4763783

Chew GL, Bleakley M, Bradley RK et al (2021) Short H2A histone variants are expressed in cancer. Nat Commun 12(1):490. https://doi.org/10.1038/s41467-020-20707-x. PMID: 33473122; PMCID: PMC7817690

Chhabra Y, Weeraratna AT (2023) Fibroblasts in cancer: unity in heterogeneity. Cell 186(8):1580–1609. https://doi.org/10.1016/j.cell.2023.03.016. PMID: 37059066; PMCID: PMC11422789

Chial H (2008) Proto-oncogenes to oncogenes to cancer. Nature Education 1(1):33

Chiang AC, Massagué J (2008) Molecular basis of metastasis. N Engl J Med 359(26):2814–2823. https://doi.org/10.1056/NEJMra0805239. PMID: 19109576; PMCID: PMC4189180

Chiappinelli KB, Strissel PL, Desrichard A et al (2015) Inhibiting DNA methylation causes an interferon response in cancer via dsRNA including endogenous retroviruses. Cell 162(5):974–986. https://doi.org/10.1016/j.cell.2015.07.011. Errata in: Cell (2016) 164(5):1073 and Cell (2017) 169(2):361. PMID: 26317466; PMCID: PMC4556003

Chiaretti S, de Curtis I (2016) Role of Liprins in the regulation of tumor cell motility and invasion. Curr Cancer Drug Targets 16(3):238–248. https://doi.org/10.2174/1568009616031602061 24103. PMID: 26882029

Chihara D, Lin R, Flowers CR et al (2022) Early drug development in solid tumours: analysis of National Cancer Institute-sponsored phase 1 trials. Lancet 400(10351):512–521. https://doi.org/10.1016/S0140-6736(22)01390-3. PMID: 35964611; PMCID: PMC9477645

Chiu HC, Chang TY, Huang CT et al (2012) EGFR and myosin II inhibitors cooperate to suppress EGFR-T790M-mutant NSCLC cells. Mol Oncol 6(3):299–310. https://doi.org/10.1016/j.molonc.2012.02.001. PMID: 22366308; PMCID: PMC5528336

Chlebowski RT, Gota CH, Chan KK et al (1982) Clinical and pharmacokinetic effects of combined warfarin and 5-fluorouracil in advanced colon cancer. Cancer Res 42(11):4827–4830. PMID: 7127318

Chlebowski RT, Blackburn GL, Thomson CA et al (2006) Dietary fat reduction and breast cancer outcome: interim efficacy results from the Women's Intervention Nutrition Study. J Natl Cancer Inst 98(24):1767–1776. https://doi.org/10.1093/jnci/djj494. PMID: 17179478

Cho WC (2017) Molecular connections of aging and cancer. Aging Dis 8(5):685–687. https://doi.org/10.14336/AD.2017.0822. PMID: 28966810; PMCID: PMC5614330

Chochinov HM, Kristjanson LJ, Breitbart W, et al (2011) Effect of dignity therapy on distress and end-of-life experience in terminally ill patients: a randomised controlled trial. Lancet Oncol 12(8):753–762. https://doi.org/10.1016/S1470-2045(11)70153-X. Epub 2011 Jul 6. PMID: 21741309; PMCID: PMC3185066

Christakoudi S, Tsilidis KK, Evangelou E et al (2021) Association of body-shape phenotypes with imaging measures of body composition in the UK Biobank cohort: relevance to colon cancer risk. BMC Cancer 21(1):1106. https://doi.org/10.1186/s12885-021-08820-6. PMID: 34654381; PMCID: PMC8518225

Chuang JS, Frentz Z, Leibler S (2019) Homeorhesis and ecological succession quantified in synthetic microbial ecosystems. Proc Natl Acad Sci USA 116(30):14852–14861. https://doi.org/10.1073/pnas.1901055116

Chung AW, Anand K, Anselme AC, et al (2021) A phase 1/2 clinical trial of the nitric oxide synthase inhibitor L-NMMA and taxane for treating chemoresistant triple-negative breast cancer. Sci Transl Med 13(624):eabj5070. https://doi.org/10.1126/scitranslmed.abj5070. Epub 2021 Dec 15. PMID: 34910551

Church G, Regis E (2012) Regenesis: how synthetic biology will reinvent nature and ourselves. Basic Books, New York

Cieślik M, Hoang SA, Baranova N et al (2013) Epigenetic coordination of signaling pathways during the epithelial-mesenchymal transition. Epigenetics Chromatin 6(1):28. https://doi.org/10.1186/1756-8935-6-28. PMID: 24004852; PMCID: PMC3847279

Cira NJ, Paull ML, Sinha S et al (2025) Structure, motion, and multiscale search of traveling networks. Nat Commun 16:1922. https://doi.org/10.1038/s41467-024-54342-7

Clancy JW, D'Souza-Schorey C (2023) Tumor-derived extracellular vesicles: multifunctional entities in the tumor microenvironment. Annu Rev Pathol 18:205–229. https://doi.org/10.1146/annurev-pathmechdis-031521-022116. Epub 2022 Oct 6. PMID: 36202098; PMCID: PMC10410237

Clark WH (1995) The nature of cancer: morphogenesis and progressive (self)-disorganization in neoplastic development and progression. Acta Oncol 34(1):3–21. https://doi.org/10.3109/02841869509093632. PMID: 7865232

Clarke MA, Rodriguez AC, Gage JC et al (2012) A large, population-based study of age-related associations between vaginal pH and human papillomavirus infection. BMC Infect Dis 12:33. https://doi.org/10.1186/1471-2334-12-33. PMID: 22316377; PMCID: PMC3292496

Classen AK (2024) Tumours form without genetic mutations. Nature 629(8012):534–535. https://doi.org/10.1038/d41586-024-01019-2. PMID: 38658714

Classen C, Butler LD, Koopman C et al (2001) Supportive-expressive group therapy and distress in patients with metastatic breast cancer: a randomized clinical intervention trial. Arch Gen Psychiat 58(5):494–501. https://doi.org/10.1001/archpsyc.58.5.494. PMID: 11343530

Claverie JM (2006) Viruses take center stage in cellular evolution. Genome Biol 7(6):110. https://doi.org/10.1186/gb-2006-7-6-110. PMID: 16787527; PMCID: PMC1779534

Clemmensen SB, Harris JR, Mengel-From J et al (2021) Familial Risk and Heritability of Hematologic Malignancies in the Nordic Twin Study of Cancer. Cancers (Basel) 13(12):3023. https://doi.org/10.3390/cancers13123023

Clifton KK, Ma CX, Fontana L, Peterson LL (2021) Intermittent fasting in the prevention and treatment of cancer. CA Cancer J Clin 71(6):527–546. https://doi.org/10.3322/caac.21694. Epub 2021 Aug 12. PMID: 34383300

Cline MJ (1987) The role of proto-oncogenes in human cancer: implications for diagnosis and treatment. Int J Radiat Oncol Biol Phys 13(9):1297–1301. https://doi.org/10.1016/0360-3016(87)90219-7. PMID: 3305447

Cohen IR, Marron A (2020) The evolution of universal adaptations of life is driven by universal properties of matter: energy, entropy, and interaction. F1000Res 9:626. https://doi.org/10.12688/f1000research.24447.3. PMID: 32802320; PMCID: PMC7416572

Cohen IR, Marron A (2023) Evolution is driven by natural autoencoding: reframing species, interaction codes, cooperation and sexual reproduction. Proc Biol Sci 290(1994):20222409. https://doi.org/10.1098/rspb.2022.2409

Cohen JD, Li L, Wang Y, et al (2018) Detection and localization of surgically resectable cancers with a multi-analyte blood test. Science 359(6378):926–930. https://doi.org/10.1126/science.aar3247. Epub 2018 Jan 18. PMID: 29348365; PMCID: PMC6080308

Cohen L, Delorme A, Cusimano A et al (2024) Examining the effects of biofield therapy through simultaneous assessment of electrophysiological and cellular outcomes. Sci Rep 14(1):29221. https://doi.org/10.1038/s41598-024-79617-3. Erratum in: Sci Rep 2025;15(1):9626. doi: 10.1038/s41598-025-93431-5. PMID: 39622875; PMCID: PMC11612308

Cole WH, Everson TC (1956) Spontaneous regression of cancer: preliminary report. Ann Surg 144(3):366–383. https://doi.org/10.1097/00000658-195609000-00007

Cole SW, Hawkley LC, Arevalo JM, Sung CY, Rose RM, Cacioppo JT (2007) Social regulation of gene expression in human leukocytes. Genome Biol 8(9):R189. https://doi.org/10.1186/gb-2007-8-9-r189

Cole SW, Nagaraja AS, Lutgendorf SK, Green PA, Sood AK (2015) Sympathetic nervous system regulation of the tumour microenvironment. Nat Rev Cancer 15(9):563–572. https://doi.org/10.1038/nrc3978

Coley WB (1893). The treatment of malignant tumors by repeated inoculations of erysipelas. With a report of ten original cases. Clin Orthop Relat Res 3–11 (1991).

Collavin L, Lunardi A, Del Sal G (2010) p53-family proteins and their regulators: hubs and spokes in tumor suppression. Cell Death Differ 17:901–911. https://doi.org/10.1038/cdd.2010.35

Colotta F, Allavena P, Sica A, Garlanda C, Mantovani A (2009) Cancer-related inflammation, the seventh hallmark of cancer: links to genetic instability. Carcinogenesis 30(7):1073–1081. https://doi.org/10.1093/carcin/bgp127

Comaills V, Castellano-Pozo M (2023) Chromosomal Instability in Genome Evolution: From Cancer to Macroevolution. Biology (Basel) 12(5):671. https://doi.org/10.3390/biology12050671

Connal S, Cameron JM, Sala A, Brennan PM, Palmer DS, Palmer JD et al (2023) Liquid biopsies: the future of cancer early detection. J Transl Med 21(1):118. https://doi.org/10.1186/s12967-023-03960-8

Connerty P, Lock RB, de Bock CE (2020) Long non-coding RNAs: major regulators of cell stress in cancer. Front Oncol 10:285. https://doi.org/10.3389/fonc.2020.00285

Conrad M (1990) The geometry of evolution. Biosystems. 24(1):61–81. https://doi.org/10.1016/0303-2647(90)90030-5

Cool CD, Kuebler WM, Bogaard HJ, Spiekerkoetter E, Nicolls MR, Voelkel NF (2020) The hallmarks of severe pulmonary arterial hypertension: the cancer hypothesis-ten years later. Am J Physiol Lung Cell Mol Physiol 318(6):L1115–L1130. https://doi.org/10.1152/ajplung.00476.2019

Cools-Lartigue J, Spicer J, Najmeh S, Ferri L (2014) Neutrophil extracellular traps in cancer progression. Cell Mol Life Sci 71(21):4179–4194. https://doi.org/10.1007/s00018-014-1683-3

Cooper SJ (2008) From Claude Bernard to Walter Cannon. Emergence of the concept of homeostasis. Appetite 51(3):419–427. https://doi.org/10.1016/j.appet.2008.06.005

Corces MR, Granja JM, Shams S, Louie BH, Seoane JA, Zhou W et al (2018) The chromatin accessibility landscape of primary human cancers. Science 362(6413):eaav1898. https://doi.org/10.1126/science.aav1898

Cordaux R, Batzer MA (2009) The impact of retrotransposons on human genome evolution. Nat Rev Genet 10(10):691–703. https://doi.org/10.1038/nrg2640

Corning PA (2023) Teleonomy in evolution: "the ghost in the machine". In: Corning PA, Kauffman SA, Noble D, Shapiro JA, Vane-Wright RI, Pross A (eds) Evolution "on purpose": teleonomy in living systems. MIT Press

Corning PA, Kline SJ (1998) Part II: Thermoeconomics and control information. Syst Res 15:453–482

Cornish-Bowden A, Cárdenas ML (2020) Contrasting theories of life: historical context, current theories. In search of an ideal theory. Biosystems 188:104063. https://doi.org/10.1016/j.biosystems.2019.104063

Cortês M, Kauffman SA, Liddle AR, Smolin L (2022) Biocosmology: biology from a cosmological perspective. arXiv:2204.09379v2. https://doi.org/10.48550/arXiv.2204.09379

Cortês M, Kauffman SA, Liddle AR, Smolin L (2024a) Biocosmology, the birth of a new science. arXiv:2204.09378v3. https://doi.org/10.48550/arXiv.2204.09378

Cortês M, Kauffman SA, Liddle AR, Smolin L (2024b) Biocosmology: biology from a cosmological perspective. arXiv:2204.09379v2. https://doi.org/10.48550/arXiv.2204.09379

Cortés-Hernández LE, Eslami SZ, Dujon AM et al (2020) Do malignant cells sleep at night? Genome Biol 21(1):276. https://doi.org/10.1186/s13059-020-02179-w

Cortés-Hernández LE, Eslami S Z, Dujon AM, Giraudeau M, Ujvari B (2024) Circulating tumor cells: from basic to translational research. Clin Chem 70(1):81–89

Cosgrove MS, Wolberger C (2005) How does the histone code work? Biochem Cell Biol 83(4):468–476. https://doi.org/10.1139/o05-137

Costanzo V, Bardelli A, Siena S, Abrignani S (2018) Exploring the links between cancer and placenta development. Open Biol 8(6):180081. https://doi.org/10.1098/rsob.180081

Costa-Silva B, Aiello NM, Ocean AJ, Singh S, Zhang H, Thakur BK et al (2015) Pancreatic cancer exosomes initiate pre-metastatic niche formation in the liver. Nat Cell Biol 17(6):816–826. https://doi.org/10.1038/ncb3169

Coudray N, Ocampo PS, Sakellaropoulos T, Narula N, Snuderl M, Fenyo D et al (2018) Classification and mutation prediction from non–small cell lung cancer histopathology images using deep learning. Nat Med 24(10):1559–1567

Creixell P, Schoof EM, Erler JT, Linding R (2012) Navigating cancer network attractors for tumor-specific therapy. Nat Biotechnol 30(9):842–848. https://doi.org/10.1038/nbt.2345

Creixell P, Schoof EM, Simpson CD, Longden J, Miller CJ et al (2015) Kinome-wide decoding of network-attacking mutations rewiring cancer signaling. Cell 163(1):202–217. https://doi.org/10.1016/j.cell.2015.08.056

Crespo J, Sun H, Welling TH, Tian Z, Zou W (2013) T cell anergy, exhaustion, senescence, and stemness in the tumor microenvironment. Curr Opin Immunol 25(2):214–221. https://doi.org/10.1016/j.coi.2012.12.003

Cristiano S, Leal A, Phallen J, Fiksel J (2017) Genome-wide cell-free DNA fragmentation in patients with cancer. Nature 570(7761):385–389. https://doi.org/10.1038/s41586-019-1272-6

Cristy M, Eckerman K (1987) Specific absorbed fractions of energy at various ages from internal photon sources. Parts I–VII. ORNL/TM-8381/V1–V7. Oak Ridge National Laboratory, Oak Ridge

Croce CM (2008) Oncogenes and cancer. N Engl J Med 358:502–511. https://doi.org/10.1056/NEJMra072367

Crook JM, O'Callaghan CJ, Duncan G, Dearnaley DP, Higano CS et al (2012) Intermittent androgen suppression for rising PSA level after radiotherapy. N Engl J Med 367(10):895–903. https://doi.org/10.1056/NEJMoa1201546

Crutchfield JP, van Nimwegen E (2002) The evolutionary unfolding of complexity. In: Landweber LF, Winfree E (eds) Evolution as computation. Springer-Verlag, Berlin

Cryan JF, O'Riordan KJ, Cowan CSM, Sandhu KV, Bastiaanssen TFS et al (2019) The microbiota-gut-brain axis. Physiol Rev 99(4):1877–2013. https://doi.org/10.1152/physrev.00018.2018

Cui Z, Willingham MC, Hicks AM, Alexander-Miller MA, Howard TD et al (2003) Spontaneous regression of advanced cancer: identification of a unique genetically determined, age-dependent trait in mice. Proc Natl Acad Sci USA 100(11):6682–6687. https://doi.org/10.1073/pnas.1031601100

Cui J, Ding R, Liu H, Ma M, Zuo R, Liu X (2024) Trends in the incidence and survival of cancer in individuals aged 55 years and older in the United States, 1975–2019. BMC Public Health 24(1):72. https://doi.org/10.1186/s12889-023-17571-x

Curth A, Peck RW, McKinney E et al (2024) Using Machine Learning to Individualize Treatment Effect Estimation: Challenges and Opportunities. Clin Pharmacol Ther. 115(4):710–719. https://doi.org/10.1002/cpt.3159

Czene K, Lichtenstein P, Hemminki K (2002) Environmental and heritable causes of cancer among 9.6 million individuals in the Swedish Family-Cancer Database. Int J Cancer 99(2):260–266. https://doi.org/10.1002/ijc.10332

D'Alessandro A, El Kasmi KC, Plecitá-Hlavatá L, Ježek P, Li M et al (2018) Hallmarks of Pulmonary Hypertension: Mesenchymal and Inflammatory Cell Metabolic Reprogramming. Antioxid Redox Signal 28(3):230–250. https://doi.org/10.1089/ars.2017.7217

D'Alterio C, Scala S, Sozzi G, Roz L, Bertolini G (2020) Paradoxical effects of chemotherapy on tumor relapse and metastasis promotion. Semin Cancer Biol 60:351–361. https://doi.org/10.1016/j.semcancer.2019.08.019

da Silva-Diz V, Lorenzo-Sanz L, Bernat-Peguera A et al (2018) Cancer cell plasticity: impact on tumor progression and therapy response. Semin Cancer Biol 53:48–58. https://doi.org/10.1016/j.semcancer.2018.08.009

Dai C, Whitesell L, Rogers AB, Lindquist S (2007) Heat shock factor 1 is a powerful multifaceted modifier of carcinogenesis. Cell 130(6):1005–1018. https://doi.org/10.1016/j.cell.2007.07.020

Dai J, Shen W, Wen W, Chang J, Wang T et al (2017) Estimation of heritability for nine common cancers using data from genome-wide association studies in Chinese population. Int J Cancer 140(2):329–336. https://doi.org/10.1002/ijc.30447

Dai J, Su Y, Zhong S, Cong L, Liu B et al (2020) Exosomes: key players in cancer and potential therapeutic strategy. Signal Transduct Target Ther 5(1):145. https://doi.org/10.1038/s41392-020-00261-0

Dailey GP, Crosby EJ, Hartman ZC (2023) Cancer vaccine strategies using self-replicating RNA viral platforms. Cancer Gene Ther 30:794–802. https://doi.org/10.1038/s41417-022-00499-6

Damasio A (2005) Descartes' error: emotion, reason, and the human brain. Penguin Books, London

Dang CV (2012) Links between metabolism and cancer. Genes Dev 26(9):877–890. https://doi.org/10.1101/gad.189365.112

Dang CV (2015) A metabolic perspective of Peto's paradox and cancer. Philos Trans R Soc Lond B Biol Sci 370(1673):20140223. https://doi.org/10.1098/rstb.2014.0223

Dang DK, Park BH (2022) Circulating tumor DNA: current challenges for clinical utility. J Clin Invest 132(12):e154941. https://doi.org/10.1172/JCI154941

Danna EA, Sinha P, Gilbert M, Clements VK, Pulaski BA, Ostrand-Rosenberg S (2004) Surgical removal of primary tumor reverses

tumor-induced immunosuppression despite the presence of metastatic disease. Cancer Res 64(6):2205–2211. https://doi.org/10.1158/0008-5472.can-03-2646
Darwin C (1859) On the origin of species. John Murray, London
Darwin C (1866) On the origin of species by means of natural selection, or the preservation of favoured races in the struggle for life, 4th edn. John Murray, London, p 153
Das SK, Eder S, Schauer S, Diwoky C, Temmel H et al (2011) Adipose triglyceride lipase contributes to cancer-associated cachexia. Science 333(6039):233–238. https://doi.org/10.1126/science.1198973
Das M, Ellies LG, Kumar D, Sauceda C, Oberg A et al (2021) Time-restricted feeding normalizes hyperinsulinemia to inhibit breast cancer in obese postmenopausal mouse models. Nat Commun 12(1):565. https://doi.org/10.1038/s41467-020-20743-7
Datta NR, Kok HP, Crezee H, Gaipl US, Bodis S (2020) Integrating loco-regional hyperthermia into the current oncology practice: SWOT and TOWS analyses. Front Oncol 10:819. https://doi.org/10.3389/fonc.2020.00819
Dauvilliers Y, Thomas F, Alix-Panabières C (2022) Dissemination of circulating tumor cells at night: role of sleep or circadian rhythm? Genome Biol 23(1):214. https://doi.org/10.1186/s13059-022-02791-y
Davidson SM, Papagiannakopoulos T, Olenchock BA, Heyman JE, Keibler MA et al (2016) Environment impacts the metabolic dependencies of Ras-driven non-small cell lung cancer. Cell Metab 23(3):517–528. https://doi.org/10.1016/j.cmet.2016.01.007
Davies P (1999) The Fifth Miracle: The Search for the Origin and Meaning of Life. Simon & Schuster, New York
Davies J, Levin M (2023) Synthetic morphology with agential materials. Nat Rev Bioeng 1:46–59. https://doi.org/10.1038/s44222-022-00001-9
Davies PC, Lineweaver CH (2011) Cancer tumors as Metazoa 1.0: tapping genes of ancient ancestors. Phys Biol 8(1):015001. https://doi.org/10.1088/1478-3975/8/1/015001
Davies PC, Demetrius L, Tuszynski JA (2011) Cancer as a dynamical phase transition. Theor Biol Med Model 8:30. https://doi.org/10.1186/1742-4682-8-30
Davies PC, Rieper E, Tuszynski JA (2013). Self-organization and entropy reduction in a living cell. Biosystems 111(1):1–10. https://doi.org/10.1016/j.biosystems.2012.10.005
Davies A, Zoubeidi A, Beltran H, Selth LA (2023) The transcriptional and epigenetic landscape of cancer cell lineage plasticity. Cancer Discov 13(8):1771–1788. https://doi.org/10.1158/2159-8290.CD-23-0225
Davila AF, Zamorano P (2013) Mitochondria and the evolutionary roots of cancer. Phys Biol 10(2):026008. https://doi.org/10.1088/1478-3975/10/2/026008
Davis FM, Stewart TA, Thompson EW, Monteith GR (2014) Targeting EMT in cancer: opportunities for pharmacological intervention. Trends Pharmacol Sci 35(9):479–488. https://doi.org/10.1016/j.tips.2014.06.006
Davis A, Gao R, Navin N (2017) Tumor evolution: Linear, branching, neutral or punctuated? Biochim Biophys Acta Rev Cancer 1867(2):151–161. https://doi.org/10.1016/j.bbcan.2017.01.003
Davis RT, Blake K, Ma D, Gabra MBI, Hernandez GA et al (2020) Transcriptional diversity and bioenergetic shift in human breast cancer metastasis revealed by single-cell RNA sequencing. Nat Cell Biol 22(3):310–320. https://doi.org/10.1038/s41556-020-0477-0
Davoli T, de Lange T (2011) The causes and consequences of polyploidy in normal development and cancer. Annu Rev Cell Dev Biol 27:585–610. https://doi.org/10.1146/annurev-cellbio-092910-154234
Davoli T, Xu AW, Mengwasser KE, Sack LM, Yoon JC et al (2013) Cumulative haploinsufficiency and triplosensitivity drive aneuploidy patterns and shape the cancer genome. Cell 155(4):948–962. https://doi.org/10.1016/j.cell.2013.10.011
de Boer RA, Meijers WC, van der Meer P, van Veldhuisen DJ (2019) Cancer and heart disease: associations and relations. Eur J Heart Fail 21(12):1515–1525. https://doi.org/10.1002/ejhf.1539
De Cabo R, Mattson MP (2019) Effects of intermittent fasting on health, aging, and disease. N Engl J Med 381(26):2541–2551. https://doi.org/10.1056/NEJMra1905136
De Carvalho AC, Kim H, Poisson LM, Winn ME, Mueller C et al (2018) Discordant inheritance of chromosomal and extrachromosomal DNA elements contributes to dynamic disease evolution in glioblastoma. Nat Genet 50(5):708–717. https://doi.org/10.1038/s41588-018-0105-0
De Coster RC, Clemens MW, Di Napoli A, Lynch EB, Bonaroti AR et al (2021) Cellular and molecular mechanisms of breast implant-associated anaplastic large cell lymphoma. Plast Reconstr Surg 147(1):30e–41e. https://doi.org/10.1097/PRS.0000000000007423
De Gregori J (2018) Adaptive oncogenesis: a new understanding of how cancer evolves within us. Harvard University Press, Cambridge, MA
De Gregori J, Pharoah P, Sasieni P, Swanton C (2020) Cancer screening, surrogates of survival, and the soma. Cancer Cell 38(4):433–437. https://doi.org/10.1016/j.ccell.2020.09.003
de Groot S, Pijl H, van der Hoeven JJM, Kroep JR (2019) Effects of short-term fasting on cancer treatment. J Exp Clin Cancer Res 38(1):209. https://doi.org/10.1186/s13046-019-1189-9
De Lange T (2005) Telomere-related genome instability in cancer. Cold Spring Harb Symp Quant Biol 70:197–204. https://doi.org/10.1101/sqb.2005.70.032
De Lellis RA, Xia L (2003) Paraneoplastic endocrine syndromes: a review. Endocr Pathol 14(4):303–317. https://doi.org/10.1385/ep:14:4:303
De Loof A (2015) From Darwin's On the Origin of Species by means of natural selection to the evolution of life with communication activity as its very essence and driving force (= mega-evolution). Funct Genom 3:153–187
De Loof A (2016) The cell's self-generated "electrome": the biophysical essence of the immaterial dimension of life? Commun Integr Biol 9(5):e1197446. https://doi.org/10.1080/19420889.2016.1197446
De Loof A (2017) The evolution of "Life": a Metadarwinian integrative approach. Commun Integr Biol 10(3):e1301335. https://doi.org/10.1080/19420889.2017.1301335
De Loof A (2023) The mega-evolution of life with its three memory systems depends on sender-receiver communication and problem-solving. A narrative review. J Physiol. https://doi.org/10.1113/JP284412
de Magalhães JP (2022) Every gene can (and possibly will) be associated with cancer. Trends Genet 38(3):216–217. https://doi.org/10.1016/j.tig.2021.09.005
de Parseval N, Heidmann T (2005) Human endogenous retroviruses: from infectious elements to human genes. Cytogenet Genome Res 110(1–4):318–332. https://doi.org/10.1159/000084964
De Pinho RA (2000) The age of cancer. Nature 408(6809):248–254. https://doi.org/10.1038/35041694
De Santis CE, Miller KD, Dale W, Mohile SG, Cohen HJ et al (2019) Cancer statistics for adults aged 85 years and older, 2019. CA Cancer J Clin 69(6):452–467. https://doi.org/10.3322/caac.21577
de Thé H (2018) Differentiation therapy revisited. Nat Rev Cancer 18(2):117–127. https://doi.org/10.1038/nrc.2017.103
de Visser KE, Joyce JA (2023) The evolving tumor microenvironment: From cancer initiation to metastatic outgrowth. Cancer Cell 41(3):374–403. https://doi.org/10.1016/j.ccell.2023.02.016
Deacon T (2010) What's missing from theories of information? In: Davies P, Gregersen NH (eds) Information and the Nature of Reality: From Physics to Metaphysics. Cambridge University Press, New York, pp 146–169
Deacon T, Srivastava A, Bacigalupi J (2014) The transition from constraint to regulation at the origin of life. Front Biosci 19:945–957

Dean DC, Shen S, Hornicek FJ, Duan Z (2018) From genomics to metabolomics: emerging metastatic biomarkers in osteosarcoma. Cancer Metastasis Rev 37(4):719–731. https://doi.org/10.1007/s10555-018-9763-8

DeBerardinis RJ, Chandel NS (2016) Fundamentals of cancer metabolism. Sci Adv 2(5):e1600200. https://doi.org/10.1126/sciadv.1600200

Deisboeck TS, Couzin ID (2009) Collective behavior in cancer cell populations. Bioessays 31(2):190–197. https://doi.org/10.1002/bies.200800084

Dekkers OM (2019) (2019) The long and winding road to causality. Eur J Epidemiol. 34(6):533–535. https://doi.org/10.1007/s10654-019-00507-4

Del Monte U (2009) Does the cell number 10(9) still really fit one gram of tumor tissue? Cell Cycle. 8(3):505–506. https://doi.org/10.4161/cc.8.3.7608

Delconte RB, Owyong M, Santosa EK et al (2024) Fasting reshapes tissue-specific niches to improve NK cell-mediated anti-tumor immunity. Immunity 57(8):1923–1938.e7. https://doi.org/10.1016/j.immuni.2024.05.021

Delisi D, Eskandari N, Gentile S (2024) Membrane potential: A new hallmark of cancer. Adv Cancer Res 164:93–110. https://doi.org/10.1016/bs.acr.2024.04.010

Deloar HM, Fujiwara T, Shidahara M et al (1998) Estimation of absorbed dose for 2-[F-18]fluoro-2-deoxy-D-glucose using whole-body positron emission tomography and magnetic resonance imaging. Eur J Nucl Med 25(6):565–574. https://doi.org/10.1007/s002590050257

Demars J, Labrune Y, Iannuccelli N et al (2022) A genome-wide epistatic network underlies the molecular architecture of continuous color variation of body extremities. Genomics 114(3):110361. https://doi.org/10.1016/j.ygeno.2022.110361

Demory Beckler M, Higginbotham JN, Franklin JL et al (2013) Proteomic analysis of exosomes from mutant KRAS colon cancer cells identifies intercellular transfer of mutant KRAS. Mol Cell Proteomics 12(2):343–355. https://doi.org/10.1074/mcp.M112.022806

Deng L, Fang H, Tso VK et al (2017) Clinical validation of a novel urine-based metabolomic test for the detection of colonic polyps on Chinese population. Int J Colorectal Dis 32(5):741–743. https://doi.org/10.1007/s00384-016-2729-9

Dennett DC (2017) From bacteria to Bach and Back: the evolution of minds. W.W. Norton & Company, New York

Derosa L, Iebba V, Silva CAC et al (2024) Cell. 187(13):3373–3389.e16. https://doi.org/10.1016/j.cell.2024.05.029

Dhamija S, Diederichs S (2016) From junk to master regulators of invasion: lncRNA functions in migration. EMT and metastasis. Int J Cancer 139(2):269–280. https://doi.org/10.1002/ijc.30039

Dhoble S, Patravale V (2019) Development of anti-angiogenic erlotinib liposomal formulation for pulmonary hypertension: a QbD approach. Drug Deliv Transl Res 9(5):980–996. https://doi.org/10.1007/s13346-019-00641-2

Di Domenico A, Wiedmer T, Marinoni I, Perren A (2017) Genetic and epigenetic drivers of neuroendocrine tumours (NET). Endocr Relat Cancer 24(9):R315–R334. https://doi.org/10.1530/ERC-17-0012

Di Gregorio E, Israel S, Staelens M et al (2022) The distinguishing electrical properties of cancer cells. Phys Life Rev 43:139–188. https://doi.org/10.1016/j.plrev.2022.09.003

Di Sario G, Rossella V, Famulari ES et al (2023) Enhancing clinical potential of liquid biopsy through a multi-omic approach: A systematic review. Front Genet 14:1152470. https://doi.org/10.3389/fgene.2023.1152470

Di Z, Garcia-Roger EM, Richmond P, et al. (2021). Transient frailty induced by cell division: observation, reasons and implications. arXiv preprint arXiv:2112.05988

Diamantopoulos MA, Tsiakanikas P, Scorilas A (2018) Non-coding RNAs: the riddle of the transcriptome and their perspectives in cancer. Ann Transl Med 6(12):241. https://doi.org/10.21037/atm.2018.06.10

Diamantopoulou Z, Castro-Giner F, Schwab FD et al (2022) The metastatic spread of breast cancer accelerates during sleep. Nature 607(7917):156–162. https://doi.org/10.1038/s41586-022-04875-y

Dias MH, Friskes A, Wang S et al (2024) Paradoxical activation of oncogenic signaling as a cancer treatment strategy. Cancer Discov 14(7):1276–1301. https://doi.org/10.1158/2159-8290.CD-23-0216

Dickerson T, Jauregui CE, Teng Y (2017) Friend or foe? Mitochondria as a pharmacological target in cancer treatment. Future Med Chem 9(18):2197–2210. https://doi.org/10.4155/fmc-2017-0110

Diehl F, Schmidt K, Choti MA et al (2007) Circulating mutant DNA to assess tumor dynamics. Nat Med 14(9):985–990. https://doi.org/10.1038/nm.1789

Dietlein F, Weghorn D, Taylor-Weiner A et al (2020) Identification of cancer driver genes based on nucleotide context. Nat Genet 52(2):208–218. https://doi.org/10.1038/s41588-019-0572-y

Dillekås H, Rogers MS, Straume O (2019) Are 90% of deaths from cancer caused by metastases? Cancer Med 8(12):5574–5576. https://doi.org/10.1002/cam4.2474

Dimarco A, Gaetani M, Orezzi P et al (1964) Daunomycin, a new antibiotic of the Rhodomycin group. Nature 201:706–707. https://doi.org/10.1038/201706a0

DiMatteo MR (2004) Social support and patient adherence to medical treatment: a meta-analysis. Health Psychol 23(2):207–218. https://doi.org/10.1037/0278-6133.23.2.207

DiNardo CD, Stein EM, de Botton S et al (2018) Durable Remissions with Ivosidenib in IDH1-Mutated Relapsed or Refractory AML. N Engl J Med 378(25):2386–2398. https://doi.org/10.1056/NEJMoa1716984

Dinarello CA (2010) Why not treat human cancer with interleukin-1 blockade? Cancer Metastasis Rev 29(2):317–329. https://doi.org/10.1007/s10555-010-9229-0

Ding XM (2014) MicroRNAs: regulators of cancer metastasis and epithelial-mesenchymal transition (EMT). Chin J Cancer 33(3):140–147. https://doi.org/10.5732/cjc.013.10094

Ding Y, Wang X, Zhang F et al (2024) The relationship between perceived social support, coping style, and the quality of life and psychological state of lung cancer patients. BMC Psychol 12(1):439. https://doi.org/10.1186/s40359-024-01927-y

Discher DE, Janmey P, Wang YL (2005) Tissue cells feel and respond to the stiffness of their substrate. Science 310(5751):1139–1143. https://doi.org/10.1126/science.1116995

Djamgoz MBA (2024) Electrical excitability of cancer cells—CELEX model updated. Cancer Metastasis Rev 43(4):1579–1591. https://doi.org/10.1007/s10555-024-10195-6

Djebali S, Davis CA, Merkel A et al (2012) Landscape of transcription in human cells. Nature 489(7414):101–108. https://doi.org/10.1038/nature11233

Dobzhansky T (1973) Nothing in biology makes sense except in the light of evolution. Am Biol Teach 35(3):125–129. https://doi.org/10.2307/4444260

Docherty AB, Harrison EM, Green CA et al (2020) Features of 20,133 UK patients in hospital with COVID-19 using the ISARIC WHO Clinical Characterisation Protocol: Prospective observational cohort study. BMJ 369:m1985. https://doi.org/10.1136/bmj.m1985

Doherty JR, Cleveland JL (2013) Targeting lactate metabolism for cancer therapeutics. J Clin Invest 123(9):3685–3692. https://doi.org/10.1172/JCI69741

Doi M (2013) What is soft matter? In: Soft Matter Physics. Oxford University Press. https://doi.org/10.1093/acprof:oso/9780199652952.003.0001

Dolan RD, McMillan DC (2020) The prevalence of cancer-associated systemic inflammation: Implications of prognostic studies using the Glasgow Prognostic Score. Crit Rev Oncol Hematol 150:102962. https://doi.org/10.1016/j.critrevonc.2020.102962

Dolberg DS, Bissell MJ (1984) Inability of Rous sarcoma virus to cause sarcomas in the avian embryo. Nature 309:552–556. https://doi.org/10.1038/309552a0

Dolce D (2023a) "Internal times" and how to second-quantize fields by means of periodic boundary conditions. Ann Phys 457:169398. https://doi.org/10.1016/j.aop.2023.169398

Dolce D (2023b) Is time a cyclic dimension? Canonical quantization implicit in classical cyclic dynamics. Ann Phys 448:169182. https://doi.org/10.1016/j.aop.2022.169182

Dominiak A, Chełstowska B, Olejarz W, Nowicka G (2020) Communication in the cancer microenvironment as a target for therapeutic interventions. Cancers (Basel) 12(5):1232. https://doi.org/10.3390/cancers12051232

Dongre A, Weinberg RA (2019) New insights into the mechanisms of epithelial-mesenchymal transition and implications for cancer. Nat Rev Mol Cell Biol 20(2):69–84. https://doi.org/10.1038/s41580-018-0080-4

Doolittle WF, Inkpen SA (2018) Processes and patterns of interaction as units of selection: An introduction to ITSNTS thinking. Proc Natl Acad Sci USA 115(16):4006–4014. https://doi.org/10.1073/pnas.1722232115

Doria-Borrell P, Pérez-García V (2024) Understanding the intersection between placental development and cancer: Lessons from the tumor suppressor BAP1. Commun Biol 7(1):1053. https://doi.org/10.1038/s42003-024-06689-2

Dossus L, Kouloura E, Biessy C, Viallon V, Siskos AP, Dimou N et al (2021) Prospective analysis of circulating metabolites and endometrial cancer risk. Gynecol Oncol 162(2):475–481. https://doi.org/10.1016/j.ygyno.2021.06.001

Douglass J, Hsiue EH, Mog BJ, Hwang MS, DiNapoli SR et al (2021) Bispecific antibodies targeting mutant RAS neoantigens. Sci Immunol 6(57):eabd5515. https://doi.org/10.1126/sciimmunol.abd5515

Downton P, Early JO, Gibbs JE (2020) Circadian rhythms in adaptive immunity. Immunology 161(4):268–277. https://doi.org/10.1111/imm.13167

Droujinine IA, Perrimon N (2013) Defining the interorgan communication network: systemic coordination of organismal cellular processes under homeostasis and localized stress. Front Cell Infect Microbiol 3:82. https://doi.org/10.3389/fcimb.2013.00082

Droujinine IA, Perrimon N (2016) Interorgan communication pathways in physiology: focus on Drosophila. Annu Rev Genet 50:539–570. https://doi.org/10.1146/annurev-genet-121415-122024

Droujinine IA, Perrimon N (2019) The multidimensional organization of interorgan communication networks. Dev Cell 50(4):395–396. https://doi.org/10.1016/j.devcel.2019.07.029

Druzd D, Matveeva O, Ince L et al (2017) Lymphocyte circadian clocks control lymph node trafficking and adaptive immune responses. Immunity 46(1):120–132. https://doi.org/10.1016/j.immuni.2016.12.011

Duesberg P, Mandrioli D, McCormack A, Nicholson JM (2011) Is carcinogenesis a form of speciation? Cell Cycle 10(13):2100–2114. https://doi.org/10.4161/cc.10.13.16352

Duim H, Otto S (2017) Towards open-ended evolution in self-replicating molecular systems. Beilstein J Org Chem 13:1189–1203. https://doi.org/10.3762/bjoc.13.118

Dujon AM, Aktipis A, Alix-Panabières C, Amend SR, Boddy AM et al (2021) Identifying key questions in the ecology and evolution of cancer. Evol Appl 14(4):877–892. https://doi.org/10.1111/eva.13190

Dujon A, Boutry J, Tissot S, et al (2022) Cancer susceptibility as a cost of reproduction and contributor to life history evolution. Deakin University. Journal contribution. https://hdl.handle.net/10536/DRO/DU:30167640

Dunican DS, McWilliam P, Tighe O et al (2002) Gene expression differences between the microsatellite instability (MIN) and chromosomal instability (CIN) phenotypes in colorectal cancer revealed by high-density cDNA array hybridization. Oncogene 21(20):3253–3257. https://doi.org/10.1038/sj.onc.1205431

Dunn TJ, Dimolareva M (2022) The effect of mindfulness-based interventions on immunity-related biomarkers: a comprehensive meta-analysis of randomised controlled trials. Clin Psychol Rev 92:102124. https://doi.org/10.1016/j.cpr.2022.102124

Dunn GP, Bruce AT, Ikeda H et al (2002) Cancer immunoediting: from immunosurveillance to tumor escape. Nat Immunol 3(11):991–998. https://doi.org/10.1038/ni1102-991

Dunn GP, Old LJ, Schreiber RD (2004) The immunobiology of cancer immunosurveillance and immunoediting. Immunity 21(2):137–148. https://doi.org/10.1016/j.immuni.2004.07.017

Duport P (2003) A database of cancer induction by low-dose radiation in mammals: Overview and initial observations. Int J Low Radiat 1:120–131

Dupré J (1998) Wittgenstein and forms of life. Philos Mag (49):24–27. https://doi.org/10.5840/tpm199849

Dupré J (2017) The metaphysics of evolution. Interface. Focus 7:20160148. https://doi.org/10.1098/rsfs.2016.0148

Dupré A, Malik HZ (2018) Inflammation and cancer: what a surgical oncologist should know. Eur J Surg Oncol 44(5):566–570. https://doi.org/10.1016/j.ejso.2018.02.209

Dustin ML (2014) The immunological synapse. Cancer. Immunol Res 2(11):1023–1033. https://doi.org/10.1158/2326-6066.CIR-14-0161

Dustin ML, Long EO (2010) Cytotoxic immunological synapses. Immunol Rev 235(1):24–34. https://doi.org/10.1111/j.0105-2896.2010.00904.x

Dutcher JM, Cole SW, Williams AC, Creswell JD (2022) Smartphone mindfulness meditation training reduces pro-inflammatory gene expression in stressed adults: a randomized controlled trial. Brain Behav Immun 103:171–177. https://doi.org/10.1016/j.bbi.2022.04.003

Dutta S, Farhadifar R, Lu W et al (2023) Self-organized intracellular twisters. bioRxiv [Preprint]. https://doi.org/10.1101/2023.04.04.534476

Dvinge H, Kim E, Abdel-Wahab O, Bradley RK (2016) RNA splicing factors as oncoproteins and tumour suppressors. Nat Rev Cancer 16(7):413–430. https://doi.org/10.1038/nrc.2016.51

Dvorak HF (1986) Tumors: wounds that do not heal. Similarities between tumor stroma generation and wound healing. N Engl J Med 315(26):1650–1659. https://doi.org/10.1056/NEJM198612253152606

Dyson F (1999) Origins of Life, 2nd edn. Cambridge University Press, Cambridge

Dzutsev A, Badger JH, Perez-Chanona E et al (2017) Microbes and cancer. Annu Rev Immunol 35:199–228. https://doi.org/10.1146/annurev-immunol-051116-052133

Ebrahim T, Ebrahim AS, Kandouz M (2024) Diversity of intercellular communication modes: a cancer biology perspective. Cells 13(6):495. https://doi.org/10.3390/cells13060495

Eder JP, Balmanoukian AS, Aggarwal C, Horn L, Patnaik A et al (2005) Viral metagenomics. Nat Rev Microbiol 3:504–510

Edwards RA, Rohwer F (2005) Viral metagenomics. Nat Rev Microbiol 3(6):504–510. https://doi.org/10.1038/nrmicro1163

Egeblad M, Nakasone ES, Werb Z (2010) Tumors as organs: complex tissues that interface with the entire organism. Dev Cell 18(6):884–901. https://doi.org/10.1016/j.devcel.2010.05.012

Eigen M (1996) Prionics or the kinetic basis of prion diseases. Biophys Chem 63(1):A1–A18. https://doi.org/10.1016/S0301-4622(96)02250-8

Eil R, Vodnala SK, Clever D et al (2016) Ionic immune suppression within the tumour microenvironment limits T cell effector function. Nature 537(7621):539–543. https://doi.org/10.1038/nature19364

Einstein A (1910) Principe de relativité. Arch Sci Phys Natur Ser 4 29(5)

Eisenhauer EA, Therasse P, Bogaerts J et al (2009) New response evaluation criteria in solid tumours: revised RECIST guideline (version 1.1). Eur J Cancer 45(2):228–247. https://doi.org/10.1016/j.ejca.2008.10.026

Eisenstein M (2025) Putting early cancer detection to the test. Nature 12(2). https://doi.org/10.1038/d41586-025-00530-4

Elaraj DM, Weinreich DM, Varghese S et al (2006) The role of interleukin 1 in growth and metastasis of human cancer xenografts. Clin Cancer Res 12(4):1088–1096. https://doi.org/10.1158/1078-0432.CCR-05-1603

Elefteriou F (2016) Role of sympathetic nerves in the establishment of metastatic breast cancer cells in bone. J Bone Oncol 5(3):132–134. https://doi.org/10.1016/j.jbo.2016.03.003

Elek G, Müller M (2024) Ervin Bauer's concept of biological thermodynamics and its different evaluations. Biosystems 235:105090. https://doi.org/10.1016/j.biosystems.2023.105090

Elenkov IJ, Wilder RL, Chrousos GP, Vizi ES (2000) The sympathetic nerve—an integrative interface between two supersystems: the brain and the immune system. Pharmacol Rev 52(4):595–638. PMID: 11121511

Elezkurtaj S, Greuel S, Ihlow J et al (2021) Causes of death and comorbidities in hospitalized patients with COVID-19. Sci Rep 11(1):4263. https://doi.org/10.1038/s41598-021-82862-5

Elia I, Haigis MC (2021) Metabolites and the tumour microenvironment: from cellular mechanisms to systemic metabolism. Nat Metab 3:21–32. https://doi.org/10.1038/s42255-020-00317-

Ellis GFR (2023) How purposeless physics underlies purposeful life. Nature 622(7982):247–249. https://doi.org/10.1038/d41586-023-03061-y

Emamalipour M, Seidi K, Zununi et al (2020) Horizontal gene transfer: from evolutionary flexibility to disease progression. Front Cell Dev Biol 8:229. https://doi.org/10.3389/fcell.2020.00229

Emami B, Lyman J, Brown A et al (1991) Tolerance of normal tissue to therapeutic irradiation. Int J Radiat Oncol Biol Phys 21(1):109–122. https://doi.org/10.1016/0360-3016(91)90171-y

Emberson JR, Kearney PM, Blackwell L, Newman C, Reith C et al (2012) Lack of effect of lowering LDL cholesterol on cancer: Meta-analysis of individual data from 175,000 people in 27 randomised trials of statin therapy. PLoS One 7(1):e29849. https://doi.org/10.1371/journal.pone.0029849

ENCODE Project Consortium (2012) An integrated encyclopedia of DNA elements in the human genome. Nature 489(7414):57–74. https://doi.org/10.1038/nature11247

Encyclopedia of Biostatistics (2005) John Wiley & Sons. https://doi.org/10.1002/0470011815.b2a03072

Endo M, Fukuda Y, Okada K et al (2023) Abscopal effect after stereotactic body radiotherapy with nivolumab for lung metastasis of head and neck cancer: A case report. Case Rep Oncol 16(1):1345–1352. https://doi.org/10.1159/000534609

Engel J, Eckel R, Halfter K et al (2022) Breast cancer: emerging principles of metastasis, adjuvant and neoadjuvant treatment from cancer registry data. J Cancer Res Clin Oncol 149(2):721–735. https://doi.org/10.1007/s00432-022-04369-4

England JL (2013) Statistical physics of self-replication. J Chem Phys 139(12):121923. https://doi.org/10.1063/1.4818538

Enriquez-Navas PM, Wojtkowiak JW, Gatenby RA (2015) Application of evolutionary principles to cancer therapy. Cancer Res 75(22):4675–4680. https://doi.org/10.1158/0008-5472.CAN-15-1337

Epstein JI, Allsbrook WC, Amin MB, Egevad LL (2005) The 2005 International Society of Urological Pathology (ISUP) consensus conference on Gleason grading of prostatic carcinoma. Am J Surg Pathol 29(9):1228–1242

Erenpreisa JA, Cragg MS, Fringes B et al (2000) Release of mitotic descendants by giant cells from irradiated Burkitt's lymphoma cell line. Cell Biol Int 24(9):635–648. https://doi.org/10.1006/cbir.2000.0558

Erklavec Zajec V, Novak U et al (2021) Dynamic multiscale metabolic network modeling of Chinese hamster ovary cell metabolism integrating N-linked glycosylation in industrial biopharmaceutical manufacturing. Biotechnol Bioeng 118(1):397–411. https://doi.org/10.1002/bit.27578

Espinosa-Carrasco G, Chiu E, Scrivo A et al (2023) Intratumoral immune triads are required for adoptive T cell therapy-mediated elimination of solid tumors. bioRxiv Preprint. https://doi.org/10.1101/2023.07.03.547423

Espinosa-Carrasco G, Chiu E, Scrivo A et al (2024) Intratumoral immune triads are required for immunotherapy-mediated elimination of solid tumors. Cancer Cell 42(7):1202–1216.e8. https://doi.org/10.1016/j.ccell.2024.05.025

Esteller M (2008) Epigenetics in cancer. N Engl J Med 358(11):1148–1159. https://doi.org/10.1056/NEJMra072067

Esteller M, Dawson MA, Kadoch C et al (2024) The epigenetic hallmarks of cancer. Cancer Discov 14(10):1783–1809. https://doi.org/10.1158/2159-8290.CD-24-0296

Esteva A, Robicquet A, Ramsundar B et al (2019) A guide to deep learning in healthcare. Nat Med 25(1):24–29. https://doi.org/10.1038/s41591-018-0316-z

Falagas ME, Zarkadoulia EA, Ioannidou EN et al (2007) The effect of psychosocial factors on breast cancer outcome: a systematic review. Breast Cancer Res 9(4):R44. https://doi.org/10.1186/bcr1744

Falanga A, Russo L, Milesi V, Vignoli A (2017) Mechanisms and risk factors of thrombosis in cancer. Crit Rev Oncol Hematol 118:79–83. https://doi.org/10.1016/j.critrevonc.2017.08.003

Falchi F, Giacomini E, Masini T et al (2017) Synthetic lethality triggered by combining olaparib with BRCA2-Rad51 disruptors. ACS Chem Biol 12(10):2491–2497. https://doi.org/10.1021/acschembio.7b00707

Falco M, Palma G, Rea D et al (2016) Tumour biomarkers: homeostasis as a novel prognostic indicator. Open Biol 6(12):160254. https://doi.org/10.1098/rsob.160254

Fan X, Alekseyenko AV, Wu J et al (2018) Human oral microbiome and prospective risk for pancreatic cancer: a population-based nested case-control study. Gut 67(1):120–127. https://doi.org/10.1136/gutjnl-2016-312580

Fan Z, Yang JY, Guo Y et al (2022) Altered levels of circulating mitochondrial DNA in elderly people with sarcopenia: association with mitochondrial impairment. Exp Gerontol 163:111802. https://doi.org/10.1016/j.exger.2022.111802

Fang X, Wei J, He X et al (2015) Landscape of dietary factors associated with risk of gastric cancer: A systematic review and dose-response meta-analysis of prospective cohort studies. Eur J Cancer 51(18):2820–2832. https://doi.org/10.1016/j.ejca.2015.09.010

Fardi M, Solali S, Farshdousti HM (2018) Epigenetic mechanisms as a new approach in cancer treatment: An updated review. Genes Dis 5(4):304–331. https://doi.org/10.1016/j.gendis.2018.05.001

Fares J, Fares MY, Khachfe HH et al (2020) Molecular principles of metastasis: a hallmark of cancer revisited. Signal Transduct Target Ther 5(1):28. https://doi.org/10.1038/s41392-020-0134-x

Faris ME, Alkawamleh DH, Madkour MI (2025) Unraveling the impact of intermittent fasting in cancer prevention, mitigation, and treatment: A narrative review. J Nutr Oncol 2:29–39. https://doi.org/10.1097/JN9.0000000000000049

Farnsworth KD, Nelson J, Gershenson C (2013) Living is information processing: from molecules to global systems. Acta Biotheor 61(2):203–222. https://doi.org/10.1007/s10441-013-9179-3

Farrokh A (2020) Constructing causal networks through regressions: a tutorial. Qual Manag Health Care 29(4):270–278. https://doi.org/10.1097/QMH.0000000000000272

Fay MP, Li F (2024) Causal interpretation of the hazard ratio in randomized clinical trials. Clin Trials. 21(5):623–635. https://doi.org/10.1177/17407745241243308

Fazilaty H, Basler K (2023) Reactivation of embryonic genetic programs in tissue regeneration and disease. Nat Genet 55(11):1792–1806. https://doi.org/10.1038/s41588-023-01526-4

Fearon ER, Vogelstein B (1990) A genetic model for colorectal tumorigenesis. Cell 61(5):759–767. https://doi.org/10.1016/0092-8674(90)90186-i

Fearon K, Strasser F, Anker SD et al (2011) Definition and classification of cancer cachexia: an international consensus. Lancet Oncol 12(5):489–495. https://doi.org/10.1016/S1470-2045(10)70218-7

Fearon KC, Glass DJ, Guttridge DC (2012) Cancer cachexia: mediators, signaling, and metabolic pathways. Cell Metab 16(2):153–166. https://doi.org/10.1016/j.cmet.2012.06.011

Feng S, Jacobsen SE, Reik W (2010) Epigenetic reprogramming in plant and animal development. Science 330(6004):622–627. https://doi.org/10.1126/science.1190614

Feng J, Song B, Zhang Y (2022) Semantic parsing of the life process by quantum biology. Prog Biophys Mol Biol 175:79–89. https://doi.org/10.1016/j.pbiomolbio.2022.09.005

Feng L, Ding R, Qu X et al (2023) BCR-ABL triggers a glucose-dependent survival program during leukemogenesis through the suppression of TXNIP. Cell Death Dis 14(4):287. https://doi.org/10.1038/s41419-023-05811-2

Ferguson LR, Chen H, Collins AR et al (2015) Genomic instability in human cancer: Molecular insights and opportunities for therapeutic attack and prevention through diet and nutrition. Semin Cancer Biol 35(Suppl):S5–S24. https://doi.org/10.1016/j.semcancer.2015.03.005

Fernandes J, Guedes PG, Lage CL et al (2012) Tumor malignancy is engaged to prokaryotic homolog toolbox. Med Hypotheses 78(4):435–441. https://doi.org/10.1016/j.mehy.2011.12.014

Fernandes MR, Aggarwal P, Costa RGF et al (2022) Targeting the gut microbiota for cancer therapy. Nat Rev Cancer 22(12):703–722. https://doi.org/10.1038/s41568-022-00513-x

Fernández de Castro I, Tenorio R, Risco C (2021) Virus factories. In: Encyclopedia of virology, pp 495–500. https://doi.org/10.1016/B978-0-12-814515-9.00001-1

Ferrara M, Samaden M, Ruggieri E, Vénéreau E (2022) Cancer cachexia as a multiorgan failure: Reconstruction of the crime scene. Front Cell Dev Biol 10:960341. https://doi.org/10.3389/fcell.2022.960341

Ferriero M, Cacciatore L, Ochoa M et al (2023) The impact of metastasectomy on survival outcomes of renal cell carcinoma: A 10-year single center experience. Cancers (Basel) 15(13):3332. https://doi.org/10.3390/cancers15133332

Feuerriegel S, Frauen D, Melnychuk V et al (2024) Causal machine learning for predicting treatment outcomes. Nat Med. 958-968. https://doi.org/10.1038/s41591-024-02902-1

Feunteun J, Ostyn P, Delaloge S (2022) Tumor cell malignancy: A complex trait built through reciprocal interactions between tumors and tissue-body system. iScience 25(5):104217. https://doi.org/10.1016/j.isci.2022.104217

Fiala C, Diamandis EP (2018) Utility of circulating tumor DNA in cancer diagnostics with emphasis on early detection. BMC Med 16(1):166. https://doi.org/10.1186/s12916-018-1157-9

Fiala C, Diamandis EP (2020) Mutations in normal tissues—some diagnostic and clinical implications. BMC Med 18(1):283. https://doi.org/10.1186/s12916-020-01763-y

Fidler IJ (2003) The pathogenesis of cancer metastasis: the 'seed and soil' hypothesis revisited. Nat Rev Cancer 3(6):453–458. https://doi.org/10.1038/nrc1098

Fidler IJ, Kripke ML (1977) Metastasis results from preexisting variant cells within a malignant tumor. Science 197(4306):893–895. https://doi.org/10.1126/science.887927. PMID: 887927

Fidler IJ, Kripke ML (2015) The challenge of targeting metastasis. Cancer Metastasis Rev 34(4):635–641. https://doi.org/10.1007/s10555-015-9586-9

Finn RS, Qin S, Ikeda M et al (2020) Atezolizumab plus bevacizumab in unresectable hepatocellular carcinoma. N Engl J Med 382(20):1894–1905. https://doi.org/10.1056/NEJMoa1915745

Fischer T, Hayn A, Mierke CT (2020) Effect of nuclear stiffness on cell mechanics and migration of human breast cancer cells. Front Cell Dev Biol 8:393. https://doi.org/10.3389/fcell.2020.00393

Fischetti M, Christiansen J (2021) A new you in 80 days. Sci Am 324(4):76

Fishbein A, Hammock BD, Serhan CN et al (2021) Carcinogenesis: Failure of resolution of inflammation? Pharmacol Ther 218:107670. https://doi.org/10.1016/j.pharmthera.2020.107670

Fisher R, Pusztai L, Swanton C (2013) Cancer heterogeneity: implications for targeted therapeutics. Br J Cancer 108:479–485. https://doi.org/10.1038/bjc.2012.581

Fitzgerald DM, Hastings PJ, Rosenberg SM (2017) Stress-induced mutagenesis: implications in cancer and drug resistance. Annu Rev Cancer Biol 1:119–140. https://doi.org/10.1146/annurev-cancerbio-050216-121919

Fjæstad KY, Rømer AMA, Goitea V et al (2022) Blockade of beta-adrenergic receptors reduces cancer growth and enhances the response to anti-CTLA4 therapy by modulating the tumor microenvironment. Oncogene 41(9):1364–1375. https://doi.org/10.1038/s41388-021-02170-0

Flavahan WA (2020) Epigenetic plasticity, selection, and tumorigenesis. Biochem Soc Trans 48(4):1609–1621. https://doi.org/10.1042/BST20191215

Flavahan WA, Gaskell E, Bernstein BE (2017) Epigenetic plasticity and the hallmarks of cancer. Science 357(6348):eaal2380. https://doi.org/10.1126/science.aal2380

Flint TR, Janowitz T, Connell CM et al (2016) Tumor-induced IL-6 reprograms host metabolism to suppress anti-tumor immunity. Cell Metab 24(5):672–684. https://doi.org/10.1016/j.cmet.2016.10.010

Floyd C, Turner FC (1941) Sarcomas at sites of subcutaneously implanted Bakelite disks in rats. J Natl Cancer Inst 2(1):81–83. https://doi.org/10.1093/jnci/2.1.81

Foglesong GD, Queen NJ, Huang W et al (2019) Enriched environment inhibits breast cancer progression in obese models with intact leptin signaling. Endocr Relat Cancer 26(5):483–495. https://doi.org/10.1530/ERC-19-0075

Fonti N, Parisi F, Lachi A et al (2024) Age at tumor diagnosis in 14,636 canine cases from the pathology-based UNIPI animal cancer registry, Italy: one size doesn't fit all. Vet Sci 11(10):485. https://doi.org/10.3390/vetsci11100485

Ford BJ (2017) Cellular intelligence: microphenomenology and the realities of being. Prog Biophys Mol Biol 131:273–287. https://doi.org/10.1016/j.pbiomolbio.2017.08.012

Forterre P (2006) The origin of viruses and their possible roles in major evolutionary transitions. Virus Res 117(1):5–16. https://doi.org/10.1016/j.virusres.2006.01.010

Forterre P (2010) Defining life: the virus viewpoint. Orig Life Evol Biosph 40(2):151–160. https://doi.org/10.1007/s11084-010-9194-1

Forterre P, Prangishvili D (2009) The great billion-year war between ribosome- and capsid-encoding organisms (cells and viruses) as the major source of evolutionary novelties. Proc N Y Acad Sci 1178:65–77

Foulds L (1958) The natural history of cancer. J Chronic Dis 8(1):2–37. https://doi.org/10.1016/0021-9681(58)90039-0

Fowler H, Belot A, Ellis L et al (2020) Comorbidity prevalence among cancer patients: a population-based cohort study of four cancers. BMC Cancer 20:2. https://doi.org/10.1186/s12885-019-6472-9

Fox Keller E (2007) A clash of two cultures. Nature 445:603. https://doi.org/10.1038/445603a

Francis N, Borniger JC (2021) Cancer as a homeostatic challenge: the role of the hypothalamus. Trends Neurosci 44(11):903–914. https://doi.org/10.1016/j.tins.2021.08.008

Freeland J (2002) Is ours the best of all possible codes? In: Landweber LF, Winfree E (eds) Evolution as computation. Springer, Berlin

Freeland S (2022) Undefining life's biochemistry: implications for abiogenesis. J R Soc Interface 19(187):20210814. https://doi.org/10.1098/rsif.2021.0814

Frieden BR, Gatenby RA (2011) Information dynamics in living systems: prokaryotes, eukaryotes, and cancer. PLoS One 6(7):e22085. https://doi.org/10.1371/journal.pone.0022085

Friedman MA, Cain DF (1990) National Cancer Institute sponsored cooperative clinical trials. Cancer 65(10 Suppl):2376–2382. https://doi.org/10.1002/1097-0142(19900515)65:10+<2376::aid-cncr2820651504>3.0.co;2-a

Friedmann-Morvinski D, Verma IM (2014) Dedifferentiation and reprogramming: origins of cancer stem cells. EMBO Rep 15(3):244–253. https://doi.org/10.1002/embr.201338254

Friston K (2010) The free-energy principle: a unified brain theory? Nat Rev Neurosci 11(2):127–138. https://doi.org/10.1038/nrn2787

Frost J (2020) Cancer's intelligence. Int J Unconv Comput 16:41–78

Fu J, Hofker M, Wijmenga C (2015) Apple or pear: size and shape matter. Cell Metab 21(4):507–508. https://doi.org/10.1016/j.cmet.2015.03.016

Fu T, Amoah K, Chan TW et al (2024) Massively parallel screen uncovers many rare 3′ UTR variants regulating mRNA abundance of cancer driver genes. Nat Commun 15(1):3335. https://doi.org/10.1038/s41467-024-46795-7

Fujii T, Mashimo M, Moriwaki Y et al (2017) Expression and function of the cholinergic system in immune cells. Front Immunol 8:1085. https://doi.org/10.3389/fimmu.2017.01085

Fulbright LE, Ellermann M, Arthur JC (2017) The microbiome and the hallmarks of cancer. PLoS Pathog 13(9):e1006480. https://doi.org/10.1371/journal.ppat.1006480

Funk L, Su KC, Ly J, Feldman D, Singh A et al (2022) The phenotypic landscape of essential human genes. Cell 185(24):4634–4653.e22. https://doi.org/10.1016/j.cell.2022.10.017

Furusawa Y, Obata Y, Fukuda S et al (2013) Commensal microbe-derived butyrate induces the differentiation of colonic regulatory T cells. Nature 504(7480):446–450. https://doi.org/10.1038/nature12721

Furuya-Kanamori L, Bell KJL, Clark J et al (2016) Prevalence of differentiated thyroid cancer in autopsy studies over six decades: a meta-analysis. J Clin Oncol 34(30):3672–3679

Fuselier C, Dumoulin A, Paré A et al (2023) Placental Galectins in cancer: why we should pay more attention. Cells 12(3):437. https://doi.org/10.3390/cells12030437

Gadadhar S, Bodakuntla S, Natarajan K, Janke C (2017) The tubulin code at a glance. J Cell Sci 130(8):1347–1353. https://doi.org/10.1242/jcs.199471

Gaddam S, Ogas O (2023) Journey of the mind: How thinking emerged from chaos. WW Norton & Company, New York

Gajewski TF, Schreiber H, Fu YX (2013) Innate and adaptive immune cells in the tumor microenvironment. Nat Immunol 14(10):1014–1022. https://doi.org/10.1038/ni.2703

Galassi C, Chan TA, Vitale I, Galluzzi L (2024) The hallmarks of cancer immune evasion. Cancer Cell 42(11):1825–1863. https://doi.org/10.1016/j.ccell.2024.09.010

Galhardo RS, Hastings PJ, Rosenberg SM (2007) Mutation as a stress response and the regulation of evolvability. Crit Rev Biochem Mol Biol 42(5):399–435. https://doi.org/10.1080/10409230701648502

Gallagher BM, Fowler JS, Gutterson NI et al (1978) Metabolic trapping as a principle of radiopharmaceutical design: some factors responsible for the biodistribution of [18-F] 2-deoxy-2-fluoro-D-glucose. J Nucl Med 19:1154–1161

Gallicchio L, Devasia TP, Tonorezos E et al (2022) Estimation of the number of individuals living with metastatic cancer in the United States. J Natl Cancer Inst 114(11):1476–1483. https://doi.org/10.1093/jnci/djac158

Gallik KL, Treffy RW, Nacke LM et al (2017) Neural crest and cancer: divergent travelers on similar paths. Mech Dev 148:89–99. https://doi.org/10.1016/j.mod.2017.08.002

Gan X, Shu Z, Wang X et al (2023) Network medicine framework reveals generic herb-symptom effectiveness of traditional Chinese medicine. Sci Adv. 9(43):eadh0215. https://doi.org/10.1126/sciadv.adh0215

Gandalovičová A, Rosel D, Fernandes M et al (2017) Migrastatics—anti-metastatic and anti-invasion drugs: promises and challenges. Trends Cancer 3(6):391–406. https://doi.org/10.1016/j.trecan.2017.04.008

Ganesh K (2023) Uncoupling metastasis from tumorigenesis. N Engl J Med 388(7):657–659. https://doi.org/10.1056/NEJMcibr2213497

Ganesh K, Massagué J (2021) Targeting metastatic cancer. Nat Med 27(1):34–44. https://doi.org/10.1038/s41591-020-01195-4

Ganesh K, Basnet H, Kaygusuz Y et al (2020) Author correction: L1CAM defines the regenerative origin of metastasis-initiating cells in colorectal cancer. Nat Cancer 1(11):1128. https://doi.org/10.1038/s43018-020-00130-3

Ganly I, Nixon IJ, Wang LY et al (2015) Survival from differentiated thyroid cancer: what has age got to do with it? Thyroid 25(10):1106–1114. https://doi.org/10.1089/thy.2015.0104

Gánti T (1987) The principle of life. OMIKK, Budapest

Gánti T (2003) The principle of life. Oxford University Press, Oxford

Ganzleben I, Neurath MF, Becker C (2021) Autophagy in cancer therapy-molecular mechanisms and current clinical advances. Cancers (Basel) 13(21):5575. https://doi.org/10.3390/cancers13215575

Gao Y, Bado I, Wang H, Zhang W et al (2019) Metastasis organotropism: redefining the congenial soil. Dev Cell 49(3):375–391. https://doi.org/10.1016/j.devcel.2019.04.012

Garcia J, Daniels J, Lee Y et al (2024) Naturally occurring T cell mutations enhance engineered T cell therapies. Nature 626(7999):626–634. https://doi.org/10.1038/s41586-024-07018-7

García-Jiménez C, García-Martínez JM, Chocarro-Calvo A, De la Vieja A (2013) A new link between diabetes and cancer: enhanced WNT/β-catenin signaling by high glucose. J Mol Endocrinol 52(1):R51–R66. https://doi.org/10.1530/JME-13-0152

Gargiulo G, Serresi M, Marine JC (2024) Cell states in cancer: drivers, passengers, and trailers. Cancer Discov 14(4):610–614. https://doi.org/10.1158/2159-8290.CD-23-1510

Garon EB, Hellmann MD, Rizvi NA et al (2019) Five-year overall survival for patients with advanced non-small-cell lung cancer treated with pembrolizumab: results from the Phase I KEYNOTE-001 study. J Clin Oncol 37(28):2518–2527. https://doi.org/10.1200/JCO.19.00934

Garraway LA, Sellers WR (2006) Lineage dependency and lineage-survival oncogenes in human cancer. Nat Rev Cancer 6(8):593–602. https://doi.org/10.1038/nrc1947

Garrett WS (2015) Cancer and the microbiota. Science 348(6230):80–86. https://doi.org/10.1126/science.aaa4972

Gatenby RA, Gillies RJ (2004) Why do cancers have high aerobic glycolysis? Nat Rev Cancer 4(11):891–899. https://doi.org/10.1038/nrc1478

Gatenby RA, Gillies RJ (2008) A microenvironmental model of carcinogenesis. Nat Rev Cancer 8(1):56–61. https://doi.org/10.1038/nrc2255

Gatenby RA, Silva AS, Gillies RJ, Frieden BR (2009) Adaptive therapy. Cancer Res 69(11):4894–4903. https://doi.org/10.1158/0008-5472.CAN-08-3658

Gaucher J, Montellier E, Sassone-Corsi P (2018) Molecular cogs: interplay between circadian clock and cell cycle. Trends Cell Biol 28(5):368–379. https://doi.org/10.1016/j.tcb.2018.01.006

Ge EJ, Bush AI, Casini A et al (2022) Connecting copper and cancer: from transition metal signalling to metalloplasia. Nat Rev Cancer 22(2):102–113. https://doi.org/10.1038/s41568-021-00417-2

Geijteman ECT, Kuip EJM, Oskam J et al (2024) Illness trajectories of incurable solid cancers. BMJ 384:e076625. https://doi.org/10.1136/bmj-2023-076625

George J, Cannon T, Lai V et al (2007) Cancer cachexia syndrome in head and neck cancer patients: Part II. Pathophysiology. Head Neck 29(5):497–507. https://doi.org/10.1002/hed.20630

George M, Smith A, Sabesan S, Ranmuthugala G (2021) Physical comorbidities and their relationship with cancer treatment and its outcomes in older adult populations: systematic review. JMIR Cancer 7(4):e26425. https://doi.org/10.2196/26425

Georgiev GY, Chatterjee A (2016) The road to a measurable quantitative understanding of self-organization and evolution. In: Jagers op Akkerhuis GAJM (ed) Evolution and transitions in complexity: general discussion. Springer

Gerasymchuk D, Hubiernatorova A, Domanskyi A (2020) MicroRNAs regulating cytoskeleton dynamics, endocytosis, and cell motility—a link between neurodegeneration and cancer? Front Neurol 11:549006. https://doi.org/10.3389/fneur.2020.549006

Gerhart J, Kirschner M (1997) Cells, embryos, and evolution: toward a cellular and developmental understanding of phenotypic variation and evolutionary adaptability. Blackwell, Malden

Gerhart J, Kirschner M (2007) The theory of facilitated variation. Proc Natl Acad Sci USA 104(Suppl 1):8582–8589. https://doi.org/10.1073/pnas.0701035104

Gerlinger M, Rowan AJ, Horswell S et al (2012) Intratumor heterogeneity and branched evolution revealed by multiregion sequencing. N Engl J Med 366(10):883–892. https://doi.org/10.1056/NEJMoa1113205

Gerstung M, Jolly C, Leshchiner I et al (2020) The evolutionary history of 2,658 cancers. Nature 578(7793):122–128. https://doi.org/10.1038/s41586-019-1907-7

Gettinger S, Horn L, Jackman D et al (2018) Five-year follow-up of nivolumab in previously treated advanced non-small-cell lung cancer: results from the CA209-003 study. J Clin Oncol 36(17):1675–1684. https://doi.org/10.1200/JCO.2017.77.0412

Geva T (2006) Magnetic resonance imaging: historical perspective. J Cardiovasc Magn Reson 8(4):573–580. https://doi.org/10.1080/10976640600755302

Ghafouri-Fard S, Khoshbakht T, Hussen BM et al (2022) A review on the role of cyclin dependent kinases in cancers. Cancer Cell Int 22:325. https://doi.org/10.1186/s12935-022-02747-z

Ghatak D, Das Ghosh D, Roychoudhury S (2021) Cancer stemness: p53 at the wheel. Front Oncol 10:604124. https://doi.org/10.3389/fonc.2020.604124

Ghețu F (2023) Can we simulate evolution? https://st.network/science/can-we-simulate-evolution.html

Ghiaseddin AP, Shin D, Melnick K, Tran DD (2020) Tumor treating fields in the management of patients with malignant gliomas. Curr Treat Options Oncol 21(9):76. https://doi.org/10.1007/s11864-020-00773-5

Giancotti FG (2013) Mechanisms governing metastatic dormancy and reactivation. Cell 155(4):750–764. https://doi.org/10.1016/j.cell.2013.10.029

Gibert MK Jr, Sarkar A, Chagari B et al (2022) Transcribed ultraconserved regions in cancer. Cells 11(10):1684. https://doi.org/10.3390/cells11101684

Gil R, Silva FJ, Peretó J, Moya A (2004) Determination of the core of a minimal bacterial gene set. Microbiol Mol Biol Rev 68(3):518–537. https://doi.org/10.1128/MMBR.68.3.518-537.2004

Gilbert DA (1968) Differentiation, oncogenesis and cellular periodicities. J Theor Biol 21(1):113–122

Gilbert SF (2024) Inter-kingdom communication and the sympoietic way of life. Front Cell Dev Biol 12:1427798. https://doi.org/10.3389/fcell.2024.1427798

Gilbert SF, Sapp J, Tauber AI (2012) A symbiotic view of life: we have never been individuals. Q Rev Biol 87(4):325–341. https://doi.org/10.1086/668166

Gillies RJ, Flowers CI, Drukteinis JS, Gatenby RA (2012) A unifying theory of carcinogenesis, and why targeted therapy doesn't work. Eur J Radiol 81(Suppl 1):S48–S50. https://doi.org/10.1016/S0720-048X(12)70018-9

Giraudeau M, Sepp T, Ujvari B et al (2018) Human activities might influence oncogenic processes in wild animal populations. Nat Ecol Evol 2(7):1065–1070. https://doi.org/10.1038/s41559-018-0558-7

Glass JI, Merryman C, Wise KS et al (2017) Minimal cells—real and imagined. Cold Spring Harb Perspect Biol 9(12):a023861. https://doi.org/10.1101/cshperspect.a023861

Glazier DS (2005) Beyond the '3/4-power law': variation in the intra- and interspecific scaling of metabolic rate in animals. Biol Rev Camb Philos Soc 80(4):611–662. https://doi.org/10.1017/S1464793105006834

Glazier DS (2022) Variable metabolic scaling breaks the law: from 'Newtonian' to 'Darwinian' approaches. Proc Biol Sci 289(1985):20221605. https://doi.org/10.1098/rspb.2022.1605

Glazier DS (2023) The relevance of time in biological scaling. Biology (Basel) 12(8):1084. https://doi.org/10.3390/biology12081084

Gleick J (1987) Chaos: Making a new science. Viking, New York

Godfrey-Smith P (2011) The evolution of the individual. The Lakatos Award Lecture, LSE, June 2011. https://www.petergodfreysmith.com/Evo_Ind_PGS_Lakatos_2011_Web.pdf

Godfrey-Smith P (2013) Darwinian individuals. In: Bouchard F, Huneman P (eds) From groups to individuals: Evolution and emerging individuality. MIT Press, Cambridge, MA, pp 17–36

Godfrey-Smith P (2021) Metazoa: Animal minds and the birth of consciousness. William Collins, London

Goff SL, Smith FO, Klapper JA et al (2010) Tumor infiltrating lymphocyte therapy for metastatic melanoma: analysis of tumors resected for TIL. J Immunother 33(8):840–847. https://doi.org/10.1097/CJI.0b013e3181f05b91

Golbidi S, Frisbee JC, Laher I (2015) Chronic stress impacts the cardiovascular system: animal models and clinical outcomes. Am J Physiol Heart Circ Physiol 308(12):H1476–H1498. https://doi.org/10.1152/ajpheart.00859.2014

Goldbeter A (1996) Biochemical oscillations and cellular rhythms. Cambridge University Press, Cambridge

Goldbeter A (2010) La vie oscilatoire: Au cœur des rythmes du vivant. Odile Jacob, Paris

Goldschmidt RB (1940) The material basis of evolution. Yale University Press, New Haven

Goldstein EB (2010) Encyclopedia of perception. Sage Publications, p 492

Gomes CC (2022) Recurrent driver mutations in benign tumors. Mutat Res Rev Mutat Res 789:108412. https://doi.org/10.1016/j.mrrev.2022.108412

Gomes AP, Blenis J (2015) A nexus for cellular homeostasis: the interplay between metabolic and signal transduction pathways. Curr Opin Biotechnol 34:110–117. https://doi.org/10.1016/j.copbio.2014.12.007

Gomes AP, Ilter D, Low V et al (2020) Age-induced accumulation of methylmalonic acid promotes tumour progression. Nature 585(7824):283–287. https://doi.org/10.1038/s41586-020-2630-0

Gomes AP, Ilter D, Low V et al (2022) Altered propionate metabolism contributes to tumour progression and aggressiveness. Nat Metab 4(4):435–443. https://doi.org/10.1038/s42255-022-00553-5

Gomez K, Miura S, Huuki LA et al (2018) Somatic evolutionary timings of driver mutations. BMC Cancer 18(1):85. https://doi.org/10.1186/s12885-017-3977-y

Gong J, Chehrazi-Raffle A, Placencio-Hickok V et al (2019) The gut microbiome and response to immune checkpoint inhibitors: preclinical and clinical strategies. Clin Transl Med 8(1):9. https://doi.org/10.1186/s40169-019-0225-x

Gong JR, Lee CK, Kim HM et al (2025) Control of cellular differentiation trajectories for cancer reversion. Adv Sci (Weinh) 12(3):e2402132. https://doi.org/10.1002/advs.202402132

Gonzalez CA, Riboli E (2006) Diet and cancer prevention: where we are, where we are going. Nutr Cancer 56(2):225–231. https://doi.org/10.1207/s15327914nc5602_14

Gonzalez-Cao M, Iduma P, Karachaliou N et al (2016) Human endogenous retroviruses and cancer. Cancer Biol Med 13(4):483–488. https://doi.org/10.20892/j.issn.2095-3941.2016.0080

Gopalakrishnan V, Helmink BA, Spencer CN et al (2018) The influence of the gut microbiome on cancer, immunity, and cancer immunotherapy. Cancer Cell 33(4):570–580. https://doi.org/10.1016/j.ccell.2018.03.015

Gorjifard S, Goldszmid RS (2016) Microbiota-myeloid cell crosstalk beyond the gut. J Leukoc Biol 100(5):865–879. https://doi.org/10.1189/jlb.3RI0516-222R

Gosain R, Gage-Bouchard E, Ambrosone C et al (2020) Stress reduction strategies in breast cancer: review of pharmacologic and non-pharmacologic based strategies. Semin Immunopathol 42(6):719–734. https://doi.org/10.1007/s00281-020-00815-y

Gostomczyk K, Marsool MDM, Tayyab H et al (2024) Targeting circulating tumor cells to prevent metastases. Hum Cell 37(1):101–120. https://doi.org/10.1007/s13577-023-00992-6

Götte M, Mohr C, Koo CY et al (2010) miR-145-dependent targeting of junctional adhesion molecule A and modulation of fascin expression are associated with reduced breast cancer cell motility and invasiveness. Oncogene 29(50):6569–6580. https://doi.org/10.1038/onc.2010.386

Goubet AG, Daillère R, Routy B et al (2018) The impact of the intestinal microbiota in therapeutic responses against cancer. C R Biol 341(5):284–289. https://doi.org/10.1016/j.crvi.2018.03.004

Gould SJ (1981) Panda's thumb: more reflections in natural history. Norton, New York

Gould HJ 3rd, Paul D (2022) Cancer as a channelopathy—appreciation of complementary pathways provides a different perspective for developing treatments. Cancers (Basel) 14(19):4627. https://doi.org/10.3390/cancers14194627

Gould MK, Munoz-Plaza CE, Hahn EE et al (2017) Comorbidity profiles and their effect on treatment selection and survival among patients with lung cancer. Ann Am Thorac Soc 14(10):1571–1580. https://doi.org/10.1513/AnnalsATS.201701-030OC

Graham D (2023) Heraclitus. Stanford Encyclopedia of philosophy. https://plato.stanford.edu/entries/heraclitus/. Accessed 30 June 2025

Grandpierre A (2008) Fundamental complexity measures of life. In: Seckbach J, Gordon R (eds) Divine Action and Natural Selection: Science, Faith and Evolution. World Scientific, Singapore

Grandpierre A, Chopra D, Kafatos M (2014) The universal principle of biology: determinism, quantum physics and spontaneity. NeuroQuantology 12(3)

Greathouse KL, White JR, Vargas AJ et al (2018) Interaction between the microbiome and TP53 in human lung cancer. Genome Biol 19(1):123. https://doi.org/10.1186/s13059-018-1501-6

Greaves M (2014) Does everyone develop covert cancer? Nat Rev Cancer 14(4):209–210. https://doi.org/10.1038/nrc3703

Greten FR, Grivennikov SI (2019) Inflammation and cancer: triggers, mechanisms, and consequences. Immunity 51(1):27–41. https://doi.org/10.1016/j.immuni.2019.06.025

Grewal S, Oosterling SJ, van Egmond M (2021) Surgery for colorectal cancer: a trigger for liver metastases development? New insights into the underlying mechanisms. Biomedicines 9(2):177. https://doi.org/10.3390/biomedicines9020177

Grippin AJ, Marconi C, Copling S et al (2025) SARS-CoV-2 mRNA vaccines sensitize tumours to immune checkpoint blockade. Nature. https://doi.org/10.1038/s41586-025-09655-y

Grivennikov SI, Karin M (2010) Dangerous liaisons: STAT3 and NF-kappaB collaboration and crosstalk in cancer. Cytokine Growth Factor Rev 21(1):11–19. https://doi.org/10.1016/j.cytogfr.2009.11.005

Grivennikov SI, Greten FR, Karin M (2010) Immunity, inflammation, and cancer. Cell 140(6):883–899. https://doi.org/10.1016/j.cell.2010.01.025

Grosberg RK, Strathmann RR (2007) The evolution of multicellularity: a minor major transition? Annu Rev Ecol Evol Syst 38:621–654. https://doi.org/10.1146/annurev.ecolsys.36.102403.114735

Grossberg S (2017) Towards solving the hard problem of consciousness: the varieties of brain resonances and the conscious experiences that they support. Neural Netw 87:38–95. https://doi.org/10.1016/j.neunet.2016.11.003

Grossberg S (2020) Developmental designs and adult functions of cortical maps in multiple modalities: perception, attention, navigation, numbers, streaming, speech, and cognition. Front Neuroinform 14:4. https://doi.org/10.3389/fninf.2020.00004

Grossberg S (2021) Conscious Mind, Resonant Brain: How Each Brain Makes a Mind. Oxford University Press, Oxford

Gruhl T, Weinert T, Rodrigues MJ et al (2023) Ultrafast structural changes direct the first molecular events of vision. Nature 615(7954):939–944. https://doi.org/10.1038/s41586-023-05863-6

Guan F, Wu X, Zhou J et al (2024) Mitochondrial transfer in tunneling nanotubes—a new target for cancer therapy. J Exp Clin Cancer Res 43:147. https://doi.org/10.1186/s13046-024-03069-w

Guay C, Regazzi R (2017) Exosomes as new players in metabolic organ cross-talk. Diabetes Obes Metab 19(Suppl 1):137–146. https://doi.org/10.1111/dom.13027

Guertin DA, Wellen KE (2023) Acetyl-CoA metabolism in cancer. Nat Rev Cancer 23(3):156–172. https://doi.org/10.1038/s41568-022-00543-5

Guida F, Tan VY, Corbin LJ et al (2021) The blood metabolome of incident kidney cancer: a case-control study nested within the MetKid consortium. PLoS Med 18(9):e1003786. https://doi.org/10.1371/journal.pmed.1003786

Guillaumond F, Bidaut G, Ouaissi M et al (2015) Cholesterol uptake disruption, in association with chemotherapy, is a promising combined metabolic therapy for pancreatic adenocarcinoma. Proc Natl Acad Sci USA 112(8):2473–2478. https://doi.org/10.1073/pnas.1421601112

Gulati GS, Sikandar SS, Wesche DJ et al (2020) Single-cell transcriptional diversity is a hallmark of developmental potential. Science 367(6476):405–411. https://doi.org/10.1126/science.aax0249

Gumilar KE, Chin Y, Ibrahim IH et al (2023) Heat shock factor 1 inhibition: a novel anti-cancer strategy with promise for precision oncology. Cancers (Basel) 15(21):5167. https://doi.org/10.3390/cancers15215167

Gumpper-Fedus K, Hart PA, Belury MA et al (2022) Altered plasma fatty acid abundance is associated with cachexia in treatment-naïve pancreatic cancer. Cells 11(5):910. https://doi.org/10.3390/cells11050910

Gunawardena J (2014) Models in biology: "accurate descriptions of our pathetic thinking". BMC Biol 12:29. https://doi.org/10.1186/1741-7007-12-29

Gundem G, Van Loo P, Kremeyer B et al (2015) The evolutionary history of lethal metastatic prostate cancer. Nature 520(7547):353–357. https://doi.org/10.1038/nature14347

Guo L, Guo N (2015) Exosomes: potent regulators of tumor malignancy and potential bio-tools in clinical application. Crit Rev Oncol Hematol 95(3):346–358. https://doi.org/10.1016/j.critrevonc.2015.04.002

Gupta PB, Kuperwasser C, Brunet JP et al (2005) The melanocyte differentiation program predisposes to metastasis after neoplastic transformation. Nat Genet 37(10):1047–1054. https://doi.org/10.1038/ng1634

Gupta SK, Luo L, Yen L (2018) RNA-mediated gene fusion in mammalian cells. Proc Natl Acad Sci U S A 115(52):E12295–E12304. https://doi.org/10.1073/pnas.1814704115

Gusev EY, Zotova NV (2019) Cellular stress and general pathological processes. Curr Pharm Des 25(3):251–297. https://doi.org/10.2174/1381612825666190319114641

Gutschner T, Diederichs S (2012) The hallmarks of cancer: a long noncoding RNA point of view. RNA Biol 9(6):703–719. https://doi.org/10.4161/rna.20481

Gvirtzman R, Livovsky DM, Tahover E et al (2021) Anemia can predict the prognosis of colorectal cancer in the pre-operative stage: a retrospective analysis. World J Surg Oncol 19(1):341. https://doi.org/10.1186/s12957-021-02452-7

Gyurkó DM, Veres DV, Módos D et al (2013) Adaptation and learning of molecular networks as a description of cancer development at the systems-level: potential use in anti-cancer therapies. Semin Cancer Biol 23(4):262–269. https://doi.org/10.1016/j.semcancer.2013.06.005

Haas MJ, Feng V, Gonzales K et al (2020) High-throughput analysis identifying drugs that reduce oxidative and ER stress in human coronary artery endothelial cells. Eur J Pharmacol 879:173119. https://doi.org/10.1016/j.ejphar.2020.173119

Hahn WC, Weinberg RA (2002) Rules for making human tumor cells. N Engl J Med 347(20):1593–1603. https://doi.org/10.1056/NEJMra021902

Hahn WC, Counter CM, Lundberg AS et al (1999) Creation of human tumour cells with defined genetic elements. Nature 400(6743):464–468. https://doi.org/10.1038/22780

Haigis KM, Cichowski K, Elledge SJ (2019) Tissue-specificity in cancer: the rule, not the exception. Science 363:1150–1151. https://doi.org/10.1126/science.aaw3472

Hajdarovic KH, Yu D, Webb AE (2022) Understanding the aging hypothalamus, one cell at a time. Trends Neurosci 45(12):942–954. https://doi.org/10.1016/j.tins.2022.10.004

Halberg F (1969) Chronobiology. Annu Rev Physiol 31:675–725. https://doi.org/10.1146/annurev.ph.31.030169.003331

Hall CAS (2004) Ecological modelling. Ecol Model 178(1–2):107–113. https://doi.org/10.1016/j.ecolmodel.2003.12.004

Hamaguchi R, Isowa M, Narui R et al (2022) Clinical review of alkalization therapy in cancer treatment. Front Oncol 12:1003588. https://doi.org/10.3389/fonc.2022.1003588

Hamann JC, Surcel A, Chen R et al (2017) Entosis is induced by glucose starvation. Cell Rep 20(1):201–210. https://doi.org/10.1016/j.celrep.2017.06.037

Hamdan F, Cerullo V (2023) Cancer immunotherapies: a hope for the uncurable? Front Mol Med 3:1140977. https://doi.org/10.3389/fmmed.2023.1140977

Hamidi H, Ivaska J (2018) Every step of the way: integrins in cancer progression and metastasis. Nat Rev Cancer 18(9):533–548. https://doi.org/10.1038/s41568-018-0038-z

Hammarlund JA, Li SY, Wu G et al (2023) Subtype-specific circadian clock dysregulation modulates breast cancer biology, invasiveness, and prognosis. bioRxiv. https://doi.org/10.1101/2023.05.17.540386

Han H, Guo W, Shi W et al (2017) Hypertension and breast cancer risk: a systematic review and meta-analysis. Sci Rep 7:44877. https://doi.org/10.1038/srep44877

Hanahan D (2022) Hallmarks of cancer: new dimensions. Cancer Discov 12(1):31–46. https://doi.org/10.1158/2159-8290.CD-21-1059

Hanahan D, Coussens LM (2012) Accessories to the crime: functions of cells recruited to the tumor microenvironment. Cancer Cell 21(3):309–322. https://doi.org/10.1016/j.ccr.2012.02.022

Hanahan D, Folkman J (1996) Patterns and emerging mechanisms of the angiogenic switch during tumorigenesis. Cell 86(3):353–364. https://doi.org/10.1016/s0092-8674(00)80108-7

Hanahan D, Monje M (2023) Cancer hallmarks intersect with neuroscience in the tumor microenvironment. Cancer Cell 41(3):573–580. https://doi.org/10.1016/j.ccell.2023.02.012

Hanahan D, Weinberg RA (2000) The hallmarks of cancer. Cell 100(1):57–70. https://doi.org/10.1016/s0092-8674(00)81683-9

Hanahan D, Weinberg RA (2011) Hallmarks of cancer: the next generation. Cell 144(5):646–674. https://doi.org/10.1016/j.cell.2011.02.013

Hanselmann RG, Welter C (2016) Origin of cancer: an information, energy, and matter disease. Front Cell Dev Biol 4:121. https://doi.org/10.3389/fcell.2016.00121

Hanson S, Thorpe G, Winstanley L et al (2020) Omega-3, omega-6 and total dietary polyunsaturated fat on cancer incidence: systematic review and meta-analysis of randomised trials. Br J Cancer 122(8):1260–1270. https://doi.org/10.1038/s41416-020-0761-6

Hao Y, Samuels Y, Li Q et al (2016) Oncogenic PIK3CA mutations reprogram glutamine metabolism in colorectal cancer. Nat Commun 7:11971. https://doi.org/10.1038/ncomms11971

Hao G, Xu ZP, Li L (2018) Manipulating extracellular tumour pH: an effective target for cancer therapy. RSC Adv 8(39):22182–22192. https://doi.org/10.1039/C8RA02095G

Haq R, Shoag J, Andreu-Perez P et al (2013) Oncogenic BRAF regulates oxidative metabolism via PGC1α and MITF. Cancer Cell 23(3):302–315. https://doi.org/10.1016/j.ccr.2013.02.003

Harding JJ, Telli M, Munster P et al (2021) A phase I dose-escalation and expansion study of telaglenastat in patients with advanced or metastatic solid tumors. Clin Cancer Res 27(18):4994–5003. https://doi.org/10.1158/1078-0432.CCR-21-1204

Harris H (1999) The Birth of a Cell. Yale University Press, New Haven

Harris JR, Hjelmborg J, Adami HO et al (2019) The Nordic Twin Study on Cancer – NorTwinCan. Twin Res Hum Genet 22(6):817–823. https://doi.org/10.1017/thg.2019.71

Hartmann K, Egberink H, Möstl K et al (2023) Feline injection-site sarcoma and other adverse reactions to vaccination in cats. Viruses 15(8):1708. https://doi.org/10.3390/v15081708

Haslem DS, Chakravarty I, Fulde G et al (2018) Precision oncology in advanced cancer patients improves overall survival with lower weekly healthcare costs. Oncotarget 9(15):12316–12322. https://doi.org/10.18632/oncotarget.24384

Hastings PJ, Slack A, Petrosino JF, Rosenberg SM (2004) Adaptive amplification and point mutation are independent mechanisms: evidence for various stress-inducible mutation mechanisms. PLoS Biol 2(12):e399. https://doi.org/10.1371/journal.pbio.0020399

Hatakeyama K, Nagashima T, Ohshima K et al (2020) Characterization of tumors with ultralow tumor mutational burden in Japanese cancer patients. Cancer Sci 111(10):3893–3901. https://doi.org/10.1111/cas.14572

Hausmann D, Hoffmann DC, Venkataramani V et al (2023) Autonomous rhythmic activity in glioma networks drives brain tumour growth. Nature 613(7942):179–186. https://doi.org/10.1038/s41586-022-05520-4

Hausser J, Alon U (2020) Tumour heterogeneity and the evolutionary trade-offs of cancer. Nat Rev Cancer 20(4):247–257. https://doi.org/10.1038/s41568-020-0241-6

Hausser J, Szekely P, Bar N et al (2019) Tumor diversity and the trade-off between universal cancer tasks. Nat Commun 10(1):5423. https://doi.org/10.1038/s41467-019-13195-1

Hay Mele B, Monticelli M, Leone S et al (2023) Oxidoreductases and metal cofactors in the functioning of the earth. Essays Biochem 67(4):653–670. https://doi.org/10.1042/EBC20230012

Hays MT, Watson EE, Thomas ST, Stabin M (2002) MIRD Dose Estimate Report No. 19: Radiation Absorbed Dose Estimates from FDG-18F. J Nucl Med 43:210–214

Hazini A, Fisher K, Seymour L (2021) Deregulation of HLA-I in cancer and its central importance for immunotherapy. J Immunother Cancer 9(8):e002899. https://doi.org/10.1136/jitc-2021-002899

He XY, Gao Y, Ng D et al (2024) Chronic stress increases metastasis via neutrophil-mediated changes to the microenvironment. Cancer Cell 42(3):474–486.e12. https://doi.org/10.1016/j.ccell.2024.01.013

Heckman TM, Best PN (2023) A global inventory of feedback. Galaxies 11:21. https://doi.org/10.3390/galaxies11010021

Hegde GV, de la Cruz C, Giltnane JM et al (2019) NRG1 is a critical regulator of differentiation in TP63-driven squamous cell carcinoma. Elife 8:e46551. https://doi.org/10.7554/eLife.46551

Heldin CH, Rubin K, Pietras K, Ostman A (2004) High interstitial fluid pressure – an obstacle in cancer therapy. Nat Rev Cancer 4(10):806–813. https://doi.org/10.1038/nrc1456

Helman E, Lawrence MS, Stewart C et al (2014) Somatic retrotransposition in human cancer revealed by whole-genome and exome sequencing. Genome Res 24(7):1053–1063. https://doi.org/10.1101/gr.163659.113

Hendrick MJ, Goldschmidt MH (1991) Do injection site reactions induce fibrosarcomas in cats? J Am Vet Med Assoc 199(8):968. PMID: 1748617

Heng HH (2016) Debating cancer the paradox in cancer research. World Scientific, Singapore

Heng HH (2019) Genome chaos: rethinking genetics, evolution and molecular medicine. Academic/Elsevier Inc

Heng DY, Wells JC, Rini BI et al (2014) Cytoreductive nephrectomy in patients with synchronous metastases from renal cell carcinoma: results from the International Metastatic Renal Cell Carcinoma Database Consortium. Eur Urol 66(4):704–710. https://doi.org/10.1016/j.eururo.2014.05.034

Heo J, Jo Y, Yoon M (2024) Synergistic effects of combined hyperthermia and electric fields treatment in non-small cell lung-cancer (NSCLC) cell lines. Clin Transl Oncol. https://doi.org/10.1007/s12094-024-03760-6

Heppner GH (1984) Tumor heterogeneity. Cancer Res 44(6):2259–2265. PMID: 6372991

Heppner GH, Shekhar M (2014) Tumor heterogeneity is fundamental to the tumor ecosystem. Oncology 28(9)

Hermida M (2016) Life on Earth is an individual. Theory Biosci 135(1–2):37–44. https://doi.org/10.1007/s12064-016-0221-2

Herms A, Jones PH (2023) Somatic mutations in normal tissues: new perspectives on early carcinogenesis. Annu Rev Cancer Biol Ser 7(1). https://doi.org/10.1146/annurev-cancerbio-061421-012447

Hernandez-Boussard T, Macklin P, Greenspan EJ et al (2021) Digital twins for predictive oncology will be a paradigm shift for precision cancer care. Nat Med. 27(12):2065–2066. https://doi.org/10.1038/s41591-021-01558-5

Hess N, Joseph JA (2025) Structured protein domains enter the spotlight: modulators of biomolecular condensate form and function. Trends Biochem Sci 50(3):206–223. https://doi.org/10.1016/j.tibs.2024.12.008

Hessvik NP, Llorente A (2018) Current knowledge on exosome biogenesis and release. Cell Mol Life Sci 75(2):193–208. https://doi.org/10.1007/s00018-017-2595-9

Heylighen F (1999) Growth of complexity. In: Heylighen F, Bollen J, Riegler A (eds) The evolution of complexity. Kluwer Academic, Dordrecht, pp 17–44

Heylighen F, Busseniers E (2023) Modeling autopoiesis and cognition with reaction networks. Biosystems 230:104937. https://doi.org/10.1016/j.biosystems.2023.104937

Hicks AM, Riedlinger G, Willingham MC et al (2006) Transferable anticancer innate immunity in spontaneous regression/complete resistance mice. Proc Natl Acad Sci U S A 103(20):7753–7758. https://doi.org/10.1073/pnas.0602382103

Higginson J, Muir CS (1977) Détermination de l'importance des facteurs environnementaux dans le cancer humain: Rôle de l'épidémiologie [Determination of the importance of environmental factors in human cancer: the role of epidemiology]. Bull Cancer 64(3):365–384. French. PMID: 922173

Hijazi MA, Gessner A, El-Najjar N (2023) Repurposing of chronically used drugs in cancer therapy: a chance to grasp. Cancers (Basel) 15(12):3199. https://doi.org/10.3390/cancers15123199

Hill AB (1965) The environment and disease: association or causation? Proc R Soc Med 58(5):295–300. https://doi.org/10.1177/003591576505800503

Himeoka Y, Kameko K (2014) Entropy production of a steady-growth cell with catalytic reactions. Phys Rev E 90(4-1):042714. https://doi.org/10.1103/PhysRevE.90.042714

Hinshaw DC, Shevde LA (2019) The tumor microenvironment innately modulates cancer progression. Cancer Res 79(18):4557–4566. https://doi.org/10.1158/0008-5472.CAN-18-3962

Hiratsuka S, Nakamura K, Iwai S et al (2002) MMP9 induction by vascular endothelial growth factor receptor-1 is involved in lung-specific metastasis. Cancer Cell 2(4):289–300. https://doi.org/10.1016/s1535-6108(02)00153-8

Hiratsuka S, Watanabe A, Sakurai Y et al (2008) The S100A8-serum amyloid A3-TLR4 paracrine cascade establishes a pre-metastatic phase. Nat Cell Biol 10(11):1349–1355. https://doi.org/10.1038/ncb1794

Hirpara A, Bloomfield M, Duesberg P (2018) Speciation theory of carcinogenesis explains karyotypic individuality and long latencies of cancers. Genes (Basel) 9(8):402. https://doi.org/10.3390/genes9080402

His M, Viallon V, Dossus L et al (2019) Prospective analysis of circulating metabolites and breast cancer in EPIC. BMC Med 17(1):178. https://doi.org/10.1186/s12916-019-1408-4

Hjelmborg J, Korhonen T, Holst K et al (2017) Lung cancer, genetic predisposition and smoking: the Nordic Twin Study of Cancer. Thorax 72:1021–1027

Ho MW (1998) Rainbow and the worm: the physics of organisms, 2nd edn. World Scientific Publishing Company, Singapore

Hoadley KA, Yau C, Wolf DM et al (2014) Multiplatform analysis of 12 cancer types reveals molecular classification within and across tissues of origin. Cell 158(4):929–944. https://doi.org/10.1016/j.cell.2014.06.049

Hoadley KA, Yau C, Hinoue T et al (2018) Cell-of-origin patterns dominate the molecular classification of 10,000 tumors from 33 types of cancer. Cell 173(2):291–304.e6. https://doi.org/10.1016/j.cell.2018.03.022

Hoang T, Kim M, Park JW et al (2023) Dysbiotic microbiome variation in colorectal cancer patients is linked to lifestyles and metabolic diseases. BMC Microbiol 23(1):33. https://doi.org/10.1186/s12866-023-02771-7

Hochhaus A, Larson RA, Guilhot F et al (2017) Long-term outcomes of imatinib treatment for chronic myeloid leukemia. N Engl J Med 376(10):917–927. https://doi.org/10.1056/NEJMoa1609324

Hoda MR, Keely SJ, Bertelsen LS et al (2007) Leptin acts as a mitogenic and antiapoptotic factor for colonic cancer cells. Br J Surg 94(3):346–354. https://doi.org/10.1002/bjs.5530

Hoekstra HE, Coyne JA (2007) The locus of evolution: evo devo and the genetics of adaptation. Evolution 61(5):995–1016. https://doi.org/10.1111/j.1558-5646.2007.00105.x

Hoeppner J, Bronsert P (2021) Metastasis and tumor cell migration of solid tumors. Cancers (Basel) 13(21):5576. https://doi.org/10.3390/cancers13215576

Hoevenaar WHM, Janssen A, Quirindongo AI et al (2020) Degree and site of chromosomal instability define its oncogenic potential. Nat Commun 11(1):1501. https://doi.org/10.1038/s41467-020-15279-9

Hoffmeyer J (1998) Surfaces inside surfaces: on the origin of agency and life. Cybern Hum Knowing 5(1):33–42

Hoffmeyer J (2008) Biosemiotics: an examination in the signs of life and life of signs. University of Scranton Press, Scranton

Hogg D, Hill A, Márquez-Rodas I, et al. (n.d.) [Entry incomplete – please provide full citation for proper formatting]

Holdhoff M, Ye X, Supko JG et al (2017) Timed sequential therapy of the selective T-type calcium channel blocker mibefradil and temozolomide in patients with recurrent high-grade gliomas. Neuro Oncol 19(6):845–852. https://doi.org/10.1093/neuonc/nox020

Holdredge C (2021) Seeing the animal whole: And why it matters. Lindisfarne Books, Great Barrington

Holmes FL (1986) Claude Bernard, the milieu intérieur, and regulatory physiology. Hist Philos Life Sci 8(1):3–25. PMID: 3534926

Holohan C, Van Schaeybroeck S, Longley DB, Johnston PG (2013) Cancer drug resistance: an evolving paradigm. Nat Rev Cancer 13(10):714–726. https://doi.org/10.1038/nrc3599

Holt-Lunstad J, Smith TB, Baker M et al (2015) Loneliness and social isolation as risk factors for mortality: a meta-analytic review. Perspect Psychol Sci 10(2):227–237. https://doi.org/10.1177/1745691614568352

Honda K, Littman DR (2016) The microbiota in adaptive immune homeostasis and disease. Nature 535(7610):75–84. https://doi.org/10.1038/nature18848

Hong D, Messier TL, Tye CE et al (2017) Runx1 stabilizes the mammary epithelial cell phenotype and prevents epithelial to mesenchymal transition. Oncotarget 8(11):17610–17627. https://doi.org/10.18632/oncotarget.15381

Hong P, Kudulaiti N, Wu S et al (2022) Tumor treating fields: a comprehensive overview of the underlying molecular mechanism. Expert Rev Mol Diagn 22(1):19–28. https://doi.org/10.1080/14737159.2022.2017283

Hooker CA (2011) Introduction to philosophy of complex systems. In: Hooker CA (ed) Philosophy of complex systems. Handbook of the philosophy of science, vol 10. North Holland/Elsevier, Amsterdam, pp 3–91

Hooker CA (2013) On the import of constraints in complex dynamical systems. Found Sci 18:757–780. https://doi.org/10.1007/s10699-012-9304-9

Hoong BYD, Gan YH, Liu H, Chen ES (2020) cGAS-STING pathway in oncogenesis and cancer therapeutics. Oncotarget 11(30):2930–2955. https://doi.org/10.18632/oncotarget.27673

Hoover G, Gilbert S, Curley O et al (2025) Nerve-to-cancer transfer of mitochondria during cancer metastasis. Nature. https://doi.org/10.1038/s41586-025-09176-8

Hordijk W, Steel M (2018) Autocatalytic networks at the basis of life's origin and organization. Life (Basel) 8(4):62. https://doi.org/10.3390/life8040062

Horlings HM, Flanagan AM, Huntsman DG (2015) Categorization of cancer through genomic complexity could guide research and management strategies. J Pathol 236(4):397–402. https://doi.org/10.1002/path.4542

Hosea R, Hillary S, Naqvi S et al (2024) The two sides of chromosomal instability: drivers and brakes in cancer. Signal Transduct Target Ther 9(1):75. https://doi.org/10.1038/s41392-024-01767-7

Hoshino A, Costa-Silva B, Shen TL et al (2015) Tumour exosome integrins determine organotropic metastasis. Nature 527(7578):329–335. https://doi.org/10.1038/nature15756

Hoshino A, Kim HS, Bojmar L et al (2020) Extracellular vesicle and particle biomarkers define multiple human cancers. Cell 182(4):1044–1061.e18. https://doi.org/10.1016/j.cell.2020.07.009

Hosios AM, Vander Heiden MG (2018) The redox requirements of proliferating mammalian cells. J Biol Chem 293(20):7490–7498. https://doi.org/10.1074/jbc.TM117.000239

Hosseini H, Obradović MMS, Hoffmann M et al (2016) Early dissemination seeds metastasis in breast cancer. Nature 540(7634):552–558. https://doi.org/10.1038/nature20785

Houghton J, Stoicov C, Nomura S et al (2004) Gastric cancer originating from bone marrow-derived cells. Science 306(5701):1568–1571. https://doi.org/10.1126/science.1099513

Howard JM, Pidgeon GP, Reynolds JV (2010) Leptin and gastro-intestinal malignancies. Obes Rev 11(12):863–874. https://doi.org/10.1111/j.1467-789X.2010.00718.x

Howe J, Rink JC, Wang B, Griffin AS (2022) Multicellularity in animals: The potential for within-organism conflict. Proc Natl Acad Sci U S A 119(32):e2120457119. https://doi.org/10.1073/pnas.2120457119

Hoyningen-Huene P (1994) Niels Bohr's argument for the irreducibility of biology to physics. In: Faye J, Folse HJ (eds) Niels Bohr and Contemporary Philosophy. Kluwer Academic Publishers, Dordrecht, pp 431–435

Hsiue EH, Wright KM, Douglass J et al (2021) Targeting a neoantigen derived from a common TP53 mutation. Science 371(6533):eabc8697. https://doi.org/10.1126/science.abc8697

Hu M, Yao J, Cai L et al (2005) Distinct epigenetic changes in the stromal cells of breast cancers. Nat Genet 37(8):899–905. https://doi.org/10.1038/ng1596

Hu Z, Ding J, Ma Z et al (2019) Quantitative evidence for early metastatic seeding in colorectal cancer. Nat Genet 51(7):1113–1122. https://doi.org/10.1038/s41588-019-0423-x

Hu Z, Li Z, Ma Z, Curtis C (2020) Multi-cancer analysis of clonality and the timing of systemic spread in paired primary tumors and metastases. Nat Genet 52(7):701–708. https://doi.org/10.1038/s41588-020-0628-z

Huang S, Ingber DE (2005) Cell tension, matrix mechanics, and cancer development. Cancer Cell 8(3):175–176. https://doi.org/10.1016/j.ccr.2005.08.009

Huang Y, Gao S, Wu S et al (2014) Multilayered molecular profiling supported the monoclonal origin of metastatic renal cell carcinoma. Int J Cancer 135(1):78–87. https://doi.org/10.1002/ijc.28654

Huang C, Wang Y, Li X et al (2020) Clinical features of patients infected with 2019 novel coronavirus in Wuhan. China. Lancet 395(10223):497–506. https://doi.org/10.1016/S0140-6736(20)30183-5

Huang B, Yan X, Li Y (2021) Cancer stem cell for tumor therapy. Cancers (Basel) 13(19):4814. https://doi.org/10.3390/cancers13194814

Huang W, Paul D, Calin GA, Bayraktar R (2023) miR-142: a master regulator in hematological malignancies and therapeutic opportunities. Cells 13(1):84. https://doi.org/10.3390/cells13010084

Huang L, Kang D, Zhao C, Liu X (2024) Correlation between surrogate endpoints and overall survival in unresectable hepatocellular carcinoma patients treated with immune checkpoint inhibitors: a systematic review and meta-analysis. Sci Rep 14(1):4327. https://doi.org/10.1038/s41598-024-54945-6

Huang S, Soto AM, Sonnenschein C (2025) The end of the genetic paradigm of cancer. PLoS Biol 23(3):e3003052. https://doi.org/10.1371/journal.pbio.3003052

Huang Z, Karaboué A, Zeng L et al (2025b) Overall survival according to time-of-day of combined immuno-chemotherapy for advanced non-small cell lung cancer: a bicentric bicontinental study. EBioMedicine 113:105607. https://doi.org/10.1016/j.ebiom.2025.105607

Hui D (2015) Prognostication of survival in patients with advanced cancer: predicting the unpredictable? Cancer Control 22(4):489–497. https://doi.org/10.1177/107327481502200415

Huo X, Zhang W, Zhao G et al (2022) FAK PROTAC inhibits ovarian tumor growth and metastasis by disrupting kinase dependent and independent pathways. Front Oncol 12:851065. https://doi.org/10.3389/fonc.2022.851065

Hurley S, Goldberg D, Von Behren J et al (2019) Chronotype and postmenopausal breast cancer risk among women in the California Teachers Study. Chronobiol Int 36(11):1504–1514. https://doi.org/10.1080/07420528.2019.1658113

Huston P, McFarlane B (2016) Health benefits of tai chi: what is the evidence? Can Fam Physician 62(11):881–890

Hutchinson L, Stenstrom B, Chen D et al (2011) Human Barrett's adenocarcinoma of the esophagus, associated myofibroblasts, and endothelium can arise from bone marrow-derived cells after allogeneic stem cell transplant. Stem Cells Dev 20(1):11–17. https://doi.org/10.1089/scd.2010.0139

Hutchison CA 3rd, Chuang RY, Noskov VN et al (2016) Design and synthesis of a minimal bacterial genome. Science 351(6280):aad6253. https://doi.org/10.1126/science.aad6253

Huysentruyt LC, Seyfried TN (2010) Perspectives on the mesenchymal origin of metastatic cancer. Cancer Metastasis Rev 29(4):695–707. https://doi.org/10.1007/s10555-010-9254-z

Iannolo G, Conticello C, Memeo L, De Maria R (2008) Apoptosis in normal and cancer stem cells. Crit Rev Oncol Hematol 66(1):42–51. https://doi.org/10.1016/j.critrevonc.2007.09.004

Ibrahim-Hashim A, Robertson-Tessi M, Enriquez-Navas PM et al (2017) Defining cancer subpopulations by adaptive strategies rather than molecular properties provides novel insights into intratumoral evolution. Cancer Res 77(9):2242–2254. https://doi.org/10.1158/0008-5472.CAN-16-2844

ICGC/TCGA Pan-Cancer Analysis of Whole Genomes Consortium (2020) Pan-cancer analysis of whole genomes. Nature 578(7793):82–93. https://doi.org/10.1038/s41586-020-1969-6

Igarashi H, Fukushi M, Nago N (2020) Oxygen use and survival in patients with advanced cancer and low oxygen saturation in home care: a preliminary retrospective cohort study. BMC Palliat Care 19(1):3. https://doi.org/10.1186/s12904-019-0511-9

Iida N, Dzutsev A, Stewart CA et al (2013) Commensal bacteria control cancer response to therapy by modulating the tumor microenvironment. Science 342(6161):967–970. https://doi.org/10.1126/science.1240527

Ilaiwy A, Ten Have GAM, Bain JR et al (2019) Identification of metabolic changes in ileum, jejunum, skeletal muscle, liver, and lung in a continuous I.V. Pseudomonas aeruginosa model of sepsis using nontargeted metabolomics analysis. Am J Pathol 189(9):1797–1813. https://doi.org/10.1016/j.ajpath.2019.05.021

Illidge TM, Cragg MS, Fringes B, Olive P, Erenpreisa JA (2000) Polyploid giant cells provide a survival mechanism for p53 mutant cells after DNA damage. Cell Biol Int 24(9):621–633. https://doi.org/10.1006/cbir.2000.0557

Inamura K, Hamada T, Bullman S et al (2022) Cancer as microenvironmental, systemic and environmental diseases: opportunity for transdisciplinary microbiomics science. Gut. https://doi.org/10.1136/gutjnl-2022-327209

Inan S, Hayran M (2019) Cell signaling pathways related to epithelial-mesenchymal transition in cancer metastasis. Crit Rev Oncog 24(1):47–54. https://doi.org/10.1615/CritRevOncog.2018029509

Ingber DE (2003a) Tensegrity I. Cell structure and hierarchical systems biology. J Cell Sci 116(7):1157–1173. https://doi.org/10.1242/jcs.00359

Ingber DE (2003b) Tensegrity II. How structural networks influence cellular information processing networks. J Cell Sci 116(8):1397–1408. https://doi.org/10.1242/jcs.00360

Ingber DE (2005) Mechanical control of tissue growth: function follows form. Proc Natl Acad Sci U S A 102(33):11571–11572. https://doi.org/10.1073/pnas.0505939102

Ingber DE (2008) Can cancer be reversed by engineering the tumor microenvironment? Semin Cancer Biol 18(5):356–364. https://doi.org/10.1016/j.semcancer.2008.03.016

Iranzo J, Martincorena I, Koonin EV (2018) Cancer-mutation network and the number and specificity of driver mutations. Proc Natl Acad Sci U S A 115(26):E6010–E6019. https://doi.org/10.1073/pnas.1803155115

Irep N, Inci K, Tokgun PE, Tokgun O (2024) Exosome inhibition improves response to first-line therapy in small cell lung cancer. J Cell Mol Med 28(4):e18138. https://doi.org/10.1111/jcmm.18138

Irigaray P, Newby JA, Clapp R et al (2007) Lifestyle-related factors and environmental agents causing cancer: an overview. Biomed Pharmacother 61(10):640–658. https://doi.org/10.1016/j.biopha.2007.10.006

Irimie AI, Zimta AA, Ciocan C et al (2018) The unforeseen non-coding RNAs in head and neck cancer. Genes (Basel) 9(3):134. https://doi.org/10.3390/genes9030134

Ishida Y (2004) The immune system as an information system. In: Ishida Y (ed) Immunity-Based Systems. Advanced Information Processing. Springer, Berlin/Heidelberg, pp 31–44. https://doi.org/10.1007/978-3-662-07863-1_3

Islami F, Goding Sauer A, Gapstur SM, Jemal A (2019) Proportion of cancer cases attributable to excess body weight by US state, 2011–2015. JAMA Oncol 5(3):384–392. https://doi.org/10.1001/jamaoncol.2018.5639

Israël M (2012) A primary cause of cancer: GABA deficiency in endocrine pancreas. Cancer Ther 8:171–183

Israël M (2014) Cancer metabolism: an alteration of the anabolic–catabolic selection switch. OA Cancer 2:1

Israël M (2019) Metabolic rewiring of stem cells and differentiated cells in cancer: the hypothetical consequences of a GABA deficiency in endocrine pancreas. J Cancer Metastasis Treat 5

Israël M (2020) The metabolic rewiring observed in cancer renders tumor cells dependent on ketone bodies and vulnerable to SCOT inhibition. Endocrinol Diabetes Metab J 4(1):1–13. https://doi.org/10.31038/EDMJ.2020412

Israel M, Schwartz L (2006) Cancer: a dysmethylation syndrome. John Libbey Eurotexte, Montrouge

Israël M, Schwartz L (2011) The metabolic advantage of tumor cells. Mol Cancer 10:70. https://doi.org/10.1186/1476-4598-10-70

Israel M, Schwartz L (2020) SCOT is a vital enzyme for tumors: with reference to Carney Triad cancers and the ketogenic diet. Trends Res 3:1000156. https://doi.org/10.15761/TR.1000156

Israël M, Berg E, Tenenbaum G (2024) Respiratory Complex II and SCOT-Oxct1 compete for succinyl-CoA in ketolytic dependent tumours. Br J Healthc Med Res 11(1):93–101. https://doi.org/10.14738/bjhmr.111.16217

Itzkowitz SH, Yio X (2004) Inflammation and cancer IV. Colorectal cancer in inflammatory bowel disease: the role of inflammation. Am J Physiol Gastrointest Liver Physiol 287(1):G7–G17. https://doi.org/10.1152/ajpgi.00079.2004

Ivanitskii GR (2010) 21st century: What is life from the perspective of physics? Phys Usp 53(4):327–356. https://doi.org/10.3367/UFNe.0180.201004a.0337

Ivanov PC, Bartsch RP (2014) Network physiology: mapping interactions between networks of physiologic networks. In: D'Agostino G, Scala A (eds) Networks of networks: the last frontier of complexity. Understanding complex systems. Springer, Cham. https://doi.org/10.1007/978-3-319-03518-5_10

Ivanova OM, Anufrieva KS, Kazakova AN et al (2023) Non-canonical functions of spliceosome components in cancer progression. Cell Death Dis 14(2):77. https://doi.org/10.1038/s41419-022-05470-9

Iwasa M, Ishihara T, Mifuji-Moroka R et al (2015) Elevation of branched-chain amino acid levels in diabetes and NAFL and changes with antidiabetic drug treatment. Obes Res Clin Pract 9(3):293–297. https://doi.org/10.1016/j.orcp.2015.01.003

Izumida Y, Yahagi N, Takeuchi Y et al (2013) Glycogen shortage during fasting triggers liver-brain-adipose neurocircuitry to facilitate fat utilization. Nat Commun 4:2316. https://doi.org/10.1038/ncomms3316

Jachowicz JW, Bing X, Pontabry J et al (2017) LINE-1 activation after fertilization regulates global chromatin accessibility in the early mouse embryo. Nat Genet 49(10):1502–1510. https://doi.org/10.1038/ng.3945

Jacquemet G, Baghirov H, Georgiadou M et al (2016) L-type calcium channels regulate filopodia stability and cancer cell invasion downstream of integrin signalling. Nat Commun 7:13297. https://doi.org/10.1038/ncomms13297

Jaeger H (2021) Toward a generalized theory comprising digital, neuromorphic, and unconventional computing. Neuromorph Comput Eng 1:012002. https://doi.org/10.1088/2634-4386/abf151

Jagers op Akkerhuis G (2016) Evolution and transitions in complexity: the science of hierarchical organization in nature. Springer, Dordrecht

Jagers op Akkerhuis G (2023) Dual closure, a new tool for recognizing thresholds between levels of organization in big history. J Big Hist 6(3):48–58. https://doi.org/10.22339/jbh.v6i3.6305

Jagers op Akkerhuis G (2024) The third law of evolution and the future of life. Springer, Dordrecht

Jain S, Gautam V, Naseem S (2011) Acute-phase proteins: as diagnostic tool. J Pharm Bioallied Sci 3(1):118–127. https://doi.org/10.4103/0975-7406.76489

Jaini S, Dadachova E (2012) FDG for therapy of metabolically active tumors. Semin Nucl Med 42(3):185–189. https://doi.org/10.1053/j.semnuclmed.2011.12.001

Jaiswal S, Ebert BL (2019) Clonal hematopoiesis in human aging and disease. Science 366(6465):eaan4673. https://doi.org/10.1126/science.aan4673

Jamal A, Arif A, Shahid MD, et al (2025) Plant secondary metabolites inhibit cancer by targeting epidermal growth factor receptor (EGFR): an updated review on their regulation and mechanisms of action. https://doi.org/10.31557/apjcb.2025.10.1.191-206

Jamal-Hanjani M, Wilson GA, McGranahan N et al (2017) Tracking the evolution of non-small-cell lung cancer. N Engl J Med 376(22):2109–2121. https://doi.org/10.1056/NEJMoa1616288

Jameera Begam A, Jubie S, Nanjan MJ (2017) Estrogen receptor agonists/antagonists in breast cancer therapy: a critical review. Bioorg Chem 71:257–274. https://doi.org/10.1016/j.bioorg.2017.02.011

Jamshidi A, Liu MC, Klein EA et al (2022) Evaluation of cell-free DNA approaches for multi-cancer early detection. Cancer Cell 40(12):1537–1549.e12. https://doi.org/10.1016/j.ccell.2022.10.022

Jansson JO, Palsdottir V, Hägg DA et al (2018) Body weight homeostat that regulates fat mass independently of leptin in rats and mice. Proc Natl Acad Sci USA 115(2):427–432. https://doi.org/10.1073/pnas.1715687114

Jassim A, Rahrmann EP, Simons BD, Gilbertson RJ (2023) Cancers make their own luck: theories of cancer origins. Nat Rev Cancer 23(10):710–724. https://doi.org/10.1038/s41568-023-00602-5

Jaynes ET (1965) Gibbs vs. Boltzmann entropies. Am J Phys 33:391–398. https://doi.org/10.1119/1.1971557

Jeffery K, Pollack R, Rovelli C (2019) On the statistical mechanics of life: Schrödinger revisited. Entropy (Basel) 21(12):1211. https://doi.org/10.3390/e21121211

Jenkins BH, Buckingham JF, Hanley CJ, Thomas GJ (2022) Targeting cancer-associated fibroblasts: challenges, opportunities and future directions. Pharmacol Ther 240:108231. https://doi.org/10.1016/j.pharmthera.2022.108231

Jenkins HA, Lloyd AL, Kembro JM, Lloyd D (2025) Time and function in the living world. Adv Microb Physiol. 86:331–375. https://doi.org/10.1016/bs.ampbs.2025.02.001

Jessy T (2011) Immunity over inability: the spontaneous regression of cancer. J Nat Sci Biol Med 2(1):43–49. https://doi.org/10.4103/0976-9668.82318

Jiang J, Srivastava S, Zhang J (2019) Starve cancer cells of glutamine: break the spell or make a hungry monster? Cancers (Basel) 11(6):804. https://doi.org/10.3390/cancers11060804

Jiang J, Li J, Zhou X, Zhao X, Huang B, Qin Y (2022) Exosomes regulate the epithelial-mesenchymal transition in cancer. Front Oncol 12:864980. https://doi.org/10.3389/fonc.2022.864980

Jiang Z, Ju YJ, Ali A et al (2023) Thinking (metastasis) outside the (primary tumor) box. Cancers (Basel) 15(22):5315. https://doi.org/10.3390/cancers15225315

Jiao Y, Lv Q (2023) Does primary tumor resection induce accelerated metastasis in breast cancer? A review. J Surg Res 283:1005–1017. https://doi.org/10.1016/j.jss.2022.11.064

Jimenez H, Wang M, Zimmerman JW et al (2019) Tumour-specific amplitude-modulated radiofrequency electromagnetic fields induce differentiation of hepatocellular carcinoma via targeting Cav3.2 T-type voltage-gated calcium channels and Ca2+ influx. EBioMedicine 44:209–224. https://doi.org/10.1016/j.ebiom.2019.05.034

Jobling P, Pundavela J, Oliveira SM et al (2015) Nerve-cancer cell cross-talk: a novel promoter of tumor progression. Cancer Res 75(9):1777–1781. https://doi.org/10.1158/0008-5472.CAN-14-3180

Johannsen W (2015) On semantic information in nature. Information 6(3):411–431. https://doi.org/10.3390/info6030411

Johnson G, Sallé A, Lorimier G et al (2008) Cancer cachexia: measured and predicted resting energy expenditures for nutritional needs evaluation. Nutrition 24(5):443–450. https://doi.org/10.1016/j.nut.2008.01.013

Johnston RJ, Comps-Agrar L, Hackney J et al (2014) The immunoreceptor TIGIT regulates antitumor and antiviral CD8(+) T cell effector function. Cancer Cell 26(6):923–937. https://doi.org/10.1016/j.ccell.2014.10.018

Jolly D, Meyer J (2009) A brief review of radiation hormesis. Australas Phys Eng Sci Med 32(4):180–187. https://doi.org/10.1007/BF03179237

José JV, Saletan EJ (1998) Classical Dynamics: A Contemporary Approach. Cambridge University Press

Joshi S, Durden DL (2019) Combinatorial approach to improve cancer immunotherapy: rational drug design strategy to simultaneously hit multiple targets to kill tumor cells and to activate the immune system. J Oncol 2019:5245034. https://doi.org/10.1155/2019/5245034

Joshi CJ, Ke W, Drangowska-Way A et al (2022) What are housekeeping genes? PLoS Comput Biol 18(7):e1010295. https://doi.org/10.1371/journal.pcbi.1010295

Joyce JA, Fearon DT (2015) T cell exclusion, immune privilege, and the tumor microenvironment. Science 348(6230):74–80. https://doi.org/10.1126/science.aaa6204

Joyce JA, Pollard JW (2009) Microenvironmental regulation of metastasis. Nat Rev Cancer 9(4):239–252. https://doi.org/10.1038/nrc2618

June CH, Sadelain M (2018) Chimeric antigen receptor therapy. N Engl J Med 379(1):64–73. https://doi.org/10.1056/NEJMra1706169

Jurj A, Zanoaga O, Braicu C et al (2020) A comprehensive picture of extracellular vesicles and their contents. Molecular transfer to cancer cells. Cancers (Basel) 12(2):298. https://doi.org/10.3390/cancers12020298

Kaadige MR, Looper RE, Kamalanaadhan S, Ayer DE (2009) Glutamine-dependent anapleurosis dictates glucose uptake and cell growth by regulating MondoA transcriptional activity. Proc Natl Acad Sci U S A 106(35):14878–14883. https://doi.org/10.1073/pnas.0901221106

Kach J, Conzen SD, Szmulewitz RZ (2015) Targeting the glucocorticoid receptor in breast and prostate cancers. Sci Transl Med 7(305):305ps19. https://doi.org/10.1126/scitranslmed.aac7531

Kakizaki M, Kuriyama S, Sone T et al (2008) Sleep duration and the risk of breast cancer: the Ohsaki Cohort Study. Br J Cancer 99(9):1502–1505. https://doi.org/10.1038/sj.bjc.6604684

Kalam F, James DL, Li YR et al (2023) (2023) Intermittent fasting interventions to leverage metabolic and circadian mechanisms for cancer treatment and supportive care outcomes. J Natl Cancer Inst Monogr 61:84–103. https://doi.org/10.1093/jncimonographs/lgad008

Kalluri R (2016a) The biology and function of exosomes in cancer. J Clin Invest 126(4):1208–1215. https://doi.org/10.1172/JCI81135

Kalluri R (2016b) The biology and function of fibroblasts in cancer. Nat Rev Cancer 16(9):582–598. https://doi.org/10.1038/nrc.2016.73

Kalluri R, LeBleu VS (2020) The biology, function, and biomedical applications of exosomes. Science 367(6478):eaau6977. https://doi.org/10.1126/science.aau6977

Kalyanaraman B, Cheng G, Hardy M et al (2018) Teaching the basics of reactive oxygen species and their relevance to cancer biology: mitochondrial reactive oxygen species detection, redox signaling, and targeted therapies. Redox Biol 15:347–362. https://doi.org/10.1016/j.redox.2017.12.012

Kamb M and Garguli S (2025) An analytic theory of creativity in convolutional diffusion models arXiv:2412.20292v2. https://doi.org/10.48550/arXiv.2412.20292

Kanarek N, Petrova B, Sabatini DM (2020) Dietary modifications for enhanced cancer therapy. Nature 579(7800):507–517. https://doi.org/10.1038/s41586-020-2124-0

Kannisto T (2017) Kant on the necessity of causal relations. Kant Studien 108(4)

Kanter NG, Cohen-Woods S, Balfour DA et al (2024) Hypothalamic-pituitary-adrenal axis dysfunction in people with cancer: a systematic review. Cancer Med 13(22):e70366. https://doi.org/10.1002/cam4.70366

Kantoff PW, Higano CS, Shore ND et al (2010) Sipuleucel-T immunotherapy for castration-resistant prostate cancer. N Engl J Med 363(5):411–422. https://doi.org/10.1056/NEJMoa1001294

Kaplan RN, Riba RD, Zacharoulis S et al (2005) VEGFR1-positive haematopoietic bone marrow progenitors initiate the pre-metastatic niche. Nature 438(7069):820–827. https://doi.org/10.1038/nature04186

Kaplan RN, Psaila B, Lyden D (2006) Bone marrow cells in the 'pre-metastatic niche': within bone and beyond. Cancer Metastasis Rev 25(4):521–529. https://doi.org/10.1007/s10555-006-9036-9

Karaboué A, Innominato PF, Wreglesworth NI et al (2024) Why does circadian timing of administration matter for immune checkpoint inhibitors' efficacy? Br J Cancer. 131(5):783–796. https://doi.org/10.1038/s41416-024-02704-9

Karadas AK, Dilmac S, Aytac G, Tanriover G (2021) Melatonin decreases metastasis, primary tumor growth and angiogenesis in a mice model of breast cancer. Hum Exp Toxicol 40(9):1545–1557. https://doi.org/10.1177/09603271211002883

Karagiannis GS, Condeelis JS, Oktay MH (2018) Chemotherapy-induced metastasis: mechanisms and translational opportunities. Clin Exp Metastasis 35(4):269–284. https://doi.org/10.1007/s10585-017-9870-x

Karampoga A, Tzaferi K, Koutsakis C et al (2022) Exosomes and the extracellular matrix: a dynamic interplay in cancer progression. Int J Dev Biol 66(1-2-3):97–102. https://doi.org/10.1387/ijdb.210120nk

Karimi, Dermani F, Datta I, Gholamzadeh Khoei S (2023) MicroRNA-452: a double-edged sword in multiple human cancers. Clin Transl Oncol 25(5):1189–1206. https://doi.org/10.1007/s12094-022-03041-0

Karin M (2006) Nuclear factor-kappaB in cancer development and progression. Nature 441(7092):431–436. https://doi.org/10.1038/nature04870

Karlsson K, Przybilla MJ, Kotler E et al (2023) Deterministic evolution and stringent selection during preneoplasia. Nature 618(7964):383–393. https://doi.org/10.1038/s41586-023-06102-8

Kartikasari AER, Huertas CS, Mitchell A, Plebanski M (2021) Tumor-induced inflammatory cytokines and the emerging diagnostic devices for cancer detection and prognosis. Front Oncol 11:692142. https://doi.org/10.3389/fonc.2021.692142

Kasas S, Ruggeri FS, Benadiba C et al (2015) Detecting nanoscale vibrations as signature of life. Proc Natl Acad Sci U S A 112(2):378–381. https://doi.org/10.1073/pnas.1415348112

Katagiri R, Goto A, Nakagawa T et al (2018) Increased levels of branched-chain amino acid associated with increased risk of pancreatic cancer in a prospective case-control study of a large cohort. Gastroenterology 155(5):1474–1482.e1. https://doi.org/10.1053/j.gastro.2018.07.033

Katayama Y, Battista M, Kao WM et al (2006) Signals from the sympathetic nervous system regulate hematopoietic stem cell egress from bone marrow. Cell 124(2):407–421. https://doi.org/10.1016/j.cell.2005.10.041

Kather JN, Halama N, Marron TU (2020) Computational pathology opens new avenues for precision oncology. Nat Rev Clin Oncol 17(11):703–715. https://doi.org/10.1038/s41571-020-0417-7

Kato S, Lippman SM, Flaherty KT, Kurzrock R (2016) The conundrum of genetic "drivers" in benign conditions. J Natl Cancer Inst 108(8):djw036. https://doi.org/10.1093/jnci/djw036

Kato Y, Kunimasa K, Sugimoto Y, Tomida A (2018) BCR-ABL tyrosine kinase inhibition induces metabolic vulnerability by preventing the integrated stress response in K562 cells. Biochem Biophys Res Commun 504(4):721–726. https://doi.org/10.1016/j.bbrc.2018.09.032

Katzmarzyk PT, Powell KE, Jakicic JM (2019) Sedentary behavior and health: update from the 2018 physical activity guidelines advisory committee. Med Sci Sports Exerc 51(6):1227–1241. https://doi.org/10.1249/MSS.0000000000001935

Kauffman SA (1971) Cellular homeostasis, epigenesis and replication in randomly aggregated macromolecular systems. J Cybern 1:71–96. https://doi.org/10.1080/01969727108545830

Kauffman SA (1993) The Origins of Order: Self-Organization and Selection in Evolution. Oxford University Press, Oxford

Kauffman S (2000) Investigations. Oxford University Press, Oxford

Kauffman S (2002) Investigations Oxford University Press

Kauffman SA (2016) Humanity in a creative universe. Oxford University Press, New York

Kauffman SA (2019) A world beyond physics: the emergence and evolution of life. Oxford University Press, New York

Kauffman S (2020) Eros and logos. Angelaki 25(3):9–23. https://doi.org/10.1080/0969725X.2020.1754011

Kauffman S (2020b) Answering Schrödinger's "what is life?". Entropy 22(8):815. https://doi.org/10.3390/e22080815

Kauffman S, Roli A (2021) The world is not a theorem. Entropy (Basel) 23(11):1467. https://doi.org/10.3390/e23111467

Kaufman CK, Mosimann C, Fan ZP et al (2016) A zebrafish melanoma model reveals emergence of neural crest identity during melanoma initiation. Science 351(6272):aad2197. https://doi.org/10.1126/science.aad2197

Kaukonen KM, Bailey M, Pilcher D, Cooper DJ, Bellomo R (2015) Systemic inflammatory response syndrome criteria in defining severe sepsis. N Engl J Med 372(17):1629–1638. https://doi.org/10.1056/NEJMoa1415236

Kawanishi S, Hiraku Y, Pinlaor S, Ma N (2006) Oxidative and nitrative DNA damage in animals and patients with inflammatory diseases in relation to inflammation-related carcinogenesis. J Biol Chem 387(4):365–372. https://doi.org/10.1515/BC.2006.049

Kawanishi S, Ohnishi S, Ma N et al (2017) Crosstalk between DNA damage and inflammation in the multiple steps of carcinogenesis. Int J Mol Sci 18(8):1808. https://doi.org/10.3390/ijms18081808

Kays JK, Shahda S, Stanley M et al (2018) Three cachexia phenotypes and the impact of fat-only loss on survival in FOLFIRINOX therapy for pancreatic cancer. J Cachexia Sarcopenia Muscle 9(4):673–684. https://doi.org/10.1002/jcsm.12307

Kazazian HH Jr, Moran JV (2017) Mobile DNA in health and disease. N Engl J Med 377(4):361–370. https://doi.org/10.1056/NEJMra1510092

Kaznatcheev A, Vander Velde R, Scott JG, Basanta D (2017) Cancer treatment scheduling and dynamic heterogeneity in social dilemmas of tumour acidity and vasculature. Br J Cancer 116(6):785–792. https://doi.org/10.1038/bjc.2017.5

Kelm JM, Pandey DS, Malin E et al (2023) PROTAC'ing oncoproteins: targeted protein degradation for cancer therapy. Mol Cancer 22:62. https://doi.org/10.1186/s12943-022-01707-5

Kempes CP, Krakauer DC (2021) The multiple paths to multiple life. J Mol Evol 89(7):415–426. https://doi.org/10.1007/s00239-021-10016-2

Key T, Spencer E (2007) Carbohydrates and cancer: an overview of the epidemiological evidence. Eur J Clin Nutr 61(S1):S112–S121. https://doi.org/10.1038/sj.ejcn.1602941

Khafizov R, Piazza E, Cui Y et al (2024) Sub-cellular imaging of the entire protein-coding human transcriptome (18933-plex) on FFPE tissue using spatial molecular imaging. bioRxiv. https://doi.org/10.1101/2024.11.27.625536

Khanmammadova N, Islam S, Sharma P, Amit M (2023) Neuro-immune interactions and immuno-oncology. Trends. Cancer 9(8):636–649. https://doi.org/10.1016/j.trecan.2023.05.002

Khorami Sarvestani S, Shojaeian S, Sarrami-Forooshani R et al (2023) Cancer is associated with the emergence of placenta-reactive autoantibodies. Biomedicines 11(2):316. https://doi.org/10.3390/biomedicines11020316

Khorana AA, Francis CW, Culakova E, Kuderer NM, Lyman GH (2007) Thromboembolism is a leading cause of death in cancer patients receiving outpatient chemotherapy. J Thromb Haemost 5(3):632–634. https://doi.org/10.1111/j.1538-7836.2007.02374.x

Kim R, Emi M, Tanabe K (2007) Cancer immunoediting from immune surveillance to immune escape. Immunology 121(1):1–14. https://doi.org/10.1111/j.1365-2567.2007.02587.x

Kim MY, Oskarsson T, Acharyya S et al (2009) Tumor self-seeding by circulating cancer cells. Cell 139(7):1315–1326. https://doi.org/10.1016/j.cell.2009.11.025

Kim SY, Wie GA, Lee WJ et al (2013) Changes in dietary intake, body weight, nutritional status, and metabolic rate in a pancreatic cancer patient. Clin Nutr Res 2(2):154–158. https://doi.org/10.7762/cnr.2013.2.2.154

Kim CS, Han KD, Choi HS et al (2020) Association of hypertension and blood pressure with kidney cancer risk: a nationwide population-based cohort study. Hypertension 75(6):1439–1446. https://doi.org/10.1161/HYPERTENSIONAHA.120.14820

Kimerling R, Lewis ET, Javier SJ, Zulman DM (2020) Opportunity or burden? A behavioral framework for patient engagement. Med Care 58(2):161–168. https://doi.org/10.1097/MLR.0000000000001240

Kimura M (1968) Evolutionary rate at the molecular level. Nature 217(5129):624–626. https://doi.org/10.1038/217624a0

Kinj R, Muggeo E, Schiappacasse L et al (2022) Stereotactic body radiation therapy in patients with oligometastatic disease: clinical state of the art and perspectives. Cancers (Basel) 14(5):1152. https://doi.org/10.3390/cancers14051152

Kinouchi K, Sassone-Corsi P (2020) Metabolic rivalry: circadian homeostasis and tumorigenesis. Nat Rev Cancer 20(11):645–661. https://doi.org/10.1038/s41568-020-0291-9

Kinzler KW, Vogelstein B (1997) Cancer-susceptibility genes. Gatekeepers and caretakers. Nature 386(6627):761–763. https://doi.org/10.1038/386761a0

Kir S, White JP, Kleiner S et al (2014) Tumour-derived PTH-related protein triggers adipose tissue browning and cancer cachexia. Nature 513(7516):100–104. https://doi.org/10.1038/nature13528

Kirkwood TB (2005) Understanding the odd science of aging. Cell 120(4):437–447. https://doi.org/10.1016/j.cell.2005.01.027

Kirschner MW, Gerhart JC (2005) The Plausibility of Life. Yale University Press, New Haven

Kirschner M, Gerhart J (2006) The plausibility of life: resolving Darwin's dilemma. Yale University Press, New Haven

Kitano H (2004) Biological robustness. Nat Rev Genet 5(11):826–837. https://doi.org/10.1038/nrg1471

Klein M (1975) François-Vincent Raspail. In: Gillispie CC (ed) Dictionary of Scientific Biography, vol 11, pp 300. Scribner, New York

Klein CA (2009) Parallel progression of primary tumours and metastases. Nat Rev Cancer 9:302–312

Klein RM, Spofford LS, Abel EV et al (2008) B-RAF regulation of Rnd3 participates in actin cytoskeletal and focal adhesion organization. Mol Biol Cell 19(2):498–508. https://doi.org/10.1091/mbc.e07-09-0895

Klein EA, Richards D, Cohn A et al (2021) Clinical validation of a targeted methylation-based multi-cancer early detection test using an independent validation set. Ann Oncol 32(9):1167–1177. https://doi.org/10.1016/j.annonc.2021.05.806

Kliemann N, Murphy N, Viallon V et al (2020) Predicted basal metabolic rate and cancer risk in the European Prospective Investigation into Cancer and Nutrition. Int J Cancer 147(3):648–661. https://doi.org/10.1002/ijc.32753

Klim J, Zielenkiewicz U, Kaczanowski S (2024) Loss-of-function mutations are main drivers of adaptations during short-term evolution. Sci Rep 14:7128. https://doi.org/10.1038/s41598-024-57694-8

Knott SRV, Wagenblast E, Khan S et al (2018) Asparagine bioavailability governs metastasis in a model of breast cancer. Nature 554(7692):378–381

Knouse KA, Wu J, Whittaker CA, Amon A (2014) Single cell sequencing reveals low levels of aneuploidy across mammalian tissues. Proc Natl Acad Sci U S A 111(37):13409–13414. https://doi.org/10.1073/pnas.1415287111

Knudson AG (2001) Two genetic hits (more or less) to cancer. Nat Rev Cancer 1:157–162

Knudson AG Jr (1971) Mutation and cancer: statistical study of retinoblastoma. Proc Natl Acad Sci USA 68(4):820–823. https://doi.org/10.1073/pnas.68.4.820

Kobets AJ, Backus R, Fluss R et al (2000) Evaluating the natural growth rate of metastatic cancer to the brain. Surg Neurol Int 11:254. https://doi.org/10.25259/SNI_291_2020

Koch J, Hau J, Pravsgaard Christensen J et al (2013) Immune cells from SR/CR mice induce the regression of established tumors in BALB/c and C57BL/6 mice. PLoS One 8(3):e59995. https://doi.org/10.1371/journal.pone.0059995

Koestler A (1969) Some general properties of self-regulating open hierarchic order (SOHO). In: Koestler A, Smythies JR (eds) Beyond Reductionism: New Perspectives in the Life Sciences. Hutchinson, London

Koestler A, Smythies JR (1971) Beyond Reductionism: New Perspectives in the Life Sciences. Houghton Mifflin Co., Boston

Koh TJ, DiPietro LA (2011) Inflammation and wound healing: the role of the macrophage. Expert Rev Mol Med 13:e23. https://doi.org/10.1017/S1462399411001943

Koh A, De Vadder F, Kovatcheva-Datchary P, Bäckhed F (2016) From dietary fiber to host physiology: short-chain fatty acids as key bacterial metabolites. Cell 165(6):1332–1345. https://doi.org/10.1016/j.cell.2016.05.041

Koh G, Degasperi A, Zou X et al (2021) Mutational signatures: emerging concepts, caveats and clinical applications. Nat Rev Cancer 21(10):619–637. https://doi.org/10.1038/s41568-021-00377-7
Koh HB, Kim HJ, Kang SW, Yoo TH (2023) Exosome-based drug delivery: translation from bench to clinic. Pharmaceutics 15(8):2042. https://doi.org/10.3390/pharmaceutics15082042
Köhler C, Nittner D, Rambow F et al (2017) Mouse cutaneous melanoma induced by mutant BRaf arises from expansion and dedifferentiation of mature pigmented melanocytes. Cell Stem Cell 21(5):679–693.e6. https://doi.org/10.1016/j.stem.2017.08.003
Kokolus KM, Zhang Y, Sivik JM et al (2017) Beta blocker use correlates with better overall survival in metastatic melanoma patients and improves the efficacy of immunotherapies in mice. Oncoimmunology 7(3):e1405205. https://doi.org/10.1080/2162402X.2017.1405205
Kolchinsky A, Wolpert DH (2018) Semantic information, autonomous agency and non-equilibrium statistical physics. Interface Focus 8(6):20180041. https://doi.org/10.1098/rsfs.2018.0041
Koltai T, Fliegel L (2024) dichloroacetate for cancer treatment: some facts and many doubts. Pharmaceuticals (Basel) 17(6):744. https://doi.org/10.3390/ph17060744
Kompaniyets L, Pennington AF, Goodman AB et al (2021) Underlying medical conditions and severe illness among 540,667 adults hospitalized with COVID-19, March 2020–March 2021. Prev Chronic Dis 18:E66. https://doi.org/10.5888/pcd18.210123
Kondepudi DK, De Bari B, Dixon JA (2020) Dissipative structures, organisms and evolution. Entropy 22(11):1305. https://doi.org/10.3390/e22111305
Koonin EV (2000) How many genes can make a cell: the minimal-gene-set concept. Annu Rev Genomics Hum Genet 1:99–116. https://doi.org/10.1146/annurev.genom.1.1.99
Kooreman NG, Kim Y, de Almeida PE et al (2018) Autologous iPSC-based vaccines elicit anti-tumor responses in vivo. Cell Stem Cell 22(4):501–513.e7. https://doi.org/10.1016/j.stem.2018.01.016
Kordes M, Larsson L, Engstrand L, Löhr JM (2021) Pancreatic cancer cachexia: three dimensions of a complex syndrome. Br J Cancer 124(10):1623–1636. https://doi.org/10.1038/s41416-021-01301-4
Korhonen T, Hjelmborg J, Harris JR et al (2022) Cancer in twin pairs discordant for smoking: the nordic twin study of cancer. Int J Cancer 151(1):33–43. https://doi.org/10.1002/ijc.33963
Kornauth C, Pemovska T, Vladimer GI et al (2022) Functional precision medicine provides clinical benefit in advanced aggressive hematologic cancers and identifies exceptional responders. Cancer Discov 12(2):372–387. https://doi.org/10.1158/2159-8290.CD-21-0538
Kornauth C, Rindler TN, Gruber AS et al (2022b) Implementation of a real-time functional precision oncology platform for advanced cancer therapy. Clin Cancer Res 28(8):1586–1599
Koshland DE Jr (2002) Special essay: The seven pillars of life. Science 295(5563):2215–2216. https://doi.org/10.1126/science.1068489
Kostic MM (2020) The second law and entropy misconceptions demystified. Entropy (Basel) 22(6):648. https://doi.org/10.3390/e22060648
Kostic AD, Chun E, Meyerson M, Garrett WS (2013) Microbes and inflammation in colorectal cancer. Cancer Immunol Res 1(3):150–157. https://doi.org/10.1158/2326-6066.CIR-13-0101
Kotas ME, Medzhitov R (2015) Homeostasis, inflammation, and disease susceptibility. Cell 160(5):816–827. https://doi.org/10.1016/j.cell.2015.02.010
Kotler S, Peterson GA, Shojaee E et al (2021) Direct observation of deterministic macroscopic entanglement. Science 372(6542):622–625. https://doi.org/10.1126/science.abf2998
Koundouros N, Poulogiannis G (2020) Reprogramming of fatty acid metabolism in cancer. Br J Cancer 122(1):4–22. https://doi.org/10.1038/s41416-019-0650-z
Koutroufinis SA (2022) Toward a logic of the organism: a process philosophical consideration. Entropy 24(1):66. https://doi.org/10.3390/e24010066
Kovács A, Guller I, Krempels K, Somlyai G (2011) Deuterium depletion may delay the progression of prostate cancer. J Cancer Ther 2:548–556. https://doi.org/10.4236/jct.2011.24075
Krajcovic M, Krishna S, Akkari L, Joyce JA, Overholtzer M (2013) mTOR regulates phagosome and entotic vacuole fission. Mol Biol Cell 24(23):3736–3745. https://doi.org/10.1091/mbc.E13-07-0408
Krakauer D, Bertschinger N, Olbrich E et al (2020) The information theory of individuality. Theory Biosci 139(2):209–223. https://doi.org/10.1007/s12064-020-00313-7
Kremer-Lecointre A, Lecointre G (2023) Démystifier le vivant: 36 métaphores à ne plus utiliser. Belin Éducation/Humensis
Krempels K, Somlyai I, Somlyai G (2008) A retrospective evaluation of the effects of deuterium depleted water consumption on 4 patients with brain metastases from lung cancer. Integr Cancer Ther 7(3):172–181. https://doi.org/10.1177/1534735408322851
Krisko A, Radman M (2013) Phenotypic and genetic consequences of protein damage. PLoS Genet 9(9):e1003810. https://doi.org/10.1371/journal.pgen.1003810
Krisko A, Radman M (2019) Protein damage, ageing and age-related diseases. Open Biol 9(3):180249. https://doi.org/10.1098/rsob.180249
Kroenke CH, Kubzansky LD, Schernhammer ES, Holmes MD, Kawachi I (2006) Social networks, social support, and survival after breast cancer diagnosis. J Clin Oncol 24(7):1105–1111. https://doi.org/10.1200/JCO.2005.04.2846
Kuczynski EA, Vermeulen PB, Pezzella F, Kerbel RS, Reynolds AR (2019) Vessel co-option in cancer. Nat Rev Clin Oncol 16(8):469–493. https://doi.org/10.1038/s41571-019-0181-9
Kühn T, Floegel A, Sookthai D et al (2016) Higher plasma levels of lysophosphatidylcholine 18:0 are related to a lower risk of common cancers in a prospective metabolomics study. BMC Med 14:13. https://doi.org/10.1186/s12916-016-0552-3
Kuhner MK, Kostadinov R, Reid BJ (2016) Limitations of the Driver/Passenger Model in Cancer Prevention. Cancer Prev Res (Phila) 9(5):335–338. https://doi.org/10.1158/1940-6207.CAPR-15-0343
Kuijk E, Blokzijl F, Jager M et al (2019) Early divergence of mutational processes in human fetal tissues. Sci Adv 5(5):eaaw1271. https://doi.org/10.1126/sciadv.aaw1271
Kull K (2023) On the concept of meaning in biology. In: Corning PA, Kauffman SA, Noble D et al (eds) Evolution "on purpose": teleonomy in living systems. MIT Press, Cambridge
Kumar S, Warrell J, Li S et al (2020) Passenger mutations in more than 2,500 cancer genomes: overall molecular functional impact and consequences. Cell 180(5):915–927.e16. https://doi.org/10.1016/j.cell.2020.01.032
Kumari N, Pullaguri N, Rath SN et al (2024) Dysregulation of calcium homeostasis in cancer and its role in chemoresistance. Cancer Drug Resist 7:11. https://doi.org/10.20517/cdr.2023.145
Kundu JK, Surh YJ (2008) Inflammation: gearing the journey to cancer. Mutat Res 659(1–2):15–30. https://doi.org/10.1016/j.mrrev.2008.03.002
Kurakin A (2005) Self-organization versus Watchmaker: stochastic dynamics of cellular organization. Biol Chem 386(3):247–254. https://doi.org/10.1515/BC.2005.030
Kutschera U (2011) From the scala naturae to the symbiogenetic and dynamic tree of life. Biol Direct. 6:33. https://doi.org/10.1186/1745-6150-6-33
Kwon M, Godinho SA, Chandhok NS et al (2008) Mechanisms to suppress multipolar divisions in cancer cells with extra centrosomes. Genes Dev 22(16):2189–2203. https://doi.org/10.1101/gad.1700908

Kypriotakis G, Vidrine DJ, Francis LE, Rose JH (2016) The longitudinal relationship between quality of life and survival in advanced stage cancer. Psychooncology 25(2):225–231. https://doi.org/10.1002/pon.3846

La Manno G, Soldatov R, Zeisel A et al (2018) RNA velocity of single cells. Nature 560(7719):494–498. https://doi.org/10.1038/s41586-018-0414-6

Labelle M, Hynes RO (2012) The initial hours of metastasis: the importance of cooperative host-tumor cell interactions during hematogenous dissemination. Cancer Discov 2(12):1091–1099. https://doi.org/10.1158/2159-8290.CD-12-0329

Labelle M, Begum S, Hynes RO (2011) Direct signaling between platelets and cancer cells induces an epithelial-mesenchymal-like transition and promotes metastasis. Cancer Cell 20(5):576–590. https://doi.org/10.1016/j.ccr.2011.09.009

Labelle M, Begum S, Hynes RO (2014) Platelets guide the formation of early metastatic niches. Proc Natl Acad Sci USA 111(30):E3053–E3061. https://doi.org/10.1073/pnas.1411082111

Lachance JC, Rodrigue S, Palsson BO (2019) Minimal cells, maximal knowledge. Elife 8:e45379. https://doi.org/10.7554/eLife.45379

Lachat C, Peixoto P, Hervouet E (2021) Epithelial to mesenchymal transition history: from embryonic development to cancers. Biomolecules 11(6):782. https://doi.org/10.3390/biom11060782

Lachmann M, Walker S (2019) What can Schrödinger's cat say about 3D printers on Mars: Life ≠ alive. Aeon Magazine.. https://aeon.co/essays/what-can-Schrödingers-cat-say-about-3d-printers-on-mars

Laconi E, Marongiu F, DeGregori J (2020) Cancer as a disease of old age: changing mutational and microenvironmental landscapes. Br J Cancer 122(7):943–952. https://doi.org/10.1038/s41416-019-0721-1

LaFave LM, Kartha VK, Ma S et al (2020) Epigenomic state transitions characterize tumor progression in mouse lung adenocarcinoma. Cancer Cell 38(2):212–228.e13. https://doi.org/10.1016/j.ccell.2020.06.006. Epub 2020 Jul 23. Erratum in: Cancer Cell. 2025;43(5):981–984. http://doi.org/10.1016/j.ccell.2025.02.025

Lagasse E, Levin M (2023) Future medicine: from molecular pathways to the collective intelligence of the body. Trends Mol Med 29(9):687–710. https://doi.org/10.1016/j.molmed.2023.06.007

Lala KN, Uller T, Feiner N et al (2024) Evolution Evolving. Yale University Press, New Haven

Laland KN, Uller T, Feldman MW et al (2015) The extended evolutionary synthesis: its structure, assumptions and predictions. Proc Biol Sci 282(1813):20151019. https://doi.org/10.1098/rspb.2015.1019

Lambert AW, Weinberg RA (2021) Linking EMT programmes to normal and neoplastic epithelial stem cells. Nat Rev Cancer 21(5):325–338. https://doi.org/10.1038/s41568-021-00332-6

Lambert AW, Pattabiraman DR, Weinberg RA (2017) Emerging biological principles of metastasis. Cell 168(4):670–691. https://doi.org/10.1016/j.cell.2016.11.037

Lander ES, Linton LM, Birren B et al (2001) Initial sequencing and analysis of the human genome. Nature 409(6822):860–921. https://doi.org/10.1038/35057062

Landré T, Karaboué A, Buchwald ZS et al (2024) Effect of immunotherapy-infusion time of day on survival of patients with advanced cancers: a study-level meta-analysis. ESMO Open 9(2):102220. https://doi.org/10.1016/j.esmoop.2023.102220

Landweber LF, Kari L (2002) Universal molecular computing in ciliates. In: Landweber LF, Winfree E (eds) Evolution as Computation. Springer, Berlin

Lane N (2009) Life ascending: the ten great inventions of evolution. W.W. Norton & Company, New York

Lane N (2015) The vital question: energy, evolution and the origins of complex life. W.W. Norton & Company, New York

Lane N (2022) Transformer: the deep chemistry of life and death. W.W. Norton & Company, New York

Lang K, Ratke J (2009) Leptin and adiponectin: new players in the field of tumor cell and leukocyte migration. Cell Commun Signal 7:27. https://doi.org/10.1186/1478-811X-7-27

Langman L, Burr HS (1947) Electrometric studies in women with malignancy of cervix uteri. Science 105(2721):209–210. https://doi.org/10.1126/science

Lanzetti L (2024) Oncometabolites at the crossroads of genetic, epigenetic and ecological alterations in cancer. Cell Death Differ 31(12):1582–1594. https://doi.org/10.1038/s41418-024-01402-6

Laplane L, Duluc D, Larmonier N et al (2018) The multiple layers of the tumor environment. Trends Cancer 4(12):802–809. https://doi.org/10.1016/j.trecan.2018.10.002

Läsche M, Emons G, Gründker C (2020) Shedding new light on cancer metabolism: a metabolic tightrope between life and death. Front Oncol 10:409. https://doi.org/10.3389/fonc.2020.00409

Läubli H, Borsig L (2010a) Selectins as mediators of lung metastasis. Cancer Microenviron 3(1):97–105. https://doi.org/10.1007/s12307-010-0043-6. PMID: 21209777

Läubli H, Borsig L (2010b) Selectins promote tumor metastasis. Semin Cancer Biol 20(3):169–177. https://doi.org/10.1016/j.semcancer.2010.04.005

Lavergne M, Janus-Bell E, Schaff M et al (2017) Platelet integrins in tumor metastasis: do they represent a therapeutic target? Cancers (Basel) 9(10):133. https://doi.org/10.3390/cancers9100133

Lavery JA, Boutros PC, Moskowitz CS et al (2025) Comorbidity in midlife and cancer outcomes. JAMA Netw Open 8(4):e253469. https://doi.org/10.1001/jamanetworkopen.2025.3469

Lawrence MS, Stojanov P, Mermel CH et al (2014) Discovery and saturation analysis of cancer genes across 21 tumour types. Nature 505(7484):495–501. https://doi.org/10.1038/nature12912

Lawrence R, Watters M, Davies CR et al (2023) Circulating tumour cells for early detection of clinically relevant cancer. Nat Rev Clin Oncol 20(7):487–500. https://doi.org/10.1038/s41571-023-00781-y

Lazar I, Clement E, Dauvillier S et al (2016) Adipocyte exosomes promote melanoma aggressiveness through fatty acid oxidation: a novel mechanism linking obesity and cancer. Cancer Res 76(14):4051–4057. https://doi.org/10.1158/0008-5472.CAN-16-0651

Lazar IM, Karcini A, Haueis JRS (2022) Mapping the cell-membrane proteome to the cancer hallmarks. bioRxiv. https://doi.org/10.1101/2022.03.18.484818

Lazebnik Y (2010) What are the hallmarks of cancer? Nat Rev Cancer 10(4):232–233. https://doi.org/10.1038/nrc2827

Le Vely B, Phan C, Berrebeh N et al (2022) Loss of cAbl tyrosine kinase in pulmonary arterial hypertension causes dysfunction of vascular endothelial cells. Am J Respir Cell Mol Biol 67(2):215–226. https://doi.org/10.1165/rcmb.2021-0332OC

Leal T, Kotecha R, Ramlau R et al (2023) Tumor treating fields therapy with standard systemic therapy versus standard systemic therapy alone in metastatic non-small-cell lung cancer following progression on or after platinum-based therapy (LUNAR): a randomised, open-label, pivotal phase 3 study. Lancet Oncol 24(9):1002–1017. https://doi.org/10.1016/S1470-2045(23)00344-3

LeCun Y, Bengio Y, Hinton G (2015) Deep learning. Nature 521:436–444. https://doi.org/10.1038/nature14539

Ledford H (2024) Why cancer risk declines sharply in old age. Nature 631(8020):261–262. https://doi.org/10.1038/d41586-024-02107-z

Lee JW (2010) Fluid and electrolyte disturbances in critically ill patients. Electrolyte Blood Press 8(2):72–81. https://doi.org/10.5049/EBP.2010.8.2.72

Lee Y (2021) Roles of circadian clocks in cancer pathogenesis and treatment. Exp Mol Med 53:1529–1538. https://doi.org/10.1038/s12276-021-00681-0

Lee MY (2022) Embryonic programs in cancer and metastasis—insights from the mammary gland. Front Cell Dev Biol 10:938625. https://doi.org/10.3389/fcell.2022.938625

Lee RC, Feinbaum RL, Ambros V (1993) The C. elegans heterochronic gene lin-4 encodes small RNAs with antisense complementarity to lin-14. Cell 75(5):843–854. https://doi.org/10.1016/0092-8674(93)90529-y

Lee L, Cheung WY, Atkinson E, Krzyzanowska MK (2011) Impact of comorbidity on chemotherapy use and outcomes in solid tumors: a systematic review. J Clin Oncol 29(1):106–117. https://doi.org/10.1200/JCO.2010.31.3049

Lee C, Raffaghello L, Brandhorst S et al (2012) Fasting cycles retard growth of tumors and sensitize a range of cancer cell types to chemotherapy. Sci Transl Med 4(124):124ra27. https://doi.org/10.1126/scitranslmed.3003293

Lee YM, Chang WC, Ma WL (2016) Hypothesis: solid tumours behave as systemic metabolic dictators. J Cell Mol Med 20(6):1076–1085. https://doi.org/10.1111/jcmm.12794

Lee SY, Ju MK, Jeon HM et al (2018) Oncogenic metabolism acts as a prerequisite step for induction of cancer metastasis and cancer stem cell phenotype. Oxid Med Cell Longev 2018:1027453. https://doi.org/10.1155/2018/1027453

Lee W, Ko SY, Mohamed MS et al (2019) Neutrophils facilitate ovarian cancer premetastatic niche formation in the omentum. J Exp Med 216(1):176–194. https://doi.org/10.1084/jem.20181170

Lee JY, Tsolis RM, Bäumler AJ (2022a) The microbiome and gut homeostasis. Science 377(6601):eabp9960. https://doi.org/10.1126/science.abp9960

Lee CH, Motzer R, Emamekhoo H et al (2022b) Telaglenastat plus everolimus in advanced renal cell carcinoma: a randomized, double-blinded, placebo-controlled, phase II ENTRATA trial. Clin Cancer Res 28(15):3248–3255. https://doi.org/10.1158/1078-0432.CCR-22-0061

Lee CY, The M, Meng C et al (2024) Illuminating phenotypic drug responses of sarcoma cells to kinase inhibitors by phosphoproteomics. Mol Syst Biol 20(1):28–55. https://doi.org/10.1038/s44320-023-00004-7

Lee-Six H, Olafsson S, Ellis P (2019) The landscape of somatic mutation in normal colorectal epithelial cells. Nature 574(7779):532–537. https://doi.org/10.1038/s41586-019-1672-7

Legátová A, Pelantová M, Rösel D et al (2023) The emerging role of microtubules in invasion plasticity. Front Oncol 13:1118171. https://doi.org/10.3389/fonc.2023.1118171

Lekic M, Kovac V, Triller N et al (2012) Outcome of small cell lung cancer (SCLC) patients with brain metastases in a routine clinical setting. Radiol Oncol 46(1):54–59. https://doi.org/10.2478/v10019-012-0007-1

Lempesis IG, Georgakopoulou VE, Papalexis P et al (2023) Role of stress in the pathogenesis of cancer (Review). Int J Oncol 63(5):124. https://doi.org/10.3892/ijo.2023.5572

Lennon AM, Buchanan AH, Kinde I et al (2020) Feasibility of blood testing combined with PET-CT to screen for cancer and guide intervention. Science 369(6499):eabb9601. https://doi.org/10.1126/science.abb9601

Leschak CJ, Eisenberger NI (2019) Two distinct immune pathways linking social relationships with health: Inflammatory and antiviral processes. Psychosom Med 81(8):711–719. https://doi.org/10.1097/PSY.0000000000000685

Letai A, Bhola P, Welm AL (2022) Functional precision oncology: Testing tumors with drugs to identify vulnerabilities and novel combinations. Cancer Cell. 40(1):26–35. https://doi.org/10.1016/j.ccell.2021.12.004

Letelier JC, Cárdenas ML, Cornish-Bowden A (2011) From L'Homme Machine to metabolic closure: steps towards understanding life. J Theor Biol 286(1):100–113. https://doi.org/10.1016/j.jtbi.2011.06.033

Letsou W (2024) The indispensable role of time in autonomous development. Biosystems 246:105340. https://doi.org/10.1016/j.biosystems.2024.105340

Letsou W, Cai L (2016) Noncommutative biology: sequential regulation of complex networks. PLoS Comput Biol 12(8):e1005089. https://doi.org/10.1371/journal.pcbi.1005089

Lévi F (2001) Circadian chronotherapy for human cancers. Lancet Oncol 2(5):307–315. https://doi.org/10.1016/S1470-2045(00)00326-0

Lévi FA, Canon C, Touitou Y et al (1998) Seasonal modulation of the circadian time structure of circulating T and natural killer lymphocyte subsets from healthy subjects. J Clin Invest 81:407–413. https://doi.org/10.1172/JCI113333

Levin M (2011) The wisdom of the body: future techniques and approaches to morphogenetic fields in regenerative medicine, developmental biology and cancer. Regen Med 6(6):667–673. https://doi.org/10.2217/rme.11.69

Levin M (2014) Molecular bioelectricity: how endogenous voltage potentials control cell behavior and instruct pattern regulation in vivo. Mol Biol Cell 25(24):3835–3850. https://doi.org/10.1091/mbc.E13-12-0708

Levin M (2019) The computational boundary of a "self": developmental bioelectricity drives multicellularity and scale-free cognition. Front Psychol 10:2688. https://doi.org/10.3389/fpsyg.2019.02688

Levin M (2021) Bioelectric signaling: reprogrammable circuits underlying embryogenesis, regeneration, and cancer. Cell 184(8):1971–1989. https://doi.org/10.1016/j.cell.2021.02.034

Levin M (2022) Technological approach to mind everywhere: an experimentally-grounded framework for understanding diverse bodies and minds. Front Syst Neurosci 16:768201. https://doi.org/10.3389/fnsys.2022.768201

Levin M (2023a) Collective intelligence of morphogenesis as a teleonomic process open access. In: Corning PA, Kauffman SA, Noble D, Shapiro JA, Vane-Wright RI, Pross A (eds) Evolution "on purpose". https://doi.org/10.7551/mitpress/14642.003.0013

Levin M (2023b) Collective intelligence of morphogenesis as a teleonomic process. In: Corning PA, Kauffman SA, Noble D, Shapiro JA, Vane-Wright RI, Pross A (eds) Evolution "On Purpose": teleonomy in living systems. MIT Press

Levin M, Martyniuk CJ (2018) The bioelectric code: an ancient computational medium for dynamic control of growth and form. Biosystems 164:76–93. https://doi.org/10.1016/j.biosystems.2017.08.009

Levin M, Selberg J, Rolandi M (2019) Endogenous bioelectrics in development, cancer, and regeneration: drugs and bioelectronic devices as electroceuticals for regenerative medicine. iScience 22:519–533. https://doi.org/10.1016/j.isci.2019.11.023

Levine AJ, Puzio-Kuter AM (2010) The control of the metabolic switch in cancers by oncogenes and tumor suppressor genes. Science 330(6009):1340–1344. https://doi.org/10.1126/science.1193494

Levine ME, Suarez JA, Brandhorst S et al (2014) Low protein intake is associated with a major reduction in IGF-1, cancer, and overall mortality in the 65 and younger but not older population. Cell Metab 19(3):407–417. https://doi.org/10.1016/j.cmet.2014.02.006

Lewontin RC (1970) The units of selection. Annu Rev Ecol Syst 1:1–18. https://doi.org/10.1146/annurev.es.01.110170.000245

Li Y (2023) Non-coding RNA performs its biological function by interacting with macromolecules. Int J Mol Sci 24(22):16246. https://doi.org/10.3390/ijms242216246

Li L, Neaves WB (2006) Normal stem cells and cancer stem cells: the niche matters. Cancer Res 66:4553–4557. https://doi.org/10.1158/0008-5472.CAN-05-3986

Li N, Zhan X (2021) Integrated genomic analysis of proteasome alterations across 11,057 patients with 33 cancer types: clinically relevant outcomes in framework of 3P medicine. EPMA J 12(4):605–627. https://doi.org/10.1007/s13167-021-00256-z

Li T, Harada M, Tamada K et al (1997) Repeated restraint stress impairs the antitumor T cell response through its suppressive effect on Th1-type CD4+ T cells. Anticancer Res 17(6D):4259–4268

Li XM, Delaunay F, Dulong S et al (2010) Cancer inhibition through circadian reprogramming of tumor transcriptome with meal timing. Cancer Res 70(8):3351–3360. https://doi.org/10.1158/0008-5472.CAN-09-4235

Li J, Bennett K, Stukalov A, Fang B et al (2013) Perturbation of the mutated EGFR interactome identifies vulnerabilities and resistance mechanisms. Mol Syst Biol 9:705. https://doi.org/10.1038/msb.2013.61

Li CH, Haider S, Shiah YJ et al (2018a) Sex differences in cancer driver genes and biomarkers. Cancer Res 78(19):5527–5537. https://doi.org/10.1158/0008-5472.CAN-18-0362

Li X, Egervari G, Wang Y et al (2018b) Regulation of chromatin and gene expression by metabolic enzymes and metabolites. Nat Rev Mol Cell Biol 19(9):563–578. https://doi.org/10.1038/s41580-018-0029-7

Li J, Xu H, Zhang L et al (2019a) Malignant ascites-derived organoid (MADO) cultures for gastric cancer in vitro modelling and drug screening. J Cancer Res Clin Oncol 145(11):2637–2647. https://doi.org/10.1007/s00432-019-03004-z

Li X, Severson E, Pignon JC et al (2019b) Comprehensive analyses of tumor immunity: implications for cancer immunotherapy. Genome Biol 20(1):309. https://doi.org/10.1186/s13059-019-1906-9

Li HY, Xie Y, Yu TT et al (2020a) Durable response to pulsatile icotinib for central nervous system metastases from EGFR-mutated non-small cell lung cancer: a case report. World J Clin Cases 8(2):370–376. https://doi.org/10.12998/wjcc.v8.i2.370

Li Z, Deng J, Sun J, Ma Y (2020b) Hyperthermia targeting the tumor microenvironment facilitates immune checkpoint inhibitors. Front Immunol 11:595207. https://doi.org/10.3389/fimmu.2020.595207

Li T, Liu J, Feng J et al (2020c) Variation in the life history strategy underlies functional diversity of tumors. Natl Sci Rev 8(2):nwaa124. https://doi.org/10.1093/nsr/nwaa124

Li M, Chen S, Zhao Y et al (2020d) Dynamic multiscale graph neural networks for 3D skeleton-based human motion prediction. In: 2020 IEEE/CVF Conference on Computer Vision and Pattern Recognition (CVPR). https://doi.org/10.1109/CVPR42600.2020.00029

Li D, Yang Y, Li Y et al (2021a) Epigenetic regulation of gene expression in response to environmental exposures: From bench to model. Sci Total Environ 776:145998. https://doi.org/10.1016/j.scitotenv.2021.145998

Li J, Duran MA, Dhanota N et al (2021b) Metastasis and immune evasion from extracellular cGAMP hydrolysis. Cancer Discov 11(5):1212–1227. https://doi.org/10.1158/2159-8290.CD-20-0387

Li Y, Zhao L, Li XF (2021c) Hypoxia and the tumor microenvironment. Technol Cancer Res Treat:15330338211036304. https://doi.org/10.1177/15330338211036304

Li R, Liu M, Yang Z, Li J et al (2022a) Proteolysis-targeting chimeras (PROTACs) in cancer therapy: present and future. Molecules 27(24):8828. https://doi.org/10.3390/molecules27248828

Li Y, Chen ZK, Duan X et al (2022b) Targeted inhibition of tumor-derived exosomes as a novel therapeutic option for cancer. Exp Mol Med 54(9):1379–1389. https://doi.org/10.1038/s12276-022-00856-3

Li L, Yang L, Yang L et al (2023) Network pharmacology: a bright guiding light on the way to explore the personalized precise medication of traditional Chinese medicine. Chin Med 18:146. https://doi.org/10.1186/s13020-023-00853-2

Li J, Chen X, Lu A, Liang C (2023a) Targeted protein degradation in cancers: orthodox PROTACs and beyond. Innovation (Camb) 4(3):100413. https://doi.org/10.1016/j.xinn.2023.100413

Li S, Pan T, Xu G et al (2023b) Deep immunophenotyping reveals clinically distinct cellular states and ecosystems in large-scale colorectal cancer. Commun Biol 6(1):785. https://doi.org/10.1038/s42003-023-05117-1

Li X, Liu K, Xing L, Rubinsky B (2023c) A review of tumor treating fields (TTFields): advancements in clinical applications and mechanistic insights. Radiol Oncol 57(3):279–291. https://doi.org/10.2478/raon-2023-0044

Li Z, Liu L, Feng C et al (2023d) LncBook 2.0: integrating human long non-coding RNAs with multi-omics annotations. Nucleic Acids Res 51(D1):D186–D191. https://doi.org/10.1093/nar/gkac999

Li X, Peng X, Li Y et al (2024) Glutamine addiction in tumor cell: oncogene regulation and clinical treatment. Cell Commun Signal 22(1):12. https://doi.org/10.1186/s12964-023-01449-x

Lichtenstein AV (2005) Cancer as a programmed death of an organism. Biochemistry (Mosc) 70(9):1055–1064. https://doi.org/10.1007/s10541-005-0224-y

Lichtenstein AV (2017) Cancer: bad luck or punishment? Biochemistry (Mosc) 82:75–80. https://doi.org/10.1134/S0006297917010084

Lichtenstein P, Holm NV, Verkasalo PK et al (2000) Environmental and heritable factors in the causation of cancer – analyses of cohorts of twins from Sweden, Denmark, and Finland. N Engl J Med 343:78–85

Lieff J (2020) The secret language of cells: what biological conversations tell us about the brain-body connection, the future of medicine, and life itself. BenBella Books

Lieffers JR, Mourtzakis M, Hall KD et al (2009) A viscerally driven cachexia syndrome in patients with advanced colorectal cancer: contributions of organ and tumor mass to whole-body energy demands. Am J Clin Nutr 89(4):1173–1179. https://doi.org/10.3945/ajcn.2008.27273

Lien EC, Vander Heiden MG (2019) A framework for examining how diet impacts tumour metabolism. Nat Rev Cancer 19(11):651–661. https://doi.org/10.1038/s41568-019-0198-5

Lim H, Moon A (2016) Inflammatory fibroblasts in cancer. Arch Pharm Res 39(8):1021–1031. https://doi.org/10.1007/s12272-016-0787-8

Lim M, Liau LM, Ashkan K et al (2021) Current state of immunotherapy for glioblastoma. Nat Rev Clin Oncol 18(7):422–434. https://doi.org/10.1038/s41571-021-00494-w

Lim L, Machingura A, Taye M et al (2025) Prognostic value of baseline EORTC QLQ-C30 scores for overall survival across 46 clinical trials covering 17 cancer types: a validation study. EClinicalMedicine 82:103153. https://doi.org/10.1016/j.eclinm.2025.103153

Lin C, Yang L, Rosenfeld MG (2012) Molecular logic underlying chromosomal translocations, random or non-random? Adv Cancer Res 113:241–279. https://doi.org/10.1016/B978-0-12-394280-7.00015-4

Lin HH, Robertson KL, Bisbee HA, Farkas ME (2020) Oncogenic and circadian effects of small molecules directly and indirectly targeting the core circadian clock. Integr Cancer Ther:1534735420924094. https://doi.org/10.1177/1534735420924094

Lineweaver CH, Davies PC, Vincent MD (2014) Targeting cancer's weaknesses (not its strengths): Therapeutic strategies suggested by the atavistic model. Bioessays 36(9):827–835. https://doi.org/10.1002/bies.201400070

Lineweaver CH, Bussey KJ, Blackburn AC et al (2021) Cancer progression as a sequence of atavistic reversions. Bioessays 43(7):e2000305. https://doi.org/10.1002/bies.202000305

Lingg L, Rottenberg S, Francica P (2022) Meiotic genes and DNA double strand break repair in cancer. Front Genet 13:831620. https://doi.org/10.3389/fgene.2022.831620

Litan A, Langhans SA (2015) Cancer as a channelopathy: ion channels and pumps in tumor development and progression. Front Cell Neurosci 9:86. https://doi.org/10.3389/fncel.2015.00086

Liu ET (2008) Functional genomics of cancer. Curr Opin Genet Dev 18(3):251–256. https://doi.org/10.1016/j.gde.2008.07.014

Liu J (2018) The dualistic origin of human tumors. Semin Cancer Biol 53:1–16. https://doi.org/10.1016/j.semcancer.2018.07.004. Epub 2018 July 21. PMID: 30040989; PMCID: PMC6553492

Liu Y, Cao X (2016) Characteristics and significance of the pre-metastatic niche. Cancer Cell 30(5):668–681. https://doi.org/10.1016/j.ccell.2016.09.011. PMID: 27846389

Liu W, Laitinen S, Khan S et al (2009) Copy number analysis indicates monoclonal origin of lethal metastatic prostate cancer. Nat Med 15(5):559–565. https://doi.org/10.1038/nm.1944

Liu Y, Gu Y, Cao X (2015) The exosomes in tumor immunity. Oncoimmunology 4(9):e1027472. https://doi.org/10.1080/2162402X.2015.1027472

Liu Y, Dou M, Song X et al (2019) The emerging role of the piRNA/piwi complex in cancer. Mol Cancer 18(1):123. https://doi.org/10.1186/s12943-019-1052-9

Liu X, Yin L, Shen S, Hou Y (2021) Inflammation and cancer: paradoxical roles in tumorigenesis and implications in immunotherapies. Genes Dis 10(1):151–164. https://doi.org/10.1016/j.gendis.2021.09.006

Liu Z, Hu M, Yang Y et al (2022) An overview of PROTACs: a promising drug discovery paradigm. Mol Biomed 3(1):46. https://doi.org/10.1186/s43556-022-00112-0

Liu F, Wu Q, Dong Z, Liu K (2023a) Integrins in cancer: emerging mechanisms and therapeutic opportunities. Pharmacol Ther 247:108458. https://doi.org/10.1016/j.pharmthera.2023

Liu C, Tang H, Hu N, Li T (2023b) Methylomics and cancer: the current state of methylation profiling and marker development for clinical care. Cancer Cell Int 23(1):242. https://doi.org/10.1186/s12935-023-03074-7

Liu F, Wu Q, Dong Z, Liu K (2023c) Integrins in cancer: emerging mechanisms and therapeutic opportunities. Pharmacol Ther 247:108458. https://doi.org/10.1016/j.pharmthera.2023.108458

Liu W, Li Y, Wu P et al (2024a) The intratumoral microbiota: a new horizon in cancer immunology. Front Cell Infect Microbiol 14:1409464. https://doi.org/10.3389/fcimb.2024.1409464

Liu M, Wen Z, Zhang T et al (2024b) The role of exosomal molecular cargo in exosome biogenesis and disease diagnosis. Front Immunol 15:1417758. https://doi.org/10.3389/fimmu.2024.1417758

Liu X, Shen B, Zhou J et al (2024c) The L-type calcium channel CaV1.3: a potential target for cancer therapy. J Cell Mol Med 28(19):e70123. https://doi.org/10.1111/jcmm.70123

Liu Z, Chen Y-C, Ao P (2024d) Entangled biphoton generation in the myelin sheath. Phys Rev E 110:024402. https://doi.org/10.1103/PhysRevE.110.024402

Liu P, Wang L, Yu H (2024e) Polyploid giant cancer cells: origin, possible pathways of formation, characteristics, and mechanisms of regulation. Front Cell Dev Biol 12:1410637. https://doi.org/10.3389/fcell.2024.1410637

Liu X, Feng C, Yan L et al (2025) Calcium channels as pharmacological targets for cancer therapy. Clin Exp Med 25(1):94. https://doi.org/10.1007/s10238-025-01632-z

Lloyd S (1991) Causality and information flow. In: Atmanspacher H, Scheingraber H (eds) Information dynamics. NATO ASI Series (Series B: Physics), vol 256. Springer, Boston. https://doi.org/10.1007/978-1-4899-2305-9_9

Lloyd D (2021) Oscillations, rhythms and synchronized time bases: the key signatures of life. In: Stefanovska A, McClintock ME (eds) Physics of biological oscillators: new insights into non-equilibrium and non-autonomous systems. Springer https://doi.org/10.1007/978-3-030-59805-1_14

Lloyd D, Aon MA, Cortassa S (2001) Why homeodynamics, not homeostasis? Scientific World J 1:133–145. https://doi.org/10.1100/tsw.2001.20

Lloyd D, Rossi RL, Rousell, MR (2008) Introduction: the temporal organisation of living systems from molecules to mind. In: Lloyd D (ed) Ultradian rhythms from molecules to mind. Springer

Loberg RD, Bradley DA, Tomlins SA et al (2007) The lethal phenotype of cancer: the molecular basis of death due to malignancy. CA Cancer J Clin 57(4):225–241. https://doi.org/10.3322/canjclin.57.4.225

Lobikin M, Chernet B, Lobo D, Levin M (2012) Resting potential, oncogene-induced tumorigenesis, and metastasis: the bioelectric basis of cancer in vivo. Phys Biol 9(6):065002. https://doi.org/10.1088/1478-3975/9/6/065002

Lo-Coco F, Avvisati G, Vignetti M et al (2013) Retinoic acid and arsenic trioxide for acute promyelocytic leukemia. N Engl J Med 369(2):111–121. https://doi.org/10.1056/NEJMoa1300874

Longo G (2023) Le cauchemar de Prométhée: Les sciences et leurs limites. PUF, Paris

Longo VD, Panda S (2016) Fasting, circadian rhythms, and time-restricted feeding in healthy lifespan. Cell Metab 23(6):1048–1059. https://doi.org/10.1016/j.cmet.2016.06.001

Longo VD, Mitteldorf J, Skulachev VP (2005) Programmed and altruistic ageing. Nat Rev Genet 6(11):866–872. https://doi.org/10.1038/nrg1706

Longo G, Montévil M, Kauffman S (2012) No entailing laws, but enablement in the evolution of the biosphere. In: Proceedings of the 14th annual conference on companion on Genetic and Evolutionary Computation (GECCO'12), pp 1379–1392. https://doi.org/10.1145/2330784.2330946

Lopez P, Newton RU, Taaffe DR et al (2022) Associations of fat and muscle mass with overall survival in men with prostate cancer: a systematic review with meta-analysis. Prostate Cancer Prostatic Dis 25(4):615–626. https://doi.org/10.1038/s41391-021-00442-0

Lopez-Camarillo C, Marchat LA, Arechaga-Ocampo E et al (2012) MetastamiRs: non-coding microRNAs driving cancer invasion and metastasis. Int J Mol Sci 13(2):1347–1379. https://doi.org/10.3390/ijms13021347

López-Novoa JM, Nieto MA (2009) Inflammation and EMT: an alliance towards organ fibrosis and cancer progression. EMBO Mol Med 1(6–7):303–314. https://doi.org/10.1002/emmm.200900043

López-Otín C, Kroemer G (2021) Hallmarks of Health. Cell 184(1):33–63. https://doi.org/10.1016/j.cell.2020.11.034. Erratum in: Cell. 2021 Apr 1;184(7):1929-1939. doi: 10.1016/j.cell.2021.03.033

López-Otín C, Blasco MA, Partridge L et al (2023a) Hallmarks of aging: an expanding universe. Cell 186(2):243–278. https://doi.org/10.1016/j.cell.2022.11.001

López-Otín C, Pietrocola F, Roiz-Valle D et al (2023b) Meta-hallmarks of aging and cancer. Cell Metab 35(1):12–35. https://doi.org/10.1016/j.cmet.2022.11.001

Lopez-Tarruella S, Escudero MJ, Pollan M et al (2019) Survival impact of primary tumor resection in de novo metastatic breast cancer patients (GEICAM/El Alamo Registry). Sci Rep 9:20081. https://doi.org/10.1038/s41598-019-55765-9

Lorenz EN (1993) The essence of chaos. University of Washington Press, Seattle

Lotka AJ (1922a) Natural selection as a physical principle. Proc Natl Acad Sci USA 8:151–154

Lotka AJ (1922b) Contribution to the energetics of evolution. Proc Natl Acad Sci USA 8:147–151

Lu Y, Chen H (2024) Deuterium-depleted water in cancer therapy: A systematic review of clinical and experimental trials. Nutrients 16(9):1397. https://doi.org/10.3390/nu16091397

Lu P, Weaver VM, Werb Z (2012) The extracellular matrix: a dynamic niche in cancer progression. J Cell Biol 196(4):395–406. https://doi.org/10.1083/jcb.201102147

Lu ZX, Huang Q, Park JW et al (2015) Transcriptome-wide landscape of pre-mRNA alternative splicing associated with metastatic colonization. Mol Cancer Res 13:305–318. https://doi.org/10.1158/1541-7786.MCR-14-0429-T

Lu Z, Zou J, Li S et al (2020) Epigenetic therapy inhibits metastases by disrupting premetastatic niches. Nature 579(7798):284–290. https://doi.org/10.1038/s41586-020-2054-x

Lu Z, Mo S, Xie D et al (2024) Polyclonal-to-monoclonal transition in colorectal precancerous evolution. Nature. https://doi.org/10.1038/s41586-024-08133-1

Luca BA, Steen CB, Matusiak M et al (2021) Atlas of clinically distinct cell states and ecosystems across human solid tumors. Cell 184(21):5482–5496.e28. https://doi.org/10.1016/j.cell.2021.09.014

Lucia U (2013) Thermodynamics and cancer stationary state. Physica A 392(17):3648–3653. https://doi.org/10.1016/j.physa.2013.04.033

Lucia U, Grisolia G, Ponzetto A, Silvagno F (2017) An engineering thermodynamic approach to select the electromagnetic wave effective on cell growth. J Theor Biol 429:181–189. https://doi.org/10.1016/j.jtbi.2017.06.029

Lucia U, Grisolia G, Ponzetto A et al (2020) Thermomagnetic resonance affects cancer growth and motility. R Soc Open Sci 7(7):200299. https://doi.org/10.1098/rsos.200299

Lucotti S, Kenific CM, Zhang H, Lyden D (2022) Extracellular vesicles and particles impact the systemic landscape of cancer. EMBO J 41(18):e109288. https://doi.org/10.15252/embj.2021109288

Lucotti S, Ogitani Y, Kenific CM et al (2025) Extracellular vesicles from the lung pro-thrombotic niche drive cancer-associated thrombosis and metastasis via integrin beta 2. Cell 188(6):1642–1661.e24. https://doi.org/10.1016/j.cell.2025.01.025

Luengo A, Li Z, Gui DY et al (2021) Increased demand for NAD+ relative to ATP drives aerobic glycolysis. Mol Cell 81(4):691–707.e6. https://doi.org/10.1016/j.molcel.2020.12.012

Luo FL (2009) Entropy production in cell and reversal of entropy flow as an anticancer therapy. Front Phys China 4(1):122–136. https://doi.org/10.1007/s11467-009-0016-1

Luo J, Margolis KL, Adami HO et al (2008) Obesity and risk of pancreatic cancer among postmenopausal women: the Women's Health Initiative (United States). Br J Cancer 99:527–531. https://doi.org/10.1038/sj.bjc.6604487

Luo J, Solimini NL, Elledge SJ (2009) Principles of cancer therapy: oncogene and non-oncogene addiction. Cell 136(5):823–837. https://doi.org/10.1016/j.cell.2009.02.024

Luo F, Wang X, Ye C, Sun H (2024) Microbial biomarkers in liquid biopsy for cancer: An overview and future directions. Cancer Control 31:10732748241292019. https://doi.org/10.1177/10732748241292019

Lutgendorf SK, Andersen BL (2015) Biobehavioral approaches to cancer progression and survival: Mechanisms and interventions. Am Psychol 70(2):186–197. https://doi.org/10.1037/a0035730

Lutgendorf SK, Lamkin DM, Jennings NB et al (2008) Biobehavioral influences on matrix metalloproteinase expression in ovarian carcinoma. Clin Cancer Res 14(21):6839–6846. https://doi.org/10.1158/1078-0432.CCR-08-0230

Lutgendorf SK, Costanzo ES, Sood AK (2012) Psychoneuroimmunology and cancer: biobehavioral influences on tumor progression. In: Segerstrom SC (ed) The Oxford handbook of psychoneuroimmunology. Oxford University Press, Oxford, pp 341–368. https://doi.org/10.1093/oxfordhb/9780195394399.013.0019

Luyendijk M, Visser O, Blommestein HM, de Hingh IHJT et al (2023) Changes in survival in de novo metastatic cancer in an era of new medicines. J Natl Cancer Inst 115(6):628–635. https://doi.org/10.1093/jnci/djad020

Lynch M (2007) The frailty of adaptive hypotheses for the origins of organismal complexity. Proc Natl Acad Sci USA 104(Suppl 1):8597–8604. https://doi.org/10.1073/pnas.0702207104

Lynch SV, Pedersen O (2016) The human intestinal microbiome in health and disease. N Engl J Med 375(24):2369–2379. https://doi.org/10.1056/NEJMra1600266

Lythgoe MP, Prasad V (2022) Repositioning canakinumab for non-small cell lung cancer—important lessons for drug repurposing in oncology. Br J Cancer 127(5):785–787. https://doi.org/10.1038/s41416-022-01893-5

Ma Y, Zhang P, Wang F et al (2010) The relationship between early embryo development and tumourigenesis. J Cell Mol Med 14(12):2697–2701. https://doi.org/10.1111/j.1582-4934.2010.01191.x

Ma F, Laster K, Dong Z (2022) The comparison of cancer gene mutation frequencies in Chinese and U.S. patient populations. Nat Commun 13(1):5651. https://doi.org/10.1038/s41467-022-33351-4

Ma L, Guo H, Zhao Y et al (2024) Liquid biopsy in cancer: current status, challenges and future prospects. Signal Transduct Target Ther 9(1):336. https://doi.org/10.1038/s41392-024-02021-w

MacArthur RH, Wilson EO (1967) The theory of island biogeography. Princeton University Press, Princeton

Maciejowski J, de Lange T (2017) Telomeres in cancer: tumour suppression and genome instability. Nat Rev Mol Cell Biol 18(3):175–186. https://doi.org/10.1038/nrm.2016.171

Mack SC, Witt H, Piro RM et al (2014) Epigenomic alterations define lethal CIMP-positive ependymomas of infancy. Nature 506(7489):445–450. https://doi.org/10.1038/nature13108

Mackie JL (1980) The cement of the universe: a study of causation. Clarendon Press, Oxford

Maddipati R, Stanger BZ (2015) Pancreatic cancer metastases harbor evidence of polyclonality. Cancer Discov 5(10):1086–1097. https://doi.org/10.1158/2159-8290.CD-15-0120

Maddocks OD, Berkers CR, Mason SM et al (2013) Serine starvation induces stress and p53-dependent metabolic remodelling in cancer cells. Nature 493(7433):542–546. https://doi.org/10.1038/nature11743

Maddocks ODK, Athineos D, Cheung EC et al (2017) Modulating the therapeutic response of tumours to dietary serine and glycine starvation. Nature 544:372–376

Magnon C (2015) Role of the autonomic nervous system in tumorigenesis and metastasis. Mol Cell Oncol 2:e975643. https://doi.org/10.4161/23723556.2014.975643

Magnon C, Hondermarck H (2023) The neural addiction of cancer. Nat Rev Cancer 23(5):317–334. https://doi.org/10.1038/s41568-023-00556-8

Magnon C, Hall SJ, Lin J et al (2013) Autonomic nerve development contributes to prostate cancer progression. Science 341:1236361. https://doi.org/10.1126/science.1236361

Magnus N, Garnier D, Meehan B et al (2014) Tissue factor expression provokes escape from tumor dormancy and leads to genomic alterations. Proc Natl Acad Sci USA 111:3544–3549. https://doi.org/10.1073/pnas.1314118111

Maia J, Caja S, Strano Moraes MC et al (2018) Exosome-Based Cell-Cell Communication in the Tumor Microenvironment. Front Cell Dev Biol 6:18. https://doi.org/10.3389/fcell.2018.00018

Maino, Vieytes CA, Taha HM, Burton-Obanla AA et al (2019) Carbohydrate nutrition and the risk of cancer. Curr Nutr Rep 8:230–239. https://doi.org/10.1007/s13668-019-0264-3

Mak JKL, McMurran CE, Kuja-Halkola R et al (2023) Clinical biomarker-based biological aging and risk of cancer in the UK Biobank. Br J Cancer 129:94–103. https://doi.org/10.1038/s41416-023-02288-w

Makarieva AM, Gorshkov VG, Li BL (2005) Energetics of the smallest: do bacteria breathe at the same rate as whales? Proc Biol Sci 272(1577):2219–2224. https://doi.org/10.1098/rspb.2005.3225

Makarieva AM, Gorshkov VG, Li BL et al (2008) Mean mass-specific metabolic rates are strikingly similar across life's major domains: evidence for life's metabolic optimum. Proc Natl Acad Sci USA 105(44):16994–16999. https://doi.org/10.1073/pnas.0802148105

Mäkelä T, Annila A (2010) Natural patterns of energy dispersal. Phys Life Rev 7:477–498. https://doi.org/10.1016/j.plrev.2010.10.001

Makhoul E, Aklinski JL, Miller J et al (2022) A Review of COVID-19 in Relation to Metabolic Syndrome: Obesity, Hypertension, Diabetes, and Dyslipidemia. Cureus 14:e27438. https://doi.org/10.7759/cureus.27438

Makinoshima H, Takita M, Matsumoto S et al (2014) Epidermal growth factor receptor (EGFR) signaling regulates global metabolic pathways in EGFR-mutated lung adenocarcinoma. J Biol Chem 289:20813–20823. https://doi.org/10.1074/jbc.M114.575464

Malakhov MM, Pan W (2025) Co-expression-wide association studies link genetically regulated interactions with complex traits. medRxiv [Preprint] 2024.10.02.24314813. https://doi.org/10.1101/2024.10.02.24314813

Malani D, Kumar A, Brück O et al (2022) Implementing a functional precision medicine tumor board for acute myeloid leukemia. Cancer Discov 12:388–401. https://doi.org/10.1158/2159-8290.CD-21-0410

Malani D, Kumar R, Samra B et al (2022b) Functional precision medicine: the next step in the evolution of individualized cancer therapy. Trends Cancer 8:42–54

Malani D, Kumar R, Samra B, Osterman T, Lehtonen R, Nykter M (2022c) Functional precision medicine: the next step in the evolution of individualized cancer therapy. Trends Cancer 8(1):42–54

Malaterre C (2024) Is Life Binary or Gradual? Life (Basel) 14:564. https://doi.org/10.3390/life14050564

Maley CC, Aktipis A, Graham TA et al (2017) Classifying the evolutionary and ecological features of neoplasms. Nat Rev Cancer 17:605–619. https://doi.org/10.1038/nrc.2017.69

Maman S, Witz IP (2018) A history of exploring cancer in context. Nat Rev Cancer 18(6):359–376. https://doi.org/10.1038/s41568-018-0006-7. PMID: 29700396

Mammoto A, Ingber DE (2009) Cytoskeletal control of growth and cell fate switching. Curr Opin Cell Biol 21(6):864–870. https://doi.org/10.1016/j.ceb.2009.08.001. PMID: 19740640

Mancusi R, Monje M (2023) The neuroscience of cancer. Nature 618:467–479. https://doi.org/10.1038/s41586-023-05968-y. PMID: 37316719

Mani K, Deng D, Lin C et al (2024) Causes of death among people living with metastatic cancer. Nat Commun 15(1):1519. https://doi.org/10.1038/s41467-024-45307-x. PMID: 38374318; PMCID: PMC10876661

Maniloff J (1996) The minimal cell genome: "on being the right size". Proc Natl Acad Sci U S A 93(19):10004–10006. https://doi.org/10.1073/pnas.93.19.10004. PMID: 8816738; PMCID: PMC38325

Manjunath Y, Porciani D, Mitchem JB et al (2020) Tumor-cell-macrophage fusion cells as liquid biomarkers and tumor enhancers in cancer. Int J Mol Sci 21(5):1872. https://doi.org/10.3390/ijms21051872. PMID: 32182935; PMCID: PMC7084898

Mann TH, Kaech SM (2019) Tick-TOX, it's time for T cell exhaustion. Nat Immunol 20:1092–1094. https://doi.org/10.1038/s41590-019-0478-y

Manning S, Danielson KM (2018) The immunomodulatory role of tumor-derived extracellular vesicles in colorectal cancer. Immunol Cell Biol. https://doi.org/10.1111/imcb.12038. PMID: 29575270

Manser RL, Dodd M, Byrnes G et al (2005) Incidental lung cancers identified at coronial autopsy: implications for overdiagnosis of lung cancer by screening. Respir Med 99(4):501–507. https://doi.org/10.1016/j.rmed.2004.08.017. PMID: 15763458

Mansour AG, Xiao R, Bergin SM et al (2021) Enriched environment enhances NK cell maturation through hypothalamic BDNF in male mice. Eur J Immunol 51(3):557–566. https://doi.org/10.1002/eji.201948358. PMID: 33169371; PMCID: PMC8212871

Mantovani G, Macciò A, Mura L et al (2000) Serum levels of leptin and proinflammatory cytokines in patients with advanced-stage cancer at different sites. J Mol Med (Berl) 78(10):554–561. https://doi.org/10.1007/s001090000137. PMID: 1119932

Mantovani A, Allavena P, Sica A, Balkwill F (2008) Cancer-related inflammation. Nature 454(7203):436–444. https://doi.org/10.1038/nature07205. PMID: 18650914

Mantovani A, Marchesi F, Malesci A et al (2017) Tumour-associated macrophages as treatment targets in oncology. Nat Rev Clin Oncol 14(7):399–416. https://doi.org/10.1038/nrclinonc.2016.217. PMID: 28117416; PMCID: PMC548060

Manzo G (2019) Similarities between embryo development and cancer process suggest new strategies for research and therapy of tumors: a new point of view. Front Cell Dev Biol 7:20. https://doi.org/10.3389/fcell.2019.00020. PMID: 30899759; PMCID: PMC6416183

Marabelle A, Fakih M, Lopez J et al (2020) Association of tumour mutational burden with outcomes in patients with advanced solid tumours treated with pembrolizumab: prospective biomarker analysis of the multicohort, open-label, phase 2 KEYNOTE-158 study. Lancet Oncol 21(10):1353–1365. https://doi.org/10.1016/S1470-2045(20)30445-9. PMID: 32919526

Marco JH, Llombart P, Romero R et al (2024) Meaning-centered psychotherapy versus cognitive behavioral therapy for cancer survivors: a randomized controlled trial. Behav Ther 55(5):1071–1083. https://doi.org/10.1016/j.beth.2024.03.005. PMID: 39174266

Margulis L (1961) On the origin of mitosing cells. J Theor Biol 14:225–274

Marijuán PC, Navarro J (2022) The biological information flow: From cell theory to a new evolutionary synthesis. Biosystems 213:104631. https://doi.org/10.1016/j.biosystems.2022.104631. PMID: 35091015

Marijuán PC, Navarro J, del Moral R (2010) On prokaryotic intelligence: strategies for sensing the environment. Biosystems 99(2):94–103. https://doi.org/10.1016/j.biosystems.2009.09.004. PMID: 19781596

Marijuán PC, del Moral R, Navarro J (2013) On eukaryotic intelligence: signaling system's guidance in the evolution of multicellular organization. Biosystems 114(1):8–24. https://doi.org/10.1016/j.biosystems.2013.06.005. PMID: 23850535

Marijuán PC, Navarro J, del Moral R (2015) How the living is in the world: An inquiry into the informational choreographies of life. Prog Biophys Mol Biol 119(3):469–480. https://doi.org/10.1016/j.pbiomolbio.2015.07.002. PMID: 2616977

Marima R, Francies FZ, Hull R, Molefi T, Oyomno M, Khanyile R et al (2021) MicroRNA and alternative mRNA splicing events in cancer drug response/resistance: potent therapeutic targets. Biomedicines 9(12):1818. https://doi.org/10.3390/biomedicines9121818. PMID: 34944633; PMCID: PMC8698559

Marinac CR, McDonnell CH, Nadauld LD et al (2025) Clinical evaluation of cancer signal origin prediction and diagnostic resolution following multi-cancer early detection testing in the PATHFINDER study. Cancer Prev Res (Phila). https://doi.org/10.1158/1940-6207.CAPR-24-0468. PMID: 4026556

Marin-Acevedo JA, Kimbrough EO, Lou Y (2021) Next generation of immune checkpoint inhibitors and beyond. J Hematol Oncol 14:45. https://doi.org/10.1186/s13045-021-01056-8

Mariscal C, Dolittle WF (2018) Life and life only: a radical alternative to life definitionism. Synthese. 197(1):2975–2989. https://doi.org/10.1007/s11229-018-1852-2

Markouli M, Strepkos D, Basdra EK et al (2021) Prominent role of histone modifications in the regulation of tumor metastasis. Int J Mol Sci 22(5):2778. https://doi.org/10.3390/ijms22052778. PMID: 33803458; PMCID: PMC7967218

Marletto C (2015) Constructor theory of life. J R Soc Interf 12(104):20141226. https://doi.org/10.1098/rsif.2014.1226. PMID: 25589566; PMCID: PMC4345487

Marletto C (2021) The science of can and can't a physicists journey through the land of counterfactuals. Penguin Books. An Imprint of Penguin Random House LLC

Marliere P (2009) The farther, the safer: a manifesto for securely navigating synthetic species away from the old living world. Syst Synth Biol 3(1–4):77–84. https://doi.org/10.1007/s11693-009-9040-9. PMID: 19816802; PMCID: PMC2759432

Marmiere M, D'Amico F, Zangrillo A, Landoni G (2021) The 5 reasons why people die of Coronavirus Disease 2019. J Cardiothorac Vasc Anesth 35(11):3454–3455. https://doi.org/10.1053/j.jvca.2021.03.045. PMID: 33966995; PMCID: PMC8012269

Marquina G, Sanchez-Ramon S, Sarnago A et al (2021) Complete spontaneous regression of lung metastases after resection of CIC-rearranged sarcoma: a case report. Case Rep Oncol 14(1):152–159. https://doi.org/10.1159/000512276. PMID: 33776697; PMCID: PMC7983645

Marshall P (2015) Evolution 2.0: breaking the deadlock between Darwin and design. BenBella Books, Inc, Dallas

Marshall P (2021) Biology transcends the limits of computation. Prog Biophys Mol Biol. 165:88–101. https://doi.org/10.1016/j.pbiomolbio.2021.04.006. PMID: 33961842

Marshall SM, Murray ARG, Cronin L (2017) A probabilistic framework for identifying biosignatures using pathway complexity. Philos Trans A Math Phys Eng Sci 375(2109):20160342. https://doi.org/10.1098/rsta.2016.0342. PMID: 29133442; PMCID: PMC5686400

Marshall SM, Mathis C, Carrick E et al (2021) Identifying molecules as biosignatures with assembly theory and mass spectrometry. Nat Commun 12(1):3033. https://doi.org/10.1038/s41467-021-23258-x. PMID: 34031398; PMCID: PMC8144626

Martin WF (2020) Older than genes: the acetyl CoA pathway and origins. Front Microbiol 11:817. https://doi.org/10.3389/fmicb.2020.00817. PMID: 32655499; PMCID: PMC7325901

Martin OA, Anderson RL (2018) Editorial: therapy-induced metastasis. Clin Exp Metastasis 35(4):219–221. https://doi.org/10.1007/s10585-018-9914-x. PMID: 29971590

Martin-Belmonte F, Perez-Moreno M (2011) Epithelial cell polarity, stem cells and cancer. Nat Rev Cancer 12(1):23–38. https://doi.org/10.1038/nrc3169. PMID: 22169974

Martincorena I (2019) Somatic mutation and clonal expansions in human tissues. Genome Med 11(1):35. https://doi.org/10.1186/s13073-019-0648-4. PMID: 31138277; PMCID: PMC6537184

Martincorena I, Campbell PJ (2015) Somatic mutation in cancer and normal cells. Science 349(6255):1483–1489. https://doi.org/10.1126/science.aab4082. PMID: 26404825

Martincorena I, Roshan A, Gerstung M et al (2015) Tumor evolution. High burden and pervasive positive selection of somatic mutations in normal human skin. Science 348(6237):880–886. https://doi.org/10.1126/science.aaa6806. PMID: 25999502; PMCID: PMC4471149

Martincorena I, Jones PH, Campbell PJ (2016) Constrained positive selection on cancer mutations in normal skin. Proc Natl Acad Sci U S A 113(9):E1128–E1129. https://doi.org/10.1073/pnas.1600910113. PMID: 26884187; PMCID: PMC4780655

Martincorena I, Raine KM, Gerstung M et al (2017) Universal patterns of selection in cancer and somatic tissues. Cell 171(5):1029–1041.e21. https://doi.org/10.1016/j.cell.2017.09.042. PMID: 29056346; PMCID: PMC5720395

Martinez Arias A (2023) The master builder: how the new science of the cell is rewriting the story of life. Basic Books

Martinez P, Sabatier JM (2025) Malignant tumors in vagal-innervated organs: exploring its homeostatic role. Cancer Lett 617:217539. https://doi.org/10.1016/j.canlet.2025.217539

Martinez-Outschoorn UE, Pavlides S, Howell A et al (2011) Stromal-epithelial metabolic coupling in cancer: integrating autophagy and metabolism in the tumor microenvironment. Int J Biochem Cell Biol 43(7):1045–1051. https://doi.org/10.1016/j.biocel.2011.01.023. PMID: 21300172; PMCID: PMC3102770

Martinez-Outschoorn UE, Lin Z, Whitaker-Menezes D et al (2012) Ketone body utilization drives tumor growth and metastasis. Cell Cycle 11(21):3964–3971. https://doi.org/10.4161/cc.22137. PMID: 23082722; PMCID: PMC3507492

Martínez-Reyes I, Chandel NS (2021) Cancer metabolism: looking forward. Nat Rev Cancer 21(10):669–680. https://doi.org/10.1038/s41568-021-00378-6. PMID: 34272515

Martins Rodrigues F, Terekhanova NV, Imbach KJ et al (2025) Precision proteogenomics reveals pan-cancer impact of germline variants. Cell 188(9):2312–2335.e26. https://doi.org/10.1016/j.cell.2025.03.026

Marusyk A, Tabassum DP, Altrock PM et al (2014) Non-cell-autonomous driving of tumour growth supports sub-clonal heterogeneity. Nature 514(7520):54–58. https://doi.org/10.1038/nature13556. PMID: 25079331; PMCID: PMC4184961

Marzese DM, Manughian-Peter AO, Orozco JIJ, Hoon DSB (2018) Alternative splicing and cancer metastasis: prognostic and therapeutic applications. Clin Exp Metastasis 35(5–6):393–402. https://doi.org/10.1007/s10585-018-9905-y. PMID: 29845349

Masri S, Cervantes M, Sassone-Corsi P (2013) The circadian clock and cell cycle: interconnected biological circuits. Curr Opin Cell Biol 25(6):730–734. https://doi.org/10.1016/j.ceb.2013.07.013. PMID: 23969329; PMCID: PMC4573394

Mathews J, Levin M (2018) The body electric 2.0: recent advances in developmental bioelectricity for regenerative and synthetic bioengineering. Curr Opin Biotechnol 52:134–144. https://doi.org/10.1016/j.copbio.2018.03.008

Mathews J, Chang AJ, Devlin L, Levin M (2023) Cellular signaling pathways as plastic, proto-cognitive systems: Implications for biomedicine. Patterns (N Y) 4(5):100737. https://doi.org/10.1016/j.patter.2023.100737. PMID: 37223267; PMCID: PMC10201306

Matias-Guiu X, Stanta G, Carneiro F et al (2020) The leading role of pathology in assessing the somatic molecular alterations of cancer: position paper of the European Society of Pathology. Virchows Arch 476(4):491–497. https://doi.org/10.1007/s00428-020-02757-0

Matson V, Fessler J, Bao R, al. (2018) The commensal microbiome is associated with anti-PD-1 efficacy in metastatic melanoma patients. Science 359(6371):104–108. https://doi.org/10.1126/science.aao3290. PMID: 29302014; PMCID: PMC6707353

Mattiuzzi C, Lippi G (2019) Current cancer epidemiology. J Epidemiol Glob Health 9(4):217–222. https://doi.org/10.2991/jegh.k.191008.001. PMID: 31854162; PMCID: PMC7310786

Mattson MP (2008) Hormesis and disease resistance: activation of cellular stress response pathways. Hum Exp Toxicol 27(2):155–162. https://doi.org/10.1177/0960327107083417. PMID: 18480142

Maturana HR, Varela FJ (1980) Autopoiesis and cognition: The realization of the living. Reidel Publishing

Maude SL, Laetsch TW, Buechner J et al (2018) Tisagenlecleucel in children and young adults with B-cell lymphoblastic leukemia. N Engl J Med 378(5):439–448. https://doi.org/10.1056/NEJMoa1709866

Mauffrey P, Tchitchek N, Barroca V et al (2019) Progenitors from the central nervous system drive neurogenesis in cancer. Nature 569(7758):672–678. https://doi.org/10.1038/s41586-019-1219-y. PMID: 31092925

Mavaddat N, Peock S, Frost D et al (2013) Cancer risks for BRCA1 and BRCA2 mutation carriers: results from prospective analysis of EMBRACE. J Natl Cancer Inst 105(11):812–822. https://doi.org/10.1093/jnci/djt095

Mayerle J, Kalthoff H, Reszka R et al (2018) Metabolic biomarker signature to differentiate pancreatic ductal adenocarcinoma from chronic pancreatitis. Gut 67(1):128–137. https://doi.org/10.1136/gutjnl-2016-312432. PMID: 28108468; PMCID: PMC5754849

Mayers JR, Vander Heiden MG (2017) Nature and nurture: what determines tumor metabolic phenotypes? Cancer Res 77(12):3131–3134. https://doi.org/10.1158/0008-5472.CAN-17-0165. PMID: 28584183

Mayers JR, Wu C, Clish CB et al (2014) Elevation of circulating branched-chain amino acids is an early event in human pancreatic adenocarcinoma development. Nat Med 20(10):1193–1198. https://doi.org/10.1038/nm.3686

Mayfield JE (2013) The engine of complexity: evolution as computation. Columbia University Press

Mayr E (1961) Cause and effect in biology. Science 134(3489):1501–1506. https://doi.org/10.1126/science.134.3489.1501. PMID: 14471768

Mazzocca A (2019) The systemic-evolutionary theory of the origin of cancer (SETOC): a new interpretative model of cancer as a complex biological system. Int J Mol Sci 20(19):4885. https://doi.org/10.3390/ijms20194885. PMID: 31581628; PMCID: PMC6801598

Mazzoccoli G (2022) Chronobiology meets quantum biology: a new paradigm overlooking the horizon? Front Physiol 13:892582. https://doi.org/10.3389/fphys.2022.892582. PMID: 35874510; PMCID: PMC9296773

Mazzone M, Dettori D, de Oliveira RL et al (2009) Heterozygous deficiency of PHD2 restores tumor oxygenation and inhibits metastasis via endothelial normalization. Cell 136(5):839–851. https://doi.org/10.1016/j.cell.2009.01.020. PMID: 19217150; PMCID: PMC4037868

Mazzone PJ, Bach PB, Carey J et al (2024) Clinical validation of a cell-free DNA fragmentome assay for augmentation of lung cancer early detection. Cancer Discov 14(11):2224–2242. https://doi.org/10.1158/2159-8290.CD-24-0519

Mbemi A, Khanna S, Njiki S, et al. PB (2020). Impact of gene-environment interactions on cancer development. Int J Environ Res Public Health, 17(21):8089. doi: https://doi.org/10.3390/ijerph17218089. PMID: 33153024; PMCID: PMC7662361

McAllister SS, Weinberg RA (2010) Tumor-host interactions: a far-reaching relationship. J Clin Oncol 28(26):4022–4028. https://doi.org/10.1200/JCO.2010.28.4257. PMID: 20644094

McAllister SS, Weinberg RA (2014) The tumour-induced systemic environment as a critical regulator of cancer progression and metastasis. Nat Cell Biol 16(8):717–727. https://doi.org/10.1038/ncb3015. PMID: 25082194; PMCID: PMC6220424

McAllister SS, Gifford AM, Greiner AL et al (2008) Systemic endocrine instigation of indolent tumor growth requires osteopontin. Cell 133(6):994–1005. https://doi.org/10.1016/j.cell.2008.04.045. PMID: 18555776; PMCID: PMC4121664

McCarthy MF (1994) Longevity effect of chromium picolinate—“Rejuvenation” of hypothalamic function. Med Hypotheses 43:253–265. https://doi.org/10.1016/0306-9877(94)90076-0

McCarty R (2016). The alarm phase and the general adaptation syndrome: two aspects of Selye's inconsistent legacy. In: Fink G (ed) Stress: Concepts, Cognition, Emotion, and Behavior. Handbook of Stress Series Volume 1. Elsevier, pp 17–23

McCorkle R, Ercolano E, Lazenby M et al (2011) Self-management: enabling and empowering patients living with cancer as a chronic illness. CA Cancer J Clin 61(1):50–62. https://doi.org/10.3322/caac.20093. PMID: 21205833; PMCID: PMC3058905

McCullough ML, Hodge RA, Campbell PT, Stevens VL, Wang Y (2021) Pre-diagnostic circulating metabolites and colorectal cancer risk in the Cancer Prevention Study-II Nutrition Cohort. Metabolites 11(3):156. https://doi.org/10.3390/metabo11030156. PMID: 33803340; PMCID: PMC8000483

McDonald OG, Li X, Saunders T et al (2017) Epigenomic reprogramming during pancreatic cancer progression links anabolic glucose metabolism to distant metastasis. Nat Genet:367–376. https://doi.org/10.1038/ng.3753

McEwen BS (2005) Stressed or stressed out: what is the difference? J Psychiatry Neurosci 30(5):315–318. PMID: 16151535; PMCID: PMC1197275

McEwen BS, Wingfield JC (2003) The concept of allostasis in biology and biomedicine. Horm Behav 43(1):2–15. https://doi.org/10.1016/s0018-506x(02)00024-7. PMID: 12614627

McFadden J, Al-Khalili J (2018) The origins of quantum biology. Proc Math Phys Eng Sci 474(2220):20180674. https://doi.org/10.1098/rspa.2018.0674. PMID: 30602940; PMCID: PMC6304024

McFarland CD, Yaglom JA, Wojtkowiak JW et al (2017) The damaging effect of passenger mutations on cancer progression. Cancer Res 77(18):4763–4772. https://doi.org/10.1158/0008-5472.CAN-15-3283-T. PMID: 28536279; PMCID: PMC5639691

McGough IJ, Vincent JP (2016) Exosomes in developmental signalling. Development 143(14):2482–2493. https://doi.org/10.1242/dev.126516. PMID: 27436038

McGranahan N, Furness AJ, Rosenthal R et al (2016) Clonal neoantigens elicit T cell immunoreactivity and sensitivity to immune checkpoint blockade. Science 351(6280):1463–1469. https://doi.org/10.1126/science.aaf1490. PMID: 26940869; PMCID: PMC4984254

McGuinn LA, Ghazarian AA, Ellison GL et al (2012) Cancer and environment: definitions and misconceptions. Environ Res 112:230–234. https://doi.org/10.1016/j.envres.2011.10.009

McMahon KW, Karunasena E, Ahuja N (2017) The roles of DNA methylation in the stages of cancer. Cancer J 23(5):257–261. https://doi.org/10.1097/PPO.0000000000000279. PMID: 28926425; PMCID: PMC5657558

McShea DW (1996) Perspective: Metazoan complexity and evolution: is there a trend? Evolution 50(2):477–492. https://doi.org/10.1111/j.1558-5646.1996.tb03861.x. PMID: 28568940

McTiernan A, Friedenreich CM, Katzmarzyk PT (2019) Physical activity in cancer prevention and survival: a systematic review. Med Sci Sports Exerc 51(6):1252–1261. https://doi.org/10.1249/MSS.0000000000001937

Meijers WC, de Boer RA (2019) Common risk factors for heart failure and cancer. Cardiovasc Res 115(5):844–853. https://doi.org/10.1093/cvr/cvz035

Meijers WC, Maglione M, Bakker SJL et al (2018) Heart failure stimulates tumor growth by circulating factors. Circulation 138(7):678–691. https://doi.org/10.1161/CIRCULATIONAHA.117.030816. PMID: 29459363

Meincke AS (2018) Autopoiesis, biological autonomy and the process view of life. Eur J Philos Sci. https://doi.org/10.1007/s13194-018-0228-2

Mejia AA, Nakamura T, Masatoshi I et al (1991) Estimation of absorbed doses in humans due to intravenous administration of fluorine-18-fluorodeoxyglucose in PET studies. J Nucl Med 32(4):699–706. PMID: 2013810

Mekler LB (1971) Hybridization of transformed cells with lymphocytes as one of the probable courses of the progression leading to the development of metastatic malignant cells. Vestn Akad Med Nauk SSSR 26:80–89

Mello T, Simeone I, Galli A (2019) Mito-nuclear communication in hepatocellular carcinoma metabolic rewiring. Cells 8(5):417. https://doi.org/10.3390/cells8050417. PMID: 31060333; PMCID: PMC6562577

Mendes C, Lemos I, Francisco I et al (2023) NSCLC presents metabolic heterogeneity, and there is still some leeway for EGF stimuli in EGFR-mutated NSCLC. Lung Cancer 182:107283. https://doi.org/10.1016/j.lungcan.2023.107283. PMID: 37379672

Mendiratta G, Ke E, Aziz M et al (2021) Cancer gene mutation frequencies for the U.S. population. Nat Commun 12(1):5961. https://doi.org/10.1038/s41467-021-26213-y. PMID: 34645806; PMCID: PMC8514428

Mendiratta G, Jones MK, Stites EC (2022) How often is each gene mutated within the cancer patient population? Mol Cell Oncol 9(1):2065176. https://doi.org/10.1080/23723556.2022.2065176. PMID: 35529901; PMCID: PMC9067461

Meng X, Schwarzenbach H, Yang Y et al (2019) Circulating mitochondrial DNA is linked to progression and prognosis of epithelial ovarian cancer. Transl Oncol 12(9):1213–1220. https://doi.org/10.1016/j.tranon.2019.05.015. PMID: 31271962; PMCID: PMC6609736

Meng W, Pan H, Sha Y et al (2024) Metabolic connectome and its role in the prediction, diagnosis, and treatment of complex diseases. Metabolites 14(2):93. https://doi.org/10.3390/metabo14020093. PMID: 38392985; PMCID: PMC10890086

Menichetti G, Barabási AL, Loscalzo J (2025) Chemical complexity of food and implications for therapeutics. N Engl J Med 392(18):1836–1845. https://doi.org/10.1056/NEJMra2413243. PMID: 40334158

Menter DG, Tucker SC, Kopetz S et al (2014) Platelets and cancer: a casual or causal relationship: revisited. Cancer Metastasis Rev 33(1):231–269. https://doi.org/10.1007/s10555-014-9498-0. PMID: 24696047; PMCID: PMC4186918

Menyhárt O, Győrffy B (2021) Multi-omics approaches in cancer research with applications in tumor subtyping, prognosis, and diagnosis. Comput Struct Biotechnol J 19:949–960. https://doi.org/10.1016/j.csbj.2021.01.009. PMID: 33613862; PMCID: PMC7868685

Meric-Bernstam F, Tannir NM et al (2022) Telaglenastat plus cabozantinib or everolimus for advanced or metastatic renal cell carcinoma: an open-label phase I trial. Clin Cancer Res 28(8):1540–1548. https://doi.org/10.1158/1078-0432.CCR-21-2972. PMID: 35140121; PMCID: PMC9164172

Merkel DE, McGuire WL (1990) Ploidy, proliferative activity and prognosis. DNA flow cytometry of solid tumors. Cancer 65(5):1194–1205. https://doi.org/10.1002/1097-0142(19900301)65:5<1194::aid-cncr2820650528>3.0.co;2-m. PMID: 2406011

Mészáros Crow E, López-Gigosos R, Mariscal-López E et al (2023) Psychosocial interventions reduce cortisol in breast cancer patients: systematic review and meta-analysis. Front Psychol 14:1148805. https://doi.org/10.3389/fpsyg.2023.1148805. PMID: 37441329; PMCID: PMC10333719

Meyer W (1931) Cancer. Its origin, its development and its self-perpetuation. The therapy of operable and inoperable cancer in the light of a systemic conception of malignancy. Paul B. Hoeber, Inc, New York

Meyer MA (2006) Positron emission tomography in cancer therapy. N Engl J Med 354:1958–1959

Meyer MA, Caday CG, Han Y et al (1996) Potential radiotherapy of human gliomas with 18F-fluorodeoxyglucose (18F-FDG). Soc Neurosci Abstr 22:948. (abstract)

Meyerhardt JA, Giovannucci EL, Holmes MD et al (2006) Physical activity and survival after colorectal cancer diagnosis. J Clin Oncol 24(22):3527–3534. https://doi.org/10.1200/JCO.2006.06.0855. PMID: 16822844

Michels EHA, Appelman B, de Brabander J et al (2023) Age-related changes in plasma biomarkers and their association with mortality in COVID-19. Eur Respir J 62(1):2300011. https://doi.org/10.1183/13993003.00011-2023. PMID: 37080568; PMCID: PMC10151455

Michod RE (2022) Multi-level selection of the individual organism. In: The Evolution of Multicellularity. CRC Press, pp 25–52

Michod RE, Nedelcu AM, Roze D (2003) Cooperation and conflict in the evolution of individuality. IV. Conflict mediation and evolvability in Volvox carteri. Biosystems 69(2–3):95–114. https://doi.org/10.1016/s0303-2647(02)00133-8. PMID: 12689724

Michod RE, Davison DR, Sanders H et al (2022) Translating research on evolutionary transitions into the teaching of biological complexity. Evolution 76(6):1124–1138. https://doi.org/10.1111/evo.14479. PMID: 35373337

Middleton JD, Stover DG, Hai T (2018) Chemotherapy-exacerbated breast cancer metastasis: a paradox explainable by dysregulated adaptive-response. Int J Mol Sci 19(11):3333. https://doi.org/10.3390/ijms19113333. PMID: 30373101; PMCID: PMC6274941

Miklos GL, Maleszka R (2001) Protein functions and biological contexts. Proteomics 1(2):169–178. https://doi.org/10.1002/1615-9861(200102)1:2<169::AID-PROT169>3.0.CO;2-C. PMID: 11680863

Miller JG (1978) Living systems. McGraw-Hill

Miller JG (1985) General living systems theory. In: Pichot P et al (eds) Biological psychiatry. Higher nervous activity. Springer, New York

Miller WB (2023) Cognition-based evolution. Natural cellular engineering and the intelligent cell. CRC Press

Miller WB Jr, Torday JS (2017) A systematic approach to cancer: evolution beyond selection. Clin Transl Med 6(1):2. https://doi.org/10.1186/s40169-016-0131-4. PMID: 28050778; PMCID: PMC5209328

Miller WB Jr, Torday JS (2018) Four domains: the fundamental unicell and post-Darwinian cognition-based evolution. Prog Biophys Mol Biol 140:49–73. https://doi.org/10.1016/j.pbiomolbio.2018.04.006. PMID: 29685747

Miller WB Jr, Reber AS, Marshall P, Baluška F (2023) Cellular and natural viral engineering in cognition-based evolution. Commun Integr Biol 16(1):2196145. https://doi.org/10.1080/19420889.2023.2196145. PMID: 37153718; PMCID: PMC10155641

Miller JL, Miller JG (1992) Greater than the sum of its parts. I. Subsystems which process both matter-energy and information. Behav Sci 37(1):1–9. https://doi.org/10.1002/bs.3830370102

Miller GE, Chen E, Zhou ES (2007) If it goes up, must it come down? Chronic stress and the hypothalamic-pituitary-adrenocortical axis in humans. Psychol Bull 133(1):25–45. https://doi.org/10.1037/0033-2909.133.1.25. PMID: 17201569

Mills JR (2017) Precision medicine—right treatment, right patient, right time, wrong approach? Clin Chem 63(4):928–929. https://doi.org/10.1373/clinchem.2016.267963. PMID: 28351859

Minea R, Helchowski C, Rubino B et al (2012) Development of a chimeric recombinant disintegrin as a cost-effective anti-cancer agent with promising translational potential. Toxicon 59(4):472–486. https://doi.org/10.1016/j.toxicon.2011.02.020. PMID: 21354198; PMCID: PMC3130806

Minger D (2010) The China study: fact or fallacy? http://rawfoodsos.com/2010/07/07/the-china-study-fact-or-fallac/

Mirica RE (2010) Deuterium-depleted water in cancer therapy. Environ Eng Manag J 9(11):1543–1545. https://doi.org/10.30638/EEMJ.2010.210

Mishra V, Mishra Y (2024) Role of gut microbiome in cancer treatment. Indian J Microbiol 64(3):1310–1325. https://doi.org/10.1007/s12088-024-01340-4. PMID: 39282183; PMCID: PMC11399371

Mistriotis A (2021) A universal model describing the structure and functions of living systems. Commun Integr Biol 14(1):27–36. https://doi.org/10.1080/19420889.2021.1887549. PMID: 33680281; PMCID: PMC7906254

Mittal D, Gubin MM, Schreiber RD, Smyth MJ (2014) New insights into cancer immunoediting and its three component phases—elimination, equilibrium and escape. Curr Opin Immunol 27:16–25. https://doi.org/10.1016/j.coi.2014.01.004. PMID: 24531241; PMCID: PMC4388310

Mittal A, Kim MS, Dunn S et al (2024) Frequently asked questions on surrogate endpoints in oncology—opportunities, pitfalls, and the way forward. EClinicalMedicine 76:102824. https://doi.org/10.1016/j.eclinm.2024.102824. PMID: 39764569; PMCID: PMC11701476

Mix LJ (2015) Defending definitions of life. Astrobiology 15(1):15–19. https://doi.org/10.1089/ast.2014.1191. PMID: 25415254

Miyamoto T, Nagao A, Okumura N, Hosaka M (2022) Effect of post-diagnosis physical activity on breast cancer recurrence: a systematic review and meta-analysis. Curr Oncol Rep 24(11):1645–1659. https://doi.org/10.1007/s11912-022-01287-z. PMID: 35947284

Moadel RM, Nguyen AV, Lin EY et al (2003) Positron emission tomography agent 2-deoxy-2-[18F]fluoro-D-glucose has a therapeutic potential in breast cancer. Breast Cancer Res 5(6):R199–

R205. https://doi.org/10.1186/bcr643. PMID: 14580255; PMCID: PMC314404
Moadel RM, Weldon RH, Katz EB et al (2005) Positherapy: targeted nuclear therapy of breast cancer with 18F-2-deoxy-2-fluoro-D-glucose. Cancer Res 65(3):698–702. PMID: 15705862
Moadel AB, Harris M, Kolidas E, Santizo R, Harris K, Paul D (2012) Psycho-educational and psycho-spiritual interventions for low-income cancer patients: results of randomized
Moazed D (2009) Small RNAs in transcriptional gene silencing and genome defence. Nature 457(7228):413–420. https://doi.org/10.1038/nature07756. PMID: 19158787; PMCID: PMC3246369
Moger-Reischer RZ, Glass JI, Wise KS et al (2023) Evolution of a minimal cell. Nature 620(7972):122–127. https://doi.org/10.1038/s41586-023-06288-x. PMID: 37407813; PMCID: PMC10396959
Mohan BP, Sivaraman L, Vijayan S, Das AG (2023) Spectrum of incidental tumors in autopsy specimens: a histopathological analysis of 797 cases. Indian J Health Sci Biomed Res 16(2):292–297. https://doi.org/10.4103/kleuhsj.kleuhsj_624_22
Mohd Khair SZN, Abd Radzak SM, Mohamed Yusoff AA (2021) The uprising of mitochondrial DNA biomarker in cancer. Dis Markers 2021:7675269. https://doi.org/10.1155/2021/7675269. PMID: 34326906; PMCID: PMC8302403
Molfino A, Iannace A, Colaiacomo MC et al (2017) Cancer anorexia: hypothalamic activity and its association with inflammation and appetite-regulating peptides in lung cancer. J Cachexia Sarcopenia Muscle 8(1):40–47. https://doi.org/10.1002/jcsm.12156
Molina JR, Sun Y, Protopopova M et al (2018) An inhibitor of oxidative phosphorylation exploits cancer vulnerability. Nat Med 24(7):1036–1046. https://doi.org/10.1038/s41591-018-0052-4
Monk M, Holding C (2001) Human embryonic genes re-expressed in cancer cells. Oncogene 20:8085–8091. https://doi.org/10.1038/sj.onc.1205088. PMID: 11781821
Monneret C (2011) Platinum anticancer drugs. From serendipity to rational design. Ann Pharm Fr 69(6):286–295. https://doi.org/10.1016/j.pharma
Montégut L, López-Otín C, Kroemer G (2024) Aging and cancer. Mol Cancer 23(1):106. https://doi.org/10.1186/s12943-024-02020-z
Montévil M (2022a) Historicity at the heart of biology. Theory Biosci 141:165–173. https://doi.org/10.1007/s12064-020-0032-8
Montévil M (2022b) Understanding living beings by analogy with computers or understanding computers as an emanation of the living. Trópoς. Rivista Di Ermeneutica e Critica Filosofica 13(2). https://doi.org/10.53136/97912218011945
Montévil M, Mossio M (2015) Biological organisation as closure of constraints. J Theor Biol 372:179–191. https://doi.org/10.1016/j.jtbi.2015.02.029
Montévil M, Mossio M (2020) The identity of organisms in scientific practice: integrating historical and relational conceptions. Front Physiol 11:611. https://doi.org/10.3389/fphys.2020.00611
Montévil M, Speroni L, Sonnenschein C, Soto AM (2016) Modeling mammary organogenesis from biological first principles: cells and their physical constraints. Prog Biophys Mol Biol 122(1):58–69. https://doi.org/10.1016/j.pbiomolbio.2016.08.004
Monti N, Verna R, Piombarolo A et al (2022) Paradoxical behavior of oncogenes undermines the somatic mutation theory. Biomolecules 12(5):662. https://doi.org/10.3390/biom12050662
Montico B, Giurato G, Pecoraro G, et al. E (2022) The pleiotropic roles of circular and long noncoding RNAs in cutaneous melanoma. Mol Oncol 16(3):565–593. https://doi.org/10.1002/1878-0261.13034
Monzel AS, Enríquez JA, Picard M (2023) Multifaceted mitochondria: moving mitochondrial science beyond function and dysfunction. Nat Metab 5(4):546–562. https://doi.org/10.1038/s42255-023-00783-1
Moody ERR, Álvarez-Carretero S, Mahendrarajah TA et al (2024) The nature of the last universal common ancestor and its impact on the early Earth system. Nat Ecol Evol 8(9):1654–1666. https://doi.org/10.1038/s41559-024-02461-1
Moore D, Walker SI, Levin M (2017) Cancer as a disorder of patterning information: computational and biophysical perspectives on the cancer problem. Conv Sci Phys Oncol 3:043001. https://doi.org/10.1088/2057-1739/aa8548
Moore L, Cagan A, Coorens THH et al (2021) The mutational landscape of human somatic and germline cells. Nature 597(7876):381–386. https://doi.org/10.1038/s41586-021-03822-7
Morales JM, Michaelian K (2020) Photon dissipation as the origin of information encoding in RNA and DNA. Entropy (Basel) 22(9):940. https://doi.org/10.3390/e22090940
Morana O, Wood W, Gregory CD (2022) The apoptosis paradox in cancer. Int J Mol Sci 23(3):1328. https://doi.org/10.3390/ijms23031328
Morange MA (2021) History of biology. Princeton University Press, Princeton
Moreno A, Mossio M (2015) Biological autonomy: a philosophical and theoretical enquiry. History, philosophy and theory of the life sciences, vol 12. Springer, Dordrecht
Morita M, Kudo K, Shima H, Tanuma N (2021) Dietary intervention as a therapeutic for cancer. Cancer Sci 112(2):498–504. https://doi.org/10.1111/cas.14777
Mormont MC, Lévi F (2003) Cancer chronotherapy: principles, applications, and perspectives. Cancer 97(1):155–169. https://doi.org/10.1002/cncr.11040
Morowitz HJ (1966) Physical background of cycles in biological systems. J Theor Biol 13:60–62
Morowitz HJ (1968) Energy flow in biology. Academic, New York
Morowitz HJ (1992) The beginnings of cellular life. Yale University Press, New Haven
Morowitz HJ, Smith E (2007) Energy flow and the organization of life. Complexity 13:51–59. https://doi.org/10.1002/cplx.20191
Morrow ZT, Powers ZM, Sauer JD (2019) Listeria monocytogenes cancer vaccines: bridging innate and adaptive immunity. Curr Clin Microbiol Rep 6(4):213–224. https://doi.org/10.1007/s40588-019-00133-4
Moschopoulos CD, Dimopoulou D, Dimopoulou A et al (2023) New insights into the fluid management in patients with septic shock. Medicina (Kaunas) 59(6):1047. https://doi.org/10.3390/medicina59061047
Moser A, Stuck AE, Silliman RA et al (2012) The eight-item modified Medical Outcomes Study Social Support Survey: psychometric evaluation showed excellent performance. J Clin Epidemiol 65(10):1107–1116. https://doi.org/10.1016/j.jclinepi.2012.04.007
Moss L (2003) What genes can't do. MIT Press, Cambridge, MA
Mossio M (2024) Organization as a scientific blind spot. In: Huneman P, Wolfe C, Reydon T (eds) Organization in biology, history, philosophy and theory of the life sciences, vol 12. Springer. https://doi.org/10.1007/978-3-031-38968-9.hal-04335078
Mossio M, Montévil M, Longo G (2016) Theoretical principles for biology: organization. Prog Biophys Mol Biol 122(1):24–35. https://doi.org/10.1016/j.pbiomolbio.2016.07.005
Motsinger-Reif AA, Reif DM, Akhtari FS et al (2024) Gene-environment interactions within a precision environmental health framework. Cell Genom 4(7):100591. https://doi.org/10.1016/j.xgen.2024.100591
Mouchemore KA, Anderson RL, Hamilton JA (2018) Neutrophils, G-CSF and their contribution to breast cancer metastasis. FEBS J 285(4):665–679. https://doi.org/10.1111/febs.14206
Mousa SA, Glinsky GV, Lin HY et al (2018) Contributions of thyroid hormone to cancer metastasis. Biomedicines 6(3):89. https://doi.org/10.3390/biomedicines6030089

Mravec B, Horvathova L, Cernackova A (2019) Hypothalamic inflammation at a crossroad of somatic diseases. Cell Mol Neurobiol 39(1):11–29. https://doi.org/10.1007/s10571-018-0631-4

Mucci LA, Hjelmborg JB, Harris JR et al (2016) Familial risk and heritability of cancer among twins in Nordic countries. JAMA 315(1):68–76. https://doi.org/10.1001/jama.2015.17703

Muenst S, Läubli H, Soysal SD et al (2016) The immune system and cancer evasion strategies: therapeutic concepts. J Intern Med 279(6):541–562. https://doi.org/10.1111/joim.12470

Muir A, Danai LV, Vander Heiden MG (2018) Microenvironmental regulation of cancer cell metabolism: implications for experimental design and translational studies. Dis Model Mech 11(8):dmm035758. https://doi.org/10.1242/dmm.035758

Mukherjee S (2011) The emperor of all Maladies. Scribner, New York

Mukherjee S (2016) The gene: an intimate history Scribner, New York

Mukherjee S (2022) The song of the cell: an exploration of medicine and the new human. Scribner, New York

Müller M, Igamberdiev AU (2024) The emergence of theoretical biology: two fundamental works of Ervin Bauer (1890–1938) in English translation. Biosystems 241:105201. https://doi.org/10.1016/j.biosystems.2024.105201

Müller MJ, Bosy-Westphal A, Heymsfield SB (2010) Is there evidence for a set point that regulates human body weight? F1000 Med Rep 2:59. https://doi.org/10.3410/M2-59. PMID: 21173874

Mun EJ, Babiker HM, Weinberg U et al (2018) Tumor-treating fields: a fourth modality in cancer treatment. Clin Cancer Res 24(2):266–275. https://doi.org/10.1158/1078-0432.CCR-17-1117

Munson JM, Shieh AC (2014) Interstitial fluid flow in cancer: implications for disease progression and treatment. Cancer Manag Res 6:317–328. https://doi.org/10.2147/CMAR.S65444

Münzberg H, Singh P, Heymsfield SB et al (2020) Recent advances in understanding the role of leptin in energy homeostasis. F1000Res 9:F1000 Faculty Rev:451. https://doi.org/10.12688/f1000research.24260.1

Muranaka H, Akinsola R, Billet S et al (2024) Glutamine supplementation as an anticancer strategy: a potential therapeutic alternative to the convention. Cancers (Basel) 16(5):1057. https://doi.org/10.3390/cancers16051057

Murgai M, Giles A, Kaplan R (2015) Physiological, tumor, and metastatic niches: opportunities and challenges for targeting the tumor microenvironment. Crit Rev Oncog 20(3–4):301–314. https://doi.org/10.1615/critrevoncog.2015013668

Mushegian AR, Koonin EV (1996) A minimal gene set for cellular life derived by comparison of complete bacterial genomes. Proc Natl Acad Sci USA 93(19):10268–10273. https://doi.org/10.1073/pnas.93.19.10268

Naci H, Zhang Y, Woloshin S et al (2024) Overall survival benefits of cancer drugs initially approved by the US Food and Drug Administration on the basis of immature survival data: a retrospective analysis. Lancet Oncol 25(6):760–769. https://doi.org/10.1016/S1470-2045(24)00152-9

Nagahata Y, Kawamoto H (2023) Evolutionary reversion in tumorigenesis. Front. Oncologia 13:1282417. https://doi.org/10.3389/fonc.2023.1282417

Naghshi S, Sadeghi O, Willett WC, Esmaillzadeh A (2020) Dietary intake of total, animal, and plant proteins and risk of all-cause, cardiovascular, and cancer mortality: systematic review and dose-response meta-analysis of prospective cohort studies. BMJ 370:m2412. https://doi.org/10.1136/bmj.m2412

Nair PR, Danilova L, Gómez-de-Mariscal E et al (2024) MLL1 regulates cytokine-driven cell migration and metastasis. Sci Adv 10(11):eadk0785. https://doi.org/10.1126/sciadv.adk0785

Naito Y, Yoshioka Y, Ochiya T (2022) Intercellular crosstalk between cancer cells and cancer-associated fibroblasts via extracellular vesicles. Cancer Cell Int 22(1):367. https://doi.org/10.1186/s12935-022-02784-8

Nakabachi A, Yamashita A, Toh H et al (2006) The 160-kilobase genome of the bacterial endosymbiont Carsonella. Science 314(5797):267. https://doi.org/10.1126/science.1134196

Nakahira K, Kyung SY, Rogers AJ et al (2013) Circulating mitochondrial DNA in patients in the ICU as a marker of mortality: derivation and validation. PLoS Med 10(12):e1001577. https://doi.org/10.1371/journal.pmed.1001577

Nakajima EC, Van Houten B (2013) Metabolic symbiosis in cancer: refocusing the Warburg lens. Mol Carcinog 52(5):329–337. https://doi.org/10.1002/mc.21863

Nakamura K, Smyth MJ (2017) Targeting cancer-related inflammation in the era of immunotherapy. Immunol Cell Biol 95(4):325–332. https://doi.org/10.1038/icb.2016.126

Nakamura H, Takada K (2021) Reactive oxygen species in cancer: current findings and future directions. Cancer Sci 112(10):3945–3952. https://doi.org/10.1111/cas.15068

Nam JW, Choi SW, You BH (2016) Incredible RNA: dual functions of coding and noncoding. Mol Cells 39(5):367–374. https://doi.org/10.14348/molcells.2016.0039

Nam AS, Chaligne R, Landau DA (2021) Integrating genetic and non-genetic determinants of cancer evolution by single-cell multi-omics. Nat Rev Genet 22(1):3–18. https://doi.org/10.1038/s41576-020-0265-5

Nardin S, Mora E, Varughese FM et al (2020) Breast cancer survivorship, quality of life, and late toxicities. Front Oncol 10:864. https://doi.org/10.3389/fonc.2020.00864

Naser R, Fakhoury I, El-Fouani A et al (2023) Role of the tumor microenvironment in cancer hallmarks and targeted therapy (Review). Int J Oncol 62(2):23. https://doi.org/10.3892/ijo.2022.5471

Nath S, Ghatak D, Das P et al (2015) Transcriptional control of mitosis: deregulation and cancer. Front Endocrinol (Lausanne) 6:60. https://doi.org/10.3389/fendo.2015.00060

Nath A, Cosgrove PA, Chang JT, Bild AH (2022) Predicting clinical response to everolimus in ER+ breast cancers using machine-learning. Front Mol Biosci 9:981962. https://doi.org/10.3389/fmolb.2022.981962

National Research Council (2011) Toward precision medicine: building a knowledge network for biomedical research and a new taxonomy of disease. Committee on a Framework for Development a New Taxonomy of Disease. The National Academy Press, Washington, DC

Naumann P, Eberlein J, Farnia B et al (2019) Cachectic body composition and inflammatory markers portend a poor prognosis in patients with locally advanced pancreatic cancer treated with chemoradiation. Cancers (Basel) 11(11):1655. https://doi.org/10.3390/cancers11111655

Naxerova K, Bult CJ, Peaston A et al (2008) Analysis of gene expression in a developmental context emphasizes distinct biological leitmotifs in human cancers. Genome Biol 9(7):R108. https://doi.org/10.1186/gb-2008-9-7-r108

Nedelcu AM (2020) The evolution of multicellularity and cancer: views and paradigms. Biochem Soc Trans 48(4):1505–1518. https://doi.org/10.1042/BST20190992

Needleman D, Dogic Z (2017) Active matter at the interface between materials science and cell biology. Nat Rev Mater 2:17048. https://doi.org/10.1038/natrevmats.2017.48

Neelapu SS, Locke FL, Bartlett NL et al (2017) Axicabtagene ciloleucel CAR T-cell therapy in refractory large B-cell lymphoma. N Engl J Med 377(26):2531–2544. https://doi.org/10.1056/NEJMoa1707447

Neisser U, Boodoo G, Bouchard TJ Jr et al (1996) Intelligence: knowns and unknowns. Am Psychol 51:77–101. https://doi.org/10.1037/0003-066X.51.2.77

Nelson M (2021) Biosphere 2's lessons about living on Earth and in space. Space Sci Technol 2021(1–4):1–11. https://doi.org/10.34133/2021/8067539

Nelson CM, Bissell MJ (2006) Of extracellular matrix, scaffolds, and signaling: tissue architecture regulates development, homeostasis, and cancer. Annu Rev Cell Dev Biol 22:287–309. https://doi.org/10.1146/annurev.cellbio.22.010305.104315

Nencioni A, Caffa I, Cortellino S et al (2018) Fasting and cancer: molecular mechanisms and clinical application. Nat Rev Cancer 18(11):707–719

Neophytou CM, Panagi M, Stylianopoulos T et al (2021) The role of tumor microenvironment in cancer metastasis: molecular mechanisms and therapeutic opportunities. Cancers (Basel) 13(9):2053. https://doi.org/10.3390/cancers13092053

Neshat MS, Raitano AB, Wang HG et al (2000) The survival function of the Bcr-Abl oncogene is mediated by Bad-dependent and -independent pathways: roles for phosphatidylinositol 3-kinase and Raf. Mol Cell Biol 20(4):1179–1186. https://doi.org/10.1128/MCB.20.4.1179-1186.2000

Netherton C, Moffat K, Brooks E et al (2007) A guide to viral inclusions, membrane rearrangements, factories, and viroplasm produced during virus replication. Adv Virus Res 70:101–182. https://doi.org/10.1016/S0065-3527(07)70004-0

Neuhaus E, Zirjacks L, Ganser K et al (2019) Alternating electric fields (TTFields) activate Cav1.2 channels in human glioblastoma cells. Cancers (Basel) 11(1):110. https://doi.org/10.3390/cancers11010110

Newman AM, Lovejoy AF, Klass DM et al (2016) Integrated digital error suppression for improved detection of circulating tumor DNA. Nat Biotechnol 34(5):547–555. https://doi.org/10.1038/nbt.3520

Ng JCM, Schooling CM (2021) Effect of basal metabolic rate on cancer: a Mendelian randomization study. Front Genet 12:735541. https://doi.org/10.3389/fgene.2021.735541

Ngoi NYL, Gallo D, Torrado C et al (2025) Synthetic lethal strategies for the development of cancer therapeutics. Nat Rev Clin Oncol 22(1):46–64. https://doi.org/10.1038/s41571-024-00966-z

Nguyen DX, Massagué J (2007) Genetic determinants of cancer metastasis. Nat Rev Genet 8(5):341–352. https://doi.org/10.1038/nrg2101

Nguyen B, Fong C, Luthra A et al (2022) Genomic characterization of metastatic patterns from prospective clinical sequencing of 25,000 patients. Cell 185(3):563–575.e11. https://doi.org/10.1016/j.cell.2022.01.003

Nicholas C, Beharry A, Bendzsak AM et al (2024) Point of care liquid biopsy for cancer treatment—early experience from a community center. Cancers (Basel) 16(14):2505. https://doi.org/10.3390/cancers16142505

Nichols CA, Gibson WJ, Brown MS et al (2020) Loss of heterozygosity of essential genes represents a widespread class of potential cancer vulnerabilities. Nat Commun 11:2517. https://doi.org/10.1038/s41467-020-16399-y

Nicholson DJ, Dupré J (2018) A manifesto for a processual philosophy of biology. In: Daniel J, Nicholson D, Dupré J (eds) Everything flows: towards a processual philosophy of biology. Oxford University Press. https://doi.org/10.1093/oso/9780198779636.001.0001

Nicholson DJ (2019) Is the cell really a machine? J Theor Biol 477:108–126. https://doi.org/10.1016/j.jtbi.2019.06.002

Nicholson JK, Holmes E, Kinross J et al (2012) Host-gut microbiota metabolic interactions. Science 336(6086):1262–1267. https://doi.org/10.1126/science.1223813

Nicolazzo C, Francescangeli F, Magri V et al (2023) Is cancer an intelligent species? Cancer Metastasis Rev 42(4):1201–1218. https://doi.org/10.1007/s10555-023-10123-0

Nidadavolu LS, Feger D, Chen D et al (2023) Associations between circulating cell-free mitochondrial DNA, inflammatory markers, and cognitive and physical outcomes in community-dwelling older adults. Immun Ageing 20(1):24. https://doi.org/10.1186/s12979-023-00342-y

Nielsen SF, Nordestgaard BG, Bojesen SE (2012) Statin use and reduced cancer-related mortality. N Engl J Med 367(19):1792–1802. https://doi.org/10.1056/NEJMoa1201735

Nik-Zainal S, Hall BA (2019) Cellular survival over genomic perfection. Science 366(6467):802–803. https://doi.org/10.1126/science.aax8046

Niu FY, Zhou Q, Yang JJ et al (2016) Distribution and prognosis of uncommon metastases from non-small cell lung cancer. BMC Cancer 16:149. https://doi.org/10.1186/s12885-016-2169-5

Niu N, Mercado-Uribe I, Liu J (2017) Dedifferentiation into blastomere-like cancer stem cells via formation of polyploid giant cancer cells. Oncogene 36(34):4887–4900. https://doi.org/10.1038/onc.2017.72

Noble D (2006) The music of life: biology beyond genes. Oxford University Press, Oxford

Noble D (2008) Claude Bernard, the first systems biologist, and the future of physiology. Exp Physiol 93(1):16–26. https://doi.org/10.1113/expphysiol.2007.038695

Noble D (2012) A theory of biological relativity: no privileged level of causation. Interface Focus 2(1):55–64. https://doi.org/10.1098/rsfs.2011.0067

Noble D (2013) A biological relativity view of the relationships between genomes and phenotypes. Prog Biophys Mol Biol 111(2–3):59–65. https://doi.org/10.1016/j.pbiomolbio.2012.09.004

Noble R, Tasaki K, Noble PJ, Noble D (2019) Biological relativity requires circular causality but not symmetry of causation: so, where, what and when are the boundaries? Front Physiol 10:827. https://doi.org/10.3389/fphys.2019.00827

Nolen SC, Evans MA, Fischer A et al (2017) Cancer incidence, prevalence and mortality in the oldest-old: a comprehensive review. Mech Ageing Dev 164:113–126. https://doi.org/10.1016/j.mad.2017.05.002

Noorani A, Li X, Goddard M et al (2020) Genomic evidence supports a clonal diaspora model for metastases of esophageal adenocarcinoma. Nat Genet 52(1):74–83. https://doi.org/10.1038/s41588-019-0551-3

Noordam L, Kaijen MEH, Bezemer K et al (2018) Low-dose cyclophosphamide depletes circulating naïve and activated regulatory T cells in malignant pleural mesothelioma patients synergistically treated with dendritic cell-based immunotherapy. Oncoimmunology 7(12):e1474318. https://doi.org/10.1080/2162402X.2018.1474318

Norcross MA (1984) A synaptic basis for T-lymphocyte activation. Ann Immunol (Paris) 135D(2):113–134. https://doi.org/10.1016/s0769-2625(84)81105-8

Nordling CO (1953) A new theory on the cancer-inducing mechanism. Br J Cancer 7:68–72

North G (2011) Third genetic code, anyone? Curr Biol 21(6):R203. https://doi.org/10.1016/j.cub.2011.02.024

Nøst TH, Alcala K, Urbarova I et al (2021) Systemic inflammation markers and cancer incidence in the UK Biobank. Eur J Epidemiol 36(8):841–848. https://doi.org/10.1007/s10654-021-00752-6

Notarangelo G, Spinelli JB, Perez EM et al (2022) Oncometabolite d-2HG alters T cell metabolism to impair CD8+ T cell function. Science 377(6614):1519–1529. https://doi.org/10.1126/science.abj5104

Novais EJ, Tran VA, Johnston SN et al (2021) Long-term treatment with senolytic drugs dasatinib and quercetin ameliorates age-dependent intervertebral disc degeneration in mice. Nat Commun 12:5213. https://doi.org/10.1038/s41467-021-25453-2

Novoa RR, Calderita G, Arranz R et al (2005) Virus factories: associations of cell organelles for viral replication and morphogenesis. Biol Cell 97:147–172. https://doi.org/10.1042/BC20040058

Nowak MA, Waclaw B (2017). Genes, environment, and "bad luck." Science, 355:1266–1267. doi: https://doi.org/10.1126/science.aam9746

Nowell PC (1976) The clonal evolution of tumor cell populations. Science 194(4260):23–28. https://doi.org/10.1126/science.959840

Noy R, Pollard JW (2014) Tumor-associated macrophages: from mechanisms to therapy. Immunity 41(1):49–61. https://doi.org/10.1016/j.immuni.2014.06.010

Nunez JJ, Leung B, Ho C et al (2023) Predicting the survival of patients with cancer from their initial oncology consultation document using natural language processing. JAMA Netw Open 6(2):e230813. https://doi.org/10.1001/jamanetworkopen.2023.0813

Nunn AVW, Guy GW, Bell JD (2022) Bioelectric fields at the beginnings of life. Bioelectricity 4(4):237–247. https://doi.org/10.1089/bioe.2022.0012

Nürnberg A, Kitzing T, Grosse R (2011) Nucleating actin for invasion. Nat Rev Cancer 11(3):177–187. https://doi.org/10.1038/nrc3003

Nurse P (2021) What is life: five great ideas in biology. W.W. Norton & Company, New York

O'Brien J, Hayder H, Zayed Y, Peng C (2018) Overview of microRNA biogenesis, mechanisms of actions, and circulation. Front Endocrinol (Lausanne) 9:402. https://doi.org/10.3389/fendo.2018.00402

O'Keeffe KP, Hong H, Strogatz SH (2017) Oscillators that sync and swarm. Nat Commun 8(1):1504. https://doi.org/10.1038/s41467-017-01190-3

Obama B (2015a) State of the Union Address. The White House, Washington, DC

Obama B (2015b) Remarks by the President on precision medicine. The White House. https://www.whitehouse.gov/the-press-office/2015/01/30/remarks-president-precision-medicine

Obenauf AC, Massagué J (2015) Surviving at a distance: organ-specific metastasis. Trends Cancer 1(1):76–91. https://doi.org/10.1016/j.trecan.2015.07.009

Ocana A, Nieto-Jiménez C, Pandiella A, Templeton AJ (2017) Neutrophils in cancer: prognostic role and therapeutic strategies. Mol Cancer 16(1):137. https://doi.org/10.1186/s12943-017-0707-7

Odum HT (1995) Self-organization and maximum empower. In: Hall CAS (ed) Maximum Power: The Ideas and Applications of H.T. Odum. Colorado University Press, Colorado

Oginuma M, Harima Y, Tarazona OA et al (2020) Intracellular pH controls WNT downstream of glycolysis in amniote embryos. Nature 584(7819):98–101. https://doi.org/10.1038/s41586-020-2428-0

Oh SY, Kim S, Keam B et al (2021) Soluble PD-L1 is a predictive and prognostic biomarker in advanced cancer patients who receive immune checkpoint blockade treatment. Sci Rep 11(1):19712. https://doi.org/10.1038/s41598-021-99311-y

Oishi Y, Manabe I (2020) Organ system crosstalk in cardiometabolic disease in the age of multimorbidity. Front Cardiovasc Med 64. doi:https://doi.org/10.3389/fcvm.2020.00064

Old LJ (2001) Cancer/testis (CT) antigens—a new link between gametogenesis and cancer. Cancer Immun 1:1. PMID: 12747762

Oliveira JC (2023) Transcribed ultraconserved regions: new regulators in cancer signaling and potential biomarkers. Genet Mol Biol 46(1 Suppl 2):e20220125. https://doi.org/10.1590/1678-4685-GMB-2022-0125

Olofsson PS, Tracey KJ (2017) Bioelectronic medicine: technology targeting molecular mechanisms for therapy. J Intern Med 282(1):3–4. https://doi.org/10.1111/joim.12624

Omar EM, Omran GA, Mustafa MF, El-Khodary NM (2022) Intermittent fasting during adjuvant chemotherapy may promote differential stress resistance in breast cancer patients. J Egypt Natl Canc Inst 34(1):38. https://doi.org/10.1186/s43046-022-00141-4

Orimo A, Weinberg RA (2006) Stromal fibroblasts in cancer: a novel tumor-promoting cell type. Cell Cycle 5(15):1597–1601. https://doi.org/10.4161/cc.5.15.3112

Orlando PA, Gatenby RA, Brown JS (2012) Cancer treatment as a game: integrating evolutionary game theory into the optimal control of chemotherapy. Phys Biol 9(6):065007. https://doi.org/10.1088/1478-3975/9/6/065007

Orme JJ, Enninga EAL, Lucien-Matteoni F et al (2020) Therapeutic plasma exchange clears circulating soluble PD-L1 and PD-L1-positive extracellular vesicles. J Immunother Cancer 8(2):e001113. https://doi.org/10.1136/jitc-2020-001113

Orr HA (2009) Fitness and its role in evolutionary genetics. Nat Rev Genet 10(8):531–539. https://doi.org/10.1038/nrg2603

Osswald M, Jung E, Sahm F et al (2015) Brain tumour cells interconnect to a functional and resistant network. Nature 528(7580):93–98. https://doi.org/10.1038/nature16071

Ostroverkhova D, Przytycka TM, Panchenko AR (2023) Cancer driver mutations: predictions and reality. Trends Mol Med 29(7):554–566. https://doi.org/10.1016/j.molmed.2023.03.007

Osum M, Kalkan R (2023) Cancer stem cells and their therapeutic usage. Adv Exp Med Biol 1436:69–85. https://doi.org/10.1007/5584_2022_758

Osuna-Marco MP, López-Barahona M, López-Ibor B, Tejera ÁM (2021) Ten reasons why people with Down syndrome are protected from the development of most solid tumors—a review. Front Genet 12:749480. https://doi.org/10.3389/fgene.2021.749480

Ottaiano A, Santorsola M, Circelli L et al (2023) Oligo-metastatic cancers: putative biomarkers, emerging challenges and new perspectives. Cancers (Basel) 15(6):1827. https://doi.org/10.3390/cancers15061827

Otto S (2022) An approach to de novo synthesis of life. Acc Chem Res 55:145–155. https://doi.org/10.1021/acs.accounts.1c00591

Ozsvari B, Sotgia F, Simmons K et al (2017) Mitoketoscins: novel mitochondrial inhibitors for targeting ketone metabolism in cancer stem cells (CSCs). Oncotarget 8(45):78340–78350. https://doi.org/10.18632/oncotarget.21259

Pacella I, Procaccini C, Focaccetti C et al (2018) Fatty acid metabolism complements glycolysis in the selective regulatory T cell expansion during tumor growth. Proc Natl Acad Sci U S A 115(28):E6546–E6555. https://doi.org/10.1073/pnas.1720113115

Paget S (1889) The distribution of secondary growths in cancer of the breast. Cancer Metastasis Rev 8:98–101

Palacios-Acedo AL, Mège D, Crescence L et al (2019) Platelets, thrombo-inflammation, and cancer: collaborating with the enemy. Front Immunol 10:1805. https://doi.org/10.3389/fimmu.2019.01805

Pan X, Zheng L (2020) Epigenetics in modulating immune functions of stromal and immune cells in the tumor microenvironment. Cell Mol Immunol 17(9):940–953. https://doi.org/10.1038/s41423-020-0505-9

Pan H, Renaud L, Chaligne R et al (2021) Discovery of candidate DNA methylation cancer driver genes. Cancer Discov 11(9):2266–2281. https://doi.org/10.1158/2159-8290.CD-20-1334

Pandey K, Zafar H (2022) Inference of cell state transitions and cell fate plasticity from single-cell with MARGARET. Nucleic Acids Res 50(15):e86. https://doi.org/10.1093/nar/gkac412

Panet F, Papakonstantinou A, Borrell M et al (2024) Use of ctDNA in early breast cancer: analytical validity and clinical potential. NPJ Breast Cancer. 10(1):50. https://doi.org/10.1038/s41523-024-00653-3

Panigrahi G, Ambs S (2021) How comorbidities shape cancer biology and survival. Trends Cancer 7(6):488–495. https://doi.org/10.1016/j.trecan.2020.12.010

Panková K, Rösel D, Novotný M (2010) Brábek J (2010) The molecular mechanisms of transition between mesenchymal and amoeboid invasiveness in tumor cells. Cell Mol Life Sci. 67(1):63–71. https://doi.org/10.1007/s00018-009-0132-1

Pannu V, Rida PC, Celik B et al (2014) Centrosome-declustering drugs mediate a two-pronged attack on interphase and mitosis in supercentrosomal cancer cells. Cell Death Dis 5(11):e1538. https://doi.org/10.1038/cddis.2014.505

Pannu V, Rida PC, Ogden A et al (2015) HSET overexpression fuels tumor progression via centrosome clustering-independent mechanisms in breast cancer patients. Oncotarget 6(8):6076–6091. https://doi.org/10.18632/oncotarget.3475

Pantel K, Alix-Panabières C (2019) Liquid biopsy and minimal residual disease – latest advances and implications for cure. Nat Rev Clin Oncol 16(7):409–424. https://doi.org/10.1038/s41571-019-0187-3

Papac RJ (1998) Spontaneous regression of cancer: possible mechanisms. In Vivo 12(6):571–578

Papier K, Atkins JR, Tong TYN et al (2024) Identifying proteomic risk factors for cancer using prospective and exome analyses of 1463 circulating proteins and risk of 19 cancers in the UK Biobank. Nat Commun 15:4010. https://doi.org/10.1038/s41467-024-48017-6

Pardoll DM (2012) The blockade of immune checkpoints in cancer immunotherapy. Nat Rev Cancer 12(4):252–264. https://doi.org/10.1038/nrc3239

Parish DC, Goyal H, Dane FC (2018) Mechanism of death: there's more to it than sudden cardiac arrest. J Thorac Dis 10(5):3081–3087. https://doi.org/10.21037/jtd.2018.04.113

Park J, Wysocki RW, Amoozgar Z et al (2016) Cancer cells induce metastasis-supporting neutrophil extracellular DNA traps. Sci Transl Med 8(361):361ra138. https://doi.org/10.1126/scitranslmed.aag1711

Park WY, Gray JM, Holewinski RJ et al (2023) Apoptosis-induced nuclear expulsion in tumor cells drives S100a4-mediated metastatic outgrowth through the RAGE pathway. Nat Cancer 4:419–435. https://doi.org/10.1038/s43018-023-00524-z

Parker M, Mohankumar KM, Punchihewa C et al (2014) C11orf95-RELA fusions drive oncogenic NF-κB signalling in ependymoma. Nature 506(7489):451–455. https://doi.org/10.1038/nature13109

Paro R, Grossniklaus U, Santoro R, Wutz A (2021) Epigenetics and metabolism. In: Paro R, Grossniklaus U, Santoro R, Wutz A (eds) Introduction to epigenetics. Springer Publishing

Parreno V, Loubiere V, Schuettengruber B et al (2024) Transient loss of Polycomb components induces an epigenetic cancer fate. Nature 629(8012):688–696. https://doi.org/10.1038/s41586-024-07328-w

Parsons BL (2008) Many different tumor types have polyclonal tumor origin: evidence and implications. Mutat Res 659(3):232–247. https://doi.org/10.1016/j.mrrev.2008.05.004

Pascal R, Pross A (2015) Stability and its manifestation in the chemical and biological worlds. Chem Commun 51:16160–16165. https://doi.org/10.1039/c5cc06260h

Pascal R, Pross A (2019) Chemistry's kinetic dimension and the physical basis for life. J Syst Chem 7:1

Pascal R, Pross A (2023) Toward the physicalization of biology: seeking the chemical origin of cognition. In: Corning PA et al (eds) Evolution "on purpose": teleonomy in living systems. MIT Press

Pasche B, Benson AB III, Kudo M, Shileds A (2025) Development and regulatory approval of a new systemic targeted therapy for advanced hepatocellular carcinoma. J Clin Oncol 43:613–613. https://doi.org/10.1200/JCO.2025.43.4_suppl.613

Pascual G, Avgustinova A, Mejetta S et al (2017) Targeting metastasis-initiating cells through the fatty acid receptor CD36. Nature 541(7635):41–45. https://doi.org/10.1038/nature20791

Pascual G, Domínguez D, Elosúa-Bayes M et al (2021) Dietary palmitic acid promotes a prometastatic memory via Schwann cells. Nature 599(7885):485–490. https://doi.org/10.1038/s41586-021-04075-0

Pastushenko I, Brisebarre A, Sifrim A et al (2018) Identification of the tumour transition states occurring during EMT. Nature 556(7702):463–468. https://doi.org/10.1038/s41586-018-0040-3

Pasut A, Matsumoto A, Clohessy JG, Pandolfi PP (2016) The pleiotropic role of non-coding genes in development and cancer. Curr Opin Cell Biol 43:104–113. https://doi.org/10.1016/j.ceb.2016.10.005

Patel SA, Kondratov RV (2021) Clock at the core of cancer development. Biology (Basel) 10(2):150. https://doi.org/10.3390/biology10020150

Patel SA, Vanharanta S (2017) Epigenetic determinants of metastasis. Mol Oncol 11(1):79–96. https://doi.org/10.1016/j.molonc.2016.09.008

Patel AV, Friedenreich CM, Moore SC et al (2019) American College of Sports Medicine roundtable report on physical activity, sedentary behavior, and cancer prevention and control. Med Sci Sports Exerc 51(11):2391–2402. https://doi.org/10.1249/MSS.0000000000002117

Patel SA, Rodrigues P, Wesolowski L, Vanharanta S (2021) Genomic control of metastasis. Br J Cancer 124(1):3–12. https://doi.org/10.1038/s41416-020-01127-6

Pătraş L, Paul D (2024) Systemic communication via extracellular vesicles in cancer: the body's wireless internet system. In: Barciszewski J (ed) Systems Biology II. Springer

Pătraş L, Shaashua L, Matei I, Lyden D (2023a) Immune determinants of the pre-metastatic niche. Cancer Cell 41(3):546–572. https://doi.org/10.1016/j.ccell.2023.02.018

Pătraş L, Paul D, Matei IR (2023b) Weaving the nest: extracellular matrix roles in pre-metastatic niche formation. Front Oncol 13:1163786. https://doi.org/10.3389/fonc.2023.1163786

Patterson E, Wall R, Fitzgerald GF et al (2012) Health implications of high dietary omega-6 polyunsaturated fatty acids. J Nutr Metab 2012:539426. https://doi.org/10.1155/2012/539426

Patterson RE, Laughlin GA, LaCroix AZ et al (2015) Intermittent fasting and human metabolic health. J Acad Nutr Diet 115(8):1203–1212. https://doi.org/10.1016/j.jand.2015.02.018

Paul D (2015) Cancer the big picture: seeing the forest beyond the trees. Oncolog-Hematolog 1:28–30

Paul D (2020) The systemic hallmarks of cancer. J Cancer Metastasis Treat 6:29. https://doi.org/10.20517/2394-4722.2020.63

Paul D, Komarova NL (2021) Multi-scale network targeting: a holistic systems-biology approach to cancer treatment. Prog Biophys Mol Biol 165:72–79. https://doi.org/10.1016/j.pbiomolbio.2021.08.004

Paul D, Nedelcu AM (2022) The underexplored links between cancer and the internal body climate: implications for cancer prevention and treatment. Front Oncol 12:1040034. https://doi.org/10.3389/fonc.2022.1040034

Paul WE, Mizuguchi J, Brown M et al (1986) Regulation of B-lymphocyte activation, proliferation, and immunoglobulin secretion. Cell Immunol 99(1):7–13. https://doi.org/10.1016/0008-8749(86)90209-1

Paul D, Gold M, Nouraddin N (2014) Prolonged survival in metastatic colorectal cancer following chemotherapy. Clin Case Rep 2(5):219–223. https://doi.org/10.1002/ccr3.109

Paul DM, Ghiuzeli CM, Rini J et al (2020a) A pilot study treatment of malignant tumors using low-dose 18F-fluorodeoxyglucose (18F-FDG). Am J Nucl Med Mol Imaging 10(6):334–341. PMID: 33329935

Paul D, Maggi P, Piero FD et al (2020b) Targeted osmotic lysis of highly invasive breast carcinomas using pulsed magnetic field stimulation of voltage-gated sodium channels and pharmacological blockade of sodium pumps. Cancers (Basel) 12(6):1420. https://doi.org/10.3390/cancers12061420

Paul S, Pearlman AH, Douglass J et al (2021) TCR β chain-directed bispecific antibodies for the treatment of T cell cancers. Sci Transl Med 13(584):eabd3595. https://doi.org/10.1126/scitranslmed.abd3595

Paull EO, Aytes A, Jones SJ et al (2021) A modular master regulator landscape controls cancer transcriptional identity. Cell 184(2):334–351.e20. https://doi.org/10.1016/j.cell.2020.11.045

Pausas JG, Bond WJ (2022) Feedbacks in ecology and evolution. Trends Ecol Evol 37(8):637–644. https://doi.org/10.1016/j.tree.2022.03.008

Pavlides S, Tsirigos A, Migneco G et al (2010) The autophagic tumor stroma model of cancer: role of oxidative stress and ketone production in fueling tumor cell metabolism. Cell Cycle 9(17):3485–3505. https://doi.org/10.4161/cc.9.17.12721

Pavlidis N, Stanta G, Audisio RA (2012) Cancer prevalence and mortality in centenarians: a systematic review. Crit Rev Oncol Hematol 83(1):145–152. https://doi.org/10.1016/j.critrevonc.2011.09.007

Pavlov VA, Tracey KJ (2012) The vagus nerve and the inflammatory reflex—linking immunity and metabolism. Nat Rev Endocrinol 8(12):743–754. https://doi.org/10.1038/nrendo.2012.189

Pavlov VA, Tracey KJ (2022) Bioelectronic medicine: preclinical insights and clinical advances. Neuron 110(21):3627–3644. https://doi.org/10.1016/j.neuron.2022.09.003

Pavlova NN, Thompson CB (2016) The emerging hallmarks of cancer metabolism. Cell Metab 23(1):27–47. https://doi.org/10.1016/j.cmet.2015.12.006

Pavlova NN, Zhu J, Thompson CB (2022) The hallmarks of cancer metabolism: still emerging. Cell Metab 34(3):355–377. https://doi.org/10.1016/j.cmet.2022.01.007

Pavlyukov MS, Yu H, Bastola S et al (2018) Apoptotic cell-derived extracellular vesicles promote malignancy of glioblastoma via intercellular transfer of splicing factors. Cancer Cell 34(1):119–135.e10. https://doi.org/10.1016/j.ccell.2018.05.012

Payen VL, Porporato PE, Baselet B, Sonveaux P (2016) Metabolic changes associated with tumor metastasis, part 1: tumor pH, glycolysis and the pentose phosphate pathway. Cell Mol Life Sci 73:1333–1348

Payne SL, Levin M, Oudin MJ (2019) Bioelectric control of metastasis in solid tumors. Bioelectricity 1(3):114–130. https://doi.org/10.1089/bioe.2019.0013

PCAWG Transcriptome Core Group, Calabrese C, Davidson NR et al (2020) Genomic basis for RNA alterations in cancer. Nature 578(7793):129–136. https://doi.org/10.1038/s41586-020-1970-0

Pearl J, Mackenzie D (2018) The book of why: the new science of cause and effect. Basic Books, New York

Pecorino L (2021) Molecular Biology of Cancer, 5th edn. Oxford University Press, Oxford

Peinado H, Rafii S, Lyden D (2008) Inflammation joins the "niche". Cancer Cell 14(5):347–349. https://doi.org/10.1016/j.ccr.2008.10.012

Peinado H, Alečković M, Lavotshkin S et al (2012) Melanoma exosomes educate bone marrow progenitor cells toward a pro-metastatic phenotype through MET. Nat Med 18(6):883–891. https://doi.org/10.1038/nm.2753

Peinado H, Zhang H, Matei IR et al (2017) Pre-metastatic niches: organ-specific homes for metastases. Nat Rev Cancer 17(5):302–317. https://doi.org/10.1038/nrc.2017.6

Peiris-Pagès M, Martinez-Outschoorn UE, Pestell RG et al (2016) Cancer stem cell metabolism. Breast Cancer Res 18(1):55. https://doi.org/10.1186/s13058-016-0712-6

Pelosof LC, Gerber DE (2010) Paraneoplastic syndromes: an approach to diagnosis and treatment. Mayo Clin Proc 85(9):838–854. https://doi.org/10.4065/mcp.2010.0099

Pendergrass JC, Targum SD, Harrison JE (2018) Cognitive Impairment Associated with Cancer: A Brief Review. Innov Clin Neurosci. 15(1-2):36–44

Peng Y, Croce CM (2016) The role of microRNAs in human cancer. Signal Transduct Target Ther 1:15004. https://doi.org/10.1038/sigtrans.2015.4

Peng X, Chen Z, Farshidfar F et al (2018) Molecular characterization and clinical relevance of metabolic expression subtypes in human cancers. Cell Rep 23(1):255–269.e4. https://doi.org/10.1016/j.celrep.2018.03.077

Peng A, Peng W, Wang R et al (2022) Regulation of 3D organization and its role in cancer biology. Front Cell Dev Biol 10:879465. https://doi.org/10.3389/fcell.2022.879465

Penrose R (2006) Before the Big Bang: An outrageous new perspective and its implications for particle physics. In: Proceedings of the EPAC 2006. Edinburgh, Scotland, pp 2759–2762

Penrose R (2012) Cycles of time: an extraordinary new view of the universe. Vintage Books, London

Penrose R (2020) Nobel lecture: Roger Penrose, nobel prize in physics 2020. Nobel Prize Committee. https://youtu.be/DpPFn0qzYT0

Perazzoli G, García-Valdeavero OM, Peña M et al (2023) Evaluating metabolite-based biomarkers for early diagnosis of pancreatic cancer: a systematic review. Metabolites 13(7):872. https://doi.org/10.3390/metabo13070872

Pérez-Fidalgo JA, Pimentel P, Caballero A et al (2011) Removal of primary tumor improves survival in metastatic breast cancer. Does timing of surgery influence outcomes? Breast 20(6):548–554. https://doi.org/10.1016/j.breast.2011.06.005

Perillo B, Di Donato M, Pezone A et al (2020) ROS in cancer therapy: the bright side of the moon. Exp Mol Med 52(2):192–203. https://doi.org/10.1038/s12276-020-0384-2

Pernigoni N, Zagato E, Calcinotto A et al (2021) Commensal bacteria promote endocrine resistance in prostate cancer through androgen biosynthesis. Science 374(6564):216–224. https://doi.org/10.1126/science.abf8403

Perrimon N, Pitsouli C, Shilo BZ (2012) Signaling mechanisms controlling cell fate and embryonic patterning. Cold Spring Harb Perspect Biol 4(8):a005975. https://doi.org/10.1101/cshperspect.a005975

Perry ACF, Asami M, Lam BYH, Yeo GSH (2023) The initiation of mammalian embryonic transcription: to begin at the beginning. Trends Cell Biol 33(5):365–373. https://doi.org/10.1016/j.tcb.2022.08.008

Pertesi M, Ekdahl L, Palm A et al (2019) Essential genes shape cancer genomes through linear limitation of homozygous deletions. Commun Biol 2:262. https://doi.org/10.1038/s42003-019-0517-0

Perunov N, Marsland RA, England JL (2016) Statistical physics of adaptation. Phys Rev X 6:021036. https://doi.org/10.1103/PhysRevX.6.021036

Pesavento PA, Agnew D, Keel MK, Woolard KD (2018) Cancer in wildlife: patterns of emergence. Nat Rev Cancer 18(10):646–661. https://doi.org/10.1038/s41568-018-0045-0

Pete G, Timár Á, Stefánsson SÖ et al (2024) Physical networks as network-of-networks. Nat Commun 15(1):4882. https://doi.org/10.1038/s41467-024-49227-8

Peters L, Venkatachalam A, Ben-Neriah Y (2024) Tissue-predisposition to cancer driver mutations. Cells 13(2):106. https://doi.org/10.3390/cells13020106

Petruzzelli M, Wagner EF (2016) Mechanisms of metabolic dysfunction in cancer-associated cachexia. Genes Dev 30(5):489–501. https://doi.org/10.1101/gad.276733.115

Petruzzelli M, Schweiger M, Schreiber R et al (2014) A switch from white to brown fat increases energy expenditure in cancer-associated cachexia. Cell Metab 20(3):433–447. https://doi.org/10.1016/j.cmet.2014.06.011

Pfeifer GP (2018) Defining driver DNA methylation changes in human cancer. Int J Mol Sci 19(4):1166. https://doi.org/10.3390/ijms19041166

Phallen J, Sausen M, Adleff V et al (2017) Direct detection of early-stage cancers using circulating tumor DNA. Sci Transl Med 9(403):eaan2415. https://doi.org/10.1126/scitranslmed.aan2415

Phan TG, Croucher PI (2020) The dormant cancer cell life cycle. Nat Rev Cancer 20:398–411. https://doi.org/10.1038/s41568-020-0263-0

Philip M, Schietinger A (2022) CD8+ T cell differentiation and dysfunction in cancer. Nat Rev Immunol 22(4):209–223. https://doi.org/10.1038/s41577-021-00574-3

Philip M, Fairchild L, Sun L et al (2017) Chromatin states define tumour-specific T cell dysfunction and reprogramming. Nature 545(7655):452–456. https://doi.org/10.1038/nature22367

Phillips KA, Sakowski JA, Trosman J et al (2014) The economic value of personalized medicine tests: what we know and what we need to know. Genet Med 16(3):251–257. https://doi.org/10.1038/gim.2013.122

Pich O, Bernard E, Zagorulya M et al (2025) Tumor-Infiltrating Clonal Hematopoiesis. N Engl J Med 392(16):1594–1608. https://doi.org/10.1056/NEJMoa2413361

Pienta KJ, Hammarlund EU, Axelrod R et al (2020) Poly-aneuploid cancer cells promote evolvability, generating lethal cancer. Evol Appl 13(7):1626–1634. https://doi.org/10.1111/eva.12929

Pienta KJ, Hammarlund EU, Austin RH et al (2022) Cancer cells employ an evolutionarily conserved polyploidization program to resist therapy. Semin Cancer Biol 81:145–159. https://doi.org/10.1016/j.semcancer.2020.11.016

Pienta KJ, Goodin PL, Amend SR (2025) Defeating lethal cancer: interrupting the ecologic and evolutionary basis of death from malignancy. CA Cancer J Clin 75(3):183–202. https://doi.org/10.3322/caac.70000

Pierce GB (1993) On the boundary between development and neoplasia. An interview with Professor G. Barry Pierce. Interview by Juan Arechaga. Int J Dev Biol 37(1):5–16. PMID: 8507570

Pierce S (2023) Life's mechanism. Life (Basel) 13(8):1750. https://doi.org/10.3390/life13081750

Piergentili R, Sechi S (2024) Non-coding RNAs of mitochondrial origin: roles in cell division and implications in cancer. Int J Mol Sci 25(13):7498. https://doi.org/10.3390/ijms25137498

Pierpont TM, Lyndaker AM, Anderson CM et al (2017) Chemotherapy-induced depletion of OCT4-positive cancer stem cells in a mouse model of malignant testicular cancer. Cell Rep 21(7):1896–1909. https://doi.org/10.1016/j.celrep.2017.10.078

Pietras K, Ostman A (2010) Hallmarks of cancer: interactions with the tumor stroma. Exp Cell Res 316(8):1324–1331. https://doi.org/10.1016/j.yexcr.2010.02.045

Piffkó A, Yang K, Panda A et al (2025) Radiation-induced amphiregulin drives tumour metastasis. Nature. https://doi.org/10.1038/s41586-025-08994-0

Pinzón N, Li B, Martinez L et al (2017) microRNA target prediction programs predict many false positives. Genome Res 27(2):234–245. https://doi.org/10.1101/gr.205146.116

Pio-Lopez L, Levin M (2023) Morphoceuticals: perspectives for discovery of drugs targeting anatomical control mechanisms in regenerative medicine, cancer and aging. Drug Discov Today 28(6):103585. https://doi.org/10.1016/j.drudis.2023.103585

Piskounova E, Viswanathan SR, Janas M et al (2008) Determinants of microRNA processing inhibition by the developmentally regulated RNA-binding protein Lin28. J Biol Chem 283(31):21310–21314. https://doi.org/10.1074/jbc.C800108200

Pitsiou G, Zarogoulidis P, Petridis D et al (2014) Inhaled tyrosine kinase inhibitors for pulmonary hypertension: a possible future treatment. Drug Des Devel Ther 8:1753–1763. https://doi.org/10.2147/DDDT.S70277

Placido D, Yuan B, Hjaltelin JX et al (2023) A deep learning algorithm to predict risk of pancreatic cancer from disease trajectories. Nat Med 29(5):1113–1122. https://doi.org/10.1038/s41591-023-02332-5

Plasschaert RN, Bartolomei MS (2014) Genomic imprinting in development, growth, behavior and stem cells. Development 141(9):1805–1813. https://doi.org/10.1242/dev.101428

Pleskova SN, Lazarenko EV, Bezrukov NA et al (2023) Differences in bacteria nanomotion profiles and neutrophil nanomotion during phagocytosis. Front Microbiol 14:1113353. https://doi.org/10.3389/fmicb.2023.1113353

Plotnikova O, Baranova A, Skoblov M (2019) Comprehensive analysis of human microRNA-mRNA interactome. Front Genet 10:933. https://doi.org/10.3389/fgene.2019.00933

Plutinsky A (2018) Explaining cancer: finding order in disorder. Oxford University Press

Podolskiy DI, Gladyshev VN (2016) Intrinsic versus extrinsic cancer risk factors and aging. Trends Mol Med 22(10):833–834. https://doi.org/10.1016/j.molmed.2016.08.001

Pohling C, Nguyen H, Chang E et al (2023) Current status of the preclinical evaluation of alternating electric fields as a form of cancer therapy. Bioelectrochemistry 149:108287. https://doi.org/10.1016/j.bioelechem.2022.108287

Polak P, Karlić R, Koren A et al (2015) Cell-of-origin chromatin organization shapes the mutational landscape of cancer. Nature 518(7539):360–364. https://doi.org/10.1038/nature14221

Pollicino F, Veronese N, Dominguez LJ et al (2023) Mediterranean diet and mitochondria: new findings. Exp Gerontol 176:112165. https://doi.org/10.1016/j.exger.2023.112165

Pollock PM, Harper UL, Hansen KS et al (2003) High frequency of BRAF mutations in nevi. Nat Genet 33(1):19–20. https://doi.org/10.1038/ng1054

Pontarotti G, Mossio M, Pocheville A (2022) The genotype–phenotype distinction: from Mendelian genetics to 21st century biology. Genetica 150(3–4):223–234. https://doi.org/10.1007/s10709-022-00159-5

Poon GYP, Watson CJ, Fisher DS, Blundell JR (2021) Synonymous mutations reveal genome-wide levels of positive selection in healthy tissues. Nat Genet 53(11):1597–1605. https://doi.org/10.1038/s41588-021-00957-1

Popa R (2004) Between necessity and probability: searching for the definition and origin of life. Springer

Popa R (2009) The complexity-independence of the origin of life. Systems science friday noon seminar series 43

Popa R (2010) Necessity, futility and the possibility of defining life are all embedded in its origin as a punctuated-gradualism. Orig Life Evol Biosph 40(2):183–190. https://doi.org/10.1007/s11084-010-9198-x

Porporato PE (2016) Understanding cachexia as a cancer metabolism syndrome. Oncogenesis 5(2):e200. https://doi.org/10.1038/oncsis.2016.3

Porporato PE, Payen VL, Pérez-Escuredo J et al (2014) A mitochondrial switch promotes tumor metastasis. Cell Rep 8(3):754–766. https://doi.org/10.1016/j.celrep.2014.06.043

Porporato PE, Payen VL, Baselet B, Sonveaux P (2016) Metabolic changes associated with tumor metastasis, part 2: mitochondria, lipid and amino acid metabolism. Cell Mol Life Sci 73:1349–1363. https://doi.org/10.1007/s00018-015-2118-3

Porta-Pardo E, Valencia A, Godzik A (2020) Understanding oncogenicity of cancer driver genes and mutations in the cancer genomics era. FEBS Lett 594(24):4233–4246. https://doi.org/10.1002/1873-3468.13781

Postow MA, Sidlow R, Hellmann MD (2018) Immune-related adverse events associated with immune checkpoint blockade. N Engl J Med 378(2):158–168. https://doi.org/10.1056/NEJMra1703481

Potter JD (2001) Morphostats: a missing concept in cancer biology. Cancer Epidemiol Biomarkers Prev 10(3):161–170. PMID: 11303583

Powell VD, Kumar N, Galecki AT et al (2022) Bad company: loneliness longitudinally predicts the symptom cluster of pain, fatigue, and depression in older adults. J Am Geriatr Soc 70(8):2225–2234. https://doi.org/10.1111/jgs.17796

Pradat Y, Viot J, Yurchenko AA et al (2023) Integrative pan-cancer genomic and transcriptomic analyses of refractory metastatic cancer. Cancer Discov 13(5):1116–1143. https://doi.org/10.1158/2159-8290.CD-22-0966

Prager BC, Xie Q, Bao S, Rich JN (2019) Cancer stem cells: the architects of the tumor ecosystem. Cell Stem Cell 24(1):41–53. https://doi.org/10.1016/j.stem.2018.12.009

Pressley M, Salvioli M, Lewis DB et al (2021) Evolutionary dynamics of treatment-induced resistance in cancer informs understanding of rapid evolution in natural systems. Front Ecol Evol 9:681121. https://doi.org/10.3389/fevo.2021.681121

Pretzsch E, Nieß H, Bösch F et al (2022) Age and metastasis—how age influences metastatic spread in cancer. Colorectal cancer as a model. Cancer Epidemiol 77:102112. https://doi.org/10.1016/j.canep.2022.102112

Prevarskaya N, Skryma R, Shuba Y (2010) Ion channels and the hallmarks of cancer. Trends Mol Med 16(3):107–121. https://doi.org/10.1016/j.molmed.2010.01.005

Prevarskaya N, Skryma R, Shuba Y (2018) Ion channels in cancer: are cancer hallmarks oncochannelopathies? Physiol Rev 98(2):559–621. https://doi.org/10.1152/physrev.00044.2016

Price JE, Naito S, Fidler IJ (1988) Growth in an organ microenvironment as a selective process in metastasis. Clin Exp Metastasis 6(1):91–102. https://doi.org/10.1007/BF01580409

Priestley P, Baber J, Lolkema MP et al (2019) Pan-cancer whole-genome analyses of metastatic solid tumours. Nature 575(7781):210–216. https://doi.org/10.1038/s41586-019-1689-y

Prigogine I, Stengers I (1984) Order out of chaos: man's new dialogue with nature. Flamingo Edition, London

Prigogine I, Stengers I (2017) Order out of chaos: man's new dialogue with nature. Verso, London

Prihantono FM (2021) Breast cancer resistance to chemotherapy: when should we suspect it and how can we prevent it? Ann Med Surg (Lond) 70:102793. https://doi.org/10.1016/j.amsu.2021.102793

Prior IA, Hood FE, Hartley JL (2020) The frequency of Ras mutations in cancer. Cancer Res 80(14):2969–2974. https://doi.org/10.1158/0008-5472.CAN-19-3682

Proença JT, Barral DC, Gordo I (2017) Commensal-to-pathogen transition: one-single transposon insertion results in two pathoadaptive traits in Escherichia coli–macrophage interaction. Sci Rep 7:4504. https://doi.org/10.1038/s41598-017-04081-1

Prokop A (2021) Towards the first principles in biology and cancer: new vistas in computational systems biology of cancer. Life (Basel) 12(1):21. https://doi.org/10.3390/life12010021

Pross A (2003) The driving force for life's emergence: kinetic and thermodynamic considerations. J Theor Biol 220(3):393–406. https://doi.org/10.1006/jtbi.2003.3178

Pross A (2011) Toward a general theory of evolution: extending Darwinian theory to inanimate matter. J Syst Chem 2:1. https://doi.org/10.1186/1759-2208-2-1

Pross A (2012) What is life?: how chemistry becomes biology. Oxford University Press, Oxford

Pross A, Khodorkovsky V (2004) Extending the concept of kinetic stability: toward a paradigm for life. J Phys Org Chem 17(4):312–316. https://doi.org/10.1002/poc.729

Pross A, Pascal R (2017) How and why kinetics, thermodynamics, and chemistry induce the logic of biological evolution. Beilstein J Org Chem 13:665–674. https://doi.org/10.3762/bjoc.13.66

Pross A, Pascal R (2023) On the emergence of autonomous chemical systems through dissipation kinetics. Life (Basel) 13(11):2171. https://doi.org/10.3390/life13112171

Pruimboom L, Muskiet FAJ (2018) Intermittent living: the use of ancient challenges as a vaccine against the deleterious effects of modern life – a hypothesis. Med Hypotheses 120:28–42. https://doi.org/10.1016/j.mehy.2018.08.002

Psaila B, Lyden D (2009) The metastatic niche: adapting the foreign soil. Nat Rev Cancer 9(4):285–293. https://doi.org/10.1038/nrc2621

Psychogios N, Hau DD, Peng J et al (2011) The human serum metabolome. PLoS One 6(2):e16957. https://doi.org/10.1371/journal.pone.0016957

Pullamsetti SS, Savai R, Seeger W, Goncharova EA (2017) Translational advances in the field of pulmonary hypertension. From cancer biology to new pulmonary arterial hypertension therapeutics. Targeting cell growth and proliferation signaling hubs. Am J Respir Crit Care Med 195(4):425–437. https://doi.org/10.1164/rccm.201606-1226PP

Pullamsetti SS, Nayakanti S, Chelladurai P et al (2020) Cancer and pulmonary hypertension: learning lessons and real-life interplay. Glob Cardiol Sci Pract 1:e202010. https://doi.org/10.21542/gcsp.2020.10

Pupo E, Avanzato D, Middonti E et al (2019) KRAS-driven metabolic rewiring reveals novel actionable targets in cancer. Front Oncol 9:848. https://doi.org/10.3389/fonc.2019.00848

Pusuluk O, Torun G, Deliduman C (2018) Quantum entanglement shared in hydrogen bonds and its usage as a resource in molecular recognition. Mod Phys Lett B 32:1850308. https://doi.org/10.1142/S0217984918503086

Qian DC, Kleber T, Brammer B et al (2021) Effect of immunotherapy time-of-day infusion on overall survival among patients with advanced melanoma in the USA (MEMOIR): a propensity score-matched analysis of a single-centre, longitudinal study. Lancet Oncol 22(12):1777–1786. https://doi.org/10.1016/S1470-2045(21)00546-5

Qiu J, Lai C, Yuan Z et al (2024) Utilizing liquid-liquid phase separation-related lncRNAs to predict the prognosis and treatment response of PCa. Discov Oncol 15(1):352. https://doi.org/10.1007/s12672-024-01226-3

Qu R, Ye F, Tu D et al (2022) Clinical features and surgical treatment of synchronous multiple primary lung adenocarcinomas with different EGFR mutations. Front Oncol 11:785777. https://doi.org/10.3389/fonc.2021.785777

Quail D, Joyce J (2013) Microenvironmental regulation of tumor progression and metastasis. Nat Med 19:1423–1437. https://doi.org/10.1038/nm.3394

Quesenberry PJ, Colvin GA, Lambert JF (2002) The new stem cell biology. Trans Am Clin Climatol Assoc 113:182–206; discussion 206–7. PMID: 12053709; PMCID: PMC2194475

Quevli N (1916) Cell intelligence, the cause of growth, heredity and instinctive actions. Colwell Press, Minneapolis

Quicke P, Sun Y, Arias-Garcia M et al (2022) Voltage imaging reveals the dynamic electrical signatures of human breast cancer cells. Commun Biol 5(1):1178. https://doi.org/10.1038/s42003-022-04077-2

Quinlan RA, Ellis RJ (2013) Chaperones: needed for both the good times and the bad times. Philos Trans R Soc Lond B Biol Sci 368(1617):20130091. https://doi.org/10.1098/rstb.2013.0091

Quinten C, Coens C, Mauer M et al (2009) Baseline quality of life as a prognostic indicator of survival: a meta-analysis of individual patient data from EORTC clinical trials. Lancet Oncol 10(9):865–871. https://doi.org/10.1016/S1470-2045(09)70200-1

Rachkovsky M, Sodi S, Chakraborty A et al (1998) Melanoma × macrophage hybrids with enhanced metastatic potential. Clin Exp Metastasis 16(4):299–312. https://doi.org/10.1023/a:1006557228604

Radha G, Lopus M (2021) The spontaneous remission of cancer: current insights and therapeutic significance. Transl Oncol 14(9):101166. https://doi.org/10.1016/j.tranon.2021.101166

Radišauskas R, Kuzmickienė I, Milinavičienė E, Everatt R (2016) Hypertension, serum lipids and cancer risk: a review of epidemiological evidence. Medicina (Kaunas) 52(2):89–98. https://doi.org/10.1016/j.medici.2016.03.002

Radkiewicz C, Järkvik Krönmark J, Adami HO, Edgren G (2022) Declining cancer incidence in the elderly: decreasing diagnostic intensity or biology? Cancer Epidemiol Biomarkers Prev 31(1):280–286. https://doi.org/10.1158/1055-9965.EPI-21-0797

Radkiewicz C, Andersson TM, Lagergren J (2023) Reranking cancer mortality using years of life lost. JNCI Cancer Spectr 7(3):pkad038. https://doi.org/10.1093/jncics/pkad038

Radman M (2019) Cellular parabiosis and the latency of age-related diseases. Open Biol 9(3):180250. https://doi.org/10.1098/rsob.180250

Rafalko JM, Kruglyak KM, McCleary-Wheeler A et al (2023) Age at cancer diagnosis by breed, weight, sex, and cancer type in a cohort of more than 3,000 dogs: determining the optimal age to initiate cancer screening in canine patients. PLoS One 18(2):e0280795. https://doi.org/10.1371/journal.pone.0280795

Rahal Z, Scheet P, Kadara H (2024) Somatic mutations in normal tissues: calm before the storm. Cancer Discov 14(4):605–609. https://doi.org/10.1158/2159-8290.CD-23-1508

Rahrmann EP, Shorthouse D, Jassim A et al (2022) The NALCN channel regulates metastasis and nonmalignant cell dissemination. Nat Genet 54(12):1827–1838. https://doi.org/10.1038/s41588-022-01182-0

Rak J (2015) Organ-seeking vesicles. Nature 527:312–314. https://doi.org/10.1038/nature15642

Rakha EA, El-Sayed ME, Lee AH et al (2008) Prognostic significance of Nottingham histologic grade in invasive breast carcinoma. J Clin Oncol 26(19):3153–3158. https://doi.org/10.1200/JCO.2007.15.5986

Rakoff-Nahoum S, Medzhitov R (2009) Toll-like receptors and cancer. Nat Rev Cancer 9(1):57–63. https://doi.org/10.1038/nrc2541

Ramalingam SS, Felip E, Goldman JW, et al (n.d.) Please confirm or complete publication details for formatting

Ramroop JR, Gerber MM, Toland AE (2019) Germline variants impact somatic events during tumorigenesis. Trends Genet 35(7):515–526. https://doi.org/10.1016/j.tig.2019.04.005

Ramya AS, Uppala D, Majumdar S et al (2015) Are salivary amylase and pH prognostic indicators of cancers? J Oral Biol Craniofac Res 5(2):81–85. https://doi.org/10.1016/j.jobcr.2015.04.002

Rando TA (2006) Stem cells, ageing and the quest for immortality. Nature 441(7097):1080–1086. https://doi.org/10.1038/nature04958

Raoult D, Forterre P (2008) Redefining viruses: lessons from Mimivirus. Nat Rev Microbiol 6:315–319. https://doi.org/10.1038/nrmicro1858

Rashevsky N (1960) Contributions to relational biology. Bull Math Biophys 22:73–84

Rashid OM, Nagahashi M, Ramachandran S et al (2013) Resection of the primary tumor improves survival in metastatic breast cancer by reducing overall tumor burden. Surgery 153(6):771–778. https://doi.org/10.1016/j.surg.2013.02.002

Raskov H, Orhan A, Gaggar S, Gögenur I (2021) Cancer-associated fibroblasts and tumor-associated macrophages in cancer and cancer immunotherapy. Front Oncol 11:668731. https://doi.org/10.3389/fonc.2021.668731

Rattan SI (2008) Hormesis in aging. Ageing Res Rev 7(1):63–78. https://doi.org/10.1016/j.arr.2007.03.002

Rayon T (2023) Cell time: how cells control developmental timetables. Sci Adv 9(10):eadh1849. https://doi.org/10.1126/sciadv.adh1849

Reber AS (2018) The First Minds: Caterpillars, Karyotes, and Consciousness. Oxford University Press

Reber AS, Baluška F (2021) Cognition in some surprising places. Biochem Biophys Res Commun 564:150–157. https://doi.org/10.1016/j.bbrc.2020.08.115

Redig AJ, McAllister SS (2013) Breast cancer as a systemic disease: a view of metastasis. J Intern Med 274(2):113–126. https://doi.org/10.1111/joim.12084

Reeves MM, Battistutta D, Capra S et al (2006) Resting energy expenditure in patients with solid tumors undergoing anticancer therapy. Nutrition 22(6):609–615. https://doi.org/10.1016/j.nut.2006.03.006

Reichert M, Bakir B, Moreira L et al (2018) Regulation of epithelial plasticity determines metastatic organotropism in pancreatic cancer. Dev Cell 45(6):696–711.e8. https://doi.org/10.1016/j.devcel.2018.05.025

Reichle A, Vogt T (2008) Systems biology: a therapeutic target for tumor therapy. Cancer Microenviron 1(1):159–170. https://doi.org/10.1007/s12307-008-0012-5

Rein M, Ben-Yacov O, Godneva A et al (2022) Effects of personalized diets by prediction of glycemic responses on glycemic control and metabolic health in newly diagnosed T2DM: a randomized dietary intervention pilot trial. BMC Med 20(1):56. https://doi.org/10.1186/s12916-022-02254-y

Reinfeld BI, Madden MZ, Wolf MM et al (2021) Cell-programmed nutrient partitioning in the tumour microenvironment. Nature 593(7858):282–288. https://doi.org/10.1038/s41586-021-03442-1

Reiter JG, Baretti M, Gerold JM et al (2019) An analysis of genetic heterogeneity in untreated cancers. Nat Rev Cancer 19(11):639–650. https://doi.org/10.1038/s41568-019-0185-x

Rekowska AK, Obuchowska K, Bartosik M et al (2023) Biomolecules involved in both metastasis and placenta accreta spectrum—does the common pathophysiological pathway exist? Cancers (Basel) 15(9):2618. https://doi.org/10.3390/cancers15092618

Repasky EA, Evans SS, Dewhirst MW (2013) Temperature matters! and why it should matter to tumor immunologists. Cancer Immunol Res 1(4):210–216. https://doi.org/10.1158/2326-6066.CIR-13-0118

Replogle JM, Zhou W, Amaro AE et al (2020) Aneuploidy increases resistance to chemotherapeutics by antagonizing cell division. Proc Natl Acad Sci U S A 117(48):30566–30576. https://doi.org/10.1073/pnas.2009506117

Rezende LFM, Sá TH, Markozannes G et al (2018) Physical activity and cancer: an umbrella review of the literature including 22 major anatomical sites and 770000 cancer cases. Br J Sports Med 52(13):826–833. https://doi.org/10.1136/bjsports-2017-098391

Reznik E, Torjani A (2024) Mechanisms of stress-attributed breast cancer incidence and progression. Cancer Causes Control 35(11):1413–1432. https://doi.org/10.1007/s10552-024-01884-2

Rheinbay E, Nielsen MM, Abascal F et al (2020) Analyses of non-coding somatic drivers in 2,658 cancer whole genomes. Nature 578(7793):102–111. https://doi.org/10.1038/s41586-020-1965-x

Ribas A, Wolchok JD (2018) Cancer immunotherapy using checkpoint blockade. Science 359(6382):1350–1355. https://doi.org/10.1126/science.aar4060

Ribatti D, Tamma R, Annese T (2020) Epithelial-mesenchymal transition in cancer: a historical overview. Transl Oncol 13(6):100773. https://doi.org/10.1016/j.tranon.2020.100773

Ricca BL, Venugopalan G, Furuta S et al (2018) Transient external force induces phenotypic reversion of malignant epithelial structures via nitric oxide signaling. Elife 7:e26161. https://doi.org/10.7554/eLife.26161

Richmond P, Roehner BM (2019) A physicist's view of the similarities and differences between tuberculosis and cancer. Physica A 534:120761. https://doi.org/10.1016/j.physa.2019.03.126

Ridker PM, Everett BM, Thuren T et al (2017) Antiinflammatory therapy with canakinumab for atherosclerotic disease. N Engl J Med 377(12):1119–1131. https://doi.org/10.1056/NEJMoa1707914

Riedl R (1978) Order in Living Organisms: A Systems Analysis of Evolution. John Wiley and Sons, New York

Riedlinger G, Adams J, Stehle JR Jr et al (2010) The spectrum of resistance in SR/CR mice: the critical role of chemoattraction in the cancer/leukocyte interaction. BMC Cancer 10:179. https://doi.org/10.1186/1471-2407-10-179

Rieppel O (2013) Biological individuals and natural kinds. Biol Theory 7:162–169. https://doi.org/10.1007/s13752-012-0051-5

Riggi N, Aguet M, Stamenkovic I (2018) Cancer metastasis: a reappraisal of its underlying mechanisms and their relevance to treatment. Annu Rev Pathol 13:117–140. https://doi.org/10.1146/annurev-pathol-020117-044127

Rivera F, Benavides M, Gallego J et al (2019) Tumor treating fields in combination with gemcitabine or gemcitabine plus nab-paclitaxel in pancreatic cancer: results of the PANOVA phase

2 study. Pancreatology 19(1):64–72. https://doi.org/10.1016/j.pan.2018.10.004
Rizaner N, Onkal R, Fraser SP et al (2016) Intracellular calcium oscillations in strongly metastatic human breast and prostate cancer cells: control by voltage-gated sodium channel activity. Eur Biophys J 45(7):735–748. https://doi.org/10.1007/s00249-016-1170-x
Rizvi NA, Hellmann MD, Snyder A et al (2015) Cancer immunology. mutational landscape determines sensitivity to PD-1 blockade in non-small cell lung cancer. Science 348(6230):124–128. https://doi.org/10.1126/science.aaa1348
Robertson MP, Joyce GF (2012) The origins of the RNA world. Cold Spring Harb Perspect Biol 4(5):a003608. https://doi.org/10.1101/cshperspect.a003608
Rock CL, Thomson C, Gansler T et al (2020) American Cancer Society guideline for diet and physical activity for cancer prevention. CA Cancer J Clin 70(4):245–271. https://doi.org/10.3322/caac.21591
Rodenfels J, Neugebauer KM, Howard J (2019) Heat oscillations driven by the embryonic cell cycle reveal the energetic costs of signaling. Dev Cell 48(5):646–658.e6. https://doi.org/10.1016/j.devcel.2018.12.024
Rodić N, Sharma R, Sharma R et al (2014) Long interspersed element-1 protein expression is a hallmark of many human cancers. Am J Pathol 184(5):1280–1286. https://doi.org/10.1016/j.ajpath.2014.01.007
Rodrigues G, Hoshino A, Kenific CM et al (2019) Tumour exosomal CEMIP protein promotes cancer cell colonization in brain metastasis. Nat Cell Biol 21(11):1403–1412. https://doi.org/10.1038/s41556-019-0404-4
Rodriguez-Martin B, Alvarez EG, Baez-Ortega A et al (2020) Pan-cancer analysis of whole genomes identifies driver rearrangements promoted by LINE-1 retrotransposition. Nat Genet 52(3):306–319. https://doi.org/10.1038/s41588-019-0562-0
Rohani N, Hao L, Alexis MS et al (2019) Acidification of tumor at stromal boundaries drives transcriptome alterations associated with aggressive phenotypes. Cancer Res 79(8):1952–1966. https://doi.org/10.1158/0008-5472.CAN-18-1604
Rolong A, Chen B, Lau KS (2021) Deciphering the cancer microenvironment from bulk data with EcoTyper. Cell 184(21):5306–5308. https://doi.org/10.1016/j.cell.2021.09.028
Romee R, Cooley S, Berrien-Elliott MM et al (2018) First-in-human phase 1 clinical study of the IL-15 superagonist complex ALT-803 to treat relapse after transplantation. Blood 131(23):2515–2527. https://doi.org/10.1182/blood-2017-12-823757
Roncati L, Figueiredo CR (2023) Editorial: Hallmark of cancer: tumor promoting inflammation. Front Oncol 13:1242407. https://doi.org/10.3389/fonc.2023.1242407
Rose JE, Levin ED (1992) Concurrent agonist-antagonist administration for the analysis and treatment of drug dependence. Pharmacol Biochem Behav 41(1):219–226. https://doi.org/10.1016/0091-3057(92)90086-u
Rosenbaum M, Knight R, Leibel RL (2015) The gut microbiota in human energy homeostasis and obesity. Trends Endocrinol Metab 26(9):493–501. https://doi.org/10.1016/j.tem.2015.07.002
Rosenberg SM, Queitsch C (2014) Medicine. Combating evolution to fight disease. Science 343(6175):1088–1089. https://doi.org/10.1126/science.1247472
Rosenberg SA, Yang JC, Sherry RM et al (2011) Durable complete responses in heavily pretreated patients with metastatic melanoma using T-cell transfer immunotherapy. Clin Cancer Res 17(13):4550–4557. https://doi.org/10.1158/1078-0432.CCR-11-0116
Rosenfeld S (2013) Are the somatic mutation and tissue organization field theories of carcinogenesis incompatible? Cancer Inform 12:221–229. https://doi.org/10.4137/CIN.S13013
Rosenthal N, Cao Z, Gundrum J et al (2020) Risk factors associated with in-hospital mortality in a US national sample of patients with COVID-19. JAMA Netw Open 3(12):e2029058. https://doi.org/10.1001/jamanetworkopen.2020.29058
Roskelley CD, Desprez PY, Bissell MJ (1994) Extracellular matrix-dependent tissue-specific gene expression in mammary epithelial cells requires both physical and biochemical signal transduction. Proc Natl Acad Sci USA 91:12378–12382
Roskelley CD, Srebrow A, Bissell MJ (1995) A hierarchy of ECM-mediated signalling regulates tissue-specific gene expression. Curr Opin Cell Biol 7:736–747
Rospars JP, Meyer-Vernet N (2021) How fast do mobile organisms respond to stimuli? Response times from bacteria to elephants and whales. Phys Biol 18(2):026002. https://doi.org/10.1088/1478-3975/abcd88
Rosslenbroich B (2023) Properties of life. Towards a theory of biological organisms. MIT Press
Rossmeislová L, Gojda J, Smolková K (2021) Pancreatic cancer: branched-chain amino acids as putative key metabolic regulators? Cancer Metastasis Rev 40(4):1115–1139. https://doi.org/10.1007/s10555-021-10016-0
Rothbart SB, Strahl BD (2014) Interpreting the language of histone and DNA modifications. Biochim Biophys Acta 1839(8):627–643. https://doi.org/10.1016/j.bbagrm.2014.03.001
Rothschild LJ, Averesch NJH, Strychalski EA et al (2024) Building synthetic cells—from the technology infrastructure to cellular entities. ACS Synth Biol 13(4):974–997. https://doi.org/10.1021/acssynbio.3c00724
Round JL, Mazmanian SK (2009) The gut microbiota shapes intestinal immune responses during health and disease. Nat Rev Immunol 9(5):313–323. https://doi.org/10.1038/nri2515
Rous P (1911) A sarcoma of the fowl transmissible by an agent separable from the tumor cells. J Exp Med 13:397–411. https://doi.org/10.1084/jem.13.4.397
Routy B, Le Chatelier E, Derosa L et al (2018) Gut microbiome influences efficacy of PD-1-based immunotherapy against epithelial tumors. Science 359(6371):91–97. https://doi.org/10.1126/science.aan3706
Rovelli C (2016) Meaning = Information + Evolution. arXiv preprint arXiv:1611.02420v1
Rovithi M, Gerritse SL, Honeywell RJ et al (2019) Phase I dose-escalation study of once weekly or once every two weeks administration of high-dose sunitinib in patients with refractory solid tumors. J Clin Oncol 37(5):411–418. https://doi.org/10.1200/JCO.18.00725
Roxburgh CS, McMillan DC (2014) Cancer and systemic inflammation: treat the tumour and treat the host. Br J Cancer 110(6):1409–1412. https://doi.org/10.1038/bjc.2014.90
Rozhok A, DeGregori J (2019) A generalized theory of age-dependent carcinogenesis. Elife 8:e39950. https://doi.org/10.7554/eLife.39950
Rozhok AI, Wahl GM, DeGregori J (2015) A critical examination of the "bad luck" explanation of cancer risk. Cancer Prev Res (Phila) 8(9):762–764. https://doi.org/10.1158/1940-6207.CAPR-15-0229
Ruan K, Fang X, Ouyang G (2009) MicroRNAs: novel regulators in the hallmarks of human cancer. Cancer Lett 285(2):116–126. https://doi.org/10.1016/j.canlet.2009.04.031
Rubio LS, Mohajan S, Gross DS (2024) Heat Shock Factor 1 forms nuclear condensates and restructures the yeast genome before activating target genes. Elife 12:RP92464. https://doi.org/10.7554/eLife.92464
Russell ES (1945) The directiveness of organic activities. Cambridge University Press
Russell MJ (2017) Life is a verb, not a noun. Geology 45:1143–1144. https://doi.org/10.1130/focus112017.1
Russell D, Peplau LA, Ferguson ML (1978) Developing a measure of loneliness. J Pers Assess 42:290–294. https://doi.org/10.1207/s15327752jpa4203_11

Russell CD, Lone NI, Baillie JK (2023) Comorbidities, multimorbidity and COVID-19. Nat Med 29(2):334–343. https://doi.org/10.1038/s41591-022-02156-9

Rutkowski MR, Conejo-Garcia JR (2015) TLR5 signaling, commensal microbiota and systemic tumor promoting inflammation: the three parcae of malignant progression. Oncoimmunology 4(8):e1021542. https://doi.org/10.1080/2162402X.2015.1021542

Saatcioglu F (2013) Regulation of gene expression by yoga, meditation and related practices: a review of recent studies. Asian J Psychiatr 6(1):74–77. https://doi.org/10.1016/j.ajp.2012.10.002

Sabelli H (2001) The co-creation hypothesis. In: Ragsdell G, Wilby J (eds) Understanding Complexity. Springer, pp 155–164

Sabelli H (2003) Mathematical development: a theory of natural creation. Kybernetes 32(5/6):752–766. https://doi.org/10.1108/03684920310477404

Sabelli H (2005) Bios: A Study of Creation (Series on Knots and Everything). World Scientific, Singapore

Sabnis AJ, Bivona T (2019) Principles of resistance to targeted cancer therapy: lessons from basic and translational cancer biology. Trends Mol Med 25(3):185–197. https://doi.org/10.1016/j.molmed.2018.12.009

Saccon TD, Nagpal R, Yadav H et al (2021) Senolytic combination of dasatinib and quercetin alleviates intestinal senescence and inflammation and modulates the gut microbiome in aged mice. J Gerontol A Biol Sci Med Sci 76(11):1895–1905. https://doi.org/10.1093/gerona/glab002

Saeed RF, Naz S, Azeem Awan U, et al. (2024) Micronutrients importance in cancer prevention—vitamins. In: Saeed RF, ul Shaheed S (eds) Cancer Prevention and Treatment Based on Lifestyles, Nutrition and Dietary Interventions in Cancer. Springer, pp 115–138

Sahai E, Astsaturov I, Cukierman E et al (2020) A framework for advancing our understanding of cancer-associated fibroblasts. Nat Rev Cancer 20(3):174–186. https://doi.org/10.1038/s41568-019-0238-1

Sahu B, Pihlajamaa P, Zhang K et al (2021) Human cell transformation by combined lineage conversion and oncogene expression. Oncogene 40:5533–5547. https://doi.org/10.1038/s41388-021-01940-0

Saini P, Courtneidge SA (2018) Tks adaptor proteins at a glance. J Cell Sci 131(1):jcs203661. https://doi.org/10.1242/jcs.203661

Saleh A, Kashir J, Thanassoulas A et al (2020) Essential role of sperm-specific PLC-Zeta in egg activation and male factor infertility: an update. Front Cell Dev Biol 8:28. https://doi.org/10.3389/fcell.2020.00028

Salgia R, Li JL, Ewaniuk DS et al (1997) BCR/ABL induces multiple abnormalities of cytoskeletal function. J Clin Invest 100(1):46–57. https://doi.org/10.1172/JCI119520

Salminen A, Kaarniranta K (2010) ER stress and hormetic regulation of the aging process. Ageing Res Rev 9(3):211–217. https://doi.org/10.1016/j.arr.2010.04.003

Saltz J, Gupta R, Hou L et al (2018) Spatial organization and molecular correlation of tumor-infiltrating lymphocytes using deep learning on pathology images. Cell Rep 23(1):181–193.e7. https://doi.org/10.1016/j.celrep.2018.03.086

Saman H, Raza SS, Uddin S, Rasul K (2020) Inducing angiogenesis, a key step in cancer vascularization, and treatment approaches. Cancers (Basel) 12(5):1172. https://doi.org/10.3390/cancers12051172

Samant H, Amiri HS, Zibari GB (2021) Addressing the worldwide hepatocellular carcinoma: epidemiology, prevention and management. J Gastrointest Oncol. 12(Suppl 2):S361–S373. https://doi.org/10.21037/jgo.2020.02.08. PMID: 34422400; PMCID: PMC8343080

Sanchez-Vega F, Mina M, Armenia J et al (2018) Oncogenic signaling pathways in The Cancer Genome Atlas. Cell 173(2):321–337.e10. https://doi.org/10.1016/j.cell.2018.03.035

Sansregret L, Swanton C (2017) The role of aneuploidy in cancer evolution. Cold Spring Harb Perspect Med 7(1):a028373. https://doi.org/10.1101/cshperspect.a028373

Santorsola M, Capuozzo M, Savarese G et al (2023) Oligo-metastatic disease in oncology: exploring the limits and the potential of genetic assessment. Genes (Basel) 14(12):2131. https://doi.org/10.3390/genes14122131

Saorin A, Di Gregorio E, Miolo G et al (2020) Emerging role of metabolomics in ovarian cancer diagnosis. Metabolites 10(10):419. https://doi.org/10.3390/metabo10100419

Sarapata EA, de Pillis LG (2014) A comparison and catalog of intrinsic tumor growth models. Bull Math Biol 76(8):2010–2024. https://doi.org/10.1007/s11538-014-9986-y

Sarfati D, Koczwara B, Jackson C (2016) The impact of comorbidity on cancer and its treatment. CA Cancer J Clin 66(4):337–350. https://doi.org/10.3322/caac.21342

Sarhadi S, Salehzadeh-Yazdi A, Damaghi M et al (2020) Omics integration analyses reveal the early evolution of malignancy in breast cancer. Cancers (Basel) 12(6):1460. https://doi.org/10.3390/cancers12061460

Sarnaik AA, Hamid O, Khushalani NI et al (2021) Lifileucel, a tumor-infiltrating lymphocyte therapy, in metastatic melanoma. J Clin Oncol 39(24):2656–2666. https://doi.org/10.1200/JCO.21.00612

Sarwar E (2023) The emerging field of precision medicine – the new paradigm for healthcare. In: Global perspectives on precision medicine: ethical, social and public health implications. Springer, pp 9–32

Sasai K, Gunji YP (2008) Heterarchy in biological systems: a logic-based dynamical model of abstract biological network derived from time-state-scale re-entrant form. Biosystems 92(2):182–188. https://doi.org/10.1016/j.biosystems.2008.02.004

Schaub J, Tang SC (2025) Beyond checkpoint inhibitors: the three generations of immunotherapy. Clin Exp Med 25(1):43. https://doi.org/10.1007/s10238-024-01546-2

Scheiermann C, Gibbs J, Ince L, Loudon A (2018) Clocking in to immunity. Nat Rev Immunol 18:423–437. https://doi.org/10.1038/s41577-018-0008-4

Schernhammer E, Bogl L, Hublin C et al (2023) The association between night shift work and breast cancer risk in the Finnish twins cohort. Eur J Epidemiol 38(5):533–543. https://doi.org/10.1007/s10654-023-00983-9

Schietinger A, Philip M, Krisnawan VE et al (2016) Tumor-specific T cell dysfunction is a dynamic antigen-driven differentiation program initiated early during tumorigenesis. Immunity 45(2):389–401. https://doi.org/10.1016/j.immuni.2016.07.011

Schiliro C, Firestein BL (2021) Mechanisms of metabolic reprogramming in cancer cells supporting enhanced growth and proliferation. Cells 10(5):1056. https://doi.org/10.3390/cells10051056

Schleidgen S, Klingler C, Bertram T et al (2013) What is personalized medicine: sharpening a vague term based on a systematic literature review. BMC Med Ethics 14:55. https://doi.org/10.1186/1472-6939-14-55

Schmidt DR, Patel R, Kirsch DG et al (2021) Metabolomics in cancer research and emerging applications in clinical oncology. CA Cancer J Clin 71(4):333–358. https://doi.org/10.3322/caac.21670

Schmitt AM, Chang HY (2016) Long noncoding RNAs in cancer pathways. Cancer Cell 29(4):452–463. https://doi.org/10.1016/j.ccell.2016.03.010

Schneider G, Schmidt-Supprian M, Rad R, Saur D (2017) Tissue-specific tumorigenesis: context matters. Nat Rev Cancer 17(4):239–253. https://doi.org/10.1038/nrc.2017.5

Schneider MA, Buzdin AA, Weber A et al (2021) Combination of antiretroviral drugs zidovudine and efavirenz impairs tumor growths in a mouse model of cancer. Viruses 13(12):2396. https://doi.org/10.3390/v13122396

Schneider A, Woortman J, Bang Jensen CB et al (2024) Therapeutic implications of phosphoproteomics in molecular cancer diagnostics. J Clin Oncol 42(16_Suppl):3068. https://doi.org/10.1200/JCO.2024.42.16_suppl.3068

Schnipper LE, Davidson NE, Wollins DS et al (2016) Updating the American Society of Clinical Oncology Value Framework: revisions and reflections in response to comments received. J Clin Oncol 34(24):2925–2934. https://doi.org/10.1200/JCO.2016.68.2518

Schöffski P, Carlino MS, Lebbé C, et al. (2025)

Schönthal AH, Swenson SD, Chen TC, Markland FS (2020) Preclinical studies of a novel snake venom-derived recombinant disintegrin with antitumor activity: a review. Biochem Pharmacol 181:114149. https://doi.org/10.1016/j.bcp.2020.114149

Schork NJ (2015) Personalized medicine: time for one-person trials. Nature 520(7549):609–611. https://doi.org/10.1038/520609a

Schrag D, Beer TM, McDonnell CH 3rd et al (2023) Blood-based tests for multicancer early detection (PATHFINDER): a prospective cohort study. Lancet 402(10409):1251–1260. https://doi.org/10.1016/S0140-6736(23)01700-2

Schreiber RD, Old LJ, Smyth MJ (2011) Cancer immunoediting: integrating immunity's roles in cancer suppression and promotion. Science 331(6024):1565–1570. https://doi.org/10.1126/science.1203486

Schrödinger E (1944) What is life? The physical aspect of the living cell with mind and matter & autobiographical sketches. Cambridge University Press, Cambridge

Schumacher B (2009) Transcription-blocking DNA damage in aging: a mechanism for hormesis. Bioessays 31(12):1347–1356. https://doi.org/10.1002/bies.200900107

Schwanhäusser B, Busse D, Li N et al (2013) Corrigendum: global quantification of mammalian gene expression control. Nature 495(7439):126–127. https://doi.org/10.1038/nature11848

Schwartsburd P (2020) A view on pathogenesis of "vicious cancer progression cycle." Front Oncol 10:690. https://doi.org/10.3389/fonc.2020.00690

Schwartz SM, Urfer SR, White M et al (2022) Lifetime prevalence of malignant and benign tumours in companion dogs: cross-sectional analysis of Dog Aging Project baseline survey. Vet Comp Oncol. 20(4):797–804. https://doi.org/10.1111/vco.12839

Schwartz PB, Nukaya M, Berres ME et al (2023a) The circadian clock is disrupted in pancreatic cancer. PLoS Genet 19(6):e1010770. https://doi.org/10.1371/journal.pgen.1010770

Schwartz L, Guais A, Israël M et al (2023b) Tumor regression with a combination of drugs interfering with the tumor metabolism: efficacy of hydroxycitrate, lipoic acid and capsaicin. Invest New Drugs 31:256–264

Schwarzenbach H, Gahan PB (2019) MicroRNA shuttle from cell-to-cell by exosomes and its impact in cancer. Noncoding RNA 5(1):28. https://doi.org/10.3390/ncrna5010028

Scott BR (2014) Radiation-hormesis phenotypes, the related mechanisms and implications for disease prevention and therapy. J Cell Commun Signal 8(4):341–352. https://doi.org/10.1007/s12079-014-0250-x

Scott AC, Dündar F, Zumbo P et al (2019) TOX is a critical regulator of tumour-specific T cell differentiation. Nature 571(7764):270–274. https://doi.org/10.1038/s41586-019-1324-y

Seca AML, Pinto DCGA (2018) Plant Secondary Metabolites as Anticancer Agents: Successes in Clinical Trials and Therapeutic Application. Int J Mol Sci. 19(1):263. https://doi.org/10.3390/ijms19010263

Segal NH, Tie J, Kopetz S et al (2024) COLUMBIA-1: a randomised study of durvalumab plus oleclumab in combination with chemotherapy and bevacizumab in metastatic microsatellite-stable colorectal cancer. Br J Cancer 131(6):1005–1013. https://doi.org/10.1038/s41416-024-02796-3

Seitz H (2024) A new perspective on microRNA-guided gene regulation specificity, and its potential generalization to transcription factors and RNA-binding proteins. Nucleic Acids Res 52(16):9360–9368. https://doi.org/10.1093/nar/gkae694

Sell MC, Ramlogan-Steel CA, Steel JC, Dhungel BP (2023) MicroRNAs in cancer metastasis: biological and therapeutic implications. Expert Rev Mol Med 25:e14. https://doi.org/10.1017/erm.2023.7

Semeradtova A, Liegertova M, Herma R et al (2025) Extracellular vesicles in cancer's communication: messages we can read and how to answer. Mol Cancer 24(1):86. https://doi.org/10.1186/s12943-025-02282-1

Senft D, Ronai ZA (2016) Regulators of mitochondrial dynamics in cancer. Curr Opin Cell Biol 39:43–52. https://doi.org/10.1016/j.ceb.2016.02.001

Sephton SE, Sapolsky RM, Kraemer HC, Spiegel D (2009) Diurnal cortisol rhythm as a predictor of breast cancer survival. J Natl Cancer Inst 92(12):994–1000. https://doi.org/10.1093/jnci/92.12.994

Sepich-Poore GD, Zitvogel L, Straussman R et al (2021) The microbiome and human cancer. Science 371(6536):eabc4552. https://doi.org/10.1126/science.abc4552

Seretis A, Cividini S, Markozannes G et al (2019) Association between blood pressure and risk of cancer development: a systematic review and meta-analysis of observational studies. Sci Rep 9(1):8565. https://doi.org/10.1038/s41598-019-45014-4

Serra M, Columbano A, Ammarah U et al (2020) Understanding metal dynamics between cancer cells and macrophages: competition or synergism? Front Oncol 10:646. https://doi.org/10.3389/fonc.2020.00646

Setiawan T, Sari IN, Wijaya YT et al (2023) Cancer cachexia: molecular mechanisms and treatment strategies. J Hematol Oncol 16(1):54. https://doi.org/10.1186/s13045-023-01454-0

Setrerrahmane S, Xu H (2017) Tumor-related interleukins: old validated targets for new anti-cancer drug development. Mol Cancer 16(1):153. https://doi.org/10.1186/s12943-017-0721-9

Sever R, Brugge JS (2015) Signal transduction in cancer. Cold Spring Harb Perspect Med 5(4):a006098. https://doi.org/10.1101/cshperspect.a006098

Seyfried TN (2012) Cancer as a Metabolic Disease: On the Origin, Management, and Prevention of Cancer. Wiley, Hoboken

Seyfried TN (2015) Cancer as a mitochondrial metabolic disease. Front Cell Dev Biol 43. https://doi.org/10.3389/fcell.2015.00043

Seyfried TN, Huysentruyt LC (2013) On the origin of cancer metastasis. Crit Rev Oncog 18(1–2):43–73. https://doi.org/10.1615/critrevoncog.v18.i1-2.40

Seyfried TN, Flores RE et al (2014) Cancer as a metabolic disease: implications for novel therapeutics. Carcinogenesis 35(3):515–527. https://doi.org/10.1093/carcin/bgt480

Seyfried TN, Yu G, Maroon JC, D'Agostino DP (2017) Press-pulse: a novel therapeutic strategy for the metabolic management of cancer. Nutr Metab (Lond) 14:19. https://doi.org/10.1186/s12986-017-0178-2

Shaashua L, Eckerling A, Israeli B et al (2020) Spontaneous regression of micro-metastases following primary tumor excision: a critical role for primary tumor secretome. BMC Biol 18(1):163. https://doi.org/10.1186/s12915-020-00893-2

Shabo I, Svanvik J, Lindström A et al (2020) Roles of cell fusion, hybridization and polyploid cell formation in cancer metastasis. World J Clin Oncol 11(3):121–135. https://doi.org/10.5306/wjco.v11.i3.121

Shafiei-Irannejad V, Samadi N, Salehi R et al (2017) New insights into antidiabetic drugs: possible applications in cancer treatment.

Chem Biol Drug Des 90(6):1056–1066. https://doi.org/10.1111/cbdd.13013

Shahid RK, Ahmed S, Le D, Yadav S (2021) Diabetes and cancer: risk, challenges, management and outcomes. Cancers (Basel) 13(22):5735. https://doi.org/10.3390/cancers13225735

Shakyawar SK, Sajja BR, Patel JC, Guda C (2024) iCluF: an unsupervised iterative cluster-fusion method for patient stratification using multiomics data. Bioinform Adv 4(1):vbae015. https://doi.org/10.1093/bioadv/vbae015

Shalapour S, Karin M (2015) Immunity, inflammation, and cancer: an eternal fight between good and evil. J Clin Invest 125(9):3347–3355. https://doi.org/10.1172/JCI80007

Shalhoub J, Sikkel MB, Davies KJ et al (2013) Systems biology of human atherosclerosis. Vasc Endovasc Surg 48(1):5–17. https://doi.org/10.1177/1538574413510628

Shan F, Somasundaram A, Bruno TC et al (2022) Therapeutic targeting of regulatory T cells in cancer. Trends Cancer 8(11):944–961. https://doi.org/10.1016/j.trecan.2022.06.008

Shannon CE (1948) A mathematical theory of communication. Bell Syst Tech J 27(3):379–423. https://doi.org/10.1002/j.1538-7305.1948.tb01338.x

Shannon CE, Weaver W (1949) The Mathematical Theory of Information. University of Illinois Press, Urbana

Shapiro JA (2005) A 21st century view of evolution: genome system architecture, repetitive DNA, and natural genetic engineering. Gene 345(1):91–100. https://doi.org/10.1016/j.gene.2004.11.020

Shapiro JA (2021) All living cells are cognitive. Biochem Biophys Res Commun 564:134–149. https://doi.org/10.1016/j.bbrc.2020.08.120

Shapiro JA (2022) Evolution: a view from the 21st century. Fortified edn. Cognition Press

Shapiro JA (2023) Evolutionary change is naturally biological and purposeful. In: Corning PA, Kauffman SA, Noble D et al (eds) Evolution "On Purpose": teleonomy in living systems. MIT Press, Cambridge

Sharma V (2016) The application of chaos theory and fractal mathematics to the study of cancer evolution: placing metabolism and immunity centre stage. Med Res Arch 4:1–12. https://doi.org/10.18103/mra.v4i6.717

Sharma P, Allison JP (2015) The future of immune checkpoint therapy. Science 348(6230):56–61. https://doi.org/10.1126/science.aaa8172

Sharma NK, Sarode SC (2023) Artificial intelligence vs. evolving super-complex tumor intelligence: critical viewpoints. Front. Artif Intell 6:1220744. https://doi.org/10.3389/frai.2023.1220744

Sharma S, Wu SY, Jimenez H et al (2019) Ca2+ and CACNA1H mediate targeted suppression of breast cancer brain metastasis by AM RF EMF. EBioMedicine 44:194–208. https://doi.org/10.1016/j.ebiom.2019.05.038

Sharma A, Czégel D, Lachmann M et al (2023) Assembly theory explains and quantifies selection and evolution. Nature 622(7982):321–328. https://doi.org/10.1038/s41586-023-06600-9

Shaul ME, Fridlender ZG (2018) Cancer-related circulating and tumor-associated neutrophils—subtypes, sources and function. FEBS J 285(23):4316–4342. https://doi.org/10.1111/febs.14524

Sheltzer JM, Blank HM, Pfau SJ et al (2011) Aneuploidy drives genomic instability in yeast. Science 333(6045):1026–1030. https://doi.org/10.1126/science.1206412

Shen L, Shi Q, Wang W (2018) Double agents: genes with both oncogenic and tumor-suppressor functions. Oncogenesis 7(3):25. https://doi.org/10.1038/s41389-018-0034-x

Shen R, Martin A, Ni A et al (2019) Harnessing clinical sequencing data for survival stratification of patients with metastatic lung adenocarcinomas JCO Precis Oncol PO.18.00307. https://doi.org/10.1200/PO.18.00307

Sherbet GV (2005) Hormonal influences on cancer progression and prognosis. Vitam Horm 71:147–200. https://doi.org/10.1016/S0083-6729(05)71006-2

Sherman MA, Yaari AU, Priebe O et al (2022) Genome-wide mapping of somatic mutation rates uncovers drivers of cancer. Nat Biotechnol 40(11):1634–1643. https://doi.org/10.1038/s41587-022-01353-8

Sheth M, Esfandiari L (2022) Bioelectric dysregulation in cancer initiation, promotion, and progression. Front Oncol 12:846917. https://doi.org/10.3389/fonc.2022.846917

Shields GS, Spahr CM, Slavich GM (2020) Psychosocial interventions and immune system function: a systematic review and meta-analysis of randomized clinical trials. JAMA Psychiatry 77(10):1031–1043. https://doi.org/10.1001/jamapsychiatry.2020.0431

Shih J, Sarmashghi S, Zhakula-Kostadinova N et al (2023) Cancer aneuploidies are shaped primarily by effects on tumour fitness. Nature 619(7971):793–800. https://doi.org/10.1038/s41586-023-06266-3

Shirakawa J, De Jesus DF, Kulkarni RN (2017) Exploring inter-organ crosstalk to uncover mechanisms that regulate β-cell function and mass. Eur J Clin Nutr 71(7):896–903. https://doi.org/10.1038/ejcn.2017.13

Shitara K, Nishikawa H (2018) Regulatory T cells: a potential target in cancer immunotherapy. Ann N Y Acad Sci 1417(1):104–115. https://doi.org/10.1111/nyas.13625

Shivappa N, Steck SE, Hurley TG et al (2014) Designing and developing a literature-derived, population-based dietary inflammatory index. Public Health Nutr 17(8):1689–1696. https://doi.org/10.1017/S1368980013002115

Shlyakhtina Y, Moran KL, Portal MM (2021) Genetic and non-genetic mechanisms underlying cancer evolution. Cancers (Basel) 13(6):1380. https://doi.org/10.3390/cancers13061380

Sholl LM, Do K, Shivdasani P et al (2016) Institutional implementation of clinical tumor profiling on an unselected cancer population. JCI Insight 1(19):e87062. https://doi.org/10.1172/jci.insight.87062

Shomar A, Barak O, Brenner N (2022) Cancer progression as a learning process. iScience 25(3):103924. https://doi.org/10.1016/j.isci.2022.103924

Shu J, Silva BVRE, Gao T et al (2017) Dynamic and modularized microRNA regulation and its implication in human cancers. Sci Rep 7(1):13356. https://doi.org/10.1038/s41598-017-13470-5

Shuldiner EG, Karmakar S, Tsai MK et al (2024) Aging represses lung tumorigenesis and alters tumor suppression. bioRxiv [Preprint]. https://doi.org/10.1101/2024.05.28.596319

Siddiqui AJ, Khan N, Fatima K et al (2023) Serum metallomics reveals insights into the associations of elements with the progression of preleukemic diseases toward acute leukemia. Biol Methods Protoc 9(1):bpad027. https://doi.org/10.1093/biomethods/bpad027

Siegel R, Ma J, Zou Z, Jemal A (2014) Cancer statistics, 2014. CA Cancer J Clin 64(1):9–29. https://doi.org/10.3322/caac.21208

Sies H, Jones DP (2020) Reactive oxygen species (ROS) as pleiotropic physiological signalling agents. Nat Rev Mol Cell Biol 21(7):363–383. https://doi.org/10.1038/s41580-020-0230-3

Silva A, Yunes JA, Cardoso BA et al (2008) PTEN posttranslational inactivation and hyperactivation of the PI3K/Akt pathway sustain primary T cell leukemia viability. J Clin Invest 118:3762–3774. https://doi.org/10.1172/JCI35131

Sincere NI, Anand K, Ashique S et al (2023) PROTACs: emerging targeted protein degradation approaches for advanced druggable strategies. Molecules 28(10):4014. https://doi.org/10.3390/molecules28104014

Singh M, Kumar V, Sehrawat N et al (2022) Current paradigms in epigenetic anticancer therapeutics and future challenges. Semin Cancer Biol 83:422–440. https://doi.org/10.1016/j.semcancer.2021.03.013

Singh A, Rappolee DA, Ruden DM (2023) Epigenetic reprogramming in mice and humans: from fertilization to primordial germ cell development. Cells 12(14):1874. https://doi.org/10.3390/cells12141874

Siravegna G, Marsoni S, Siena S et al (2017) Integrating liquid biopsies into the management of cancer. Nat Rev Clin Oncol 14(9):531–548. https://doi.org/10.1038/nrclinonc.2017

Siva S, MacManus MP, Martin RF, Martin OA (2015) Abscopal effects of radiation therapy: a clinical review for the radiobiologist. Cancer Lett 356(1):82–90. https://doi.org/10.1016/j.canlet.2013.09.018

Siversten I, Dahstrom AW (1921) Relation of muscular activity to carcinoma: a preliminary report. J Cancer Res 6(4):365–378. https://doi.org/10.1158/jcr.1921.365

Skulachev VP (1999) Phenoptosis: programmed death of an organism. Biochemistry (Mosc) 64(12):1418–1426

Skulachev VP (2012) What is "phenoptosis" and how to fight it? Biochemistry (Mosc) 77(7):689–706. https://doi.org/10.1134/S0006297912070012

Skulachev MV, Skulachev VP (2014) New data on programmed aging – slow phenoptosis. Biochemistry (Mosc) 79(10):977–993. https://doi.org/10.1134/S0006297914100010

Slack FJ, Chinnaiyan AM (2019) The role of non-coding RNAs in oncology. Cell 179(5):1033–1055. https://doi.org/10.1016/j.cell.2019.10.017

Slijepcevic P (2021) Principles of information processing and natural learning in biological systems. J Gen Philos Sci 52:227–245. https://doi.org/10.1007/s10838-019-09471-9

Slocombe L, Al-Khalili JS, Sacchi M (2021) Quantum and classical effects in DNA point mutations: Watson–Crick tautomerism in AT and GC base pairs. Phys Chem Chem Phys 23:4141–4150. https://doi.org/10.1039/D0CP05781A

Slocombe L, Winokan M, Al-Khalili J, Sacchi M (2022) Proton transfer during DNA strand separation as a source of mutagenic guanine–cytosine tautomers. Commun Chem 5(1):144. https://doi.org/10.1038/s42004-022-00760-x

Slocombe L, Winokan M, Al-Khalili J, Sacchi M (2023) Quantum tunnelling effects in the guanine–thymine wobble misincorporation via tautomerism. J Phys Chem Lett 14(1):9–15. https://doi.org/10.1021/acs.jpclett.2c03171

Slominski RM, Raman C, Chen JY, Slominski AT (2023) How cancer hijacks the body's homeostasis through the neuroendocrine system. Trends Neurosci 46(4):263–275. https://doi.org/10.1016/j.tins.2023.01.003

Smith JM (1986) The problems of biology. Oxford University Press, Oxford

Smith JM (1989) Evolutionary genetics. Oxford University Press, Oxford

Smith SE (2017) Organisms as persisters. Philos Theory Pract Biol 9(14)

Smith E, Morowitz HJ (2016) The origin and nature of life on earth: the emergence of the fourth geosphere. Cambridge University Press

Smith JC, Sheltzer JM (2018) Systematic identification of mutations and copy number alterations associated with cancer patient prognosis. Elife 7:e39217. https://doi.org/10.7554/eLife.39217

Smith JC, Sheltzer JM (2022) Genome-wide identification and analysis of prognostic features in human cancers. Cell Rep 38(13):110569. https://doi.org/10.1016/j.celrep.2022.110569

Smith AL, Whitehall JC, Bradshaw C et al (2020) Age-associated mitochondrial DNA mutations cause metabolic remodelling that contributes to accelerated intestinal tumorigenesis. Nat Cancer 1(10):976–989. https://doi.org/10.1038/s43018-020-00112-5

Smithers DW (1962) Cancer: an attack on cytologism. Lancet 279(7228):493–499. https://doi.org/10.1016/S0140-6736(62)91475-7

Smolarz B, Durczyński A, Romanowicz H et al (2022) miRNAs in cancer (review of literature). Int J Mol Sci 23(5):2805. https://doi.org/10.3390/ijms23052805

Smolin L (2023) Temporal naturalism. In: Rovelli C, Lestienne R, Harris PA (eds) Time and science – volume 1: the metaphysics of time and its evolution. WSPC (Europe)

Sobczuk P, Szczylik C, Porta C, Czarnecka AM (2017) Renin angiotensin system deregulation as renal cancer risk factor. Oncol Lett 14(5):5059–5068. https://doi.org/10.3892/ol.2017.6826

Sökeland G, Schumacher U (2019) The functional role of integrins during intra- and extravasation within the metastatic cascade. Mol Cancer 18(1):12. https://doi.org/10.1186/s12943-018-0937-3

Somasundaram K, Gupta B, Jain N, Jana S (2022) LncRNAs divide and rule: the master regulators of phase separation. Front Genet 13:930792. https://doi.org/10.3389/fgene.2022.930792

Somlyai G et al (1993) Naturally occurring deuterium is essential for the normal growth rate of cells. FEBS Lett 317(1–2):1–4. https://doi.org/10.1016/0014-5793(93)81479-J

Somlyai G, Kovács BZ, Papp A, Somlyai I (2023) A preliminary study indicating improvement in the median survival time of glioblastoma multiforme patients by the application of deuterium depletion in combination with conventional therapy. Biomedicines 11(7):1989. https://doi.org/10.3390/biomedicines11071989

Somlyai G, Papp A, Somlyai I et al (2025) Real-world data confirm that the integration of deuterium depletion into conventional cancer therapy multiplies the survival probability of patients. Biomedicines 13(4):876. https://doi.org/10.3390/biomedicines13040876

Sommer F, Bäckhed F (2013) The gut microbiota—masters of host development and physiology. Nat Rev Microbiol 11(4):227–238. https://doi.org/10.1038/nrmicro2974

Song M, Giovannucci E (2016) Preventable incidence and mortality of carcinoma associated with lifestyle factors among White adults in the United States. JAMA Oncol 2(9):1154–1161. https://doi.org/10.1001/jamaoncol.2016.0843

Song W, Wu J, Ge G, Lin Q (2008) Two domains of the epidermal growth factor receptor are involved in cytoskeletal interactions. Biochem Biophys Res Commun 370(4):589–593. https://doi.org/10.1016/j.bbrc.2008.03.138

Song M, Vogelstein B, Giovannucci EL et al (2018) Cancer prevention: molecular and epidemiologic consensus. Science 361(6409):1317–1318. https://doi.org/10.1126/science.aau3830

Song KX, Wang JX, Huang D (2023) Therapy-induced senescent tumor cells in cancer relapse. J Natl Cancer Cent 3(4):273–278. https://doi.org/10.1016/j.jncc.2023.09.001

Sonnenschein C, Soto AM (1998) The Society of Cells. Garland Science

Sonnenschein C, Soto AM (2016) Carcinogenesis explained within the context of a theory of organisms. Prog Biophys Mol Biol 122(1):70–76. https://doi.org/10.1016/j.pbiomolbio.2016.07.004

Soria G, Polo SE, Almouzni G (2012) Prime, repair, restore: the active role of chromatin in the DNA damage response. Mol Cell 46(6):722–734. https://doi.org/10.1016/j.molcel.2012.06.002

Sorkin R (2000) A historical perspective on cancer. arXiv:physics/0011002. https://doi.org/10.48550/arXiv.physics/0011002

Soto AM, Sonnenschein C (2011) The tissue organization field theory of cancer: a testable replacement for the somatic mutation theory. Bioessays 33(5):332–340. https://doi.org/10.1002/bies.201100025

Sottoriva A, Kang H, Ma Z et al (2015) A big bang model of human colorectal tumor growth. Nat Genet 47(3):209–216. https://doi.org/10.1038/ng.3214

Sottoriva A, Barnes CP, Graham TA (2017) Catch my drift? Making sense of genomic intra-tumour heterogeneity. Biochim Biophys Acta Rev Cancer 1867 2:95–100. https://doi.org/10.1016/j.bbcan.2016.12.003

Sou IF, Hamer G, Tee WW et al (2023) Cancer and meiotic gene expression: two sides of the same coin? Curr Top Dev Biol 151:43–68. https://doi.org/10.1016/bs.ctdb.2022.06.002

Soucy SM, Huang J, Gogarten JP (2015) Horizontal gene transfer: building the web of life. Nat Rev Genet 16(8):472–482. https://doi.org/10.1038/nrg3962

Southam CM, Ehrlich J (1943) Effects of extract of western bed-cedar heart-wood on certain wood-decaying fungi in culture. Phytopathology 33(6):517–524

Sowinski DR, Carroll-Nellenback J, Markwick RN, Piñero J, Gleiser M, Kolchinsky A et al (2023) Semantic information in a model of resource gathering agents. PRX Life 1:023003. https://doi.org/10.1103/PRXLife.1.023003

Spencer VA, Costes S, Inman JL et al (2011) Depletion of nuclear actin is a key mediator of quiescence in epithelial cells. J Cell Sci 124(1):123–132. https://doi.org/10.1242/jcs.073197

Sporn MB (1991) Carcinogenesis and cancer: different perspectives on the same disease. Cancer Res 51(23 Pt 1):6215–6218

Sporn MB (1996) The war on cancer. Lancet 347(9012):1377–1381. https://doi.org/10.1016/s0140-6736(96)91015-6

Srivastava R (2019) The role of proton transfer on mutations. Front Chem 7:536. https://doi.org/10.3389/fchem.2019.00536

Stadler ZK, Maio A, Chakravarty D (2021) Therapeutic implications of germline testing in patients with advanced cancers. J Clin Oncol 39(24):2698–2709. https://doi.org/10.1200/JCO.20.03661

Staff (1954) Science: the fleeting flesh. Time Magazine (Oct 11) https://time.com/archive/6869550/science-the-fleeting-flesh/

Stan MC, Paul D (2024) Diabetes and cancer: a twisted bond. Oncol Rev 18:1354549. https://doi.org/10.3389/or.2024.1354549

Stanková K, Brown JS, Dalton WS, Gatenby RA (2019) Optimizing cancer treatment using game theory: a review. JAMA Oncol 5(1):96–103. https://doi.org/10.1001/jamaoncol.2018.3395

Stanley RF, Abdel-Wahab O (2022) Dysregulation and therapeutic targeting of RNA splicing in cancer. Nat Cancer 3(5):536–546. https://doi.org/10.1038/s43018-022-00384-z

Statello L, Guo CJ, Chen LL, Huarte M (2021) Gene regulation by long non-coding RNAs and its biological functions. Nat Rev Mol Cell Biol 22(2):96–118. https://doi.org/10.1038/s41580-020-00315-9

Steen CB, Luca BA, Esfahani MS, Azizi A, Sworder BJ, Nabet BY et al (2021) The landscape of tumor cell states and ecosystems in diffuse large B cell lymphoma. Cancer Cell 39(10):1422–1437.e10. https://doi.org/10.1016/j.ccell.2021.08.011

Stehelin D, Varmus HE, Bishop JM, Vogt PK (1976) DNA related to the transforming gene(s) of avian sarcoma viruses is present in normal avian DNA. Nature 260(5547):170–173. https://doi.org/10.1038/260170a0

Stein M, Lin H, Jeyamohan C et al (2010) Targeting tumor metabolism with 2-deoxyglucose in patients with castrate-resistant prostate cancer and advanced malignancies. Prostate 70(13):1388–1394. https://doi.org/10.1002/pros.21172

Stein EM, DiNardo CD, Fathi AT et al (2019) Molecular remission and response patterns in patients with mutant-IDH2 acute myeloid leukemia treated with enasidenib. Blood 133(7):676–687. https://doi.org/10.1182/blood-2018-08-869008

Stephen W (2009) The evolution of evolution: Darwin, Enlightenment and Scotland. Luath Press

Stephens PJ, Greenman CD, Fu B et al (2011) Massive genomic rearrangement acquired in a single catastrophic event during cancer development 144(1):27–40. https://doi.org/10.1016/j.cell.2010.11.055

Sterling P (2018) Predictive regulation and human design. eLife 7:e36133. https://doi.org/10.7554/eLife.36133

Sterling P (2020) What is health? MIT Press

Sterling P, Eyer J (1988) Allostasis: a new paradigm to explain arousal pathology. In: Fisher S, Reason JT (eds) Handbook of life stress, cognition, and health. Wiley, Chichester

Stern DL, Orgogozo V (2008) The loci of evolution: how predictable is genetic evolution? Evolution 62(9):2155–2177. https://doi.org/10.1111/j.1558-5646.2008.00450.x

Stewart TA, Yapa KT, Monteith GR (2015) Altered calcium signaling in cancer cells. Biochim Biophys Acta 1848(10 Pt B):2502–2511. https://doi.org/10.1016/j.bbamem.2014.08.016

Stone JR, Yang S (2006) Hydrogen peroxide: a signaling messenger. Antioxid Redox Signal 8:243–270. https://doi.org/10.1089/ars.2006.8.243

Stone J, Mitrofanis J, Johnstone DM et al (2018) Acquired resilience: an evolved system of tissue protection in mammals. Dose Response 16(4):1559325818803428. https://doi.org/10.1177/1559325818803428

Stopsack KH, Whittaker CA, Gerke TA et al (2019) Aneuploidy drives lethal progression in prostate cancer. Proc Natl Acad Sci USA 116(23):11390–11395. https://doi.org/10.1073/pnas.1902645116

Strahl BD, Allis CD (2000) The language of covalent histone modifications. Nature 403(6765):41–45. https://doi.org/10.1038/47412

Strassler M (2024) Waves in impossible sea. Basic Books

Strauss JD, Pursell ZF (2023) Replication DNA polymerases, genome instability and cancer therapies. NAR. Cancer 5(3):zcad033. https://doi.org/10.1093/narcan/zcad033

Strauss B, Bertolaso, Ernberg I, Bissell MJ (2021) Rethinking cancer a new paradigm for the posgenomics era. Edited by Bernhard Strauss, Marta Bertolaso, Ingemear Ernberg and Mina J. Bissell The MIT Press

Street K, Risso D, Fletcher RB et al (2018) Slingshot: cell lineage and pseudotime inference for single-cell transcriptomics. BMC Genomics 19(1):477. https://doi.org/10.1186/s12864-018-4772-0

Stupar P, Podolski-Renić A, Villalba MI et al (2021) Nano-motion analysis for rapid and label-free assessing of cancer cell sensitivity to chemotherapeutics. Medicina (Kaunas) 57(5):446. https://doi.org/10.3390/medicina57050446

Su SC, Hsieh MJ, Yang WE et al (2017) Cancer metastasis: mechanisms of inhibition by melatonin. J Pineal Res 62(1):e12370. https://doi.org/10.1111/jpi.12370

Su JX, Li SJ, Zhou XF et al (2023) Chemotherapy-induced metastasis: molecular mechanisms and clinical therapies. Acta Pharmacol Sin 44(9):1725–1736. https://doi.org/10.1038/s41401-023-01093-8

Su J, Song Y, Zhu Z et al (2024) Cell-cell communication: new insights and clinical implications. Signal Transduct Target Ther 9(1):196. https://doi.org/10.1038/s41392-024-01888-z

Su K, Zeng D, Zhang W, et al. (2025) Integrating cancer medicine into metabolic rhythms. Trends Endocrinol Metab S1043-2760(25)00053-0. doi: https://doi.org/10.1016/j.tem.2025.03.007

Suárez-González J, Martínez-Laperche C, Kwon M et al (2018) Donor cell-derived hematologic neoplasms after hematopoietic stem cell transplantation: a systematic review. Biol Blood Marrow Transplant 24(7):1505–1513. https://doi.org/10.1016/j.bbmt.2018.01.033

Sud A, Kinnersley B, Houlston RS (2017) Genome-wide association studies of cancer: current insights and future perspectives. Nat Rev Cancer 17(11):692–704. https://doi.org/10.1038/nrc.2017.82

Sulli G, Lam MTY, Panda S (2019) Interplay between circadian clock and cancer: new frontiers for cancer treatment. Trends Cancer 5(8):475–494. https://doi.org/10.1016/j.trecan.2019.07.002

Sullivan MR, Danai LV, Lewis CA et al (2019) Quantification of microenvironmental metabolites in murine cancers reveals determinants of tumor nutrient availability. Elife 8:e44235. https://doi.org/10.7554/eLife.44235

Sun ZC, Ge JL, Guo B et al (2016) Extremely low frequency electromagnetic fields facilitate vesicle endocytosis by increasing presynaptic calcium channel expression at a central synapse. Sci Rep 6:21774. https://doi.org/10.1038/srep21774

Sun X, Zhan L, Chen Y et al (2018) Increased mtDNA copy number promotes cancer progression by enhancing mitochondrial oxidative phosphorylation in microsatellite-stable colorectal cancer. Signal Transduct Target Ther 3:8. https://doi.org/10.1038/s41392-018-0011-z

Sun S, Hong J, You E et al (2023a) Cancer cells co-evolve with retrotransposons to mitigate viral mimicry. bioRxiv 2023.05.19.541456. https://doi.org/10.1101/2023.05.19.541456

Sun Z, Fan J, Dang Y et al (2023b) Enhancer in cancer pathogenesis and treatment. Genet Mol Biol 46(3):e20220313. https://doi.org/10.1590/1678-4685-GMB-2022-0313
Sun H, Xia T, Ma S et al (2025) Intercellular communication is crucial in the regulation of healthy aging via exosomes. Pharmacol Res 212:107591. https://doi.org/10.1016/j.phrs.2025.107591
Sundelacruz S, Levin M, Kaplan DL et al (2013) Depolarization alters phenotype, maintains plasticity of predifferentiated mesenchymal stem cells. Tissue Eng Part A 19(17–18):1889–1908. https://doi.org/10.1089/ten.tea.2012.0425.rev
Sung H, Ferlay J, Siegel RL et al (2021) Global cancer statistics 2020: GLOBOCAN estimates of incidence and mortality worldwide for 36 cancers in 185 countries. CA Cancer J Clin 71(3):209–249. https://doi.org/10.3322/caac.21660
Surh YJ (2003) Cancer chemoprevention with dietary phytochemicals. Nat Rev Cancer 3(10):768–780. https://doi.org/10.1038/nrc1189
Sutton EF, Beyl R, Early KS et al (2018) Early time-restricted feeding improves insulin sensitivity, blood pressure, and oxidative stress even without weight loss in men with prediabetes. Cell Metab 27(6):1212–1221.e3. https://doi.org/10.1016/j.cmet.2018.04.010
Swamy K (2022) Stereotactic body radiotherapy immunological planning—a review with a proposed theoretical model. Front Oncol 12:729250. https://doi.org/10.3389/fonc.2022.729250
Swanton C (2012) Intratumor heterogeneity: evolution through space and time. Cancer Res 72(19):4875–4882. https://doi.org/10.1158/0008-5472.CAN-12-2217
Swanton C, Bernard E, Abbosh C et al (2024) Embracing cancer complexity: Hallmarks of systemic disease. Cell 187(7):1589–1616. https://doi.org/10.1016/j.cell.2024.02.009
Switzer B, Puzanov I, Gandhi S et al (2024) Targeting beta-adrenergic receptor pathways in melanoma: how stress modulates oncogenic immunity. Melanoma Res 34(2):89–95. https://doi.org/10.1097/CMR.0000000000000943
Syroeshkin AV, Antipova NV, Zlatska AV et al (2018) The effect of the deuterium depleted water on the biological activity of the eukaryotic cells. J Trace Elem Med Biol 50:629. https://doi.org/10.1016/j.jtemb.2018.05.004
Syvanen M, Kado CI (eds) (1998) Horizontal gene transfer. Chapman & Hall, London
Szabo S (1998) Hans Selye and the development of the stress concept: Special reference to gastroduodenal ulcerogenesis. Ann N Y Acad Sci 851:19–27. https://doi.org/10.1111/j.1749-6632.1998.tb08972.x
Szczepanek J, Tretyn A (2023) MicroRNA-mediated regulation of histone-modifying enzymes in cancer: Mechanisms and therapeutic implications. Biomolecules 13(11):1590. https://doi.org/10.3390/biom13111590
Szostak JW (2003) Functional information: Molecular messages. Nature 423(6941):689. https://doi.org/10.1038/423689a
Szturz P, Vermorken JB (2022) Steering decision making by terminology: oligometastatic versus argometastatic. Br J Cancer 127(4):587–591. https://doi.org/10.1038/s41416-022-01879-3
Szulcek R, Grune J, Kuebler WM (2020) The updated cancer paradigm of PAH: recognizing complexity. Am J Physiol Lung Cell Mol Physiol 318(6):L1111–L1114. https://doi.org/10.1152/ajplung.00159.2020
Tachè Y (2014) Hans Selye and the stress response: From "the first mediator" to the identification of the hypothalamic corticotropin-releasing factor. Ideggyogy Sz 67(3–4):95–98. PMID: 26118248
Tafani M, Sansone L, Limana F et al (2016) The interplay of reactive oxygen species, hypoxia, inflammation, and sirtuins in cancer initiation and progression. Oxid Med Cell Longev 2016:3907147. https://doi.org/10.1155/2016/3907147
Tai YL, Chen KC, Hsieh JT, Shen TL (2018) Exosomes in cancer development and clinical applications. Cancer Sci 109(8):2364–2374. https://doi.org/10.1111/cas.13697
Tajada S, Villalobos C (2020) Calcium permeable channels in cancer hallmarks. Front Pharmacol 11:968. https://doi.org/10.3389/fphar.2020.00968
Tamura R, Tanaka T, Yamamoto Y et al (2018) Dual role of macrophage in tumor immunity. Immunotherapy 10(10):899–909. https://doi.org/10.2217/imt-2018-0006
Tang S, Li S, Tang B et al (2023) Hormetic and synergistic effects of cancer treatments revealed by modelling combinations of radio- or chemotherapy with immunotherapy. BMC Cancer 23(1):1040. https://doi.org/10.1186/s12885-023-11542-6
Tanner K, Mori H, Mroue R et al (2012) Coherent angular motion in the establishment of multicellular architecture of glandular tissues. Proc Natl Acad Sci U S A 109(6):1973–1978. https://doi.org/10.1073/pnas.1119578109
Tannir NM, Agarwal N, Porta C et al (2022) Efficacy and safety of Telaglenastat plus Cabozantinib vs placebo plus Cabozantinib in patients with advanced renal cell carcinoma: The CANTATA randomized clinical trial. JAMA Oncol 8(10):1411–1418. https://doi.org/10.1001/jamaoncol.2022.3511
Tao H, O'Neil A, Choi Y et al (2020) Pre- and post-diagnosis diabetes as a risk factor for all-cause and cancer-specific mortality in breast, prostate, and colorectal cancer survivors: a prospective cohort study. Front Endocrinol (Lausanne) 11:60. https://doi.org/10.3389/fendo.2020.00060
Tarrado-Castellarnau M, de Atauri P, Cascante M (2016) Oncogenic regulation of tumor metabolic reprogramming. Oncotarget 7(38):62726–62753. https://doi.org/10.18632/oncotarget.10911
Tatsumi S, Miyagawa A, Kaneko I et al (2016) Regulation of renal phosphate handling: inter-organ communication in health and disease. J Bone Miner Metab 34(1):1–10. https://doi.org/10.1007/s00774-015-0705-z
Taylor T, Bedau M, Channon A et al (2016) Open-ended evolution: perspectives from the OEE workshop in York. Artif Life 22(3):408–423. https://doi.org/10.1162/ARTL_a_00210
Taylor AM, Shih J, Ha G et al (2018) Genomic and functional approaches to understanding cancer aneuploidy. Cancer Cell 33(4):676–689.e3. https://doi.org/10.1016/j.ccell.2018.03.007
Taylor AMR, Rothblum-Oviatt C, Ellis NA et al (2019) Chromosome instability syndromes. Nat Rev Dis Primers 5(1):64. https://doi.org/10.1038/s41572-019-0113-0
Taylor SR, Falcone JN, Cantley LC, Goncalves MD (2022) Developing dietary interventions as therapy for cancer. Nat Rev Cancer 22(8):452–466. https://doi.org/10.1038/s41568-022-00485-y
Tchkonia T, Thomou T, Zhu Y et al (2013) Mechanisms and metabolic implications of regional differences among fat depots. Cell Metab 17(5):644–656. https://doi.org/10.1016/j.cmet.2013.03.008
Tegmark M (2020) Importance of quantum decoherence in brain processes. Phys Rev E 61:4194. https://doi.org/10.1103/PhysRevE.61.4194
Temel JS, Greer JA, Muzikansky A et al (2010) Early palliative care for patients with metastatic non-small-cell lung cancer. N Engl J Med 363(8):733–742. https://doi.org/10.1056/NEJMoa1000678
Ter Stege JA, Woerdeman LAE, Kieffer JM et al (2024) Efficacy of a decision aid in breast cancer patients considering immediate reconstruction: Results of a randomized controlled trial. Plast Reconstr Surg 154(4):706–722. https://doi.org/10.1097/PRS.0000000000011100
Terlizzi M, Casolaro V, Pinto A, Sorrentino R (2014) Inflammasome: cancer's friend or foe? Pharmacol Ther 143(1):24–33. https://doi.org/10.1016/j.pharmthera.2014.02.002
Terracciano D, Terreri S, de Nigris F et al (1868) (2017) The role of a new class of long noncoding RNAs transcribed from ultraconserved regions in cancer. Biochim Biophys Acta Rev Cancer 2:449–455. https://doi.org/10.1016/j.bbcan.2017.09.001
Thaxton CB, Bradley WL, Olsen RL et al (2020) The mystery of life's origin. Discovery Institute Press

Thibaudeau TA, Smith DM (2019) A practical review of proteasome pharmacology. Pharmacol Rev 71(2):170–197. https://doi.org/10.1124/pr.117.015370

Thiery JP, Acloque H, Huang RY, Nieto MA (2009) Epithelial-mesenchymal transitions in development and disease. Cell 139(5):871–890. https://doi.org/10.1016/j.cell.2009.11.007

Thims L (2007) Human chemistry, vol 1. LuLu, Morrisville

Thom R (1975) Structural stability and morphogenesis: an outline of a general theory of models, reading. W. A. Benjamin, Inc., Reading

Thom R (2009) Prédire n'est pas expliquer. Flammarion, Paris

Thomas L (1959) Discussion of cellular and humoral aspects of hypersensitive states. In: Lawrence HS (ed) Cellular and humoral aspects of hypersensitive states. Hoeber-Harper, New York

Thomas L (1982) On immunosurveillance in human cancer. Yale J Biol Med 5(3–4):329–333. PMID: 6758376; PMCID: PMC2596448

Thomas ET, Del Mar C, Glasziou P et al (2017) Prevalence of incidental breast cancer and precursor lesions in autopsy studies: a systematic review and meta-analysis. BMC Cancer 17(1):808. https://doi.org/10.1186/s12885-017-3808-0

Thomas F, DeGregori J, Marusyk A et al (2024) A new perspective on tumor progression: evolution via selection for function. Evol Med Public Health 12(1):172–177. https://doi.org/10.1093/emph/eoae021

Thompson DW (1917) On growth and form. Cambridge University Press

Thurner S, Klimek P, Hanel R (2018) Preface. In: Introduction to the theory of complex systems. Oxford University Press. https://doi.org/10.1093/oso/9780198821939.001.0001

Timp W, Feinberg AP (2013) Cancer as a dysregulated epigenome allowing cellular growth advantage at the expense of the host. Nat Rev Cancer 13(7):497–510. https://doi.org/10.1038/nrc3486

Tinetti ME, Naik AD, Dodson JA (2016) Moving from disease-centered to patient goals-directed care for patients with multiple chronic conditions: patient value-based care. JAMA Cardiol 1(1):9–10. https://doi.org/10.1001/jamacardio.2015.0248

Tiwari S, Sapkota N, Han Z (2022) Effect of fasting on cancer: a narrative review of scientific evidence. Cancer Sci 113(10):3291–3302. https://doi.org/10.1111/cas.15492

Todor RD, Bratucu G, Moga MA et al (2021) Int J Environ Res Public Health. 18(4):1721. https://doi.org/10.3390/ijerph18041721

Tohme S, Simmons RL, Tsung A (2017) Surgery for cancer: a trigger for metastases. Cancer Res 77(7):1548–1552. https://doi.org/10.1158/0008-5472.CAN-16-1536

Tokizane K, Imai SI (2025) Inter-organ communication is a critical machinery to regulate metabolism and aging. Trends Endocrinol Metab. https://doi.org/10.1016/j.tem.2024.11.013

Tomasetti C (2019) Mutated clones are the new normal. Science 364(6444):938–939. https://doi.org/10.1126/science.aax5525

Tomasetti C, Vogelstein B (2015) Cancer etiology. Variation in cancer risk among tissues can be explained by the number of stem cell divisions. Science 347(6217):78–81. https://doi.org/10.1126/science.1260825

Tomasetti C, Vogelstein B, Parmigiani G (2013) Half or more of the somatic mutations in cancers of self-renewing tissues originate prior to tumor initiation. Proc Natl Acad Sci USA 110(6):1999–2004. https://doi.org/10.1073/pnas.1221068110

Tomasetti C, Marchionni L, Nowak MA et al (2015) Only three driver gene mutations are required for the development of lung and colorectal cancers. Proc Natl Acad Sci USA 112:118–123. https://doi.org/10.1073/pnas.1421839112

Tomasetti C, Li L, Vogelstein B (2017) Stem cell divisions, somatic mutations, cancer etiology, and cancer prevention. Science 355(6331):1330–1334. https://doi.org/10.1126/science.aaf9011

Tominaga N, Kosaka N, Ono M et al (2015) Brain metastatic cancer cells release microRNA-181c-containing extracellular vesicles capable of destructing blood-brain barrier. Nat Commun 6:6716. https://doi.org/10.1038/ncomms7716

Tonkin-Reeves A, Giuliani CM, Price JT (2023) Inhibition of autophagy; an opportunity for the treatment of cancer resistance. Front Cell Dev Biol 11:1177440. https://doi.org/10.3389/fcell.2023.1177440

Tononi G (2004) An information integration theory of consciousness. BMC Neurosci 5:42. https://doi.org/10.1186/1471-2202-5-42

Topalian SL, Hodi FS, Brahmer JR et al (2012) Safety, activity, and immune correlates of anti-PD-1 antibody in cancer. N Engl J Med 366(26):2443–2454. https://doi.org/10.1056/NEJMoa1200690

Topol EJ (2019) High-performance medicine: the convergence of human and artificial intelligence. Nat Med 25:44–56. https://doi.org/10.1038/s41591-018-0300-7

Topper MJ, Vaz M, Chiappinelli KB et al (2017) Epigenetic therapy ties MYC depletion to reversing immune evasion and treating lung cancer. Cell 171(6):1284–1300.e21. https://doi.org/10.1016/j.cell.2017.10.022

Torday JS (2014) On the evolution of development. Trends Dev Biol 8:17–37. PMID: 25729239; PMCID: PMC4339279

Torday JS (2015) The cell as the mechanistic basis for evolution. Wiley Interdiscip Rev Syst Biol Med 7(5):275–284. https://doi.org/10.1002/wsbm.1305

Torday JS (2024) The synchronic, diachronic cell as the holism of consciousness. Prog Biophys Mol Biol 188:19–23. https://doi.org/10.1016/j.pbiomolbio.2024.02.006

Torday J (2025) Consciousness-Based Evolution. CRC Press

Torday JS, Rehan VK (2017) Evolution, the logic of biology. Wiley

Torra OS, Loeb KR (2011) Donor cell–derived leukemia and myelodysplastic neoplasm: unique forms of leukemia. Am J Clin Pathol 135(4):501–504. https://doi.org/10.1309/AJCPXW8DKEG5QMTB

Torres CM, Biran A, Burney MJ et al (2016) The linker histone H1.0 generates epigenetic and functional intratumor heterogeneity. Science 353(6307):aaf1644. https://doi.org/10.1126/science.aaf1644

Torti D, Trusolino L (2011) Oncogene addiction as a foundational rationale for targeted anti-cancer therapy: promises and perils. EMBO Mol Med 3(11):623–636. https://doi.org/10.1002/emmm.201100176

Touyz R, Briones A (2011) Reactive oxygen species and vascular biology: implications in human hypertension. Hypertens Res 34:5–14. https://doi.org/10.1038/hr.2010.201

Tracey KJ (2002) The inflammatory reflex. Nature 420(6917):853–859. https://doi.org/10.1038/nature01321

Tracey K (2025) The great nerve: the new science of the vagus nerve and how to harness its healing reflexes. Avery

Trachootham D et al (2009) Targeting cancer cells by ROS-mediated mechanisms: a radical therapeutic approach? Nat Rev Drug Discov 8(7):579–591. https://doi.org/10.1038/nrd2803

Trefil J, Morowitz HJ, Smith E (2009) The origin of life: a case is made for the descent of electrons. Am Sci 97(3):206–213

Trevors JT, Saier MH Jr (2010) Three laws of biology. Water Air Soil Pollut 205(Suppl 1):S87–S89. https://doi.org/10.1007/s11270-009-0069-y

Trewavas A (1999) Le calcium, c'est la vie: calcium makes waves. Plant Physiol 120(1):1–6. https://doi.org/10.1104/pp.120.1.1

Trewavas AJ, Baluška F (2011) The ubiquity of consciousness. EMBO Rep 12:1221–1225. https://doi.org/10.1038/embor.2011.218

Trifonov EN (2011) Vocabulary of definitions of life suggests a definition. J Biomol Struct Dyn 29(2):259–266. https://doi.org/10.1080/073911011010524992

Trigos AS, Pearson RB, Papenfuss AT, Goode DL (2017) Altered interactions between unicellular and multicellular genes drive hallmarks of transformation in a diverse range of solid tumors. Proc Natl Acad Sci USA 114(24):6406–6411. https://doi.org/10.1073/pnas.1617743114

Trigos AS, Pearson RB, Papenfuss AT, Goode DL (2018) How the evolution of multicellularity set the stage for cancer. Br J Cancer 118(2):145–152. https://doi.org/10.1038/bjc.2017.398

Trigos AS, Bongiovanni F, Zhang Y et al (2024) Disruption of metazoan gene regulatory networks in cancer alters the balance of co-expression between genes of unicellular and multicellular origins. Genome Biol 25(1):110. https://doi.org/10.1186/s13059-024-03247-1

Tsoli M, Robertson G (2013) Cancer cachexia: malignant inflammation, tumorkines, and metabolic mayhem. Trends Endocrinol Metab 24(4):174–183. https://doi.org/10.1016/j.tem.2012.10.006

Tsugane S, Sasazuki S (2007) Diet and the risk of gastric cancer: review of epidemiological evidence. Gastric Cancer 10(2):75–83. https://doi.org/10.1007/s10120-007-0420-0

Tubio JMC, Li Y, Ju YS et al (2014) Mobile DNA in cancer: extensive transduction of nonrepetitive DNA mediated by L1 retrotransposition in cancer genomes. Science 345(6196):1251343. https://doi.org/10.1126/science.1251343

Tufail M, Jiang CH, Li N (2024) Altered metabolism in cancer: insights into energy pathways and therapeutic targets. Mol Cancer 23(1):21. https://doi.org/10.1186/s12943-024-02119-3

Tumeh PC, Harview CL, Yearley JH et al (2014) PD-1 blockade induces responses by inhibiting adaptive immune resistance. Nature 515(7528):568–571. https://doi.org/10.1038/nature13954

Tung JC, Barnes JM, Desai SR et al (2015) Tumor mechanics and metabolic dysfunction. Free Radic Biol Med 79:269–280. https://doi.org/10.1016/j.freeradbiomed.2014.11.020

Tuomisto AE, Mäkinen MJ, Väyrynen JP (2019) Systemic inflammation in colorectal cancer: underlying factors, effects, and prognostic significance. World J Gastroenterol 25(31):4383–4404. https://doi.org/10.3748/wjg.v25.i31.4383

Turajlic S, Xu H, Litchfield K et al (2018) Tracking cancer evolution reveals constrained routes to metastases: TRACERx Renal. Cell 173(3):581–594.e12. https://doi.org/10.1016/j.cell.2018.03.057

Turcotte A, Achi S, Mamlouk O, Mandayam S (2022) Electrolyte disturbances in cancer patients. Curr Opin Nephrol Hypertens 31(5):425–434. https://doi.org/10.1097/MNH.0000000000000819

Turizo-Smith AD, Córdoba-Hernandez S, Mejía-Guarnizo LV et al (2024) Inflammation and cancer: friend or foe? Front Pharmacol 15:1385479. https://doi.org/10.3389/fphar.2024.1385479

Turley SJ, Cremasco V, Astarita JL (2015) Immunological hallmarks of stromal cells in the tumour microenvironment. Nat Rev Immunol 15(11):669–682. https://doi.org/10.1038/nri3902

Turner KM, Deshpande V, Beyter D et al (2017) Extrachromosomal oncogene amplification drives tumour evolution and genetic heterogeneity. Nature 543(7643):122–125. https://doi.org/10.1038/nature21356

Tuszynski JA, Costa F (2022) Low-energy amplitude-modulated radiofrequency electromagnetic fields as a systemic treatment for cancer: review and proposed mechanisms of action. Front Med Technol 4:869155. https://doi.org/10.3389/fmedt.2022.869155

Udrişte O (1978) Gena Ancestrală şi Originea Cancerului (in Romanian). Editura Ştiinţifică şi Enciclopedică, Bucharest

Udrişte O (1982) Gena Marker Ancestrală şi Originea Cancerului (in Romanian). Editura Medicală, Bucharest

Ugai T, Sasamoto N, Lee HY et al (2022) Is early-onset cancer an emerging global epidemic? Current evidence and future implications. Nat Rev Clin Oncol 19(10):656–673. https://doi.org/10.1038/s41571-022-00672-8

UK Biobank Whole-Genome Sequencing Consortium (2025) Whole-genome sequencing of 490,640 UK Biobank participants. Nature.:2025. https://doi.org/10.1038/s41586-025-09272-9.

Uller T (2023) Agency, goal orientation, and evolutionary explanations. In: Corning PA, Kauffman SA, Noble D, Shapiro JA, Vane-Wright RI, Pross A (eds) Evolution "On Purpose": Teleonomy in Living Systems. MIT Press

Unger JM, Shulman LN, Facktor MA et al (2024) National estimates of the participation of patients with cancer in clinical research studies based on Commission on Cancer Accreditation Data. J Clin Oncol 42(18):2139–2148. https://doi.org/10.1200/JCO.23.01030

Urzúa-Traslaviña CG, van Lieshout T, Boulogne F et al (2024) Co-expression in tissue-specific gene networks links genes in cancer-susceptibility loci to known somatic driver genes. BMC Med Genomics 17(1):186. https://doi.org/10.1186/s12920-024-01941-4

Utzeri VJ, Ribani A, Schiavo G, Fontanesi L (2021) Describing variability in the tyrosinase (TYR) gene, the albino coat color locus, in domestic and wild European rabbits. Ital J Anim Sci 20(1):181–187. https://doi.org/10.1080/1828051X.2021.1877574

Valcz G, Újvári B, Buzás EI et al (2022) Small extracellular vesicle DNA-mediated horizontal gene transfer as a driving force for tumor evolution: facts and riddles. Front Oncol 12:945376. https://doi.org/10.3389/fonc.2022.945376

Valente LJ, Tarangelo A, Li AM et al (2020) p53 deficiency triggers dysregulation of diverse cellular processes in physiological oxygen. J Cell Biol 219(11):e201908212. https://doi.org/10.1083/jcb.201908212

Valero-Bover D, Monterde D, Carot-Sans G et al (2023) Is age the most important risk factor in COVID-19 patients? The relevance of comorbidity burden: a retrospective analysis of 10,551 hospitalizations. Clin Epidemiol 15:811–825. https://doi.org/10.2147/CLEP.S408510

Valiant L (2013) Probably approximately correct: nature's algorithms for learning and prospering in a complex world. Basic Books

Valiathan MS (2009) An Ayurvedic way of life. Curr Sci 96(9):1186–1192

van den Berg YW, Osanto S, Reitsma PH, Versteeg HH (2012) The relationship between tissue factor and cancer progression: insights from bench and bedside. Blood 119(4):924–932. https://doi.org/10.1182/blood-2011-06-317685

van den Brink GR, Hardwick JC (2006) Hedgehog Wnt interaction in colorectal cancer. Gut 55(7):912–914. https://doi.org/10.1136/gut.2005.085902

van den Brink GR, Hardwick JC, Tytgat GN et al (2001) Sonic hedgehog regulates gastric gland morphogenesis in man and mouse. Gastroenterology 121(2):317–328. https://doi.org/10.1053/gast.2001.26261

van der Burg SH, Arens R, Ossendorp F et al (2016) Vaccines for established cancer: overcoming the challenges posed by immune evasion. Nat Rev Cancer 16(4):219–233. https://doi.org/10.1038/nrc.2016.16

van der Laak J, Litjens G, Ciompi F (2021) Deep learning in histopathology: the path to the clinic. Nat Med 27(5):775–784. https://doi.org/10.1038/s41591-021-01343-4

van Der Merwe R (2024) Stuart Kauffman's metaphysics of the adjacent possible: a critique. Interdiscip Sci Rev 48(1). https://doi.org/10.1080/03080188.2022.2125614

van der Ryn S, Cowan S (1997) Ecological Design. Island Press

van der Schaar M (2025) Genies: the future of AI agents. Accessed April 10, 2025. https://www.vanderschaar-lab.com/genies-the-future-of-ai-agents/

van der Schaar M, Peck R, McKinney E et al. (2025) Revolutionizing Clinical Trials: A Manifesto for AI-Driven Transformation arXiv:2506.09102v1 https://doi.org/10.48550/arXiv.2506.09102

van Herck Y, Feyaerts A, Alibhai S et al (2021) Is cancer biology different in older patients? Lancet Healthy Longev 2(10):e663–e677. https://doi.org/10.1016/S2666-7568(21)00179-3

van Roosbroeck K, Calin GA (2017) Cancer hallmarks and microRNAs: the therapeutic connection. Adv Cancer Res 135:119–149. https://doi.org/10.1016/bs.acr.2017.06.002

van Tongelen A, Loriot A, De Smet C (2017) Oncogenic roles of DNA hypomethylation through the activation of cancer-germline genes. Cancer Lett 396:130–137. https://doi.org/10.1016/j.canlet.2017.03.029

Vander Heiden MG, DeBerardinis RJ (2017) Understanding the intersections between metabolism and cancer biology. Cell 168(4):657–669. https://doi.org/10.1016/j.cell.2016.12.039

Vander, Heiden MG, Cantley LC, Thompson CB (2009) Understanding the Warburg effect: the metabolic requirements of cell proliferation. Science 324(5930):1029–1033. https://doi.org/10.1126/science.1160809

Varela F, Maturana H, Uribe R (1974) Autopoiesis: the organization of living systems, its characterization and a model. BioSystems 5:187–195

Various Authors (2015) Letters on cancer risk (reply to Tomasetti and Vogelstein). Science 347:727–730

Varmus H (1989) https://www.nobelprize.org/prizes/medicine/1989/varmus/speech/

Vasudevan A, Baruah PS, Smith JC et al (2020) Single-chromosomal gains can function as metastasis suppressors and promoters in colon cancer. Dev Cell 52(4):413–428.e6. https://doi.org/10.1016/j.devcel.2020.01.034

Vaupel P, Mayer A (2012) Availability, not respiratory capacity governs oxygen consumption of solid tumors. Int J Biochem Cell Biol 44(9):1477–1481. https://doi.org/10.1016/j.biocel.2012.05.019

Vaux DL, Korsmeyer SJ (1999) Cell death in development. Cell 96(2):245–254. https://doi.org/10.1016/s0092-8674(00)80564-4

Vázquez-Fernández E, Young HS, Requena JR, Wille H (2017) The structure of mammalian prions and their aggregates. Int Rev Cell Mol Biol 329:277–301. https://doi.org/10.1016/bs.ircmb.2016.08.013

Venditti S, Verdone L, Reale A et al (2020) Molecules of silence: effects of meditation on gene expression and epigenetics. Front Psychol 11:1767. https://doi.org/10.3389/fpsyg.2020.01767

Venerin AA, Venerina YA, Alexandrov YI (2020) Cell functioning in norm and pathology in terms of the activity paradigm: oncogenesis. Med Hypotheses 144:110240. https://doi.org/10.1016/j.mehy.2020.110240

Venturelli L, Harrold ZR, Murray AE et al (2021) Nanomechanical biosensing for fast and reliable detection of viability and susceptibility of microorganisms. Sens Actuators B Chem 348:130650. https://doi.org/10.1016/j.snb.2021.130650

Verhaak RGW, Bafna V, Mischel PS (2019) Extrachromosomal oncogene amplification in tumour pathogenesis and evolution. Nat Rev Cancer 19(5):283–288. https://doi.org/10.1038/s41568-019-0128-6

Verhey KJ, Gaertig J (2007) The tubulin code. Cell Cycle 6(17):2152–2160. https://doi.org/10.4161/cc.6.17.4633

Verlande A, Masri S (2019) Circadian Clocks and Cancer: Timekeeping Governs Cellular Metabolism. Trends Endocrinol Metab 30(7):445–458. https://doi.org/10.1016/j.tem.2019.05.001

Verma P, Cohen SM (2015) miR-965 controls cell proliferation and migration during tissue morphogenesis in the Drosophila abdomen. Elife 4:e07389. https://doi.org/10.7554/eLife.07389

Verschuren EW (2021) Niche reconstruction to revert or transcend the cancer state. In: Bertolaso M (ed) Rethinking cancer: a new paradigm for the postgenomics era. MIT Press, Cambridge

Vesely MD, Schreiber RD (2013) Cancer immunoediting: antigens, mechanisms, and implications to cancer immunotherapy. Ann N Y Acad Sci 1284(1):1–5. https://doi.org/10.1111/nyas.12105

Vetsigian K, Woese C, Goldenfeld N (2006) Collective evolution and the genetic code. Proc Natl Acad Sci U S A 103(28):10696–10701. https://doi.org/10.1073/pnas.0603780103

Viaud S, Saccheri F, Mignot G et al (2013) The intestinal microbiota modulates the anticancer immune effects of cyclophosphamide. Science 342(6161):971–976. https://doi.org/10.1126/science.1240537

Victorson D, Murphy K, Benedict C et al (2020) A randomized pilot study of mindfulness-based stress reduction in a young adult cancer sample: feasibility, acceptability, and changes in patient-reported outcomes. Psychooncology 29(5):841–850. https://doi.org/10.1002/pon.5355

Vidaurre T, Wilkerson J, Simon R et al (2009) Stable disease is not preferentially observed with targeted therapies and as currently defined has limited value in drug development. Cancer J 15(5):366–373. https://doi.org/10.1097/PPO.0b013e3181b9d37b

Vihervaara A, Duarte FM, Lis JT (2018) Molecular mechanisms driving transcriptional stress responses. Nat Rev Genet 19(6):385–397. https://doi.org/10.1038/s41576-018-0001-6

Vincent MD (2010) The animal within: carcinogenesis and the clonal evolution of cancer cells are speciation events sensu stricto. Evolution 64(4):1173–1183. https://doi.org/10.1111/j.1558-5646.2009.00942.x

Vincent MD (2011) Cancer: beyond speciation. Adv Cancer Res 112:283–350. https://doi.org/10.1016/B978-0-12-387688-1.00010-7

Vincent M (2012) Cancer: a de-repression of a default survival program common to all cells? A life-history perspective on the nature of cancer. Bioessays 34(1):72–82. https://doi.org/10.1002/bies.201100049

Vincent MD (2017) Cancer: Towards a general theory of the target: All successful cancer therapies, actual or potential, are reducible to either (or both) of two fundamental strategies. Bioessays 39(9). https://doi.org/10.1002/bies.201700059. PMID: 28833403

Vinnitsky VB (1989) Oncogerminative hypothesis of tumor growth [Russian]. Exp Oncol 21:239–242

Vinnitsky VB (1993) Oncogerminative hypothesis of tumor formation. Med Hypotheses 40(1):19–27. https://doi.org/10.1016/0306-9877(93)90191-r

Vinnitsky V (2014) The development of a malignant tumor is due to a desperate asexual self-cloning process in which cancer stem cells develop the ability to mimic the genetic program of germline cells. Intrinsically Disord Proteins 2(1):e29997. https://doi.org/10.4161/idp.29997

Vinogradov AE, Anatskaya OV (2023) Systemic alterations of cancer cells and their boost by polyploidization: Unicellular Attractor (UCA) model. Int J Mol Sci 24(7):6196. https://doi.org/10.3390/ijms24076196

Vitto VAM, Bianchin S, Zolondick AA et al (2022) Molecular mechanisms of autophagy in cancer development, progression, and therapy. Biomedicine 10(7):1596. https://doi.org/10.3390/biomedicines10071596

Vodovotz Y, An G, Androulakis IP (2013) A systems engineering perspective on homeostasis and disease. Front Bioeng Biotechnol 1:6. https://doi.org/10.3389/fbioe.2013.00006

Vogelstein B, Kinzler KW (1993) The multistep nature of cancer. Trends Genet9(4):138–141.https://doi.org/10.1016/0168-9525(93)90209-z

Vogelstein B, Kinzler K (1998) The genetic basis of human cancer. McGraw-Hill, New York

Vogelstein B, Kinzler KW (2004) Cancer genes and the pathways they control. Nat Med 10(8):789–799. https://doi.org/10.1038/nm1087

Vogelstein B, Kinzler KW (2015) The path to cancer—three strikes and you're out. N Engl J Med 373(20):1895–1898. https://doi.org/10.1056/NEJMp1508811

Vogelstein B, Papadopoulos N, Velculescu VE et al (2013) Cancer genome landscapes. Science 339(6127):1546–1558. https://doi.org/10.1126/science.1235122

Vogenberg FR, Isaacson Barash C, Pursel M (2010) Personalized medicine: part 1: evolution and development into theranostics. P T 35(10):560–576

Vogler M, Braun Y, Smith VM et al (2025) The BCL2 family: from apoptosis mechanisms to new advances in targeted therapy. Signal Transduct Target Ther 10(1):91. https://doi.org/10.1038/s41392-025-02176-0

Vollhardt LT (1991) Psychoneuroimmunology: a literature review. Am J Orthopsychiatry 61(1):35–47. https://doi.org/10.1037/h0079226

Von Behren J, Goldberg D, Hurley S et al (2024) Prospective analysis of sleep characteristics, chronotype, and risk of breast cancer in the California teachers study. Cancer Causes Control 35(4):597–604. https://doi.org/10.1007/s10552-023-01817-5

Von Bertalanffy L (1952) Problems of life. Harper and Brothers, New York

von Neumann J (1951) The general and logical theory of automata. In: Jeffress LA (ed) The Hixon Symposium. Wiley, New York

von Neumann J (1966) In: Burks A (ed) The theory of self-reproducing Automata. University of Illinois Press

Voutsadakis IA (2017) Proteasome expression and activity in cancer and cancer stem cells. Tumour Biol 39(3):1010428317692248. https://doi.org/10.1177/1010428317692248

Vrinzen CEJ, Delfgou L, Stadhouders N et al (2023) A systematic review and multilevel regression analysis reveals the comorbidity prevalence in cancer. Cancer Res 83(7):1147–1157. https://doi.org/10.1158/0008-5472.CAN-22-1336

Vyas S, Zaganjor E, Haigis MC (2016) Mitochondria and cancer. Cell 166(3):555–566. https://doi.org/10.1016/j.cell.2016.07.002

Waddington CH (1957) The strategy of the genes: a discussion of some aspects of theoretical biology. Allen & Unwin, London

Wade R, Nevitt S, Liu Y et al (2025) Multi-cancer early detection tests for general population screening: a systematic literature review. Health Technol Assess. 29(2):1–105. https://doi.org/10.3310/DLMT1294

Wagner GP, Kshitiz DA, Levchenko A (2022) The coevolution of placentation and cancer. Annu Rev Anim Biosci 10:259–279. https://doi.org/10.1146/annurev-animal-020420-031544

Wainwright EN, Scaffidi P (2017) Epigenetics and cancer stem cells: unleashing, hijacking, and restricting cellular plasticity. Trends Cancer 3(5):372–386. https://doi.org/10.1016/j.trecan.2017.04.004

Walford GA, Davis J, Warner AS et al (2013) Branched chain and aromatic amino acids change acutely following two medical therapies for type 2 diabetes mellitus. Metabolism 62(12):1772–1778. https://doi.org/10.1016/j.metabol.2013.07.003

Walker SI (2024) Life as no one knows it: the physics of life's emergence. Riverhead Books, New York

Walker C, Burggren W (2020) Remodeling the epigenome and (epi) cytoskeleton: a new paradigm for co-regulation by methylation. J Exp Biol 223(13):jeb220632. https://doi.org/10.1242/jeb.220632

Walker S, Cronin L (2023) Time is an object. Aeon, published May 19:2023. https://aeon.co/essays/time-is-not-an-illusion-its-an-object-with-physical-size

Walker SI, Davies PC (2012) The algorithmic origins of life. J R Soc Interface 10(79):20120869. https://doi.org/10.1098/rsif.2012.0869

Wallace DC (2011) Bioenergetic origins of complexity and disease. Cold Spring Harb Symp Quant Biol 76:1–16. https://doi.org/10.1101/sqb.2011.76.010462

Walther-Antonio M, Schulze-Makuch D (2023) The hypothesis of a "living pulse" in cells. Life (Basel) 13(7):1506. https://doi.org/10.3390/life13071506

Wan JCM, Massie C, Garcia-Corbacho J et al (2017) Liquid biopsies come of age: towards implementation of circulating tumour DNA. Nat Rev Cancer. 17(4):223–238. https://doi.org/10.1038/nrc.2017.7

Wan L, Liu Q, Liang D et al (2021) Circulating tumor cell and metabolites as novel biomarkers for early-stage lung cancer diagnosis. Front Oncol 11:630672. https://doi.org/10.3389/fonc.2021.630672

Wan Y, Liu J, Mai Y et al (2024) Current advances and future trends of hormesis in disease. npj. Aging 10:26. https://doi.org/10.1038/s41514-024-00155-3

Wang XF, Chen G (2003) Complex networks: small-world, scale-free and beyond. IEEE Circuits and Systems Magazine 3(1):6–20. https://doi.org/10.1109/MCAS.2003.1228503

Wang ZY, Chen Z (2008) Acute promyelocytic leukemia: from highly fatal to highly curable. Blood 111(5):2505–2515. https://doi.org/10.1182/blood-2007-07-102798

Wang H, Tso V, Wong C et al (2014) Development and validation of a highly sensitive urine-based test to identify patients with colonic adenomatous polyps. Clin Transl Gastroenterol 5(3):e54. https://doi.org/10.1038/ctg.2014.2

Wang C, Gu Y, Zhang K et al (2016) Systematic identification of genes with a cancer-testis expression pattern in 19 cancer types. Nat Commun 7:10499. https://doi.org/10.1038/ncomms10499

Wang G, Yuan R, Zhu X, Ao P (2018) Endogenous molecular-cellular network cancer theory: a systems biology approach. Methods Mol Biol 1702:215–245. https://doi.org/10.1007/978-1-4939-7456-6_11

Wang C, Jin S, Xu S, Cao S (2020a) High systemic immune-inflammation index (SII) represents an unfavorable prognostic factor for small cell lung cancer treated with etoposide and platinum-based chemotherapy. Lung 198(2):405–414. https://doi.org/10.1007/s00408-020-00333-6

Wang YP, Li JT, Qu J, Yin M, Lei QY (2020b) Metabolite sensing and signaling in cancer. J Biol Chem 295(33):11938–11946. https://doi.org/10.1074/jbc.REV119.007624

Wang G, Zhang H, Lyden D (2021) Tumour-regulated anorexia preceding cachexia. Nat Cell Biol 23(2):111–113. https://doi.org/10.1038/s41556-021-00635-8

Wang EJ et al (2022) Alterations of cytoskeleton networks in cell fate determination and cancer development. Biomolecules 12(12):1862. https://doi.org/10.3390/biom12121862

Wang G et al (2023a) Tumour extracellular vesicles and particles induce liver metabolic dysfunction. Nature 618(7964):374–382. https://doi.org/10.1038/s41586-023-06114-4

Wang HM, Zhang CY, Peng KC et al (2023b) Using patient-derived organoids to predict locally advanced or metastatic lung cancer tumor response: a real-world study. Cell Rep Med 4(2):100911. https://doi.org/10.1016/j.xcrm.2022.100911

Wang K, Lu Y, Li H et al (2024a) Role of long non-coding RNAs in metabolic reprogramming of gastrointestinal cancer cells. Cancer Cell Int 24(1):15. https://doi.org/10.1186/s12935-023-03194-0

Wang Z, Wang M, Lin M et al (2024b) East meets west: integrating Yin-Yang theory with immunology teaching. Front Immunol 15:1441863. https://doi.org/10.3389/fimmu.2024.1441863

Wang S, Wang J, Chen Z et al (2024c) Targeting M2-like tumor-associated macrophages is a potential therapeutic approach to overcome antitumor drug resistance. NPJ Precis Onc 8:31. https://doi.org/10.1038/s41698-024-00522-z

Wang Y, Jia J, Wang F et al (2024d) Pre-metastatic niche: formation, characteristics and therapeutic implication. Signal Transduct Target Ther 9(1):236. https://doi.org/10.1038/s41392-024-01937-7

Wang C, Qiu X, Yang X et al (2024e) Factors influencing social isolation among cancer patients: a systematic review. Healthcare (Basel) 12(10):1042. https://doi.org/10.3390/healthcare12101042

Wang Y, Narasimamurthy R, Qu M et al (2024f) Circadian regulation of cancer stem cells and the tumor microenvironment during metastasis. Nat Cancer 5(4):546–556. https://doi.org/10.1038/s43018-024-00759-4

Wang N-Y, Wang LK, Tsao H-P (2024g) The Yin/yang properties of cancers and correlation with infectious diseases, autoimmune diseases, and allergies: treatment with small bidirectional molecules. In: Wang LK (ed) Evolutionary progress in science, technology, engineering, arts and mathematics, No. 8A. Lenox Institute Press

Wang Z, Wang Q, Qin F, Chen J (2024h) Exosomes: a promising avenue for cancer diagnosis beyond treatment. Front Cell Dev Biol 12:1344705. https://doi.org/10.3389/fcell.2024.1344705

Wang N-Y, Wang LK, Tsao H-P (2024i) The yin/yang properties of cancers and correlation with infectious diseases, autoimmune diseases, and allergies: treatment with small bidirectional molecules.

In: Wang LK (ed) Evolutionary progress in science, technology, engineering, arts and mathematics, No. 8A. Lenox Institute Press, MA
Wang JF, Wang MC, Jiang LL, Lin NM (2025a) The neuroscience in breast cancer: current insights and clinical opportunities. Heliyon 11(3):e42293. https://doi.org/10.1016/j.heliyon.2025.e42293
Wang J, Zeng Z, Li Z et al (2025b) The clinical application of artificial intelligence in cancer precision treatment. J Transl Med 23:120. https://doi.org/10.1186/s12967-025-06139-5
Wangsa D, Braun R, Schiefer M et al (2018) The evolution of single cell-derived colorectal cancer cell lines is dominated by the continued selection of tumor-specific genomic imbalances, despite random chromosomal instability. Carcinogenesis 39(8):993–1005. https://doi.org/10.1093/carcin/bgy068
Warburg O (1924) Über den Stoffwechsel der Carcinomzelle. Die Naturwissenschaften 12(50):1131–1137
Warburg O (1956) On the origin of cancer cells. Science 123(3191):309–314. https://doi.org/10.1126/science.123.3191.309
Warburg O, Posener K, Negelein E (1924) Über den Stoffwechsel der Tumoren. Biochem Z 152:309
Ward PS, Thompson CB (2012) Metabolic reprogramming: a cancer hallmark even Warburg did not anticipate. Cancer Cell 21(3):297–308. https://doi.org/10.1016/j.ccr.2012.02.014
Ward C, Meehan J, Gray ME et al (2020) The impact of tumour pH on cancer progression: strategies for clinical intervention. Explor Target Antitumor Ther 1(2):71–100. https://doi.org/10.37349/etat.2020.00005
Warner TFCS (1975) Cell hybridization: an explanation for the phenotypic diversity of certain tumours. Med Hypoth 1:51–57. https://doi.org/10.1016/0306-9877(75)90042-0
Watkins TBK, Lim EL, Petkovic M et al (2020) Pervasive chromosomal instability and karyotype order in tumour evolution. Nature 587(7832):126–132. https://doi.org/10.1038/s41586-020-2698-6
Watson OJ, Barnsley G, Toor J et al (2022) Global impact of the first year of COVID-19 vaccination: a mathematical modelling study. Lancet Infect Dis 22(9):1293–1302. https://doi.org/10.1016/S1473-3099(22)00320-6
Wattanathamsan O, Pongrakhananon V (2022) Emerging role of microtubule-associated proteins on cancer metastasis. Front Pharmacol 13:935493. https://doi.org/10.3389/fphar.2022.935493
Weaver VM, Petersen OW, Wang F et al (1997) Reversion of the malignant phenotype of human breast cells in three-dimensional culture and in vivo by integrin blocking antibodies. J Cell Biol 137(1):231–245. https://doi.org/10.1083/jcb.137.1.231
Weber BH (2009) Embracing the biosemiotic perspective. Biosemiotics 2:367–375. https://doi.org/10.1007/s12304-009-9065-1
Weberpals J, Feuerriegel S, van der Schaar M et al (2025) Opportunities for causal machine learning in precision oncology. NEJM 2(8). https://doi.org/10.1056/AIp2500277
Weeden CE, Hill W, Lim EL et al (2023) Impact of risk factors on early cancer evolution. Cell 186(8):1541–1563. https://doi.org/10.1016/j.cell.2023.03.013
Weedon-Fekjaer H, Lindqvist BH, Vatten LJ et al (2008) Breast cancer tumor growth estimated through mammography screening data. Breast Cancer Res 10(3):R41. https://doi.org/10.1186/bcr2092
Wei J, Sun J, Xu H et al (2015) Effects of extremely low frequency electromagnetic fields on intracellular calcium transients in cardiomyocytes. Electromagn Biol Med 34(1):77–84. https://doi.org/10.3109/15368378.2014.881744
Wei AL, Li M, Li GQ et al (2020a) Oral microbiome and pancreatic cancer. World J Gastroenterol 26(48):7679–7692. https://doi.org/10.3748/wjg.v26.i48.7679
Wei Q, Qian Y, Yu J, Wong CC (2020b) Metabolic rewiring in the promotion of cancer metastasis: mechanisms and therapeutic implications. Oncogene 39(39):6139–6156. https://doi.org/10.1038/s41388-020-01432-7
Wei L, Niraula D, Gates EDH et al (2023) Artificial intelligence (AI) and machine learning (ML) in precision oncology: a review on enhancing discoverability through multiomics integration. Br J Radiol 96(1150):20230211. https://doi.org/10.1259/bjr.20230211
Weibel ER (1994) Design of biological organisms and fractal geometry. In: Nonnenmacher TF, Losa GA, Weibel ER (eds) Fractals in Biology and Medicine. Springer, pp 1–17
Weichselbaum RR, Liang H et al (2017) Radiotherapy and immunotherapy: a beneficial liaison? Nat Rev Clin Oncol 14(6):365–379. https://doi.org/10.1038/nrclinonc.2016.211
Weikert S, Boeing H, Pischon T et al (2008) Blood pressure and risk of renal cell carcinoma in the European prospective investigation into cancer and nutrition. Am J Epidemiol 167(4):438–446. https://doi.org/10.1093/aje/kwm321
Weilandt DR, Masid M, Hatzimanikatis V (2021) Thermodynamics of metabolic pathways. In: Metabolic engineering: concepts and applications, vol 13b, chapter 7. https://doi.org/10.1002/9783527823468.ch7
Weinberg R (1999) One renegade cell: how cancer begins. Basic Books
Weinberg R (2006) The biology of cancer. Garland Science
Weinberg RA (2014) Coming full circle—from endless complexity to simplicity and back again. Cell 157(1):267–271. https://doi.org/10.1016/j.cell.2014.03.004
Weinstein IB (2000) Disorders in cell circuitry during multistage carcinogenesis: the role of homeostasis. Carcinogenesis 21(5):857–864. https://doi.org/10.1093/carcin/21.5.857
Weinstein IB (2002) Cancer. Addiction to oncogenes—the Achilles heal of cancer. Science 297(5578):63–64. https://doi.org/10.1126/science.1073096
Weinstein IB, Joe AK (2006) Mechanisms of disease: oncogene addiction-a rationale for molecular targeting in cancer therapy. Nat Clin Pract Oncol 3(8):448–457. https://doi.org/10.1038/ncponc0558
Weiss MC, Preiner M, Xavier JC et al (2018) The last universal common ancestor between ancient Earth chemistry and the onset of genetics. PLoS Genet 14(8):e1007518. https://doi.org/10.1371/journal.pgen.1007518
Welch DR, Hurst DR (2019) Defining the hallmarks of metastasis. Cancer Res 79(12):3011–3027. https://doi.org/10.1158/0008-5472.CAN-19-0458
Wells J (2014) Membrane patterns carry ontogenetic information that is specified independently of DNA. BIO-Complexity 2014(2):1–28
West J, You L, Zhang J, Gatenby RA et al (2020) Towards Multidrug Adaptive Therapy. Cancer Res. 80(7):1578–1589. https://doi.org/10.1158/0008-5472.CAN-19-2669
West J, Schenck RO, Gatenbee C et al (2021) Normal tissue architecture determines the evolutionary course of cancer. Nat Commun 12(1):2060. https://doi.org/10.1038/s41467-021-22123-1
Wheeler JA (1989) Information, physics, quantum: the search for links. In: Proceedings III international symposium on foundations of quantum mechanics. Tokyo, pp 354–358
Wheeler DL, Iida M, Dunn EF (2009) The role of Src in solid tumors. Oncologist 14(7):667–678. https://doi.org/10.1634/theoncologist.2009-0009
Whitehead AN (1929) Process and reality. an essay in cosmology. Gifford lectures delivered in the University of Edinburgh During the Session 1927–1928. Macmillan/Cambridge University Press, New York/Cambridge UK
Whiteside TL (2015) The role of regulatory T cells in cancer immunology. Immunotargets Ther 4:159–171. https://doi.org/10.2147/ITT.S55415

Whiteside TL (2016) Exosomes and tumor-mediated immune suppression. J Clin Invest 126(4):1216–1223. https://doi.org/10.1172/JCI81136

Wicken JS (1980) A thermodynamic theory of evolution. J Theor Biol 87(1):9–23. https://doi.org/10.1016/0022-5193(80)90216-7

Wicken JS (1989) Evolution and thermodynamics: the new paradigm. Syst Res 6(3):181–186

Widman AJ, Shah M, Frydendahl A et al (2024) Ultrasensitive plasma-based monitoring of tumor burden using machine-learning-guided signal enrichment. Nat Med 30(6):1655–1666. https://doi.org/10.1038/s41591-024-03040-4

Wike-Hooley JL, Haveman J, Reinhold HS (1984) The relevance of tumour pH for the treatment of malignant disease. Radiother Oncol 2:343–366. https://doi.org/10.1016/S0167-8140(84)80077-8

Wild CP (2005) Complementing the genome with an "exposome": the outstanding challenge of environmental exposure measurement in molecular epidemiology. Cancer Epidemiol Biomarkers Prev 14(8):1847–1850. https://doi.org/10.1158/1055-9965.EPI-05-0456

Wild CP, Weiderpass E, Stewart BW (eds) (2020) World cancer report: cancer research for cancer prevention. International Agency for Research on Cancer, Lyon. Available from: https://www.ncbi.nlm.nih.gov/books/NBK606505/

Wilson BE, Jacob S, Yap ML et al (2019) Estimates of global chemotherapy demands and corresponding physician workforce requirements for 2018 and 2040: a population-based study. Lancet Oncol 20(6):769–780. https://doi.org/10.1016/S1470-2045(19)30163-9

Wirbel J, Pyl PT, Kartal E et al (2019) Meta-analysis of fecal metagenomes reveals global microbial signatures that are specific for colorectal cancer. Nat Med 25(4):679–689. https://doi.org/10.1038/s41591-019-0406-6

Wishart DS (2019) Metabolomics for investigating physiological and pathophysiological processes. Physiol Rev 99(4):1819–1875. https://doi.org/10.1152/physrev.00035.2018

Witek Janusek L, Tell D, Mathews HL (2019) Mindfulness based stress reduction provides psychological benefit and restores immune function of women newly diagnosed with breast cancer: a randomized trial with active control. Brain Behav Immun 80:358–373. https://doi.org/10.1016/j.bbi.2019.04.012

Witt H, Mack SC, Ryzhova M et al (2011) Delineation of two clinically and molecularly distinct subgroups of posterior fossa ependymoma. Cancer Cell 20(2):143–157. https://doi.org/10.1016/j.ccr.2011.07.007

Wittenbrink N, Ananthasubramaniam B, Münch M et al (2018) (2018) High-accuracy determination of internal circadian time from a single blood sample. J Clin Invest. 128(9):3826–3839. https://doi.org/10.1172/JCI120874

Witzany G (2010) Biocommunication and natural genome editing. World J Biol Chem 1(11):348–352

Witzany G (2020) What is life? Front Astron Space Sci 7:1–13. https://doi.org/10.3389/fspas.2020.00001

Witzany G, Baluška F (2012) Turing: a formal clash of codes. Nature 483(7391):541. https://doi.org/10.1038/483541c

Wodarz D, Newell AC, Komarova NL (2018) Passenger mutations can accelerate tumour suppressor gene inactivation in cancer evolution. J R Soc Interface 15(143):20170967. https://doi.org/10.1098/rsif.2017.0967

Woese CR (2002) On the evolution of cells. Proc Natl Acad Sci U S A 99(13):8742–8747. https://doi.org/10.1073/pnas.132266999

Woese CR (2004) A new biology for a new century. Microbiol Mol Biol Rev 68(2):173–186. https://doi.org/10.1128/MMBR.68.2.173-186.2004

Wolchok JD, Chiarion-Sileni V, Gonzalez R et al (2022) Long-term outcomes with Nivolumab plus Ipilimumab or Nivolumab alone versus Ipilimumab in patients with advanced melanoma. J Clin Oncol 40(2):127–137. https://doi.org/10.1200/JCO.21.02229

Wolchok JD, Chiarion-Sileni V, Rutkowski P et al (2025) Final, 10-year outcomes with Nivolumab plus Ipilimumab in advanced melanoma. N Engl J Med 392(1):11–22. https://doi.org/10.1056/NEJMoa2407417

Wolf DM, Segawa M, Kondadi AK et al (2019) Individual cristae within the same mitochondrion display different membrane potentials and are functionally independent. EMBO J 38(22):e101056. https://doi.org/10.15252/embj.2018101056

Wolfram S (2023) Computational foundations for the second law of thermodynamics. In: Stephen Wolfram Writings. https://writings.stephenwolfram.com/2023/02/computational-foundations-for-the-second-law-of-thermodynamics

Wong SH, Kwong TNY, Chow TC et al (2017) Quantitation of faecal Fusobacterium improves faecal immunochemical test in detecting advanced colorectal neoplasia. Gut 66(8):1441–1448. https://doi.org/10.1136/gutjnl-2016-312766

Wong ML, Cleland CE, Arend D Jr et al (2023) On the roles of function and selection in evolving systems. Proc Natl Acad Sci U S A 120(43):e2310223120. https://doi.org/10.1073/pnas.2310223120

Wong J, Muralidhar R, Wang L, Huang CC (2024) Epigenetic modifications of cfDNA in liquid biopsy for the cancer care continuum. Biomed J 100718. https://doi.org/10.1016/j.bj.2024.100718

Wood LD, Parsons DW, Jones S et al (2007) The genomic landscapes of human breast and colorectal cancers. Science 318(5853):1108–1113. https://doi.org/10.1126/science.1145720

Wortzel I, Dror S, Kenific CM, Lyden D (2019) Exosome-mediated metastasis: communication from a distance. Dev Cell 49(3):347–360. https://doi.org/10.1016/j.devcel.2019.04.011

Woting A, Blaut M (2016) The intestinal microbiota in metabolic disease. Nutrients 8(4):202. https://doi.org/10.3390/nu8040202

Wright NA, Poulsom R (2012) Omnis cellula e cellula revisited: cell biology as the foundation of pathology. J Pathol 226(2):145–147. https://doi.org/10.1002/path.3030

Wu S, Hannun Y (2016) The importance of extrinsic factors in the development of cancers. Mol Cell Oncol 3(3):e1143079. https://doi.org/10.1080/23723556.2016.1143079

Wu M, Pastor-Pareja JC, Xu T (2010) Interaction between Ras(V12) and scribbled clones induces tumour growth and invasion. Nature 463(7280):545–548. https://doi.org/10.1038/nature08702

Wu S, Powers S, Zhu W, Hannun YA (2016) Substantial contribution of extrinsic risk factors to cancer development. Nature 529(7584):43–47. https://doi.org/10.1038/nature16166

Wu KT, Hishamunda JB, Chen DT et al (2017) Transition from turbulent to coherent flows in confined three-dimensional active fluids. Science 355(6331):eaal1979. https://doi.org/10.1126/science.aal1979

Wu J, Hu S, Zhang L et al (2020) Tumor circulome in the liquid biopsies for cancer diagnosis and prognosis. Theranostics. 10(10):4544–4556. https://doi.org/10.7150/thno.40532

Wulczyn E, Steiner DF, Moran M et al (2021) Interpretable survival prediction for colorectal cancer using deep learning. NPJ Digit Med 4(1):71. https://doi.org/10.1038/s41746-021-00439-5

Wurzer H, Hoffmann C, Al Absi A, Thomas C (2019) Actin cytoskeleton straddling the immunological synapse between cytotoxic lymphocytes and cancer cells. Cells 8(5):463. https://doi.org/10.3390/cells8050463

Xavier CPR, Caires HR, Barbosa MAG et al (2020) The role of extracellular vesicles in the hallmarks of cancer and drug resistance. Cells 9(5):1141. https://doi.org/10.3390/cells9051141

Xavier JC, Hordijk W, Kauffman S et al (2020b) Autocatalytic chemical networks at the origin of metabolism. Proc Biol Sci 287(1922):20192377. https://doi.org/10.1098/rspb.2019.2377

Xia M, Zhang Y, Jin K et al (2019) Communication between mitochondria and other organelles: a brand-new perspective on mitochondria in cancer. Cell Biosci 9:27. https://doi.org/10.1186/s13578-019-0289-8

Xiao S, Zhou L (2017) Gastric cancer: Metabolic and metabolomics perspectives (Review). Int J Oncol 51(1):5–17. https://doi.org/10.3892/ijo.2017.4000

Xiao R, Bergin SM, Huang W et al (2016) Environmental and genetic activation of hypothalamic BDNF modulates T-cell immunity to exert an anticancer phenotype. Cancer Immunol Res 4(6):488–497. https://doi.org/10.1158/2326-6066.CIR-15-0297

Xiao A, Feng Y, Yin B et al (2025) Pancreatic cancer cachexia: a systemic consequence of multi-organ interactions. hLife. https://doi.org/10.1016/j.hlife.2025.05.002. (in press)

Xiong J (2024) AI-generated hallmarks of aging and cancer: a computational approach using causal emergence and dependency networks. BioRxiv. https://doi.org/10.1101/2024.08.28.610200

Xu XG, Bednarz B, Paganetti H (2008) A review of dosimetry studies on external-beam radiation treatment with respect to second cancer induction. Phys Med Biol 53(13):R193–R241. https://doi.org/10.1088/0031-9155/53/13/R01

Xu M, Zhang T, Xia R et al (2022) Targeting the tumor stroma for cancer therapy. Mol Cancer 21(1):208. https://doi.org/10.1186/s12943-022-01670-1

Xu X, Zeng C, Qing B et al (2024) Development of a urine-based metabolomics approach for multi-cancer screening and tumor origin prediction. Front Immunol 15:1449103. https://doi.org/10.3389/fimmu.2024.1449103

Xuan K, Zhao T, Sun C et al (2021) The association between hypertension and colorectal cancer: a meta-analysis of observational studies. Eur J Cancer Prev 30(1):84–96. https://doi.org/10.1097/CEJ.0000000000000578

Xue AG, Chan M, Gujral TS (2020) Pan-cancer analysis of the developmental pathways reveals non-canonical Wnt signaling as a driver of mesenchymal-type tumors. Transl Res 224:1–15. https://doi.org/10.1016/j.trsl.2020.06.003

Yachida S, Jones S, Bozic I et al (2010) Distant metastasis occurs late during the genetic evolution of pancreatic cancer. Nature 467(7319):1114–1117. https://doi.org/10.1038/nature09515

Yaeger LS (2009) How evolution guides complexity. HFSP J 3(5):328–339. https://doi.org/10.2976/1.3233712

Yaeger R, Chatila WK, Lipsyc MD et al (2018) Clinical sequencing defines the genomic landscape of metastatic colorectal cancer. Cancer Cell 33(1):125–136.e3. https://doi.org/10.1016/j.ccell.2017.12.004

Yafremava LS, Wielgos M, Thomas S et al (2013) A general framework of persistence strategies for biological systems helps explain domains of life. Front Genet 4:16. https://doi.org/10.3389/fgene.2013.00016

Yaglova NV, Obernikhin SS, Yaglov VV, Nazimova SV (2020) Time-dependent effect of deuterium depletion on tumor growth and metastasis. Russ Open Med J 9(2):e0210. https://doi.org/10.15275/rusomj.2020.0210

Yamaguchi H, Condeelis J (2007) Regulation of the actin cytoskeleton in cancer cell migration and invasion. Biochim Biophys Acta 1773(5):642–652. https://doi.org/10.1016/j.bbamcr.2006.07.001

Yamamoto N, Yang M, Jiang P, Xu M et al (2003) Determination of clonality of metastasis by cell-specific color-coded fluorescent-protein imaging. Cancer Res 63(22):7785–7790

Yamamoto A, Doak AE, Cheung KJ (2023) Orchestration of collective migration and metastasis by tumor cell clusters. Annu Rev Pathol 18:231–256. https://doi.org/10.1146/annurev-pathmechdis-031521-023557

Yamashita K, Miyamoto A, Hama N et al (2015) Survival impact of pulmonary metastasis as recurrence of pancreatic ductal adenocarcinoma. Dig Surg 32(6):464–471. https://doi.org/10.1159/000439545

Yan H, Bu P (2021) Non-coding RNA in cancer. Essays Biochem 65(4):625–639. https://doi.org/10.1042/EBC20200032

Yan L, Sundaram S, Mehus AA, Picklo MJ (2019) Time-restricted feeding attenuates high-fat diet-enhanced spontaneous metastasis of Lewis lung carcinoma in mice. Anticancer Res 39(4):1739–1748. https://doi.org/10.21873/anticanres.13280

Yang M, Brackenbury WJ (2013) Membrane potential and cancer progression. Front Physiol 4:185. https://doi.org/10.3389/fphys.2013.00185

Yang YA, Yu J (2013) EZH2, an epigenetic driver of prostate cancer. Protein Cell 4(5):331–341. https://doi.org/10.1007/s13238-013-2093-2

Yang G, Wang Y, Wu Y et al (2015) The road to effective tobacco control in China. Lancet 385(9972):1019–1028. https://doi.org/10.1016/S0140-6736(15)60174-X

Yang N, Zhu S, Lv X et al (2018) MicroRNAs: pleiotropic regulators in the tumor microenvironment. Front Immunol 9:2491. https://doi.org/10.3389/fimmu.2018.02491

Yang P, Jiang Y, Rhea PR et al (2019) Human biofield therapy and the growth of mouse lung carcinoma. Integr Cancer Ther 18:1534735419840797. https://doi.org/10.1177/1534735419840797

Yang J, Wang L, Jia R (2020a) Role of de novo cholesterol synthesis enzymes in cancer. J Cancer 11(7):1761–1767. https://doi.org/10.7150/jca.38598

Yang E, Wang X, Gong Z et al (2020b) Exosome-mediated metabolic reprogramming: the emerging role in tumor microenvironment remodeling and its influence on cancer progression. Signal Transduct Target Ther 5(1):242. https://doi.org/10.1038/s41392-020-00359-5

Yang H, Wang J, Huang G (2022) Small extracellular vesicles in metabolic remodeling of tumor cells: cargos and translational application. Front Pharmacol 13:1009952. https://doi.org/10.3389/fphar.2022.1009952

Yang MQ, Zhang SL, Sun L et al (2024a) Targeting mitochondria: restoring the antitumor efficacy of exhausted T cells. Mol Cancer 260. https://doi.org/10.1186/s12943-024-02175-9

Yang L, Tang L, Min Q et al (2024b) Emerging role of RNA modification and long noncoding RNA interaction in cancer. Cancer Gene Ther 31:816–830. https://doi.org/10.1038/s41417-024-00734-2

Yang J, Zhou F, Luo X et al (2025) Enhancer reprogramming: critical roles in cancer and promising therapeutic strategies. Cell Death Discov 11:84. https://doi.org/10.1038/s41420-025-02366-3

Yankaskas CL, Bera K, Stoletov K et al (2021) The fluid shear stress sensor TRPM7 regulates tumor cell intravasation. Sci Adv 7(28):eabh3457. https://doi.org/10.1126/sciadv.abh3457

Yao H, Fan C, Lu Y et al (2020) Alteration of gut microbiota affects expression of adiponectin and resistin through modifying DNA methylation in high-fat diet-induced obese mice. Genes Nutr 15(1):12. https://doi.org/10.1186/s12263-020-00671-3

Yarchoan M, Hopkins A, Jaffee EM (2017) Tumor mutational burden and response rate to PD-1 inhibition. N Engl J Med 377(25):2500–2501. https://doi.org/10.1056/NEJMc1713444

Yates FE (1992) Fractal applications in biology: Scaling time in biochemical networks: (Numerical methods). Methods Enzymol 219:636–676

Yates FE (1994) Order and complexity in dynamical systems: Homeodynamics as a generalized mechanics for biology. Math Comput Model 19:49–74. https://doi.org/10.1016/0895-7177(94)90189-9

Yates FE, Yates LB (2008) Ultradian rhythms as the dynamic signatures of life. In: Lloyd D, Rossi EL (eds) Ultradian rhythms from molecules to mind, a new vision of life. Springer

Ye J, Medzhitov R (2019) Control strategies in systemic metabolism. Nat Metab 1(10):947–957. https://doi.org/10.1038/s42255-019-0118-8

Ye H, Yu CH, Li L et al (2012) Meta-analysis of human colorectal cancer transcriptome. Int J Colorectal Dis 27(8):1125–1128. https://doi.org/10.1007/s00384-011-1371-9

Ye H, Adane B, Khan N et al (2018) Subversion of systemic glucose metabolism as a mechanism to support the growth of leukemia cells. Cancer Cell 34(4):659–673.e6. https://doi.org/10.1016/j.ccell.2018.08.016

Yeung CC, Dondelinger F, Schoof EM et al (2022) Circadian regulation of protein cargo in extracellular vesicles. Sci Adv 8(14):eabc9061. https://doi.org/10.1126/sciadv.abc9061

Yin Y, Cai X, Chen X et al (2014) Tumor-secreted miR-214 induces regulatory T cells: a major link between immune evasion and tumor growth. Cell Res 24(10):1164–1180. https://doi.org/10.1038/cr.2014.121

Yizhak K, Aguet F, Kim J et al (2019) RNA sequence analysis reveals macroscopic somatic clonal expansion across normal tissues. Science 364(6444):eaaw0726. https://doi.org/10.1126/science.aaw0726

Yoon J, Davtyan C, van der Schaar M (2017) Discovery and clinical decision support for personalized healthcare IEEE. J Biomed Health Inform 21(4):1133–1145. https://doi.org/10.1109/JBHI.2016.2574857

Yoshimasu T, Ohashi T, Oura S et al (2015) A theoretical model for the hormetic dose-response curve for anticancer agents. Anticancer Res 35(11):5851–5855. PMID: 26504007

Youssef KK, Narwade N, Arcas A et al (2024) Two distinct epithelial-to-mesenchymal transition programs control invasion and inflammation in segregated tumor cell populations. Nat Cancer 5(11):1660–1680. https://doi.org/10.1038/s43018-024-00839-5

Yu X, Li S (2017) Non-metabolic functions of glycolytic enzymes in tumorigenesis. Oncogene 36(19):2629–2636. https://doi.org/10.1038/onc.2016.410

Yu J, Feng Q, Wong SH et al (2017a) Metagenomic analysis of faecal microbiome as a tool towards targeted non-invasive biomarkers for colorectal cancer. Gut 66(1):70–78. https://doi.org/10.1136/gutjnl-2015-309800

Yu H, Xia H, Tang Q et al (2017b) Acetylcholine acts through M3 muscarinic receptor to activate the EGFR signaling and promotes gastric cancer cell proliferation. Sci Rep 7:40802. https://doi.org/10.1038/srep40802

Yu X, Ma R, Wu Y et al (2018) Reciprocal regulation of metabolic reprogramming and epigenetic modifications in cancer. Front Genet 9:394. https://doi.org/10.3389/fgene.2018.00394

Yu B, Zanetti KA, Temprosa M et al (2019) The Consortium of Metabolomics Studies (COMETS): Metabolomics in 47 prospective cohort studies. Am J Epidemiol 188(6):991–1012. https://doi.org/10.1093/aje/kwz028

Yu W, Hurley J, Roberts D et al (2021) Exosome-based liquid biopsies in cancer: opportunities and challenges. Ann Oncol 32(4):466–477. https://doi.org/10.1016/j.annonc.2021.01.074

Yu D, Li Y, Wang M et al (2022) Exosomes as a new frontier of cancer liquid biopsy. Mol Cancer 21:56. https://doi.org/10.1186/s12943-022-01509-9

Yu X, Zhao H, Wang R et al (2024) Cancer epigenetics: from laboratory studies and clinical trials to precision medicine. Cell Death Discov 10(1):28. https://doi.org/10.1038/s41420-024-01803-z

Yuan S, Almagro J, Fuchs E (2024) Beyond genetics: driving cancer with the tumour microenvironment behind the wheel. Nat Rev Cancer 24(4):274–286. https://doi.org/10.1038/s41568-023-00660-9

Yun J, Rago C, Cheong I et al (2009) Glucose deprivation contributes to the development of KRAS pathway mutations in tumor cells. Science 325(5947):1555–1559. https://doi.org/10.1126/science.1174229

Zack TI, Schumacher SE, Carter SL et al (2013) Pan-cancer patterns of somatic copy number alteration. Nat Genet 45(10):1134–1140. https://doi.org/10.1038/ng.2760

Zacksenhaus E, Egan SE (2021) Progression to metastasis of solid cancer. Cancers (Basel) 13(4):717. https://doi.org/10.3390/cancers13040717

Zafereo M, Wang RJ, Busaidy NL et al (2025) Neoadjuvant pembrolizumab in combination with dabrafenib and trametinib (DTP) for BRAF V600E-mutated anaplastic thyroid cancer (BRAFm-ATC): a multicenter phase 2 trial. J Clin Oncol 43(16_suppl):Abstract 6008

Zajicek G (1978) Cancer as a systemic disease. Med Hypotheses 4(3):193–207. https://doi.org/10.1016/0306-9877(78)90002-6

Zajicek G (1995) Streaming organism. Med Hypotheses 45(4):403–407. https://doi.org/10.1016/0306-9877(95)90105-1

Zajicek G (1996) A new cancer hypothesis. Med Hypotheses 47(2):111–115. https://doi.org/10.1016/s0306-9877(96)90449-1. PMID: 8869926

Zaki NF, Sabri YM, Farouk O et al (2020) Depressive symptoms, sleep profiles and serum melatonin levels in a sample of breast cancer patients. Nat Sci Sleep 12:135–149. https://doi.org/10.2147/NSS.S206768

Zamarron BF, Chen W (2011) Dual roles of immune cells and their factors in cancer development and progression. Int J Biol Sci 7(5):651–658. https://doi.org/10.7150/ijbs.7.651

Zambalde EP, Adamoski D, Gradia DF et al (2022) Transcribed ultraconserved regions are associated with clinicopathological features in breast cancer. Biomolecules 12(2):214. https://doi.org/10.3390/biom12020214

Zanotelli MR, Rahman-Zaman A, VanderBurgh JA et al (2019) Energetic costs regulated by cell mechanics and confinement are predictive of migration path during decision-making. Nat Commun 10:4185. https://doi.org/10.1038/s41467-019-12020-2

Zareifar P, Ahmed HM, Ghaderi P et al (2024) miR-142-3p/5p role in cancer: from epigenetic regulation to immunomodulation. Cell Biochem Funct 42(2):e3931. https://doi.org/10.1002/cbf.3931

Zarkavelis G, Amylidi AL, Verbaanderd C et al (2023) Off-label despite high-level evidence: a clinical practice review of commonly used off-patent cancer medicines. ESMO Open 100604. https://doi.org/10.1016/j.esmoop.2022.100604

Zebrowska A, Skowronek A, Wojakowska A et al (2019) Metabolome of exosomes: focus on vesicles released by cancer cells and present in human body fluids. Int J Mol Sci 20(14):3461. https://doi.org/10.3390/ijms20143461

Zeevi D, Korem T, Zmora N et al (2015) Personalized nutrition by prediction of glycemic responses. Cell 163(5):1079–1094. https://doi.org/10.1016/j.cell.2015.11.001

Zeida A, Trujillo M, Ferrer-Sueta G et al (2019) Catalysis of peroxide reduction by fast reacting protein thiols. Chem Rev 119(19):10829–10855. https://doi.org/10.1021/acs.chemrev.9b00371

Zeleznik OA, Eliassen AH, Kraft P et al (2020) A prospective analysis of circulating plasma metabolites associated with ovarian cancer risk. Cancer Res 80(6):1357–1367. https://doi.org/10.1158/0008-5472.CAN-19-2567

Zhang D, Frenette PS (1999) Cross talk between neutrophils and the microbiota. Blood 133(20):2168–2177. https://doi.org/10.1182/blood-2018-11-844555

Zhang L, Yu D (2019) Exosomes in cancer development, metastasis, and immunity. Biochim Biophys Acta Rev Cancer 1871(2):455–468. https://doi.org/10.1016/j.bbcan.2019.04.004

Zhang L, Zhang Y (2015) Tunneling nanotubes between rat primary astrocytes and C6 glioma cells alter proliferation potential of gli-

oma cells. Neurosci Bull 31(3):371–378. https://doi.org/10.1007/s12264-014-1522-4
Zhang P, Meng X, Li D et al (2018) Commensal homeostasis of gut microbiota-host for the impact of obesity. Front Physiol 8:1122. https://doi.org/10.3389/fphys.2017.01122
Zhang R, Zhang F, Sun Z et al (2019) LINE-1 retrotransposition promotes the development and progression of lung squamous cell carcinoma by disrupting the tumor-suppressor gene FGGY. Cancer Res 79(17):4453–4465. https://doi.org/10.1158/0008-5472.CAN-19-0076
Zhang X, Wang J, Zubarev RA (2020a) Slight deuterium enrichment in water acts as an antioxidant: is deuterium a cell growth regulator? Mol Cell Proteomics 19(11):1790–1804. https://doi.org/10.1074/mcp.RA120.002231
Zhang Y, Guoqiang L, Sun M, Lu X (2020b) Targeting and exploitation of tumor-associated neutrophils to enhance immunotherapy and drug delivery for cancer treatment. Cancer Biol Med 17(1):32–43. https://doi.org/10.20892/j.issn.2095-3941.2019.0372
Zhang Y, Wang J, Wang X et al (2020c) The autophagic-lysosomal and ubiquitin proteasome systems are simultaneously activated in the skeletal muscle of gastric cancer patients with cachexia. Am J Clin Nutr 111(3):570–579. https://doi.org/10.1093/ajcn/nqz347
Zhang L, Xu F, Li S et al (2021a) Influence of fluid balance on the prognosis of patients with sepsis. BMC Anesthesiol 21(1):269. https://doi.org/10.1186/s12871-021-01489-1
Zhang Y, Qian J, Gu C, Yang Y (2021b) Alternative splicing and cancer: a systematic review. Signal Transduct Target Ther 6(1):78. https://doi.org/10.1038/s41392-021-00486-7
Zhang J, Cunningham J, Brown J, Gatenby R (2022a) Evolution-based mathematical models significantly prolong response to abiraterone in metastatic castrate-resistant prostate cancer and identify strategies to further improve outcomes. Elife 11:e76284. https://doi.org/10.7554/eLife.76284
Zhang M, Zhang Y, Sun Y et al (2022b) Intratumoral microbiota impacts the first-line treatment efficacy and survival in non-small cell lung cancer patients free of lung infection. J Healthc Eng 2022:5466853. https://doi.org/10.1155/2022/5466853
Zhang M, Liu Y, Shi L et al (2022c) Neural stemness unifies cell tumorigenicity and pluripotent differentiation potential. J Biol Chem 298(7):102106. https://doi.org/10.1016/j.jbc.2022.102106
Zhang Y, He J, Jin J, Ren C (2022d) Recent advances in the application of metallomics in diagnosis and prognosis of human cancer. Metallomics 14(7):mfac037. https://doi.org/10.1093/mtomcs/mfac037
Zhang B, Zhao D, Chen F et al (2023a) Acquired miR-142 deficit in leukemic stem cells suffices to drive chronic myeloid leukemia into blast crisis. Nat Commun 14(1):5325. https://doi.org/10.1038/s41467-023-41167-z
Zhang H, Yu X, Ye J et al (2023b) Systematic investigation of mitochondrial transfer between cancer cells and T cells at single-cell resolution. Cancer Cell 41(10):1788–1802.e10. https://doi.org/10.1016/j.ccell.2023.09.003
Zhang M, Lin Y, Chen R et al (2023c) Ghost messages: cell death signals spread. Cell Commun Signal 21(1):6. https://doi.org/10.1186/s12964-022-01004-0
Zhang C, Qin C, Dewanjee S et al (2024) Tumor-derived small extracellular vesicles in cancer invasion and metastasis: molecular mechanisms, and clinical significance. Mol Cancer 23(1):18. https://doi.org/10.1186/s12943-024-01932-0
Zhang C, Wang Y, Cheng L et al (2025) Gut microbiota in colorectal cancer: a review of its influence on tumor immune surveillance and therapeutic response. Front Oncol 15:1557959. https://doi.org/10.3389/fonc.2025.1557959
Zhao CM, Hayakawa Y, Kodama Y et al (2014) Denervation suppresses gastric tumorigenesis. Sci Transl Med 6(250):250ra115. https://doi.org/10.1126/scitranslmed.3009569
Zhao ZM, Zhao B, Bai Y et al (2016) Early and multiple origins of metastatic lineages within primary tumors. Proc Natl Acad Sci USA 113(8):2140–2145. https://doi.org/10.1073/pnas.1525677113
Zhao ZD, Yang WZ, Gao C et al (2017) A hypothalamic circuit that controls body temperature. Proc Natl Acad Sci U S A 114(8):2042–2047. https://doi.org/10.1073/pnas.1616255114
Zhao Y, Oreskovic E, Zhang Q et al (2021) Transposon-triggered innate immune response confers cancer resistance to the blind mole rat. Nat Immunol 22(10):1219–1230. https://doi.org/10.1038/s41590-021-01027-8
Zhao G, Tan Y, Cardenas H et al (2022) Ovarian cancer cell fate regulation by the dynamics between saturated and unsaturated fatty acids. Proc Natl Acad Sci U S A 119(41):e2203480119. https://doi.org/10.1073/pnas.2203480119
Zhao J, Xu L, Sun J et al (2023) Global trends in incidence, death, burden and risk factors of early-onset cancer from 1990 to 2019. BMJ Oncol 2(1):e000049. https://doi.org/10.1136/bmjonc-2023-000049
Zheng J (2012) Energy metabolism of cancer: glycolysis versus oxidative phosphorylation (Review). Oncol Lett 4(6):1151–1157. https://doi.org/10.3892/ol.2012.928
Zheng J, Gao P (2019) Toward normalization of the tumor microenvironment for cancer therapy. Integr Cancer Ther 18:1534735419862352. https://doi.org/10.1177/1534735419862352
Zheng Z, Jia S, Shao C et al (2020) Irradiation induces cancer lung metastasis through activation of the cGAS-STING-CCL5 pathway in mesenchymal stromal cells. Cell Death Dis 11(5):326. https://doi.org/10.1038/s41419-020-2546-5
Zhi X, Wu F, Qian J et al (2025) Nociceptive neurons promote gastric tumour progression via a CGRP-RAMP1 axis. Nature 640(8059):802–810. https://doi.org/10.1038/s41586-025-08591-1
Zhong W, Sun T (2023) Editorial: Epithelial-mesenchymal transition (EMT) as a therapeutic target in cancer, Volume II. Front Oncol 13:1218855. https://doi.org/10.3389/fonc.2023.1218855
Zhou W, Fong MY, Min Y et al (2014) Cancer-secreted miR-105 destroys vascular endothelial barriers to promote metastasis. Cancer Cell 25(4):501–515. https://doi.org/10.1016/j.ccr.2014.03.007
Zhou JX, Cisneros L, Knijnenburg T et al (2018) Phylostratigraphic analysis of tumor and developmental transcriptomes. Converg Sci Phys Oncol 4:025002. https://doi.org/10.1088/2057-1739/aaa0ec
Zhou N, Xu Y, Huang Y et al (2025) Comprehensive genomic profiling of Chinese lung cancer characterizes germline-somatic mutation interactions influencing cancer risk. J Transl Med 23(1):199. https://doi.org/10.1186/s12967-025-06096-z
Zhu W, Huang L, Li Y et al (2012) Exosomes derived from human bone marrow mesenchymal stem cells promote tumor growth in vivo. Cancer Lett 315(1):28–37. https://doi.org/10.1016/j.canlet.2011.10.002
Zhu Y, Tchkonia T, Pirtskhalava T et al (2015) The Achilles' heel of senescent cells: from transcriptome to senolytic drugs. Aging Cell 14:644–658. https://doi.org/10.1111/acel.12344
Zhu Y, Du Y, Liu H et al (2016) Study of efficacy and safety of pulsatile administration of high-dose gefitinib or erlotinib for advanced non-small cell lung cancer patients with secondary drug resistance: a single center, single arm, phase II clinical trial. Thorac Cancer 7(6):663–669. https://doi.org/10.1111/1759-7714.12384

Zhu L, Ploessl K, Zhou R et al (2017) Metabolic imaging of glutamine in cancer. J Nucl Med 58(4):533–537. https://doi.org/10.2967/jnumed.116.182345

Zhu B, Dacso CC, O'Malley BW (2018) Unveiling "Musica Universalis" of the cell: a brief history of biological 12-hour rhythms. J Endocr Soc 2(7):727–752. https://doi.org/10.1210/js.2018-00113

Zhu S, Wang J, Zellmer L et al (2022) Mutation or not, what directly establishes a neoplastic state, namely cellular immortality and autonomy, still remains unknown and should be prioritized in our research. J Cancer 13(9):2810–2843. https://doi.org/10.7150/jca.72628

Zhu Y, Zheng Y, Dai R, Gu X (2024) Crosstalk between circadian rhythm dysregulation and tumorigenesis, tumor metabolism and tumor immune response. Aging Dis 16(4):2073–2099. https://doi.org/10.14336/AD.2024.0533

Zhuravlev YV, Avetisov VA (2006) The definition of life in the context of its origin. Biogeosciences 3:281–291 www.biogeosciences.net/3/281/2006/

Zijlstra A, Lewis J, Degryse B et al (2008) The inhibition of tumor cell intravasation and subsequent metastasis via regulation of in vivo tumor cell motility by the tetraspanin CD151. Cancer Cell 13(3):221–234. https://doi.org/10.1016/j.ccr.2008.01.031

Zong Y, Wang X, Cui B et al (2023) Decoding the regulatory roles of non-coding RNAs in cellular metabolism and disease. Mol Ther 31(6):1562–1576. https://doi.org/10.1016/j.ymthe.2023.04.012

zur Hausen H (1991) Viruses in human cancers. Science 254(5035):1167–1173. https://doi.org/10.1126/science.1659743

Zviran A, Schulman RC, Shah M et al (2020) Genome-wide cell-free DNA mutational integration enables ultra-sensitive cancer monitoring. Nat Med 26(7):1114–1124. https://doi.org/10.1038/s41591-020-0915-3

Index

D. Paul, *Cancer: An Integrative Approach*, https://doi.org/10.1007/978-3-032-08808-6

F

G

H

MIX
Papier aus verantwortungsvollen Quellen
Paper from responsible sources
FSC® C105338

If you have any concerns about our products,
you can contact us on
ProductSafety@springernature.com

In case Publisher is established outside the EU,
the EU authorized representative is:
Springer Nature Customer Service Center GmbH
Europaplatz 3, 69115 Heidelberg, Germany

Printed by Libri Plureos GmbH
in Hamburg, Germany